ANDREW ELBY

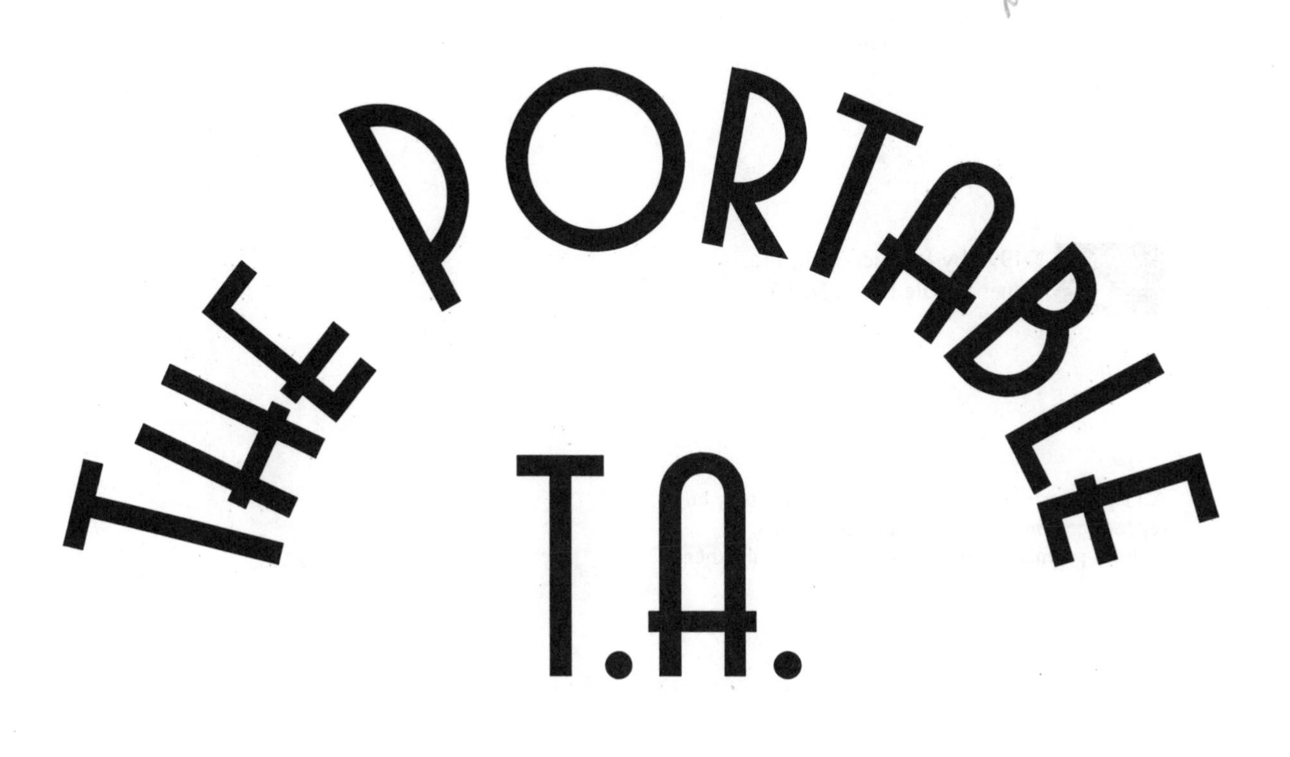

A PHYSICS PROBLEM SOLVING GUIDE
VOLUME I

PRENTICE HALL Upper Saddle River, NJ 07458

Executive Editor: *Alison Reeves*
Production Editor: *Mindy De Palma*
Special Projects Manager: *Barbara A. Murray*
Manufacturing Buyer: *Ben Smith*

Upper Saddle River, NJ 07458

Printed in the United States of America

10 9

ISBN 0-13-231713-3

Prentice-Hall International (UK) Limited, *London*
Prentice-Hall of Australia Pty. Limited, *Sydney*
Prentice-Hall Canada, Inc., *Toronto*
Prentice-Hall Hispanoamericana, S.A., *Mexico*
Prentice-Hall of India Private Limited, *New Delhi*
Prentice-Hall of Japan, Inc., *Tokyo*
Prentice-Hall Asia Pte. Ltd., *Singapore*
Editora Prentice-Hall do Brasil, Ltda., *Rio de Janeiro*

Contents

Preface

In my college physics courses, I always hated the scarcity of feedback. After doing a problem at the end of a chapter, I could look in the back of the book to see if I got the wrong answer. But I never knew whether a particular wrong answer stemmed from a careless error or from a lack of understanding. The chapters themselves contained worked-out sample problems; but they always "tested" a topic I had just read about thirty seconds ago. So, of course I knew what problem-solving technique to use. The solution sets we received after handing in a homework assignment usually did not materialize until days later, and often failed to discuss in depth the relevant intuitions and problem-solving techniques. Besides, by then, we had rushed on to new topics.

When I started teaching first-semester physics, I realized I was not alone. My students also wanted more opportunities to test their level of knowledge, and to learn physics by *practicing* it. So, I started writing a collection of problems with detailed answers. Unfortunately, students found some of the problems unenlightening and several of the answers unhelpful. In other words, some of my stuff flopped. So, I revised many of the questions and answers, deleted some problems entirely, and added new ones. From working closely with students, I also pinpointed trouble spots, and added extra problems to address them. Each semester, students gave me new ideas, and I revised accordingly. My study guide resulted from two years of this cycle. I hope you benefit from the feedback it provides, as much as I benefited from the feedback I received while writing it. In any case, I would appreciate more comments and suggestions. Do not hesitate to e-mail me at elby@physics.berkeley.edu.

I would like to thank Richard Dalven, Toni Torres, Cynthia Hess, Hope Jahren, and Bruce Birkett for much-needed encouragement; William R. Frazer for encouragement and problem suggestions; Ray Henderson at Prentice Hall for giving this book a chance; and most of all, my students. When it comes to feedback, they rule.

I dedicate this book to Diana Perry, who is goddess-like in all respects.

May 1995

Andrew Elby
University of California, Berkeley

Introduction

Hi, I am Andy, a graduate student and teaching assistant at the University of California, Berkeley. Before describing how to use these practice problems effectively, I should warn you how not to use them. Do not read this study guide the way you would read a regular book. Students who simply read the questions and then read the answers do not learn the material well. The concepts and problem-solving skills just will not sink in, and will not become integrated with each other or with your prior knowledge, unless you attempt to solve the problems yourself before reading the answers. Maybe this was not true in high school, but believe me (and believe my students), it is true now. Below, I will discuss how my students have used this study guide effectively, and I will tell you about the associated computer simulations. But first, let me address some common questions.

◆ ***Why should I bust my back carrying around a whole book of nothing but practice problems?***

In college-level physics, you cannot learn the concepts solely by hearing and reading about them. I learned this the hard way. In high school, I got A's just by taking notes and studying them before the test. But in college, my old study strategy bombed. Why?

Physics, like driving, can be learned only by *doing* it. After listening to a lecture and reading the corresponding textbook chapter, you may think to yourself, "That is not so hard–I understand it fine." But then when you try to *apply* those concepts on a hard problem, you discover that you cannot do it. (This happened to me all the time.) I hear some students say, "I understand the concepts, but cannot do the problems." This really means that the student understands the concepts as well as they can be understood from reading and listening. But you can acquire a *deeper* understanding by solving lots of practice problems. This deeper understanding makes the physics more interesting and also enables you to score better on exams.

◆ ***O.K., so I should do practice problems. But will not the homework problems be enough?***

No. Let us face it: When homework deadline approaches, many people scour the textbook looking for relevant formulas, and then plug them in. Or they "help" each other (if it is allowed). Or they play around with numbers until finding a way to reproduce the answer at the back of the textbook. For these reasons, some students "ace" the homework but bomb the tests.

I hope you will do the homework assignments properly, by which I mean the following: Try to understand every formula and concept, instead of just "chugging and plugging." If you work together, make sure you can also address each problem by yourself. And finally, spread out your work over a few days. Your brain synthesizes concepts more effectively if you do not "overload" by cramming the whole assignment into one night.

But even if you do the homework assignments properly, you will still need more practice problems. At least, I did. Here is why: First, 7 to 11 problems a week do not provide enough practice, especially if you want to increase your speed and accuracy. Second, some homework problems are too plug 'n' chug. Third, many homework problems address one very specific topic; and you can easily figure out what that topic is by seeing to what textbook section the homework problem corresponds. By contrast, an exam question tends to involve multiple concepts, and does not come with a "label" telling you what kind of problem it is. For these reasons, I strongly advise you to do your homework problems properly, but also to do most of my practice problems.

◆ ***What is special about your practice problems?***

I have written the problems specifically to give you practice applying the central concepts, the ones most likely to show up on tests. Taking the suggestions of my students, I have modified many of these problems over the past two years, added new ones, and cut out the unhelpful ones.

Equally important, my answers discuss *in detail* the relevant concepts and problem-solving skills. The solutions do not assume that you already understand the material perfectly. Instead, they try to explain the relevant material intuitively. And I have made it easy to "test" yourself: About a quarter of the problems in this book are practice exam questions. See Chapters 23-32.

◆ ***All right, enough advertising. How can I use your practice problems most effectively?***

Find the study guide chapters corresponding to the material covered on this week's (or last week's) homework assignment. For each problem,

- Read the question carefully, making sure you understand what is being asked and what is given.
- Instead of jumping right to a formula, visualize what is happening, sketching pictures or graphs to organize your thoughts.
- Try to formulate a problem-solving strategy. Only after thinking the problem through should you dive into formulas.
- Even if you cannot complete the problem, go as far as you can. Once you have *actively* tried to solve it yourself, the material you learn from reading the answer will "stick"; you will be able to apply it on future problems.

Given all this, here is how I recommend using these chapters.

(1) Work through each problem in the way just described.
(2) A few weeks later, go back to the beginning of the study guide chapter and re-work each problem, skipping the algebra to save time. (Most students do this the week before a test.) If you can solve each problem from scratch, then you understand the material well.
(3) A few days before the test, do the relevant "review problem set" and practice midterms or practice final exams. (See the table of contents.)

◆ ***Should I do your practice problems before or after completing the corresponding homework assignment?***

My students have found that both ways work fine. The problems range in difficulty from medium-easy to very hard. Some of them will help you with your homework. The hardest ones are probably best to postpone until after you have learned the basics and tried the homework problems. In any case, what you should NOT do is rifle through these practice problems the night before a homework assignment deadline, searching for useful formulas and hints. As noted earlier, anything you "learn" in this manner will not stick.

◆ ***What is up with the computer simulations?***

Paul Manly and I have written Interactive Physics™ computer simulations to accompany many of these practice problems. They can help you visualize complicated physical processes. They also let you "play around" with various physical situations, by inviting you to adjust parameters such as mass, initial velocity, and so on. You can develop your physical intuitions by asking yourself a "what if" question, making your own prediction, and then testing it out on the computer. For instance, how does a projectile's trajectory change if there is wind resistance? How does the oscillation frequency of a mass on a spring depend on the size of the oscillations?

These simulations are available *free* over the Internet, though you will need to buy the Interactive Physics™ program in order to run them. Your campus book store should carry the Interactive Physics™ Student Edition, in both Macintosh and Windows versions. If not, the store can order it from Prentice Hall. The Mac version order number is ISBN 013-101874-4. The Windows version order number is ISBN 013-101866-3. The simulations should run on either platform.

To get the simulations, you (or a friend) will need access to the Internet, and a World Wide Web browser such as Mosaic or Netscape. Your school's computer support office can tell you how to get Mosaic or Netscape. They are free. For up-to-date directions on how to download the simulations, finger elby@physics.berkeley.edu. (On a UNIX system connected to the Internet, you can do this by typing "finger elby@physics.berkeley.edu".) Or, using a World Wide Web browser, go to the Berkeley Physics Department WWW home page, at "http://www.physics.berkeley.edu/". Click on the *Berkeley Physics Problem Solving Guide*, and follow the directions.

◆ ***What about units and significant digits?***

Units. In most cases, I use standard SI units (meters, seconds, kilograms, etc.). Sometimes I use other metric units such as centimeters and grams, and occasionally you will see feet and miles. (No "slugs," though.) Although units are important, this study guide focuses on the central physical concepts. For this reason, few of the problems involve unit conversions.

Significant digits. When an integer ends with zero, it is not clear how many digits are significant. For instance "100 meters" could contain either one, two, or three significant digits. In ambiguous cases such as these, the number of significant digits "follows" the other quantities listed in the problem. For instance, if I say a car travels 100 meters in 5.0 seconds, then the "100 meters" contains two significant digits. But if the car travels 100 meters in 5.00 seconds, then the "100 meters" contains three significant digits.

During the "intermediate steps" of calculations, I usually retain more digits than are significant. But then I round to the correct number of significant digits in the final answer. An example will illustrate what I mean. If a car travels 100 meters in 6.0 seconds, how far will it travel in 11 seconds, assuming constant velocity throughout? To answer this, I will first calculate the car's velocity as an intermediate step: $v = \frac{\Delta x}{\Delta t} = \frac{(100\ \text{m})}{(6.0\ \text{s})} = 16.7\ \text{m/s}$. Notice that I have kept an "extra" significant digit. But now when I use that velocity to calculate D, the distance the car moves in 11 seconds, I will "round" to the correct number of significant digits in this problem, which is two: $D = vt = (16.7\ \text{m/s})(11\ \text{s}) = 180$ meters, not "184 meters."

◆ ***Will this study guide improve my love life?***

I guarantee that doing these practice problems may improve your romantic success by *up to* 400%. (Individual results may vary.)

Calculus Review

1 CHAPTER

◆ *Please skim through the Introduction before getting started. It contains hints on how to use this study guide most efficiently.*

QUESTION 1-1

A biologist studies a species of bacteria called *FAST*. Her research reveals that FAST bacteria reproduce prodigiously. In fact, if a bacteria culture contains 10 FAST bacteria at time $t = 0$, the number of FAST bacteria in the culture at later time t is $N(t) = 10 + 6t + 2t^2$, where t is expressed in seconds. (By contrast, most bacteria grow exponentially.)

(a) How many FAST bacteria occupy the culture at time $t = 10$ s? At time $t = 11$ s? From these answers, calculate how many new bacteria were born during the interval between $t = 10$ and $t = 11$ s. Assume no bacteria died.

(b) Now use calculus to find the birth rate of bacteria (i.e., the rate at which new bacteria get born) at time $t = 10$ s. It may help you to sketch a rough graph of N vs. t.

(c) Explain, conceptually, why your answers to parts (a) and (b) are close but not exactly the same.

ANSWER 1-1

Before jumping to formulas, you should graph, diagram, or otherwise visually represent the situation. Especially in difficult physics problems, making a sketch will help you picture what is going on. Here, we can graph N vs. t. The graph tells us how many FAST bacteria occupy the culture at different times. From $N(t) = 10 + 6t + 2t^2$, we see that N increases as time passes. The $+2t^2$ term indicates that the graph is a parabola "facing up." So, a rough sketch looks like this.

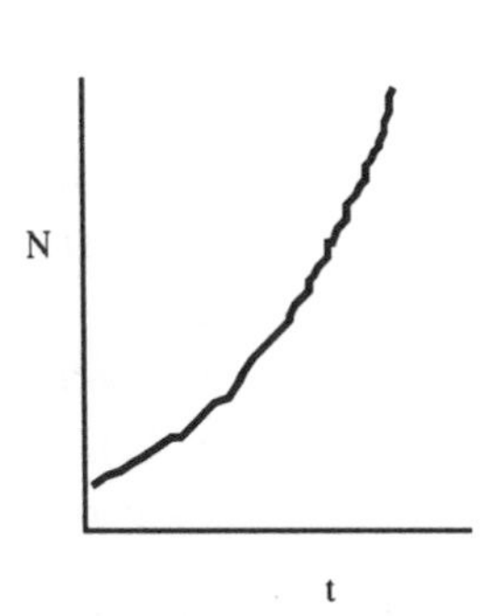

You could, of course, generate a more precise graph by plotting points. But that is unnecessary, because we are just getting a rough idea of the graph's shape.

(a) The given formula tells us the number of bacteria at time t. At $t = 10$ s, $N(10) = 10 + 6(10) + 2(10)^2 = 270$. A second later, at $t = 11$ s, the bacteria population has swelled to $N(11) = 10 + 6(11) + 2(11)^2 = 318$. So, between $t = 10$ and $t = 11$ s, the bacteria population changes by $\Delta N = N(11) - N(10) = 48$. In words, 48 bacteria were born during the one-second interval between times $t = 10$ s and $t = 11$ s.

(b) We need to calculate the rate at which new bacteria get born, at time $t = 10$ s. This "birth rate" specifies how quickly the bacteria population increases, i.e., how fast N increases with time. This rate of increase is given by the slope of the N vs. t graph. You can see from the slope that, at earlier times, the birth rate is comparatively small; N increases slowly. But later on, bacteria get born at a tremendous rate. (If you feel uncomfortable with this kind of reasoning, please see your instructor to get "caught up." A good intuitive understanding of slopes is essential for physics, right from the start.)

Here is the upshot: The birth rate at $t = 10$ s is given by the slope of the N vs. t graph at time $t = 10$ s. To calculate that slope, first take the derivative of N with respect to t: $\frac{dN}{dt} = 0 + 6 + 2(2t) = 6 + 4t$. That is the birth rate at arbitrary time t. So, at time $t = 10$ seconds, the bacterial birth rate is $6 + 4(10) = 46$ bacteria per second.

By the way, notice that $\frac{dN}{dt} = 6 + 4t$ gets bigger as time passes. This agrees with the qualitative conclusion reached by graphing.

(c) After explaining why the answers to parts (a) and (b) are close, I will spell out why they do not agree exactly.

In part (a), we found that during the one-second time interval between $t = 10$ and $t = 11$ s, 48 new bacteria were born. In part (b), we used calculus to find that, at time $t = 10$ seconds, new bacteria were born at a rate of 46 bacteria per second. Part (b) told us that, at a certain moment during the time interval between $t = 10$ and $t = 11$, bacteria were getting born at a rate of 46 per second. Therefore, over that one-second interval, we expect approximately 46 new bacteria to appear. That is what we found in part (a).

I can not emphasize enough, however, that we should not expect parts (a) and (b) to agree *exactly*. Here is why: In part (a), we found the *average* birth rate between $t = 10$ s and $t = 11$ s. In part (b), we found the *instantaneous* birth rate at $t = 10$ s, by which I mean the birth rate *right at that moment*. But the instantaneous birth rate changes between $t = 10$ and $t = 11$ s. For instance, plugging $t = 10.25$ s into $\frac{dN}{dt} = 6 + 4t$ gives us the birth rate, at $t = 10.25$ s, as 47 bacteria per second. By equivalent reasoning, the birth rate at $t = 11$ s is 50 bacteria per second. So, between $t = 10$ and $t = 11$ s, the instantaneous birth rate increases–i.e., the N vs. t graph gets steeper. Bacteria keep getting born at a faster and faster rate. That is why the number of bacteria born during that one-second interval, 48, slightly exceeds the birth rate at $t = 10$ s, which is 46 per second. The *average* rate at which they are getting born, 48 per second, is higher than the instantaneous birth rate at $t = 10$ s, but lower than the instantaneous birth rate at $t = 11$ s.

In general, the average birth rate over an interval does not equal the instantaneous birth rate at some moment during that interval. But for small time intervals, we expect close agreement. This crucial distinction between average and instantaneous rates becomes important in your textbook's discussion of velocity and acceleration.

QUESTION 1-2

Consider a water reservoir. The reservoir gradually loses water, because people insist on showering, washing clothes, etc. The "use rate" R is the rate at which consumers use the water.

(a) (*Easier*) On September 10, between 1 p.m. and 2 p.m., R is 5 gallons per second. How much water was drained from the reservoir between 1 p.m. and 1:30 p.m.?

(b) (*Harder*) On September 11, R gradually decreases from 5 gallons per second to 4 gallons per second between 1 p.m. and 2 p.m., because a big industrial user gradually shuts down. In the following graph, the thick line is R.

Although the graph is "choppy," you may approximate it as a straight line. How much water got taken from the reservoir between 1 p.m. and 2 p.m.?

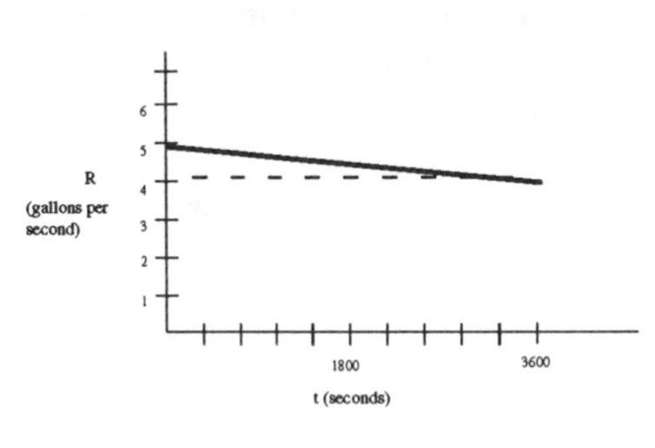

(c) At midnight, when everyone is asleep, a big industrial water user suddenly becomes active. Let w denote the amount of water taken from the reservoir (in gallons), with $w = 0$ at midnight. It turns out that $w = 2t^2$, with t expressed in seconds. So for instance, during the first 5 seconds after midnight, a total of $(2)(5^2) = 50$ gallons of water were removed from the reservoir. Given this information, calculate R, the use rate, at $t = 10$ s.

ANSWER 1-2

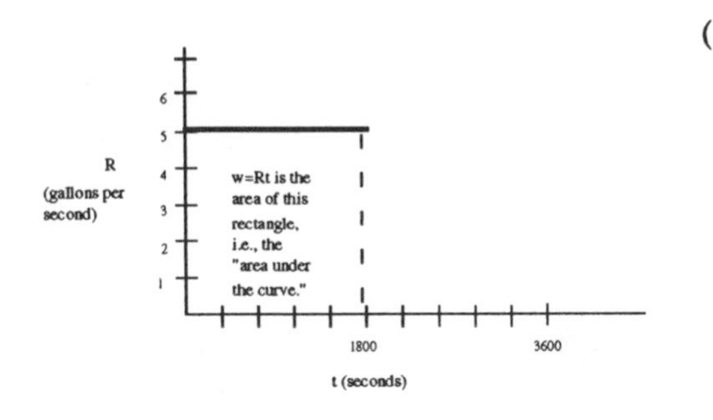

(a) Each second, the reservoir loses 5 gallons of water. Let w denote the total amount of water taken from the reservoir after 1 p.m. So, one second after 1 p.m., $w = 5$ gallons. Two seconds after 1 p.m., $w = 10$ gallons. And so on. From these examples, you can see that *for constant R*, "amount equals rate times time": $w = Rt$. Notice that $w = Rt$ is the area under the curve.

Here, we want to know w at time 1:30 p.m., which is 30 minutes $= 1800$ seconds after 1 p.m. So, $w = Rt = (5\text{ gallons / seconds})(1800\text{ seconds}) = 9000\text{ gallons}$.

(b) You might be tempted to again try $w = Rt$, with $R = 5$ gallons per second and $t = 3600$ s. But R does not equal 5 gallons/second during the whole hour; it gradually decreases. We cannot use the simple formula $w = Rt$, because R is not constant; it changes with time. However, if we are careful, we *can* use a modified version of that formula. Let Δw denote the amount of water used during a very small time interval Δt. So, $\Delta w = R\Delta t$, where R is the use rate *during that time interval*. For instance, at time $t = 1800$ s, the use rate is $R = 4.5$ gallons per second. Therefore, during the time interval $\Delta t = 0.1$ s between $t = 1800$ s and $t = 1800.1$ s, the amount of water used is almost exactly $\Delta w = R\Delta t = (4.5\text{ gallons / s})(0.1\text{ s}) = 0.45\text{ gallons}$.

Now suppose we divide up the entire time between 1 p.m. and 2 p.m. into small time intervals, as represented in this graph.

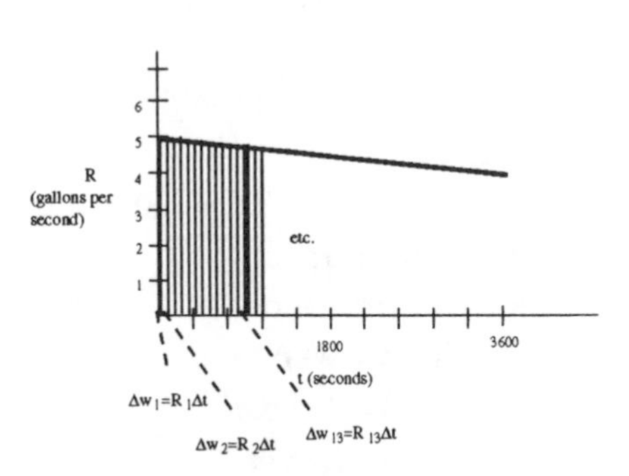

The amount of water used during the first time interval, Δw_1, is given by $R_1\Delta t_1$. Notice that the use rate R_1 during that interval is about $R_1 = 5$ gallons per second. But $R_1\Delta t_1$ is the area of the first (left-most) rectangle. Similarly, the amount of water used during the second time interval, Δw_2, is the area of the second rectangle. (The second rectangle is slightly smaller than the first, because R_2 is slightly less than 5.) The total amount of water used between 1 p.m. and 2 p.m. is the sum of all the little "bits" of water used during each time interval: $w = \Delta w_1 + \Delta w_2 + \Delta w_3 + \ldots$. This sum of Δw corresponds to the sum of all the areas of the thin rectangles. But that sum is the total area under the R vs. t curve.

In summary, to find w, we must calculate the area under the R vs. t curve. Normally, we would need to integrate. But here, the graph is simple enough that you can figure out the area using high-school geometry.

$$\text{(total area under curve)} = \text{(area of rectangle)} + \text{(area of triangle)}.$$

The rectangle's area is $(4 \text{ gallons / second})(3600 \text{ seconds}) = 14,400 \text{ gallons}$. The triangle's area is $\frac{1}{2}(\text{base})(\text{height}) = \frac{1}{2}(3600 \text{ seconds})(1 \text{ gallon / second}) = 1800 \text{ gallons}$. So,

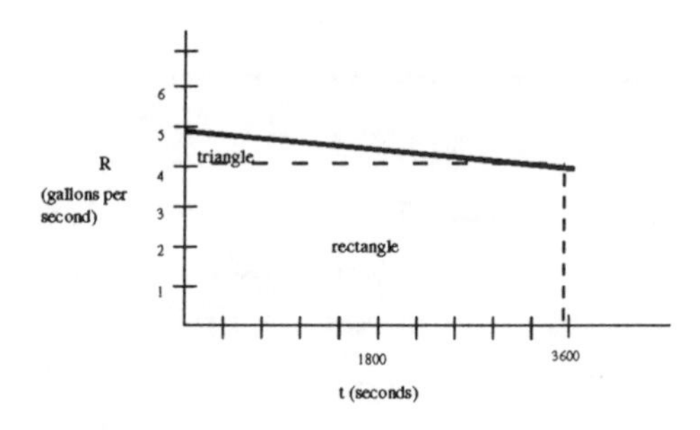

$$w = \text{(total area under curve)}$$
$$= 14,400 \text{ gallons} + 1800 \text{ gallons} = 16,200 \text{ gallons}.$$

Do we get the same answer by integrating? Of course! By "trial and error" or by using $y = mx + b$ reasoning, you can see that $R(t) = 5 - \frac{t}{3600}$.

In the limit at Δt approaches 0, $\Delta w = R\Delta t$ "transforms" into $dw = Rdt$. To add up all those infinitesimal bits of water, integrate over dw:

$$w = \int dw = \int_0^{3600} Rdt = \int_0^{3600}\left(5 - \frac{t}{3600}\right)dt = \left[5t - \frac{t_2}{7200}\right]_0^{3600} = 18000 - 1800 = 16200 \text{ gallons},$$

the same answer obtained using geometry to calculate the area under the curve.

In this particular problem, you can easily "read off" from the graph that the *average* use rate between 1 p.m. and 2 p.m. is $\overline{R} = 4.5$ gallons per second. (A bar on top of a symbol denotes the average value of that quantity.) So, the amount of water used is

$$w = \overline{R}t = (4.5 \text{ gallons / second})(3600 \text{ seconds}) = 16200 \text{ gallons},$$

again in agreement with the above methods. But integrating works even when you cannot "guess" the average rate.

(c) As always, visually represent the information in some way. Here, we can sketch w vs. t.

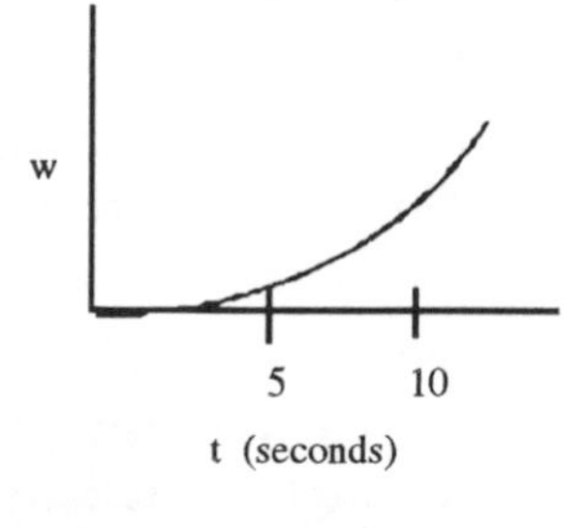

Here is a common line of reasoning: At time $t = 10$ s, the total amount of water that has been used is $w(10) = 2t^2 = 2(10)^2 = 200$ gallons. This water removal occurred over 10 seconds. So, the rate of water usage is $\overline{R} = \frac{w}{t} = 200 \text{ gallons / 10 seconds} = 20 \text{ gallons / second}$.

This answer is the *average* rate at which water gets used between $t = 0$ and $t = 10$ s. But we are looking for the *instantaneous* usage rate *right at* $t = 10$ s. R is the slope of the w vs. t graph; a steep slope means lots of water gets used over a short time interval. We need to find the slope of the graph right at $t = 10$ s. Notice that the slope gets steeper as time passes. Physically, this means water gets used more rapidly at $t = 10$ s than it does at $t = 5$ s.

To find the slope, differentiate w with respect to t:

$$R = \frac{dw}{dt} = \frac{d(2t^2)}{dt} = 2\frac{d(t^2)}{dt} = (2)(2t) = 4t \text{ gallons per second.}$$

Therefore, at time $t = 10$ s, the slope is $R = (4)(10) = 40$ gallons / second. This is twice the *average* usage rate between $t = 0$ and $t = 10$, $\overline{R} = 20$ gallons / second.

But that is not mysterious. During the first few seconds, water gets used at a low rate. For instance, at $t = 1$ s, R is only 4 gallons/second. As time passes, the rate gradually increases to $R(10) = 40$ gallons / second. So, we expect the *average* use rate between $t = 0$ and $t = 10$ s to lie somewhere between those two extremes. In this case, $\overline{R}$ happens to fall right in the middle.

QUESTION 1-3

A traffic engineer studies traffic flow patterns, in order to decide how many new roads to build. Specifically, she is interested in how many cars drive into Middletown during morning rush hour, from 7 a.m. through 8 a.m. She graphs her data. The graph shows the number of cars entering Middletown per minute. This "car entrance rate" is called R. Notice that time $t = 0$ corresponds to 7 a.m., time $t = 1$ minute corresponds to 7:01 a.m., and so on.

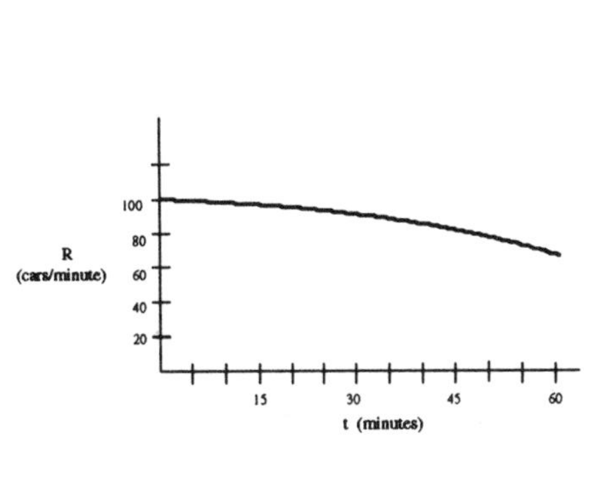

After studying the graph, the engineer notices that R is approximated by $R(t) = 100 - (.01)t^2$, with t given in minutes.

(a) If you had to drive into Middletown somewhere between 7 a.m. and 8 a.m., when would be the best time for you to arrive? Can you decide based on the above graph?

(b) How many cars drive into Middletown between 7 a.m. and 8 a.m.

ANSWER 1-3

(a) At first glance, you might reason as follows: From the graph of R vs. t, the number of cars per minute entering Middletown gradually decreases between 7 a.m. and 8 a.m. In other words, traffic apparently gets lighter. Therefore, you would want to arrive as close to 8 a.m. as possible.

But on second thought, we cannot conclude that traffic gets lighter near 8 a.m. Sure, fewer cars per minute enter the city at that hour, and this could be because fewer cars are on the road. But it could also be because the cars move really slowly–a bumper-to-bumper traffic jam. In that case, you would rather reach Middletown earlier.

So, from the R vs. t graph alone, you cannot make an informed decision about when to drive in. In general, it is important to realize what graphs can *and cannot* tell you.

(b) Let N denote the total number of cars that enter Middletown between 7 a.m. and 8 a.m. If the rate at which cars enter Middletown were constant, N would be easy calculate; we would just use $N = Rt$. But here, R is not constant. It gradually decreases. Still, N is given by the area under the R vs. t curve, as explained in Question 1-2(b) above. Here, we cannot use geometry to find the area. We have no choice but to integrate.

Let dN denote the (tiny) number of cars entering Middletown during a small time interval dt. By filling in the area under the curve with tall, thin rectangles (as in Question 1-2(b)), you can see that $dN = Rdt$. So,

$$N = \int dN = \int_0^{60} R dt = \int_0^{60} (100-.01t^2) dt$$

$$= 100t - (.01)\left(\frac{1}{3}t^3\right)\Big]_0^{60}$$

$$= \left[100(60) - (.01)\left(\frac{1}{3}60^3\right)\right] - \left[100(0) - (.01)\left(\frac{1}{3}0^3\right)\right]$$

$$= 5280 \text{ cars}.$$

That is how many cars entered Middletown between 7 a.m. ($t = 0$) and 8 a.m. ($t = 60$ minutes).

If you sometimes have trouble deciding when to use an integral and when to use a derivative, ask your instructor for more problems of this sort.

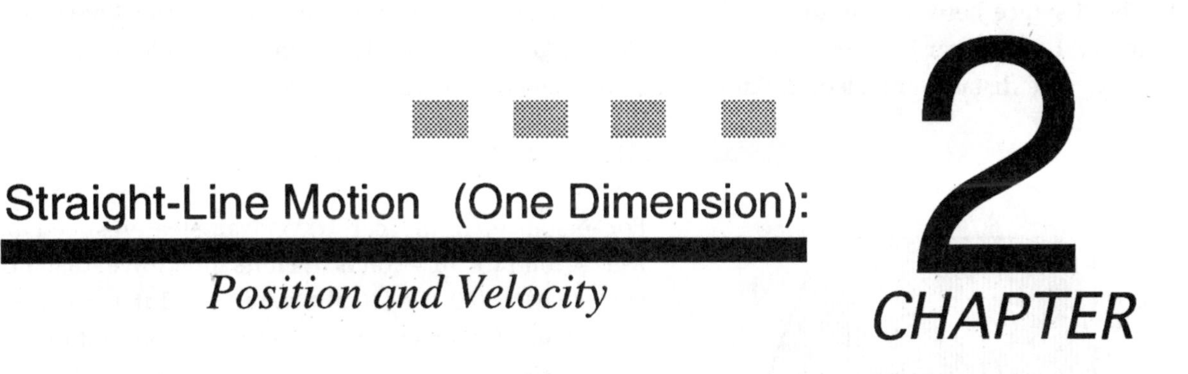

Straight-Line Motion (One Dimension): *Position and Velocity*

2 CHAPTER

- *Before getting started, please read the Introduction, which gives hints on how to use this study guide most efficiently.*
- *This chapter illustrates how to use basic calculus concepts to address kinematics problems. If you have not yet learned derivatives and integrals, skip to Chapter 3, which contains mostly non-calculus problems.*

QUESTION 2-1

Consider a car traveling along a straight highway in New Jersey. The car is initially stopped at a toll booth. But then, at time $t = 0$, the car starts pulling away from the toll booth. Between $t = 0$ and $t = 10$ seconds, the car's velocity, in feet per second, is given by $v = 10t$ (from $t = 0$ to $t = 10$ s). Then, for the next minute and a half, the car's velocity is $v = 100$ feet/second (from $t = 10$ to $t = 100$ s). At this point, the car slows down, because it is approaching another toll booth. Its velocity in feet per second is $v = 100 - 5(t - 100)$ (from $t = 100$ to $t = 120$ s). At $t = 120$ seconds, the car is stopped at the second toll booth.

(a) Sketch a rough graph of v vs. t for the car between $t = 0$ and $t = 120$ seconds. (You should do this even if not explicitly asked.)

(b) How far apart are the two toll booths?

ANSWER 2-1

(a) Between $t = 0$ and $t = 10$ s, the velocity steadily increases from 0 to 100 feet/second. On the v vs. t graph, this corresponds to an upward-sloped straight line. Between $t = 10$ and $t = 100$ s, the velocity is constant at 100 feet/second. This corresponds to a straight horizontal line on the v vs. t graph, because v neither increases nor decreases. Then, between $t = 100$ and $t = 120$ s, the velocity gradually decreases from $v = 100$ feet/s (at $t = 100$ s) to $v = 0$ (at $t = 120$ s). To prove this to yourself, you can plot points. Putting all this together, we get the following graph. To picture this process, see the computer simulation.

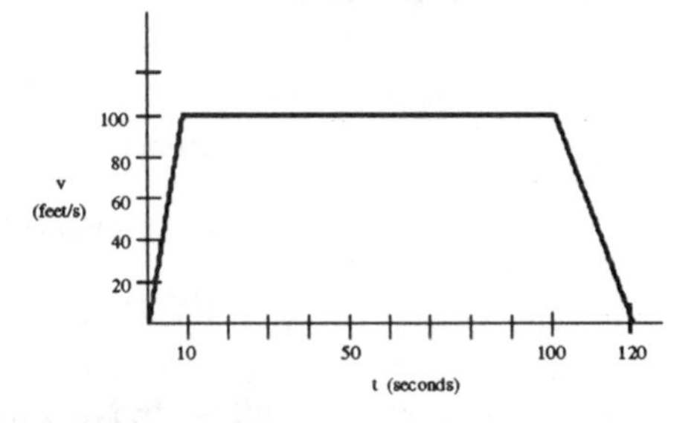

(b) The distance between the toll booths is the distance traveled by the car during the 120-second interval. If the car traveled at a constant rate (i.e., constant velocity), then this problem would be easy: The distance traveled, Δx during the time interval Δt would be given by

$$\Delta x = v\Delta t \text{ [for constant } v\text{]}. \tag{1}$$

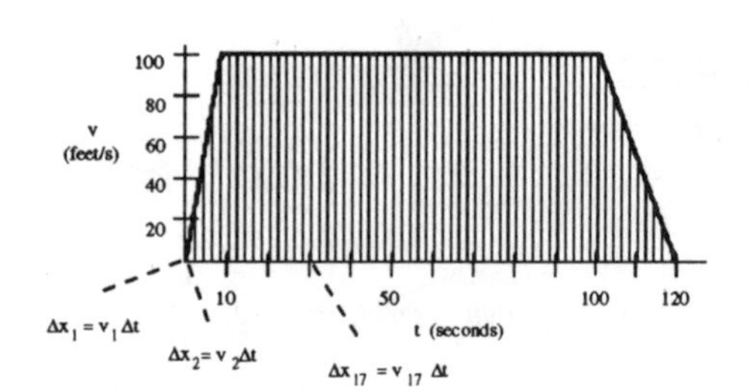

For instance, if the car travels at steady speed 80 feet/second for 10 seconds, then its total travel distance is $\Delta x = v\Delta t = (80 \text{ ft/s})(10 \text{ s}) = 800$ feet. But we cannot use formula (1) directly, because v is not constant for the whole trip. It is constant only between $t = 10$ and $t = 100$ s.

To understand the problem-solving technique needed here, divide up the entire journey between $t = 0$ and $t = 120$ s into small time intervals, as represented in this graph.

Notice that the velocity during time interval 2 differs from the velocity during interval 17. But *within* a given time interval, the velocity is nearly constant, especially if we use tiny intervals. For instance, over the course of a .001-second interval, the velocity does not have time to change very much.

The distance Δx_1 traveled during the first time interval is given by $\Delta x_1 = v\Delta t_1$. But $v\Delta t_1$ is the area of the first (left-most) rectangle. (On the graph, the first "rectangle" does not look much like a rectangle, because I did not draw the horizontal "top" of it.) Similarly, the distance Δx_2 traveled during the second time interval is given by the area of the second rectangle. And so on. So, the total distance traveled between $t = 0$ and $t = 120$, $\Delta x = \Delta x_1 + \Delta x_2 + \Delta x_3 + \ldots$, is the sum of the areas of all the little rectangles. But the sum of those rectangles is just the total area under the v vs. t curve.

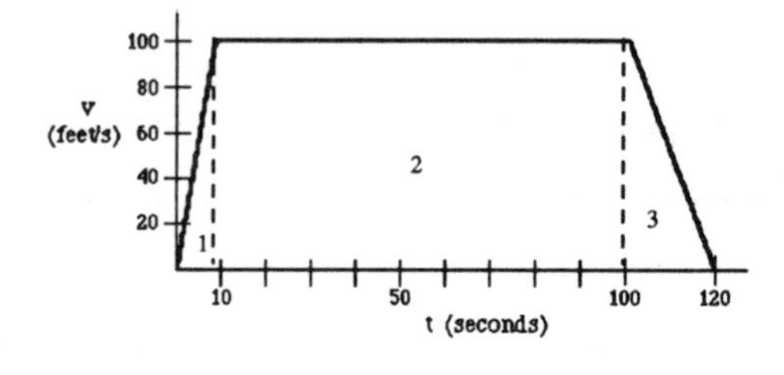

In summary, to find the total distance Δx covered by the car, we must calculate the area under the v vs. t curve. In general, you would need to integrate, finding $\int v dt$. But here, we can find the area using simple geometry. The total area can be divided into a big rectangle and two small triangles.

The area of a rectangle is $A = (\text{base})(\text{height})$, while the area of a triangle is $A = \frac{1}{2}(\text{base})(\text{height})$. So,

$$\begin{aligned}\Delta x &= \text{total area under } v \text{ vs. } t \text{ curve} \\ &= \text{area } 1 + \text{area } 2 + \text{area } 3 \\ &= \frac{1}{2}(10 \text{ s})(100 \text{ ft/s}) + (90 \text{ s})(100 \text{ ft/s})\frac{1}{2}(20 \text{ s})(100 \text{ ft/s}) \\ &= 10500 \text{ ft.},\end{aligned}$$

which is about two miles. In New Jersey, that sounds about right.

For the purposes of illustration, let us also use calculus to find the area under the curve. In the limit as Δt approaches zero, $\Delta x = v\Delta t$ "transforms" into $dx = vdt$. To get the total Δx, add up all the dx's by integrating: $\Delta x = \int dx = \int vdt$. Because the mathematical curve "changes shape" at $t = 10$ s and again at $t = 100$ s, we must break up the integral into three parts. Between $t = 0$ and

$t = 10$ s, $v = 10t$, and therefore we must integrate $10t$ between $t = 0$ and $t = 10$ s. (This integral is the area of the first triangle, area 1.) To that, we must add the integral of $v = 100$ evaluated between $t = 10$ and $t = 100$ s. (This integral is the big rectangle's area.) Finally, we must add in the integral of $v = 100 - 5(t - 100)$ evaluated between $t = 100$ and $t = 120$ s. In symbols, this three-part sum can be written

$$\Delta x = (\text{Total area under curve}) = \int_{t=0}^{t=10} (10t)dt + \int_{t=10}^{t=100} (100)dt + \int_{t=100}^{t=120} (100 - 5[t - 100])dt.$$

The first integral evaluates to $10(\frac{1}{2}t^2)|_0^{10} = 5(10^2 - 0^2) = 500$ feet. The second integral evaluates to $100t|_{10}^{100} = 100(100 - 10) = 9000$ feet. The third integral comes to $(600t - \frac{5}{2}t^2)|_{100}^{120} = \ldots = 1000$ feet. So, the total distance is $(500 + 9000 + 1000) = 10500$ feet, the same answer obtained above using geometry. The math confirms what the computer simulation shows us: The car covers most of its total distance while moving at 100 ft/s, between $t = 10$ and $t = 100$ seconds.

QUESTION 2-2

The graph represents the position of a bicyclist who, starting from rest at $t = 0$, travels along a straight path. Between $t = 0$ and $t = 9$ s, the mathematical equation describing her position is $x = t^2$, with x in feet and t in seconds.

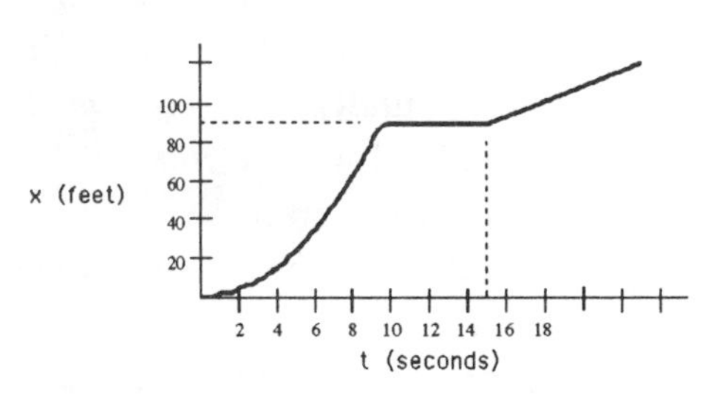

(a) Make a rough sketch of the bicycle's velocity vs. time.

(b) What is the cyclist's exact velocity at time $t = 3$ seconds? At time $t = 6$ s?

(c) What total distance did the cyclist travel between $t = 0$ and $t = 15$ s?

(d) Let me now add some additional information about the cyclist's trip. While riding along the path, she suddenly noticed a skunk ahead of her. But a few seconds later, the skunk wandered back into the woods. From looking at the x vs. t graph, guess when the cyclist first saw the skunk, and when the skunk was safely out of the way.

ANSWER 2-2

(a) Before diving into math, it is good to get a qualitative "feel" for what is going on. That way, you will be more likely to set up the graphs and mathematics correctly. See the computer simulation for extra help. For now, you can qualitatively understand what is happening by studying the x vs. t graph. Velocity is the rate at which the bike's position changes with time. In non-calculus language, $v = \frac{\Delta x}{\Delta t}$. But $\frac{\Delta x}{\Delta t}$ is the "rise over run" of the x vs. t graph, which is the *slope*. So, velocity is the slope of the position vs. time graph.

Between $t = 0$ and $t = 9$ s, the x vs. t graph goes "up," indicating that the bicycle moves forward. In addition, the slope increases. This means the bicycle moves faster and faster.

Between $t = 9$ and $t = 10$ s, the bike continues to move forward (i.e., x is still increasing), but at a slower and slower rate. In other words, the bike suddenly slows down between $t = 9$ and $t = 10$ s; the cyclist must have slammed on her brakes. Her velocity decreases to zero.

Between $t = 10$ and $t = 15$, the bike's position stays the same; it is not moving. Therefore, the velocity is zero.

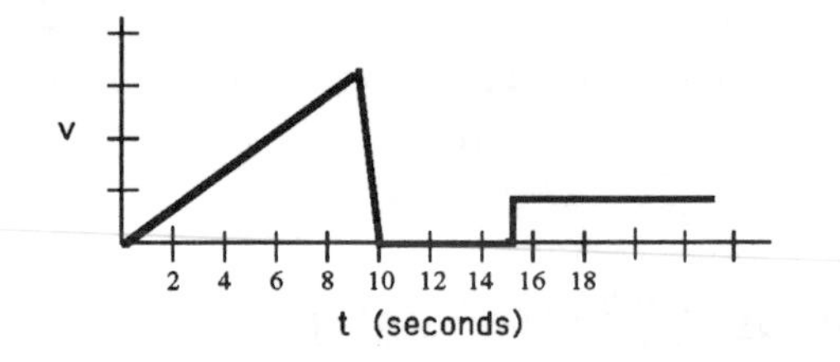

Finally, after $t = 15$ s, the bike moves forward again. From the graph, we see that the bike's position increases steadily, i.e., at a constant rate. Put another way, the slope of the x vs. t graph is constant. Hence, the velocity is constant.

Putting together all these pieces gives us the following v vs. t graph.

Actually, the "corners" should be rounded, because the cyclist's speed cannot change so abruptly.

(b) As noted above, velocity corresponds to the slope of the x vs. t graph. Between $t = 0$ and $t = 9$ s, we know $x = t^2$. To find the slope of x vs. t, differentiate x with respect to t: $v = \frac{dx}{dt} = 2t$. Therefore, at time $t = 3$ seconds, her velocity is $v = 2(3) = 6$ feet per second. At time $t = 6$ s, the velocity is $v = 2(6) = 12$ feet per second. These answers confirm our graphical intuition that the bike's velocity gradually increases between $t = 0$ and $t = 9$ s.

(c) A common mistake is to find the area under the x vs. t curve. You should not do this, because the x vs. t curve *directly* tells you the position at any time. From reading the graph, you can see that at $t = 15$ s, the cyclist occupies position $x = 90$ feet. She started off at $x_0 = 0$. Therefore, between $t = 0$ and $t = 15$ s, she travels a distance of $\Delta x = x - x_0 = 90$ feet. As we have seen before, distance is the area under the *velocity* vs. time curve. So, you could find Δx using the v vs. t graph drawn in part (a). But why calculate that nasty area when you can directly "read off" the bike's position from the x vs. t graph?

(d) As noted above, the cyclist suddenly slammed on her brakes at $t = 9$ seconds. We can infer that she first saw the skunk right before $t = 9$ s. The cyclist did not move between $t = 10$ and $t = 15$ s, but started moving forward again at $t = 15$ s. We can infer that the skunk was safely out of the way by $t = 15$ s. See the computer simulation.

QUESTION 2-3

An astronaut standing on Mars throws a ball straight up. The ball leaves his hand at time $t = 0$, with initial speed 20 m/s. Due to gravity, the ball gradually slows down. In fact, for reasons we will soon study in detail, this ball's velocity, as a function of time, is approximately given by $v = 20 - 4t$, with t in seconds and v in meters per second.

(a) At what time does the ball reach its maximum height off the ground?

(b) What is the ball's maximum height? In other words, how far up does the ball travel from the astronaut's hand?

(c) At time $t = 7$ seconds, what is the ball's velocity? Physically, what does the minus sign in your answer mean?

ANSWER 2-3

Again, you should picture what is going on qualitatively before solving numerically. Throw your pencil straight up into the air. While rising, it gradually slows down. In other words, it travels upward at a smaller and smaller rate. Eventually, just for a moment, the pencil comes to a stop in mid-air. Then it starts to fall back down, slowly at first, and then getting faster and faster. Notice that the "turning point" is where the pencil reaches its maximum height. So, simply by watching the pencil, we can see that the point of maximum height is the turning point, i.e., the point at which $v = 0$. To observe this whole process in slower motion, see the computer simulation.

(a) As just noted, the ball reaches its maximum height when $v = 0$. Set $v = 20 - 4t$ equal to zero, and solve for t to get $t = 5$ s.

Let us see how that answer relates to your mathematical techniques for finding maxima and minima. As you learned in calculus, if we graph y vs. t, where y is the ball's vertical height, then the maximum (or minimum) value of y occurs where the y vs. t graph has zero slope–i.e., where $\frac{dy}{dt} = 0$. To graph y vs. t, we must find y as a function of time. As shown in Question 2-1 above, Δy is given by the area under the v vs. t curve. To find this area, we must integrate v:

$$y(t) = \text{area under } v \text{ vs. } t \text{ curve between 0 and } t$$
$$= \int_0^t v\,dt = \int_0^t (20 - 4t)\,dt = 20t - 2t^2.$$

So, $y = 20t - 2t^2$. Here is what that curve looks like.

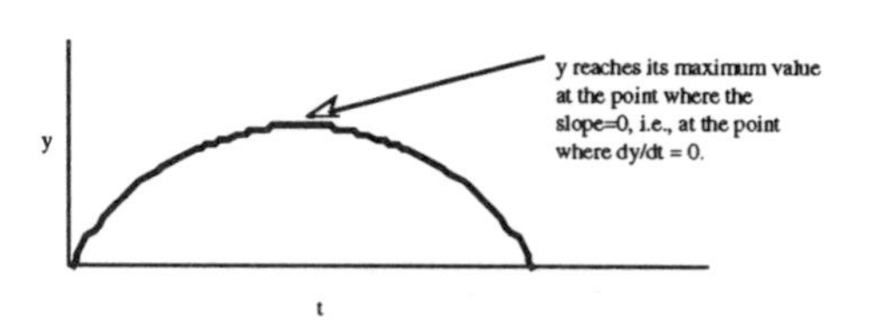

Be careful not to misread this graph as indicating that the ball arcs. It travels straight up and down. The parabola describes how the ball rises and falls as time passes.

To find the t at which the height is maximized, we must find at what time the slope (i.e., the derivative) equals zero. But the derivative $\frac{dy}{dt}$ is the ball's velocity! Taking the derivative of $y = 20t - 2t^2$ gives us $v = \frac{dy}{dt} = 20 - 4t$, right back where we started. So, "finding the derivative of y with respect to t and setting it equal to zero" is just another way of saying, "set the velocity equal to zero." In other words, the ball's height is maximized when its velocity equals zero. We just arrived at this conclusion by purely *mathematical* reasoning. And it is the same conclusion we reached using the *physical* insight obtained from throwing a pencil. The maximum height occurs when $\frac{dy}{dt} = v = 20 - 4t = 0$, as we saw above. So, $t = 5$ s.

(b) In part (a), we found that the ball reaches its maximum height at $t = 5$ s. So here, we need to find $y(5)$, the ball's height at time $t = 5$ s.

As shown in part (a), since $v = \frac{dy}{dt}$, we have $\Delta y = \int v\,dt$. Above, we did the integral and got $y(t) = 20t - 2t^2$. Plug in $t = 5$ s to obtain $y(5) = 20(5) - 2(5)^2 = 50$ meters.

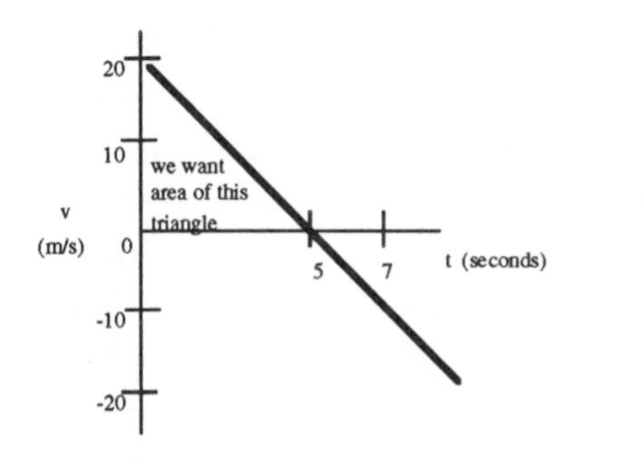

Graphically, this integral corresponds to the area under the v vs. t curve, from $t = 0$ to $t = 5$ s. The curve is a straight line that "starts" at $v = 20$ m/s and has slope -4 m/s^2. (Remember $y = mx + b$?)

The triangle's area is $\frac{1}{2}$(base)(height) $= \frac{1}{2}(5\text{ s})(20\text{ m/s}) = 50$ meters, in agreement with the above integral. That is how high the ball travels upwards from the astronaut's hand.

(c) Simply by substituting $t = 7$ s into $v = 20 - 4t$, we get $v(7) = -8$ m/s. The minus sign indicates that the ball's

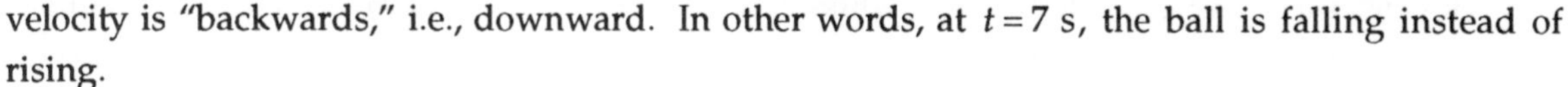

velocity is "backwards," i.e., downward. In other words, at $t = 7$ s, the ball is falling instead of rising.

This negative answer fits in with our part (a) and part (b) answers. In part (a), we found that the ball reaches its maximum height at $t = 5$ s. So, between $t = 0$ and $t = 5$ s, the ball rises. But after $t = 5$ s, the ball starts to fall back down. At $t = 7$ s, it is falling at a rate of 8 meters per second. The v vs. t graph in part (b) confirms this result.

QUESTION 2-4

A hiker walks up a treacherous mountain path at constant speed 2 miles per hour, and returns down the same path at 3 mph. The hike ends when she reaches her starting point.

(a) Calculate the average *velocity* for the round trip.

(b) Calculate the average *speed* for the round trip.

(c) If the round trip took 5 hours, then how long is the stretch of path along which the hiker walked?

ANSWER 2-4

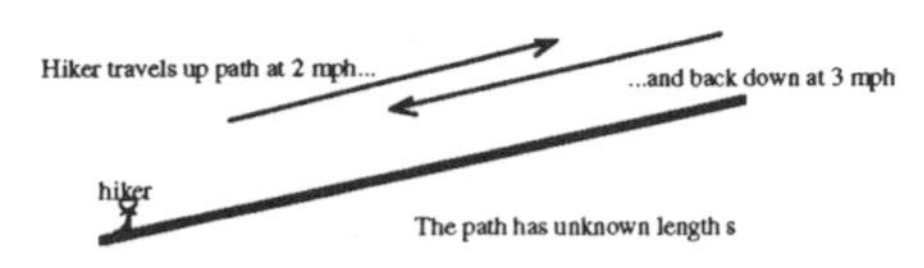

(a) Recall the difference between instantaneous and average velocity. Instantaneous velocity, $v = \frac{dx}{dt}$, tells us how fast and in what direction something is moving at a particular moment. Average velocity, $\bar{v} = \frac{\Delta x}{\Delta t}$, is the total displacement over time, i.e., how fast *on average* the hiker moves.

To better understand this distinction, consider a car driving through traffic along a straight highway. Let us say the car moves forward a distance of 80 miles in 2 hours. So, the car's *average* velocity was $\bar{v} = \frac{\Delta x}{\Delta t} = 40$ mph. But this does not mean the car was *always* moving at 40 mph. Sometimes, when traffic was light, the car moved faster, say 60 mph. At other times, when traffic was heavy, the car crawled along at only 10 or 20 mph. On average, the car's velocity was 40 mph. But its instantaneous velocity was sometimes higher and sometimes lower than that.

Now let us consider the problem at hand. We are trying to find $\bar{v} = \frac{\Delta x}{\Delta t}$, where $\Delta x = x - x_0$, the final position minus the initial position. The most common mistake is to set Δx equal to the distance covered by the hiker, which is twice the length of the path. But the hiker travels down the path as far as she travels up the path. Her final position *is* her initial position: $x = x_0$. Therefore, since her net displacement is $\Delta x = x - x_0 = 0$, her average *velocity* is $\bar{v} = \frac{\Delta x}{\Delta t} = 0$.

Seems weird? Think of it this way: The hiker travels up the hill with a certain positive velocity. Coming back down the hill, she moves "backwards" compared to her initial direction of motion, and hence her velocity is negative. So, the hiker's velocity is sometimes positive and sometimes negative. When we take the appropriate weighted average, these positive and negative contributions cancel out. (This is because the hiker spends more time going up than she spends coming down, as I will now explain.)

(b) The average speed is $\overline{\text{speed}} = \frac{\text{distance covered}}{\Delta t}$. The distance covered differs from Δx, the hiker's displacement. Specifically, displacement is how far the hiker ends up from where she started. But distance covered is how far the hiker moves, i.e., how many miles register on the hiker's "odometer." In this problem, since the hiker ends up where she started, $\Delta x = 0$. But the distance covered is twice the length of the path. When calculating speeds, there is no difference between "positive distance" and "negative distance." All motion "counts" as positive.

You may think this problem cannot be solved, because you do not know the path length. Fortunately, you do not need this information. *Here is a general problem-solving hint: If you are missing an important physical quantity that you cannot calculate, just give it a name. For instance, let s denote the length of the path. You can solve the problem using s. If you are lucky, all factors of "s" will cancel out of your final answer. Then, it will not matter that you do not know s.*

A reasonable guess for $\overline{\text{speed}}$ is 2.5 mph, since the hiker travels half her total distance at 2 mph and half her total distance at 3 mph. This intuitive guess would be right if the hiker spent half her *time* traveling at 2 mph and half her *time* traveling at 3 mph. But the "up time" t_{up} and the "down time" t_{down} *are not equal.* To see why, imagine yourself slowly climbing a flight of stairs, and then running back down. Your climb time, t_{up}, is bigger than your return time, t_{down}. Sure, you cover the same *distance* going up as you cover going down. But while climbing, you cover that distance more slowly; and while returning, you cover that same distance more quickly. Similarly, because the hiker goes slower while climbing the hill, she spends more time going up than she spends coming back down. Specifically, the hiker spends more time traveling at 2 mph than she spends traveling at 3 mph. Therefore, intuitively speaking, her average speed is closer to 2 mph than it is to 3 mph. Now I will calculate the exact number.

As just mentioned, the hiker covers a distance $2s$, where s denotes the length of the path. So,

$$\overline{\text{speed}} = \frac{\text{distance covered}}{\Delta t} = \frac{2s}{\Delta t}, \tag{1}$$

where the total travel time Δt is $\Delta t = t_{\text{up}} + t_{\text{down}}$. Since the hiker goes up the mountain a distance s at speed $v_{\text{up}} = 2$ mph, we have $v_{\text{up}} = 2\text{ mph} = \frac{\Delta x_{\text{up}}}{\Delta t_{\text{up}}} = \frac{s}{t_{\text{up}}}$. Solving for t_{up} gives $t_{\text{up}} = \frac{s}{v_{\text{up}}} = \frac{s}{2\text{ mph}}$. By similar reasoning, $t_{\text{down}} = \frac{s}{3\text{ mph}}$. So, the total travel time is

$$\begin{aligned}\Delta t &= t_{\text{up}} + t_{\text{down}} \\ &= \frac{s}{2\text{ mph}} + \frac{s}{3\text{ mph}} \\ &= s\left\{\frac{5}{6\text{ mph}}\right\}.\end{aligned} \tag{2}$$

Substituting this into Eq. (1) gives

$$\overline{\text{speed}} = \frac{\text{distance covered}}{\Delta t} = \frac{2s}{\Delta t} = \frac{2s}{s\left\{\frac{5}{6\text{ mph}}\right\}} = 2.4\text{ mph}.$$

Notice that the s factors canceled. We did not need to know the path's length to solve this problem.

By the way, $\overline{\text{speed}} = 2.4$ mph confirms the intuitive guess that the average speed is closer to 2 mph than it is to 3 mph, because the hiker spends more time traveling at the slower speed.

(c) As just shown, by considering t_{up} and t_{down}, we get Eq. (2): $\Delta t = s\left\{\frac{5}{6\text{ mph}}\right\}$, where s is the path length, and $\Delta t = t_{\text{up}} + t_{\text{down}}$ is the round trip travel time. Plugging in $\Delta t = 5$ hours, and solving for s, gives $s = (\frac{6}{5}\text{ mph})\Delta t = (\frac{6}{5}\text{ mph})(5\text{ hours}) = 6$ miles. Let us see how all this fits together. The hiker travels up the 6mile path at speed 2 mph. So, the hiker takes $t_{\text{up}} = 3$ hours to complete that leg of the trip. Then the hiker travels 6 miles down the path at 3 mph. Therefore, that leg of the trip takes only $t_{\text{down}} = 2$ hours. We now see that the hiker spends 50% more time going up than she spends going down. That is why the average speed is closer to 2 mph than it is to 3 mph. In fact, let us recalculate by taking the weighted average of 2 mph and 3 mph. Since the hiker spends 3 hours at 2 mph and only 2 hours at 3 mph, the relevant "weighting factors" are 3 and 2, respectively:

$$\overline{\text{speed}} = \frac{v_{\text{up}}t_{\text{up}} + v_{\text{down}}t_{\text{down}}}{t_{\text{down}}} = \frac{(2\text{ mph})(3\text{ hours}) + (3\text{ mph})(2\text{ hours})}{5\text{ hours}} = 2.4\text{ mph},$$

in agreement with our earlier result.

Although you can take weighted averages to find $\overline{\text{speed}}$ or $\bar{v}$, it is usually easier to invoke the "shortcut," $\overline{\text{speed}} = \frac{\text{distance covered}}{\Delta t}$ and $\bar{v} = \frac{\Delta x}{\Delta t}$. By playing around with examples such as this problem, you can convince yourself that the shortcut and the weighted average always yield the same answer. Indeed, they are really different ways of saying the same thing. To see why, notice that the numerator of $\frac{v_{\text{up}}t_{\text{up}} + v_{\text{down}}t_{\text{down}}}{t_{\text{total}}}$ is equal to $\Delta x_{\text{up}} + \Delta x_{\text{down}}$, the total distance covered.

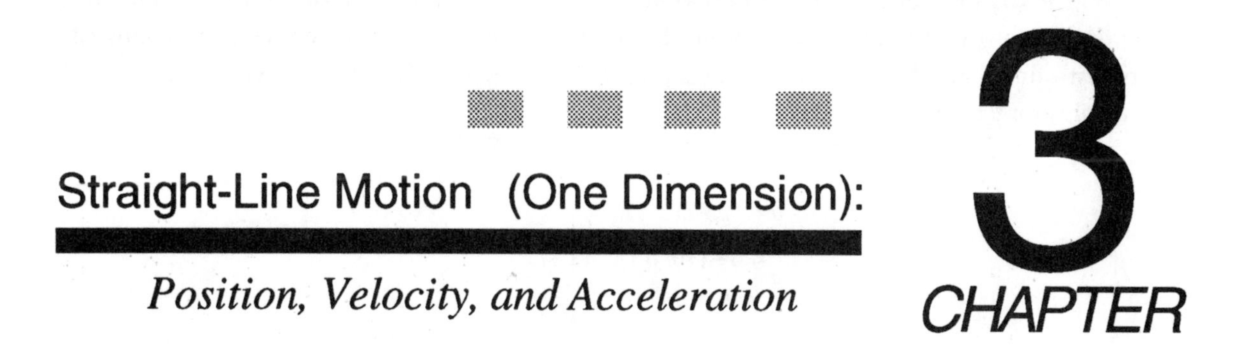

Straight-Line Motion (One Dimension): Position, Velocity, and Acceleration

CHAPTER 3

◆ *If you have not done so, please read the Introduction for hints on how to use these problems effectively.*

QUESTION 3-1

A brick of mass 1.0 kg is dropped from a height of 80 meters. Neglect wind resistance. To simplify your calculations, let us say the acceleration due to gravity near the Earth's surface is $a = 10\ \mathrm{m/s^2}$. (Actually, it is $9.8\ \mathrm{m/s^2}$, but we need not be exact here.)

(a) For how much time does the brick fall before hitting the ground?
(b) How fast is the brick traveling just before it hits the ground?
(c) From the definition of velocity, $\frac{\Delta x}{\Delta t}$, you can calculate the brick's velocity during its fall. Does the v calculated in this way agree with your part (b) answer? If not, explain the apparent contradiction.

ANSWER 3-1

(a) First, an intuitive guess. Eighty meters is approximately 250 feet, about the 25th story of a building. How much time does a brick dropped from the 25th story take to fall? A few seconds, I would guess. Certainly not less than a second and not greater than 10 seconds.

An object undergoing constant acceleration covers distance $\Delta x = x - x_0$, given by $x = x_o + v_0 t + \frac{1}{2}at^2$, where v_0 is the object's initial velocity. This is one of the three basic kinematic formulas you will use repeatedly. The other two are $v = v_0 + at$, and $v^2 = v_0^2 + 2a\Delta x$. The additional kinematic formulas listed in the book all "come from" these three. I recommend using only the basic three, to avoid becoming confused by a flood of symbols. **These kinematic formulas apply only when the acceleration is constant.**

Here, the brick is initially motionless; $v_0 = 0$. Also, we know $a = 10\ \mathrm{m/s^2}$ downward and $\Delta x = x - x_0 = 80$ m downward. So, picking the downward direction as positive, and solving for t, gives

$$t = \sqrt{\frac{2(x - x_0)}{a}} = \sqrt{\frac{2(80\ \mathrm{m})}{10\ \mathrm{m/s^2}}} = 4.0\ \mathrm{s},$$

in line with the intuitive guess.

(b) Before using $v = v_0 + at$, let me show where that formula comes from. By definition, acceleration (in one dimension) is the rate at which an object speeds up or slows down, i.e., the change in velocity with time. So, if acceleration is constant, $a = \frac{\Delta v}{\Delta t} = \frac{(v - v_0)}{t}$. Multiply this equation through by t and solve for v to get $v = v_0 + at$.

In part (a), we found that the brick falls for $t = 4$ seconds. So

$$\begin{aligned} v &= v_0 + at \\ &= 0 + (10\ \text{m/s}^2)(4\ \text{s}) \\ &= 40\ \text{m/s}. \end{aligned}$$

You could also obtain this answer using $v^2 = v_0^2 + 2a\Delta x$, with $a = 10\ \text{m/s}^2$ and $\Delta x = 80$ m.

(c) Substituting in $\Delta x = 80$ meters and $\Delta t = 4$ seconds (from part a), we get $v = \frac{\Delta x}{\Delta t} = 20$ m/s. This disagrees with our part (b) answer, $v = 40$ m/s. To reconcile this apparent discrepancy, we must look more closely at the concepts behind the formulas.

In part (b), we calculated the brick's *instantaneous* final velocity, i.e., how fast the brick is falling just at the moment it crashes into the ground. By contrast, the total displacement over time, $\overline{v} = \frac{\Delta x}{\Delta t} = 20$ m/s is the brick's *average* velocity during its fall.

Why is the final velocity bigger than the average velocity? Because the brick gradually speeds up (accelerates) during its fall. It starts out falling slowly, but then gets faster and faster.

Here, the brick starts at 0 m/s, and gradually speeds up to 40 m/s right before crashing. Therefore, it is no surprise that the average velocity, $\overline{v} = \frac{\Delta x}{\Delta t}$, is 20 m/s, midway between 0 m/s and 40 m/s. (*Important note:* Only when the acceleration is constant must the average velocity be midway between the initial and final velocity.)

QUESTION 3-2

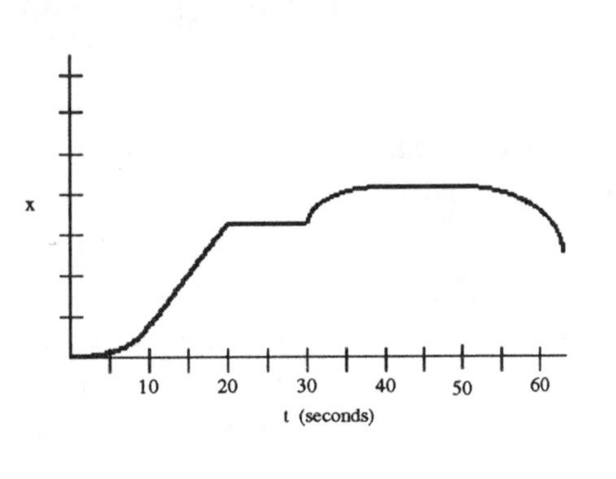

A driver is stuck in traffic. Bored stiff, he decides to graph his position (as indicated by the markers posted along the edge of the highway) versus time. Over a certain 1-minute period, his graph looks like this:

I have deliberately omitted the units and numbers for x, because this is a qualitative problem.

(a) Sketch a rough graph of the driver's velocity vs. time for that 1-minute interval.

(b) Now sketch the driver's acceleration vs. time.

(c) Approximately when did the driver shift into reverse?

ANSWER 3-1

As we have seen, an object's instantaneous velocity is the *rate* at which its position changes in time. This rate corresponds to the slope of the x vs. t graph. Similarly, an object's acceleration is the *rate* at which its velocity changes, i.e., how quickly the object speeds up or slows down. So, acceleration is given by the slope of the v vs. t graph.

(a) Between $t = 0$ and $t = 10$ s, the car moves forward, i.e., x gets bigger. So, the velocity is positive in that region. Furthermore, the velocity gradually increases, as indicated by the increasing slope of the x vs t curve. The car gets faster and faster. An increasing velocity corresponds to an upward-sloped line on the v vs. t graph. From the given information, we cannot tell whether v is represented by a straight line or by a curve. For simplicity, I will draw a line. But you would not lose points for drawing a mild curve.

By contrast, between $t = 10$ and $t = 20$ s, the car's position increases at a steady rate; the velocity is constant. This corresponds to a horizontal line on the v vs. t graph.

Between $t = 20$ and $t = 30$ s, the car's position does not change; it just sits there motionless, stuck in traffic. Therefore, $v = 0$ during that interval.

At $t = 30$ s, the car suddenly lurches forward at high velocity, but then gradually slows down, reaching $v = 0$ at $t = 40$ s. I could tell this by looking at the slope of the x vs. t curve. So, between $t = 30$ and $t = 40$ s, the car's velocity is always positive (forward); but the velocity starts high and ends low, corresponding to a downward-sloped line.

Between $t = 40$ and $t = 50$ s, the car is once again stopped ($v = 0$).

Then, at $t = 50$ s, it starts moving *backwards* (i.e., x decreases). The velocity is negative. Furthermore, the negative velocity gets bigger and bigger, since the car goes backwards at a higher and higher rate, as indicated by the steepening slope of the x vs. t graph. Again, we cannot know whether to draw a downward-sloped line or a downward-sloped curve.

Putting all this together, we get the following velocity vs. time graph.

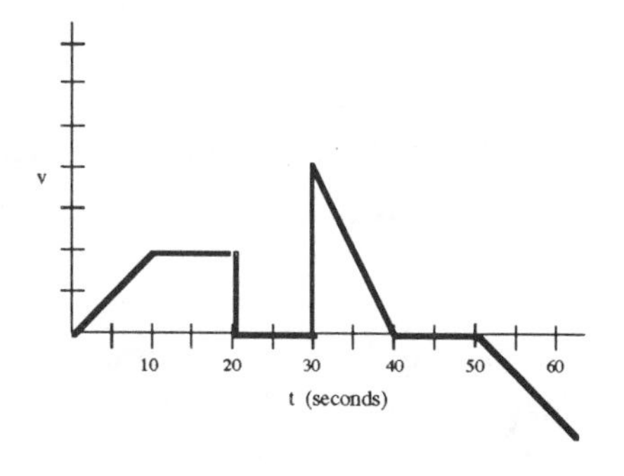

The computer simulation can help you visualize how the real-life motion relates to the graphs.

(b) To sketch this graph, we can work from either the x vs. t graph given in the problem, or the v vs. t graph drawn in part (a). It is easier to use v vs. t, because acceleration is the rate of change of velocity, i.e., the slope of v vs. t.

Between $t = 0$ and $t = 10$ s, the car's velocity increases steadily (or nearly steadily). So, the acceleration is *constant* (or nearly constant) and positive.

But between $t = 10$ and $t = 20$ s, the velocity is constant; the car neither speeds up nor slows down. So, $a = 0$.

Right at $t = 20$ s, the car suddenly slows down to zero. In other words, the velocity suddenly *decreases*. This corresponds to a negative acceleration. Notice that the entire deceleration takes place during a very short time interval near $t = 20$ s. But between $t = 20.5$ and $t = 30$ s, the car does not move; $v = 0$. Because the car neither speeds up nor slows down, $a = 0$ during that interval.

At $t = 30$ s, the velocity suddenly gets large (big positive acceleration), but then gradually decreases. (Mathematically, the slope of v vs. t is negative.) This gradual decrease in velocity corresponds to a negative acceleration.

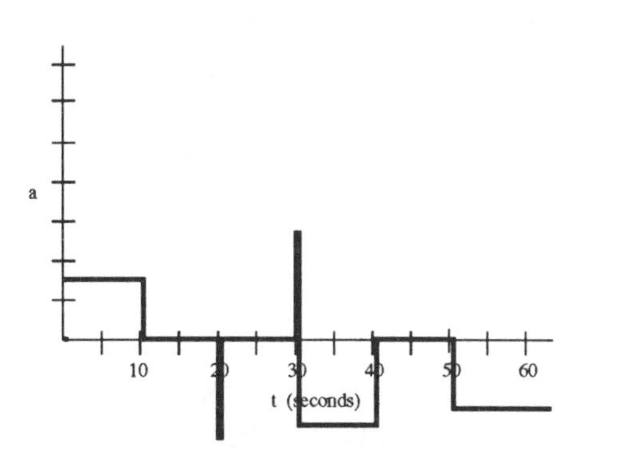

At $t = 40$ s, the car stops and stays motionless until $t = 50$ s; so once again, $v = 0$ and $a = 0$.

Between $t = 50$ and $t = 60$ s, the velocity gradually gets more and more negative, i.e., the car speeds up in the backwards direction. Mathematically, the slope of the v vs. t graph is negative. So the acceleration is constant (or nearly constant) and negative.

Putting all this together generates this acceleration vs. time graph.

If you did not get the "kinks" at $t = 20$ and $t = 30$ s, that is o.k.

(c) The car started going backwards at $t = 50$ s. Between $t = 40$ and $t = 50$ s, the car was stopped. So, the driver must have shifted into reverse somewhere between $t = 40$ and $t = 50$ s. He decided to get into the shoulder and back up to the last exit.

QUESTION 3-3

While driving down the highway at 30 m/s (which is about 72 miles per hour) John spots a police car, sirens flashing. At exactly 1:00 p.m., when the police car is also traveling at 30 m/s and is 100 meters behind John's car, John floors the gas pedal, giving his car an acceleration of 2 m/s^2. John keeps the pedal floored for 5 seconds. During this time, the police car continues moving at 30 m/s, and the officer radios for backup.

(a) How fast is John's car moving at the end of those 5 seconds?

(b) At the end of those 5 seconds, how much distance is between John's car and the police car?

(c) At the end of those 5 seconds, John, in a fit of lawfulness, hits the brake (lightly). His car slows down at a constant rate of 1.5 m/s^2. At what time will John's car come to a complete halt?

(d) Make a rough (non-numerical) sketch of John's position vs. time, starting from 1:00 p.m. and ending when his car comes to rest.

*DO **NOT** TRY THIS EXPERIMENT AT HOME. (My lawyer made me say that.)*

ANSWER 3-3

Before diving into formulas, "think through" the series of events. A diagram or graph helps. For instance, in the following x vs. t graph, $t = 0$ corresponds to 1:00 p.m. The solid line is John's car, and the dotted line is the police car. At $t = 0$, John is ahead of the police car by 100 meters. Since the police car moves at a constant velocity, its x vs. t graph is an upward-sloped straight line. By contrast, John gradually speeds up. So his x vs. t graph is an upward (parabolic) curve. John's curve has a steeper and steeper slope, corresponding to a larger and larger velocity. In part (b), we will find D, the distance between the police car and John at $t = 5$ seconds. This graph does not represent what happens after John hits the brake.

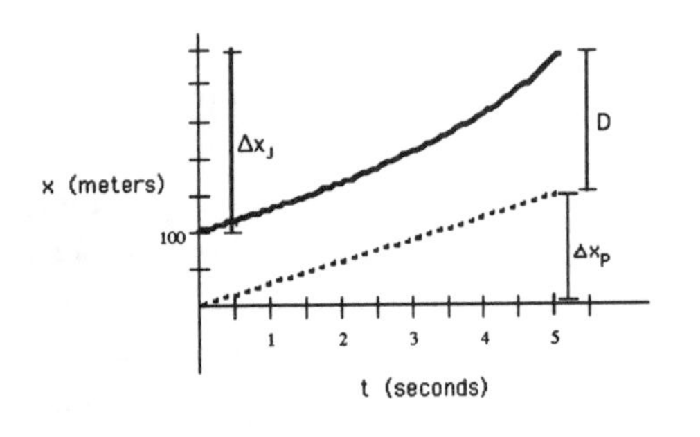

(a) During those 5 seconds, John's car accelerates by $a = 2$ m/s per second. His initial speed at $t = 0$ is $v_0 = 30$ m/s. Therefore we can solve for his velocity at $t = 5$ s using one of our basic constant-acceleration kinematic formulas:

$$\begin{aligned} v &= v_0 + at \\ &= 30 \text{ m/s} + (2 \text{ m/s})(5 \text{ s}) \\ &= 40 \text{ m/s}. \end{aligned}$$

You could have jumped to this answer almost intuitively: Since John speeds up by 2 m/s each second, over the course of 5 seconds he speeds up by 10 m/s. Therefore, since he started at 30 m/s, he ends up (at $t = 5$) with velocity $v = 40$ m/s.

(b) We can easily use the displacement formula,

$$x = x_0 + v_0 t + \frac{1}{2}at^2, \tag{1}$$

to find how far John's car travels during those 5 seconds. But that will not fully answer the question, because the police car *also* moves. Furthermore, John started with a 100 meter "head start." Somehow, we must sort out all that information. The computer simulation can help you visualize everything.

I like to sketch graphs. On the one above, I have labeled some distances. We are trying to find D, the distance between John and the police car at $t = 5$. Over those 5 seconds, Δx_{p} is the distance traveled by the police car, while Δx_{J} is the distance traveled by John. Notice that John's position at $t = 5$ is $x_{\text{J}} = 100 + \Delta x_{\text{J}}$, because John's initial position was $x_0 = 100$ meters.

By looking at the graph, you can see that

$$D = x_{\text{J}} - x_{\text{p}}, \tag{2}$$

where x_{J} and x_{p} are the positions at time $t = 5$ s. So, let us solve for D by separately finding x_{J} and x_{p}, and then subtracting.

I will start with John. Eq. (1), with $x_0 = 100$ m, $t = 5$ s, $a_{\text{J}} = 2$ m/s^2, and $v_{0\text{J}} = 30$ m/s, gives

$$\begin{aligned} x_{\text{J}} &= x_{0\text{J}} + v_{0\text{J}} t + \frac{1}{2}a_{\text{J}} t^2 \\ &= 100 \text{ m} + (30 \text{ m/s})(5 \text{ s}) + \frac{1}{2}(2 \text{ m/s}^2)(5 \text{ s})^2 \\ &= 275 \text{ meters.} \end{aligned}$$

Finding x_{p} is easier, since the police car does not accelerate $(a_{\text{p}} = 0)$:

$$\begin{aligned} x_{\text{p}} &= x_{0\text{p}} + v_{0\text{p}} t + \frac{1}{2}a_{\text{p}} t^2 \\ &= 0 + (30 \text{ m/s})(5 \text{ s}) + 0 \\ &= 150 \text{ meters.} \end{aligned}$$

Notice that when $a = 0$, this kinematic equation reduces to $\Delta x = vt$.

Substitute these positions into Eq. (2) to get

$$\begin{aligned} D &= x_{\text{J}} - x_{\text{p}} \\ &= 275 \text{ meters} - 150 \text{ meters} \\ &= 125 \text{ meters.} \end{aligned}$$

Remember that John started off with a 100 meter head start. (Those hundred meters are "built into" x_{J}.) By accelerating for 5 seconds, John "gains on" the police car by only 25 meters. That is why he decides to give himself up.

(c) As calculated in part (a), John is moving at 40 m/s (in the "positive" direction) right when he hits the brake. In other words, his "initial" velocity as he starts to slow down is $v_0 = 40$ m/s. The initial velocity in this part of the problem is the *final* velocity from part (a).

You are asked how much time the car takes to stop. In other words, at what time will the velocity be $v = 0$? Well, the acceleration during this process is $a = -1.5\ \text{m/s}^2$. The minus sign indicates that the acceleration points "backwards" compared to the positive direction. Since the car continues moving in the positive (forward) direction while braking, this backwards acceleration makes the car slow down. For instance, after 1 second of slowing down, the car is going 38.5 m/s. After 2 seconds of slowing down, it is going 37 m/s. And so on. We could "count down" like this all the way to 0 m/s. Equivalently, you can start with the constant-acceleration kinematic formula $v = v_0 + at$. Solve for t, then substitute in $v = 0$, $v_0 = 40\ \text{m/s}$, and $a = -1.5\ \text{m/s}^2$, to get $t = \frac{v - v_0}{a} = \frac{0 - 40\ \text{m/s}}{-1.5\ \text{m/s}^2} = 27\ \text{s}$. Recall that John floored the gas pedal for 5 seconds starting at 1:00 p.m. Then, he hit the brake and started to slow down. We have just calculated that he slowed down for 27 seconds before stopping. Therefore, his car stops at $5 + 27 = 32$ seconds past 1:00 p.m.

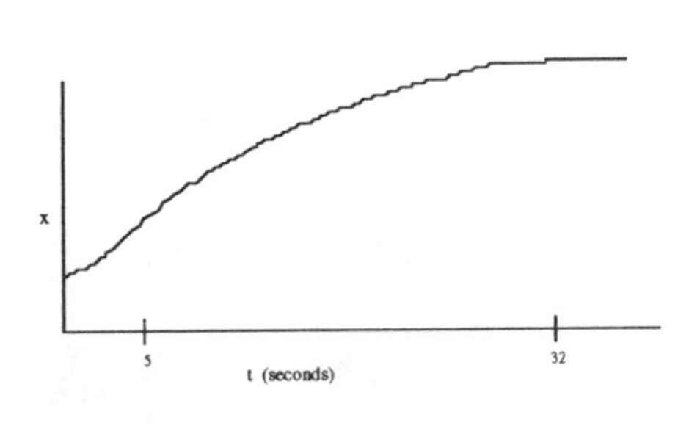

(d) You might think that John's position vs. time graph starts "coming down" after $t = 5$ s. That would be the case if he started moving backwards. But he does not move backwards. He keeps traveling forward, though at a slower and slower rate. Therefore, although his x vs. t graph keeps going "up," the *slope* gradually gets smaller and smaller, until it hits zero when John stops entirely. Above, I already sketched a magnified view of the first 5 seconds of this curve.

QUESTION 3-4

Stuck in traffic, Leticia decides to graph her motion. But instead of graphing her position vs. time, Leticia sketches her velocity vs. time. Over a 1-minute interval, her graph looks like this.

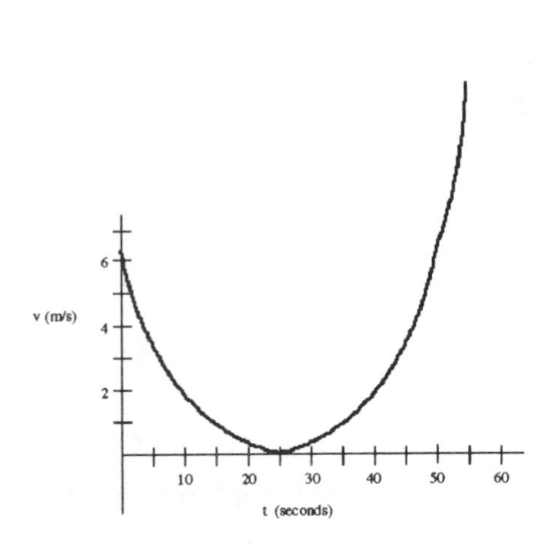

Leticia discovers that her graph approximately fits the formula $v = .01(t^2 - 50t + 625)$, where t is in seconds and v is in meters per second.

(a) Without performing any calculations, make a rough, non-numerical graph of Leticia's position vs. time.

(b) Same as (*a*), but now sketch her acceleration vs. time.

(c) Find a formula for $x(t)$, Leticia's position at any given time, and compare that result to your graph in part (a). You may assume that her initial position is $x_0 = 0$.

(d) Find $a(t)$, Leticia's acceleration at any time. Does this formula "agree" with your graph from part (b)?

(e) What was Leticia's average velocity between $t = 0$ and $t = 60.0$ s?

(f) What was Leticia's instantaneous velocity at $t = 30.0$ s? Does this equal your answer to part (e)? Intuitively, why or why not?

ANSWER 3-4

As you can see from the v vs. t graph, the velocity neither steadily increases nor steadily decreases; the slope is not constant. Since the slope of v vs. t is acceleration, the acceleration is not constant. *Consequently, we cannot use those constant-acceleration kinematic formulas.* Instead, we must use graphs and calculus.

(a) You might have drawn the x vs. t graph as initially "coming down" (i.e., x getting smaller). But during the whole trip, Leticia's velocity is either positive or zero. She is always moving forward (or not moving). A decreasing velocity does *not* mean Leticia moves backward. It means that she travels forward at a slower and slower rate (i.e., she slows down). Only a negative velocity implies backwards motion.

Between $t=0$ and $t=25$ s, Leticia's velocity decreases. So, the slope of her x vs. t graph decreases, hitting zero right at $t=25$ s. But then, between $t=25$ and $t=60$, her velocity increases. In other words, her position increases at a faster and faster rate. This corresponds, in the x vs. t graph, to an increasing slope. Putting all this together gives the following graph.

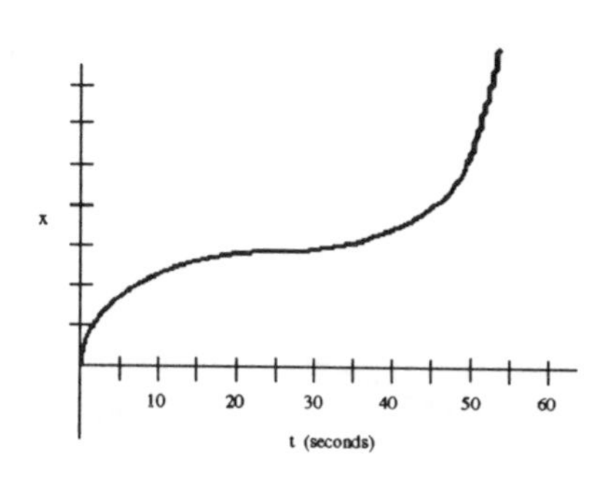

(b) Acceleration is the *rate* at which velocity changes with time. So, Leticia's acceleration corresponds to the slope of her v vs. t graph. Between $t=0$ and $t=25$ s, Leticia's velocity decreases; she moves forward but slows down. This corresponds to a negative acceleration. Notice that Leticia's *rate* of slowing down gets smaller and smaller; the v vs. t graph gradually gets less steep between $t=0$ and $t=25$ s. For instance, between $t=0$ and $t=5$ s, Leticia slows down by a lot more than she does between $t=15$ and $t=20$ s. So, between $t=0$ and $t=25$ s, Leticia's acceleration starts off large and negative, but gradually gets "smaller"and "smaller" (i.e., closer to zero). Right at $t=25$ s, the slope of the v vs. t graph is zero; Leticia is neither speeding up nor slowing down. So, right at $t=25$ s, $a=0$.

Then, between $t=25$ and $t=60$ s, Leticia speeds up while moving forward. This corresponds to a positive acceleration. Furthermore, she speeds up at a faster and faster rate, as indicated by the steepening slope of the v vs. t curve. So, between $t=25$ and $t=60$ s, the acceleration increases.

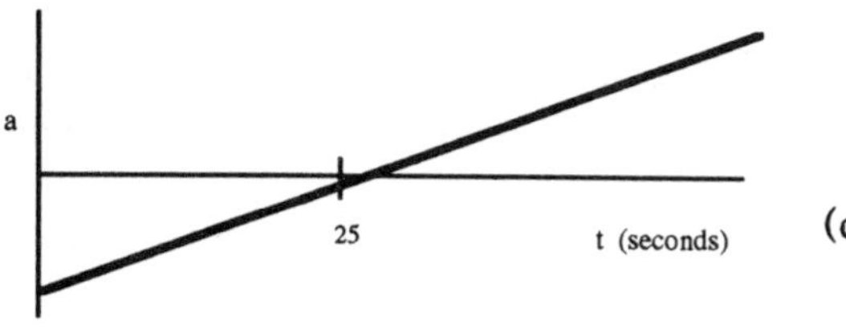

From the graph, we cannot tell the exact shape of the a vs. t curve. I will draw line segments, but mild curves are also acceptable. Putting all this together gives thi graph.

(c) As shown in Question 2-1, the displacement $\Delta x = x - x_0$ corresponds to the area under the v vs. t curve. Since Leticia starts off at $x_0 = 0$, her displacement Δx equals her position $x(t)$.

Here, we cannot find the area under the curve using simple geometry. We must integrate:

$$x(t) = \text{Area under } v \text{ vs. } t \text{ curve between } t=0 \text{ and } t=t$$
$$= \int_0^t v\,dt = \int_0^t .01(t^2 - 50t + 625)dt$$
$$= .01\left[\frac{t^3}{3} - (50)\frac{t^2}{2} + 625t\right]_0^t$$
$$= .01\left[\frac{t^3}{3} - (50)\frac{t^2}{2} + 625t\right].$$

This is a cubic polynomial, the graph of which indeed resembles the qualitative curve we drew in part (a). So, our qualitative graph "agrees" with the rigorous calculus, as far as I can tell.

(d) As mentioned in part (b), acceleration is the slope (rate of change) of the v vs. t curve, calculated by differentiating v with respect to i:

$$a = \frac{dv}{dt} = .01\frac{d(t^2 - 50t + 625)}{dt} = .01(2t - 50 + 0) = .02t - 0.5.$$

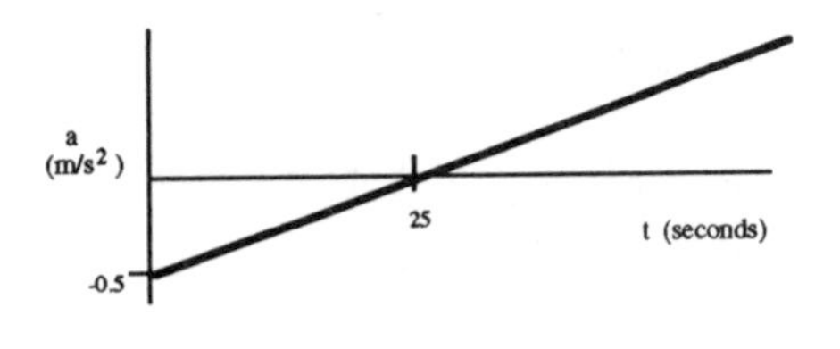

This equation describes a straight line with slope .02 that intercepts the vertical axis at $a = -0.5$. (Remember $y = mx + b$?) So, here is the graph of $a = .02t - 0.5$.

Notice that the curve intercepts the horizontal axis at $t = 25$ s, since $t = 25$ s is when $a = (.02t - 0.5)$ hits zero. This matches our intuitive graph from part (b). Our intuitions often tell us the relevant physical information, while calculus fills in the details.

(e) By looking at the v vs. t graph and plugging numbers into the given formula for v, we see that Leticia's velocity ranges from 6.25 m/s (at $t = 0$) to 0 m/s (at $t = 25$) to 12.25 m/s (at $t = 60$). But we can quickly find her average velocity just by looking at her total displacement for the whole trip: $\bar{v} = \frac{\Delta x}{\Delta t} = \frac{x_{\text{final}} - x_0}{\Delta t}$. Her initial position is $x_0 = 0$, while her final position is x at $t = 60$ s. In part (c), we found an expression for $x(t)$. Substitute $t = 60$ s into that expression to get

$$x_{\text{final}} = x(60) = .01\left[\frac{60^3}{3} - (50)\frac{60^2}{2} + 625(60)\right] = 195 \text{ meters.}$$

So, in 60 seconds, Leticia travels 195 meters. Therefore, her *average* velocity is $\bar{v} = \frac{\Delta x}{\Delta t} = \frac{195 \text{ m}}{60 \text{ s}} = 3.25 \text{ m/s}$. Sometimes, Leticia travels faster than $\bar{v}$, and sometimes she moves slower. But *on average*, she covers 3.25 meters each second.

(f) The problem tells us that Leticia's velocity is $v = .01(t^2 - 50t + 625)$. Substitute in $t = 30.0$ s to get

$$v(30) = .01[30^2 - 50(30) + 625] = .0250 \text{ m/s},$$

an inch per second. This is much smaller than her average velocity of 3.25 m/s found in part (e). But that is not surprising. During her journey, Leticia slows down and then speeds up. Time $t = 30$ s is one of her slow times. By contrast, at $t = 5$ and $t = 55$ s, she moves much faster than her average speed of 3.25 m/s. When all the slow and fast portions of her journey get averaged together, the average velocity comes out to be 3.25 m/s, as calculated in part (e).

QUESTION 3-5

Jill is riding a bicycle along a straight road at constant speed v_1. Her friend, Anita, is driving a car along the same road, at constant speed v_2, where $v_2 > v_1$. Initially, Anita is way behind Jill.

At time $t = 0$, however, Anita sees Jill and therefore presses on the accelerator in order to catch up quickly. Her acceleration is a_2. At time $t = 0$, Anita was a distance L behind Jill. While Anita is accelerating, Jill continues to cycle with constant velocity.

(a) On the same set of axes, make rough sketches of Anita's position vs. time and Jill's position vs time. Use a solid line (or curve) for Anita and a dashed line (or curve) for Jill.

(b) At what time does Anita catch up to Jill? In other words, at what time is Anita's car right next to Jill's bicycle? Express your answer in terms of L, v_1, v_2, and a_2.

(c) How fast is Anita's car moving when she catches up to Jill? Express your answer in terms of the quantities listed above, and/or your answer to part (a).

(d) Would Anita have caught up to Jill even if she had not accelerated? If so, would the "new" answers to parts (b) and (c) be bigger or smaller than the answers you obtained above?

ANSWER 3-5

(a) Since Jill cycles with constant velocity, her position steadily increases. This corresponds to an upward-sloped line on the x vs. t graph. Furthermore, she starts out a distance L ahead of Anita. So, her graph "starts" at $x = L$, not at the origin.

Since Anita accelerates, her velocity increases. In other words, her position increases not at a steady rate, but at an ever-increasing rate (i.e., increasing slope). Therefore, her x vs. t graph curves upward.

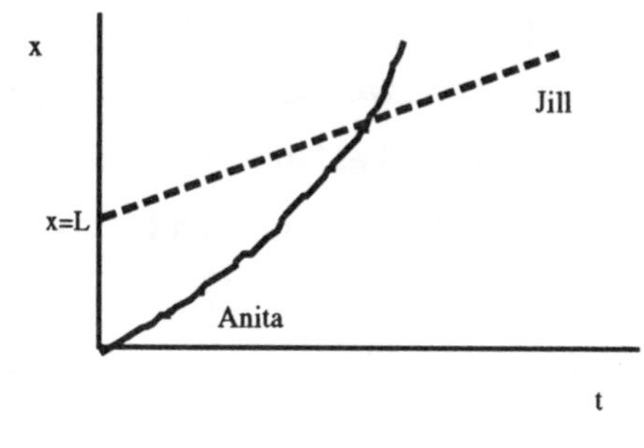

Notice that the curves intersect. The intersection point is where Jill and Anita have the same position, i.e., where Anita catches up to Jill.

(b) The hardest part of this problem is seeing how to get started. Here is the trick: When Anita catches up to Jill, Anita and Jill have the same position. See the graph from part (a). Let x_1 and x_2 denote Jill's and Anita's position, respectively. Since Anita and Jill have the same position when they meet ($x_1 = x_2$), we can use the **"meeting strategy"**:

(1) Write x_1 and x_2 as functions of time, using kinematics and whatever other tools you need.

(2) Set $x_1(t) = x_2(t)$, and solve for t. The t you obtain is the time at which Anita and Jill have the same position, i.e., the "meeting time."

(3) Substitute that meeting time into kinematic formulas to solve for any "meeting" positions or velocities you wish to know. We will use this step in part (b).

Now I will implement the strategy:

Step 1: Write the positions as functions of time

Let us say that Anita's position at time $t = 0$ is $x_0 = 0$. Then, since Jill starts off with a head start of distance L, Jill's initial position at time $t = 0$ is $x_0 = L$. Applying the displacement equation to Jill's motion gives us

Jill
$$x_1 = x_0 + v_0 t + \frac{1}{2} a_1 t^2$$
$$= L + v_1 t,$$

since Jill does not accelerate; her velocity is constant.

By similar reasoning, Anita's position as a function of time is

Anita
$$x_2 = x_0 + v_0 t + \frac{1}{2} a_2 t^2$$
$$= 0 + v_2 t + \frac{1}{2} a_2 t^2,$$

since Anita's initial velocity at $t = 0$ is v_2.

Now that we know Anita and Jill's positions as functions of time, we can proceed to *Step 2: Equate the positions and solve for the meeting time*

$$x_1 = x_2$$
$$L + v_1 t = v_2 t + \frac{1}{2} a_2 t^2.$$

Solve this equation for t, in order to find the time at which Anita and Jill occupy the same position, i.e., the time at which Anita catches up to Jill. To do so, we must solve a quadratic equation, which I will first express in standard form: $\frac{1}{2} a_2 t^2 + (v_2 - v_1)t - L = 0$. The quadratic formula gives us $t = \frac{v_1 - v_2 \pm \sqrt{(v_2 - v_1)^2 - 4(\frac{1}{2} a_2)(-L)}}{a_2}$. Which root is the physically correct answer? Well, we know that t must be positive. And since $v_2 > v_1$, we know $v_1 - v_2$ is negative. Therefore, for t to be positive, we must add (instead of subtract) the square root term. So,

$$t_{\text{meet}} = \frac{v_1 - v_2 \pm \sqrt{(v_2 - v_1)^2 - 4(\frac{1}{2} a_2)(-L)}}{a_2}.$$

Let us check this answer against our intuitions. If Anita accelerates very quickly, then we expect her to catch Jill sooner. Our answer for the meeting time t confirms this; a bigger a_2 produces a smaller t, because the a_2 factor in the denominator beats out the square root of a_2 in the numerator. To see all this physically, check out the computer simulation.

(c) *Step 3: Substitute that meeting time into kinematic formulas to find what you want*

At $t = 0$, Anita starts out with initial speed $v_0 = v_2$. She then gradually speeds up (with acceleration a_2) until time t_{meet} meet from step 2, at which point she catches Jill. So, using $v = v_0 + at$, we can find Anita's speed at time t_{meet}: $v = v_2 + a_2 t_{\text{meet}}$. You may leave your answer in this form, without plugging in that messy expression for t_{meet} obtained in part (a). *Indeed, on a test, many instructors allow you to express answers to later parts of a problem in terms of your answers to earlier parts, even if you did not get the answer to the earlier part. This can save precious time. Check with your instructor about this.*

(d) Intuitively, since Anita was driving faster than Jill, she would have caught up eventually, even without accelerating. See the simulation. The acceleration just helps Anita catch up more quickly. In other words, with no acceleration, the "meeting time" would have been bigger than the answer you obtained above.

You can see this by redrawing your graph from part (a), "turning off" Anita's acceleration. The thin line represents Anita's motion when she accelerates, and the thick line represents her motion when she remains at constant speed v_2. Since $v_2 > v_1$, Anita catches Jill even without accelerating, though it takes her longer.

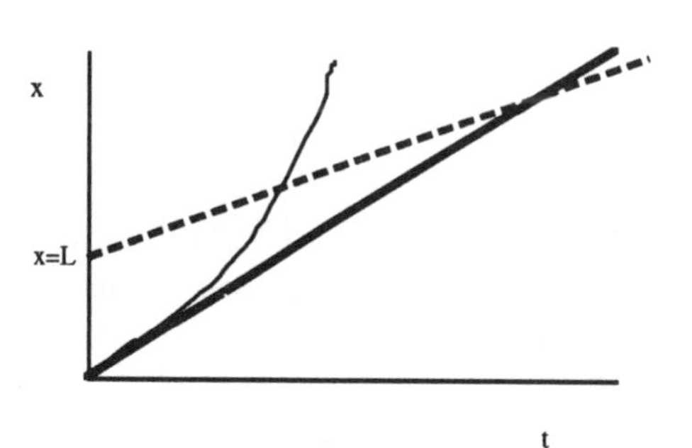

By the way, if you substitute $a_2 = 0$ into your part (a) expression for t_{meet}, both the numerator and denominator become 0. You have to evaluate this expression using L'Hopital's rule to take the limit as $a_2 \to 0$.

With no acceleration, Anita's "meeting speed" would have been her initial speed, v_2.

QUESTION 3-6

Here is a legal question. Suppose Jan gets hit by a car, driven by Mike. According to the law, if the accident resulted partly from Mike's breaking the speed limit, then Mike is considered more negligent than if he had not been speeding. This negligence usually translates into higher cash awards for the victim.

Here is the evidence on which all the witnesses agree: Mike was driving along when suddenly Jan walked into the road. Mike slammed on his high-tech brakes. The skid marks from this braking process were measured to be 60.0 meters long.

Furthermore, a "test" car just like Mike's was driven along the same road at 20.0 m/s, at which point the driver slammed on the brakes. The deceleration was approximately constant, and the resulting skid marks were 44.4 meters long.

What Jan and Mike disagree about is how fast Mike was traveling before he hit his brakes. The speed limit on that road is 50 miles per hour. (Recall that $1 \text{ m/s} \approx 2.4$ miles per hour.)

Suppose you are Jan's lawyer. How do you argue?

ANSWER 3-6

Intuitively, the faster Mike is cruising before he slams the brakes, the farther he skids. Below, we will see that the skid distance is proportional to the *square* of the initial velocity. For instance, if you slam the brakes while going 60 mph, you will skid four times as far as you would have at 30 mph.

Jan's lawyer must prove that the length of Mike's skid implies that his initial speed broke the speed limit. To do so, she should present the following argument to the judge and jury:

"Your honor . . . My argument will consist of two parts. First, by considering the 'test' car, I will establish that the braking acceleration of Mike's car is -4.5 m/s^2, where the minus sign indicates 'backwards.' Then, using that acceleration, I will show that Mike must have been exceeding the speed limit when he slammed on his brakes.

Consider exhibit A, the test car. Since it is just like Mike's car, and since both cars were driven on the same road under the same conditions, we can reasonably assume that both cars experience the same deceleration when the brakes are slammed. Since the deceleration is constant, we can use our constant-acceleration kinematic formulas from physics. Specifically, we can start with exhibit (1),

$$v^2 = v_0^2 + 2a\Delta x \qquad (2)$$

For the test car, the final velocity (when it stops) is $v = 0$. We also know the test car's initial velocity and skid length Δx. Solve exhibit (1) for acceleration, and plug in the known facts, to get

$$\text{TEST CAR} \qquad a = \frac{v^2 - v_0^2}{2\Delta x} = \frac{0 - (20 \text{ m/s}^2)}{2(44.4 \text{ m})} = -4.50 \text{ m/s}^2.$$

As just noted, this was also the acceleration of Mike's car on that fateful day he plowed into my client. Again start with exhibit (1); but now I am talking about Mike's car, not the test car. In Mike's case, the final velocity was again $v = 0$, but the skid distance was $\Delta x = 60.0 \text{ m}$. Solve exhibit (1) for v_0, and substitute in these facts, to get

$$\text{MIKE'S CAR} \quad v_0 = \sqrt{v^2 - 2a\Delta x} = \sqrt{0 - 2(-4.50 \text{ m/s}^2)(60 \text{ m})} = 23.2 \text{ m/s}.$$

Converting this to mph shows that Mike was initially traveling at

$$(32.2 \text{ m/s})\frac{2.4 \text{ miles per hour}}{1 \text{ m/s}} = 55.8 \text{ miles per hour},$$

which exceeds the speed limit of 50 mph.

Your honor, if you remain unconvinced, the accompanying computer simulation will prove beyond reasonable doubt that if Mike's initial speed had been $50 \text{ mph} = 20.8 \text{ m/s}$, then he would have stopped in well *under* 60 meters.

Although I am sure Mike did not intentionally ram his car into my client, her injury resulted in part from his lawless negligence. He should be required to pay all medical expenses, as well as hefty punitive damages."

QUESTION 3-7

Two long-separated friends, June and Bill, spot each other in a bus terminal from a distance of $D = 25$ meters apart. Starting from rest, they run toward each other. Bill accelerates at a constant rate of $a_B = 1.5 \text{ m/s}^2$, while June accelerates at a constant rate of $a_J = 1.2 \text{ m/s}^2$. How far from June's initial position do they meet?

ANSWER 3-7

You can use the standard "meeting" strategy, which relies on the following physical insight: When two objects meet (or crash or whatever), they have the same position. Therefore,

(1) Using kinematics or whatever tools you have available, write x_1 and x_2 (the positions of object 1 and object 2) as functions of time.

(2) The meeting occurs when $x_1 = x_2$. So, set $x_1(t) = x_2(t)$, and solve for t, the time at which the meeting occurs.

(3) Use that meeting time t to find whatever you need.

In this problem, the most common mistake involves minus signs. To avoid this error, sketch what is happening. You will notice that the two friends accelerate in *opposite* directions. So, one of the accelerations must get a minus sign. I will arbitrarily give the minus sign to Bill; but you get the same answer by giving the minus sign to June.

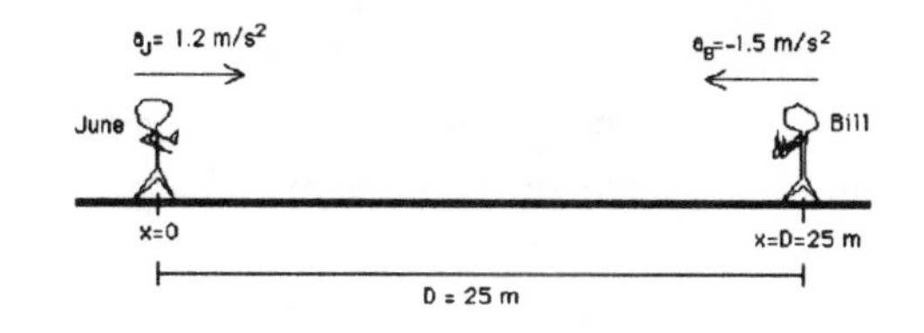

I have chosen the rightward direction as positive. Let us implement the three-step strategy:

Step 1: Write the relevant positions as functions of time.

Since both people run with constant acceleration in one dimension, their positions are given by the kinematic formula $x = x_0 + v_0 t + \frac{1}{2}at^2$.

Let us start with June. From the diagram, her initial position is $x_0 = 0$. Since she starts from rest, $v_0 = 0$. So, her position at arbitrary time t is

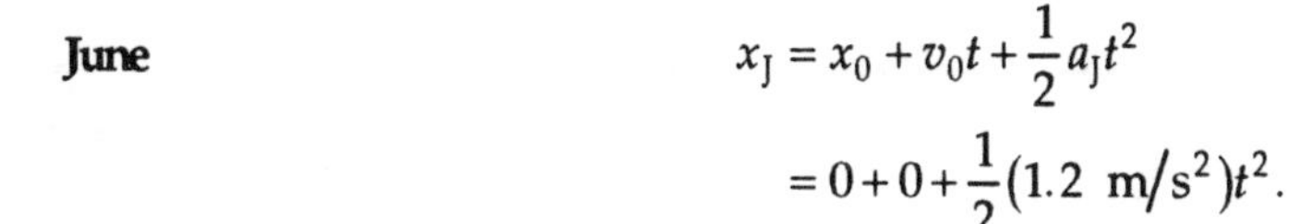

June

$$\begin{aligned} x_J &= x_0 + v_0 t + \frac{1}{2}a_J t^2 \\ &= 0 + 0 + \frac{1}{2}(1.2\ \text{m/s}^2)t^2. \end{aligned}$$

Bill also starts with initial speed $v_0 = 0$. But his initial position is $x_0 = D = 25$ m, and his acceleration is *backwards*, $a_B = -1.5\ \text{m/s}^2$. So,

Bill

$$\begin{aligned} x_B &= x_0 + v_0 t + \frac{1}{2}a_B t^2 \\ &= 25\ \text{m} + 0 + \frac{1}{2}(-1.5\ \text{m/s}^2)t^2. \end{aligned}$$

Step 2: Equate those positions and solve for t.

$$\begin{aligned} x_J &= x_B \\ \frac{1}{2}(1.2\ \text{m/s}^2)t^2 &= 25\ \text{m} + \frac{1}{2}(-1.5\ \text{m/s}^2)t^2. \end{aligned} \qquad (1)$$

Solve for t to get $t_{\text{meet}} = 4.3$ s. That is how much time June and Bill take to reach each other.

Step 3: Use that "meeting time" to solve for the quantity of interest.

We are looking for the distance from June's initial position at which the meeting occurs. Since June started at $x_0 = 0$, this distance is just x_J, June's position at the meeting time. So, substitute $t_{\text{meet}} = 4.3$ s into the above expression for June's position to get

$$\begin{aligned} x_J &= \frac{1}{2}(1.2\ \text{m/s}^2)t_{\text{meet}}^2 \\ &= \frac{1}{2}(1.2\ \text{m/s}^2)(4.3\ \text{s})^2 \\ &= 11\ \text{m}. \end{aligned}$$

Notice that the meeting occurs closer to June's initial position than it does to Bill's initial position, because she accelerates at a slower rate.

Many students solve this problem by setting $\Delta x_J + \Delta x_B = 25$ meters. This is equivalent to my method, as you can confirm by starting with Eq. (1) and adding $\frac{1}{2}(1.5\ \text{m/s}^2)t^2$ to both sides. I recommend using the meeting strategy, because it is more general. For instance, only the meeting strategy works with Jill and Anita in Question 3-5 above.

QUESTION 3-8

When Jan floors the gas pedal in her car, it accelerates at 1.5 m/s^2. When she slams the brakes, the car decelerates at 2.5 m/s^2.

Starting from rest at time $t = 0$, Jan floors the gas pedal for 5 seconds, at which point she slams the brakes. The car soon skids to rest.

(a) Sketch a rough graph of Jan's position vs. time and velocity vs. time. The graphs need not be numerically exact.

(b) When the car finally stops, how far is it from its starting point?

ANSWER 3-8

(a) As always, start by visualizing the motion. For the first five seconds, Jan speeds up at constant acceleration. In other words, her velocity increases at a steady rate; she gradually gets faster and faster. This corresponds to an upward-sloped straight line on a v vs. t graph. But then, when she slams the brakes, the car slows down at a steady rate. This corresponds to a downward-sloped line on a v vs. t graph. So, sketching v vs. t will be easier than sketching x vs. t. We might as well start with velocity vs. time.

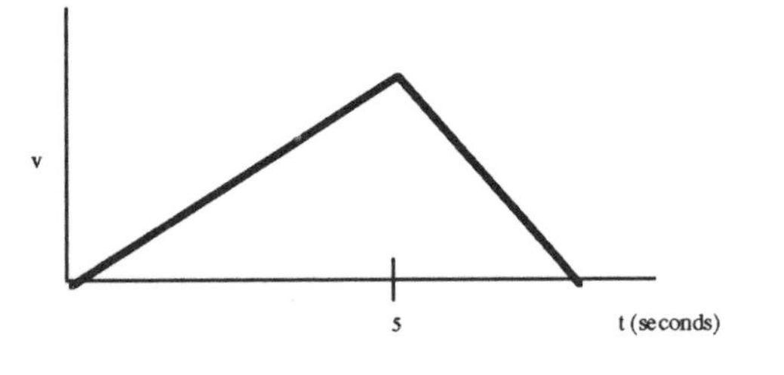

Given this v vs. t graph, you might be tempted to draw an x vs. t graph in which x increases for the first five seconds, and then decreases. But when Jan slams the brakes, the car does not go backwards. It keeps going forward, though at a slower and slower rate. Throughout the trip, x keeps increasing; the x vs. t graph always "goes up." The issue is, at what *rate* does x go up?

Since velocity is the rate at which position increases–i.e., the slope of x vs. t–we see that for the first five seconds, x increases at a higher and higher rate. The slope of the x vs. t graph increases for the first five seconds. But after $t = 5$ s, the slope of x vs. t decreases. As noted above, Jan keeps moving forward, but at a slower and slower rate (slope). Eventually, when the car stops, x stops increasing at all. This does not mean $x = 0$. Rather, it means that x stays constant at its final value, corresponding to a flat line on the x vs. t graph.

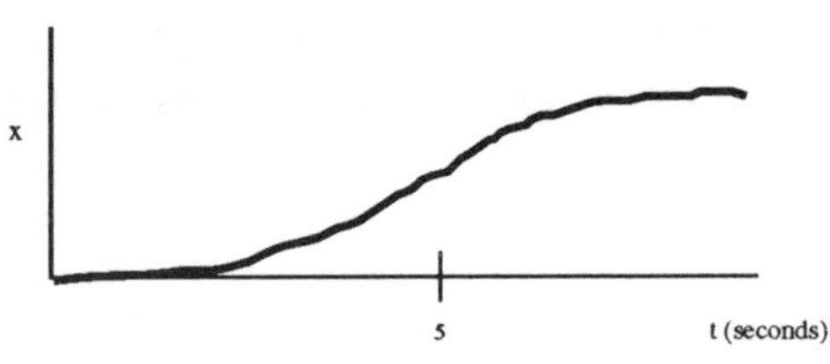

(b) A common mistake is to try to solve the whole problem in one step, using one equation. But remember, those kinematic equations apply only when the acceleration is constant. Here, the acceleration *changes* at $t = 5$ s.

Nonetheless, the acceleration between $t = 0$ and $t = 5$ s is constant at $a_1 = 1.5 \text{ m/s}^2$. And between $t = 5$ s and the "stop" time, the acceleration is constant at $a_2 = -2.5 \text{ m/s}^2$, where the minus sign indicates "backwards" compared to the positive (forward) direction. So, we can use constant-acceleration kinematic formulas if we divide the problem into two parts: The "speed-up" leg of the journey–call it "leg 1"–and the "braking" leg of the journey (call it "leg 2"). The above graphs, especially v vs. t, strongly suggest that it is "natural" to break up the overall trip into these two legs.

So, here is a strategy. First, we can figure out how far the car goes during leg 1, i.e., between $t = 0$ and $t = 5$ s. Call that distance Δx_1. Then we can find the additional distance the car travels during leg 2 (while the brakes are slammed after $t = 5$ s). Call that distance Δx_2. The total travel distance is $x_{\text{total}} = \Delta x_1 + \Delta x_2$.

Finding Δx_1 is not too hard. It is the distance covered in $t = 5$ s at acceleration $a_1 = 1.5 \text{ m/s}^2$, starting from rest ($v_0 = 0$). So, we can use

$$\begin{aligned}\Delta x_1 &= v_0 t + \frac{1}{2}at^2 \\ &= 0 + \frac{1}{2}(1.5 \text{ m/s}^2)(5 \text{ s})^2 \\ &= 18.8 \text{ m.}\end{aligned}$$

Finding Δx_2, the distance traveled during leg 2, is harder. Since we do not know the time the car takes to stop, we cannot immediately reuse the above formula. Let us step back and figure out what we do know about leg 2. The car's final velocity at the end of that segment is $v_f = 0$. But we do not know the car's initial speed. Remember, for leg 2 of the journey, the initial speed is not $v_0 = 0$; rather, "v_0" is the velocity at time $t = 5$ s, right when the brakes are first slammed. In other words, the "initial" velocity for leg 2 of Jan's trip is the *final* velocity from leg 1, namely, her speed after accelerating for 5 seconds. And once we find that "initial" velocity for leg 2, we can use a kinematic equation to calculate the distance covered during that leg.

Since we know Jan's acceleration during leg 1, and the time for which she accelerates, we can find her speed at the end of those 5 seconds:

$$\begin{aligned}v_{0 \text{ for leg } 2} = v_{f \text{ for leg } 1} &= v_0 + a_1 t \\ &= 0 + (1.5 \text{ m/s}^2)(5 \text{ s}) \\ &= 7.5 \text{ m/s.}\end{aligned}$$

We now know Jan's initial and final velocities during leg 2, as well as her acceleration during that leg. Therefore, without bothering to find the time the car takes to stop, we can immediately use the "time-independent" kinematic formula $v^2 = v_0^2 + 2a\Delta x$. Solve for Δx to get

$$\Delta x_2 = \frac{v_{f \text{ for leg } 2}^2 - v_{0 \text{ for leg } 2}^2}{2a_2} = \frac{0 - (7.5 \text{ m/s})^2}{2(-2.5 \text{ m/s}^2)} = 11.2 \text{ m.}$$

That is how far the car moved while braking. So, the total travel distance is

$$\begin{aligned}x_{\text{total}} &= \Delta x_1 + \Delta x_2 \\ &= 18.8 \text{ m} + 11.2 \text{ m} \\ &= 30 \text{ m.}\end{aligned}$$

v
area of this triangle is $\Delta x_1 = 18.8$ m.
This area is only $\Delta x_2 = 11.2$ m
5
t (seconds)

According to these equations, the car traveled a greater distance while speeding up (leg 1) than it traveled while braking (leg 2). Do our rough graphs from part (a) confirm this conclusion? Well, the distance traveled corresponds to the area under the v vs. t curve. And you can see that the area under the 1st five seconds of that curve is greater than the area under the 2nd five seconds.

QUESTION 3-9

A Red Cross helicopter delivers crates of medicine to a war-torn area. Because of the hilly terrain, the helicopter cannot safely land. Instead, the helicopter must hover above the ground, while someone drops the crates onto the ground.

The crates are designed to withstand "impact velocities" of up to 8.0 m/s. You can approximate the acceleration due to gravity as $a = 10\ \text{m/s}^2$.

(a) Assuming the crates are dropped from rest, what is the highest the helicopter can be off the ground, such that the crates do not break when they land?

(b) The helicopter pilot accidentally starts ascending before the last crate has been dropped. As a result, the last crate is "dropped" when the helicopter is 9.0 meters off the ground and rising at 3.0 m/s. How much time passes between when that crate gets dropped and when it lands?

(c) How fast is the crate from part (b) moving right before it hits the ground?

ANSWER 3-9

(a) The higher the helicopter hovers off the ground, the more time the crate falls before landing, and hence, the higher its velocity upon impact with the ground. We want the helicopter to hover at a height such that the crates barely do not break, i.e., so that crates strike the ground at $v = 8.0$ m/s.

I see two efficient ways to solve. Using $v_0 = 0$, $v = 8.0$ m/s, and $a = 10\ \text{m/s}^2$, you could invoke $v = v_0 + at$ to find the time for which the crate falls. Then you could substitute that "fall time" into $x = x_0 + v_0 t + \frac{1}{2}at^2$, to find the distance through which the crate can safely fall.

To reach the solution even more quickly, however, you can avoid worrying about t by whipping out the time-independent kinematic equation, $v^2 = v_0^2 + 2a\Delta x$. Solve for Δx to get

$$\Delta x = \frac{v^2 - v_0^2}{2a} = \frac{(8.0\ \text{m/s})^2 - 0}{2(10\ \text{m/s}^2)} = 3.2\ \text{m},$$

which is about 10 feet. A skilled pilot could easily hover at that height. Notice that I implicitly picked the downward direction as positive, since I plugged in positive numbers for both the gravitational acceleration and the final velocity.

(b) Here, you must have the following physical insight: When the crate is "dropped" from a helicopter moving upward at speed 3.0 m/s, the crate has initial speed $v_0 = 3.0$ m/s, not $v_0 = 0$. In general, an object "dropped" from a moving vehicle acquires the initial speed of the vehicle. To see this, drop a small rock out of a moving car's window (when no one is around). You will notice that the rock does not simply hit the road and stop. Instead, it bounces forward. This proves that the rock had some forward initial speed, even though you "dropped" it. It acquired that initial speed from your hand. Remember, if you and your hand are riding in a car traveling at 30 mph, then your hand "automatically" has 30 mph of speed, because it is getting "carried along" by the car. When you "drop" the rock, your hand is moving forward at 30 mph. Therefore, the rock essentially gets "thrown forward" at 30 mph. Intuitively, getting "carried along" at 30 mph and getting "thrown" at 30 mph might seem different. But as far as the rock's concerned, those are really two different ways of saying the same thing.

In the rising helicopter problem, when the relief worker drops the crate from "rest," her hands are really moving upward at 3.0 m/s, carried along by the helicopter. So, the crate is essentially "thrown upward" at $v_0 = 3.0$ m/s.

I see two ways to solve this problem. The long way involves (i) finding how much time the rising crate takes to reach its peak, and then (ii) finding the time it takes to fall from its peak to the ground. The total flight time is the sum of the rise time and the fall time. Let me demonstrate this method. Then I will show you a shortcut.

The long way. Let t_{up} denote the time the crate takes to reach its peak. By throwing something straight up, you can confirm that the peak is where the object's velocity momentarily reaches 0. Intuitively, the peak is where the object is moving neither upward nor downward, for an instant. So, if we are considering only the upward part of the crate's flight, the "final" velocity is $v = 0$. If we say the initial velocity is $v_0 = +3.0$ m/s, then we have implicitly chosen upward as the positive direction. (So, I have "reversed" the sign convention used in part (a).) Since gravity pulls things downward, the gravitational acceleration is therefore negative: $a = -10 \text{ m/s}^2$. Starting with $v = v_0 + at$, we can immediately solve for t_{up} to get

$$t_{up} = \frac{v - v_0}{a} = \frac{0 - 3.0 \text{ m/s}}{-10 \text{ m/s}^2} = 0.30 \text{ s}.$$

To complete the problem, we need to find t_{down}, the time the crate takes to fall from its peak to the ground. Since we are now considering only the crate's fall, its "initial" velocity is the velocity at the peak: $v_0 = 0$. Because we do not know the crate's speed when it lands, we cannot immediately recycle the formula just used to find t_{up}. Intuitively, the time it takes the crate to fall depends on how high it starts. So, we need to find the crate's peak height, x_{peak}. Given that height, we can then find the fall time, t_{down}.

How can we find the peak height? Well, the crate was "thrown" upward at $v_0 = 3.0$ m/s from initial distance $x = 9.0$ meters off the ground. And above, we found the time it takes to reach the peak. So, we can find the peak height using the kinematic formula that relates distance to time. Remember, since I have picked upward as my positive direction, the acceleration is negative:

From helicopter to peak

$$\begin{aligned} x_{peak} &= x_0 v_0 t_{up} + \frac{1}{2}at_{up}^2 \\ &= 9.0 \text{ m} + (3.0 \text{ m/s})(0.30 \text{ s}) + \frac{1}{2}(-10 \text{ m/s}^2)(0.30 \text{ s})^2 \\ &= 9.45 \text{ m}. \end{aligned}$$

Now that we know the height from which the crate falls, we can reuse that same formula to solve for t_{down}. Starting at rest from initial height $x_0 = x_{peak} = 9.45$ meters off the ground, the crate falls to final height $x = 0$ in (unknown) time t_{down}. So,

From peak to ground

$$\begin{aligned} x &= x_0 + v_0 t_{down} + \frac{1}{2}at_{down}^2 \\ 0 &= 9.45 \text{ m} + 0 + \frac{1}{2}(-10 \text{ m/s}^2)t_{down}^2. \end{aligned}$$

Solve for t_{down} to get

$$t_{down} = \sqrt{\frac{2(x - x_0)}{a}} = \sqrt{\frac{2(0 - 9.45\text{ m})}{-10\text{ m/s}^2}} = 1.4\text{ s}.$$

At this point, we have found how much time the crate takes to reach its peak, and then how much time it takes to fall from the peak to the ground. The total flight time is

$$t_{total} = t_{up} + t_{down} = 0.30\text{ s} + 1.4\text{ s} = 1.7\text{ s}.$$

Now that I have solved this problem the "long way" by separately considering the upward and downward parts of the crate's journey, let me show you a shortcut. Even though the crate does not go "directly" down from the helicopter to the ground, we can still write kinematic equations in which "initial" refers to the crate when it is released from the helicopter, and "final" refers to the crate at the ground. This is "legal," because the crate's acceleration is the same for the entire trip. The time t in that equation will be the total flight time. Initially (at the helicopter), the crate's position is $x_0 = 9.0\text{ m}$, and its velocity is $v_0 = 3.0\text{ m/s}$. The final position, when it reaches the ground, is $x = 0$. So, we can write

From helicopter to ground

$$x = x_0 + v_0 t + \frac{1}{2}at^2$$

$$0 = 9.0\text{ m} + (3.0\text{ m/s})t + \frac{1}{2}(-10\text{ m/s}^2)t^2.$$

We can immediately solve this quadratic equation for t:

$$t = \frac{-3.0\text{ m/s} \pm \sqrt{(3.0\text{ m/s})^2 - 4(-5\text{ m/s}^2)(9.0\text{ m})}}{2(-5\text{ m/s}^2)} = -1.1\text{ s or } +1.7\text{ s}.$$

Since the time of flight cannot be negative, the physically correct answer is $t = 1.7\text{ s}$, in agreement with the answer obtained the "long way."

(c) As in part (b), you can solve this the "long way" by separately considering the crate's upward and downward motion. Or you can treat the crate's entire motion in one equation.

Working the long way, you would first figure out the crate's height at its peak. We already did this in part (b), and found $x_{peak} = 9.45\text{ m}$. At its peak, the crate is momentarily motionless. So, from its peak, the crate falls through distance 9.45 m. You could use a kinematic equation to find how fast it is moving after falling through that distance.

Instead of finishing the problem by that method, however, let me demonstrate the short cut. As emphasized in part (b), we are "allowed" to do this because the crate's acceleration while it rises equals its acceleration while it falls. (While rising, the crate slows down; and while falling, it speeds up. Both of these motions correspond to downward acceleration.) So for instance, we can write

From helicopter to ground

$$v = v_0 + at$$
$$= 3.0\text{ m/s} - (10\text{ m/s}^2)(1.7\text{ s})$$
$$= -14\text{ m/s},$$

where I substituted in the total flight time from part (b). The minus sign indicates that the crate's final velocity is downward. (If you wrote "$v = 14$ m/s," that is fine.)

By the way, you could also get this answer using $v^2 = v_0^2 + 2a\Delta x$. The key is to realize that $\Delta x = -9.0$ m. Remember, Δx is not the distance traveled by the crate. Rather, it is the "displacement," by which I mean the difference between the crate's initial and final positions. So, even though the crate goes up 0.45 meters and then falls down 9.45 meters, its total displacement is still only −9.0 meters, instead of 9.9 meters, because it ends up 9.0 meters below where it started. Similarly, if you walk north for a mile and then south for a mile, your displacement is $\Delta x = 0$, not $\Delta x = 2$ miles, because you end up right where you started.

QUESTION 3-10

Two rocks are dropped from rest from the top of a high tower. Rock 1 is dropped exactly one second before rock 2 is dropped. Assume negligible wind resistance throughout.

(a) At the moment rock 2 gets dropped, what is the distance between the two rocks?

(b) On a single set of axes, sketch the velocity vs. time of rock 1 with a solid line or curve, and the velocity vs. time of rock 2 with a dashed line or curve. On your graphs, let downward velocities count as "positive." Do these graphs ever cross?

(c) *Without* writing any formulas, figure out whether the distance between the two rocks increases, decreases, or stays the same as they fall. After the rocks have been falling for a few seconds, is the distance between them greater than or less than (or equal to) the answer you found in part (a)? Do not get a number; I just want a qualitative answer, and your reasoning behind it. Hint: Look at your part (b) graph, but be careful not to misinterpret it.

(d) Rock 1 hits the ground at exactly 11:00 a.m. At what time will rock 2 hit the ground? Again, answer without plugging in any formulas.

ANSWER 3-10

(a) When rock 2 gets dropped, rock 1 has been falling for exactly one second. So, this problem is really asking how far a rock falls in one second. Since we know the rock was dropped from rest ($v_0 = 0$) and accelerates downward at $a = g = 9.8\ \text{m/s}^2$, we can immediately use

$$\begin{aligned} x &= x_0 + v_0 t + \frac{1}{2}at^2 \\ &= 0 + 0 + \frac{1}{2}(9.8\ \text{m/s}^2)(1.0\ \text{s})^2 \\ &= 4.9\ \text{m}. \end{aligned}$$

Notice that I picked downward as my positive direction; that is why I used $a = +9.8\ \text{m/s}^2$ instead of $a = -9.8\ \text{m/s}^2$.

(b) Both rocks accelerate downward at the same constant rate. In other words, both rocks steadily speed up at the same rate. So, both rocks' graphs are upward-sloped straight lines with the same slope. Remember, acceleration is the rate of change of velocity, i.e., the slope of the velocity vs. time graph. Since the rocks have the same acceleration, their v vs. t graphs automatically have the same slope.

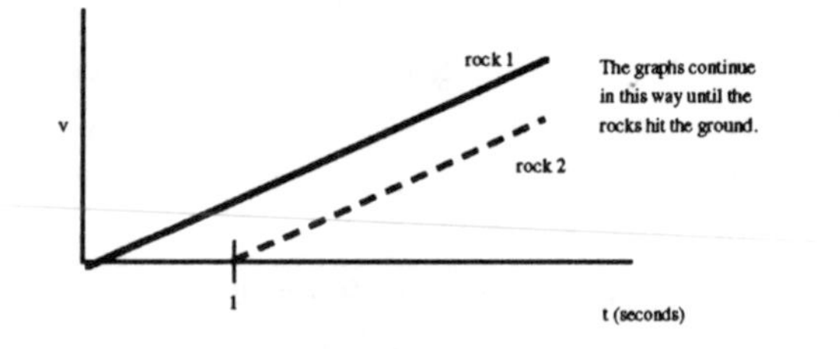

So, the dashed and solid graphs differ in only one way: The dashed (rock 2) graph "starts a second later" than the solid (rock 1) graph starts. In other words, at time $t = 1$ second after rock 1 gets dropped, rock 2 still has no velocity. In mathematics lingo, the rock 2 graph is "shifted along the time axis" compared to the rock 1 graph.

Notice that, since the graphs have the same slope, they are parallel. Therefore, they never cross (until rock 1 lands, at least). So, rock 1's velocity graph is always "above" rock 2's velocity graph. Physically, this means that rock 1 *always* travels faster than rock 2, until it lands.

(c) It would be easy to look at the graph and say "Oh, the distance between the rocks is always the same." But this graph shows velocity vs. time, not distance vs time. The graph *really* tells us that rock 1's velocity is always greater than rock 2's velocity by the same amount.

Does this mean the rocks stay the same distance apart? No! Since rock 1 is always moving faster than rock 2, it is always "gaining" on rock 2. So, rock 1 gets farther and farther ahead of rock 2. The distance between the rocks keeps increasing.

If this does not make intuitive sense, imagine two joggers. Jogger 1 starts ahead of jogger 2. And furthermore, jogger 1 is faster. Therefore, jogger 1 gets more and more ahead of jogger 2. If you were jogger 2, you would see jogger 1 "increasing her lead" on you. The same thing happens with the falling rocks.

We can confirm this conclusion by sketching x vs. t graphs for the rocks. Since both rocks speed up, the slopes of their x vs. t graphs increase. This yields an upward curve. (Remember, I picked downward as the positive direction. So, these upward curves indicate downward-moving rocks.) Crucially, as we saw above in the v vs. t graphs, both rocks are described by the *same* curve, except that rock 2 starts a second later.

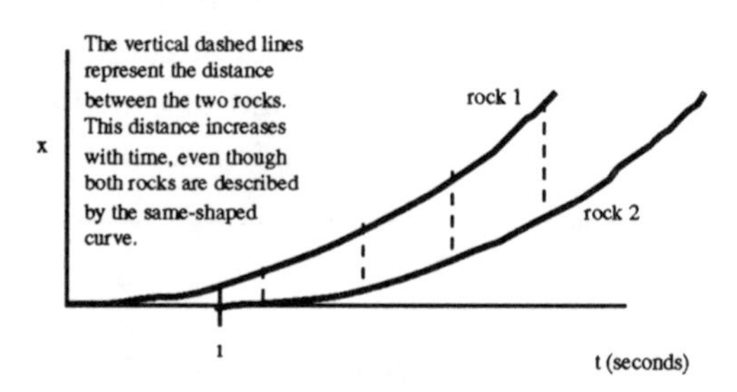

(d) You can avoid turning this into a big ugly formula plug-fest by thinking intuitively. As the above graphs show, rock 2's motion is exactly like rock 1's motion, except it is a second later. Since both rocks get released from rest from the same height, they both fall for the same amount of time before reaching the ground. So for instance, if rock 1 takes five seconds to reach the ground, then so does rock 2. Therefore, since rock 2 was released a second later than rock 1, it also lands a second later than rock 1.

You might have mistrusted this answer on the following grounds: As shown in part (c), rock 1's "lead" over rock 2 keeps increasing. So, when rock 1 lands, it is *way* ahead of rock 2. In other words, when rock 1 lands, rock 2 still has a long distance to travel before landing–so long that rock 2 might take more than one second to cover that distance.

This argument contains an element of truth. When rock 1 lands, rock 2 indeed has a long distance to travel before landing. But remember: throughout the fall, both rocks speed up. By the time rock 1 lands, both rocks are moving quite fast. When rock 1 lands, even though rock 2 still has a long distance to travel before landing, it covers that distance quickly. In fact, rock 2 covers that distance in exactly 1 second. It lands one second after rock 1 landed.

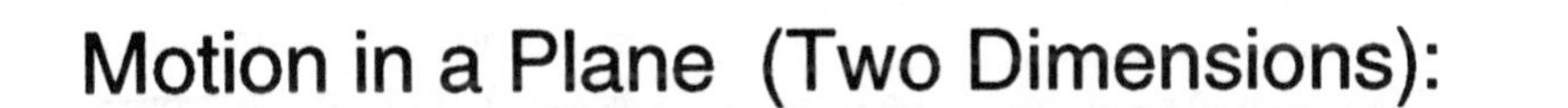

Motion in a Plane (Two Dimensions): Projectiles

4 CHAPTER

QUESTION 4-1

(This problem reviews vectors.)

Phil and Cindy are avid hikers. At 2 p.m., they both start from the same place, but walk along different paths. Specifically, Phil walks at a speed of 3 miles per hour for 2 hours, in the northeast direction (i.e., at a 45° angle east of north), and then he stops to prepare a campsite, forgetting all about Cindy. While Phil was walking, Cindy hiked at 3.5 miles per hour for 2 hours, due west. At 4 p.m., Cindy realizes that Phil has flaked out, and that it is going to be dark in 3 hours. Guessing the location of Phil's campsite (based on a previous discussion with him), she decides to walk there. For safety reasons, she must reach his campsite by 7 p.m. But Cindy does not want to walk any faster than needed to reach Phil's campsite in time.

At what speed, and in which direction, should Cindy walk to reach Phil's campsite?

ANSWER 4-1

Notation note: Boldface symbols represent vectors, while plain letters denote regular numbers. For instance, $\mathbf{v}_0$ is the magnitude and direction of initial velocity, while v_0 is the initial speed (without any direction information).

As always in physics problems, and especially in ones involving multiple directions, sketch what is going on. Phil walks a certain distance northeast, while Cindy walks a certain distance west. Since their accelerations equal 0, the kinematic formula $x = x_0 + v_0 t + \frac{1}{2}at^2$ reduces to $x = x_0 + v_0 t$, which can be rewritten as the constant-velocity formula $\Delta x = v\Delta t$. Phil covers

$$\Delta x_P = v_P \Delta t = (3 \text{ miles/hour})(2 \text{ hours}) = 6 \text{ miles},$$

while Cindy covers

$$\Delta x_C = v_C \Delta t = (3.5 \text{ miles/hour})(2 \text{ hours}) = 7 \text{ miles}.$$

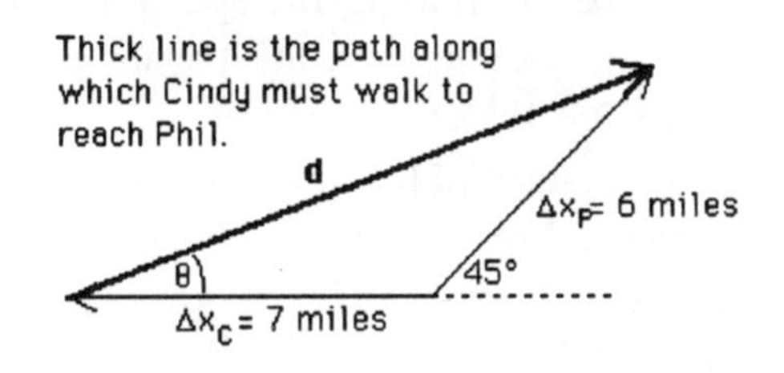

Here is a vector diagram showing what has happened so far. To reach Phil's campsite most efficiently, Cindy must travel along the thick line, which I will call **d**. Using trigonometry, we can figure out the magnitude and direction of $\mathbf{d} = d_x\hat{\mathbf{j}} + d_y\hat{\mathbf{j}}$, where $\hat{\mathbf{i}}$ and $\hat{\mathbf{j}}$ are unit vectors in the eastward and northward directions. Once we do so, we can calculate how fast Cindy must walk to cover distance d before nightfall.

To solve for **d**, we can work directly with the vectors drawn above, *or* we can work with vector components. I will now find **d** using both methods.

Method 1: Vector components.

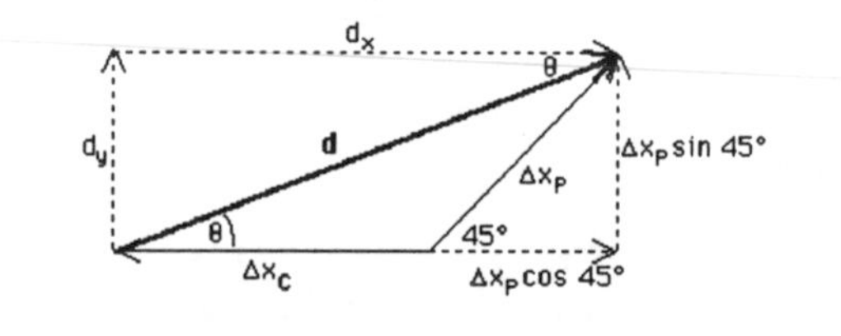

In this method, we break up all relevant vectors into their components. I have let the x-direction be east, and the y-direction be north. Notice that Cindy's displacement vector $\Delta \mathbf{x}_C$ need not be broken up, because it points entirely in the negative x-direction, with no y "contribution."

Our "components" strategy is to find d_x and d_y separately. Then, we can add those components vectorially, using Pythagoras' theorem, to get the **d** vector.

From the diagram, you can see that

$$\begin{aligned} d_x &= \Delta x_C + \Delta x_P \cos 45° \\ &= 7 \text{ miles} + (6 \text{ miles}) \cos 45° \\ &= 11.2 \text{ miles.} \end{aligned}$$

You can also see that the y distance separating Phil from Cindy equals the y-component of Phil's walk, since Cindy hiked solely in the x-direction:

$$\begin{aligned} d_y &= \Delta x_P \sin 45° \\ &= (6 \text{ miles}) \sin 45° \\ &= 4.2 \text{ miles.} \end{aligned}$$

Now that we know d_x and d_y, the components of **d**, we can use Pythagoras' theorem to calculate the length of **d**:

$$d = \sqrt{d_x^2 + d_y^2} = \sqrt{(11.2 \text{ miles})^2 + (4.2 \text{ miles})^2} = 12.0 \text{ miles.}$$

And we can calculate the angle θ in the above diagram using trigonometry. For instance, $\tan\theta = \frac{d_y}{d_x}$. Take the inverse tangent (i.e., the arctangent) of both sides to get

$$\theta = \tan^{-1}\left(\frac{d_y}{d_x}\right) = \tan^{-1}\left(\frac{4.2 \text{ miles}}{11.2 \text{ miles}}\right) = 20.6°.$$

Before finding the speed at which Cindy must walk to cover distance $d = 12$ miles in time $t = 3$ hours, let me solve for **d** again, using a different method of vector analysis.

Method 2: Whole vectors.

Instead of breaking vectors up into components, we will work directly with the "whole vectors."

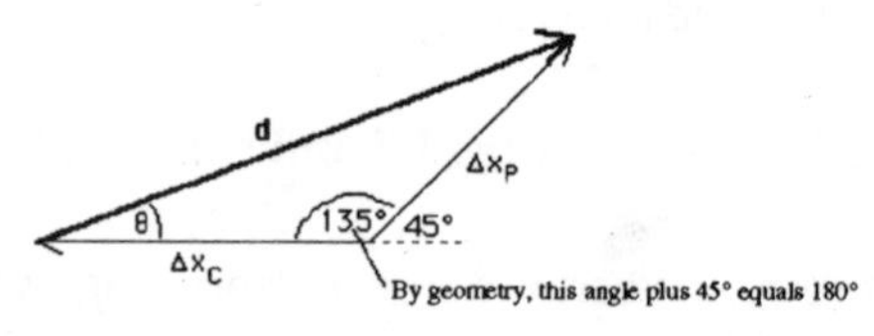

The trigonometric "tools" at our disposal include the law of sines and the law of cosines. According to the law of cosines, d is given by

$$d^2 = (\Delta x_P)^2 + (\Delta x_C)^2 - 2(\Delta x_P)(\Delta x_C)\cos(135°).$$

Substituting in $\Delta x_P = 6$and $\Delta x_C = 7$ gives us

$$d = \sqrt{(6 \text{ miles})^2 + (7 \text{ miles})^2 - 2(6 \text{ miles})(7 \text{ miles})\cos 135°}$$
$$= 12 \text{ miles},$$

the same answer obtained above using vector components. To find θ, use the law of sines: $\frac{\sin 135°}{d} = \frac{\sin \theta}{\Delta x_P}$. Solving for θ eventually gives $20.6°$, the same answer obtained above using components.

So, Cindy must walk about 12 miles, at a direction $20.7°$ north of east. She has three hours to complete the hike. Therefore, her average speed must be

$$\overline{v} = \frac{\Delta x}{\Delta t} = \frac{12 \text{ miles}}{3 \text{ hours}} = 4 \text{ miles per hour.}$$

In this problem, using vector components was no easier than using whole vectors. **In general, however, vector components are easier.**

QUESTION 4-2

A powerful cannon shoots cannon balls at 100 m/s. When I align the cannon at an angle θ to the ground and fire the cannon, the ball takes 1.0 second to reach a height of 50 meters above the nozzle of the cannon. To simplify calculations, let us say the acceleration due to gravity is 10 m/s^2.

(a) What is θ, in degrees?

(b) One second after it is fired, how far away from the cannon nozzle is the ball?

(c) How fast is the cannon ball moving one second after it is fired?

ANSWER 4-2

As always, you should sketch what is going on. The colored-in circle represents the ball at $t = 1$ s. The other circle represents the ball just as it leaves the cannon nozzle. Following the usual convention, I have let the x-direction and y-direction be horizontal and vertical, respectively. Let us say the the ball is fired from $x_0 = y_0 = 0$. At $t = 1$ s, the ball reaches height $y = 50$ meters. (We are not told the horizontal position, x, at $t = 1$ s.) We need to find the cannon's firing angle θ, i.e., the angle of the initial velocity v_0.

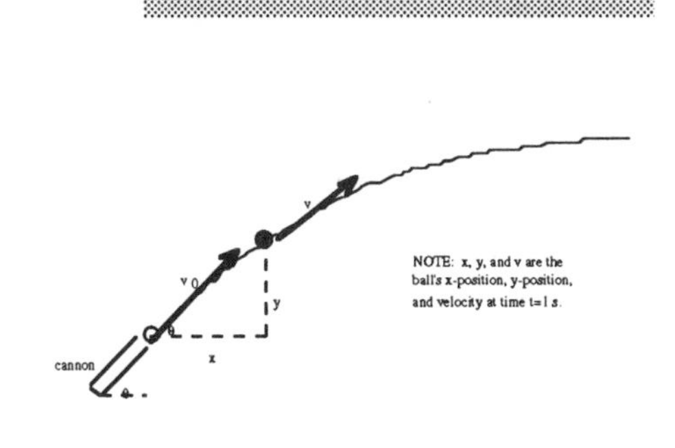

(a) After it is fired, the ball undergoes constant acceleration (due to gravity). Therefore, we can use our constant-acceleration kinematic formulas, *provided* we remember that positions, velocities, and accelerations are *vectors*. In other words, we have to worry about directions.

A useful formula here, for instance, is $\mathbf{r} = \mathbf{r}_0 + \mathbf{v}_0 t + \frac{1}{2}\mathbf{a}t^2$. In projectile motion problems, it is difficult–and sometimes confusing–to use this equation directly. Instead, you should break it up into its horizontal and vertical components:

$$x = x_0 + v_{0x}t + \frac{1}{2}a_x t^2. \tag{1a}$$

$$y = y_0 + v_{0y}t + \frac{1}{2}a_y t^2. \tag{1b}$$

Let us review what the symbols in formulas (1a) and (1b) mean. x is the ball's horizontal position, while y is its vertical position (height). Similarly, a_y is the ball's acceleration in the vertical direction, i.e., the rate at which the ball's *vertical* velocity speeds up or slows down. Here, gravity accelerates the ball *downward* at a rate of $g = 10$ m/s per second. So, if we call the upward direction positive, $a_y = -g = -10\ \text{m/s}^2$. By contrast, a_x is the ball's horizontal acceleration, the rate at which its sideways motion speeds up or slows down. Since gravity does not push or pull the ball sideways, and since we can neglect air resistance, the ball's horizontal motion neither increases nor decreases. In other words, the ball's horizontal velocity is constant. So $a_x = 0$. *This is always true in standard projectile motion problem.* The computer simulation helps to demonstrate this fact. Notice in particular that $a_x = 0$ does *not* mean the ball's horizontal motion dies away as the ball rises or falls. The ball "keeps" all of its sideways velocity until it lands. Put another way, when the ball gains (or loses) vertical velocity, it does not do so "at the expense" of sideways velocity. **The sideways and vertical motion are independent. That's why you can use separate equations, such as (1a) and (1b), to deal with the horizontal and vertical aspects of the ball's motion.**

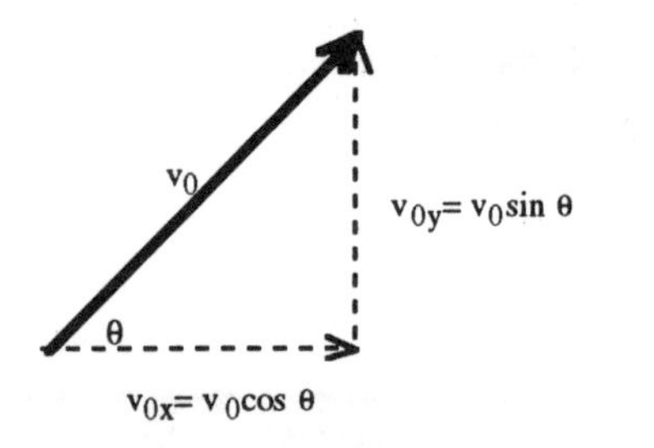

Similarly, v_{0x} and v_{0y} are the horizontal and vertical "parts" of the ball's initial velocity. Since the ball was fired at $v_0 = 100$ m/s at angle θ, the ball's total initial velocity is a vector of length 100, oriented at angle θ.

I have broken up the initial velocity into components. Remember, we are solving for θ.

Now that we have pictured what is going on, let us find θ by considering the horizontal motion, the vertical motion, or both. Well, the horizontal displacement Eq. (1a) will not help, because we do not know x at $t = 1$ s. On the other hand, we *do* know that at $t = 1$ s, $y = 50$ m. So, let us try Eq. (1b):

$$y = y_0 + v_{0y}t + \frac{1}{2}a_y t^2.$$

$$y = y_0 + (v_0 \sin\theta) + \frac{1}{2}(-g)t^2.$$

The only unknown is $\sin\theta$. Solve for it to get

$$\sin\theta = \frac{y - y_0 + \frac{1}{2}gt^2}{v_0 t} = \frac{50\ \text{m} - 0 + \frac{1}{2}(10\ \text{m/s}^2)(1\ \text{s})^2}{(100\ \text{m/s})(1\ \text{s})} = 0.55,$$

and hence $\theta = \sin^{-1}(.55) = 33°$.

(b) Let d denote the ball's total distance from the cannon at $t = 1$ s, and look at the diagram from two pages ago. You can see that

$$d^2 = x^2 + y^2. \tag{2}$$

The problem tells us that $y = 50$ m. So, to solve Eq. (2) for d, we need to know x.

You might think that we can solve for x using $\tan\theta = \frac{y}{x}$, since we know $y = 50$ m and we just solved for θ in part (a). To see why this does not work, you must carefully picture the ball's motion between $t = 0$ and $t = 1$ s. I will "magnify" part of my previous diagram.

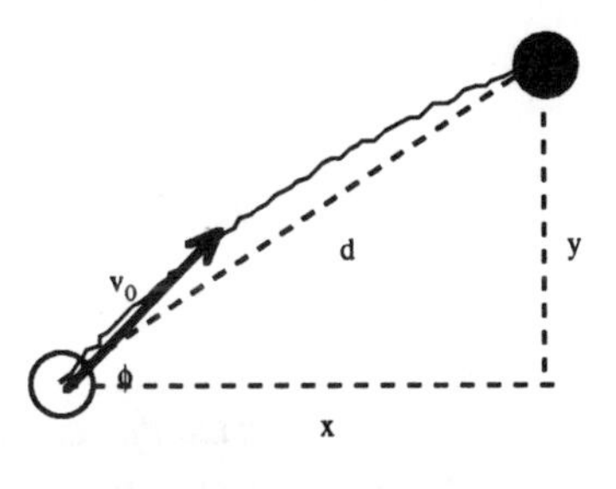

I have drawn in **d**, the straight line connecting the ball's initial position to its position at $t = 1$ s. That straight line makes angle ϕ with the horizontal. It is true that $\tan\phi = \frac{y}{x}$. *But ϕ is not the angle we found in part (a).* In part (a), we found the angle θ that $\mathbf{v}_0$ makes with the horizontal.

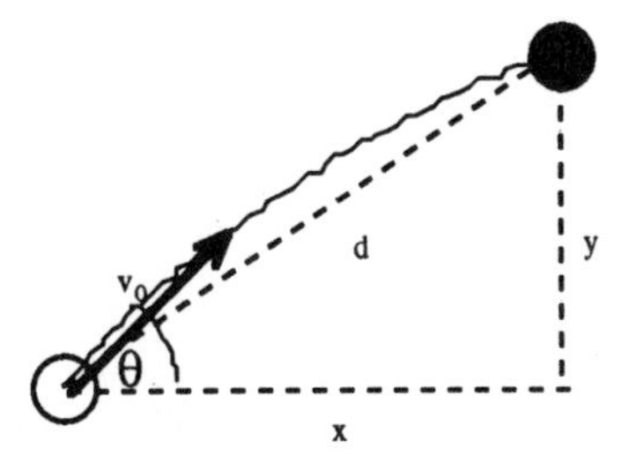

I cannot emphasize enough that θ is bigger than ϕ, because the ball does not travel along the straight line **d**. Instead, it arcs. So, $\tan\theta$ does not equal $\frac{y}{x}$. We must find a different way to calculate x.

Let us try Eq. (1a). As discussed above, $a_x = 0$; the ball's horizontal motion is steady. And from the vector diagram of the initial velocity, $v_{0x} = v_0 \cos\theta$. Substituting all that into Eq. (1a) gives

$$\begin{aligned} x &= x_0 + v_{0x}t + \frac{1}{2}a_x t^2. \\ &= 0 + (100 \text{ m/s})(\cos 33°)(1 \text{ s}) + 0 \\ &= 83 \text{ m}. \end{aligned}$$

In summary, after 1 second, the ball has traveled $y = 50$ meters up and $x = 83$ meters sideways. Therefore, from Pythagoras' theorem, the ball's total distance from the cannon at $t = 1$ s is

$$d = \sqrt{x^2 + y^2} = \sqrt{(83.5 \text{ m})^2 + (50 \text{ m})^2} = 97.3 \text{ m}.$$

(c) You might be tempted to use the formula $\mathbf{v} = \mathbf{v}_0 + \mathbf{a}t$ directly, by plugging in $v_0 = 100$ m/s and $a = -10$ m/s^2. But you cannot, because the formula contains vectors. If all the vectors "point" forwards and backwards along the *same* direction, then you can use such formulas directly. But $\mathbf{v}_0$ points at an angle, while the gravitational acceleration **a** points straight down. So, you cannot just "add" $\mathbf{v}_0$ to $\mathbf{a}t$. That would be like saying a person who walks three meters east and then four meters north ends up $3 + 4 = 7$ from where she started.

When a formula contains vectors that are not "lined up," you can break that formula into vector components. In this case, $\mathbf{v} = \mathbf{v}_0 + \mathbf{a}t$ factors into

$$\begin{aligned} v_x &= v_{0x} + a_x t, \\ y_y &= v_{0y} + a_y t. \end{aligned}$$

Using these formulas, we can *separately* solve for v_x and v_y, the ball's sideways and upward speed at $t = 1$ s. Then we can obtain the total speed by *vectorially* adding those components.

Let us start with v_x. As discussed in parts (a) and (b), $a_x = 0$. Therefore,

$$\begin{aligned} v_x &= v_{0x} + a_x t \\ &= v_0 \cos\theta + 0 \\ &= (100 \text{ m/s})\cos 33^\circ \\ &= 84 \text{ m/s}, \end{aligned}$$

the same horizontal speed it started with. (The ball "keeps" all its horizontal motion.) We can also solve for for v_y:

$$\begin{aligned} v_y &= v_{0y} + a_y t \\ &= v_0 \sin\theta - gt \\ &= (100 \text{ m/s})\sin 33^\circ - (10 \text{ m/s}^2)(1 \text{ s}) \\ &= 44 \text{ m/s}. \end{aligned}$$

So, the total velocity at $t = 1$ s is the following vector.

As you can confirm by looking at my first diagram in this problem, the angle of this velocity is less than the angle of $\mathbf{v}_0$, because the ball arcs down. The magnitude of this velocity, i.e., the ball's speed, is

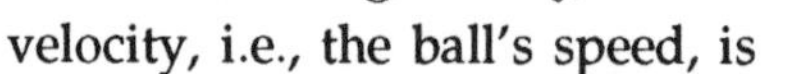

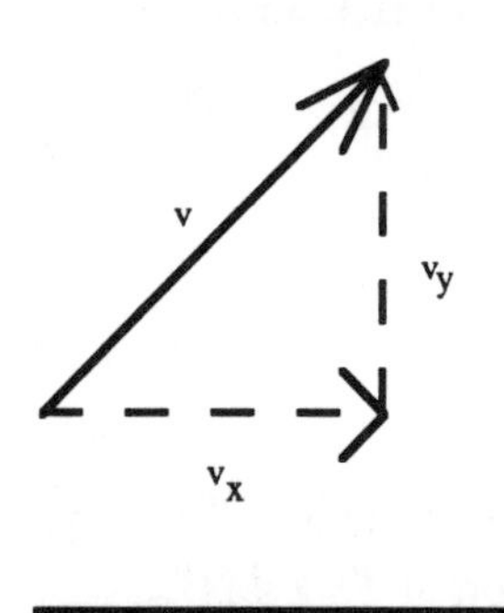

$$v = \sqrt{v_x^2 + v_y^2} = \sqrt{(84 \text{ m/s})^2 + (44 \text{ m/s})^2} = 95 \text{ m/s}.$$

The ball has slowed down from its initial speed of 100 m/s. That is because, while the ball rises, gravity slows down the upward part of its motion.

QUESTION 4-3

A cannon fires a cannon ball from ground level with speed v_0. The cannon makes an angle θ with the ground. The acceleration due to gravity is g.

(a) Neglecting air resistance, how far from the cannon does the ball land? Answer in terms of v_0, θ, and g. Do not plug in numbers. In this problem, you are essentially *deriving* a useful formula about projectile motion. Do not look up the "range" formula in the book. Instead, derive that formula.

(b) Give an intuitive explanation of why the cannon ball travels the farthest when fired at 45°. Do not mathematically derive that result; just explain in words why it makes sense. Experimenting with the computer simulation might help.

ANSWER 4-3

(a) As always, sketch what is happening.

As usual in projectile motion problems (and in all two-dimensional motion problems), it is useful to "divide up" the motion into its horizontal and vertical components. We are trying to find x, the ball's horizontal distance covered in flight. (Let $x_0 = 0$.) The ball experiences no horizontally-directed pushes once it is in mid-air. So, the x-component of its velocity, v_x, is constant. We can easily write a formula for x as a function of time. Therefore, to complete the problem, we need to calculate how much time the ball spends in the air. To figure this out, we can consider the vertical aspect of the ball's motion, because we know the ball lands at the same vertical height from which it was fired. I will explain this more fully later on. The point is, to solve this problem, **we need to separately consider both the x- and the y-components of the ball's motion. That is true in most projectile motion problems.**

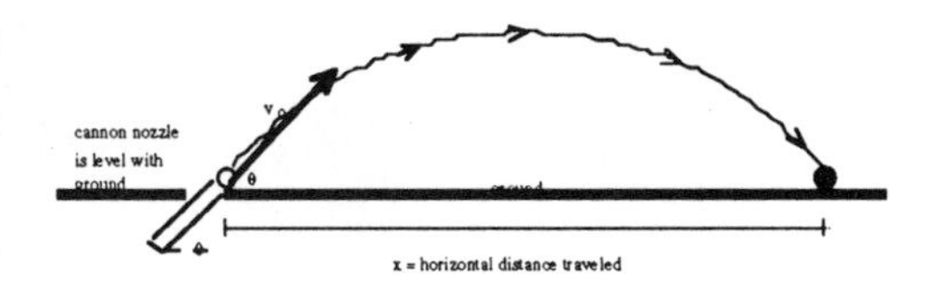

Taking the horizontal and vertical components of $\mathbf{r} = \mathbf{r}_0 + \mathbf{v}_0 t + \frac{1}{2}\mathbf{a}t^2$ yields

$$x = x_0 v_{0x} t + \frac{1}{2} a_x t^2. \tag{1a}$$

$$y = y_0 v_{0y} t + \frac{1}{2} a_y t^2. \tag{1b}$$

In this problem, t is the time spent by the ball in the air before it lands. Because the ball experiences no sideways-directed pushes or pulls, $a_x = 0$, and hence Eq. (1a) reduces to $x = x_0 + v_{0x} t$, which is just another way of writing the constant-velocity formula $v_x = \frac{\Delta x}{\Delta t}$.

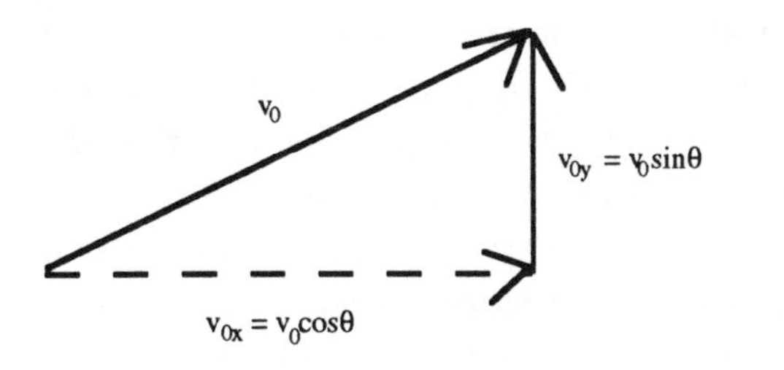

To solve this equation for x, we must find t and v_{0x}. Can we do so? Well, to get v_{0x}, break up the initial velocity into components.

To find t, consider the ball's vertical motion. If I pick upward as positive, the vertical acceleration due to gravity is $a_y = -g$. The ball starts off at ground level, $y_0 = 0$. So, to solve Eq. (1b) for t, we just need y, the ball's final height. But the final height is $y = 0$, because the ball ends up at ground level, the same height from which it started. Substituting all this information into Eq. (1b) gives

$$y = y_0 + v_{0y} t + \frac{1}{2} a_y t^2.$$

$$0 = 0 + (v_0 \sin\theta) - \frac{1}{2} g t^2.$$

Dividing through by t, and then isolating the remaining factor of t, gives $t = \frac{2v_0 \sin\theta}{g}$. Before continuing, let us check this formula against intuition. We expect t to be biggest when θ is near $90°$, and smallest when θ is near $0°$. The above formula confirms this intuition, because $\sin 90° = 1$ and $\sin 0° = 0$. Furthermore, it makes sense that the flight time is proportional to v_0, the firing speed.

Substitute this t into Eq. (1a) to get

$$x = x_0 + v_{0x}t + \frac{1}{2}a_x t^2.$$
$$= 0 + (v_0 \cos\theta)\left(\frac{2v_0 \sin\theta}{g}\right) + 0$$
$$= \frac{2v_0^2 \sin\theta\cos\theta}{g}$$
$$= \frac{v_0^2 \sin 2\theta}{g}.$$

In the last step of algebra, I invoked the trig identity $\sin 2\theta = 2\sin\theta\cos\theta$.

To arrive at this formula, we used the fact that $y = y_0$. So, ***this "range" formula applies only when the ball lands at the same height from which it was fired.***

Some people solve this problem by calculating Δx from the firing point to the peak of the trajectory, and then multiplying by two. (Notice the symmetry of the trajectory; the ball covers the same distance "going up" as it does "coming down.") This method works fine. But remember, if the ball lands at a different height from which it was fired, the trajectory will not be symmetrical. In those cases, it is quicker to treat the "whole trajectory" at once instead of dividing it into pieces.

(b) Later in this chapter, I will review how to solve maximization/minimization problems. Here, let us focus on intuitions. Why is 45° the "best" angle to fire the ball? By playing around with the simulation, or by throwing things, you can see the following: If the firing angle is too low, then the ball lands quickly; it does not spend enough time in the air to cover lots of distance. By contrast, if the firing angle is too high, the ball spends plenty of time in the air. But since only a small percentage of its overall motion is horizontal, the ball does not fly *forward* very far; most of its motion is up and down. To travel forward a long way, the ball must "compromise" between having lots of forward velocity but spending too little time aloft, and spending lots of time aloft but having too little forward motion. The best "compromise" angle turns out to be 45°. *This is true only when the ball lands at the same height from which it was fired.*

QUESTION 4-4

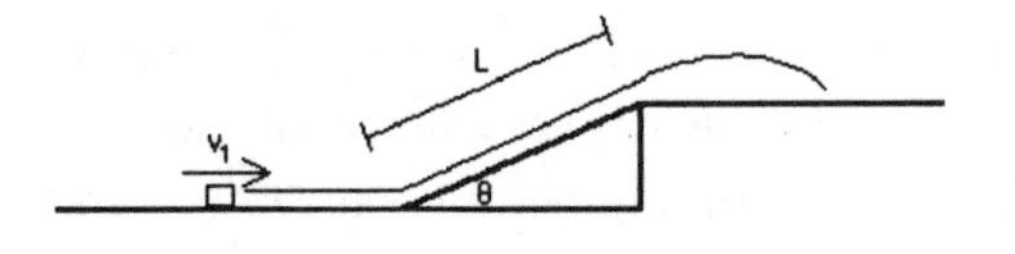

A block slides frictionlessly towards a ramp at speed $v_1 = 15$ m/s. The block slides onto the ramp without "bumping" or otherwise slowing down. The block slides up the ramp and then shoots into the air. Eventually the block lands on a plateau that is level with the top of the ramp.

The ramp is $L = 12.5$ meters long and makes a $\theta = 30°$ angle with the floor. While on the ramp, the block decelerates, and the *magnitude* of the deceleration is $a = 5$ m/s^2.

How far from the top of the ramp does the block land (on the plateau)? Express your answer in terms of v_1, L, θ, and a; plug in numbers only after you obtain the answer in terms of these symbols. (Even when you are not explicitly required to do so, it is best to solve problems symbolically, and plug in numbers at the end. That way, you can more easily catch algebra errors and conceptual mistakes.)

ANSWER 4-4

This is a multi-step problem. We must picture the physical situation carefully in order to formulate a problem-solving strategy.

The block slides up and then shoots off the ramp. How far the block travels in mid-air depends, intuitively, on the angle of the ramp and on the block's speed when it leaves the ramp. But as the block slides up the ramp, it slows down from its initial speed v_1.

Nonetheless, if we know the block's velocity when it flies off the ramp, then we have a standard projectile motion problem. In fact, consider the previous question in this reader: A ball, fired out of a cannon, lands at the same height from which it was fired. In this problem, the block shoots off the ramp, and then lands at the same height from which it was "fired."

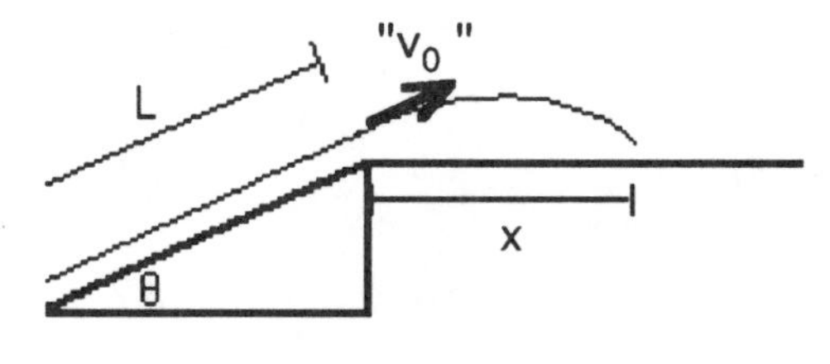

Here is my point: If the block shoots off the ramp at angle θ and with speed v_0, it travels exactly as far (in mid-air) as a cannon ball fired at angle θ and with speed v_0. The block does not care whether it was fired from a cannon or from a ramp. It just knows how fast and in what direction it was "fired." In the previous problem, we figured out how far such a cannon ball goes. In this problem, we can simply rederive–or if we are pressed for time, recycle–that result.

So, the only *new* work we must do in this problem is to find the speed v_0 with which the block exits the ramp. (Do not confuse this ramp-exit speed v_0 with the block's initial speed on the floor, v_1.) In summary: First, we can use kinematic reasoning to figure out how fast the block shoots off the ramp. (We already know the "firing" angle, θ.) Then, we can reuse the projectile-motion reasoning of the previous problem to calculate how much horizontal distance the block covers in the air.

I will now implement this strategy. Given the block's speed v_1 at the bottom of the ramp, how can we calculate its speed v_0 at the top of the ramp? We can use *one-dimensional* kinematics. You might think it necessary to break up the block's motion on the ramp into x- and y-components. But that is unnecessary, for the following reason: While on the ramp, the block's velocity and acceleration vectors all "point" forwards and backwards along the same line. Specifically, the velocity and acceleration are parallel to the ramp's surface.

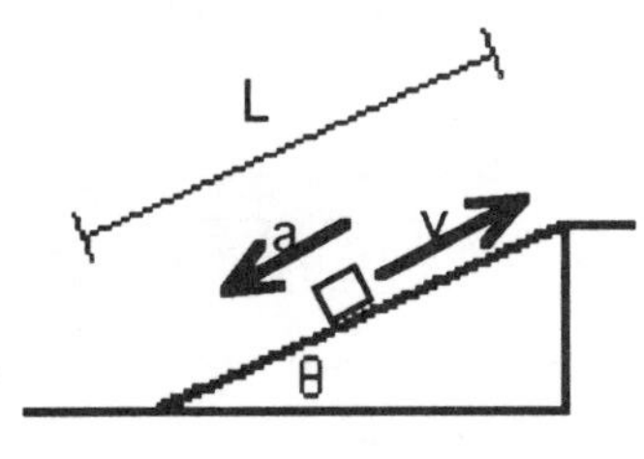

Therefore, you can say that the "x-direction" is along the ramp, and use one-dimensional kinematic formulas. Only when the vectors in your formulas are *not* parallel must you worry about components. (For instance, in projectile problems, the acceleration always points down, but the velocity usually points at some angle. Therefore, you must use vector components.)

In this problem, the acceleration is down-the-ramp; the block slows down. The most efficient kinematic equation to use is $v^2 = v_i^2 + 2a\Delta x$. As just noted, Δx denotes the displacement along the ramp, *not* the horizontal displacement. The block's initial velocity at the bottom of the ramp is $v_i = v_1$. Its "final" velocity, by which I mean the velocity at the top of the ramp, is v_0. (Remember, the block's "final" velocity in this stage of the problem is its initial velocity when it becomes a projectile.) The displacement along the ramp is $\Delta x = L$. And the acceleration is $-a$, since the acceleration is backwards compared to the positive (up-the-ramp) direction, and the given "a" is a positive number. Substituting all this in, and solving for v_0, yields $v_0 = \sqrt{v_1^2 + 2(-a)L}$. By the way, you could also obtain v_0 by using $\Delta x = L = v_1 t - \frac{1}{2}at^2$ to figure out how much time the block spends on the ramp, and then using $v = v_1 - at$ to calculate the speed with which it leaves the ramp.

Now we know the block's "initial" velocity as it becomes a projectile. In Question 4-3 above, we found that a projectile fired at velocity v_0 and angle θ, when allowed to land at the same height from which it was "fired," travels a horizontal distance $x = \frac{v_0^2 \sin 2\theta}{g}$. For practice, you should rederive this formula and check it against my derivation in Answer 4-3. In any case, by substituting in the expression just obtained for v_0, we get

$$\begin{aligned} x &= \frac{(v_1^2 - 2aL)\sin 2\theta}{g} \\ &= \frac{[(15 \text{ m/s})^2 - 2(5 \text{ m/s}^2)(12.5 \text{ m})]\sin 60°}{10 \text{ m/s}^2} \\ &= 8.7 \text{ m.} \end{aligned}$$

This example illustrates two important problem-solving skills. First, you must learn to break up complicated problems into smaller pieces. Here, we divided the problem into a block-sliding-up-a-ramp question and a projectile question. Second, you must recognize when a problem, or part of a problem, is exactly like something you already know (or can easily figure out). Here, you had to realize that the block shot off the ramp behaves exactly like a ball shot out of a cannon.

QUESTION 4-5

A flare is shot horizontally from a flare gun located 60.0 m above the ground. The flare's speed as it leaves the gun is 200 m/s. Neglect wind resistance.

(a) What horizontal distance does the flare travel before striking the ground?

(b) What is its speed immediately before striking the ground?

(c) If the projectile were simply dropped from 60.0 m, instead of fired horizontally from that height, how much time would it take to reach the ground? How does this compare with the time it took to hit the ground in part (a)? Why?

ANSWER 4-5

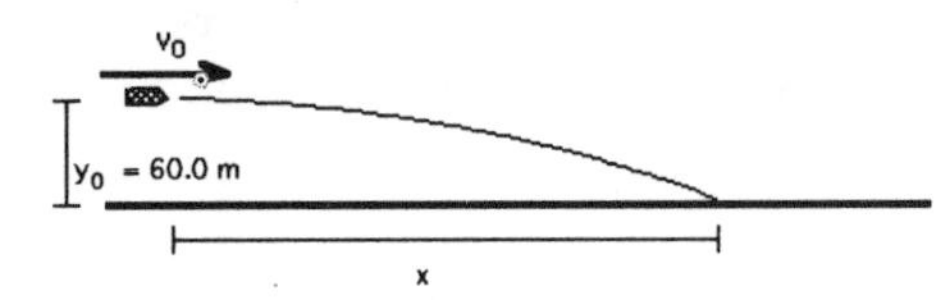

As always, you should start not by plugging in formulas, but by sketching what is going on.

(a) and (c) The horizontal and vertical aspects of a projectile's motion are independent. How much time a projectile takes to "fall" through a given height depends *only* on its *vertical* acceleration and its initial *vertical* velocity. The *horizontal* motion (if any) cannot affect the "fall time." In this problem, since the flare gets fired horizontally, it has no initial vertical velocity: $v_{0y} = 0$. If it were dropped, it would also have $v_{0y} = 0$. So, in both cases, the flare hits the ground at the same time. See the computer simulation.

We can reach this conclusion mathematically by starting with one of our standard vector constant-acceleration formulas, $\mathbf{r} = \mathbf{r}_0 + \mathbf{v}_0 t + \frac{1}{2}\mathbf{a}t^2$. The vertical part of this formula is

$$y = y_0 + v_{oy}t + \frac{1}{2}a_y t^2. \tag{1}$$

Do we have enough information to solve for t? Well, from the diagram, $y_0 = 60.0$ m and $y = 0$. (I have chosen the upward direction as positive.) The acceleration due to gravity is $a_y = -g = -9.80\ \text{m/s}^2$, where the minus sign indicates "downward." And as noted above, the flare initially has only horizontal velocity, no upward or downward velocity: $v_{0y} = 0$. So, we can solve Eq. (1) for t. And our answer for t in no way depends on v_{0x} or on any other aspect of the flare's horizontal motion. Indeed, since $v_{0y} = 0$, the algebra gives

$$t = \sqrt{\frac{2(y - y_0)}{a_y}} = \sqrt{\frac{2(0 - 60.0\ \text{m})}{-9.8\ \text{m/s}^2}} = 3.50\ \text{s}.$$

That is how long the projectile takes to hit the ground, whether or not it has horizontal motion.

Now that we know the time of flight, we can solve for x using the horizontal part of $\mathbf{r} = \mathbf{r}_0 + \mathbf{v}_0 t + \frac{1}{2}\mathbf{a}t^2$, namely $x = x_0 + v_{0x}t + \frac{1}{2}a_x t^2$. We know that $v_{0x} = v_0 = 200$ m/s, since the initial velocity is entirely horizontal. What about a_x? For any projectile in mid-air, $a_x = 0$ if we neglect wind resistance, because nothing makes the projectile's *sideways* motion speed up or slow down. In other words, a projectile's horizontal velocity is constant. With this information, we can solve for x:

$$\begin{aligned} x &= x_0 + v_{0x}t + \frac{1}{2}a_x t^2 \\ &= 0 + (200\ \text{m/s})(3.50\ \text{s}) + 0 \\ &= 700\ \text{m}, \end{aligned}$$

which is slightly under half a mile.

As usual in projectile motion problems, you could solve only by separately considering both the horizontal and the vertical aspects of the motion.

(b) You may be tempted to use $\mathbf{v} = \mathbf{v}_0 + \mathbf{a}t$ directly, since you know the flight time, the acceleration, and the initial velocity. But you *cannot* use that equation directly, because the vectors are not all parallel. The initial velocity $\mathbf{v}_0$ is horizontal, while the acceleration (due to gravity) is downwards.

Since the relevant vectors point in different directions, you must divide them into components. Breaking $\mathbf{v} = \mathbf{v}_0 + \mathbf{a}t$ into its vertical and horizontal components gives us

$$v_y = v_{0y} + a_y t, \tag{2a}$$

$$v_x = v_{0x} + a_x t = v_{0x} \qquad [\text{since } a_x = 0]. \tag{2b}$$

After separately solving for v_x and v_y, we can vectorially add those components to find the projectile's total speed right before it hits the ground.

I will first solve for v_y. From part (a), the flare's initial *vertical* velocity is $v_{0y} = 0$, and its vertical acceleration is due to gravity, $a_y = -g = -9.80\ \text{m/s}^2$. Substituting this in, along with the "travel time" of $t = 3.50$ s, gives

$$\begin{aligned} v_y &= v_{0y} + a_y t \\ &= 0 + (-9.80 \text{ m/s}^2)(3.50 \text{ s}) \\ &= -34.4 \text{ m/s}. \end{aligned}$$

Since I chose upward as positive, the minus sign indicates a downward velocity.

The projectile, however, also lands with a horizontal velocity $v_x = v_{0x} = 200$ m/s. I just used Eq. (2b).

The projectile's total velocity is the vector sum of v_x and v_y.

$v_x = 200$ m/s

$v_y = -34.3$ m/s

The speed, i.e., the magnitude of that velocity, is

$$v = \sqrt{v_x^2 + v_y^2} = \sqrt{(200 \text{ m/s})^2 + (-34.3 \text{ m/s})^2} = 203 \text{ m/s}.$$

Intuitively, the flare keeps all its horizontal speed, but also gains some vertical speed while falling. You might be surprised that the final velocity is only 3 m/s greater than the initial velocity, given that the flare gains 34.4 m/s of vertical speed. But as the above velocity vector diagram shows, the hypotenuse is only slightly longer than the horizontal "side."

QUESTION 4-6

(Extra hard problem. SKIP IT unless your teacher specifically says to solve it.)

A puck on a frictionless frozen pond is sliding northward at speed 20 m/s. At time $t = 0$, an easterly wind starts to blow, and keeps blowing until time $t = 5.0$ seconds. Between $t = 0$ and $t = 5$ s, the wind causes the puck to accelerate. The *eastward* acceleration, in meters per second per second, is given by

$$a = 1 - 0.2t \qquad [\text{between } t = 0 \text{ and } t = 5 \text{ s}]$$

At time $t = 5$ seconds, what is the puck's speed, and in what direction is it going? Specify the direction as an angle east of north.

ANSWER 4-6

In this problem, the acceleration is not constant. It changes with time. Therefore, **you cannot use constant-acceleration kinematic formulas**. Fortunately, old-style graphical reasoning works (as does calculus), if we remember that positions, velocities, and accelerations are vectors.

Let us think things through intuitively. The eastward wind gives the puck some eastward velocity. But the puck also has some initial northward velocity. As we have seen, the different components of an object's motion are independent. The puck's eastward velocity does not come "at the expense" of its northward velocity. Instead, the puck keeps all its northward velocity. Therefore, we expect its final velocity to be partly northward and partly eastward. More precisely, the final velocity $\mathbf{v}$ at $t = 5$ s is the sum of the (northward) initial velocity $\mathbf{v}_0$ and the (eastward) change in velocity $\Delta\mathbf{v}$ due to the wind:

$$\mathbf{v} = \mathbf{v}_0 + \mathbf{v}_{\text{due to wind}} \tag{1}$$

where every quantity in Eq. (1) is a vector. I have picked eastward as the x-direction.

We know $v_0 = 20$ m/s. Therefore, to find **v**, we must calculate $v_x = v_{\text{due to wind}}$, the eastward velocity. If the acceleration were constant, we could just use $v_x = v_{0x} + a_x t$. But the acceleration gradually decreases between $t = 0$ and $t = 5$. How can we figure out v_x? With graphs and/or calculus! Here is the eastward acceleration vs. time, a graph of the equation $a = 1 - 0.2t$.

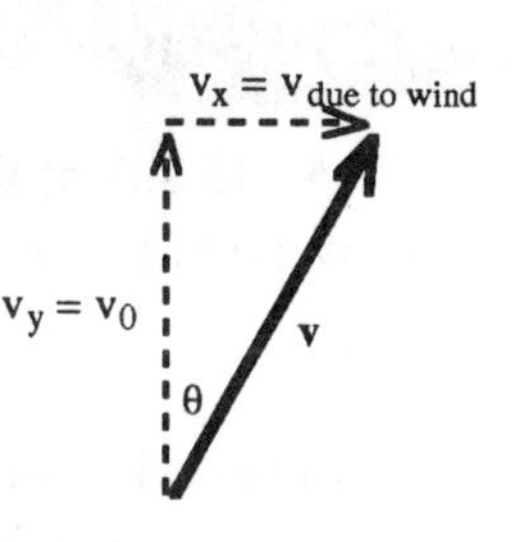

Just as the area under the v_x vs. t curve gives you Δx, the area under an a_x vs. t curve gives you Δv_x, the change in the puck's eastward velocity.

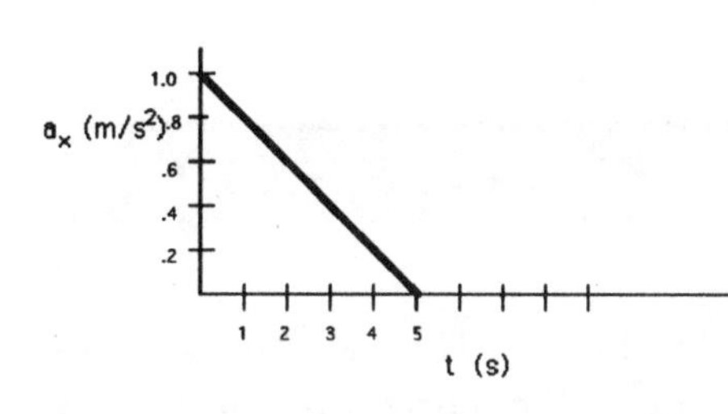

This is precisely what we are trying to find. (Since the puck's initial eastward velocity is $v_{0x} = 0$, we have $\Delta v_x = v_x - v_{0x} = v_x$.)

Calculus confirms this conclusion. By definition, $a_x = \frac{dv_x}{dt}$. So, $dv_x = a_x dt$, and therefore

$$\Delta v_x = \int dv_x = \int a_x dt.$$

The integral $\int a_x dt$ is the area under the a_x vs. t curve.

In this problem, you need no calculus to find that area, because it is just a triangle:

$$\begin{aligned}\Delta v_x &= \text{area under the } a_x \text{ vs. } t \text{ curve}\\ &= \frac{1}{2}(\text{base})(\text{height})\\ &= \frac{1}{2}(5\text{ s})(1.0\text{ m/s}^2)\\ &= 2.5\text{ m/s},\end{aligned}$$

the same answer you would get by evaluating $\int_0^5 (1-.2t)dt$.

So, $v_x = v_{\text{due to wind}} = 2.5$ m/s. Therefore, from Eq. (1) and the accompanying vector diagram, we get

$$\begin{aligned}v &= \sqrt{v_x^2 + v_y^2}\\ &= \sqrt{v_{\text{due to wind}}^2 + v_0^2}\\ &= \sqrt{(2.5\text{ m/s})^2 + (20\text{ m/s})^2}\\ &= 20.2\text{ m/s}.\end{aligned}$$

To find the direction of that velocity, i.e., to find θ in the diagram accompanying Eq. (1), we must use trig. For instance, $\tan\theta = \frac{v_x}{v_y} = \frac{2.5\text{ m/s}}{20\text{ m/s}}$, and hence

$$\theta = \tan^{-1}\left(\frac{v_x}{v_y}\right) = \tan^{-1}\left(\frac{2.5\text{ m/s}}{20\text{ m/s}}\right) = 7.1°.$$

The puck's final direction of motion is $7.1°$ east of north.

QUESTION 4-7

A kid sits at the bottom of a steep hill that makes a 30° angle to the horizontal, and throws rocks up the hill, releasing them from essentially ground level.

(a) Suppose a rock is thrown at angle θ_0 to the horizontal, where $\theta_0 > 30°$. If the rock is thrown at speed v_0, how far does it land from the kid? Express your answer in terms of θ_0, v_0, g, and 30°. (I recommend drawing a picture to get started, as always.)

(b) *(Extra hard problem. Ask your instructor whether you should do it.)*

At what angle θ should the rock be thrown so as to minimize its speed when it lands? It is o.k. if you do not complete all the math. But set things up.

ANSWER 4-7

(a) Start with a picture. The thick line is the hill.

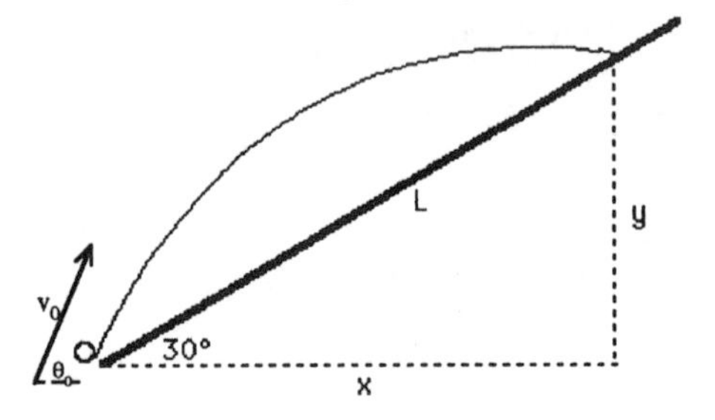

We are trying to solve for L. This is hard, because usually we solve for x or y, the horizontal or vertical position at a given time. To address this problem, you must realize that the horizontal and vertical distance traveled by the rock are

$$x = L\cos 30°$$
$$y = L\sin 30°,$$

as indicated on the diagram. So, we can generate equations containing L by writing our usual equations for x and y.

Set $x_0 = y_0 = 0$. The initial horizontal and vertical velocities are found in the usual way.

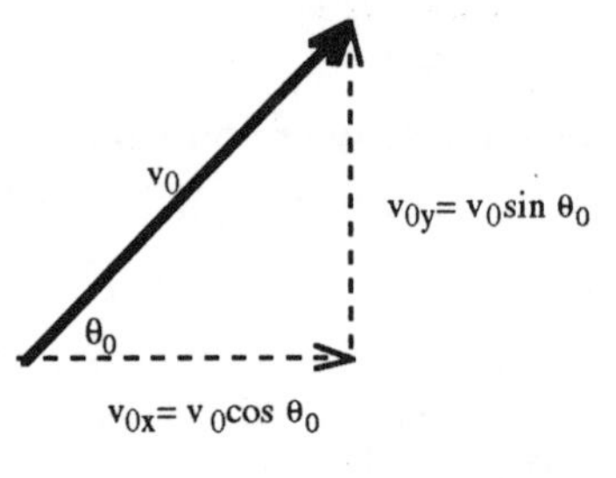

Keep in mind that the "throwing" angle θ_0 *differs from the hill angle of* 30°. As usual in projectile problems, the sideways acceleration is $a_x = 0$, while the vertical acceleration is due to gravity, $a_y = -g$. Putting all this together, we get

$$x = x_0 + v_{0x}t + \frac{1}{2}a_x t^2$$
$$L\cos 30° = 0 + (v_0 \cos\theta_0)t + 0,$$

which simplifies to

$$L\cos 30° = (v_0 \cos\theta_0)t. \tag{1a}$$

This equation contains two unknowns, L and t. We need more information. So, consider the vertical motion:

$$y = y_0 + v_{0y}t + \frac{1}{2}a_y t^2$$
$$L\sin 30° = 0 + (v_0 \sin\theta_0)t - \frac{1}{2}gt^2. \tag{1b}$$

Equations (1a) and (1b) are two equations in two unknowns, L and t. Therefore, we can solve for L. But before doing so, let me discuss some of the intuitions underlying my problem-solving strategy. From the hill diagram, we know that at time t, the rock has traveled a horizontal distance x and a vertical distance y that must be related by $\frac{y}{x} = \tan 30°$. So, we could write equations for $x(t)$ and $y(t)$, and then set $\frac{y}{x} = \tan 30°$. In essence, that is what I have done. To see this, divide Eq. (1b) by Eq. (1a). The result is $\tan 30° = \frac{\text{expression for } y}{\text{expression for } x}$.

I will now complete the algebra. *No more physics thinking will happen for a while. I am just solving two equations in two unknowns.*

Algebra starts here. Divide Eq. (1b) by Eq. (1a), as suggested above. Dividing the left-hand sides gives us $\frac{L\sin 30°}{L\cos 30°} = \tan 30°$. Dividing the right-hand sides yields

$$\frac{(v_0 \sin\theta_0)t - \frac{1}{2}gt^2}{(v_0 \cos\theta_0)t} = \tan\theta_0 - \frac{g}{2v_0\cos\theta_0}t.$$

So, equating the divided left-hand sides to the divided right-hand sides yields

$$\tan 30° = \tan\theta_0 - \frac{g}{2v_0\cos\theta_0}t.$$

Isolate t to get $t = \frac{2v_0\cos\theta_0}{g}(\tan\theta_0 - \tan 30°)$. Now that we know the flight time, we can substitute it into Eq. (1a) or (1b) to find L. I will use Eq. (1a). Solve for L, and substitute in this t, to get

$$\begin{aligned} L &= \frac{1}{\cos 30°}(v_0\cos\theta)t \\ &= \frac{1}{\cos 30°}(v_0\cos\theta_0)\frac{2v_0\cos\theta_0}{g}(\tan\theta_0 - \tan 30°) \\ &= \frac{2v_0^2\cos^2\theta_0}{g\cos 30°}(\tan\theta_0 - \tan 30°). \end{aligned}$$

End of algebra.

Some instructors give lots of partial credit for setting up the equations correctly, even if you make a careless error while completing the algebra. Ask your instructor for her or his policy.

In summary, by considering the horizontal and vertical aspects of the rock's motion, I generated equations that we could solve for L. I had to take advantage of the fact that $x = L\cos 30°$ and $y = L\sin 30°$.

(b) This is a classic minimization problem. The following **maximization/minimization strategy** applies to all such problems, not just to projectiles:

(1) Figure out what variable you are trying to maximize or minimize. Then figure out which variable you "alter" in order to achieve maximization/minimization. This parameter is called the "free variable."

(2) Write the variable you are trying to maximize or minimize as a function of the free variable. The equation may also contain constants, *but it may not contain other unknowns besides the free variable and variable you are maximizing or minimizing.*

(3) Using your equation from step (2), take the derivative of the variable you are maximizing/minimizing *with respect to the free variable.* Set the derivative equal to 0, and solve for the free variable.

To see where this strategy comes from, consider the following graph of \$ vs. t, where \$ is the amount of money in my bank account.

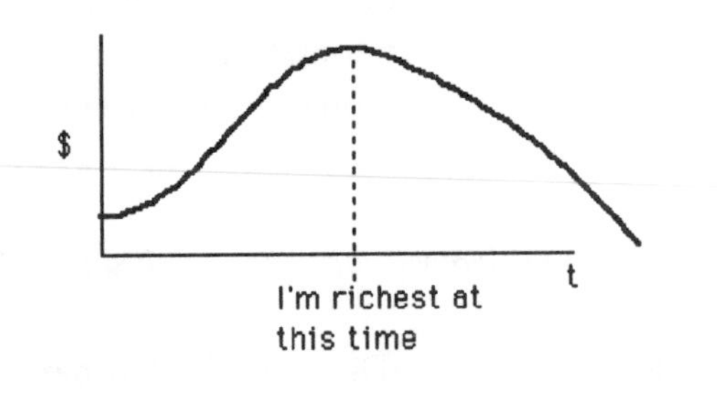

At what time did I have the most money? In other words, when was my money maximized? (The variable we can "adjust" here is t.) From the graph, we see that \$ is maximized when the slope is 0. But the slope of the \$ vs. t graph is the derivative of \$ with respect to t. So, to find the time at which I had the most money, set $\frac{d\$}{dt}=0$, and solve for t. This summarizes the strategy outlined above.

Now let us turn to the problem at hand. It looks like we need to write the rock's landing speed as a function of the throw angle θ. But we can avoid this difficult step, using the following physical insight: For a given initial speed v_0, the farther the rock lands up the hill, the slower it is moving when it lands. In other words, the throw angle that *minimizes* v is also the throw angle that *maximizes* L. Intuitively, the rock loses speed as it rises; so, the farther it has to rise before landing, the more speed it loses in mid-air.

If you do not buy this intuition, I can prove it mathematically, by writing an expression for v in terms of L. You will see that v is smallest when L is biggest. The rock's horizontal velocity is constant, because $a_x=0$. So, its horizontal landing velocity is $v_x=v_{0x}=v_0\cos\theta$. And its vertical landing velocity can be found from $v_y^2=v_{0y}^2+2a_y\Delta y=v_0^2\sin^2\theta-2g(L\sin 30°)$. Pythagorasing together these velocity components gives

$$v=\sqrt{v_x^2+v_y^2}=\sqrt{v_0^2\cos^2\theta+v_0^2\sin^2\theta-2gL\sin 30°}=\sqrt{v_0^2-2gL\sin 30°}.$$

From this equation, we immediately see that v is smallest when L is largest, other things being equal. In summary, I have proven both intuitively and mathematically that, for given initial speed, the rock lands with less speed if it lands farther up the hill. Therefore, maximizing L is the same thing as minimizing v.

So, to solve this problem, we can either minimize v or maximize L. I will choose to maximize L, not because it is an intrinsically "better" way to proceed, but simply because we already found an expression for L in part (a) above.

To get an intuitive feel for which throw angle θ produces the biggest L, experiment with the computer simulation. Notice in particular that 45° is *not* the best angle, because the rock lands too quickly. Nor is 30°+45°= 75° the best angle, because the rock "wastes" too much of its motion going up and down instead of sideways. *45° maximizes the range of a projectile only if it lands at the same height from which it was thrown.*

I will now implement the three-step maximization strategy.

Step 1: Figure out what is being maximized and what is the free variable.

We are trying to maximize L. To do so, we can adjust the "throw" angle θ. So, L is the variable we are maximizing, and θ is the free variable.

Step 2: Write equation for maximized variable in terms of free variable.

We already did this in part (a): $L=\frac{2v_0^2\cos^2\theta}{g\cos 30°}(\tan\theta-\tan 30°)$. But before continuing, we had better double check that the variable we are maximizing (L) and the free variable (θ) are the only unknowns. Yes they are; everything else is given.

Step 3: Differentiate with respect to free variable, and set equal to 0.

The calculus and algebra take a long time. On a test, some (but not all) instructors would give partial credit for writing, "now set $\frac{dL}{d\theta}$ equal to 0, and solve for θ." For the record, let me carry the calculation through:

Math starts here.

$$\begin{aligned}0 &= \frac{dL}{d\theta} \\ &= \frac{2v_0^2}{g\cos 30°}\left[2\cos\theta(-\sin\theta)(\tan\theta - \tan 30°) + \cos^2\theta\left(\frac{1}{\cos^2\theta}\right)\right] \\ &= \frac{2v_0^2}{g\cos 30°}[-2\sin^2\theta + 2\sin\theta\cos\theta\tan 30° + 1].\end{aligned}$$

So far, I have used nothing more than the chain rule. Now for some trig identities. First, $2\sin\theta\cos\theta = \sin 2\theta$. Second, $1 = \cos^2\theta + \sin^2\theta$. So, the equation becomes

$$\begin{aligned}0 &= \frac{2v_0^2}{g\cos 30°}[-2\sin^2\theta + \sin 2\theta\tan 30° + \cos^2\theta + \sin^2\theta] \\ &= \frac{2v_0^2}{g\cos 30°}[-\sin^2\theta + \sin 2\theta\tan 30° + \cos^2\theta] \\ &= \frac{2v_0^2}{g\cos 30°}[\cos 2\theta + \sin 2\theta\tan 30°].\end{aligned}$$

where in the last step I invoked the identity $-\sin^2\theta + \cos^2\theta = \cos 2\theta$. So, we have $\cos 2\theta + \sin 2\theta\tan 30° = 0$. Therefore, $(\tan 30°)\sin 2\theta = -\cos 2\theta$. Divide through by $\cos 2\theta$ and by $\tan 30°$ to get $\tan 2\theta = \frac{1}{\tan 30°}$. Taking the inverse tangent of both sides, and then dividing through by 2, gives $\theta = \frac{1}{2}\tan^{-1}(\frac{1}{\tan 30°}) = 60°$.

End of math.

This is the "best" throwing angle, i.e., the angle that maximizes the distance the rock travels up the hill, and therefore minimizes its landing speed. You can confirm this with the computer simulation. By repeating this derivation with an arbitrary hill angle, you can confirm that the "best" angle is always midway between the hill angle and 90°. So for instance, if the hill angle were 80°, the "best" throw angle would by 85°. And if the hill angle were 0°–if the "hill" were just a flat plain–then the best throw angle would be 45°. We have seen that result before.

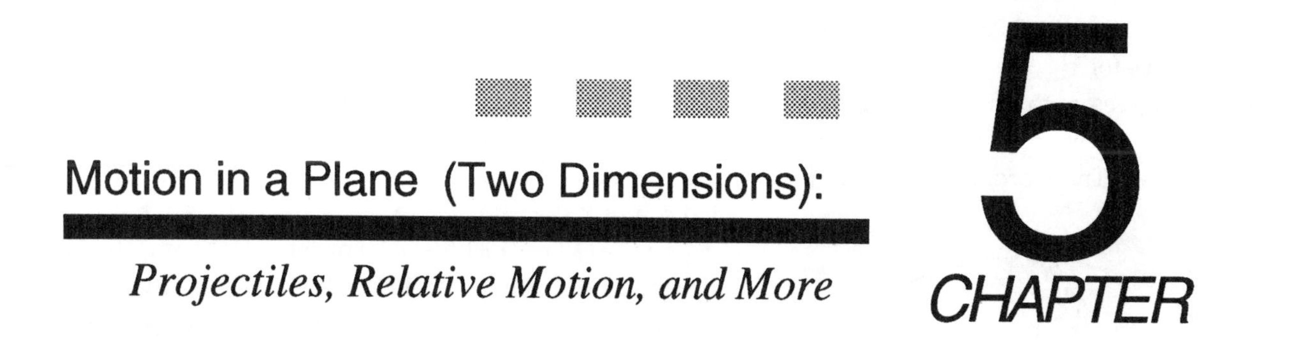

QUESTION 5-1

A child named Kim is sitting in a train, playing with a toy ball. Kim is so absorbed that she does not realize the train is moving forward with speed 20 m/s. Kim thinks the train is still stopped in the station. According to Kim, she is throwing her ball straight up, letting it fall back down, and then catching it at the same place she threw it. Kim notices that exactly 1.5 seconds elapse between her throw and her catch.

The train's windows are big and surprisingly clean. Standing on the ground at the side of the railroad tracks is Joe. As the train speeds by, Joe watches Kim throw and catch her ball.

For simplicity, let us say the acceleration due to gravity is $10\ \mathrm{m/s^2}$.

(a) Suppose Kim is a physics prodigy. According to Kim, how fast does she throw her ball upward?

(b) According to Joe, Kim's ball does not travel straight up and down. Sketch the trajectory of Kim's ball as seen by Joe. According to Joe, what is the distance d between where Kim throws her ball and where she catches it?

(c) According to Joe, how fast was Kim's ball moving when it left her hand?

(d) Who is "right" about the initial velocity of the ball, Kim or Joe?

(e) According to Kim, what peak height does the ball reach above her hand? According to Joe, what peak height does the ball reach above Kim's hand?

ANSWER 5-1

(a) As the computer simulation shows, Kim "sees" the ball go straight up and down. (She also sees everything outside the train moving "backwards" at 20 m/s, but that is irrelevant here.) Therefore, to calculate the ball's initial velocity, she uses a standard one-dimensional constant-acceleration formula,

$$y = y_0 + v_{0y}t + \frac{1}{2}a_y t^2, \tag{1}$$

According to Kim, she needs only this vertical formula, because the ball has no horizontal motion.

Kim wants to find v_{0y}, the speed with which she threw the ball straight up. Let us say $y_0 = 0$. Then, at $t = 1.5$ seconds, $y = y_0 = 0$, because she catches the ball at the same height from which she threw it. If we call the up direction positive, then the acceleration due to gravity is $a_y = -g = -10\ \mathrm{m/s^2}$. Substituting all this into Eq. (1) gives $0 = 0 + v_{0y}(1.5\ \mathrm{s}) + \frac{1}{2}(-10\ \mathrm{m/s^2})(1.5\ \mathrm{s})^2$. Solving for v_{0y} gives us $v_{0y} = 7.5\ \mathrm{m/s}$.

You could also have solved this by realizing that the ball reaches its peak midway through its trip, at time $t = 0.75$ s. At the peak, $v_y = 0$, as I will discuss below. So, you could use $v_y = v_{0y} + a_y t$ to solve for v_{0y}.

(b) Joe agrees with Kim that the ball stays aloft for $t = 1.5$ seconds. But according to Joe, Kim is carried forward at the train's constant speed of 20 m/s. The computer simulation shows this clearly. Therefore, according to Joe, during the 1.5-second flight time of the ball, Kim moves forward a distance $\Delta x = v_x t = (20\text{ m/s})(1.5\text{ s}) = 30$ meters. From Joe's perspective, Kim catches the ball a distance $d = \Delta x = 30$ meters from where she threw it. Joe sees the ball was thrown at a forward angle, not straight up.

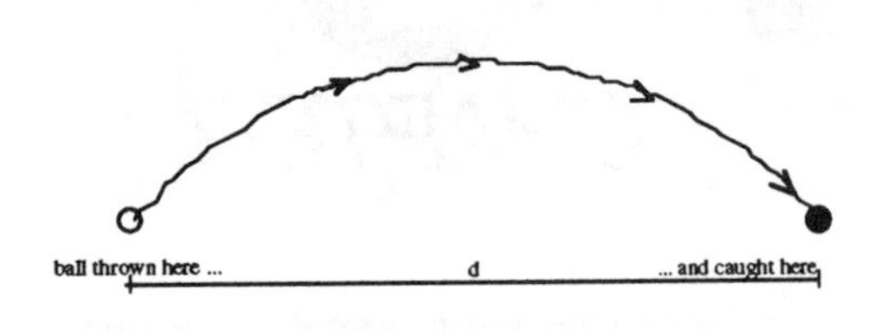

(c) As just noted, according to Joe, Kim threw the ball both upward and forward. So Joe describes the ball's trajectory using the vector formula $\mathbf{r} = \mathbf{r}_0 + \mathbf{v}_0 t + \frac{1}{2}\mathbf{a}t^2$, the vertical part of which is Eq. (1) above:

$$y = y_0 + v_{0y}t + \frac{1}{2}a_y t^2. \tag{1}$$

So, Joe agrees with Kim about all *vertical* aspects of the ball's motion. For instance, he agrees that $y = y_0 = 0$, and he agrees that the vertical acceleration is $a_y = -g = -10\text{ m/s}^2$. Less obviously, he agrees that the *vertical* part of the ball's initial velocity is $v_{0y} = 7.5\text{ m/s}$, as found in part (a). (Physically, he agrees with Kim about how much *upward* motion she gives the ball.) What Kim and Joe *disagree* about are the horizontal aspects of the ball's motion. Kim thinks the ball has no sideways motion. But according to Joe, when Kim throws the ball "up," the ball also acquires forward motion, namely the forward motion of the train. Think of it this way: When Kim releases the ball, her hand–like the rest of her–is moving horizontally at 20 m/s. This horizontal motion gets imparted to the ball. The ball cannot tell whether Kim's hand "willfully" moves sideways at 20 m/s or passively gets carried along at 20 m/s. Either way, the ball receives 20 m/s of horizontal velocity. In summary, Joe says the ball's initial *horizontal* velocity is not 0, but $v_{0x} = 20\text{ m/s}$, the forward velocity of the train. Therefore, says Joe, the ball leaves Kim's hand with a total initial velocity given by the vector sum of $v_{0y} = 7.5\text{ m/s}$ and $v_{0x} = 20\text{ m/s}$.

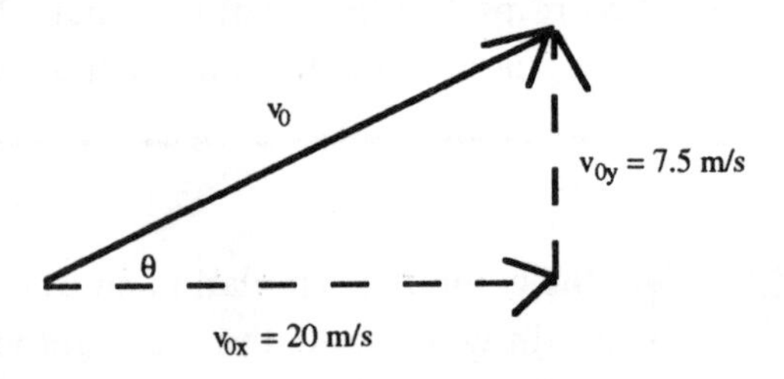

So, Joe "sees" the total initial speed as

$$v_0 = \sqrt{v_{0x}^2 + v_{0y}^2} = \sqrt{(20\text{ m/s})^2 + (7.5\text{ m/s})^2} = 21\text{ m/s}.$$

(d) You might be tempted to say Joe is correct, because he is not moving. On the other hand, you might say Kim is correct, since she is the one throwing the ball.

According to the principle of *relativity*, however, both Kim and Joe are equally "right." Neither frame of reference is special. Therefore, we cannot say Kim "really is" moving and Joe really is standing still. We can say only that Joe is motionless with respect to the Earth, while

Kim is not. (Of course, Kim is motionless with respect to the train, while Joe is not.) The frame of reference "attached" to the Earth is not special, as Martians would quickly point out.

Of course, it is usually convenient to take the Earth as our "fixed" frame of reference. But convenience is different from fundamental *correctness*.

Joe and Kim's reference frames are equally good for the following reason: Both of them can predict and explain the ball's motion using the same set of physical laws. Sure, Joe has to consider horizontal motion; but he and Kim use the same arsenal of constant-acceleration kinematic equations. Similarly, he and Kim could both use Newton's laws, conservation of energy, and all the other physical laws you will learn this semester.

By contrast, if you were riding a merry-go-round and threw a ball straight up, you would not see the ball as moving either straight up and down or along a parabolic path. From your perspective, it would follow some weird curve. (Try it!) The ball's motion would seem to violate the laws of physics, at least until you take your rotation into account. This is because a rotating reference frame is "illegal"; it is not "covered" by the principle of relativity. Although Joe's and Kim's reference frames are both equally good, your rotating frame is not.

(O.K., the Earth is rotating, and therefore our usual frame of reference is, strictly speaking, "illegal." But Earth's rotation rate is slow enough that we can neglect it in most circumstances.)

I am getting too theoretical. If you find this interesting, talk with your instructor or with an advanced physics student. They will have a lot to say.

(e) According to Kim, the ball at its peak is motionless: $v_{y\,\text{peak}} = 0$. Using that fact, she can invoke the constant-acceleration formula

$$v_{y\,\text{peak}}^2 = v_{0y}^2 + 2a_y\Delta y. \quad (2)$$

Kim says $v_{0y} = 7.5 \text{ m/s}$, and $a_y = -g = -10 \text{ m/s}^2$. Solving for Δy gives

$$\Delta y = \frac{v_{y\,\text{peak}}^2 - v_{0y}^2}{2a_y} = \frac{0-(7.5 \text{ m/s})^2}{2(-10 \text{ m/s}^2)} = 2.8 \text{ meters}.$$

This is the ball's peak height above her hand, according to Kim. She could get the same answer by first calculating how much time the ball takes to reach its peak (or realizing by symmetry that it takes half the total flight time), and then using $y = y_0 + v_{0y}t + \frac{1}{2}a_y t^2$.

According to Joe, however, the ball is moving at its peak. But, as you can see from the computer simulation or from throwing your pencil at an angle, **a projectile at its peak has only horizontal motion. Its *vertical* velocity at the peak is $v_y = 0$.**

So, Joe agrees with Kim that $v_{y\,\text{peak}} = 0$. He also agrees with Kim about v_{0y} and a_y–indeed, about all *vertical* aspects of the ball's motion. Therefore, he would use Eq. (2) and get the same answer for Δy that Kim got. This is not surprising. Imagine yourself in Joe's shoes. You would not disagree with a passenger in the train about how high an object is off the train floor. You would argue with them only about horizontal motion issues.

QUESTION 5-2

A medieval army attacks a castle with tall walls, 20 meters high. The army's cannon is entrenched exactly 30 meters from the castle. The Head Knight decides to fire cannon balls *over* the castle wall. Specifically, the Head Knight wants the cannon ball to reach the peak of its trajectory (i.e., its highest point) when it is directly over the wall. (This is not to say that the ball "skims" the top of the wall. The ball clears the wall, and its peak is supposed to be directly over the wall.)

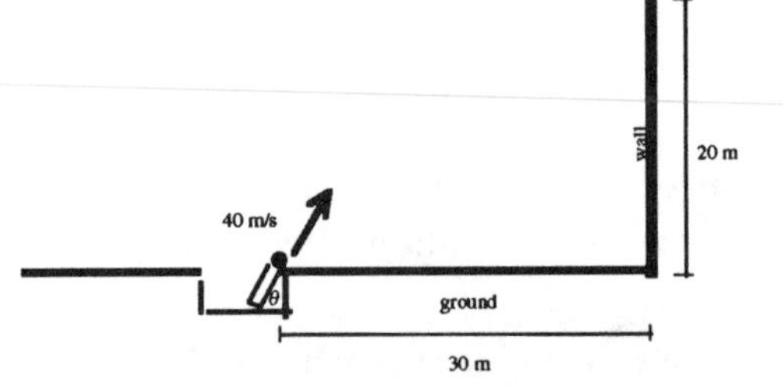

The cannon fires balls at 40 m/s, from ground level. For calculational simplicity, let us say the acceleration due to gravity is 10 m/s^2.

(a) *(Tricky problem!)* At what angle θ to the horizontal should the Head Knight align the cannon?

(b) What peak height does the ball reach when fired at that angle?

(c) If the Head Knight screws up and aligns the cannon at a $\theta = 30°$ angle to the ground, the cannon ball hits the wall. How high off the ground does the ball hit the wall?

(a) When a projectile reaches its peak, its *vertical* velocity is $v_y = 0$. Here is a way of understanding why: Immediately before reaching its peak, the projectile travels sideways and upward. Immediately after reaching its peak, the projectile travels sideways and downward. The peak is where it "turns around" from upward to downward motion. Hence, at the peak, its motion is purely sideways. We want this point to occur when the ball is directly over the castle wall, which is a horizontal distance $x = 30$ m from the cannon. So, in essence, the problem asks you to find the firing angle θ such that the ball has vertical velocity $v_y = 0$ at the moment it has traveled a horizontal distance $x = 30$ m.

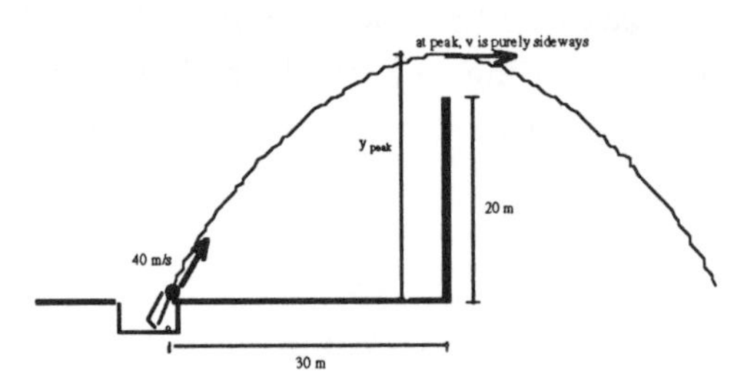

This suggests a strategy. By taking advantage of the fact that is the ball's vertical velocity at the peak is $v_y = 0$, we can write an expression for the time it takes the ball to reach the peak. Then we can substitute that time into a formula for x, and set $x = 30$ meters. With luck, we can solve the resulting equation for the firing angle θ.

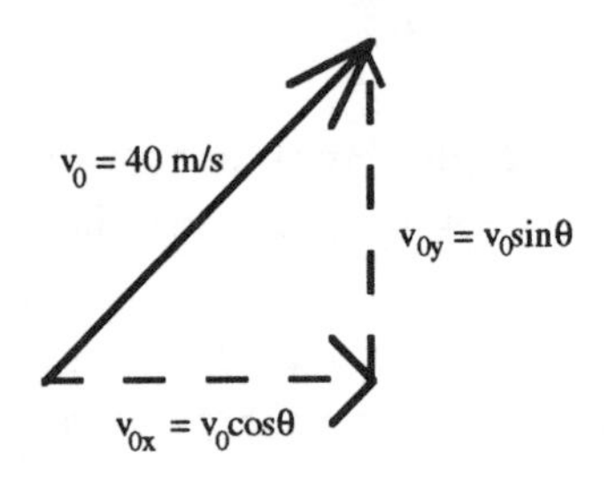

Since we want to relate time to the peak vertical velocity, let's start with the vertical component of the vector constant-acceleration formula, $\mathbf{v} = \mathbf{v}_0 + \mathbf{a}t$. We know $a_y = -g = -10$ m/s^2. With a velocity vector diagram, we can easily find v_{0y}.

Finally, as just noted, when $t = t_{\text{peak}}$, $v_y = 0$. So, $v_y = v_{0y} + a_y t$

$$0 = v_0 \sin\theta - g t_{\text{peak}}. \tag{1}$$

We cannot yet solve for θ, because this equation contains a second unknown, t_{peak}. We need more information. Of course, we can solve for t_{peak} in terms of θ, or vice versa. And we know that when $t = t_{\text{peak}}$, the ball has moved a horizontal distance $x = 30$ m from the cannon. So, start with the horizontal component of $\mathbf{r} = \mathbf{r}_0 + \mathbf{v}_0 t + \frac{1}{2}\mathbf{a}t^2$:

$$\begin{aligned} x &= x_0 + v_{0x}t + \frac{1}{2}a_x t^2 \\ &= 0 + (v_0 \cos\theta)t_{\text{peak}} + 0. \end{aligned} \tag{2}$$

(Remember, $a_x = 0$ for a projectile, because no force speeds up or slows down its horizontal motion.)

We know $x = 30$ m. So, at this point,we have two equations in two unknowns, θ and t_{peak}. We are done with the physics. It is just a matter of algebra to solve Eqs. (1) and (2) for θ.

Algebra starts here. Solving Eq. (1) for t_{peak} yields $t_{\text{peak}} = \frac{(v_0 \sin\theta)}{g}$. Substitute that into Eq. (2), and remember the trig identity $\sin\theta\cos\theta = \frac{1}{2}\sin 2\theta$, to get

$$\begin{aligned} x &= (v_0 \cos\theta)t_{\text{peak}} \\ &= (v_0 \cos\theta)\frac{v_0 \sin\theta}{g} \\ &= \frac{v_0^2 \sin 2\theta}{2g}. \end{aligned}$$

Now isolate $\sin 2\theta$ to get $\sin 2\theta = \frac{2gx}{v_0^2}$. Take the arcsine of both sides, and then divide by 2:

$$\theta = \frac{1}{2}\sin^{-1}\left(\frac{2gx}{v_0^2}\right) = \frac{1}{2}\sin^{-1}\left(\frac{2(10\ \text{m/s}^2)(30\ \text{m})}{(40\ \text{m/s})^2}\right) = \frac{1}{2}\sin^{-1}(0.375).$$

At this point, your calculator may screw you up. It probably says that $\sin^{-1}(.375) = 22°$, in which case θ would be only $11°$. But this makes no sense. If fired at $11°$, the ball does not clear the wall! You should be able to see this intuitively by looking at the diagram.

To resolve this dilemma, you must remember that arcsines, like square roots, have more than one "answer." Here is why. In general, $\sin\phi = \sin(180° - \phi)$. It follows that if $\sin^{-1}(\text{blah}) = \phi$, then $\sin^{-1}(\text{blah})$ *also* equals $180° - \phi$. Here, since $\sin^{-1}(.375) = 22°$, we also have $\sin^{-1}(.375) = 180° - 22° = 158°$. So, the physically correct firing angle is

$$\theta = \frac{1}{2}\sin^{-1}(.375) = \frac{1}{2}(158°) = 79°.$$

With many instructors, missing this bit of mathematical erudition would cost you few points on a test. Here at Berkeley, we give lots of partial credit for setting up the equations correctly. Find out your instructor's policy on this issue.

End of algebra.

The cannon must point nearly straight up, as you can confirm with the computer simulation.

(b) Now that we know the firing angle, we can immediately solve a kinematic formula for y_{peak}. I see two good options, both of which invoke only the *vertical* component of the relevant equation.

(i) We could solve Eq. (1) or (2) for t_{peak}, and then plug it into $y = y_0 + v_{0y}t + \frac{1}{2}a_y t^2$.

(ii) We could use $v_y^2 = v_{oy}^2 + 2a_y\Delta y$, since we know $v_y = 0$ at the peak.

I will use method (ii). Remembering that $\Delta y = y - y_0$, and setting $y_0 = 0$, we can immediately solve for $y = y_{\text{peak}}$ to get

$$y_{\text{peak}} = \frac{v_y^2 - v_{0y}^2}{2a_y} = \frac{0 - v_0^2 \sin^2\theta}{2(-g)} = \frac{0 - (40\ \text{m/s})^2 \sin^2 79°}{2(-10\ \text{m/s}^2)} = 77\ \text{m}.$$

So, the ball clears the top of the wall comfortably.

(c) You might think to use $\theta = 30°$ in the above expression for y_{peak}. But when fired at 30°, the ball slams into the wall *before* reaching its peak. The computer simulation confirms that the ball is still rising when it crashes. So, we are not trying to find its peak height.

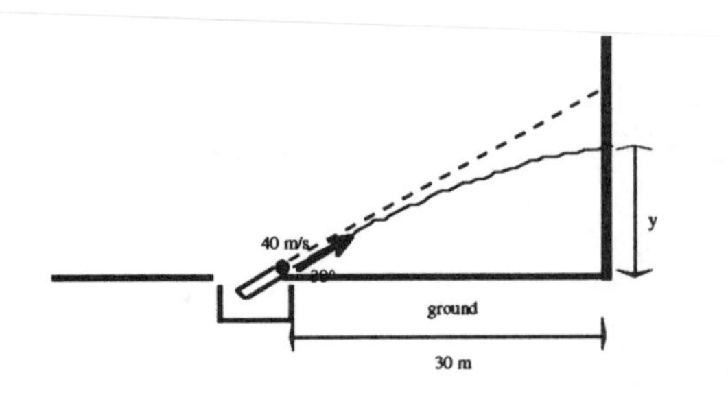

What ARE we trying to find? A diagram can help. I have drawn in the straight dashed line to emphasize that the ball does *not* travel along that path. If it did, we could quickly solve for y using $\tan 30° = \frac{y}{x}$. But the y you would get using that formula is bigger than the y we are seeking here, because the ball arcs below that "hypotenuse."

Fortunately, the picture shows us that we are looking for the ball's height y after it has traveled a horizontal distance $x = 30$ m. This suggests a strategy. By considering the ball's horizontal motion, we can find t, the time it takes to crash into the wall. Then we can substitute that t into a formula for y. Once again, we will need to consider both the horizontal and vertical motion.

Let us start with the horizontal motion, to get the "flight time." Since we know $x = 30$ m, start with $x = x_0 + v_{0x}t + \frac{1}{2}a_x t^2$. Since no horizontal forces act on the ball, $a_x = 0$. From the usual velocity vector diagram, $v_{0x} = v_0 \cos 30°$. So, solving for t gives

$$t = \frac{x - x_0}{v_0 \cos\theta} = \frac{30\text{ m} - 0}{(40\text{ m/s})\cos 30°} = 0.87\text{ s}.$$

Now that we know the flight time, we can again use the vertical position formula

$$\begin{aligned} y &= y_0 + v_{0y}t + \frac{1}{2}a_y t^2 \\ &= 0 + (40\text{ m/s})(\sin 30°)(.87\text{ s}) + \frac{1}{2}(-10\text{ m/s}^2)(.87\text{ s})^2 \\ &= 14\text{ meters,} \end{aligned}$$

which is 6 meters short of clearing the wall.

QUESTION 5-3

(Tricky problem.)

Consider a charged particle on a frictionless tabletop. The table is placed in an electric field. As a result, the particle accelerates, and the acceleration is constant.

Let the x-direction be eastward and the y-direction be northward along the tabletop. When the particle, initially at $\mathbf{r} = 0$, is released from rest and allowed to move freely for 4 seconds, its position in centimeters (cm) at $t = 4$ s is observed to be $\mathbf{r} = 2\hat{\imath} + 3\hat{\jmath}$. Remember, $\hat{\imath}$ and $\hat{\jmath}$ are unit vectors in the x-direction and y-direction, respectively.

Suppose we now take the same particle, put it at $\mathbf{r} = 0$, and give it an initial velocity of 4 cm/s in the y-direction at time $t = 0$. What will be the particle's position 4 seconds later? Express your answer as a vector, using $\hat{\imath}$, $\hat{\jmath}$ notation.

ANSWER 5-3

In this problem, a particle undergoing constant acceleration is given an initial velocity, in cm/s, of $\mathbf{v}_0 = 4\hat{\mathbf{j}}$. We also know the initial position, $\mathbf{r}_0 = 0$, and we want to know the particle's position at $t = 4$ s. So, we can try the constant-acceleration *vector* equation

$$\mathbf{r} = \mathbf{r}_0 + \mathbf{v}_0 t + \frac{1}{2}\mathbf{a}t^2, \tag{1}$$

which we can break into components. But to solve Eq. (1) for position, we need to know the acceleration. Fortunately, we are *told* that when the particle starts from rest, it reaches a known position in known time. Using that information, we can figure out the acceleration.

In summary: By considering the first "experiment," when the particle started from rest, we will figure out the acceleration caused by the electric field. Then we will use that acceleration to consider the second "experiment," in which the particle gets some initial velocity. The particle's acceleration is the same in both experiments, because it "feels" the same electric field (and hence, the same force) both times.

Experiment #1: Particle initially at rest. Solve for acceleration.

Although you can address this problem using whole vectors, because the particle travels along a straight line, I recommend using vector components. (The components will help us deal more easily with experiment #2.) I will separately solve for a_x and a_y, the eastward and northward parts of the overall acceleration. To do so, factor Eq. (1) into its components.

$$x = x_0 + v_{0x}t + \frac{1}{2}a_x t^2 \tag{1a}$$

$$y = y_0 + v_{0y}t + \frac{1}{2}a_y t^2. \tag{1b}$$

The independence of the x-motion and y-motion is not something special about projectile motion. You can separately consider the x- and y-motion in all kinds of problems. The particle under consideration was initially at rest ($v_{0x} = v_{0y} = 0$) at position $x_0 = y_0 = 0$. At $t = 4$ s, it reaches position $\mathbf{r} = 2\hat{\mathbf{i}} + 3\hat{\mathbf{j}}$, which is another way of saying $x = 2$ cm and $y = 3$ cm:

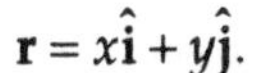

$$\mathbf{r} = x\hat{\mathbf{i}} + y\hat{\mathbf{j}}.$$

Solving Eqs. (1a) and (1b) for a_x and a_y gives

$$a_x = \frac{2(x - x_0 - v_{0x}t)}{t^2} = \frac{2(2\text{ cm} - 0 - 0)}{(4\text{ s})^2} = 0.25\text{ cm/s}^2.$$

$$a_y = \frac{2(y - y_0 - v_{0y}t)}{t^2} = \frac{2(3\text{ cm} - 0 - 0)}{(4\text{ s})^2} = 0.375\text{ cm/s}^2.$$

As demonstrated by the computer simulation, when the particle is released from rest, this acceleration makes it move in a straight line along the direction of the acceleration.

$$a_x = \frac{dv_x}{dt} = 2$$
$$a_y = \frac{dv_y}{dt} = 1,$$

in units of m/s per second. So, the total acceleration is

$$\mathbf{a} = a_x\hat{\mathbf{i}} + a_y\hat{\mathbf{j}}$$
$$= 2\hat{\mathbf{i}} + 1\hat{\mathbf{j}}$$

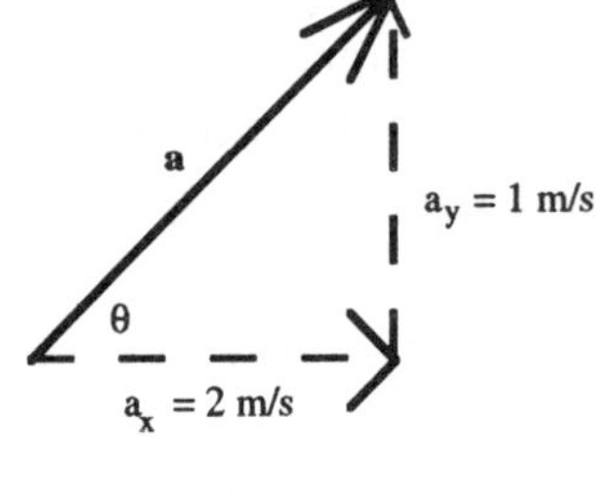

Notice that the acceleration happens to be constant; it does not change with time! But unlike in a projectile problem, this constant acceleration is not downwards.

The magnitude of this acceleration is $a = \sqrt{a_x^2 + a_y^2} = \sqrt{5}\ \text{m/s}^2$, and the angle is

$$\phi = \tan^{-1}\left(\frac{a_y}{a_x}\right) = \tan^{-1}\left(\frac{1\ \text{m/s}^2}{2\ \text{m/s}^2}\right) = 27°.$$

(c) To solve this, you must picture the situation physically. When the particle moves due east, it (momentarily) has no northward or southward motion. In other words, at that instant, $v_y = 0$. So, this problem asks for the particle's speed when $v_y = 0$.

This suggests a strategy. First, we can find the time at which $v_y = 0$. Then, we can put that time into our part (a) equation for v_x. This gives us the particle's *total* speed at that moment, because $v_y = 0$.

Since $v_y = -4 + t$, we can set $0 = v_y = -4 + t$, and hence $t = 4$ s is the moment when the particle moves due east. The particle's eastward speed at that moment is

$$v_x = 3 + 2t$$
$$= 3 + 2(4)$$
$$= 11\ \text{m/s}.$$

QUESTION 5-5

Consider a sailboat moving smoothly through the ocean at steady speed $V = 5.0$ m/s. The ship travels parallel to the shoreline. So, an observer on the shore gets a "side view" of the ship. A sailor sits in the crow's nest, a small basket attached to the top of the mast. The mast of this ship is vertical, and $L = 6.0$ meters long.

The sailor accidentally drops her watch. Assume the watch was dropped from right next to the top of the mast. It falls until hitting the deck of the ship near the bottom of the mast. Neglect wind resistance.

(a) Does the watch land in front of, behind, or right next to the mast? Explain your answer.

(b) How much time, after it is dropped, does the watch take to hit the deck?

(c) According to an observer on the shore, how much horizontal distance does the watch cover while falling?

ANSWER 5-5

This problem requires you to picture projectile motion as seen by two different observers–someone on the boat, and someone on the shore.

To develop a "feel" for what is happening, imagine yourself riding in a train, like Kim from Question 5-1. When you drop a watch, it falls onto your foot. To you, the watch appears to fall straight down. But according to Sam, who is sitting in the station watching you through the train window, the watch does not fall straight down. Sam says the watch also moves forward. In fact, the watch moves forward at exactly the rate your foot moves forward, namely the rate of the train. If you are having trouble picturing this, look at 5-1 above.

Here is the crucial physical insight: When you "drop" the watch, its initial velocity is not zero, according to Sam. Here is what Sam claims: "Let us say the train chugs along at 50 mph. Then, you are also moving at 50 mph, carried along by the train. Therefore, just as you 'drop' the watch, your hand is moving forward at 50 mph. In other words, you 'throw' the watch forward at 50 mph. Since your foot also moves forward at 50 mph, the watch has no forward or backward motion *relative to your foot*. Therefore, as the watch falls, it 'stays even' with your foot. Hence, it hits your foot, instead of landing in front of or behind your foot."

In summary: According to you, the watch falls straight down. According to Sam, the watch moves forward while falling down. Who is "right"? As discussed in Question 5-1(d), the principle of relativity insists that *both* people are equally "right." Neither perspective is more valid, because both people can use the laws of physics to make accurate predictions. For instance, you claim that the watch falls straight down onto your foot, while Sam claims that it falls "forward" onto your (forward-moving) foot. But everyone agrees with the painfully-accurate prediction that the watch smashes your foot!

Given all this, let us address the ship problem.

(a) Before visualizing the train example just discussed, you might have thought the watch lands behind the mast, because the watch falls straight down while the mast keeps moving forward. This reasoning would work if the watch were dropped from rest *relative to the shore*. But the "dropped" watch actually moves forward at $V = 5$ m/s relative to the shore, because the sailor's hand was moving at that speed (relative to the shore) when she dropped it. In other words, according to Sam on the shore, the watch was "thrown" forward at 5 m/s.

So, according to Sam, the falling watch has $V = 5$ m/s of forward (horizontal) motion. Does this forward motion make the watch land in front of the mast? No, because the mast–like everything on the ship–*also* moves forward at $V = 5$ m/s. Although the watch moves forward, so does the ship, at precisely the same rate. Therefore, the watch does not move forward or backward *relative to the mast*. As the watch falls, it moves forward a certain distance. But during that "fall time," the mast moves forward by the same distance. So, the watch lands next to the mast.

People on the ship make the same prediction; but they say the watch falls straight down.

(b) Let us exploit the *independence* of the horizontal and vertical aspects of an object's motion. If the watch has no initial *vertical* velocity, then how much time it takes to fall through a given *vertical* distance does *not* depend on whether the watch also moves *horizontally*. To see this, put a ball on the edge of a table. At the same moment you knock the ball sideways off the table, drop a second ball from the same height. If you get the timing right, both balls hit the floor simultaneously.

"best" firing angle, for *any* initial velocity. Mathematically speaking, v_0 is "independent" from the firing angle θ; we can adjust θ without changing v_0. When some quantity is "independent" of the free variable in this sense, you can treat it as a known constant, even if you do not happen to know its value.

Is the flight time t also independent of θ? No! When you adjust θ, you unavoidably change t. For instance, by increasing the firing angle, you automatically increase how much time the arrow spends in the arrow. So, you cannot "pretend" the flight time is a constant. It is another unknown. Since the equation obtained in step B must contain no unknowns *except* the free variable and the variable we are maximizing, we must rid Eq. (1) of t.

To do so, consider the arrow's vertical motion. It starts and lands at height $y = y_0 = 0$, after being fired with initial vertical speed $v_0 \sin\theta$, and accelerating at rate $a_y = -g$. So, we have

$$y = y_0 + v_{0y}t + \frac{1}{2}a_y t^2$$
$$0 = 0 + (v_0 \sin\theta)t + \frac{1}{2}(-g)t^2.$$

Solve for t to get $t = \frac{2v_0 \sin\theta}{g}$, and substitute it into Eq. (2) to get

$$\begin{aligned} R &= (v_0 \cos\theta)\frac{2v_0 \sin\theta}{g} \\ &= \frac{v_0^2 \sin 2\theta}{g}. \end{aligned} \tag{2}$$

I just used the trig identity $\sin 2\theta = 2\sin\theta\cos\theta$.

At this point, we have expressed the variable we are maximizing, R, in terms of the free variable, θ; and the equation contains no other unknowns. (Remember, we can pretend v_0 is a given constant, because it is independent of the firing angle.) So, we are done with step B. Now we can proceed to the last step of this subproblem, which is to find the "best" firing angle when the arrow lands at the same height from which it was fired.

Step C: Set derivative equal to zero, and solve for the free variable.

Take the derivative of the variable we are maximizing with respect to the free variable, and set it equal to 0. Solve the resulting equation for the free variable: $0 = \frac{dR}{d\theta} = \frac{v_0^2}{g}2\cos 2\theta$. Since $\cos\phi = 0$ when $\phi = 90°$, it follows that $2\theta = 90°$, and hence $\theta = 45°$. That firing angle maximizes the horizontal range. We are done with subproblem 1.

Subproblem 2: *Find the arrows' initial speed*

We are told that $R_{max} = 500$ m. In other words, when the arrow gets fired at 45°, it covers a horizontal distance of $x = 500$ m. From this information, we must extract the arrows' initial speed.

Well, Eq. (2) above relates the horizontal range to the initial velocity: $R = \frac{v_0^2 \sin 2\theta}{g}$. *If you did not just read through subproblem 1, go over the derivation contained in step B of that subproblem. As you will see, this range formula applies* ***only*** *when the projectile lands at the same height from which it was fired.*

So, to find v_0, algebraically manipulate that range formula, and set $R = 500$ m and $\theta = 45°$:

$$v_0 = \sqrt{\frac{Rg}{\sin 2\theta}} = \sqrt{\frac{(500\text{ m})(9.8\text{ m/s}^2)}{\sin 2(45°)}} = 70\text{ m/s}.$$

Subproblem 3: *Find the range of the arrows when fired at 70°.*

Once again, we can use the horizontal range formula derived in subproblem 1 step B, namely $R = \frac{v_0^2 \sin 2\theta}{g}$. Substitute in $\theta = 70°$ and $v_0 = 70$ m/s to get

$$R = \frac{v_0^2 \sin 2\theta}{g} = \frac{(70 \text{ m/s})^2 \sin 2(70°)}{9.8 \text{ m/s}^2} = 320 \text{ m}.$$

Notice that the arrow travels much farther, namely 500 m, when fired at the "best" angle, 45°.

QUESTION 5-7

An astronaut, Dudley J. Spacewalker, loves to play golf, though he is not very good. On a level fairway on Earth, when he hits a golf ball as hard as he can at a $\theta = 30°$ angle to the ground, it lands a distance $D = 100$ meters away from him. The acceleration due to gravity near the Earth's surface is $g = 10$ m/s^2, and you may neglect wind resistance.

Now suppose Spacewalker plays golf on the Moon, where gravity is six times weaker than Earthly gravity. Furthermore, suppose he hits the ball off the edge of a moon cliff, so that the ball lands a distance $H = 20$ m below its starting point.

(a) If Spacewalker again hits the golf ball as hard as he can at a $\theta = 30°$ angle to the horizontal, how much horizontal distance does the ball cover before landing? You may assume that his space suit does not hamper his swing. Answer in terms of D, g, θ, and H. Then, obtain a numerical answer.

(b) In 30 seconds or fewer, estimate how far Spacewalker's ball would have gone if it had landed at the same height from which it was hit. Do not use formulas: You have only 24 seconds left!

ANSWER 5-7

(a) I will start with a picture of Spacewalker's moon shot. We want x. As you can see intuitively (or by writing formulas for x), we cannot find x until we know v_0. Since Spacewalker hits the ball equally hard on the Earth and Moon, the ball has the same initial velocity both times. Therefore, we can solve this problem in two parts: (1) Using Earthly projectile reasoning, find the initial velocity with which Spacewalker hits a golf ball, and then (2) Use that initial velocity to calculate how far the ball travels on the moon.

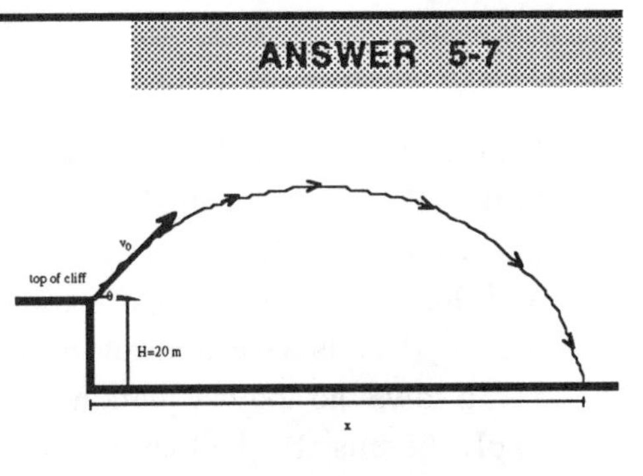

Subproblem 1: Find v_0 using "Earthly" reasoning.

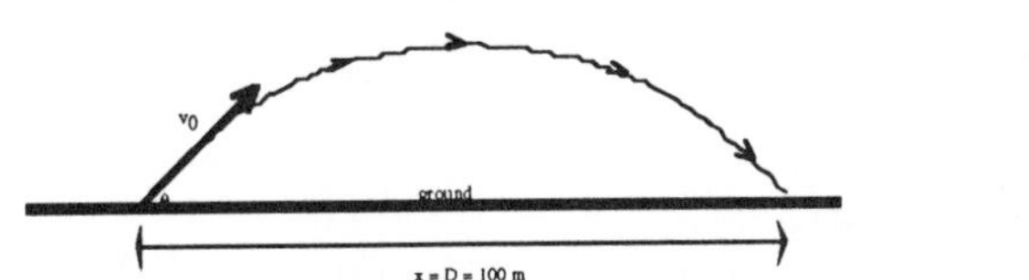

On the Earth, Spacewalker's shot looks like this figure.

We have dealt with this situation before. But instead of just invoking the "range" formula, let me briefly show where it comes from. We are trying to find v_0, and we know x and θ. As always, break up the initial velocity vector into components.

So, let us start with

$$x = x_0 + v_{0x}t + \frac{1}{2}a_x t^2 \tag{1}$$

$$D = 0 + (v_0 \cos\theta)t + 0,$$

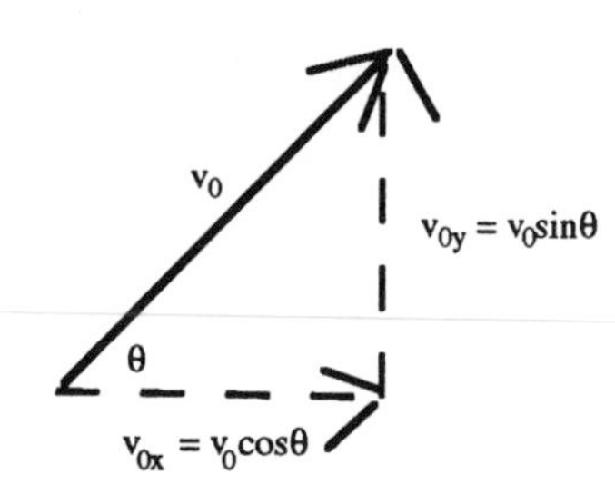

since the ball experiences no horizontal pushes once it is in mid-air ($a_x = 0$). To solve for v_0, we need to know the Earthly flight time t. Well, the ball lands at the same height from which it was fired: $y = y_0 = 0$. So, we can use

$$\begin{aligned} y &= y_0 + v_{0y}t + \frac{1}{2}a_y t^2 \\ 0 &= 0 + (v_0 \sin\theta)t + \frac{1}{2}(-g)t^2. \end{aligned} \tag{2}$$

Now that we have two equations in two unknowns (v_0 and t), we can solve for v_0. After a few steps of algebra, you get

$$v_0 = \sqrt{\frac{Dg}{2\sin\theta\cos\theta}} = \sqrt{\frac{(100\text{ m})(10\text{ m/s}^2)}{2\sin 30^\circ \cos 30^\circ}} = 34\text{ m/s}.$$

In summary: By considering Spacewalker's golf shot *on Earth*, we found the speed with which he hits golf balls. Now we can use that v_0 to calculate how far his moon shot travels.

Subproblem 2: Find x on Moon.

YOU CANNOT SIMPLY USE THE 'RANGE' FORMULA, BECAUSE THE BALL LANDS AT A DIFFERENT HEIGHT FROM WHICH IT WAS FIRED.

Since we are solving for x, let us again start with

$$\begin{aligned} x &= x_0 + v_{0x}t + \frac{1}{2}a_x t^2 \\ &= 0 + (v_0\cos\theta)t + 0. \end{aligned}$$

On the Moon, as on the Earth, $a_x = 0$, because the projectile "feels" no horizontal pushes or pulls; it keeps all of its horizontal motion. But on the Moon, the vertical acceleration a_y is less than its Earthly value. The ball feels a less intense downward pull. Intuitively, this means the ball stays aloft longer. If we let "g_m" denote the acceleration due to gravity near the Moon's surface, then $g_m = \frac{g}{6}$ That is what it means to say gravity is six times weaker on the Moon.

To solve the above equation for x, we need to know the flight time t. Notice that Eq. (2) does not apply, because the ball lands below its starting point. Picking the upward direction as positive, I will say the "final" height is $y = 0$ while the initial height is $y_0 = H = 20$ m. (You could equally well say $y_0 = 0$ and $y = -H = -20$ m.) So, we have

$$\begin{aligned} y &= y_0 + v_{0y}t + \frac{1}{2}a_y t^2 \\ 0 &= H + (v_0\sin\theta)t + \frac{1}{2}\left(-\frac{1}{6}g\right)t^2. \end{aligned}$$

The only unknown is this equation is t. So, we can solve for t, and then substitute it into the above formula for x. We are done with the physics; only algebra remains.

Algebra starts here. Unfortunately, we must solve a quadratic equation to find t. Using the quadratic formula, I get

$$t = \frac{-v_0 \sin\theta \pm \sqrt{(v_0 \sin\theta)^2 - 4\left(-\frac{g}{12}\right)(H)}}{2\left(-\frac{g}{12}\right)}$$

$$= \frac{-(34\ \text{m/s})\sin 30° \pm \sqrt{[(34\ \text{m/s})\sin 30°]^2 - 4\left(-\frac{10\ \text{m/s}^2}{12}\right)(20\ \text{m})}}{2\left(-\frac{10\ \text{m/s}^2}{12}\right)}$$

$$= 21.5\ \text{s or } -1.1\ \text{s}.$$

Since the flight time must be positive, it is $t = 21.5$ s, much longer than an Earthly golf ball would ever stay aloft.

Substitute that t into the above expression for x to get

$$\begin{aligned} x &= (v_0 \cos\theta)t \\ &= (34\ \text{m/s})(\cos 30°)(21.5\ \text{s}) \\ &= 630\ \text{m}, \end{aligned}$$

which is nearly half a mile. **End of algebra.**

(b) Intuitively, if gravity is six times weaker, we expect the ball to travel six times farther, other things being equal. This ends up being correct, as you can confirm with the "range" formula. If Spacewalker's moon shot had landed at the same height from which it was "fired," it would have traveled $x = 6D = 600$ m.

But Spacewalker's moon shot lands below its starting point. Therefore, the ball gets to stay aloft longer than if it had landed at the same height from which it was hit. In other words, by landing below its starting point, the ball acquires some "extra" flight time. During this extra flight time, the ball continues moving forward. That is why Spacewalker's shot travels forward 630 m instead of 600 m.

Forces and Newton's Laws: *The Main Ideas*

CHAPTER 6

QUESTION 6-1

A T-1200 engine can propel a snowmobile with a certain force. With a T-1200 engine installed, my snowmobile can accelerate from 0 to 10 m/sin 5.0 seconds. But Jen's snowmobile, with the same engine installed, accelerates from 0 to 10 m/s in 10 seconds.

In this problem, to keep the math simple, neglect wind resistance and friction.

(a) When I say the "mass of a snowmobile," I mean the total mass including the driver and engine. What can you say about the mass of my snowmobile versus the mass of Jen's snowmobile? Which snowmobile is more massive, and by what ratio (i.e., three times as massive, 1.5 times as massive, or what)? Use your intuitions; do not plug in any formulas to solve this part. See if you can answer in under 30 seconds.

(b) My snowmobile engine breaks down. Jen offers to tow my snowmobile with her snowmobile, by attaching them with a strong, light cord. Jen's snowmobile has a T-1200 engine. While she is towing my snowmobile (with me riding in it), what is her snowmobile's acceleration?

ANSWER 6-1

(a) Jen's snowmobile accelerates more slowly than mine, given the same push. Even if you did not know Newton's 2nd law, you would know intuitively that heavier objects are harder to push. So, Jen's snowmobile is more massive.

By how much? Well, for a given push, her acceleration is only half of mine. Remember, acceleration is the rate at which an object speeds up or slows down. Here, I speed up from 0 to 10 m/s twice as quickly as Jen does. So, my acceleration is twice her acceleration, given the same push. Intuition suggests, and Newton's 2nd law confirms, that her snowmobile must be twice as massive.

(b) You might think we lack sufficient information to solve this part. After all, when Jen's snowmobile tows mine, the force of a single T-1200 engine must acceleration two masses–my snowmobile and Jen's snowmobile. Indeed, you can think of Jen's engine as pulling a "single" combined mass of $M_{\text{Jen}} + m_{\text{me}}$, where M_{Jen} and m_{me} are the snowmobiles' masses (including drivers and engines). When both snowmobiles are attached, they move "together," i.e., they share the same acceleration. I will call that acceleration a_{tow}. Newton's 2nd law applied to the two-snowmobile system tells us that $F_{\text{T-1200}} = (M_{\text{Jen}} + m_{\text{me}})a_{\text{tow}}$, and hence

$$a_{\text{tow}} = \frac{F_{\text{T-1200}}}{M_{\text{Jen}} + m_{\text{me}}}, \tag{1}$$

where $F_{\text{T-1200}}$ is the force generated by a single T-1200 engine. The trouble is, we do not know $F_{\text{T-1200}}$, M_{Jen}, or m_{me}. We need more information.

Well, we know from part (a) that $M_{\text{Jen}} = 2m_{\text{me}}$. That gets rid of one of the unknowns in Eq. (1):

$$\begin{aligned} a_{\text{tow}} &= \frac{F_{\text{T-1200}}}{M_{\text{Jen}} + m_{\text{me}}} \\ &= \frac{F_{\text{T-1200}}}{2m_{\text{me}} + m_{\text{me}}} \\ &= \frac{1}{3}\left(\frac{F_{\text{T-1200}}}{m_{\text{me}}}\right). \end{aligned} \tag{1'}$$

At this point, you can solve for a_{tow} by noticing that the ratio $\frac{F_{\text{T-1200}}}{m_{\text{me}}}$ on the right-hand side of Eq. (1′) is the acceleration of my snowmobile (alone) when the engine works. Applying Newton's 2nd law to my snowmobile alone yields $F_{\text{T-1200}} = m_{\text{me}}a_{\text{me}}$, and hence $a_{\text{me}} = \frac{F_{\text{T-1200}}}{m_{\text{me}}}$. So, Eq. (1′) becomes

$$\begin{aligned} a_{\text{tow}} &= \frac{1}{3}\left(\frac{F_{\text{T-1200}}}{m_{\text{me}}}\right) \\ &= \frac{1}{3}a_{\text{me}}. \end{aligned} \tag{1''}$$

The tied-together snowmobiles accelerate one third as quickly as my snowmobile alone, because the combined system is three times heavier. Indeed, using the intuitive reasoning of part (a), you could have jumped right to this conclusion, without formally using Newton's 2nd law. In any case, from the given kinematic information, we can find a_{me} and substitute it into Eq. (1″). Starting from $v = v_0 + at$ or from the definition of acceleration as change in velocity per change in time, we get $a_{\text{me}} = \frac{v - v_0}{t} = \frac{10 \text{ m/s} - 0}{5 \text{ s}} = 2.0 \text{ m/s}^2$, and hence $a_{\text{tow}} = \frac{1}{3}a_{\text{me}} = \frac{1}{3}(2 \text{ m/s}^2) = 0.67 \text{ m/s}^2$.

QUESTION 6-2

An electric field exerts a constant force on a charged bead of mass m, which slides frictionlessly on top of a table. The force pushes the bead to the left.

Before the electric field was turned on, the bead was moving leftward at speed v_0. At time $t = 0$, the electric field is turned on, and it stays on until time t_1. At t_1, the bead is observed to be moving leftward at speed $3v_0$.

Express your answers in terms of the quantities listed in the problem (m, v_0, and t_1).

(a) How fast was the bead moving at time $t = \frac{t_1}{4}$?

(b) What force does the electric field exert on the bead?

(c) At time t_1, the electric field is suddenly "reversed." In other words, it now creates an electric force equal in strength but opposite in direction to the original electric force. For how long must the "reversed" field be left on before the bead comes to a complete stop?

(d) Qualitatively, if the electric field is never turned off, what happens to the bead after it momentarily comes to rest?

ANSWER 6-2

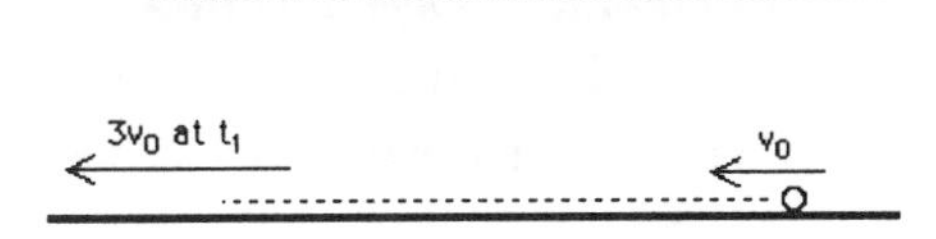

Draw a picture to help yourself visualize the problem. The computer simulation can also help.

(a) The electric force accelerates the bead leftward, causing it to speed up. Before using formulas, let us think intuitively about what is happening. Over time t_1, the bead speeds up from v_0 to $3v_0$, at a constant rate (i.e., constant acceleration). The question is, how fast is the bead moving one quarter of the way into this acceleration, at time $\frac{t_1}{4}$? Intuitively, since the bead steadily speeds up between $t = 0$ and $t = t_1$, at time $t = \frac{t_1}{4}$ it should have completed one quarter of its total velocity increase. In other words, its speed should be one quarter of the way from v_0 to $3v_0$.

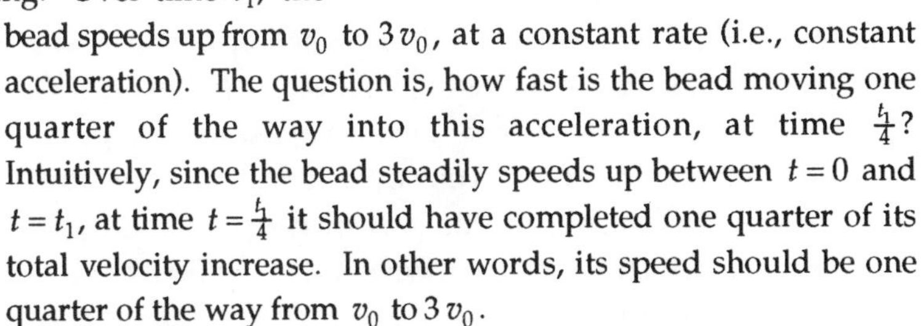

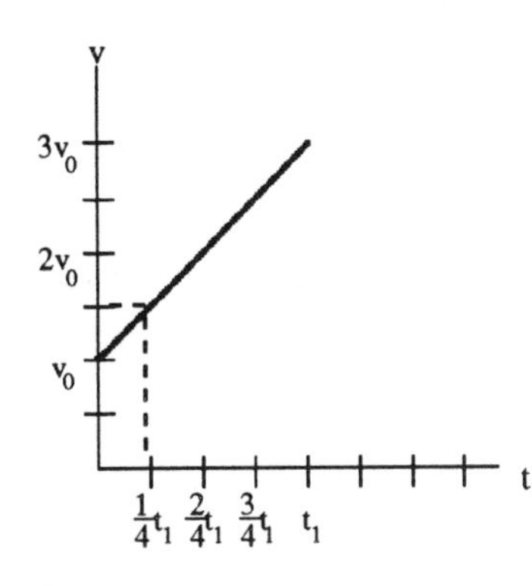

Let us see if a graph agrees with this intuitive conclusion. You can see that at $t = \frac{t_1}{4}$, the bead's speed is $\frac{3v_0}{2}$, exactly one quarter of the way from v_0 to $3v_0$.

Now let us see if equations give the same answer. Let v_1 denote the bead's speed at time $\frac{t_1}{4}$. Since everything happens along a straight line, we need not worry about vector components. At time $\frac{t_1}{4}$, the velocity is

$$v_1 = v_0 + a\left(\frac{1}{4}t_1\right). \qquad (1)$$

To solve Eq. (1) for v_1, we need to know the bead's acceleration, a. How can we find it? Well, we cannot use Newton's 2nd law, $a = \frac{F_{net}}{m}$, because we do not know the force. We will have to use other information.

At time t_1, the bead's "final" speed is $v = 3v_0$. So, starting again with $v = v_0 + at$, we get $3v_0 = v_0 + at_1$. Solving for a gives $a = \frac{3v_0 - v_0}{t_1} = \frac{2v_0}{t_1}$. Put that acceleration back into Eq. (1) to get

$$\begin{aligned} v_1 &= v_0 + a\left(\frac{1}{4}t_1\right) \\ &= v_0 + \frac{2v_0}{t_1}\left(\frac{1}{4}t_1\right) \\ &= v_0 + \frac{1}{2}v_0 \\ &= \frac{3}{2}v_0, \end{aligned}$$

the same answer obtained above using intuitive reasoning.

(b) Since we just found the bead's acceleration, $a = \frac{2v_0}{t_1}$, it is easy to find the net force using Newton's 2nd law, $F_{net} = ma$. Because the electric force is the *only* force acting on the bead along its direction of motion, $F_{net} = F_{elec}$: $F_{elec} = F_{net} = ma = m\frac{2v_0}{t_1}$.

(c) While under the influence of the "reversed" electric field, the bead gets pushed rightward instead of leftward. Does this mean the bead starts moving rightward immediately after time t_1? No. The rightward net force causes a rightward acceleration, which is a rate of change of velocity. When the leftward-moving bead experiences a rightward acceleration, it gradually slows down. It keeps moving leftward, at a slower and slower rate, until it stops entirely. We are looking for the time t at which this happens.

The reversed electric field creates a force–and therefore an acceleration–of the same "strength" (magnitude) as before. So, if we pick leftward as our positive direction, the new acceleration is $a = -\frac{2v_0}{t_1}$, where the minus sign indicates "rightward."

To find the "stop" time t, we can again use $v_{\text{final}} = v_{\text{initial}} + at$, if we are careful. In this part of the problem, the "final" velocity is $v = 0$, since we are calculating how long the bead takes to stop. The initial velocity, in this part of the problem, is the bead's speed *when the electric field gets reversed*, which happens at time t_1. So, our new v_{initial} is $3v_0$, the "final" velocity from parts (a) and (b). Start with $v_{\text{final}} = v_{\text{initial}} + at$, and solve for t to get

$$t = \frac{v_{\text{final}} - v_{\text{initial}}}{a} = \frac{0 - 3v_0}{-\frac{2v_0}{t_1}} = \frac{3}{2}t_1.$$

That is how long the bead takes to stop, after the electric field is reversed.

(d) If the electric field stays on after the bead stops, the bead starts to slide rightward (backwards), gradually getting faster and faster.

You have seen analogous situations before. Throw your pencil straight up. While in mid-air, the pencil always feels a downward force–and hence a downward acceleration–due to gravity. So, why does the pencil keep rising for a while? *Not* because an upward force continues pushing it up after it leaves your hand. What keeps the pencil moving upward is its own natural "tendency" to stay in motion, i.e., its inertia. But as it rises, the downward acceleration gradually decreases its velocity, until $v = 0$ at the peak. Then, the downward acceleration makes the pencil fall back down, getting faster and faster. Analogously, the leftward-moving bead, when subjected to a rightward force, gradually slows down, stops, and then speeds up in the opposite direction.

QUESTION 6-3

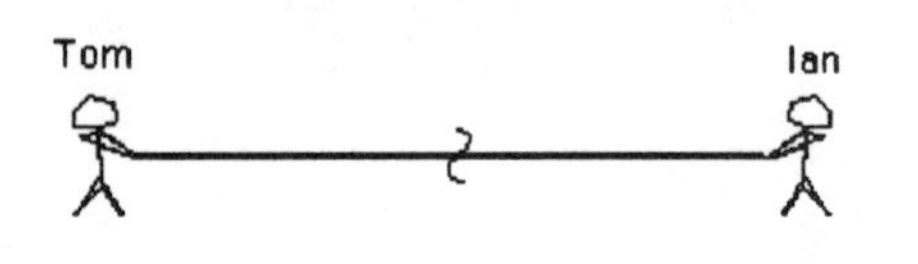

Tom and Ian procrastinate by playing tug-of-war. Tom pulls on one end of the rope, while Ian pulls on the other end. A ribbon is tied to the middle of the rope. Starting at time $t = 0$, Tom pulls with a horizontal force of 200.0 newtons, while Ian pulls with a horizontal force of 200.2 newtons. The rope and ribbon together have a total mass of 4.00 kg. At time $t = 0$, when the game begins, the ribbon is motionless right in the middle, at $x = 0$. Let the rightward direction count as "positive."

(a) Where will the ribbon be at time $t = 4$ seconds?

(b) At time $t = 4$ s, Tom, realizing he is about to lose, lets go of his end of the rope. Describe qualitatively what happens.

ANSWER 6-3

As always with multiple force problems, start by drawing a "free-body" force diagram, showing all the forces acting upon the object of interest. Here, things get confusing, because the players exert forces on the rope, and the rope exerts forces on the players. You have to be careful not to mix up these forces. In part (a), we care only about how the rope moves. Therefore, you should draw a free-body diagram showing the forces acting *on the rope.* In your diagram, *do not include any forces exerted by the rope on other objects; include only forces exerted by other objects on the rope.* Here, the forces on the rope are those exerted by Tom and Ian.

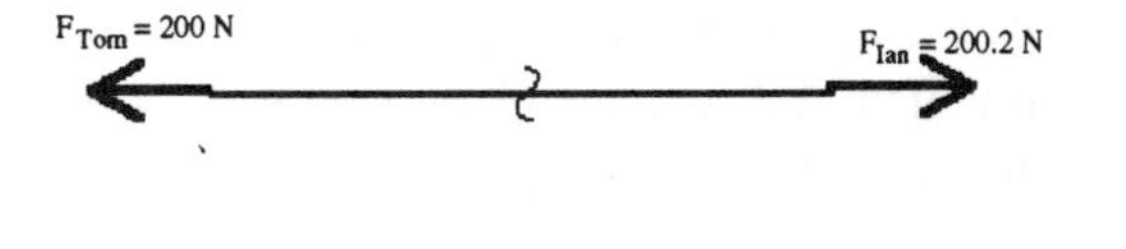

On the diagram, I have omitted the downward force of gravity, because we are interested only in the sideways motion of the rope. (Remember the main "lesson" learned from projectile motion: We can treat the x- and y-component of an object's motion separately.) Notice that Tom's force points in the negative (leftward) direction. The total force on the rope is the sum of the individual forces:

$$F_{net} = F_{Ian} - F_{Tom} = 200.2 \text{ newtons} - 200.0 \text{ newtons} = 0.2 \text{ newtons}.$$

In words, the rope "feels" an overall rightward pull of only 0.2 N.

(a) We are asked for the ribbon's position at time $t = 4$ seconds. Since a constant net force acts on the rope, it undergoes a constant acceleration. So, we can use our usual constant-acceleration kinematic formulas to calculate how far the rope (and therefore the ribbon) moves during those 4 seconds.

First, we need to calculate the acceleration, using Newton's 2nd law:

$$\sum F = ma$$
$$F_{Ian} - F_{Tom} = ma$$
$$200.2 \text{ N} - 200 \text{ N} = (4 \text{ kg})a.$$

Solve for acceleration to get $a = 0.05 \text{ m/s}^2$. Now that we know the acceleration, we can substitute it into an old kinematic formula for position. The ribbon starts off motionless ($v_0 = 0$) right in the middle ($x_0 = 0$), and gets tugged for $t = 4$ s. So, we can use the displacement formula

$$\begin{aligned} x &= x_0 + v_0 t + \frac{1}{2}at^2 \\ &= 0 + 0 + \frac{1}{2}(.05 \text{ m/s}^2)(4 \text{ s})^2 \\ &= 0.4 \text{ meters,} \end{aligned}$$

which is a little over a foot. Perhaps Tom gave up too quickly.

By playing tug of war, you can see that what "matters" is not your force or your opponent's force. What matters is the *net* force, i.e., the difference between the two forces.

(b) Intuitively, you can see that Ian will topple over or otherwise "fall" rightward. Let us see why. While Tom pulls on the rope, the rope is taut. In other words, it carries a "tension." Tension is the force exerted *by* a rope on other objects. This tension tries to pull Ian leftward. But by "digging in" and leaning back, Ian also ensures that a rightward force acts on him. When Tom lets go, the rope goes slack, losing its tension. Therefore, the leftward force on Ian suddenly disappears. All that remains is the rightward force. That is why Ian topples rightward.

QUESTION 6-4

Experimental high-energy physicists love to crash particles into each other at outrageously high speeds. During such a collision, new kinds of particles are created–particles that exist for only a few billionths of a second before decaying into other kinds of particles. Physicists study the properties of these short-lived particles.

A *linear accelerator* produceshigh-speed electrons that crash into a target, thereby producing the new particles. A simplified linear accelerator works as follows: An electron enters a "push" region in which high-voltage plates create an electric field. The electric field exerts a constant force on the electron. After the electron leaves a push region, it enters a "float" region where no forces act on it until it reaches the next "push" region. (For simplicity, we can neglect the gravitational influences on the electron.) The inside of the linear accelerator contains almost no air or stray particles. A typical linear accelerator is over half a mile long and contains dozens of "push" regions. The diagram shows just a piece of an accelerator.

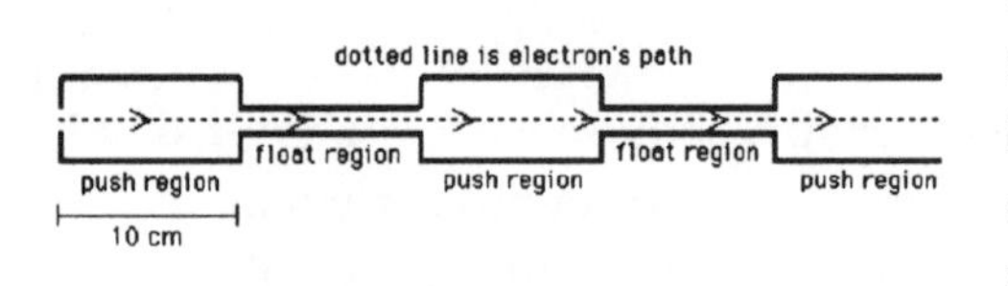

(a) While in a float region, does the electron speed up, slow down, or stay at the same speed? Explain.

(b) Suppose an electron enters the first "push" region traveling at 50.0 centimeters per second (cm/s). Inside any push region, the electric field exerts a force of 1.00×10^{-22} dynes (i.e., $\text{grams} \cdot \text{cm/s}^2$) on the electron. The push region is 10 centimeters long. How fast will the electron be travelling when it leaves the push region? An electron's mass is about 1.00×10^{-27} grams.

(c) Let Δv denote the change in the electron's speed in the first push region. When the electron enters the second push region, it speeds up by an amount I will call $\Delta v'$. Without doing any math, figure out whether $\Delta v'$ is bigger than, smaller than, or the same size as Δv. Justify your answer in words. This is tricky, so try to picture things carefully.

ANSWER 6-4

(a) You might think that the electron in the float region slows down, because no force pushes it forward. But remember, an object does not need a force to *keep* it moving. It needs a force only to *get* it moving in the first place, or to change its motion. In the absence of forces, an object's motion does not die away. Instead, the object "keeps" all its motion. For instance, the electron in the float region floats along at constant velocity until reaching the next push region. It neither speeds up nor slows down.

Let me reexpress this argument more mathematically using Newton's 2nd law, $F_{\text{net}} = ma$. In the float region, the electron experiences no forces. So, $F_{\text{net}} = 0$, and hence the acceleration is $a = 0$. Since acceleration is the rate of change of velocity, the velocity does not change.

(b) We are solving for v, the electron's velocity when it leaves the first push region. Since the electron in the push region experiences a constant force and therefore a constant acceleration, we can use one of our old constant-acceleration kinematic formulas. But first, let us find the acceleration using Newton's 2nd law:

$$a = \frac{F_{\text{net}}}{m} = \frac{10^{-22} \text{ dynes}}{10^{-27} \text{ grams}} = 10^5 \text{ cm/s}^2 .$$

To decide which kinematic equation works best, let us review what we know about the electron. The initial velocity is $v_0 = 50$ cm/s. We also know the distance over which the electron gets accelerated; the push region is $\Delta x = x - x_0 = 10$ cm long. Since all the relevant velocities and acceleration point along the same line, we need not use vectors. We can use the one-dimensional constant-acceleration formula $v^2 = v_0^2 + 2a\Delta x$. Solve for v to get

$$\begin{aligned} v &= \sqrt{v_0^2 + 2a\Delta x} \\ &= \sqrt{(50 \text{ cm/s})^2 + 2(10^5 \text{ cm/s}^2)(10 \text{ cm})} \\ &= 1410 \text{ cm/s}, \end{aligned}$$

which is over 30 miles per hour.

(c) We just found that in the first push region, the electron speeds up by

$$\Delta v = v - v_0 = 1420 \text{ cm/s} - 50 \text{ cm/s} = 1370 \text{ cm/s}.$$

In the second push region, does the electron speed up by an additional 1370 centimeters per second? Or does it speed up by less than that amount, or by more than that amount?

In each push region, the electron experiences the same force, and therefore undergoes the same acceleration. You therefore might jump to the conclusion that the electron speeds up by the same amount in each push region. But this conclusion fails to take into account a crucial fact: The electron spends more time in the first push region than it spends in the second push region. Let me first explain why these times differ. Then I will explain why it matters.

The electron, when it enters the second push region, is already traveling very fast–at $v = 1420$ cm/s, to be exact. Therefore, the electron whizzes through the second push region quickly. It spends hardly any time in that region. By contrast, when it entered the first push region, the electron was travelling slowly, only 50 cm/s. Therefore, the electron spends more time in the first push region than it does in the second.

Let me bolster this argument. In the first push region, the electron speeds up from 50 cm/s to 1420 cm/s. So, its average speed in that region is less than 1420 cm/s. (Indeed, since the acceleration is constant, its average speed is midway between 50 and 1420 cm/s.) Because the electron enters the second push region at 1420 cm/s, and leaves that region with even higher velocity, its average velocity in the second push region is greater than 1420 cm/s. So, the electron speeds through the second push region more quickly than it speeds through the first. Therefore, it spends more time in the first push region.

Why does this matter? Because an object's change in velocity Δv depends not just on the acceleration, but on the time over which the acceleration gets to "act." For instance, suppose you accelerate a desk by pushing it across the floor. If you accelerate it for two seconds, it ends up moving faster than if you would accelerated it for just one second. In formulas, $\Delta v = a\Delta t$; change in velocity depends on acceleration *and* time.

Since the electron spends more time in push region 1, it gets accelerated over a longer time interval, and therefore gains more velocity: $\Delta v = \Delta v'$.

You may have answered this question incorrectly if you reasoned that the electron exits the second push region moving faster than when it exited the first push region. This is correct. Indeed, it turns out that the electron leaves the second push region with speed $v = 2000$ cm/s. But its *change* in velocity in push region 2 is only $\Delta v' = 2000 \text{ cm/s} - 1420 \text{ cm/s} = 580 \text{ cm/s}$, while its *change* in velocity in push region 1 is $\Delta v = 1420 \text{ cm/s} - 50 \text{ cm/s} = 1370 \text{ cm/s}$. Do not confuse changes in velocity with final velocity.

QUESTION 6-5

A stage light is suspended above the stage by two ropes, as drawn below. The stage light weighs 200 newtons. This means that gravity exerts a force of 200 newtons on the light. What is the tension in rope 1? In other words, what force does rope 1 exert on the light?

ANSWER 6-5

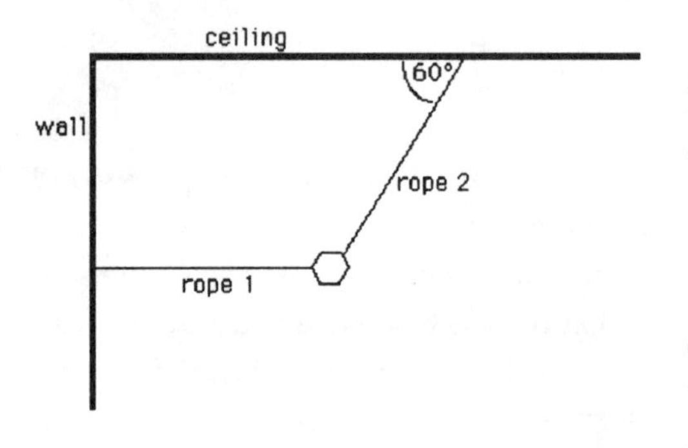

The stage light stays motionless. It never speeds up or slows down. So, the light's acceleration is $\mathbf{a} = 0$. Newton's 2nd law immediately implies that the *net* force on the light is $\mathbf{F}_{\text{net}} = 0$.

That net force consists of 3 forces: Gravity pulling the light down, rope 1 pulling the light to the left, and rope 2 pulling the light upwards and to the right. (*A rope always pulls an object along the direction in which the rope is stretched.*) To keep track of forces, always draw a force diagram.

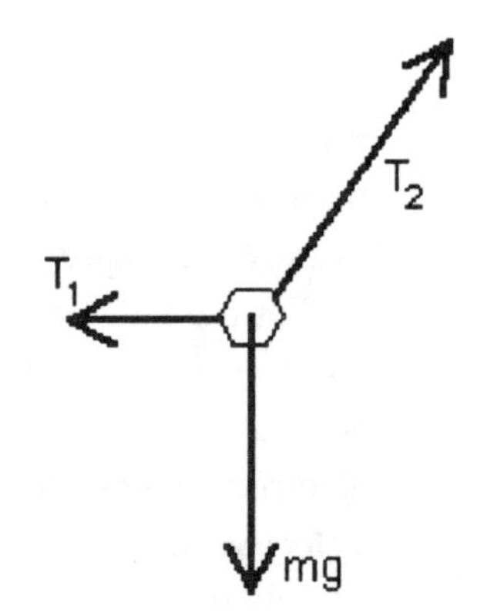

Since $\mathbf{F}_{\text{net}} = 0$, you might be tempted to write $F_{\text{net}} = T_1 + T_2 + mg = 0$. But since these forces do not all point along the same line, you cannot add and subtract them like numbers. You must use vectors.

As always, you can deal with the vectors by breaking them into components and then *separately* dealing with the x- and y-components of the motion. If we choose the x-direction as horizontal and the y-direction as vertical, then the only vector that needs to be broken up is $\mathbf{T}_2$.

The dotted arrows are not separate forces acting on the stage light. They are the horizontal and vertical components of $\mathbf{T}_2$.

Instead of working directly with the vector equation $F_{\text{net}} = ma$, we can break it into components:

$$F_{\text{net}\,x} = \sum F_x = ma_x \tag{1a}$$

$$F_{\text{net}\,y} = \sum F_y = ma_y. \tag{1b}$$

$\sum F_x$ is the sum of the forces in the x-direction. **Whether or not an object is moving, it is best to factor Newton's 2nd law into components.**

A careful free-body diagram, with forces broken up into components, makes Eqs. (1a) and (1b) easy to use. Let us choose rightward as our positive x-direction. From the diagram, we see that the only x-directed forces are $T_2 \cos 60°$ and $-T_1$. (The minus sign indicates that $\mathbf{T}_1$ points "backwards.") We also know that the stage light's horizontal acceleration is $a_x = 0$, because it never moves. Substituting all this information into Eq. (1a) gives

$$\sum F_x = ma_x$$
$$T_2 \cos 60° - T_1 = 0. \qquad (2)$$

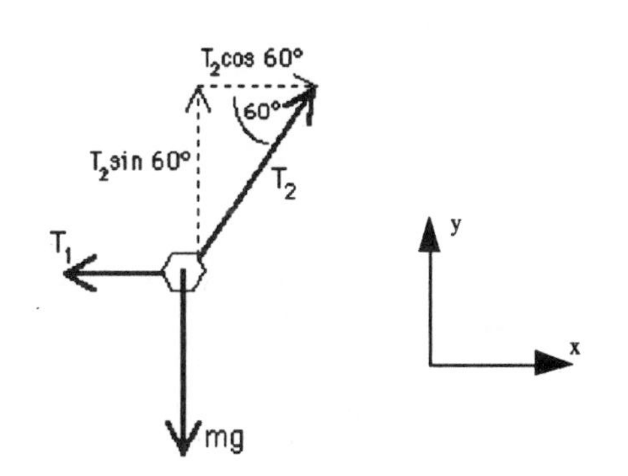

Since Eq. (2) contains two unknowns (T_1 and T_2), we cannot yet solve for T_1. To get more information, consider the vertical (y-directed) forces. I will choose the upward direction as positive. From the diagram and Eq. (1b), I get

$$\sum F_y = ma_y$$
$$T_2 \cos 60° - mg = 0. \qquad (3)$$

Since we know $mg = 200$ N, we have two equations in the two unknowns, T_1 and T_2. Therefore, we are done with the physics; it is just a matter of algebra to solve for T_1.

Algebra starts here. From Eq. (3), we get $T_2 = \frac{mg}{\sin 60°}$. Now solve Eq. (2) for T_1, and set $T_2 = \frac{mg}{\sin 60°}$, to get

$$\begin{aligned} T_1 &= T_2 \cos 60° \\ &= \frac{mg}{\sin 60°} \cos 60° \\ &= mg \cot 60° \\ &= (200 \text{ N}) \cot 60° \\ &= 115 \text{ N}. \end{aligned}$$

End of algebra.

Let me reiterate the strategy I just used, because it is the way you should approach almost all force problems:

1. *Draw a careful force diagram, breaking forces into components if necessary.*
2. *Use the vector components of Newton's 2nd law,* $\sum F_x = ma_x$ *and/or* $\sum F_y = ma_y$. *Either you will use the known accelerations to solve for some forces, or you will use some known forces to solve for acceleration.*

QUESTION 6-6

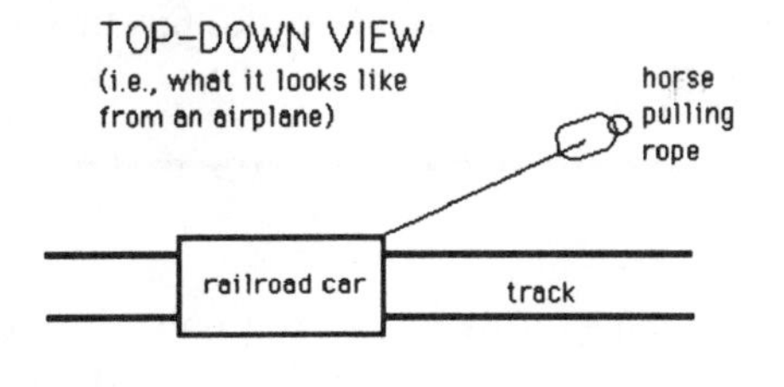

A small railroad car sits on an east-to-west track. Let the eastward direction be the positive x-direction. The car has mass $M = 7500$ kg. A horse pulls on a rope attached to the car. The rope makes an angle $\theta = 20°$ with the track. The horse pulls with force $P = 6500$ N. When the car is pulled, friction in the wheels, air resistance, etc., act as a "resistive" force that tends to keep the car from moving forward. Let $\mathbf{F}_r$ denote this resistive force.

As the horse pulls, the car's acceleration is constant, at least for the first several seconds. It is observed that if the car starts from rest, then after being pulled for time $t_1 = 5.0$ s, the car has speed $v_1 = 0.60$ m/s.

What is the magnitude and direction of the resistive force $\mathbf{F}_r$?

ANSWER 6-6

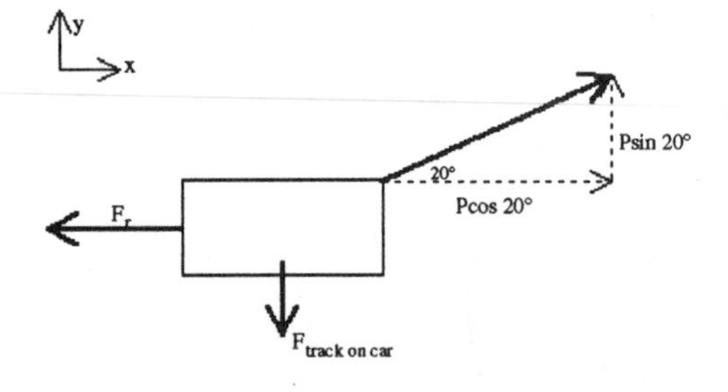

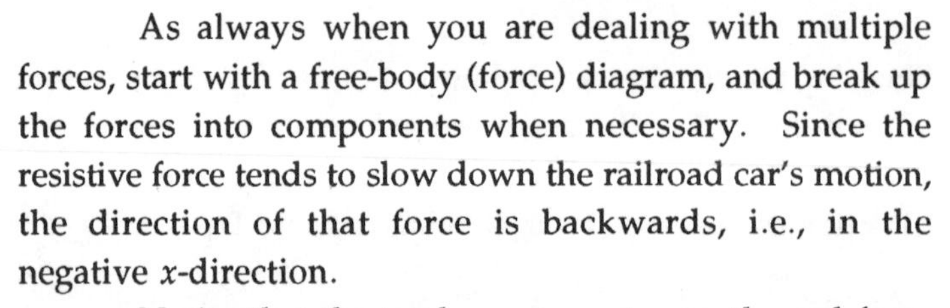

As always when you are dealing with multiple forces, start with a free-body (force) diagram, and break up the forces into components when necessary. Since the resistive force tends to slow down the railroad car's motion, the direction of that force is backwards, i.e., in the negative x-direction.

Notice that the track must exert a southward force on the car, or else the northward component of the horse's force, $P\sin 20°$, would pull the car off the track. But you could get away with forgetting this force, because it turns out not to matter. Only the x-directed forces will enter into our reasoning. But for the record, my free-body diagram omits two forces: Gravity, pulling the railroad car "into the page"; and the normal force, pushing the car "out of the page." Gravity and the normal force cancel. I will discuss normal forces later in this study guide.

We are trying to solve for F_r. So, let us start with Newton's 2nd law, concentrating on the x-directed forces only:

$$\sum F_x = Ma_x.$$
$$P\cos\theta - F_r = Ma_x. \qquad (1)$$

We are given P, M and θ. So, if we can just find a_x, then we can easily solve Eq. (1) for F_r.

How can we obtain this acceleration, i.e., the rate at which the train speeds up? Well, we are told that the train's velocity increases from $v_0 = 0$ to $v_x = 0.60$ m/s in time $t_1 = 5.0$ s. So, we can use $v_x = v_{0x} + a_x t$. Solving for a_x gives us

$$a_x = \frac{v_x - v_{0x}}{t} = \frac{v_1 - 0}{t_1} = \frac{0.60\ \text{m/s} - 0}{5.0\ \text{s}} 0.12\ \text{m/s}^2.$$

Now solve Eq. (1) for F_r, and set $a_x = 0.12\ \text{m/s}^2$, $P = 6500$ N, $\theta = 20°$, and $M = 7500$ kg.

$$\begin{aligned} F_r &= P\cos\theta - Ma_x \\ &= (6500\ \text{N})\cos 20° - (7500\ \text{kg})(0.12\ \text{m/s}^2) \\ &= 5200\ \text{N}. \end{aligned}$$

This problem again illustrates the general strategy of first drawing a careful free-body diagram with forces broken into components, and then using $\sum F_x = ma_x$ and/or $\sum F_y = ma_y$.

QUESTION 6-7

When a tennis ball falls through the air, wind resistance pushes upward on the ball. In fact, the upward force due to wind resistance is $F = kv$, where v is the ball's downward velocity and k is a proportionality constant.

When a tennis gets dropped from an airplane, it accelerates downward. But as its velocity increases, the upward force on it due to wind resistance gets bigger. Eventually, the upward wind resistance force becomes as big as the downward gravitational force.

(a) Suppose the tennis ball is dropped from rest from way high up. Without doing any math, draw a rough sketch of the ball's velocity vs. time. Pick downward as the positive direction.

(b) A tennis ball has mass $m = 0.060$ kg. The measured terminal velocity of a tennis ball is about 100 mph, which "translates" to 40 m/s. What is the proportionality constant k for a tennis ball? Be sure to get the units right.

(c) At the moment the ball is traveling at *half* its terminal velocity, what is its downward acceleration?

ANSWER 6-7

(a) To get a feel for what is happening, stand on a table, lightly wad up a piece of paper, and drop it. Also, experiment with the computer simulation, which allows you to change the air resistance to see how that affects the motion. The wad (or tennis ball) speeds up until reaching a maximum speed, at which point it just floats down at that speed. So, the velocity starts at 0 and ends at some "terminal" speed v_t. But this does not tell us what "shape" to draw the graph. Is it a straight line, a concave curve, or a convex curve?

Well, the slope of the v vs. t curve–i.e., the rate of change of velocity–is acceleration. So, by considering the ball's acceleration, we can gain insight into the slope of the v vs. t graph. This should help us choose the right "shape" for v vs. t.

To figure out what is going on with acceleration, consider the forces on the ball.

The downward force of gravity is constant. But the upward force increases as the ball gets faster, since that force is proportional to v. So, as the ball speeds up, the *net* downward force gets smaller and smaller. Eventually, when kv gets as big as mg, the net force reaches 0. In summary, the net downward force, and hence the downward acceleration, decreases as the ball speeds up. The ball keeps speeding up, but at a smaller and smaller rate of change. For instance, maybe during its first second of falling, the ball speeds up from 0 to 9 m/s. But maybe during its 2nd second of falling, it only speeds up from 9 m/sto 15 m/s. And during its third second of falling, it speeds up from 15 m/s to 21 m/s. It keeps speeding up, but at a slower and slower rate. In graphical terms, the acceleration–the slope of v vs. t–gets smaller and smaller. So, the graph must look something like the following.

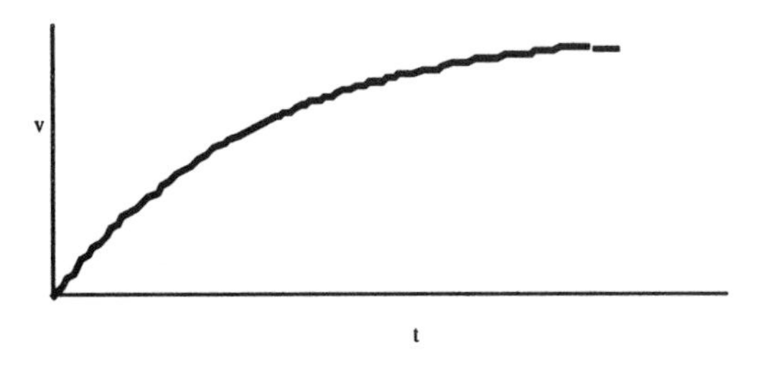

I cannot emphasize enough that although the *acceleration* (i.e., the slope of v vs. t) gets smaller until eventually reaching 0, the velocity itself increases the whole time, until reaching terminal velocity. A common mistake is to make the v vs, t graph "come down," which corresponds to a decreasing velocity. *But the ball never slows down. It just speeds up at a smaller and smaller rate of change, until it eventually stops speeding up at all.* When it stops speeding up, it keeps falling with constant speed.

(b) At its terminal velocity v_t, the ball no longer speeds up. Its acceleration becomes $a_y = 0$. Therefore, the net force on it must be zero. So, from the above force diagram, we get

$$\sum F_y = ma_y$$
$$mg - kv_t = 0.$$

I used the terminal velocity v_t, because only when the ball reaches that velocity does its acceleration vanish. We can immediately solve for k to get $k = \frac{mg}{v_t} = \frac{(0.060 \text{ kg})(9.8 \text{ m/s}^2)}{40 \text{ m/s}} = 0.015 \text{ kg/s}$. Weird units. I have no intuitive explanation for why the units are kg/s.

(c) Here, we can again use Newton's 2nd law, $\sum F_y = ma_y$. But this time, the velocity is $\frac{v_t}{2}$, and the acceleration is not zero. (You can see from the slope of the above v vs. t graph that the acceleration does not hit zero until the ball reaches terminal velocity.)

$$\sum F_y = ma_y$$
$$mg - k\frac{v_t}{2} = ma_y.$$

Solve for a_y to get

$$a_y = \frac{mg - k\frac{v_t}{2}}{m}$$
$$= \frac{(0.060 \text{ kg})(9.8 \text{ m/s}^2) - (0.015 \text{ kg/s})\frac{(40 \text{ m/s})}{2}}{0.060 \text{ kg}}$$
$$= 4.9 \text{ m/s}^2,$$

exactly half of g. Intuitively, when $v = \frac{v_t}{2}$, the upward force is half as big as mg. (By contrast, when $v = v_t$, the upward force is every bit as big as mg.) So, when $v = \frac{v_t}{2}$, wind resistance cancels half of gravity's downward force. As a result, the ball accelerates downward at half the "free-fall" acceleration, g.

Forces and Newton's Laws: More Core Concepts

CHAPTER 7

QUESTION 7-1

A small bungee cord, when stretched a distance s, exerts a force $F_{\text{bungee}} = Ks^2$, where K is a constant. One end of the cord is attached to the ceiling. On the other end is attached a potted plant, which has mass m (including the pot, the hooks, and the soil).

(a) How far below the ceiling will the plant hang motionless? Answer in terms of m, K, and any other constants you need.

(b) When a bungee cord gets hot, it becomes "looser," by which I mean easier to stretch. How would this loosening be reflected in the constant K? In other words, is "K" for a cold bungee cord bigger or smaller than "K" for a hot bungee cord? If we hang the plant in a hot greenhouse, will it hang "higher" or "lower" than it did in part (a)?

(c) Now suppose we take the original bungee cord and attach one end to a wall and the other end to a slippery brick of mass M, so that the bungee cord is stretched horizontally. The brick is pulled to a distance L from the wall and released from rest. Pulled by the cord, the brick slides frictionlessly along the grease-coated floor until slamming into the wall.

How could you figure out the time it takes for the brick to reach the wall (in terms of M, L, and K)? Set up the equations you need, but d onot solve them.

ANSWER 7-1

(a) Recall that s is the "stretch distance" of the bungee cord. By making a sketch, you can see that s is the distance by which the plant hangs below the ceiling. So, we are solving for s.

As always, start with a free-body (force) diagram showing all the forces acting on the object of interest, which here is the plant.

The plant hangs motionless in the air; it has no acceleration. So, the net force on the plant is $F_{\text{net}} = 0$. In other words, the upward pull of the bungee cord must balance the downward force due to gravity:

$$\begin{aligned}\sum F_y &= ma_y \\ Ks^2 - mg &= 0.\end{aligned} \qquad (1)$$

Intuitively, the plant stretches the bungee cord just far enough so that the bungee cord pulls the plant upward just as hard as gravity pulls the plant downward. Solve Eq. (1) for s to get $s = \sqrt{\frac{mg}{K}}$.

(b) A bungee cord stretched a distance x exerts a force $F = Kx^2$. Suppose we stretch a looser and a stiffer bungee cord by the same distance s. Then the looser cord exerts a weaker force:

$$F_{\text{looser cord}} < F_{\text{stiffer cord}}$$
$$(K_{\text{looser cord}})s < (K_{\text{stiffer cord}})s.$$

So, $K_{\text{looser cord}}$ must be smaller than $K_{\text{stiffer cord}}$. Since hotter bungees get looser, a hot bungee cord has a smaller K than a cold bungee cord.

Intuitively, the looser the bungee cord, the farther down from the ceiling the plant hangs. Therefore, the plant hangs "lower" in the greenhouse. To see if this intuition agrees with our answer to part (a), imagine plugging a bigger K and a smaller K into $s = \sqrt{\frac{mg}{K}}$. As just shown, the bigger K corresponds to a stiffer cord, which means the plant hangs less far below the ceiling. This conclusion agrees with the formula $s = \sqrt{\frac{mg}{K}}$; a bigger (stiffer) K leads to a smaller s.

(c) This is a trick question, designed to illustrate in which situations your old constant-acceleration kinematic formulas do not apply. Let me trace the standard way people solve this problem, and then show why it does not work.

As always, start with a force diagram, and then use it to write Newton's 2nd law in the x-direction and/or the y-direction. *(This part of the reasoning is correct.)*

Gravity and the normal force cancel, leaving us with horizontal motion only:

$$\sum F_x = Ma_x$$
$$KL^2 = Ma_x \tag{2}$$

At this point, we can easily solve for a_x to get $a_x = \frac{KL^2}{M}$. *So far, so good.* But now, you might think to substitute that acceleration into a kinematic formula to solve for the time the brick takes to travel a horizontal distance $x = L$. For instance, you could try

NOT VALID
$$x = x_0 + v_{0x}t + \frac{1}{2}a_x t^2$$
$$L = 0 + 0 + \frac{1}{2}\left(\frac{KL^2}{M}\right)t^2$$

Solving this equation for t gives the **wrong answer**. Those old kinematic formulas apply only when the acceleration is constant. But here, the acceleration keeps changing. Intuitively, as the brick gets closer to the wall, the bungee cord becomes less and less stretched, and therefore pulls less and less hard on the brick. Mathematically, $F_{\text{bungee}} = Kx^2$, where x is the brick's distance from the wall. As x gets smaller, F_{bungee} gets smaller. Since $\sum F_x = F_{\text{bungee}} = Ma_x$, it follows that as x gets smaller, the acceleration decreases. This does not mean the brick slows down; it means the brick speeds up at a lesser and lesser rate, just like the tennis ball falling with wind resistance. In any case, since the brick's acceleration is not constant, we must find another way to solve for the time it takes to reach the wall.

Later in this course, you will learn conservation laws, such as conservation of energy. These laws allow you to solve problems such as this even when the forces and accelerations are non-constant. For now, let us just recall that acceleration is the change of velocity with time, while velocity is the change of position with time:

$$a = \frac{dv}{dt} = \frac{d(\frac{dx}{dt})}{dt} = \frac{d^2x}{dt^2}.$$

From the above force reasoning, applied not just at $x = L$ but at *any* x, we get

$$\sum F_x = Ma_x$$

$$Kx^2 = M\frac{d^2x}{dt^2}.$$

This is a second-order differential equation, which you will learn to solve in an advanced math class. For now, the point of this problem is to set up the differential equation correctly, and even more important, to realize that you cannot use your old constant-acceleration kinematic formulas.

QUESTION 7-2

A modern sculptor wants to hang her sculpture, which weighs 100 newtons, by two symmetrically-placed thin cords.

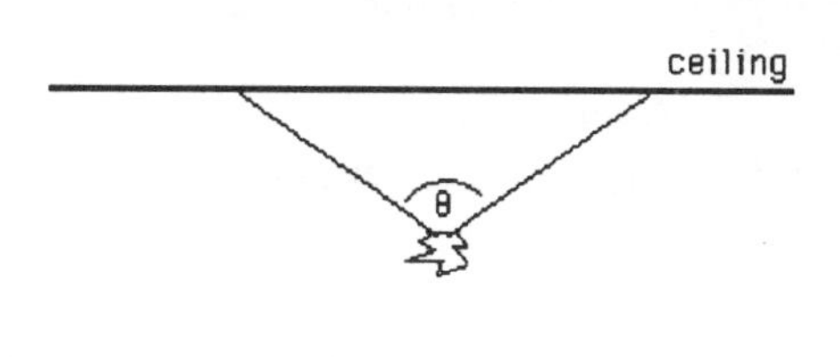

She uses thin fish-line cords because they are less visually distracting. For this same reason, she wants the angle between the cords, θ, to be as large as possible. Unfortunately, the tension in each thin cord must not exceed 80 newtons, or the cord will break.

What angle θ should the sculptor put between the cords when she hangs the sculpture?

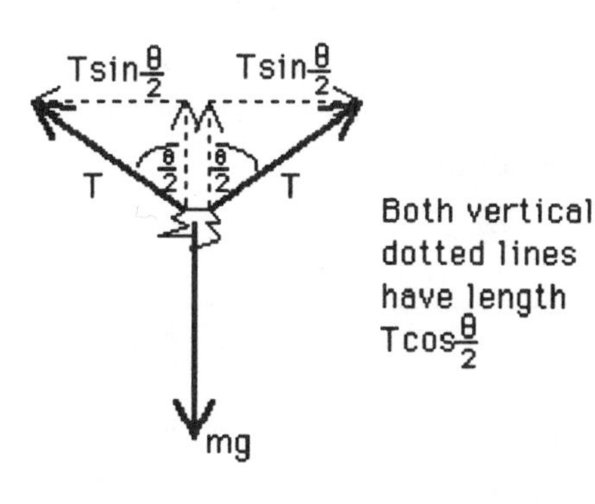

We must find the greatest angle between the cords such that the tension in each cord is 80 newtons or less. We will see below that, as the angle θ gets bigger, the tension in each cord increases. So, to find the biggest "safe" angle, we can see what angle θ corresponds to setting the tension in each cord equal to 80 N. Intuitively, because the cords are symmetrically placed, they both bear the same brunt, i.e., they both exert the same force (tension) on the sculpture. But those forces point in different directions.

As always, start with a free-body diagram, and break up the force vectors into components.

The tension in each cord is $T = 80$ newtons, and sculpture's weight is $mg = 100$ N. (By definition, "weight" is the gravitational force on an object.) Since the total angle between the cords is θ, the angle between a given cord and an imaginary vertical line is $\frac{\theta}{2}$.

I will pick the upward and rightward directions as positive. Because the sculpture hangs motionless, its net acceleration is $a_x = a_y = 0$. Therefore, the net force on the sculpture, in both the x- and y-directions, is zero. From the diagram, we get

$$\sum F_x = ma_x$$
$$T\sin\left(\frac{1}{2}\theta\right) - T\sin\left(\frac{1}{2}\theta\right) = 0 \tag{1}$$

$$\sum F_y = ma_y$$
$$T\cos\left(\frac{1}{2}\theta\right) + T\cos\left(\frac{1}{2}\theta\right) - mg = 0. \tag{2}$$

Eq. (1) confirms our guess that both cords exert the same tension. If you had not realized this, and labeled those tensions T_1 and T_2, then Eq. (1) would have proven that $T_1 = T_2$.

Let us now focus on Eq. (2). Since we know $T = 80$ N and $mg = 100$ N, θ is the only unknown. Let us solve for it.

Algebra starts here. Starting with Eq. (2), add mg to both sides to get $2T\cos(\frac{1}{2}\theta) = mg$. Divide through by $2T$, and then take the inverse cosine of both sides, to get $\frac{1}{2}\theta = \cos^{-1}\left(\frac{mg}{2T}\right)$. Finally, multiply through by 2 to get $\theta = 2\cos^{-1}\left(\frac{mg}{2T}\right) = 2\cos^{-1}\left[\frac{100\text{ N}}{2(80\text{ N})}\right] = 103°$, a bit more than a right angle.

QUESTION 7-3

A charged bead of mass $m = .0010$ kg travels rightward into a region containing an electric field. Since this experiment takes place in outer space, you can neglect the gravitational force on the bead. The region containing the electric field is between two metal plates. In other words, the particle experiences an electric force only while between the two plates. The field exerts an electric force of $F_e = .0020$ newtons *in the y-direction*, i.e., at right angles to the initial velocity of the bead.

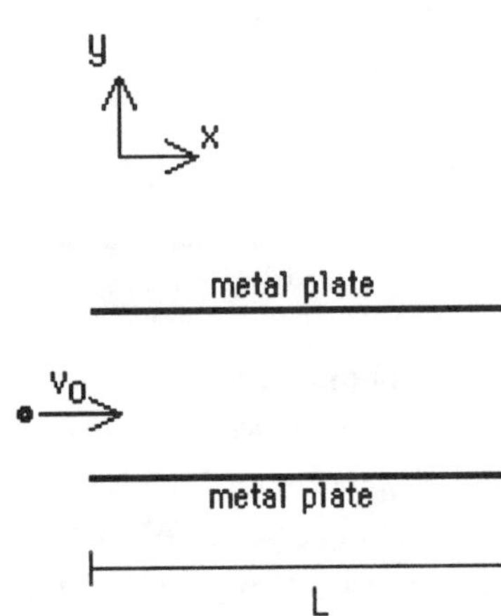

Before entering the electric field, the bead's velocity was $v_0 = 2.0$ m/s rightward. The metal plates are $L = 1.0$ meter "wide," as shown in the diagram.

What is the bead's velocity (speed and direction) when it leaves the electric field? In other words, how fast is the bead going, and in what direction, when it emerges from the right side of the electric field? Plug in numbers only after obtaining a symbolic answer in terms of m, F_e, v_0, and L.

ANSWER 7-3

Before playing around with formulas, let us picture the situation physically. The bead enters the electric field traveling rightward. While between the metal plates, the bead gets accelerated "upward," and therefore gains upward velocity. What about the bead's rightward velocity? You might think it slows down, because the bead's upward velocity comes "at the expense" of its horizontal speed, or because the rightward motion dies away. But no horizontal force acts on the bead. Therefore, it experiences no horizontal acceleration. Therefore, it is rightward velocity stays constant. In brief, *the y-forces on the bead in no way affect its x-motion.* The x-motion and y-motion are independent; you can address them separately in your problem-solving. This is a general result, applicable in all situations. You have seen this before. The horizontal component of a cannon ball's velocity stays constant, because no horizontal force acts on the projectile. This is not something special about projectiles. In general, if only y-directed forces act on an object, its x-component of velocity stays constant, since $a_x = 0$. And if only x-directed forces act on an object, its y-component of velocity stays constant.

In summary, while between the plates, the bead keeps its horizontal velocity, but also gradually gains vertical velocity. So, the bead starts curving upward parabolically, just as a horizontally-thrown ball arcs downward due to gravity. See the computer simulation, which lets you control the strength of the electric field (and hence the upward force). Here is a picture of the particle's motion.

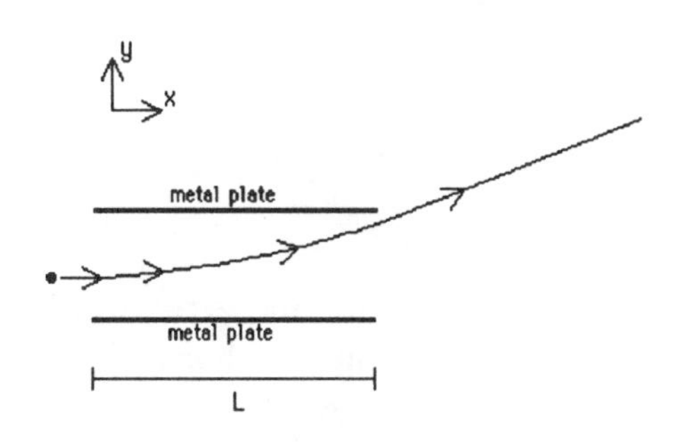

After it leaves the electric field, the particle travels in a straight line at constant speed. That is because no force, and hence no acceleration "acts on" on it; the bead no longer speeds up, slows down, or changes direction.

Now that we have figured out what is happening qualitatively, we can formulate a plan. The usual force problem strategy works fine:

(1) Free-body diagram.

(2) Newton's 2nd law, in x- and y-direction. In this case, we can use those equations to find a_x and a_y, the horizontal and vertical acceleration.

(3) Kinematics to find what we want, in this case the final velocity. As always, consider the x- and y-motion separately.

Step 1: Free-body diagram

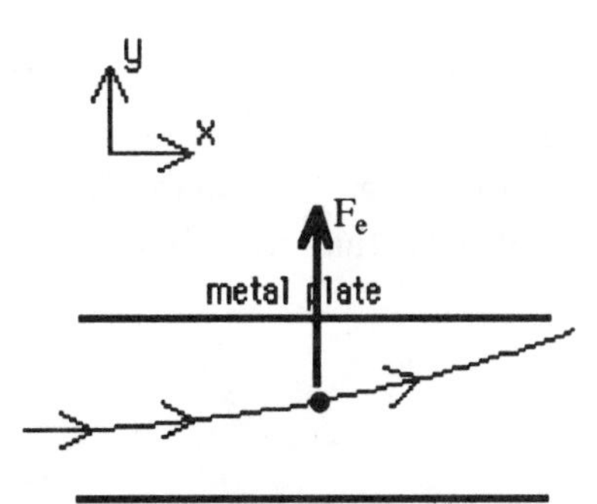

You might wonder how the particle keeps moving (partly) sideways even though the force on it points upward. Once an object is in motion, it does not "need" a force to keep moving. It needs a force only to *change* its motion somehow, either in speed or in direction (or both). As noted above, while between the plates, the bead keeps its sideways motion but also starts to acquire upward motion because of the electric force.

Step 2: Newton's laws

From the diagram, $\sum F_x = 0 = ma_x$, and hence $a_x = 0$. This restates our qualitative conclusion that the bead's *sideways* velocity neither speeds up nor slows down. But in the vertical direction,

$$\sum F_y = ma_y,$$
$$F_e = ma_y,$$

and hence $a_y = \frac{F_e}{m}$.

Step 3: Kinematics

As usual, it is easiest to deal separately with the x- and y-components of motion. We are trying to find the total final velocity **v** when the bead leaves the electric field. To do so, we can separately find v_x and v_y, the bead's final horizontal and vertical velocity. Then, we can vectorially add those components to obtain the total final velocity.

To solve for v_x and v_y, break up the vector formula $\mathbf{v} = \mathbf{v}_0 + \mathbf{a}t$ into its horizontal and vertical parts. As noted above, since $a_x = 0$, the bead's sideways velocity stays at its initial value:

$$\begin{aligned} v_x &= v_{0x} + a_x t \\ &= v_0 + 0 \\ &= 2.0 \text{ m/s}. \end{aligned} \tag{1}$$

Considering the vertical motion, we see that the particle initially has no vertical velocity; all of it is initial motion is horizontal. So, $v_{0y} = 0$. We also know the vertical acceleration, a_y, from step 2.

$$\begin{aligned} v_y &= v_{0y} + a_y t. \\ &= 0 + \frac{F_e}{m} t. \end{aligned} \tag{2}$$

To solve Eq. (2) for v_y, we need to know t, the time it takes the bead to pass between the plates. What information can help us find t? While between the plates, the bead travels a sideways distance $x - x_0 = L = 1 \text{ meter}$. So, to find t, we can try the horizontal part of the vector formula $\mathbf{r} = \mathbf{r}_0 + \mathbf{v}_0 t + \frac{1}{2}\mathbf{a}t^2$, namely

$$x = x_0 + v_{0x}t + \frac{1}{2}a_x t^2. \tag{3}$$

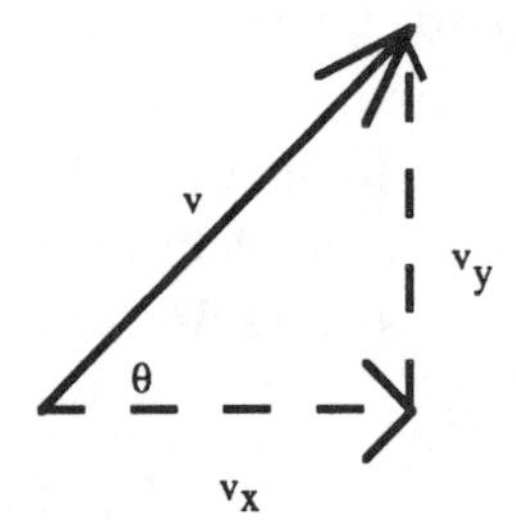

Set $x_0 = 0$. We know $x = L = 1 \text{ m}$. From Eq. (1), we also know $v_{0x} = v_0 = 2 \text{ m/s}$, and $a_x = 0$. So, we can immediately solve Eq. (3) for t to get $t = \frac{x - x_0}{v_{0x}} = \frac{L}{v_0} = \frac{1.0 \text{ m}}{2.0 \text{ m/s}} = 0.50$ seconds. That is how long the bead spends accelerating upward in the electric field. Substituting that time into Eq. (2) gives

$$\begin{aligned} v_y &= v_{0y} + a_y t \\ &= 0 + \frac{F_e}{m}\left(\frac{L}{v_0}\right) \\ &= 0 + (2.0 \text{ m/s}^2)(0.50 \text{ s}) \\ &= 1.0 \text{ m/s}. \end{aligned}$$

So, the bead's total final velocity is the vector sum of $v_y = 1.0 \text{ m/s}$ and $v_x = v_0 = 2.0 \text{ m/s}$.

The magnitude of this velocity vector is

$$v = \sqrt{v_x^2 + v_y^2} = \sqrt{(2.0 \text{ m/s})^2 + (1.0 \text{ m/s})^2} = 2.2 \text{ m/s}.$$

The angle θ specifies the direction of the bead's final velocity. Since $\tan\theta = \frac{v_y}{v_x}$, we get

$$\theta = \tan^{-1}\left(\frac{v_y}{v_x}\right) = \tan^{-1}\left(\frac{1.0 \text{ m/s}}{2.0 \text{ m/s}}\right) = 27°.$$

QUESTION 7-4

A physics teaching assistant (T.A.) wants to demonstrate constantly-accelerated motion by letting an essentially frictionless ice cube slide down an inclined plane. The ramp is $L = 1.5$ meters long. The T.A. can control the angle θ that the plane makes with the floor. The T.A. plans to release the ice cube from rest at the top of the ramp. In order for students to see everything clearly, the T.A. wants the ice cube to take exactly $t = 2.0$ seconds to slide all the way down the ramp.

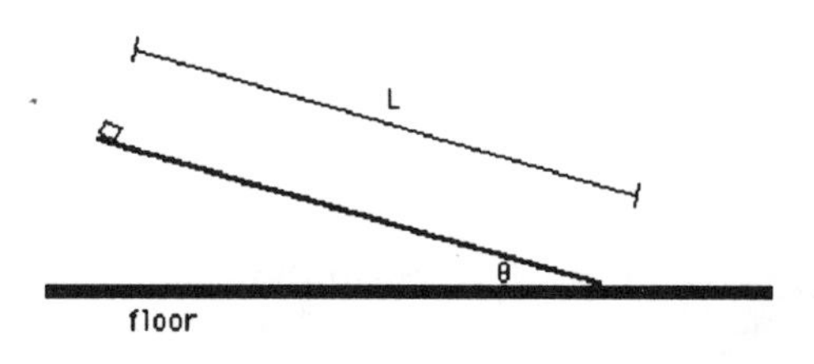

What angle θ should the ramp make with the floor? Express your answer symbolically in terms of L, t, and g before plugging in numbers.

ANSWER 7-4

Instead of flailing around with formulas, let us first formulate a strategy by picturing what is going on. The ice cube must take $t = 2.0$ seconds to slide down the $L = 1.5$ m ramp. If the T.A. picks a ramp angle θ that is too large, the ice cube speeds up too quickly, reaching the bottom of the ramp before 2 seconds have elapsed. On the other hand, if the T.A. picks too small a ramp angle, the ice cube accelerates so slowly that it does not reach the bottom in time. The T.A. needs to pick the ramp angle that produces the "correct" acceleration, i.e., the acceleration that makes the ice cube reach the end of the ramp in exactly 2 seconds. To get a feel for which angle generates this "correct" acceleration, experiment with the computer simulation.

This description suggests a two-part strategy:

(A) Figure out what acceleration the ice cube must have to slide down the ramp in exactly 2 seconds, and then

(B) Figure out what ramp angle produces the desired acceleration.

This is just a backwards version of the usual force problem-solving strategy. Normally, we would use forces to figure out the acceleration, and then substitute the acceleration into kinematic formulas to calculate the travel time t. Here, we are given t, and we must use kinematic equations to find the corresponding acceleration; then we must find what ramp angle generates the "right" force to produce that acceleration. So, we will work through the same steps as always, though perhaps in a different order.

You might be worried that I have not supplied the ice cube's mass, m. As you will see, m cancels out of your final answer.

In this problem, as in most ramp problems, pick your coordinate axes parallel and perpendicular to the surface of the ramp, instead of horizontal and vertical. In other words, "rotate" your coordinate axes. I will let the x-direction be parallel to the ramp, and the y-direction be perpendicular to the ramp.

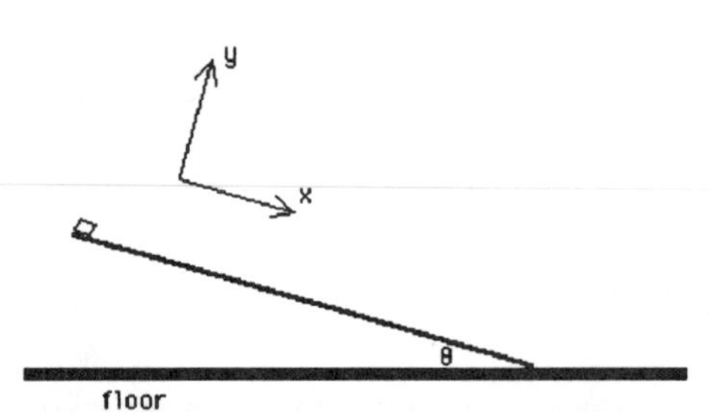

Let us do step (A) first. The ice cube, starting from rest, must speed up with enough acceleration so that in time $t = 2$ seconds, it covers a distance

$$\Delta x = x - x_0 = L = 1.5 \text{ meters}.$$

We can start with the displacement formula

$$\begin{aligned} x &= x_0 + v_{0x}t + \frac{1}{2}a_x t^2. \\ L &= 0 + 0 + \frac{1}{2}a_x t^2. \end{aligned} \tag{1}$$

Solve for a_x, the acceleration along the direction of motion, to get $a_x = \frac{2L}{t^2} = \frac{2(1.5\text{ m})}{(2\text{ s})^2} = 0.75\ \text{m/s}^2$. This is the acceleration we want the ice cube to have.

Now we can tackle step (B). To figure out what ramp angle produces an acceleration $a_x = 0.75\ \text{m/s}^2$, we must draw a free-body diagram, breaking up the force vectors into their x- and y-components. Then, Newton's 2nd law can relate those forces to the acceleration we desire.

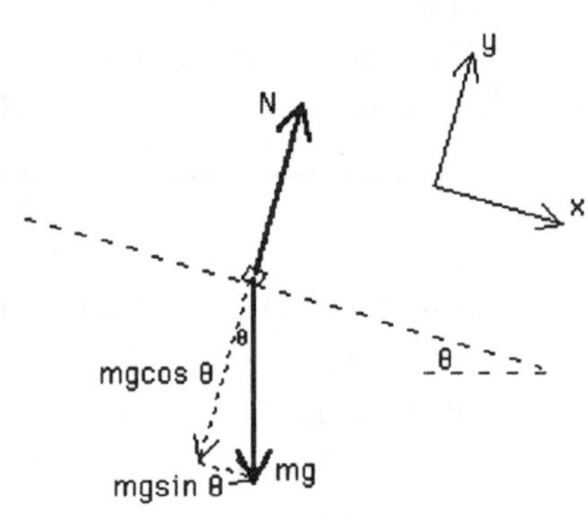

N is the "normal" force exerted by the ramp on the ice cube. *"Normal" forces always act at right angles to surface on which the object slides or rests.* So, the normal force points in the y-direction, perpendicular ("normal") to the ramp.

I have broken the gravitational force vector into its components parallel and perpendicular to the ramp. Using geometry, you can prove that the two θ's in the diagram are indeed the same angle. Physically, $mg\sin\theta$ is the "part" of gravity pushing the ice cube along the ramp, while $mg\cos\theta$ is the part of gravity tending to pull the ice cube "into" the surface of the ramp.

Now we can use the vector components of Newton's 2nd law, $\mathbf{F}_{\text{net}} = m\mathbf{a}$, to solve for the ramp angle θ needed to make the block slide with the desired acceleration of $a_x = \frac{2L}{t^2} = 0.75\ \text{m/s}^2$. (Remember, that is the acceleration needed to make the ice cube reach the bottom of the ramp in $t = 2$ seconds.) Usually, we would need to consider both the x- and y-directed forces. But here, we can extract all the information we need from the x-directed forces:

$$\begin{aligned} \sum F_x &= ma_x \\ mg\sin\theta &= m\frac{2L}{t^2} \\ g\sin\theta &= \frac{2L}{t^2} = 0.75\ \text{m/s}^2. \end{aligned} \tag{2}$$

Notice that the ice cube's mass canceled out; the ice cube's mass does not "matter." Solving Eq. (2) for θ yields

$$\theta = \sin^{-1}\left(\frac{a_x}{g}\right) = \sin^{-1}\left(\frac{\frac{2L}{t^2}}{g}\right) = \sin^{-1}\left(\frac{0.75\ \text{m/s}^2}{9.8\ \text{m/s}^2}\right) = 4.4°.$$

So, the T.A. must slope the ramp *very* gently. The computer simulation confirms this.

QUESTION 7-5

Sharon pushes a tall box across the floor. Her arms, and therefore her push force of $P = 400$ N, make a $\theta = 30°$ angle with the horizontal. Since the box rests on ball bearings, it is essentially frictionless. Sick of her job, Sharon decides to give the box a brief shove, lasting only $t = 0.20$ seconds. Find the velocity of the box at the completion of Sharon's shove, assuming

(a) the box's weight is 490 N

(b) the box's weight is 122.5 N.

In both cases, the box starts from rest.

Hint: Does the box stay on the floor?

ANSWER 7-5

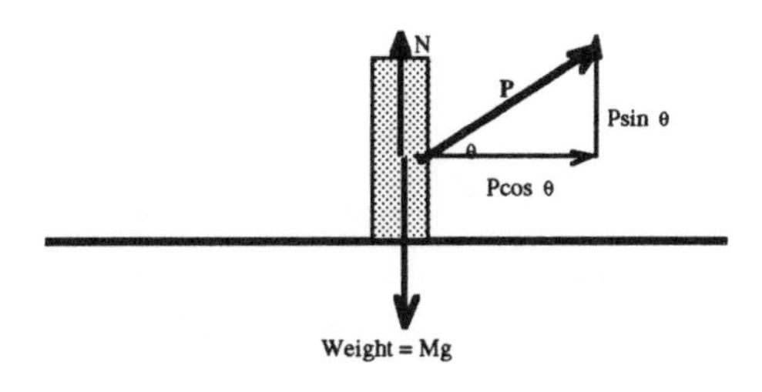

As always, start by drawing a free-body diagram, and divide up the force vectors into components when necessary. Since the box sits on a flat floor, it makes sense to choose the usual horizontal and vertical coordinate axes.

N denotes the "normal" force exerted by the floor on the box. A normal force always points perpendicular to the surface on which the object sits or slides. By definition, "weight" is the gravitational force on an object. So, near the surface of the Earth, Weight $= mg$.

(a) In this part of the problem, the box's weight is $mg = 490$ N. That is bigger than the upward component of the push force: $P_y = P\sin\theta = (400\text{ N})\sin 30° = 200$ N. Therefore, the box stays on the floor. Its only acceleration is horizontal: $a_y = 0$.

By the way, the normal force pushes as hard as needed to keep the box from falling down through the floor. Usually, the normal force would have to counteract the box's whole weight. But here, the normal force gets "help" from the upward part of Sharon's push. The normal force supplies just enough upward force so that N together with $P\sin\theta$ cancels gravity.

To figure out the final horizontal velocity of the box, we must find its horizontal acceleration, and then use kinematics. To find a_x, use Newton's 2nd law, looking specifically at the x-directed forces. Letting rightward count as the "positive" direction, I get $\sum F_x = P\cos\theta = ma_x$. Before we can solve for a_x, we need to know the box's mass. Well, since Weight $= mg$, we have $m = \frac{\text{Weight}}{g} = (490\text{ N})(9.8\ \text{m/s}^2) = 50$ kg.. So, from Newton's 2nd law,

$$a_x = \frac{P\cos\theta}{m} = \frac{(400\text{ N})\cos 30°}{50\text{ kg}} = 6.9\ \text{m/s}^2.$$

Now that we know the horizontal acceleration, we can find the box's speed after time $t = 0.20$ seconds using

$$\begin{aligned} v_x &= v_{0x} + a_x t \\ &= 0 + (6.9 \text{ m/s})(0.20 \text{ s}) \\ &= 1.4 \text{ m/s}. \end{aligned}$$

It has only x-directed velocity, because it stays on the floor.

(b) In this part of the problem, something unexpected happens. The box weighs only 122.5 newtons. But as we saw above, the upward part of Sharon's push is bigger: $P \sin\theta = 200$ newtons. Therefore, the box gets thrown off the floor. *Consequently, the floor can no longer exert forces on the box.* There is no longer a normal force. We must modify the free-body diagram.

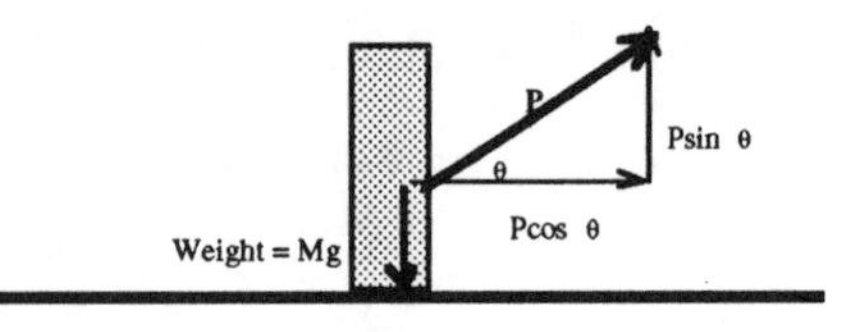

So now, the box accelerates both rightward and upward. As always, you can deal separately with the x- and y-components of motion. Specifically, we can use the x- and y-components of Newton's 2nd law to separately calculate a_x and a_y, the horizontal and vertical acceleration. Then we can use those acceleration components to find v_x and v_y, the box's horizontal and vertical velocity at $t = 0.20$ s. Finally, we can vectorially add those velocity components to obtain the magnitude and direction of the total velocity.

Since Weight $= mg$, the box's mass is $m = \frac{\text{Weight}}{g}(122.5 \text{ N})(9.8 \text{ m/s}^2) = 12.5 \text{ kg}$. From the modified free-body diagram,

$$\begin{aligned} \sum F_x &= ma_x \\ P\cos\theta &= ma_x, \end{aligned}$$

and hence $a_x = \frac{P\cos\theta}{m} = \frac{(400 \text{ N})\cos 30^\circ}{12.5 \text{ kg}} = 28 \text{ m/s}^2$. By equivalent reasoning,

$$\begin{aligned} \sum F_x &= ma_y \\ P\sin\theta - mg &= ma_y, \end{aligned}$$

and hence $a_y = \frac{P\sin\theta - mg}{m} = \frac{(400 \text{ N})\sin 30^\circ - (12.5 \text{ kg})(9.8 \text{ m/s}^2)}{12.5 \text{ kg}} = 6.2 \text{ m/s}^2$.

Now that we know the horizontal and vertical components of the box's acceleration, we can combine them in a vector diagram to obtain the total acceleration. Or we can work with them separately to get v_x and v_y. I will choose the second option:

$$\begin{aligned} v_x &= v_{0x} + a_x t \\ &= 0 + (28 \text{ m/s}^2)(0.20 \text{ s}) \\ &= 5.6 \text{ m/s}. \end{aligned}$$

$$\begin{aligned} v_y &= v_{0y} + a_y t \\ &= 0 + (6.2 \text{ m/s}^2)(0.20 \text{ s}) \\ &= 1.2 \text{ m/s}. \end{aligned}$$

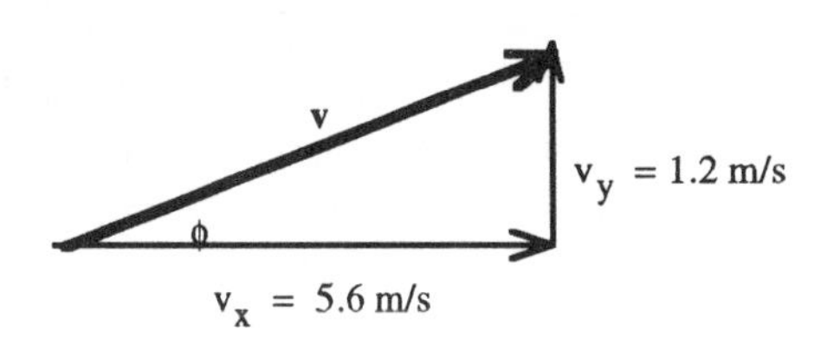

Let me represent these velocity components in a vector diagram.

Notice that the angle ϕ in this diagram differs from $\theta = 30°$. In fact, since $\tan\phi = \frac{v_y}{v_x}$,

$$\phi = \tan^{-1}\left(\frac{v_y}{v_x}\right) = \tan^{-1}\left(\frac{1.2 \text{ m/s}}{5.6 \text{ m/s}}\right) = 12°.$$

Remember, when asked for a vector quantity such as velocity or acceleration, you must solve for direction (angle) as well as magnitude. The magnitude of this velocity (i.e., the box's speed) is

$$v = \sqrt{v_x^2 + v_y^2} = \sqrt{(5.6 \text{ m/s})^2 + (1.2 \text{ m/s})^2} = 5.7 \text{ m/s}.$$

QUESTION 7-6

Two astronauts, Joe and Moe, take an untethered space walk. They decide to have an outer-space tug of war, using a light rope. In their space suits, Joe has mass 100 kg, whereas Moe has mass 110 kg. The loser is the first astronaut to cross an imaginary line midway between them.

(a) Draw a force (free-body) diagram for each astronaut after they start tug-of-warring. Of all the forces you have drawn, which is biggest and which is second biggest? Explain why.

(b) Moe will definitely win. Explain why. Do not just say "'cause he is heavier."

(c) Miffed at his loss, Joe challenges Moe to a tug of war on Earth. Draw the force diagram for each astronaut during the tug of war.

(d) On Earth, it is possible (though difficult) for Joe to win. What could Joe do to increase his chances of winning? Relate your answer to relevant physical principles.

ANSWER 7-6

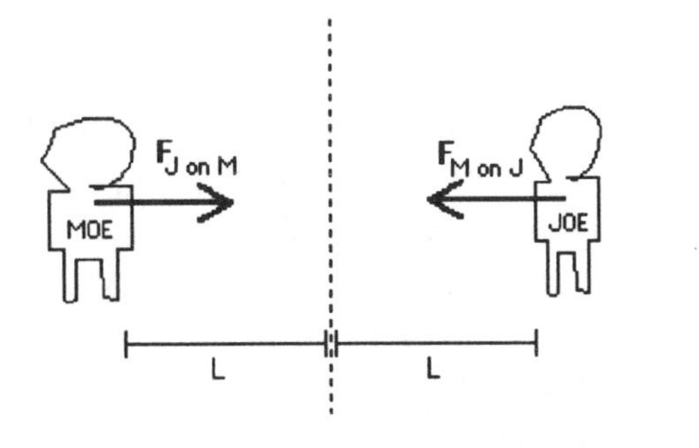

(a) We can neglect the small gravitational forces exerted on the astronauts by the Earth, the Moon, etc. The only "big" forces experienced by the astronauts are the forces they exert on each other, via the rope.

You might think that the force on Moe is bigger, because $F = ma$, and Moe has more mass. But it is misleading to think that a force *is* a mass times acceleration. A force is a push. For instance, you can apply the same push to a bowling ball and a tennis ball. The force on the bowling ball is not bigger just because it is heavier. Instead, both forces are the same; and since $F = ma$, the bowling ball accelerates more slowly. The equation $F = ma$ tells us that a force (push) *causes* a mass to accelerate. So, we cannot say the force on Moe is greater just because he is more massive. Which force is greater depends entirely on who is getting pulled harder.

On the other hand, you might think that Moe can pull harder on Joe than vice versa. By Newton's 3rd law, however, the force Moe exerts on Joe must be equal in magnitude and opposite in direction to the force Joe exerts on Moe: $\mathbf{F}_{\text{J on M}} = -\mathbf{F}_{\text{M on J}}$. However hard Joe gets pulled leftward, that is how hard Moe gets pulled rightward. Ironically, the harder Moe pulls on the rope, the more quickly he shoots rightward, even if Joe just holds his end of the rope. To see why, imagine yourself on roller skates, playing tug of war with a tree. The harder you pull on your end of the rope, the more quickly you accelerate toward the tree. In real-life tug of war, your sneakers help to hold you in place; you might not notice that the harder you pull on the rope, the harder it "pulls back" on you. But the pull-back effect is always there, even if your opponent is a person instead of a tree.

If this does not make sense, consider the tension in the rope. Whatever tension Moe creates by pulling on his end of the rope, that tension acts not only on Joe, but also on Moe. To see why, think of the rope as a flexible bungee cord. By pulling on his end, Moe stretches the cord, which then tries to "unstretch," pulling Moe rightward. This reasoning applies to any rope, not just one that stretches by a visible amount, because all ropes stretch a little.

(b) Let L denote the distance from either astronaut to the imaginary line midway between them. Whichever astronaut covers that distance first loses. Intuitively, whichever astronaut has higher acceleration will cover distance L first, thereby losing the tug of war. (You can confirm this intuition by playing around with $x = x_0 + v_0t + \frac{1}{2}at^2$.)

At first glance, you might guess that the astronaut who experiences the biggest force will lose. But as discussed in part (a), both astronauts "feel" the same force. Does this mean they both cover distance L in the same time? No. Joe is lighter. Therefore, his acceleration is higher. To see this intuitively, push a tennis ball and a bowling ball with the same force. The tennis ball speeds up faster. Similarly, since Joe and Moe experience the same force, but Joe has less mass, Joe accelerates faster, and therefore reaches the imaginary line first. In formulas,

$$a_M = \frac{F_{J\text{ on }M}}{110\text{ kg}},$$

$$a_J = \frac{F_{M\text{ on }J}}{100\text{ kg}}.$$

Since both forces are the same by Newton's 3rd law, $a_J > a_M$. (If you are not convinced, plug in a number, say 200 newtons, for the force.) Joe speeds up more quickly, and therefore he loses.

(c) N denotes the normal force exerted by the ground on the astronaut's shoes. Roughly speaking, the normal force "holds you up." Instead of writing $F_{M\text{ on }J}$ and $F_{J\text{ on }M}$, you could have called both those forces "T" for tension.

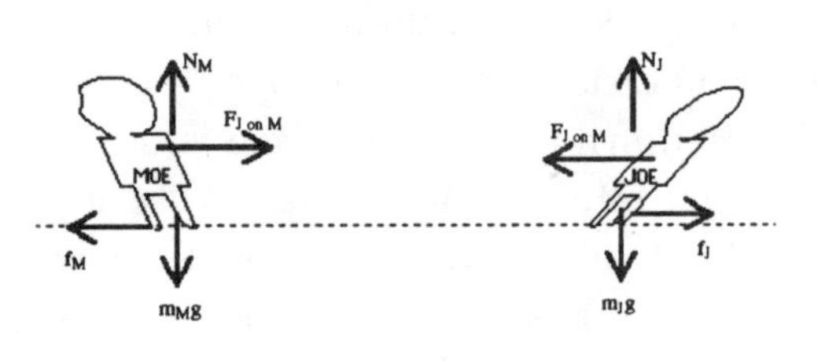

The frictional force f, like the normal force, is exerted by the ground on the people's shoes.

A common mistake is to draw f_M as pointing towards Joe, and f_J as pointing towards Moe. Here is the intuition: When Joe leans back, digs in, and pushes his shoes against the ground with all his might, he is pushing *forward*, i.e., towards Moe. That intuition is correct, and shows that Joe is "pushing the ground" toward Moe. In other words, the force exerted *by* Joe *on* the ground is forward, toward Moe. But when drawing a force diagram of Joe, we do not care about the forces he exerts on other objects. We care only about the forces exerted by other objects on him. As just noted, Joe pushes forward on the ground. Therefore, by Newton's 3rd law, the ground pushes backward against him. That *backward* force is friction.

Here is another way to see why the frictional forces point "backward." If Joe and Moe play tug of war on ice, they *would* fly toward each other, just like in outer space. Friction is what prevents them from hurtling toward each other. In other words, friction must "resist" the astronauts' tendencies to get pulled toward each other. Therefore, the frictional forces must point backward, i.e., away from each other.

This stuff is extremely subtle, and hard to explain in writing. Please see your instructor if you are confused.

(d) As shown above, Joe cannot win simply by pulling harder on the rope than Moe pulls; whatever extra tension Joe creates in the rope pulls on Joe as well as Moe. That was the main point of part (a). Of course, if Joe catches Moe off balance and "jerks" the rope, he might make Moe fall over. But if Moe stays "leaned back," Joe's frantic tugs end up affecting Joe just as much as they affect Moe.

Joe will win if he can generate a frictional force bigger than the tension in the rope (i.e., f_J must exceed $F_{M\text{ on }J}$). In that case, Joe accelerates rightward, pulling Moe with him. To generate a big f_J, Joe could wear shoes with excellent traction. As I will explain in Chapter 8, $F_J = \mu N_J$, where μ is the "coefficient of friction" between Joe's shoes and the ground. Good traction corresponds to high μ. Leaning back also helps Joe to generate a good contact between his shoes and ground, and ensures that he will not be caught off balance if Moe suddenly jerks the rope. Nonetheless, Joe remains at a disadvantage. From the force diagrams, the vertical forces cancel: $N_J = m_J g$, and $N_M = m_M g$. So, since Moe is heavier, $N_M > N_J$. Therefore, because $f = \mu N$, Moe generates a bigger frictional force, other things being equal. As just noted, Joe can make sure other things *are not* equal by using high-traction shoes, so that his μ exceeds Moe's μ.

Forces and Newton's Laws:

Applications, with Introduction to Multi-Block Problems

8

CHAPTER

QUESTION 8-1

A suitcase of mass $m = 40$ kg is pulled along a nearly frictionless floor. The strap, which makes a $\theta = 30°$ with the floor, pulls with force $F_p = 25$ newtons.

strap

(a) If the suitcase starts at rest, how fast will it be going after it is been pulled a distance $d = 3.0$ meters?

(b) Suppose the strap is now pulled as hard as it can be pulled such that the suitcase does not come off the floor. Starting from rest, the suitcase is pulled for time $t = 0.30$ s. How far does the suitcase travel in that time?

ANSWER 8-1

(a) We can use the standard force problem-solving strategy:

(1) Draw a free-body diagram, with forces broken into components where necessary.

(2) Write down Newton's 2nd law in the x- and y-direction. If you know the forces, you can solve for acceleration; and vice versa.

(3) If you obtained the acceleration in step 2, substitute it into kinematic formulas to find positions, velocities, and times.

I will now implement the strategy:

Step 1: Free-body diagram.

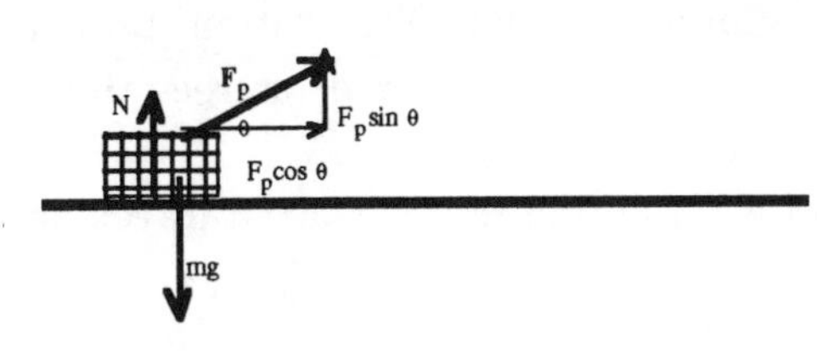

N, the "normal" force exerted by the floor on the suitcase, points perpendicular (i.e, "normal") to the surface on which the suitcase slides. Since the vertical component of the strap's pull,

$$F_p \sin\theta = (25 \text{ newtons})\sin 30° = 12.5 \text{ newtons},$$

is less than the suitcase's weight, the case stays on the floor.

Step 2: Newton's 2nd law.

To find a_x, we must consider the horizontal forces acting on the suitcase. The sideways part of the strap's pull is the only one. So, $\sum F_x = F_p \cos 30° = ma_x$. Solving for a_x gives $a_x = \frac{F_p \cos 30°}{m} = \frac{(25 \text{ N})\cos 30°}{40 \text{ kg}} = 0.54 \text{ m/s}^2$, rightward along the floor.

Step 3: Kinematics to find the quantity of interest.

We want to know the speed v after the suitcase has slid through distance $\Delta x = d = 3$ meters, with acceleration $a = 0.54 \text{ m/s}^2$. So, we can use the (one-dimensional) "time-independent" kinematic formula $v^2 = v_0^2 + 2a\Delta x$. This gives

$$v = \sqrt{v_0^2 + 2a\Delta x} = \sqrt{0 + 2ad} = \sqrt{2(0.54 \text{ m/s}^2)(3.0 \text{ m})} = 1.8 \text{ m/s}.$$

(b) Let $F_{p\,\max}$ denote the biggest force the strap can exert such that the suitcase stays on the floor. Given $F_{p\,\max}$, we can use it to find the suitcase's new acceleration. Then, we can use kinematics to find how far the suitcase travels in a given time. So, the "meat" of this problem is finding $F_{p\,\max}$.

If the upward part of the strap's pull exceeds the downward force of gravity, the suitcase comes off the floor. But what if we pull the strap just a little more gently, so that the suitcase *barely* stays in contact with the floor. Since the suitcase barely skims the floor, it does not press down very hard against the floor, and the floor does not press hard against it. In other words, the normal force is small. Indeed, if the suitcase *barely* stays on the floor, the normal force is only $N = 0.00001$ newtons. That is, $N \approx 0$. **This is a general insight you will use throughout this course. When an object barely stays in contact with a surface, or barely starts to lose contact with the surface, the normal force is $N \approx 0$.**

But even though $N \approx 0$, the suitcase does not jump off the floor. It moves horizontally, skimming along the floor. So, $a_y = 0$. It follows that the upward part of the strap's pull must exactly cancel the downward force of gravity. To see this mathematically, start with

$$\begin{aligned} \sum F_y &= ma_y \\ N + F_p \sin 30° - mg &= ma_y \\ 0 + F_{p\,\max} \sin 30° - mg &= ma_y \quad [\text{since } N \approx 0 \text{ when } F_p = F_{p\,\max}] \\ F_{p\,\max} \sin 30° - mg &= 0 \quad [a_y = 0, \text{ because motion is horizontal}]. \end{aligned}$$

Therefore, $F_{p\,\max} \sin 30° = mg$; the upward pull of the strap cancels gravity. Solve for $F_{p\,\max}$ to get

$$F_{p\,\max} = \frac{mg}{\sin 30°} = 2mg = 2(40 \text{ kg})(9.8 \text{ m/s}^2) = 780 \text{ N}.$$

At this point, we can recycle the reasoning of part (a). A component of $F_{p\,\max}$ provides the only horizontal force. So, $\sum F_x = F_{p\,\max} \cos 30° = ma_x$. Solve for the acceleration to get $a_x = \frac{F_{p\,\max} \cos 30°}{m} = \frac{(780 \text{ N}) \cos 30°}{40 \text{ kg}} = 17 \text{ m/s}^2$, which is nearly twice the acceleration of a free-falling object. (The person is flinging the suitcase; soon it will be out of her arm's reach.)

We now know the suitcase's (constant) acceleration, and want to find the distance it travels in time $t = 0.30$ s. So, we can use

$$\begin{aligned} x &= x_0 + v_0 t + \frac{1}{2}at^2 \\ &= 0 + 0 + \frac{1}{2}(17 \text{ m/s}^2)(0.30 \text{ s})^2 \\ &= 0.76 \text{ m} \end{aligned}$$

QUESTION 8-2

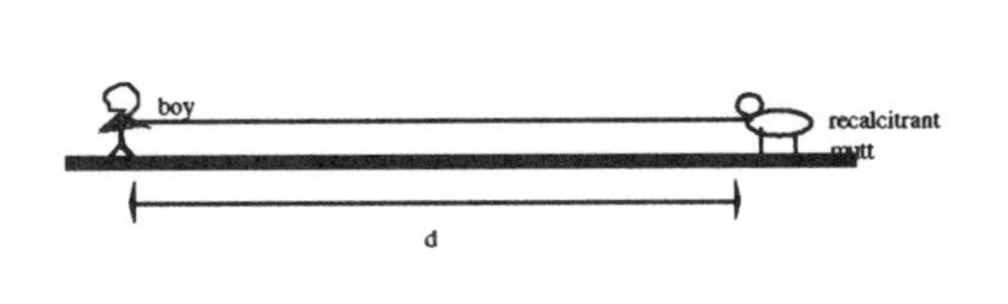

A boy of mass $M = 32$ kg is on the nearly frictionless ice-covered surface of a sidewalk with his dog, who has mass $m = 8.0$ kg. The boy and his dog are initially a distance $d = 4.0$ m apart. By pulling on the leash (a long rope), the boy exerts a constant force $F = 5.0$ N on the dog. As a result, the dog slides toward him. Remember, the boy and dog are standing on frictionless ice.

How fast is the dog moving when the dog and boy meet?

ANSWER 8-2

You might think the boy stays still while the dog slides toward him. But as Moe and Joe proved in Chapter 7, when the boy pulls on the rope, the rope pulls back on him. As a result, the boy slides toward the dog as the dog slides toward the boy. The boy and dog meet somewhere "in the middle."

You should recognize this as a standard "meeting" problem. When the dog and boy meet, they share the same position x. So, we can use the meeting strategy introduced in Question 3-5:

(1) Write x_B and x_d, the position of the boy and dog, as functions of time, using kinematics, Newton's laws, and whatever other tools you have available.

(2) Set $x_B = x_d$, and solve the resulting equation for t, the time at which the boy and dog meet.

(3) Once you know the meeting time, use kinematic formulas to find whatever you need.

Step 1: Write x_B and x_d as functions of time.

We can try using $x = x_0 + v_0 t + \frac{1}{2}at^2$ for the boy and dog. So, we need their accelerations. Since we know something about the relevant forces, I will draw force diagrams and use Newton's 2nd law.

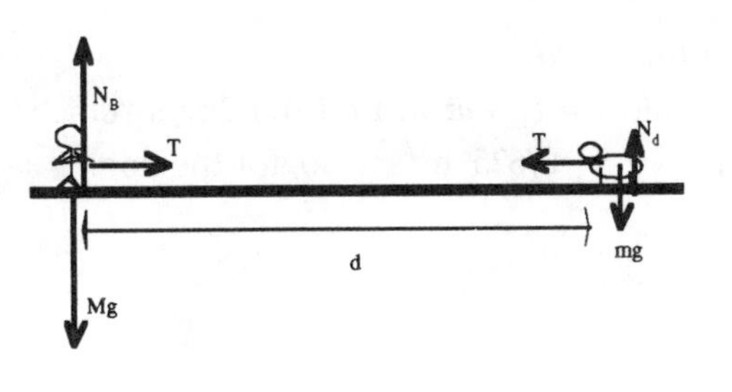

We are told the force on the dog, but not the force on the boy. Are we stuck? No! **By Newton's 3rd law, the force exerted by the boy on the dog must be equal in magnitude and opposite in direction to the force exerted by the dog on the boy.** Let me call that force T for tension.

In this problem, we care only about the horizontal motion. (In any case, the vertical forces cancel.) Let me choose rightward as positive. Then, the net force on the dog is negative, and hence its acceleration is negative:

$$\text{Dog: } \sum F_x = -T = ma_d \quad \Rightarrow \quad a_d = -\frac{T}{m} = -\frac{5.0 \text{ N}}{8.0 \text{ kg}} = -0.625 \text{ m/s}^2.$$

$$\text{Boy: } \sum F_x = T = Ma_B \quad \Rightarrow \quad a_B = \frac{T}{m} = \frac{5.0 \text{ N}}{32 \text{ kg}} = 0.156 \text{ m/s}^2.$$

Now that we know the accelerations, we can use $x = x_0 + v_0 t + \frac{1}{2}at^2$ to write expressions for x_B and x_d. Let us call the boy's initial position $x_{0B} = 0$. His initial velocity is $v_{0B} = 0$. So,

$$\text{Boy:} \quad x_B = x_{0B} + v_{0B}t + \frac{1}{2}a_B t^2,$$
$$= 0 + 0 + \frac{1}{2}\frac{T}{M}t^2.$$

Since I set $x_{0B} = 0$ for the boy, I must say that the dog's initial position is $x_{0d} = d$. (See the diagram given in the question.) The dog has no initial speed. And its acceleration is leftward, i.e., negative:

$$\text{Dog:} \quad x_d = x_{0d} + v_{0d}t + \frac{1}{2}a_d t^2,$$
$$= d + 0 + \frac{1}{2}\left(-\frac{T}{m}\right)t^2.$$

Step 2: Set $x_B = x_d$ and solve for t.

The "meeting point" is where $x_B = x_d$, i.e., where the dog and boy have the same position. Using the expressions just obtained for x_B and x_d, I get

$$x_B = x_d$$
$$\frac{1}{2}\frac{T}{M}t^2 = d - \frac{1}{2}\frac{T}{m}t^2.$$

Solve for t to get

$$t = \sqrt{\frac{d}{\frac{1}{2}\left(\frac{T}{M} + \frac{T}{m}\right)}} = \sqrt{\frac{4.0 \text{ m}}{\frac{1}{2}\left(\frac{5.0 \text{ N}}{32 \text{ kg}} + \frac{5.0 \text{ N}}{8.0 \text{ kg}}\right)}} = 3.2 \text{ s},$$

which sounds reasonable. That is how much time elapses, after the boy starts pulling, before the boy and dog meet.

Step 3: Substitute the meeting time into kinematic formulas to find what you need

(a) Now that we know the meeting time t, we can substitute it into $v = v_0 + at$ to find the dog's "crash" velocity. Above, we found the dog's acceleration to be $a_d = -\frac{T}{m} = -0.625 \text{ m/s}^2$. So, for the dog,

$$v = v_{0d} + a_d t$$
$$= 0 - (0.625 \text{ m/s}^2)(3.2 \text{ s})$$
$$= -2.0 \text{ m/s},$$

where the minus sign indicates leftward.

By the way, when you plug the meeting time back into our above expression for x_B, you get $x_B = 0.80$ m. So, the boy and dog meet after the boy has slid 0.8 m and the dog has slid 3.2 m (since they started 4 meters apart). The dog slides four times as far as the boy slides. That makes sense, because the dog is four times lighter, and hence its acceleration is four times larger.

QUESTION 8-3

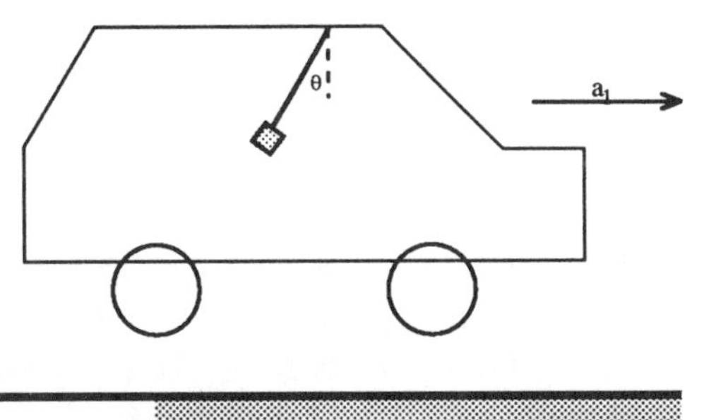

Hanging from the ceiling of my 1970's retro car is a pair of fuzzy dice. Consider one of them. It hangs from a very light string. Suppose my car accelerates forward with acceleration a_1. As a result, the die swings back, eventually settling at a certain angle. What angle θ does the string make with the vertical? Express your answer in terms of a_1 and g.

ANSWER 8-3

Let us picture what is happening. If the car were sitting still or moving at constant speed, the die would hang straight down. (Try this with your shoe the next time you are riding in a car at steady speed!) But here, the car *accelerates* forward. As a result, the passengers and the die feel "thrown back." Does this mean some force pushes the die backward? No. The die "tries" to sit still, while the car accelerates forward. So, the die moves backward relative to the car. Think of it this way: The die does not get thrown toward the back of the car. Instead, the back of the car accelerates forward toward the die.

But the die does not keep moving backwards relative to the car. After it "settles down," the die gets dragged along with the car. Since the car accelerates forward, so does the die. Indeed, the die's acceleration must equal the car's acceleration, or else the die would get left behind. Since the die accelerates forward, some force must be pushing it forward, according to Newton's 2nd law. On your force diagram, you might be tempted to call this force "*ma*." But remember, acceleration is not a "kind" of force, like gravity or tension. Acceleration is a change in velocity that *results* from forces (pushes) such as gravity and tension. So, although "mass times acceleration" should appear in your reasoning (on the right-hand side of $F = ma$), it should *not* appear on your force diagram. Given all this, the question remains, What force pushes the die forward so that it "keeps up" with the car's acceleration?

To answer this, draw a force diagram. (Indeed, even if you did not think through these qualitative considerations, you should have started off by drawing the free-body diagram.) There is no normal force, because the floor does not touch on the die. Tension always points along the direction in which the rope is stretched.

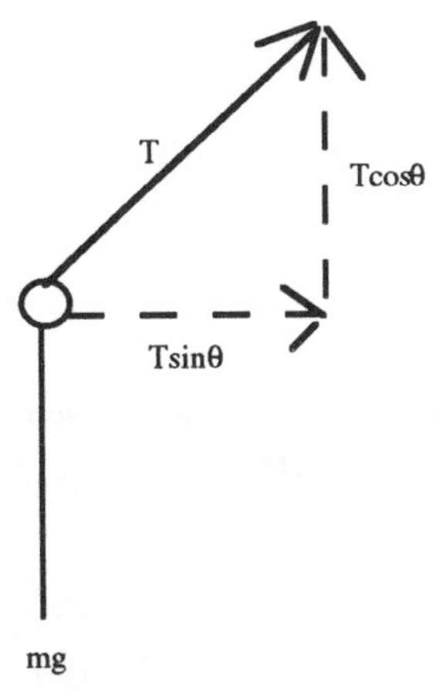

We now see what is "supplying" the forward force on the die. It is the horizontal component of the tension. In summary: When the car starts accelerating, the die gets "thrown back" more and more. (Actually, as we saw above, the car gets thrown forward instead of the die's getting thrown back; but let me speak in rough terms.) The die eventually reaches a throw-back angle θ such that the horizontal component of tension supplies a big enough forward force to make the die "keep up" with the car. That is the angle at which the die settles. After the die settles, it has no vertical motion: $a_y = 0$. And its horizontal acceleration "matches" the car's: $a_x = a_1$. **Whenever one object is motionless with respect to another object, those two objects must "share" the same velocity and acceleration.**

How can we solve for θ? Well, we know the horizontal and vertical accelerations. As always in force problems, we can try Newton's 2nd law in the x- and y-directions. Reading the forces off the free-body diagram, I get

$$\sum F_x = ma_x$$
$$T \sin\theta = ma_1 \qquad [\text{since } a_x = a_1] \tag{1}$$

$$\sum F_y = ma_y$$
$$T \cos\theta - mg = 0 \qquad [\text{since } a_y = 0] \tag{2}$$

At this point, we can solve for θ, because we have two equations in two unknowns (θ and T). So, we are done with the physics. All that remains is algebra.

Algebra starts here. To complete the algebra quickly, first rewrite Eq. (2) as $T\cos\theta = mg$. So, we have

$$T \sin\theta = ma_1 \tag{1}$$

$$T \cos\theta = mg. \tag{2}$$

Now for a trick: Divide Eq. (1) by Eq. (2), remembering that $\tan\theta = \frac{\sin\theta}{\cos\theta}$. When we take this quotient, the T's and m's cancel, leaving us with $\tan\theta = \frac{a_1}{g}$. So, $\theta = \tan^{-1}\left(\frac{a_1}{g}\right)$.
End of algebra.

Physically, this formula means that when the car accelerates more rapidly (bigger a_1), the die swings back further, as we intuitively expect.

QUESTION 8-4

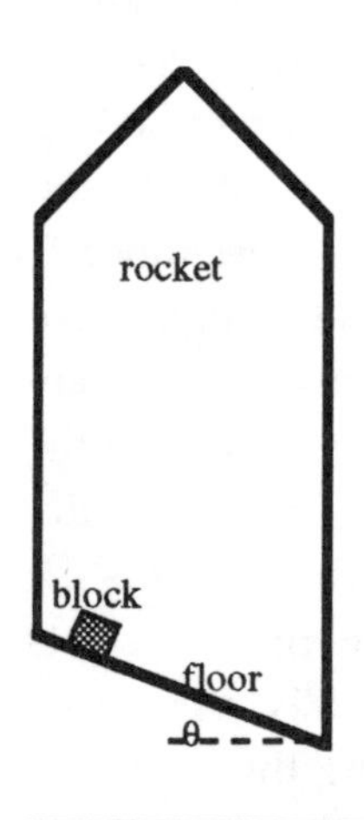

Consider a poorly constructed rocket ship, whose floor makes an angle θ with the horizontal. A block of mass m slides frictionlessly down the floor. Find the block's acceleration *relative to the rocket's floor* in the following situations:

(a) Rocket sitting still on the launch pad. (Scaffolding keeps the rocket pointing up, so that the floor is tilted at angle θ.)

(b) Rocket ascending at constant speed v.

(c) Rocket riding sideways in a truck at constant speed v. (The rocket still points up.)

(d) Rocket descends with acceleration A. (*Extra hard. Do this part only if your instructor says to. The same goes for part e.*)

(e) In part (d), what is the normal force on the block? Is it bigger or smaller than the normal force in part (a)? Explain why, intuitively.

ANSWER 8-4

In all parts of this problem, the basic free-body diagram looks like the following figure. The acceleration A, when it exists, is not a force, and is therefore not included on the free-body diagram. The acceleration is a *result* of the forces already drawn, not a "new" force. More on this later.

In all parts of this problem, use the usual strategy: (1) Draw a free-body diagram, with vectors broken into components; then (2) Write Newton's 2nd law in component form (x-direction and y-direction separately).

(a) When the rocket sits still, it is convenient to pick your coordinate axes parallel and perpendicular to the floor's surface. Break up the force vectors into components and add them to the full body diagram.

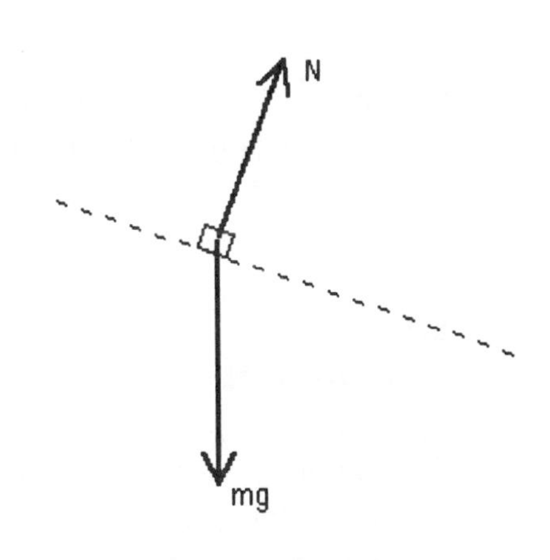

Since the block neither jumps off the floor nor burrows into the floor, the block has no y-ward motion: ay=0. (Using this fact, you could solve for the normal force, if you wanted to, bysetting $\sum F_y = N - mg\cos\theta = 0$.) All of the block's acceleration is x-ward, along the floor. So, Newton's 2nd law yields $\sum F_x = mg\sin\theta = ma_x$, and hence $a_x = g\sin\theta$. So, a steeper floor (i.e., a bigger θ) produces a larger acceleration, as you intuitively know.

(b) and (c) You might think that when the rocket moves, it affects the rate at which the block speeds up as it slides along the floor. But the block's motion down the floor remains unaffected when the rocket floor moves at *constant speed*. Any constant-speed motion adds nothing to the block's *acceleration*. In other words, the block accelerates down the floor *as if* the floor were sitting still: $a = g\sin\theta$.

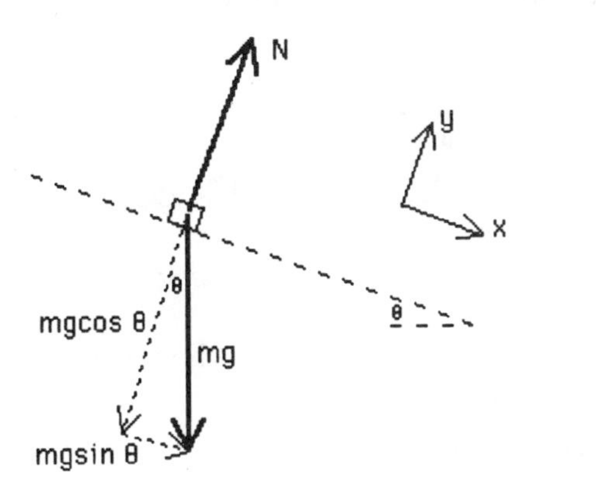

"But wait! The floor's *not* sitting still. It is moving upward with the rocket." True enough. But the block is also traveling upward at this same constant velocity. Therefore, when the block is released, it is not moving *relative to* the floor, and the floor is not moving relative to the block. So, from the block's point of view, the floor starts "at rest." As we saw in Question 5-1(d) with Kim on the train, the block's perspective (i.e., its frame of reference) is just as valid as anyone else's.

If this seems implausible, imagine yourself riding in a train at constant speed. If you take a ball and start playing "catch" with yourself or with another person, you cannot "tell" that the train is moving. The ball behaves the same way it normally does; its acceleration is no different. The same phenomenon applies on a rocket at constant speed. (If your experiences in elevators do not confirm this, it is because elevators spend much of their time speeding up and slowing down–that is, accelerating–instead of moving at constant velocity.) The point is that an object's acceleration cannot be affected by making that object "ride along" at constant speed in a rocket, elevator, train, car, or anything else.

Still not convinced? Then remember that the Earth whips around the Sun at 70,000 mph. That motion in no way affects how a block accelerates down a tilted Earthly floor.

(d) A downward-moving rocket that is speeding up has a downward acceleration. Let me choose upward as my positive direction. Then, the rocket's acceleration is $-A$. (If you choose downward as the positive direction, the rocket's acceleration is $+A$.)

The floor's acceleration is $-A$. Therefore, if the block were sitting still on the floor, its acceleration would also be $-A$. But the block is not just sitting on the floor. It is sliding down the floor. As a result, it has additional downwards and sideways acceleration, besides $-A$. Let $\mathbf{a}_{\text{rel}}$ denote the block's acceleration relative to the floor. Then the block's total acceleration, which I will call $\mathbf{a}_{\text{tot}}$, is given by

$$\mathbf{a}_{\text{tot}} = \mathbf{a}_{\text{rel}} + \mathbf{A}_{\text{rocket}}, \tag{1}$$

where $\mathbf{A}_{\text{rocket}}$ is a downward acceleration vector of magnitude A. In words, the block's total acceleration is its acceleration relative to the floor, plus the floor's acceleration.

If Eq. (1) does not make sense, consider a velocity example. Suppose a train chugs eastward at 20 m/s. Inside the train, a kid runs forward (relative to the train) at speed 3 m/s. How fast is the kid moving relative to the ground? You can see intuitively that the kid's total speed relative to the ground is $v_{\text{tot}} = 23$ m/s. How did you get that answer? You took the kid's velocity relative to the train, $v_{\text{rel}} = 3$ m/s, and added it to the train's velocity, $v_{\text{train}} = 20$ m/s. That is, you used $v_{\text{tot}} = v_{\text{rel}} + v_{\text{train}}$. Eq. (1) is analogous. Indeed, you could derive Eq. (1) from $v_{\text{tot}} = v_{\text{rel}} + v_{\text{train}}$ by differentiating both sides with respect to t, and substituting a rocket for the train.

Not all the vectors in Eq. (1) are parallel. For instance, $\mathbf{A}_{\text{rocket}}$ points downward, while $\mathbf{a}_{\text{rel}}$, the block's acceleration relative to the floor, points along the floor's surface. Since it is almost always easiest to deal separately with the x- and y-components of an object's motion, we should break up Eq. (1) into vector components.

Usually in an "inclined plane" problem, I would tilt my x- and y-directions to be parallel and perpendicular to the incline. That is what we did in part (a). But here, since the floor undergoes a *downward* acceleration, it is more convenient to say the y-direction is vertical (i.e., up-and-down), and the x-direction is sideways (horizontal). *Indeed, whenever a problem is set inside an accelerating train, rocket, elevator, or whatever, it is convenient to choose your coordinate axes to be parallel and perpendicular to the direction of that "imposed" acceleration.* So, I will say "up" is the positive y-direction, and rightward is the positive x-direction. You get the same final answer no matter what coordinate axes you choose; a "good" choice just simplifies the math.

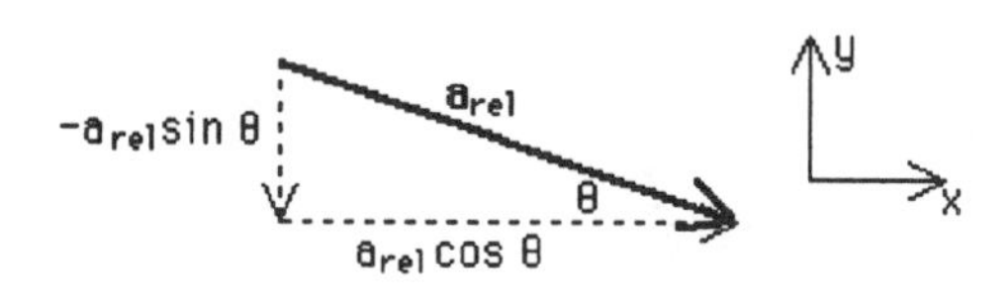

The minus sign on the vertical component indicates "downward." With this choice of coordinates, Eq. (1) breaks up into

$$a_{x\,\text{tot}} = a_{\text{rel}} \cos\theta \tag{2a}$$

$$a_{y\,\text{tot}} = -a_{\text{rel}} \sin\theta - A. \tag{2b}$$

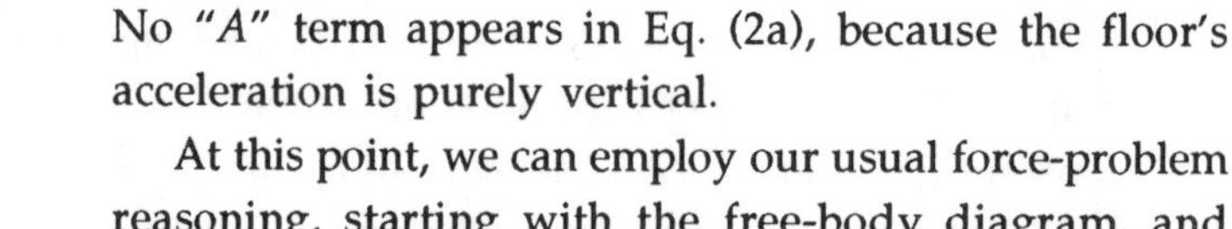

No "A" term appears in Eq. (2a), because the floor's acceleration is purely vertical.

At this point, we can employ our usual force-problem reasoning, starting with the free-body diagram, and breaking up the force vectors along these new coordinate axes. *Remember that the rocket's acceleration A does not get drawn in the free-body diagram, because that acceleration is not a new "kind" of force, but is simply a result of the forces already drawn.*

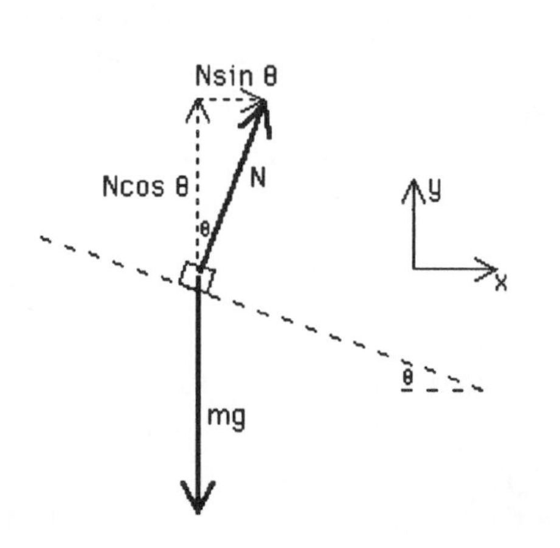

I cannot stress enough that this is the same free-body diagram we used in part (a). All that is changed is the coordinate axes. In these "new" coordinates, we had to break up the normal force instead of gravity.

Now we can use Newton's 2nd law in the usual way. **Key point: In Newton's 2nd law, always use the total acceleration, NOT the relative acceleration.** Looking at the horizontal and vertical forces, and invoking the above expressions for $a_{x\text{ tot}}$ and $a_{y\text{ tot}}$, gives

$$\sum F_x = ma_{x\text{ tot}} \qquad (3a)$$
$$N\sin\theta = m(a_{\text{rel}}\cos\theta)$$

$$\sum F_y = ma_{y\text{ tot}} \qquad (3b)$$
$$N\cos\theta - mg = m(-a_{\text{rel}}\sin\theta - A)$$

We are trying to solve for $\mathbf{a}_{\text{rel}}$. Fortunately, we now have two equations in the two unknowns, $\mathbf{a}_{\text{rel}}$ and N. So, it is just a matter of algebra to solve for $\mathbf{a}_{\text{rel}}$.

Before completing the math, let me summarize what modifications I added to the usual force problem-solving strategy. Besides making a "clever" choice of coordinate axes, all I did was write an expression for the block's total acceleration in terms of its acceleration relative to the floor. I then substituted this total acceleration into Newton's 2nd law in the usual way. So, although they look tremendously difficult, force problems set in accelerating vehicles are not really so bad. You just have to take the vehicle's acceleration into account when writing the total acceleration.

I will now complete the math.

Algebra starts here. Solving Eq. (3a) for N yields $N = \frac{ma_{\text{rel}}\cos\theta}{\sin\theta}$. Substitute that expression into Eq. (3b) to get $\frac{ma_{\text{rel}}\cos\theta}{\sin\theta}\cos\theta - mg = m(-a_{\text{rel}}\sin\theta - A)$. Now we just have to isolate $\mathbf{a}_{\text{rel}}$. To do so, first multiply both sides through by $\sin\theta$. Then add $ma_{\text{rel}}\sin^2\theta$ to both sides, to get $ma_{\text{rel}}\sin^2\theta + ma_{\text{rel}}\cos^2\theta - mg\sin\theta = -mA\sin\theta$. Use the trig identity $\sin^2\theta + \cos^2\theta = 1$, and then add $mg\sin\theta$ to both sides, to get $ma_{\text{rel}} = mg\sin\theta - mA\sin\theta$. (That is one of the few trig identities you need to know.) Dividing through by m gives us, finally, $a_{\text{rel}} = (g - A)\sin\theta$.
End of algebra.

So, when the rocket accelerates downward, the block accelerates down the floor more slowly than usual. (Recall from part (a) that $a = g\sin\theta$ for a non-accelerating floor.) To help yourself visualize this, experiment with the computer simulation, which lets you adjust A.

(e) In the midst of the part (d) algebra, I obtained an expression for N: $N = \frac{ma_{\text{rel}}\cos\theta}{\sin\theta}$. Later on, we found that $a_{\text{rel}} = (g - A)\sin\theta$. So,

$$\begin{aligned} N &= \frac{ma_{\text{rel}}\cos\theta}{\sin\theta} \\ &= \frac{m[(g - A)\sin\theta]\cos\theta}{\sin\theta} \\ &= m(g - A)\cos\theta. \end{aligned}$$

Let is see what this answer means physically. In part (a), the normal force turns out to be $N = mg\cos\theta$. Here, N is smaller. In words, the floor does not push as hard against the block, or vice versa. This makes sense. When you stand inside a downward-accelerating elevator, the floor partially "rushes out from under your feet," the therefore pushes less hard against your feet than happens when the elevator sits still (or moves at constant velocity). For this reason, you feel "lighter" than usual. By contrast, in an upward-accelerating elevator, the floor pushes harder on your feet (or on the block) than would otherwise be the case.

QUESTION 8-5

Hard problem, but worth a shot.

The windshield of your car makes an angle θ with the horizontal. You are cruising down the highway when suddenly, the truck in front of you kicks up a huge frictionless ruby, which happens to bounce off your hood up onto your windshield, and does not bounce off. Thinking fast, you realize that the ruby will slide down your windshield (and out of your life) unless you accelerate forward. But you desperately want the ruby to stay in place, neither sliding up nor sliding down your windshield, at least until you can get off the highway. You estimate that the ruby feels a backward force due to wind resistance equal to one half the ruby's weight.

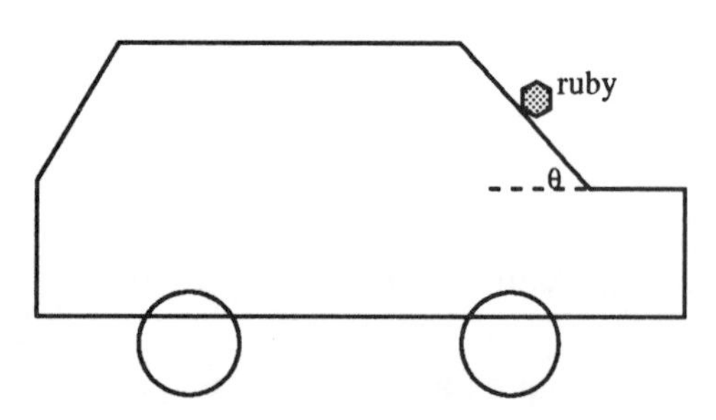

What must be your car's acceleration A to ensure that the ruby stays in place. Answer in terms of θ and any other constants you need.

ANSWER 8-5

Let us start by visualizing. The car accelerates rightward with some acceleration A. If A is too small, then the ruby slides down the windshield. And if A is too big, the windshield "slides out from under" the ruby, i.e., the ruby slides up the windshield.

We need to find the A such that the ruby does not slide up or down the windshield. It is hard to know where to begin. Therefore, I will try to get my bearings by drawing a force diagram for the ruby. Since the windshield gets accelerated rightward, I will choose the horizontal and vertical directions as my coordinate axes, instead of using the usual "tilted" coordinate axes. Do not forget about the backward wind resistance force equal to half the ruby's weight: $F_{\text{drag}} = \frac{mg}{2}$.

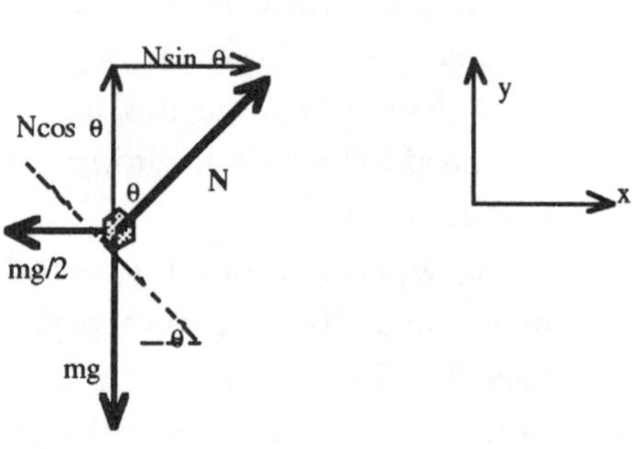

The diagram does not include the engine's force acting on the car. It includes only the forces acting *directly* on ruby.

How can we proceed? Well, from experience, we know that Newton's 2nd law will probably be helpful. As we saw in Question 8-4(d), Newton's 2nd law requires us to consider the ruby's *total* acceleration. In general, the ruby's *total* acceleration $\mathbf{a}_{\text{tot}}$ would consist of two parts: the acceleration of the windshield "carrying along" the ruby, plus the ruby's acceleration relative to the windshield: $\mathbf{a}_{\text{tot}} = \mathbf{a}_{\text{rel}} + \mathbf{A}_{\text{windshield}}$. (Look over Question 8-4(d) if it is unclear where this formula comes from.) But the whole point of this problem is that the ruby does not slide up or down the windshield. Its acceleration relative to the windshield is $\mathbf{a}_{\text{rel}} = 0$. The ruby just "rides" the windshield, sharing its acceleration:

$$\begin{aligned}\mathbf{a}_{\text{tot}} &= \mathbf{a}_{\text{rel}} + \mathbf{A}_{\text{windshield}} \\ &= 0 + \mathbf{A}_{\text{windshield}}.\end{aligned}$$

This simplifies matters greatly, because the windshield's acceleration–and hence, the ruby's acceleration–is purely horizontal. So, the ruby's total horizontal acceleration is $a_{x\,\text{tot}} = A$, while its vertical acceleration is $a_{y\,\text{tot}} = 0$.

Given these accelerations and the above force diagram, we can extract useful information from the x- and y-components of Newton's 2nd law. Remembering that the x- and y-directions are horizontal and vertical, I get

$$\sum F_y = ma_{y\,\text{tot}} \qquad (1)$$
$$N\cos\theta - mg = 0$$

$$\sum F_x = ma_{x\,\text{tot}} \qquad (2)$$
$$N\sin\theta - \frac{1}{2}mg = mA$$

These two equations give us enough information to solve for A and N, the two unknowns. We are done with the physics.

Algebra starts here. Solve Eq. (1) for N to get $N = \frac{mg}{\cos\theta}$, and substitute that expression into Eq. (2) to get $\frac{mg}{\cos\theta}\sin\theta - \frac{1}{2}mg = mA$. Divide through by m, and remember that $\tan\theta = \frac{\sin\theta}{\cos\theta}$, to get $A = g(\tan\theta - \frac{1}{2})$.

End of algebra.

Notice that the normal force found above is bigger than N would be if the windshield were sitting still. (Here, $N = \frac{mg}{\cos\theta}$, which is larger than the "usual" value, $N = mg\cos\theta$.) This is because, when the windshield gets accelerated horizontally, the surface of the windshield gets pressed harder against the ruby.

QUESTION 8-6

A block of mass $m_1 = 2.0$ kg on a frictionless ramp of angle $\theta = 30°$ is connected via an essentially massless cord to a second block, of mass $m_2 = 3.0$ kg. The cord passes over a massless, frictionless pulley. The blocks are released from rest.

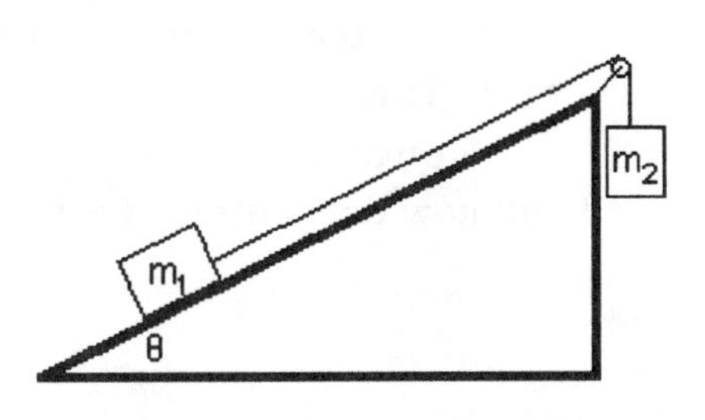

(a) How much time does it take block 2 to fall through height $h = 0.50$ meters?

(b) How much force does the cord exert on block 1?

ANSWER 8-6

(a) This is a classic "multi-block problem," by which I mean a problem in which multiple objects all move in tandem, because they are attached by ropes, pushing on each other directly, or otherwise constrained to move in sync. By thinking qualitatively, let us first gain some physical insights into what is happening. Then we can use those insights to formulate a strategy with which you can address *all* multi-block problems.

Physical insight 1: The two blocks "move together." Since block 2 drags block 1, those two blocks must move at the same speed. Therefore, if block 2 speeds up, so does block 1. In other words, *the blocks must "share" the same acceleration:* $a_1 = a_2$.

Physical insight 2: The tension in the rope acts on both blocks; and *the tension at both ends of the rope is the same.* The tension not only pulls block 1 up the ramp, but also pulls upward on block 2, thereby slowing the fall of block 2. So, block 2 falls downward with some acceleration *less than* g.

Physical insight 3: We cannot just write down an expression for that tension. It is an unknown. For instance, you might guess that the tension is $T = m_2 g$. But if that were true, the forces on block 2 would cancel out, and block 2 would not move! Since block 2 accelerates down, it must be the case that $m_2 g$ overcomes the upward tension on block 2. So, T is less than $m_2 g$. Alternatively, you might think that the tension equals the force with which gravity pulls block 1 down the ramp: $T = m_1 g \sin 30°$. But again, if that were the case, the forces on block 1 would cancel, and hence block 1 would not move. Since block 1 accelerates up the ramp, the tension must overcome $m_1 g \sin 30°$. So, T is greater than $m_1 g \sin 30°$. And we cannot just guess how much greater.

Given these insights, we can now formulate two different strategies for attacking such problems. One method involves treating the whole multi-block system as "one big mass," and looking at the external forces acting on this mass: $F_{\text{external}} = M_{\text{total}} a$. If your instructor teaches this method, by all means use it. But this method is easy to misapply, and in some cases gets you to the answer more slowly than my "multi-block" strategy. In addition, although my multi-block strategy is "clunky," it is relatively easy to apply once you get used to it. Here it is:

MULTI-BLOCK STRATEGY

(1) **Draw a free-body diagram *for each mass individually*,** including only the forces acting *directly* on that mass. For instance, $m_2 g$ does not act directly on block 1. It acts indirectly on block 1 via the tension. The tension is what *directly* pulls block 1 up the ramp.

(2) **Write Newton's 2nd law along the direction of motion for each mass individually.** Remember, the blocks all share the same acceleration. I will call it a.

Make sure that your choice of "positive" direction for each block reflects the fact that the blocks all move together. For instance, block 1 moves up the ramp when block 2 moves downward. Therefore, if we call "up the ramp" the positive direction for block 1, we are *committed* to calling downward positive for block 2.

(3) **Use the equations generated in step 2 to solve for the acceleration.**

(4) **Then, use that acceleration in kinematic equations or in Newton's 2nd law to solve for the quantity of interest.**

I will now demonstrate this strategy.

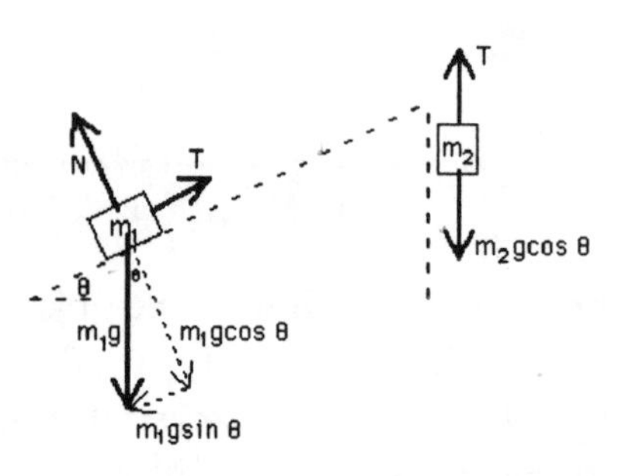

Step 1: Free-body diagram for each mass.

Notice that, since the same rope acts on both blocks, the tension T is the same on both blocks. A normal force always points perpendicular to the surface exerting that force. Divide vectors up into their components parallel and perpendicular to the direction of motion, to simplify step 2 below.

Step 2: Newton's 2nd law for each mass individually.

At this point, we should decide which direction counts as positive. To make things easier, let us choose the direction of motion. So, up-the-ramp is positive for block 1, while downward is positive for block 2. I cannot emphasize enough that our choice of positive directions must reflect the fact that the blocks move together.

Write Newton's 2nd law *along the direction of motion* for each block. For block 1, that is parallel to the ramp. For block 2, that is vertical, with downward as positive. From the force diagrams, I get

$$\text{BLOCK 1:} \quad \sum F_{1,x} = m_1 a_1 \qquad (1)$$
$$T - m_1 g \sin\theta = m_1 a,$$

$$\text{BLOCK 2:} \quad \sum F_{2,x} = m_2 a_2 \qquad (2)$$
$$m_2 g - T = m_2 a.$$

Notice that I set $a_1 = a_2$ and called that acceleration "a," because both blocks share the same acceleration.

We now have two equations in two unknowns, a and T. Only algebra remains.

Step 3: Algebra to solve for acceleration

To find the acceleration quickly, use the algebraic shortcut of adding Eq. (1) to Eq. (2). Do not worry; it is o.k. to add x-forces to y-forces (for a change!), because the pulley "twists" the direction of the forces so that they effectively all act along a straight line. When you add the equations, the T in Eq. (1) cancel the $-T$ in Eq. (2), leaving us with $-m_1 g \sin\theta + m_2 g = (m_1 + m_2)a$. Solving for a yields

$$\begin{aligned} a &= g\left(\frac{m_2 - m_1 \sin\theta}{m_1 + m_2}\right) \\ &= (9.8 \text{ m/s}^2)\frac{3 \text{ kg} - (2 \text{ kg})\sin 30°}{2 \text{ kg} + 3 \text{ kg}} \\ &= 3.9 \text{ m/s}^2, \end{aligned}$$

two fifths the "regular" acceleration due to gravity. That makes sense, because the upward tension on block 2 prevents it from free-falling with full acceleration g.

Step 4: Kinematics or Newton's 2nd law to find what you want.

In part (a), we want to know how much time block 2 takes to fall through height h. Since we know the constant acceleration, we can use $y = y_0 + v_{0y}t + \frac{1}{2}a_y t^2$. I will continue using downward as my positive direction. So, if we say the block's initial position is $y_0 = 0$, then it is final position is $y = +h$. It is initial velocity is $v_{0y} = 0$, and we just found its acceleration to be $a_y = 3.9 \text{ m/s}^2$. So, we can immediately solve for t to get $t = \sqrt{\frac{2(y-y_0)}{a_y}} = \sqrt{\frac{2(h-0)}{a_y}} = \sqrt{\frac{2(0.50 \text{ m}-0)}{3.9 \text{ m/s}^2}} = .51$ s.

(b) We want to know the tension in the rope. This is a continuation of step 4. Just take the acceleration found in step 3, and substitute it into Eq. (1) or Eq. (2), the Newton's 2nd law equations we obtained above. I will use Eq. (2). Solving for T gives

$$\begin{aligned} T &= m_2(g - a) \\ &= (3 \text{ kg})(9.8 \text{ m/s}^2 - 3.9 \text{ m/s}^2) \\ &= 18 \text{ N}. \end{aligned}$$

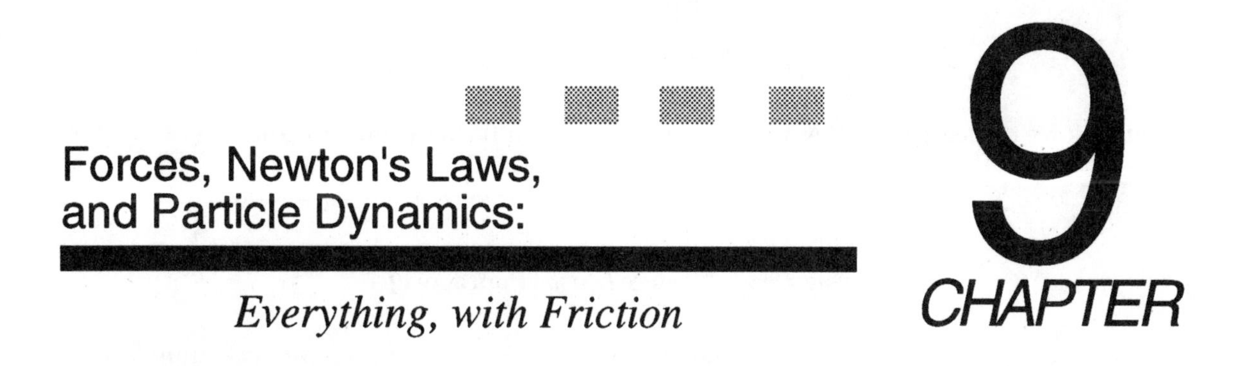

Forces, Newton's Laws, and Particle Dynamics: Everything, with Friction

9 CHAPTER

QUESTION 9-1

A desk of mass $m = 100$ kg is pushed across the floor by Kim, with force $F_{\text{Kim}} = 200$ newtons, directed horizontally. Starting from rest at time $t = 0$, the desk accelerates forward. At $t = 3.0$ s, the desk has speed $v = 1.5$ m / s.

What is the coefficient of kinetic friction, μ, between the floor and the desk?

ANSWER 9-1

Before answering the question, I would like to review the "theory" behind the coefficient of friction. Rub your hands together. It is easy when you press them together lightly. But when you press your hands together hard, it becomes more difficult to rub them back and forth; friction gets bigger. The more your hands "dig into" each other, the harder it gets to slide them over each other. Similarly, the harder the desk and floor dig into each other, the more friction resists Kim's push. The "hardness" with which two objects dig into each other is the normal force N. So, I have just made it plausible that the frictional force f is proportional to the normal force between the two surfaces: $f \sim N$.

Now imagine two identical desks, one on pavement and the other on ice. The desk on ice is easier to push. Why? Not because the normal forces differ. It is because pavement is rougher, more "frictionful," than ice. Let μ denote this "roughness" constant. We see that $\mu_{\text{pavement}} > \mu_{\text{ice}}$. Indeed these roughness constants are defined to be proportional to the frictional force: $f \sim \mu$. In summary, the force of friction is proportional to the "roughness" of the surface over which the object slides, and also proportional to the normal force exerted by the surface on the object. Putting all this together gives

$$f = \mu N, \tag{1}$$

the basic equation relating the frictional force to the normal force acting on that object, for kinetic friction. (More on static friction later.)

Given this relation, we can address the problem at hand. We are asked for μ between the floor and the desk. To calculate μ, we can figure out f and N, and then substitute those values into Eq. (1). Since we are looking at forces, let us use the usual Newton's-law reasoning, starting with a force diagram.

Notice that friction "resists" Kim's push. Since the desk neither falls through the floor nor jumps off the floor, the vertical acceleration is $a_y = 0$. Therefore,

$$\sum F_y = ma_y$$
$$N - mg = 0.$$

So, the normal force cancels gravity: $N = mg$. Now let us look at the horizontal forces:

$$\sum F_x = ma_x$$
$$F_{\text{Kim}} - \mu N = ma_x \qquad (2)$$
$$F_{\text{Kim}} - \mu mg = ma_x \qquad \text{[since } N = mg \text{ from above].}$$

Although we know F_{Kim}, m, and g, we still cannot solve Eq. (2) for μ, because it contains another unknown, namely a_x. Fortunately, we can quickly calculate a_x from the given kinematic information. The box speeds up from $v_0 = 0$ to $v = 1.5\text{ m/s}$ in time $t = 3\text{ s}$. You may be able to see intuitively that the acceleration is $a_x = 0.50\text{ m/s}$ per second If not, start with the one-dimensional kinematic formula $v = v_0 + at$, and solve for a to get $a_x = \frac{v - v_0}{t} = \frac{1.5\text{ m/s}-0}{3.0\text{ s}} = 0.50\text{ m/s}^2$. Now we can solve Eq. (2) for μ, and set $a_x = 0.50\text{ m/s}^2$, $m = 100\text{ kg}$, $F_{\text{Kim}} = 200N$, and $g = 9.8\text{ m/s}^2$, to get

$$\mu = \frac{F_{\text{Kim}} - ma_x}{\text{mg}}$$
$$= \frac{200 - N - (100\text{ kg})(.50\text{ m/s}^2)}{(100\text{ kg})(9.8\text{ m/s}^2)}$$
$$= 0.15,$$

which is quite slippery. Remember, $\mu = 0$ corresponds to extremely slippery, and $\mu = 1$ corresponds to extremely rough.

QUESTION 9-2

Four crates of hockey equipment are connected by massless ropes as drawn below, on an ice rink. All the ice is essentially frictionless, except for the dirty ice under crate 1. So, the coefficient of kinetic friction between crate 1 (of mass m_1) and the ice is μ. Someone sitting off the ice pulls rightward on rope 4, with force P. The crates have mass m_1, m_2, m_3, and m_4, respectively. What is the tension in rope 3? Answer in terms of the given constants.

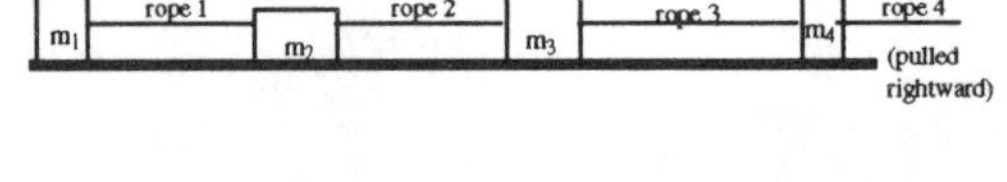
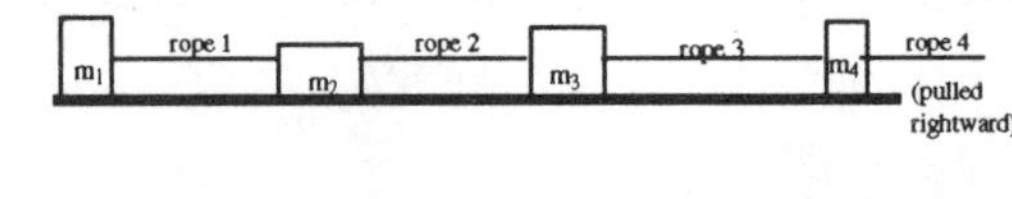

ANSWER 9-2

This is another classic "multi-block" problem. You will not be able to solve for the tension in rope 3 simply by drawing a force diagram for m_3, because the resulting Newton's law equation will contain too many unknowns. You must consider each crate. Let me reiterate the strategy introduced in Question 8-6.

(1) Draw a free-body diagram *for each mass individually*, including only the forces acting *directly* on that mass. For instance, neither P nor friction acts directly on crate 3. Therefore, those forces do not get included in the free-body diagram of crate 3.

(2) Write Newton's 2nd law along the direction of motion for each mass individually. Since the crates all move together, they share the same acceleration a.

(3) Use the equations generated in step 2 to solve for the acceleration.

(4) Then, use that acceleration in kinematic equations or in Newton's 2nd law to solve for the quantity of interest.

Let us do it.

Step 1: Separate force diagram for each crate

The trickiest part of this problem is sorting out the tensions. The tensions at both ends of a single rope are the same. For instance, rope 2 exerts a forward force on m_2 and a backward force on m_3; and those two forces are equal. But different ropes carry different tensions. For instance, the tension in rope 1 differs from the tension in rope 2. In general, $T_1 \neq T_2 \neq T_3$:

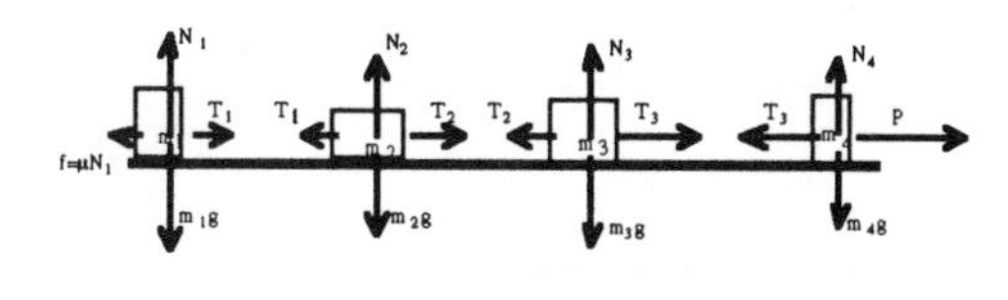

Step 2: Newton's 2nd law for each crate.

Since none of the crates undergoes vertical motion, the upward normal force cancels the downward gravitational force on each crate. We must take advantage of this fact with crate 1, because we need to know f. Well, since the y-forces on crate 1 cancel, $N_1 = m_1 g$, and hence $f = \mu m_1 g$.

I will let the rightward direction be positive for all four crates. From the free-body diagrams,

$$\begin{aligned}
\text{Crate 1:} \quad & \sum F_{1,x} = T_1 - \mu m_1 g = m_1 a, \\
\text{Crate 2:} \quad & \sum F_{2,x} = T_2 - T_1 = m_2 a, \\
\text{Crate 3:} \quad & \sum F_{3,x} = T_3 - T_2 = m_3 a, \\
\text{Crate 4:} \quad & \sum F_{4,x} = P - T_3 = m_4 a,
\end{aligned}$$

At this point, we have four equations in four unknowns (a, T_1, T_2, and T_3). So, only algebra remains.

Step 3: Solve for acceleration.

Even though you are ultimately solving for T_3, it is usually fastest to first solve for acceleration. You can often do this by adding the equations generated in step 2. When we do so here, the T_1 in Eq. (1) cancels the $-T_1$ in Eq. (2). The T_2 in Eq. (2) cancels the $-T_2$ in Eq. (3). And so on. Therefore, we are left with $F_p - \mu m_1 g = (m_1 + m_2 + m_3 + m_4)a$, and hence $a = \frac{F_p - \mu m_1 g}{m_1 + m_2 + m_3 + m_4}$. You would get this answer for acceleration more quickly by treating the four crates as one big blob of mass $M = (m_1 + m_2 + m_3 + m_4)$, acted on by an external force $F_{\text{ext}} = F_p - \mu m_1 g$. But even so, you cannot immediately solve for T_3 without the above Newton's 2nd law equations for the individual crates. That is one of the reasons I recommend the multi-block strategy.

Step 4: Kinematics or Newton's 2nd law to obtain what you want.

To obtain T_3 most efficiently, solve the crate 4 equation for T_3 to get

$$\begin{aligned}
T_3 &= F_p - m_4 a \\
&= F_p - m_4 \frac{F_p - \mu m_1 g}{m_1 + m_2 + m_3 + m_4}.
\end{aligned}$$

QUESTION 9-3

A block slides down a ramp angled at $\theta = 30°$ to the floor. The ramp is $D = 1.5$ meters long. The coefficient of friction (both kinetic and static) is $\mu = 0.20$.

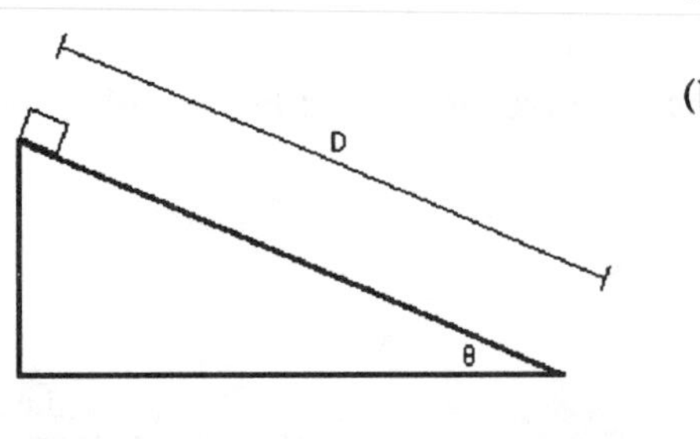

(a) If the block is released from rest from the top of the ramp, how much time does it take to reach the bottom?

(b) Suppose the block is now coated with resin to make it less slippery. We want to slap just enough resin onto the block so that it does not slide down the ramp. What is the smallest value of μ_s, the coefficient of static friction, such that the block will not slide down the ramp? Do not look up a formula in the book; *derive* the relevant formula.

ANSWER 9-3

(a) Let us first picture what is happening. Gravity, or more accurately the component of gravity parallel to the ramp, pulls the block down the ramp. But friction resists the block's motion, thereby reducing the acceleration. Consequently, the block takes more time to reach the bottom. To see how the coefficient of friction affects the travel time, experiment with the computer simulation by changing the coefficient of kinetic friction.

We can use the standard force-problem strategy. (1) Draw a free-body diagram, with force vectors broken into components when necessary, (2) Write Newton's 2nd law in the x- and/or y-direction, and solve for the acceleration along the direction of motion; then (3) Use that acceleration in kinematic formulas to calculate the quantity of interest.

You may feel stuck because I have not given you the block's mass. **As always, if you are not given a quantity that you need in your equations, just make up a symbol. For instance, you can call the block's mass "*m*." Then, work the problem through, and hope that the *m*'s cancel out of your final answer.**

I will now implement the force-problem strategy.

Step 1: Free-body diagram.

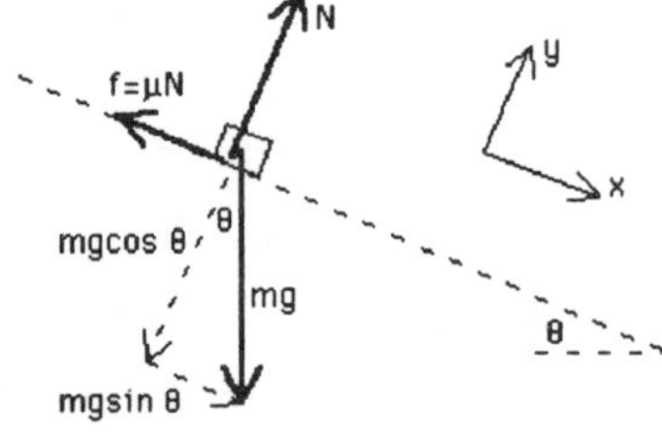

Recall that in standard ramp problems (when the ramp itself is not moving), it is convenient to choose "tilted" coordinate axes, with one of the axes parallel to the ramp.

Step 2: Newton's 2nd law.

We are trying to find the acceleration along the direction of motion, a_x. So, let us look at the x-directed forces:

$$\sum F_x = ma_x$$
$$mg\sin\theta - \mu N = ma_x. \tag{1}$$

To solve for a_x, we need to know N, the normal force. Since N is a y-directed force, let us consider Newton's 2nd law in the y-direction. Because the block moves along the surface of the ramp (in the x-direction only), the y-acceleration is $a_y = 0$. So,

$$\sum F_y = ma_y$$
$$N - mg\cos\theta = 0, \tag{2}$$

and hence $N = mg\cos\theta$. Substitute that expression for N into Eq. (1) to get

$$mg\sin\theta - \mu mg\cos\theta = ma_x.$$

Notice that the m's cancel, as we desired, giving us

$$\begin{aligned} a_x &= g(\sin\theta - \mu\cos\theta) \\ &= (9.8\ \text{m/s}^2)(\sin 30° - .2\cos 30°) \\ &= 3.2\ \text{m/s}^2. \end{aligned}$$

Step 3: Kinematics.

Now that we know the block's constant acceleration, we can use kinematics to find how much time the block takes to travel from $x_0 = 0$ to $x = D = 1.5$ m. (Remember, the x-direction here is parallel to the ramp, not horizontal. I have chosen down-the-ramp as my positive direction.) So, let us start with the displacement equation $x = x_0 + v_0 t + \frac{1}{2}a_x t^2$, with $v_{0x} = 0$. Solving for t gives us $t = \sqrt{\frac{2(x-x_0)}{a_x}} = \sqrt{\frac{2D}{a_x}} = \sqrt{\frac{2(1.5\text{ m})}{3.2\text{ m/s}^2}} = 0.97$ s about one second. The computer simulation can help you develop a feel for how the coefficient of friction affects the motion.

(b) Intuitively, the block stays in place if the frictional force is big enough to cancel the component of gravity pulling the block down the ramp. This is another way of saying the net x-directed force is $\sum F_x = 0$, and hence, $a_x = 0$. Therefore, we can repeat the reasoning of step 2 above, except now let us set $a_x = 0$ so that all the forces cancel:

$$\begin{aligned} \sum F_x &= ma_x \\ mg\sin\theta - \mu_s N &= 0 \\ mg\sin\theta - \mu_s mg\cos\theta &= 0. \end{aligned} \tag{3}$$

In the last step of algebra, I used $N = mg\cos\theta$, which we found above by noticing that the y-forces cancel; see Eq. (2).

When we solve Eq. (3) for the coefficient of friction, m and g both cancel out, leaving us with $\mu_s = \frac{\sin\theta}{\cos\theta} = \tan\theta = \tan 30° = .57$. If the coefficient of friction is this big or bigger, then friction cancels out the relevant component of gravity, and the block does not move. Confirm this with the computer simulation.

Now actually, I have just used some fishy reasoning. I assumed that the force due to *static* friction is $f_s = \mu_s N$. This is not always true. Let me take a theoretical digression to show why f_s is less than $\mu_s N$ in most cases. Then I will show why you could set $f_s = \mu_s N$ in this *particular* problem.

Theoretical digression: When does $f_s = \mu_s N$?

Kinetic friction always equals $\mu_k N$. But the force of static friction can be less than $\mu_s N$. Here is why:

Imagine putting an eraser on your desk. What is the force of friction on the eraser? Zero, because the desk is not tilted, and therefore no friction is "needed" to hold the eraser in place. (If there *were* a frictional force, it would be the *only* horizontal force, and therefore the eraser would accelerate across the desk. This does not happen.) So, in that case, $f_s = 0$, which is much less than $\mu_s N$.

Now you start tilting the desk by lifting one end. As a result, friction must counteract the component of gravity that pulls the eraser along the desk. As shown above, f_s must equal $mg\sin\theta$, to hold the eraser still. So, as the tilt angle gets bigger and bigger, friction must increase, in order to hold the eraser in place. This goes to show that static friction can "adjust" itself, getting larger or smaller as necessary to hold an object still.

Of course, if you tilt the desk too steeply, friction cannot hold the eraser any more. Although static friction is adjustable, its *maximum* possible value is $f_{s\,\max} = \mu_s N$.

In this particular problem, a certain amount of friction, $f_s = mg\sin\theta$, is needed to hold the block in place. For the sake of argument, let us say f_s works out to be 5.7 newtons, and the normal force works out to be 10 newtons. So, if the coefficient of static friction were 0.7, then the block would stay in place, because $f_s = 10$ newtons is less than

$$f_{s\,\max} = \mu_s N = (0.7)(10 \text{ newtons}) = 7 \text{ newtons}.$$

Things also "work" if $\mu_s = 0.6$. But if $\mu_s = 0.5$, then the biggest frictional force the ramp can generate is $f_{s\,\max} = \mu_s N = (0.5)(10 \text{ newtons}) = 5$ newtons, less than the 5.7 newtons needed to hold the block in place. We see that $\mu_s = 0.57$–the μ_s for which f_s equals $\mu_s N$–is the *smallest* coefficient of static friction that keeps the block in place. For this reason, whenever you are trying to make μ_s as small as possible, you can set $f_s = \mu_s N$.

QUESTION 9-4

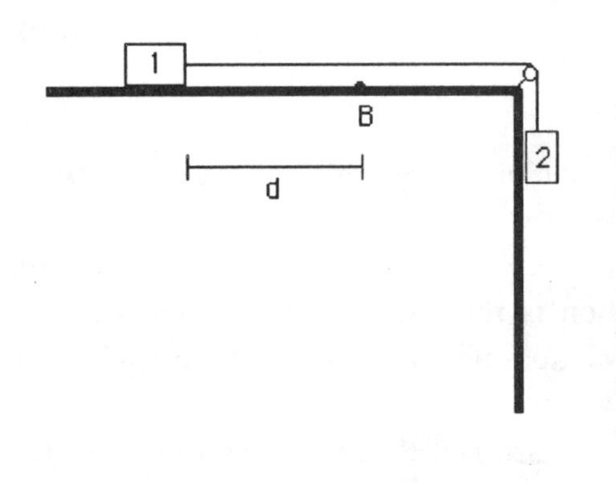

Consider the following pulley configuration.

The string connecting the blocks is essentially massless, as is the pulley. The whole pulley system works frictionlessly. But the coefficient of kinetic friction between block 1 and the table, μ, is not negligible.

At time $t = 0$, the blocks are released from rest from the configuration shown. The blocks accelerate. How fast will block 1 be moving when it reaches point B, which is a distance d from the starting point of block 1? Answer in terms of m_1, m_2, g, d, and μ.

ANSWER 9-4

We are trying to find block 1's speed when it has traveled distance d. As we have seen before, you can use force reasoning to obtain the acceleration, which you can then use in kinematic formulas. Indeed, this is a classic multi-block problem. Recall the strategy from Chapter 8:

(1) Draw a free-body diagram for each separate mass.

(2) Write Newton's 2nd law along the direction of motion for each mass separately. Because the masses all move together, they "share" the same acceleration a.

(3) Solve your equations for the "shared" acceleration of the blocks, and then

(4) Use that acceleration in kinematic equations or in Newton's 2nd law to figure out the quantity of interest.

The only "twist" here is friction.

Step 1: Free-body diagrams

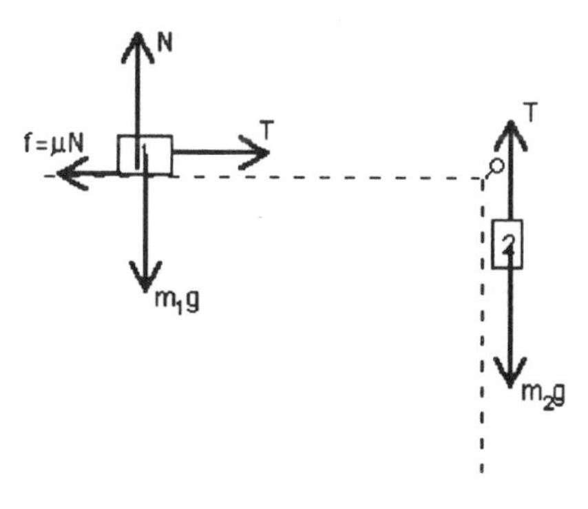

Recall that the tension at both ends of a rope is the same, even if the rope passes over a massless pulley. That is why the " T" acting on block 1 is the same T acting on block2. (If the pulley has mass, things get more complicated.)

You might be tempted to write $T = m_2g$. But if that were the case, block 2 would experience no net force; it would just hang there motionless. Gravity on block 2 must overcome T, to pull that block downward. So, for now, tension is an unknown.

Step 2: Newton's law for each block separately.

I will first consider block 1. Let the positive direction be the direction of motion. Newton's 2nd law (along the direction of motion) tells us that

$$\text{BLOCK 1, } x\text{-direction} \quad \sum F_{1,x} = m_1a$$
$$T - \mu N = m_1a,$$

where a is the shared acceleration of the blocks. For block 1, the acceleration is x-ward, while for block 2, the acceleration is y-ward. But this does not affect your multi-block strategy. The pulley effectively "twists" the y-ward acceleration of block 2 into an x-ward acceleration of block 1.

We can solve for the unknown normal force N by considering the vertical forces on block 1. Since block 1 has no vertical motion, $a_{1y} = 0$. So,

$$\text{BLOCK 1, } y\text{-direction} \quad \sum F_{1,y} = m_1a_{1y}$$
$$N - m_1g = 0,$$

and hence $N = m_1g$. Substitute that expression for N into the above $\sum F_{1,x}$ equation to get

$$T - \mu m_1 g = m_1 a. \tag{1}$$

Now let us consider block 2. Since we chose rightward as positive for block 1, we *must* pick downward as positive for block 2. Why? When block 1 moves rightward, block 2 moves downward. Our choices of positive directions must reflect the fact that the blocks move together in this way. So, for block 2,

$$\sum F_{2,y} = m_2a$$
$$m_2g - T = m_2a. \tag{2}$$

Step 3: Solve for acceleration.

Eqs. (1) and (2) give us enough information to solve for a and T, the two unknowns. We want a. Complete the algebra quickly by adding the two equations. The $+T$ in Eq. (1) cancels the $-T$ in Eq. (2), leaving us with $-\mu m_1g + m_2g = (m_1 + m_2)a$. Solve for a to get $a = g\frac{m_2 - \mu m_1}{m_1 + m_2}$.

Step 4: Kinematics.

Given this acceleration, we can now use kinematic reasoning to figure out the speed v of block 1 after it has traveled $\Delta x = d$. Since we also know the initial velocity is $v_0 = 0$, the best one-dimensional kinematic formula to use is $v^2 = v_0^2 + 2a\Delta x$. We immediately get

$$v = \sqrt{v_0^2 + 2a\Delta x} = \sqrt{0 + 2g\frac{m_2 - \mu m_1}{m_1 + m_2}d}.$$

QUESTION 9-5

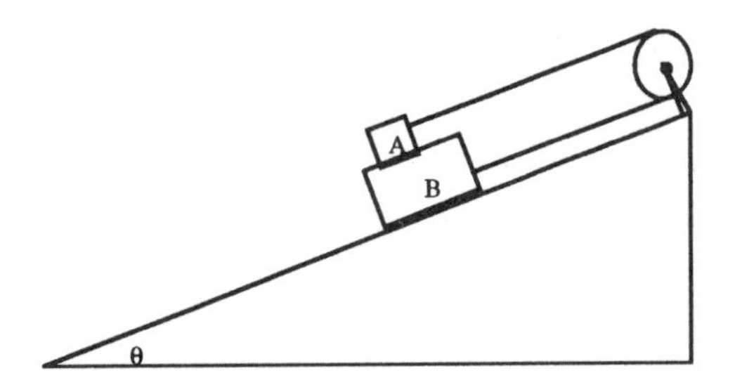

Blocks A and B sit on top of each other on a ramp as drawn below. They are attached by a massless rope that is slung over a massless, frictionless pulley. Both "segments" of rope are stretched parallel to the surface of the ramp, as drawn below. Block B is heavier than block A.

Block A is so slippery that it slides frictionlessly on block B. But the coefficient of kinetic friction between block B and the ramp is μ, which is non-negligible.

(a) Assuming block B accelerates down the ramp after it is released, what is its acceleration?

(b) What is the tension in the rope?

ANSWER 9-5

We can follow the usual multi-block strategy, reiterated at the beginning of Question 9-3. Here, it is difficult to get the normal and gravitational forces sorted out properly in your free-body diagram. But once you sort them out, everything becomes straightforward.

Step 1: Separate free-body diagram for each block.

Let us start with block A. It is "supported" by block B. In other words, block B exerts an upward-angled normal force on block A. I will call this force N_{BA}. Crucially, the ramp does not *directly* push on block A. So, in your force diagram for block A, you should not include a normal force due to the ramp.

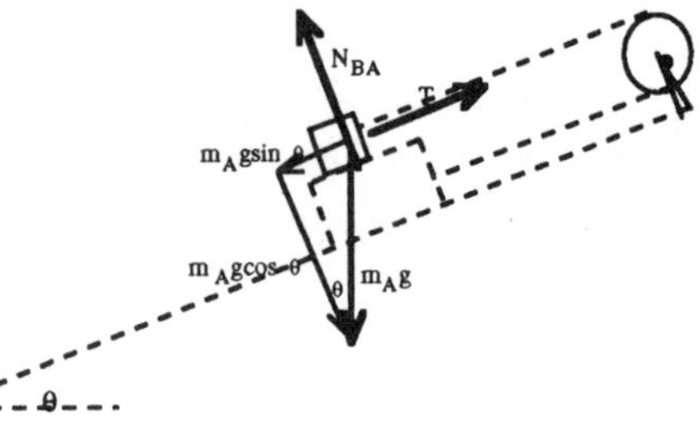

Of course, block A also feels a downward gravitational pull, and the rope's tension pulling it along block B.

Now let us consider the forces on block B. By Newton's 3rd law, since block B exerts a normal force on block A, block A must exert a normal force of equal magnitude on block B. That is, block A exerts a downward-angle normal force of magnitude N_{BA} on block B. This makes physical sense; since block A rests on block B, block A pushes down on block B, just as you are pushing down on your chair right now. But just as block B does not push block A straight up, block A does not push block B straight down. See the diagram below.

Block A does not exert the only normal ("contact") force felt by block B. Block B also feels an upward-angled normal force due to the ramp. I will call that normal force N_r.

Clearly the rope's tension pulls on block B. But what about gravity? You might be tempted to say that the gravitational force on block B is $(m_A + m_B)g$, because block A must get "built in." That is a good intuition. But remember, your force diagram should already include the downward-angled normal force that block A exerts on block B. In other words, your force diagram for block B *already takes into account* the downward push due to block A. So, it would be redundant to also include block A as part of block B's weight. The gravitational force acting *directly* on block B is $m_B g$.

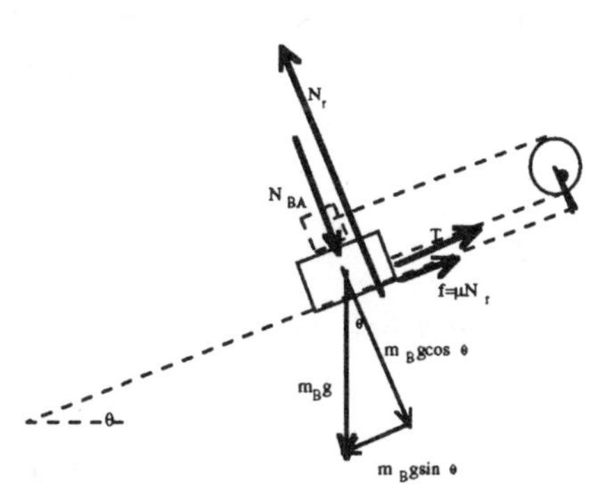

Crucially, block B gets pulled down the ramp with force $m_B g \sin\theta$, *not* with force $(m_B + m_A)\sin\theta$. This is because blocks A and B do not stay stuck together as one big blob; instead, block B slides down the ramp while block A gets dragged up the ramp.

Since friction resists block B's motion down the ramp, friction points up the ramp. Because friction acts between the ramp and block B, the correct normal force to use in "$f = \mu N$" is N_r. If there were friction between blocks A and B, we would also have to include a frictional force proportional to N_{BA}.

Because these force diagrams were so hard to draw correctly, let me address some common mistakes.

Mistake 1: Misapplied Newton's 3rd law

Many people draw two N_{BA} arrows coming out of the same block. According to Newton's 3rd law, if A exerts a force on B, then B exerts a "reaction" force on A. The reaction force never acts on the same body that the "original" force acts on. Rather, the reaction force acts on the "other" body. Here, the "action" force N_{BA} acts at an upward angle on block A, while the "reaction" force N_{BA} acts at a downward angle on block B. (You can switch which force you call the "action" and which you call the "reaction." The key is to avoid drawing both those forces as acting on the same block.)

Mistake 2: Direction of force by A on B

A related mistake is to say block A pushes block B straight downward. It is true that gravity pulls block A straight down; that fact is reflected on the force diagram *for block* A. But normal forces always act perpendicular to the plane of the touching surfaces.

Mistake 3: Two upward-angled forces on block A or block B

Many people say that block A experiences a normal force, as well as a force due to block B. This is redundant; the normal force on block A is the force due to block B. (Intuitively, block B supports block A.) There is no *separate* force on A due to B, besides the normal force N_{BA}.

Step 2: Separate Newton's 2nd law equation for each block.

Now that we have accurate force diagrams, we can proceed to Newton's 2nd law. As always, we are interested in the forces and acceleration along the direction of motion. For block A, I will choose up-the-ramp as my positive x-direction. This *commits* me to calling down-the-ramp positive for block B. Why? Because our choice of positives must reflect the fact that when the blocks "move together" in this problem, they slide in opposite directions. It might seem strange that the positive direction for block A ends up being the negative direction for block B. But it is less strange if you keep in mind that when block B drags block A, the blocks move "oppositely."

From the above force diagrams, looking only at x-directed forces (i.e., force components parallel to the ramp), and using the consistent choice of positive directions just discussed, I get

$$\text{BLOCK A:} \qquad \sum F_x = T - m_A g \sin\theta = m_A a \qquad (1)$$

$$\text{BLOCK B:} \qquad \sum F_x = m_B g \sin\theta - T - \mu N_r = m_B a. \tag{2}$$

Right now, these two equations contain three unknowns: $T, a,$ and N_r. But we can solve for N_r by considering the y-directed forces, i.e., the forces perpendicular to the incline.

Looking at the y-directed forces on block B, I get

$$\text{BLOCK B:} \qquad \sum F_y = m_B a_y$$
$$N_r - N_{BA} - m_B g \cos\theta = 0.$$

We know $a_y = 0$, because the blocks move only in the x-direction. Before we can solve this equation for N_r, we need to know N_{BA}. To get it, look at block A:

$$\text{BLOCK A:} \qquad \sum F_y = m_A a_y$$
$$N_{BA} - m_A g \cos\theta = 0.$$

Solve for N_{BA} to get $N_{BA} = m_A g \cos\theta$. Substitute that into the block B $\sum F_y$ equation, and solve for N_r, to get

$$\begin{aligned} N_r &= N_{BA} + m_B g \cos\theta \\ &= m_A g \cos\theta + m_B g \cos\theta \\ &= (m_A + m_B) g \cos\theta. \end{aligned}$$

Now we can substitute this into Eq. (2). Let me rewrite Eqs. (1) and (2) for easy reference:

$$T - m_A g \sin\theta = m_A a \tag{1}$$

$$m_B g \sin\theta - T - \mu(m_A + m_B) g \cos\theta = m_B a. \tag{2}$$

At this point, we have two equations in two unknowns, T and a. It is just a matter of algebra to solve.

Step 3: Algebra to solve for acceleration

(a) To obtain the acceleration quickly, add Eq. (1) to Eq. (2). The $+T$ in Eq. (1) cancels the $-T$ in Eq. (2), leaving us with $-m_A g \sin\theta + m_B g \sin\theta - \mu(m_A + m_B) g \cos\theta = (m_A + m_B) a$, the same equation you would get by considering the two-block system as a whole. Divide through by $(m_A + m_B)$ to get

$$a = \frac{(m_B - m_A) g \sin\theta - \mu(m_B + m_A) g \cos\theta}{m_B + m_A}.$$

Step 4: Newton's 2nd law or kinematics to find whatever you need

(b) Now we can immediately solve for the tension T by substituting the acceleration back into Eq. (1) or Eq. (2). Either way, you will get the same answer. For instance, Eq. (1) yields

$$\begin{aligned} T &= m_A a + m_A g \sin\theta \\ &= m_A \frac{(m_B - m_A) g \sin\theta - \mu(m_B + m_A) g \cos\theta}{m_B + m_A} + m_A g \sin\theta. \end{aligned}$$

Before closing, let me discuss some common errors:

Mistake 1: Fundamental misapplication of Newton's law

According to Newton's 2nd law, the force on an object equals the mass of that object times the acceleration of that object. So, if we are talking about block B, Newton's 2nd law says $F_{\text{net on B}} = m_B a$.. If we are talking about the combined two-block system composed of A and B, Newton's law says $F_{\text{net on A on B}} = (m_B + m_A)a$.

Many people incorrectly try to solve this problem by setting $F_{\text{net on B}}$ equal to $(m_B + m_A)a$, or by setting $F_{\text{net on A on B}}$ equal to $m_B a$.

Mistake 2: Setting $T = m_A g \sin\theta$

If this equality holds, the forces on block A cancel out. But the forces on block A *cannot* cancel, because if they do, then block A will not accelerate. In other words, since block A gets accelerated up the ramp, the net force on it cannot be 0. The net force on block A is $T - m_A g \sin\theta = m_A a$, not $T - m_A g \sin\theta = 0$.

Mistake 3: Setting $f = \mu m_B g \cos\theta$

As derived above, the normal force exerted by the inclined plane on block B is $(m_A + m_B) g \cos\theta$, not $m_B g \cos\theta$. Intuitively, this is because the inclined plane must "support" both block B and block A, at least indirectly.

Mistake 4: Setting $T = T_A + T_B$.

If we let T_A and T_B denote the rope's force on block A and block B, some people think the "total" tension in the rope is the sum of these two. But actually, the tension in the rope "spreads out" to both ends of the rope. For instance, if the rope is stretched so that it exerts 3 newtons of force on block A, then it also exerts 3 newtons of force on block B. The tension, in that case, is 3 newtons: $T = T_A = T_B$.

QUESTION 9-6

Masses $m_1 = 1.0$ kg and $m_2 = 3.0$ kg are connected by a taut rope. Mass m_2 is just over the edge of a ramp inclined at an angle $\theta = 30°$. Both masses have a coefficient of kinetic friction $\mu_k = 0.20$ with the surface. At time $t = 0$, the system is given an initial speed $v_0 = 0.80$ m/s, which starts mass m_2 down the ramp.

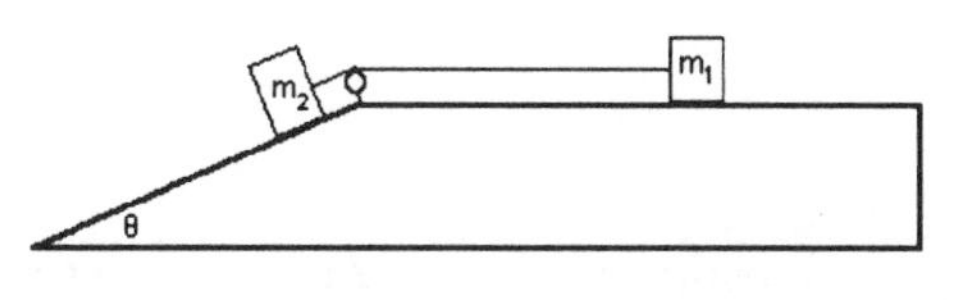

(a) Draw the force diagram for each mass.

(b) How fast are the blocks sliding after 1.0 second? (Assume that m_2 has not yet reached the bottom of the ramp, and m_1 has not yet crashed into the pulley.)

ANSWER 9-6

We can again apply the multi-block strategy.

(a) *Step 1: Force diagrams*

The blocks' initial speed does not appear in the force diagrams. Include forces only. The force that gave the blocks their initial speed no longer acts on the blocks.

The rope's tension is the same on each block. But the gravitational, normal, and frictional forces on block 1 differ from those acting on block 2. Notice also that friction resists the motion of each block, and therefore points "backwards" in both cases.

(b) *Step 2: Newton's law for each mass separately*

I will choose the down-the-ramp direction as positive for block 2. This commits me to choosing leftward as positive for block 1. Why? Because when block 2 slides down the ramp, block 1 is dragged leftward; and our choice of "positives" must be consistent.

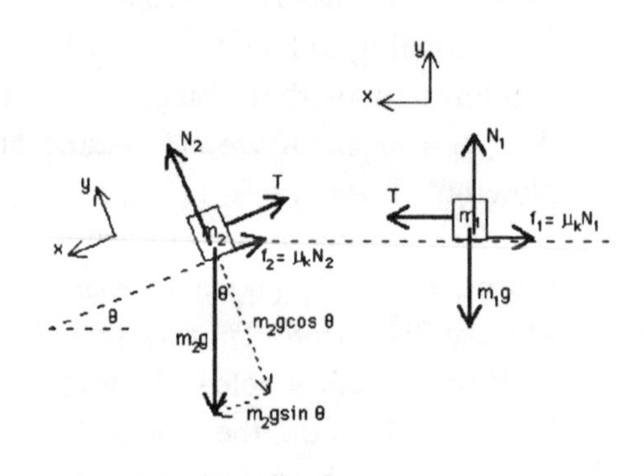

Because the blocks move together, they "share" the same acceleration, a.

Looking at the forces along the direction of motion (i.e., sideways for block 1, along the ramp for block 2), I get

$$\text{BLOCK 1, } x\text{-direction: } \sum F_x = T - \mu_k N_1 = m_1 a$$

$$\text{BLOCK 2, } x\text{-direction: } \sum F_x = m_2 g \sin\theta - \mu_k N_2 - T = m_2 a.$$

We can re-express the normal forces in terms of given constants by considering the forces along the direction perpendicular to the direction of motion. These forces must cancel out.

$$\text{BLOCK 1, } y\text{-direction: } \sum F_y = N_1 - m_1 g = m_1 a_y = 0$$

$$\text{BLOCK 2, } y\text{-direction: } \sum F_y = N_2 - m_2 g \cos\theta = m_2 a_y = 0,$$

from which we get

$$N_1 = m_1 g$$
$$N_2 = m_2 g \cos\theta.$$

Substitute those normal forces into the $\sum F_x$ equations above to get

$$T - \mu_k m_A g = m_1 a \qquad (1)$$

$$m_2 g \sin\theta - \mu_k m_2 g \cos\theta - T = m_2 a. \qquad (2)$$

Step 3: Solve for acceleration

To complete the algebra quickly, add Eq. (1) to Eq. (2). The T's cancel, leaving us with

$$-\mu_k m_1 g + m_2 g \sin\theta - \mu_k m_2 g \cos\theta = (m_1 + m_2)a.$$

Divide through by $m_1 + m_2$ to get

$$a = \frac{m_2 g \sin\theta - \mu_k g(m_1 + m_2 \cos\theta)}{m_1 + m_2}.$$

$$= \frac{(3.0\text{ kg})(9.8\text{ m/s}^2)\sin 30° - (.20)(9.8\text{ m/s}^2)(1\text{ kg} + [3.0\text{ kg}]\cos 30°)}{1\text{ kg} + 3\text{ kg}}$$

$$= 1.9\text{ m/s}^2.$$

Step 4: Kinematics or Newton's 2nd law to solve for what we want

If asked for the tension in the rope, we could plug this acceleration back into Eq. (1) or Eq. (2). But instead, we are asked for the blocks' speed at time $t = 1.0$ s. We know the acceleration, along with the initial speed ($v_0 = 0.8$ m/s). So, we can use the one-dimensional kinematic formula

$$\begin{aligned} v &= v_0 + at \\ &= 0.80 \text{ m/s} + (1.9 \text{ m/s}^2)(1.0 \text{ s}) \\ &= 2.7 \text{ m/s}. \end{aligned}$$

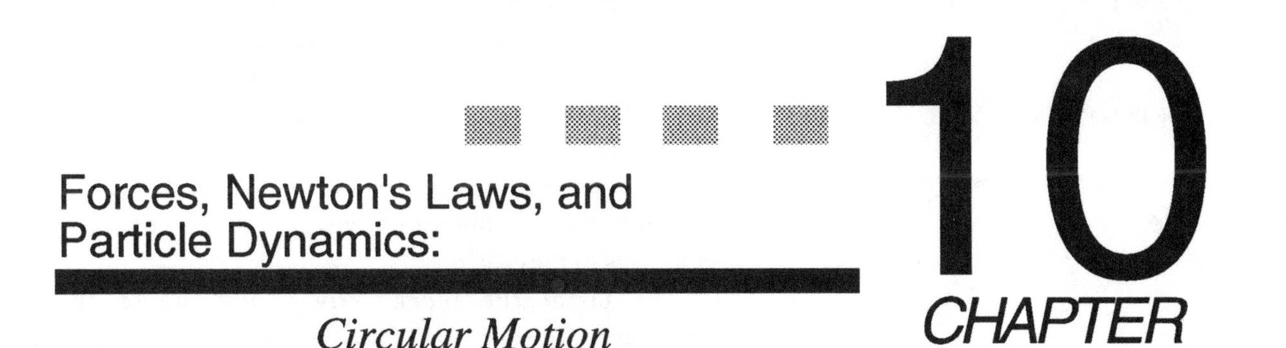

Forces, Newton's Laws, and Particle Dynamics:

Circular Motion

10 CHAPTER

QUESTION 10-1

One end of a massless rope of length s is nailed to a tabletop. The other end of the rope is attached to a frictionless block of mass m. Someone gives the block a brief push, and then lets go. As a result, the block slides around in a circle. The tension in the rope is measured to be T.

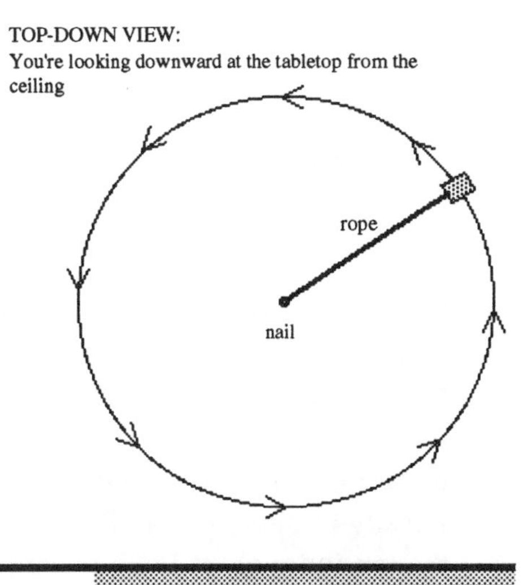

(a) Let P denote the block's period of revolution, i.e., how much time the block takes to complete one full circle. (Normally, we would use "T" to denote the period; but we have already used "T" to denote tension.) Solve for P in terms of s, m, and T.

Hint: Solve for the block's velocity, and then figure out how v relates to P.

(b) While the block is moving in a circle, what is the normal force exerted by the table on the block?

ANSWER 10-1

(a and b) This is a standard circular motion problem. Before I outline the general strategy, let us picture what is going on.

The rope, by pulling inward on the block, makes the block move in a circle. Once it gets started, the block continues moving in a circle even if no one pushes the block along its direction motion; in the absence of the friction, the block does not slow down. So, the force exerted by the rope does not speed up or slow down the block. Rather, the force continuously changes the *direction* of the block's motion, so that it moves in a circle instead of a straight line. (If someone suddenly cuts the rope, the block slides off in a straight line.)

Here is my point: When an object moves in a circle at constant speed, it is accelerating, because its velocity changes. You might think the velocity is constant, because the block neither speeds up nor slows down. But remember, velocity is a vector. It has a direction. When an object's direction of motion changes, its velocity changes. Intuitively speaking, you must accelerate an object, by pushing or pulling on it, to change its direction of motion. For an object moving in a circle, this push must be inward towards the center of the circle. So, while the object's velocity is along ("tangential to") to the circle, its acceleration is radially directed, toward the center.

TOP-DOWN VIEW:

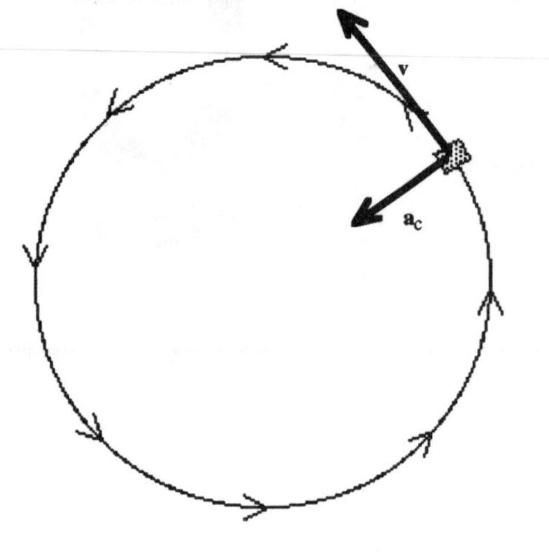

This so-called "centripetal" acceleration keeps changing the direction of the velocity, so that the block continues to trace a circle.

The faster the block moves, the bigger this acceleration must be, and hence the harder the rope must pull. You can feel this intuitively by swinging a ball (or shoe) around on a string.

As shown in the textbook, the magnitude of this radial ("centripetal") acceleration is $a_c = \frac{v^2}{r}$, where r is the radius of the circle traced out by the object. Remember, this acceleration points *radially*. Therefore, Newton's 2nd law implies that the net *radial* force on the object must equal ma_c:

$$\sum F_{\text{radial}} = ma_{\text{radial}} = m\frac{v^2}{r}.$$

Some textbooks call $\sum F_{\text{radial}}$ the "centripetal force" and abbreviate it F_c. I must emphasize that the "centripetal" force is not some new kind of force acting on the object. Instead, it is just a shorthand way of denoting the"net radial force" supplied by the already-existing forces. For instance, in this problem, the tension pulls radially inward.

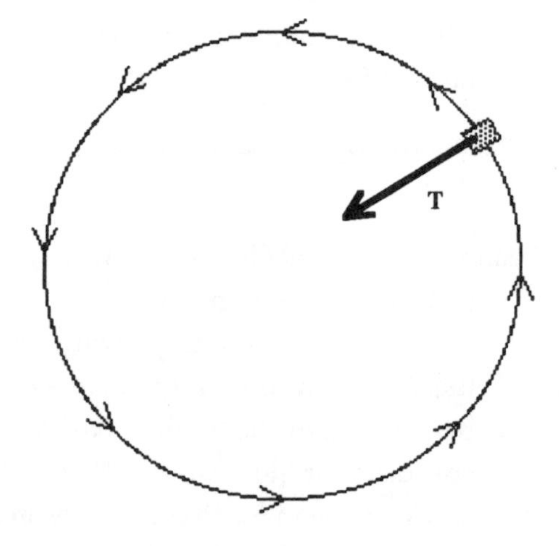

"F_c" does not appear in the force diagram, because it is just a fancy way of saying the net radial force. Here, the "centripetal" force is provided by the tension in the rope.

These considerations suggest a problem-solving hint that applies whenever an object undergoes circular motion:

Circular motion problem-solving hint #1:

Use Newton's 2nd law in the radial direction, i.e., set $\sum F_{\text{radial}} = ma_{\text{radial}} = \frac{mv^2}{r}$. *This usually gives you useful information.*

We now see that most circular motion problems are nothing more than another application of Newton's 2nd law.

(a) In this particular problem, we can use Newton's 2nd law to obtain the block's speed:

$$\sum F_{\text{radial}} = ma_{\text{radial}}$$
$$T = m\frac{v^2}{r}, \tag{1}$$

and hence $v = \sqrt{\frac{Tr}{m}} = \sqrt{\frac{Ts}{m}}$, since the block traces a circle of radius $r = s$. But we are not asked for the velocity. We want the period P. So, we must figure out how v relates to P.

Intuitively, the faster the block travels, the less time it takes to complete a circle. So, a high v corresponds to a small P, and vice versa. We can relate speed to period more precisely. In time P, the block completes one full circle. The circumference of a circle is $2\pi r$. So, in time P, the block covers distance $2\pi r$. But since the speed is constant, we have speed $= \frac{\text{distance covered}}{\text{time}} = \frac{2\pi r}{P}$. As intuitively expected, v is inversely proportional to P. You need not memorize this formula; it is easy to derive.

Solve this relation for P to get $P = \frac{2\pi r}{v}$. Remembering that $r = s$ in this case, and substituting in our expression for v from Eq. (1) above, yields

$$P = \frac{2\pi s}{v}$$
$$= \frac{2\pi s}{\sqrt{\frac{Ts}{m}}}$$
$$= 2\pi\sqrt{\frac{ms}{T}},$$

which, has the correct dimensions (units), as you can confirm.

(b) So far, we have considered only the forces and accelerations in the radial direction. But part (b) asks for the normal force N. On the above force diagram, we cannot even *see* N, because it points "out of the page." So, let us redraw the force diagram from another vantage point.

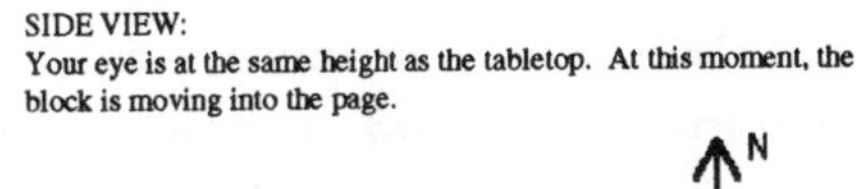

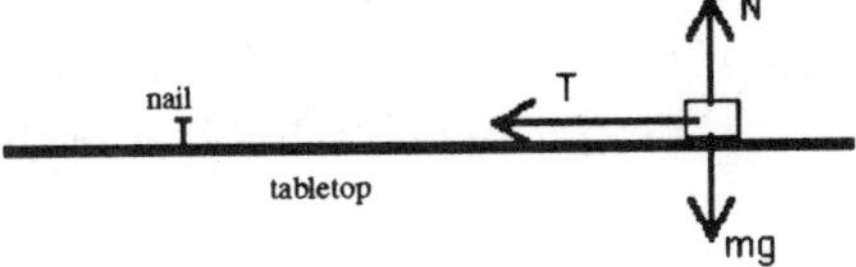

The block moves entirely in the plane of the tabletop. It does not move vertically. Therefore, its vertical acceleration is $a_y = 0$: The vertical forces must cancel:

$$\sum F_y = ma_y$$
$$N - mg = 0,$$

and hence $N = mg$.

This illustrates another general circular motion problem solving hint.

Circular motion problem-solving hint #2:

When an object traces a circle, all of its motion is in the plane of the circle. Therefore, the forces *perpendicular* to the plane of the circle must cancel out.

Again, this is just an application of Newton's 2nd law. You have seen it before, for example, in the context of blocks sliding down ramps. Since the block's motion is entirely along the ramp's surface, the forces perpendicular to the ramp's surface cancel out.

QUESTION 10-2

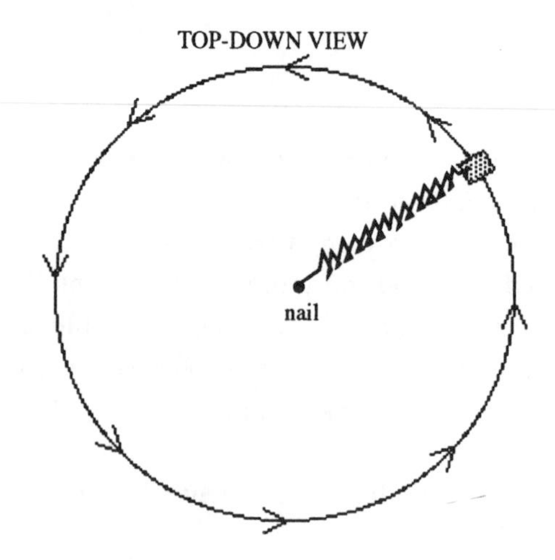

The force exerted by a spring is $F_s = -kx$, where k is a "stiffness constant," and x is how far the spring is stretched from its equilibrium length (i.e., its natural length, when nothing pulls on it). The negative sign indicates that a stretched spring "wants" to get shorter.

Consider a spring with spring constant k and equilibrium length L_0. One end of this spring is nailed to a tabletop. The other end of the spring is attached to a frictionless block of mass m. A playful physicist sets the block in motion in such a way that the block starts moving in a circle around the nail at constant speed v.

How far from the nail is the block?

ANSWER 10-2

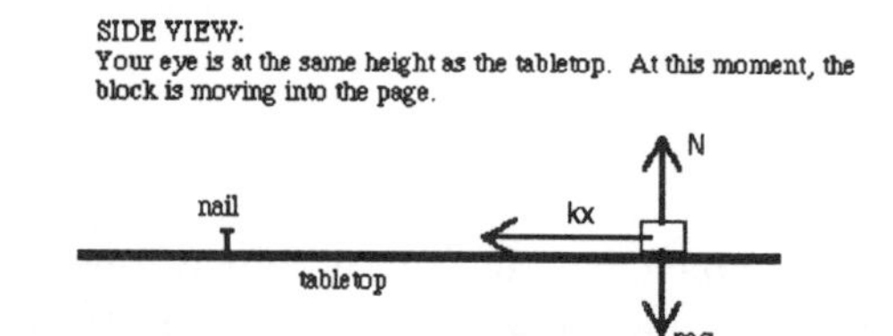

Use the circular motion reasoning described in Question 9-1. Besides gravity and the normal force, which cancel each other, the only force acting on the block comes from the spring.

Remember, the minus sign in $F_s = -kx$ indicates that a stretched spring wants to unstretch. In other words, the stretched spring pulls the block toward the nail. So, I have built the minus sign into my force diagram by letting kx point inward (toward the nail) instead of outward. It would redundant to include that negative sign "again" by writing "$-kx$" in the force diagram.

From Newton's 2nd law applied to circular motion, we get

$$\sum F_{\text{radial}} = ma_{\text{radial}}$$

$$kx = m\frac{v^2}{r}, \tag{1}$$

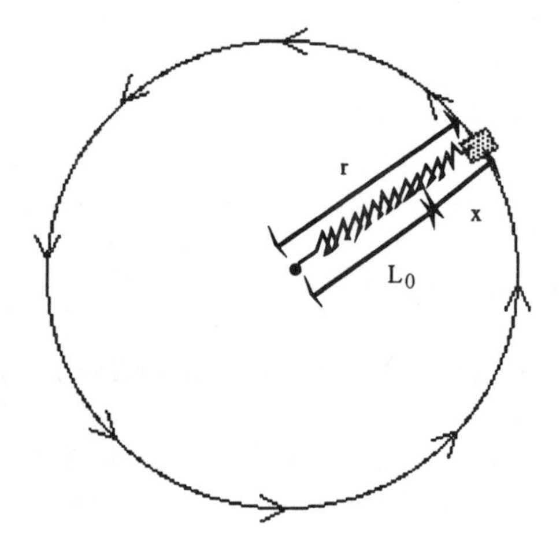

where r is the radius of the circle. We are solving for r, because the circle's radius is the block's distance from the nail.

Crucially, x does not equal r, because x is how the far the spring is stretched *from its equilibrium length*. For instance, suppose the spring is naturally $L_0 = 0.5$ meters long, and the block in this problem traces a circle of radius $r = 0.6$ meters. Then the spring is stretched by only $x = 0.1$ meters from its equilibrium length. From this hypothetical example, we see that $x = r - L_0$. The previous diagram confirms this relationship between x and r.

Again, we see that $r = L_0 + x$, and hence

$$x = r - L_0. \qquad (2)$$

Now we can solve Eqs. (1) and (2) for r.

Algebra starts here. Substitute Eq. (2) into Eq. (1) to get $k(r - L_0) = m\frac{v^2}{r}$. Multiply both sides by r, and then subtract mv^2 from both sides. This leaves us with a quadratic equation in standard form: $kr^2 - kL_0 r - mv^2 = 0$. The quadratic formula yields $r = \frac{kL_0 \pm \sqrt{(kL_0)^2 - 4(k)(-mv^2)}}{2k} = \frac{kL_0 \pm \sqrt{(kL_0)^2 - 4(k)(mv^2)}}{2k}$. The radius of a circle cannot be negative. So, the physically correct answer for r is $r = \frac{kL_0 \pm \sqrt{(kL_0)^2 + 4kmv^2}}{2k}$.

QUESTION 10-3

Consider an airplane flying in a circle of diameter D. The circle is horizontal, by which I mean the plane of the circle is parallel to the ground. In other words, as the airplane circles around, its height never changes. It is like when you are in a holding pattern over Newark Airport, and you desperately want to land because you are going to get sick, but the airplane just keeps circling and circling...but I digress.

The airplane's speed, v, is constant.

According to fluid dynamics, air pressing against the bottom of the airplane's wings exerts a "lift" force F_L. The direction of this lift force is always perpendicular to the wings. For instance, consider the following diagram, in which the airplane is seen from behind. The airplane is "banking." The lift force is perpendicular to the wings, as shown.

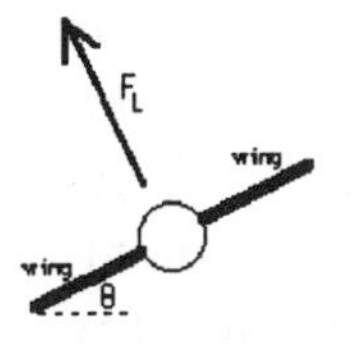

BACK VIEW
(i.e., you're looking at the airplane from behind. The airplane is flying "into the page")

The airplane has mass M. The airplane's engines exert a forward "thrust" force on the plane F_t, while air resistance exerts a backwards "drag" force F_d. By "forwards" and "backwards," I mean with respect to the airplane's direction of motion.

(a) Is the thrust force F_t bigger, smaller, or the same magnitude as the drag force F_d? Explain your answer.

(b) At what angle θ must the airplane be banked in order to continue flying in its present circular path? Express your answer in terms of M, v, and D. (Note: The algebra is difficult.)

ANSWER 10-3

(a) Here is a "top-down" view, with the thrust and drag forces drawn in. I have not included any of the other forces, because they all point perpendicular to F_t and F_d, and therefore play no role in this part of the problem.

You may think that $F_t > F_d$, because the airplane keeps moving forward. By the same reasoning, you would think that when I push a book across my desk with constant speed, my "push" force is bigger than the frictional "drag" force. But that is not true. When I push my book at constant speed, it is neither speeding up nor slowing down–i.e., it is not accelerating. Therefore, the net force on it is zero. The "push" force exactly equals the "drag" force.

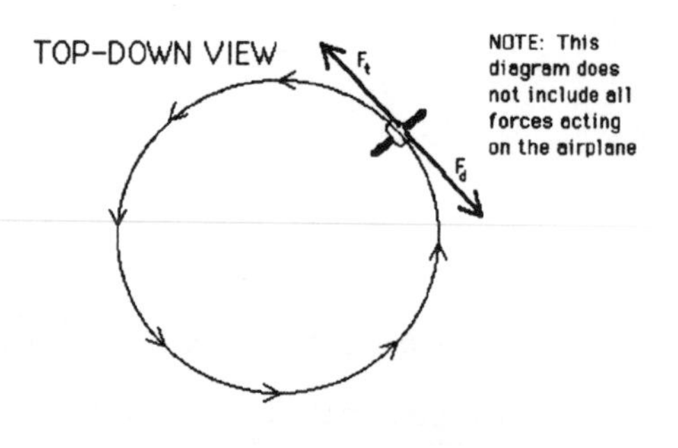

This seems less strange if you remember that when I first start pushing the book, speeding it up from 0 mph to its final velocity, I am accelerating the book. The net force is not zero during the acceleration–my push force exceeds friction. But once the book reaches its final speed, and just keeps moving at constant velocity, it no longer "needs" a net force to keep it moving. It is own inertia "wants" it to keep moving. (Think of a hockey puck sliding on frictionless ice.) Once in motion, the book stays in motion at constant speed, unless a net force speeds it up or slows it down. So, my pushforce needs to be just strong enough to cancel out friction, ensuring a zero net force.

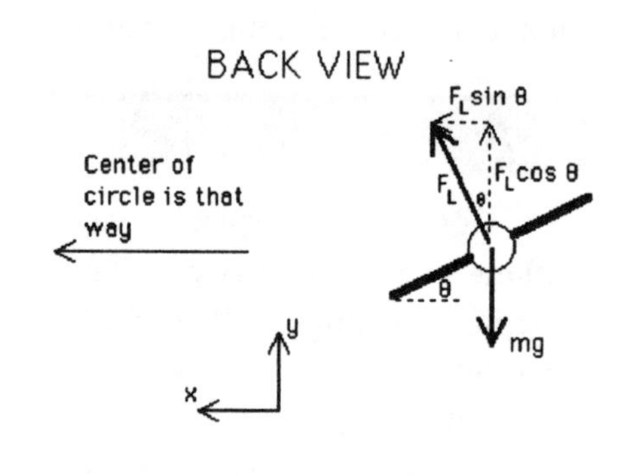

Similarly, in this problem, the airplane neither speeds up nor slows down. Therefore, its acceleration *along the direction of motion* is 0. So, the net forward force on the airplane is 0. Therefore, F_d exactly cancels F_t; those forces are equal. Of course, the airplane has a *radial* (centripetal) acceleration directed towards the center of the circle. But that acceleration is not caused by F_t or F_d, neither of which points in the radial direction.

(b) As always, start with a force diagram. The best "view" is from behind the airplane.

I have omitted the thrust and drag forces, which point into and out of the page, respectively. As just shown, those forces cancel.

Circular motion problem solving hint #1 advises us to write Newton's 2nd law for the radial direction, taking advantage of the fact that an object in circular motion has radial (centripetal) acceleration $a_{\text{radial}} = \frac{v^2}{r}$. Here, the radial direction is x-ward (horizontal), because the circle traced out by the airplane is parallel to the ground. So, we get

$$\sum F_{\text{radial}} = ma_{\text{radial}}$$
$$F_L \sin\theta = M\frac{v^2}{r}. \tag{1}$$

We are trying to solve for the banking angle θ. Unfortunately, Eq. (1) contains a second unknown, the lift force F_L. (We *are* given M and v; and the radius is half the diameter, $r = \frac{D}{2}$.) We need another equation.

To get it, use circular motion problem solving hint #2: The net force perpendicular to the plane of motion must be zero. Here, the airplane flies in a horizontal circle. It never changes altitude. That is, the airplane has no vertical motion, and hence $a_y = 0$. It follows that the vertical forces cancel:

$$\sum F_y = Ma_y$$
$$F_L \cos\theta - Mg = 0,$$

and hence

$$F_L \cos\theta = Mg. \qquad (2)$$

At this point, we have two equations in the two unknowns, F_L and θ. Therefore, only algebra remains.

Algebra starts here: To finish quickly, divide Eq. (1) by Eq. (2). The F_L's (and also the M's) cancel out, leaving us with $\tan\theta = \frac{v^2}{rg}$. Set $r = \frac{D}{2}$, and take the arctangent of both sides, to get $\theta = \tan^{-1}\left(\frac{2v^2}{Dg}\right)$.

End of algebra.

By playing with this expression for θ, you can confirm your experience that a larger velocity corresponds to a larger ("steeper") banking angle θ, for a fixed diameter.

This problem emphasizes, again, that circular motion problems are just like other force problems. You must draw a free-body diagram and consider the x- and y-components. The only difference is that you should always choose your x-direction or y-direction to be radial, towards the center of the circle. Let's say you choose x-ward as radial. Then, you know that $a_{radial} = a_x = \frac{v^2}{r}$.

QUESTION 10-4

(Very hard. Skip this until you are confident with circular motion. If you solved the previous problem, give this one a go, and see how much progress you can make.)

A toy car of mass $M = 0.40$ kg travels along a circular banked road, with banking angle $\theta = 30°$. The circle has radius $r = 1.0$ m. Nina wants to make the car move as fast as possible. But if it goes too fast, it "flies off" the road. Therefore, Nina cuts a 1-meter piece of fishline (i.e., an ultra-thin cord), attaches one end to the car, and attaches the other end to a nail stuck in the center of the circle. See the drawing below. Unfortunately, the fishline is so thin and brittle that it cannot withstand a tension greater than $T_{max} = 5.0$ N. If the tension gets any bigger than that, the fishline breaks.

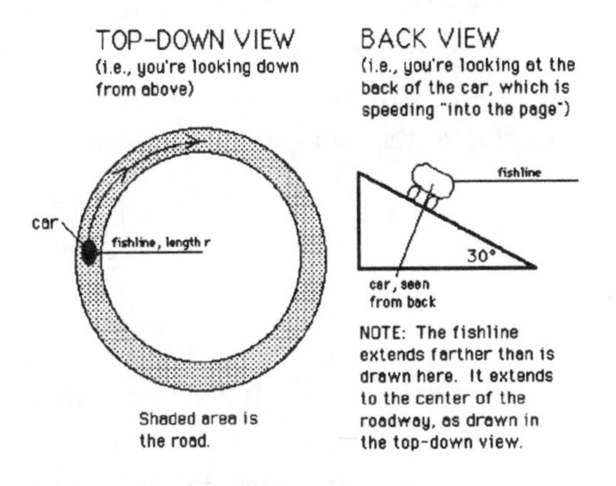

What is the maximum speed the car can travel around the track without breaking the fishline?

ANSWER 10-4

Intuitively speaking, the faster the car goes, the more tension builds up in the fishline. We want the car to go as fast as possible. So, the tension in the fishline will be T_{max}. Since we are relating the car's speed v to forces acting on the car, let us use circular motion problem solving hint #1: Draw a force diagram, and then consider Newton's 2nd law in the radial direction. Do not forget the normal force. Here, $\theta = 30°$.

I have chosen the x-direction as towards the center of the circle (horizontal), and the y-direction as vertical, instead of picking "tilted" coordinates. Why? Because we want to consider the radial forces and acceleration, so that we can use $a_{radial} = \frac{v^2}{r}$. The radial direction is towards the center of the circle, which is located at the end of the fishline not attached to the car. Since the fishline is

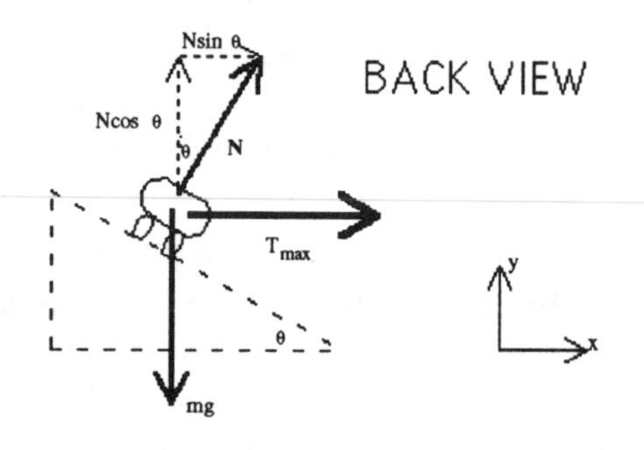

stretched horizontally, the radial direction is horizontal. To see this, notice that the car does not move up or down. The circle it traces is parallel to the ground, i.e., horizontal. (If you hold a circle (e.g., a plate) parallel to the ground, you will see that the radial direction is horizontal.)

Newton's 2nd law in the radial direction gives

$$\sum F_{\text{radial}} = ma_{\text{radial}}$$
$$T_{\max} + N\sin\theta = M\frac{v^2}{r}. \tag{1}$$

We cannot yet solve for v, because Eq. (1) contains a second unknown, namely the normal force N. As always, to extract more information, consider the y-directed forces. Since the car's motion lies entirely in a horizontal plane–that is, since the car does not move up or down–its vertical acceleration is $a_y = 0$. Therefore,

$$\sum F_y = ma_y$$
$$N\cos\theta - Mg = 0. \tag{2}$$

Now we have two equations in the two unknowns, v and N. No more physical thinking is needed.

Algebra starts here. To complete the algebra efficiently, first solve Eq. (2) for N to get $N = \frac{Mg}{\cos\theta}$. Now substitute that expression for N into Eq. (1) to get $T_{\max} + \frac{Mg}{\cos\theta}\sin\theta = M\frac{v^2}{r}$. Solve for v, and plug in the numbers provided in the problem, to get

$$\begin{aligned} v &= \sqrt{gr\tan\theta + \frac{T_{\max}r}{M}} \\ &= \sqrt{(9.8\ \text{m/s}^2)(1.0\ \text{m})\tan 30° + \frac{(5.0\ \text{N})(1.0\ \text{m})}{0.40\ \text{kg}}} \\ &= 4.3\ \text{m/s}. \end{aligned}$$

If the car goes any faster, the fishline will break.

QUESTION 10-5

A block of mass m hangs from one end of a light string, the other end of which is attached to the ceiling. The string has length L.

Obviously, if we just let the block hang, the tension in the string will be mg, to balance the downward force of gravity . But suppose we displace the block as shown. When released, the block swings down. Let point P denote the "bottom" of the block's swing-path.

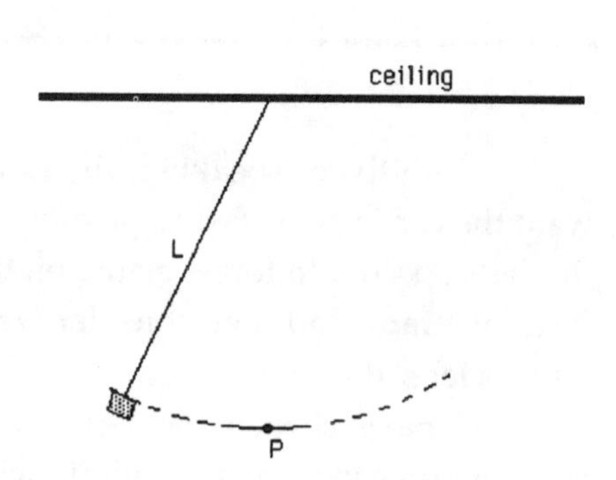

If we release the block from the position shown, its speed when it passes through point P is v.

(a) At the moment the block passes through point P, is the tension in the string bigger than, smaller than, or the same as mg? Answer this question in two minutes or fewer, using purely intuitive reasoning. Do not do math until part (b).

(b) At the moment the block passes through point P, what is the tension in the string? Answer in terms of m, g, L, and v.

ANSWER 10-5

(a) When the block hangs motionless, the upward force exerted by the string must "fight" the downward force of gravity. When the block swings, the string must continue to fight gravity, and must ALSO keep the block moving along its circular path. So, the string pulls harder when the block swings; $T > mg$.

Let me flesh out this argument. Pretend that the experiment takes place in outer space, where gravity is negligible. If we let the block "hang," then the tension in the string is $T = 0$, because the block just floats there. But if the block is made to "swing," then the string must exert enough tension to produce the "centripetal" acceleration needed to keep the block moving in a circle. This outer-space example shows that the tension is greater when the block swings.

Does this conclusion hold on Earth, too? Well, as noted above, the string must fight gravity whether or not the block swings. In other words, the tension in the string "devoted" to fighting gravity is the same whether or not the block moves. (This tension is mg.) But when the block swings, there is also *additional* tension in the string "devoted" to keeping the block moving in a circle.

(b) Since the block traces out a circular path, we can use circular motion reasoning, by which I mean Newton's 2nd law applied to the radial direction. As always in force problems, start with a force diagram. Here, the free-body diagram-body should depict the block at the moment it reaches point P.

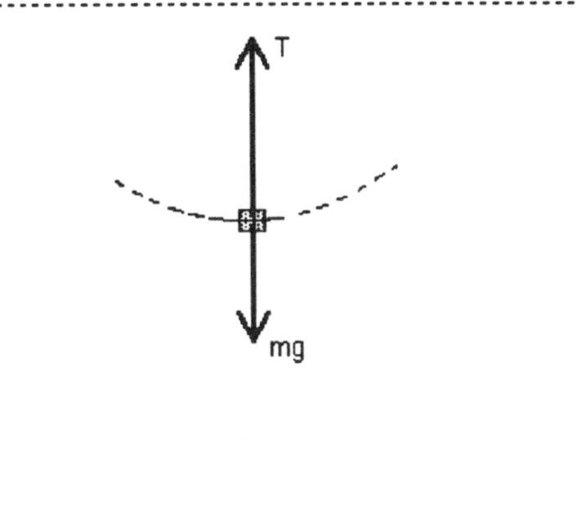

At point P, the rope is vertical, and hence tension points upward.

The center of the circle is where the rope touches the ceiling, directly above point P. Hence, the radial direction is upward. (This would not be true if the block were not at point P.) So,

$$\sum F_{\text{radial}} = ma_{\text{radial}}$$

$$T - mg = m\frac{v^2}{r} \qquad (1)$$

$$= m\frac{v^2}{L},$$

since the block traces a circle whose radius is the string's length: $r = L$. We can immediately solve for the tension to get $T = mg + m\frac{v^2}{L}$. Let us hook up this mathematical answer with the intuitive reasoning of part (a). The tension in the string must fight gravity, and must also keep the block moving circularly. The amount of tension devoted to counteracting gravity is mg, the block's weight. The amount of tension devoted to keeping the block moving in a circle is $\frac{mv^2}{L}$. So, the total tension in the string is the sum of mg and $\frac{mv^2}{L}$.

11 CHAPTER

Work and Energy:

With Core Concepts of Potential Energy and Energy Conservation

QUESTION 11-1

Starting from rest, a block of mass $m = 5.0$ kg slides down a frictionless ramp. The ramp has a vertical height of $H = 1.0$ meter and makes a $\theta = 30°$ angle with the floor. After leaving the ramp, the block slides along the floor. But the floor is not frictionless. In fact, the floor's coefficient of kinetic friction is $\mu = 0.20$.

H

θ

B

NOTE: Point B is the bottom of the ramp.

(a) What is the block's speed at point B? Solve this *without* using any kinematic formulas. (I want you to learn the "shortcut" techniques.)

(b) How much work does the floor do on the block as the block slows down and eventually comes to rest?

(c) How far along the floor does the block slide before coming to rest?

(d) How would your answers to parts (b) and (c) change if the floor's coefficient of friction were $\mu = .40$ instead of $\mu = .20$? Do not solve numerically; just describe qualitatively how things would be different.

ANSWER 11-1

(a) I know you could treat this as a standard force problem: Draw a force diagram, find the net force parallel to the ramp, use $\sum F_x = ma_x$ to solve for the acceleration, and then invoke kinematics to find the velocity at the bottom of the ramp. In this chapter, however, I want you to become comfortable with alternative problem-solving techniques that, in many cases, provide a substantial shortcut. The new techniques are (1) Work and energy, and (2) Conservation of energy. Usually, when you have a choice, conservation of energy is the quickest, easiest way to solve a problem. When it is not applicable, the work-energy theorem can save the day.

Because many of you are probably focusing only on work-energy stuff this week, and saving energy conservation until next week, I will not yet introduce energy conservation. Instead, I will use work-energy considerations. But after you learn about energy conservation, come back and redo this problem.

You can solve with the work-energy theorem. Before introducing this useful result, let me review the underlying concepts.

Kinetic energy is an object's energy of motion, which you can think of as the object's ability to knock something over. So for instance, a bowling ball rolling at 10 mph has much more kinetic energy than a tennis ball at that same speed. So, kinetic energy increases with mass. Velocity also matters; a bowling ball at 20 mph knocks down more pins than a bowling ball at 10 mph. So, we expect kinetic energy to increase with velocity. It is not quite intuitive, however, that kinetic energy increases with the square of velocity. See your textbook for a deeper explanation. The result is Kinetic energy $= K = \frac{1}{2}mv^2$.

Now let us talk about work. Roughly speaking, it is the "effort" exerted to *successfully* move an object. For instance, if you push really hard on a wall, you exert no work, because you do not *succeed* in moving the wall.

Intuitively, it is clear that work should be proportional to the force doing the work. But distance also matters. For instance, pushing a desk ten meters across the floor requires more effort than pushing it two meters. So, work is proportional to distance. Given all this, you might think the formula for work is $W = F\Delta x$. But one complication remains. To see why, imagine pushing the desk horizontally vs. pushing at a downward angle.

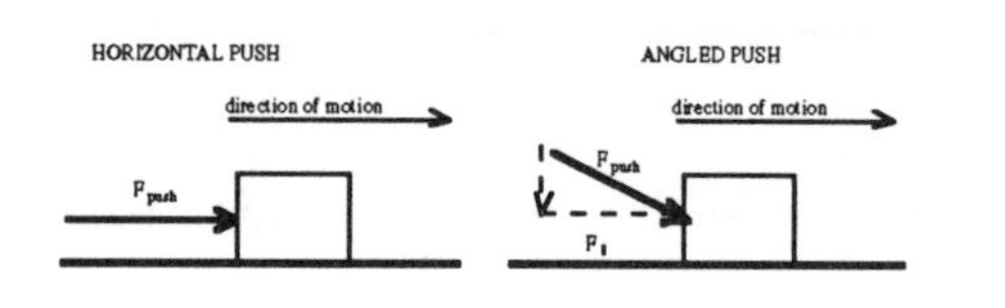

When you push horizontally, *all* of your push gets "devoted" to moving the desk across the floor. But when you push at a downward angle, the downward component of your push gets "wasted" jamming the desk into the floor. It is "better" to push the desk horizontally. Because the desk moves horizontally, only the horizontal component of your push actually helps to move it across the floor. In other words, *only the component of the force parallel to the direction of motion does work.* If we call this parallel component of the force "$F_{\parallel}$," then we can write Work $= W = F_{\parallel}\Delta x$, where Δx is the distance over which the force gets exerted. (If the force is not constant, you must use $W = \int F_{\parallel}dx$, as I will explain later in this chapter.) Your textbook probably writes this as $W = \mathbf{F}\cdot\Delta\mathbf{x}$, a dot product. The dot product is just a fancy way of "picking off" the components of two vectors that are parallel to each other. See your textbook for details. So, $F_{\parallel}\Delta x$ and $\mathbf{F}\cdot\Delta\mathbf{x}$ are the same thing.

Now that I have reviewed work and kinetic energy, I can state the result we will use to solve this problem:

$$\textbf{Work-energy theorem:}\quad W_{\text{net}} = \Delta K.$$

In words, when work gets done on an object, the object speeds up or slows down; and its *change* in kinetic energy ($K_f - K_0$) equals the net work done. See your textbook for a proof.

Let us apply the work-energy theorem to the problem at hand. We want to know the block's speed v_B at point B. Well, at the top of the ramp, the block was motionless, and therefore had no kinetic energy: $K_0 = 0$. Its (as-yet-unknown) kinetic energy at point B is $K_B = \frac{1}{2}mv_B^2$. To find the net work, we must consider the forces acting on the block.

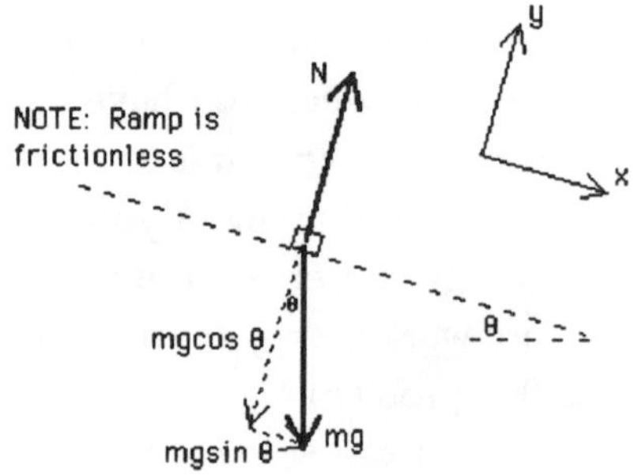

Remember, only the force components *along the direction of motion* do any work. Here, the net force along the direction of motion is $F_{\parallel \text{net}} = mg\sin\theta$. So, to calculate the work done on the block as it slides down the ramp, we just have to multiply $F_{\parallel \text{net}}$ by the length of the ramp, Δx.

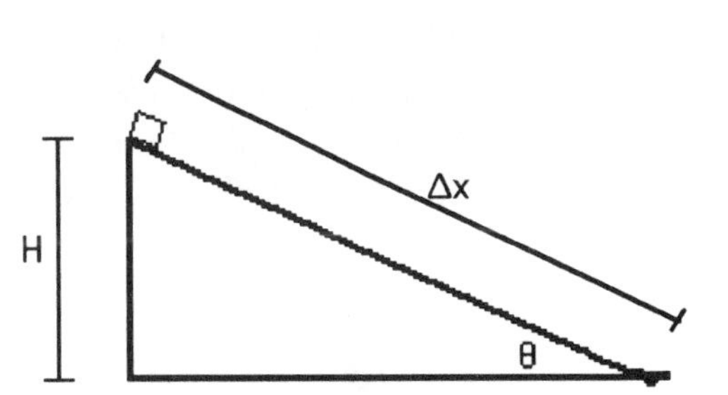

From trig, $H = \Delta x \sin\theta$, and hence $\Delta x = \frac{H}{\sin\theta}$. Putting all this together gives

$$\begin{aligned} W_{\text{net}} &= \Delta K \\ F_{\parallel \text{net}}\Delta x &= K_B - K_0 \\ (mg\sin\theta)\frac{H}{\sin\theta} &= \frac{1}{2}mv_B^2 - 0. \end{aligned}$$

We can solve this for v_B. Notice that $\sin\theta$ cancels out, leaving us with

$$v_B = \sqrt{2gH} = \sqrt{2(9.8\ \text{m/s}^2)(1.0\ \text{m})} = 4.4\ \text{m/s},$$

the same answer you would get using Newton's 2nd law and kinematics.

You might be surprised that the block's speed at point B does not depend on the ramp angle θ. A steep or shallow ramp of the same height both give the block the same final speed. This result will seem more sensible in light of energy conservation. For now, let me provide a rough intuition. If the ramp is steep, then the block accelerates more quickly; but it does not accelerate for much time, because it reaches the bottom quickly. So, a high acceleration acts for a short time. By contrast, if the ramp is shallow, then the block experiences a small acceleration; but it spends more time on the ramp before reaching the bottom. So, a small acceleration acts for a long time. It turns out that the big, short-acting acceleration yields the same final velocity as does the small, long-acting acceleration.

(b) We can again use the work-energy theorem as a shortcut, instead of Newton's 2nd law and kinematics. The only force doing work on the block is friction, i.e., the force due to the floor. (Gravity and the normal force cannot do work, because they point perpendicular to the direction of motion. Besides, they cancel out.) In this part of the problem, we are considering the block's slide across the floor. So, its "initial" speed is v_B, and its final speed is $v = 0$ when it comes to rest. Substituting all this into the work-energy theorem gives

$$\begin{aligned} W_{\text{net}} &= \Delta K \\ W_{\text{by floor}} &= K_f - K_B \\ &= 0 - \frac{1}{2}mv_B^2 \\ &= -\frac{1}{2}m(2gH) \\ &= -mgH, \end{aligned}$$

which comes out to be $mgH = -(5\ \text{kg})(9.8\ \text{m/s}^2)(1\ \text{m}) = -49$ joules. The work is negative because it slows the block down instead of speeding it up. In other words, the floor "takes" kinetic energy from the block instead of giving the block energy. *Notice that the work-energy theorem allowed us to avoid worrying about the forces and distances involved*. It was not necessary to solve part (c) before solving part (b), because the work-energy theorem is a great shortcut.

This answer makes intuitive sense. As we saw in part (a), gravity does $(mg\sin\theta)(\frac{H}{\sin\theta}) = mgH$ of work on the block as it slides down the ramp, thereby giving the block kinetic energy. The floor's friction then "steals" all that kinetic energy. So, the floor must do an equal amount of negative work.

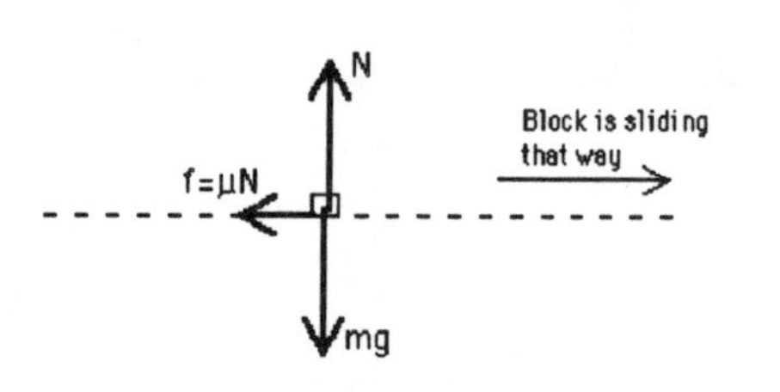

(c) Now that we know how much work the floor does on the block, we can use the definition of work to find how far the block slid. Start with a force diagram.

Since the normal force cancels gravity, $N = mg$. So, the force exerted by the floor along the direction of motion is $F_{\parallel\,\text{floor}} = -\mu N = -\mu mg$. The minus sign is crucial. It indicates that $F_{\parallel}$ points backwards compared to the direction of motion, and therefore slows the block down instead of speeding it up. Let s denote the distance the block slides. From the definition of work, and using our above result that $W_{\text{by floor}} = -mgH$, I get

$$W_{\text{by floor}} = (F_{\parallel\,\text{floor}})s$$
$$-mgH = (-\mu mg)s.$$

Solve for s to get $s = \frac{H}{\mu} = \frac{1.0\text{ m}}{0.20} = 5.0\text{ m}$. In words, the distance the block slides across the floor is proportional to the height of the ramp, but inversely proportional to the "roughness" of the floor. This makes sense.

Of course, you get the same answer for s using old-style force-and-acceleration reasoning. But these new-fangled energy considerations get the job done even when the forces and accelerations are not constant, in which case you cannot use your old kinematic formulas.

(d) In part (b), we saw that in order to take away the block's kinetic energy, the floor must "take away" the work done on the block as it slid down the ramp. That work is $mgH = 49$ joules. No matter what the coefficient of friction equals, the floor must do that same amount of work. Your part (b) answer does not "care" about μ.

So, the floor will definitely do the same amount of work as above. But if μ gets doubled, then the floor exerts *twice* as much force. Therefore, to perform the same total work, the force needs to act over only *half* the distance. Our answer to part (c), $s = \frac{H}{\mu}$, confirms this intuitive conclusion: If we double μ, the slide distance s gets cut in half.

QUESTION 11-2

A bookshelf, of mass $m = 40$ kg, slides down a moving van's loading ramp, which is $s = 5.0$ m long and $h = 2.0$ m high. Jo pushes on the bookshelf, resisting its motion, so that it slides down at constant velocity (instead of careening out of control). The direction of Jo's push is parallel to the ramp's surface. The coefficient of kinetic friction between the bookshelf and the ramp is $\mu_k = 0.25$. Solve for

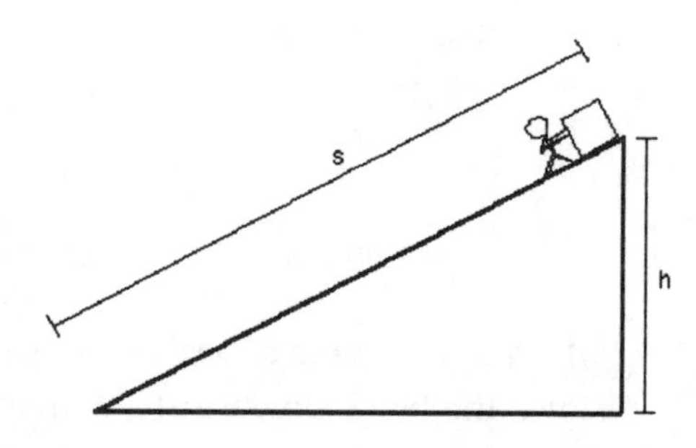

(a) the work done by Jo on the bookshelf,

(b) the work done by gravity on the bookshelf,

(c) the work done by the normal force on the bookshelf, and
(d) the net work done on the bookshelf
as the bookshelf slides all the way down the ramp. Express all your answers in terms of symbols before plugging in numbers.

ANSWER 11-2

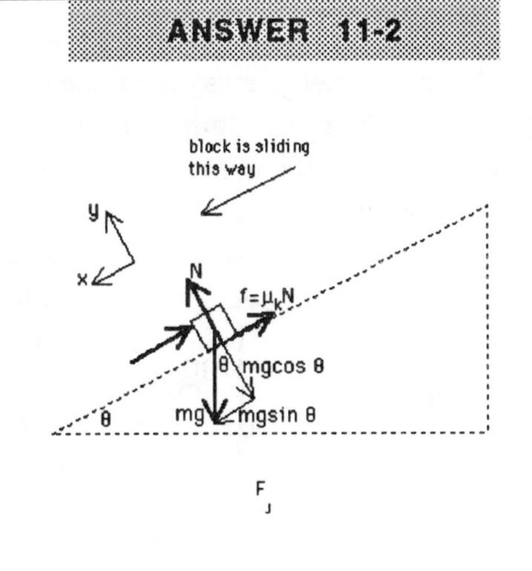

This problem requires us to distinguish between the work done by a *single* force and the *net* work done by all the forces. Because we must sort out the work done by various forces, let us start by sketching a free-body diagram. F_J denotes the force exerted by Jo. Friction points up the ramp, because it "resists" the bookshelf's motion down the ramp. Also, by looking at the diagram given in the problem, you can see that $\sin\theta = \frac{h}{s}$, and hence $\theta = \sin^{-1}(\frac{h}{s}) = \sin^{-1}(\frac{2.0\text{ m}}{5.0\text{ m}}) = 23.6°$.

(a) Since the bookshelf slides at constant speed, its kinetic energy does not change: $\Delta K = 0$. Therefore, according to the work-energy theorem, "the work is zero." This is true: But the work that appears in the the work-energy theorem is *not* the work due to Jo, or the work due to gravity, or the work due to the normal force or friction. It is the *net* work done by all those forces–i.e., it is the work done by the *net* force. So, we can use the work-energy theorem in part (d). But not here.

Here, we must resort to old-fashioned force-and-acceleration reasoning. Once we calculate Jo's force, we can multiply it by Δx to find the work she exerts.

Let the x-direction be parallel to the ramp. We know $a_x = 0$, because the bookshelf neither speeds up nor slows down; its velocity is constant. Remembering that the frictional force is $f = \mu_k N$, we get

$$\begin{aligned} \sum F_x &= ma_x \\ mg\sin\theta - \mu_k N - F_J &= 0. \end{aligned} \qquad (1)$$

In order to solve for F_J, we need to know the normal force N. By the usual reasoning,

$$\begin{aligned} \sum F_y &= ma_y \\ N - mg\cos\theta &= 0, \end{aligned}$$

and hence $N = mg\cos\theta$. Substitute this into Eq. (1), and solve for F_J, to get

$$\begin{aligned} F_J &= mg\sin\theta - \mu_k mg\cos\theta \\ &= mg(\sin\theta - \mu_k\cos\theta). \end{aligned}$$

Now that we know Jo's force, we can calculate her work. The work done *by Jo* on the bookshelf, when the bookshelf travels distance s, is $W_J = \mathbf{F}_J \cdot \mathbf{s} = F_{\parallel J} s$, where $F_{\parallel J}$ is the component of Jo's force along the direction of motion. Here, $\mathbf{F}_J$ points up the ramp, while the displacement vector $\mathbf{s}$ points down the ramp (because the bookshelf moves down the ramp). So, $F_{\parallel J} = -F_J$, because Jo's force points "opposite" to the direction of motion. Hence,

$$\begin{aligned} W_J &= F_{\parallel J} s \\ &= -mg(\sin\theta - \mu_k \cos\theta)s \\ &= -(40 \text{ kg})(9.8 \text{ m/s}^2)(\sin 23.6° - .25\cos 23.6°)(5 \text{ m}) \\ &= -340 \text{ J}. \end{aligned}$$

The minus sign indicates that Jo's push tends to resist rather than help the bookshelf's motion. Jo takes energy away from the bookshelf instead of giving it energy.

The most common error in this problem is to calculate the net work. The *net* work is the work done by the net force, not the work done by one of the individual forces. Be careful to make this distinction.

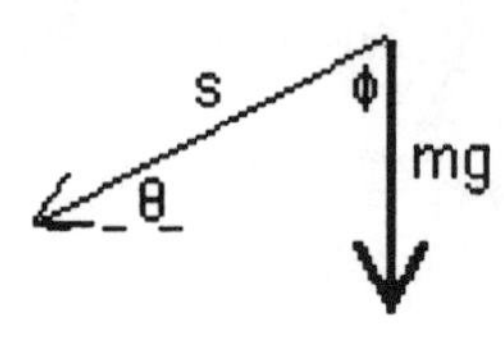

(b) Mathematically speaking, I see two ways of thinking about this, either of which is valid. Gravity is a vector of magnitude mg pointing straight down, while **s** points along the ramp.

The angle between these two vectors is $\phi = 90° - \theta$, where $\theta = 23.6°$ is the ramp angle. Hence, $\phi = 90° - 25° = 66.4°$. So, using the dot product formula for work, we get

$$\begin{aligned} W_{\text{grav}} &= \mathbf{F}_{\text{grav}} \cdot \mathbf{s} \\ &= (mg)s\cos\theta \\ &(40 \text{ kg})(9.8 \text{ m/s}^2)(5.0 \text{ m})\cos(66.4°) \\ &= 780 \text{ J}. \end{aligned}$$

Although this method works fine, you could accidentally use θ instead of ϕ as the angle in the dot-product formula. To avoid this mistake, I recommend writing the dot product in the form $W_{\text{grav}} = \mathbf{F}_{\text{grav}} \cdot \mathbf{s} = F_{\parallel \text{ grav}} s$, where $F_{\parallel \text{ grav}}$ is the component of gravity along the direction of motion. As shown in the free-body diagram, the component of gravity parallel to the surface of the ramp is $F_{\parallel \text{ grav}} = mg\sin\theta$, where θ is the ramp angle. Intuitively, this is the "part" of gravity pulling the bookshelf down the ramp. So, the work done by gravity is

$$\begin{aligned} W_{\text{grav}} &= F_{\parallel \text{ grav}} s \\ &= (mg\sin\theta)s, \end{aligned}$$

which is equivalent to the above expression for W_{grav}, because $\sin\theta = \cos\phi$. So, these two "different" techniques for calculating work give the same answer, because they are really two different ways of saying the same thing: $F_{\parallel}\Delta x$ is just another way of writing the dot product. The math works out because $\sin\theta = \cos(90 - \theta) = \cos\phi$.

(c) A common mistake is to write $W_N = Ns = (mg\cos\theta)s$. But remember, work gets performed only by the component of a force *parallel* to the direction of motion. The normal force points entirely *perpendicular* to the bookshelf's direction of motion. Roughly speaking, the normal force neither helps nor hinders the bookshelf's motion down the ramp. So, the work done by the normal force is $W_N = 0$.

Of course, the normal force helps to "create" the frictional force, which *does* hinder the bookshelf's motion. But that hindrance gets reflected in the work done by friction, not in the work done *directly* by the normal force.

(d) The net work is the sum of the work done by all the forces. So here, $W_{net} = W_J + W_{grav} + W_N + W_{fric}$, where W_{fric} is the work done by friction. We have already found W_J, W_{grav}, and W_N in parts (a), (b), and (c). So, we could complete the problem by finding W_{fric}, and then adding together all these individual works.

But there is shortcut. Another (equivalent) way to write the net work is $W_{net} = F_{\parallel\,net}\Delta x$. You can confirm that these two ways of writing W_{net} are really two different ways of saying the same thing. In any case, as discussed in part (a), the net force on the bookshelf must be zero, because it does not accelerate. Since $F_{\parallel\,net} = 0$, it follows that $W_{net} = 0$. Physically, the positive work performed by gravity gets "canceled" by the negative work done by Jo and friction.

The work-energy theorem confirms this conclusion. Since the bookshelf neither speeds up nor slows down, its kinetic energy does not change: $\Delta K = 0$. So, by the work-energy theorem, $W_{net} = 0$.

QUESTION 11-3

A ball slides off a cliff as shown. It is not rolling or otherwise spinning.

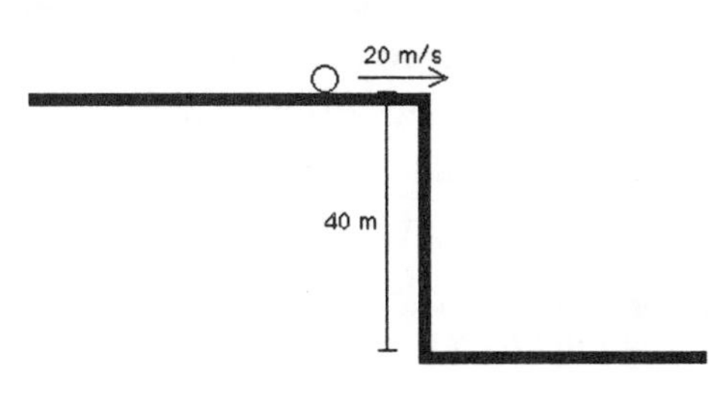

The ball's horizontal speed as it leaves the cliff is 20 m/s. The distance from the top of the cliff to the ground is 40 meters.

How fast is the ball moving just before it hits the ground? HINT: You can solve quickly using energy considerations. Take a shortcut around your old-style projectile reasoning.

ANSWER 11-3

You could solve the "old way" with projectile motion reasoning. Using kinematic formulas, you would separately solve for the x- and y-components of the final velocity. Then you would vectorially add those components, using Pythagoras' theorem, to find v. But it is quicker to use conservation of energy.

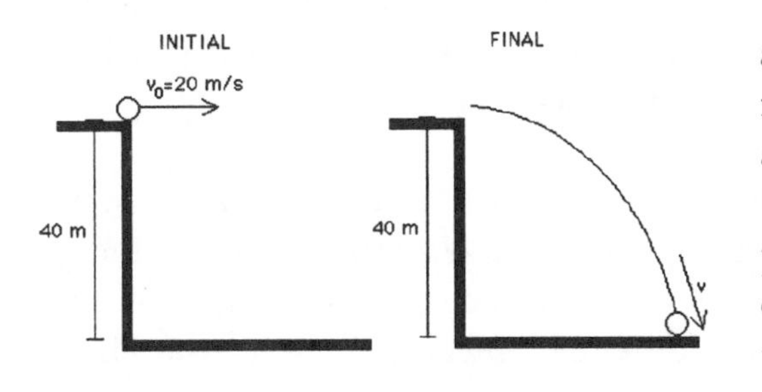

When using energy conservation, it is almost always helpful to draw a picture corresponding to each relevant "stage" of the process. Here, the relevant stages are "initial," when the ball shoots off the cliff, and "final," when the ball is about to land. Of course, in this problem, you do not really need to doodle, because everything's easy to visualize. Still, I want you to get in the habit of drawing pictures, with velocity vectors drawn and labeled, because in more complicated problems, these drawings will help you organize your thoughts and avoid making careless errors.

The ball leaves the cliff with both kinetic energy and gravitational potential energy. As it falls, it loses gravitational potential energy but gains kinetic energy. The "lost" potential energy must equal the "gained" kinetic energy, so that the total energy remains the same.

In this particular problem, you could solve by setting the loss in potential energy equal to the gain kinetic energy: $\Delta K = -\Delta U$. But that strategy will not work if three or more kinds of energy are involved. I recommend writing conservation of energy in the form $E_1 = E_2$, where E_1 and E_2 are the *total* energy at "stage 1" and "stage 2." Here, the two stages are "initial" and "final." So, we can start with

$$E_i = E_f, \tag{1}$$

Until the ball slams into the ground and heats up, the only kinds of energy it carries are kinetic (K) and gravitational potential (U). So Eq. (1), "written out," is

$$K_i + U_i = K_f + U_f. \tag{1'}$$

The initial kinetic energy is $K_i = \frac{1}{2}mv_0^2$, where $v_0 = 20$ m/s. The final kinetic energy is $\frac{1}{2}mv^2$, where v is what we are trying to find. **When calculating energies, we need not worry about directions, because energy does not have a direction.** It is a scalar, not a vector. This means you do not have to worry about vector components. Always use the "whole" velocity.

What about potential energy? As I will explain below, the gravitational potential energy stored in an object is $U = mgh$, where h is height. We can arbitrarily pick at what point $h = 0$. Let us say the ground is where $h = 0$. Then the ball's final potential energy is $U_f = 0$, while its initial potential energy is $U_i = mgH$. Substituting all this into Eq. (1') gives

$$K_i + U_i = K_f + U_f.$$
$$\frac{1}{2}mv_0^2 + mgH = \frac{1}{2}mv^2 + 0.$$

The m's cancel. Solving for v, and setting $v_0 = 20$ m/s and $H = 40$ meters, gives

$$v = \sqrt{v_0^2 + 2gh} = \sqrt{(20 \text{ m/s})^2 + 2(9.8 \text{ m/s}^2)(40 \text{ meters})} = 34 \text{ m/s}.$$

Here, energy conservation got us to the answer more quickly than projectile motion reasoning would have. Energy conservation usually supplies a shortcut. But not always. For instance, if I had asked you where the ball lands, you would have needed to use projectile (kinematic) reasoning.

Theoretical digression: What is potential energy?

Potential energy is the energy stored up in an object because work has been done on it, when that work does not get "devoted to" kinetic energy. For instance, suppose I expend 20 joules of work lifting a box from the ground onto a desk. The box now has 20 joules of potential energy stored in it. We can "release" that potential energy by nudging the box off the desk; the potential converts into kinetic energy as it falls. By *definition*, potential energy at a given point is the amount of work it took to put the object at that point.

What does this have to do with mgh? Well, suppose you lift an object of mass m straight up. Your upward force must (at least) cancel the downward pull of gravity, mg. So, you are lifting with force $F = mg$. If you raise the box through vertical distance $\Delta y = h$, then the work you have done is $F\Delta y = (mg)h$.

Now that I have shown where the formula $U_{grav} = mgh$ comes from, you can just use that expression without rederiving it each time. But keep the derivation in mind, because you can invoke similar reasoning to derive the potential energy stored in a spring, a bungee cord, a galaxy, etc. See your textbook for a full discussion of the connection between work and potential energy, and the interesting issue of "path independence."

QUESTION 11-4

A block of mass m slides frictionlessly at speed v_0 towards a frictionless ramp, which is nailed down. The block is sliding rightward. The ramp angle is θ.

(I have drawn a "rounded bottom" on the ramp to emphasize that the block smoothly enters and exits the ramp, without bouncing).

Here, v_0 is small enough that the block does not make it all the way up the ramp. Instead, it slides about halfway up the ramp, and then slides back down. Eventually it slides off the ramp and travels leftward along the floor with unknown speed v.

What is v? Express your answer in terms of v_0, m, and θ.

ANSWER 11-4

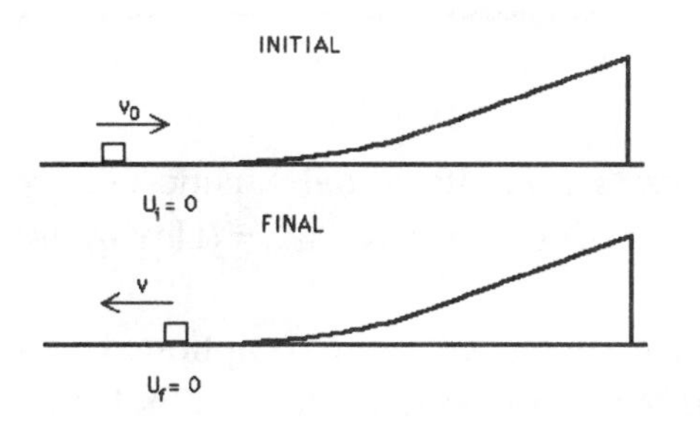

You can jump right to the answer using energy conservation. Initially, as the block slides rightward towards the ramp, all of its energy is kinetic. As it slides up the ramp, the block's kinetic energy converts into gravitational potential energy; the block slows down, but rises higher and higher off the floor.

But then, as the block slides back down the ramp, all the potential energy turns back into kinetic energy. When the block again reaches the floor and slides leftward, all of its energy is kinetic. No energy gets dissipated as heat or sound. So, the block's initial and final energy are entirely kinetic.

By conservation of energy,

$$E_i = E_f,$$

$$K_i + U_i = K_f + U_f.$$

$$\frac{1}{2}mv_0^2 + 0 = \frac{1}{2}mv^2 + 0.$$

Therefore, $v = |v_0|$. The block comes off the ramp exactly as fast as it entered the ramp, though in the opposite direction.

To solve this, you did not need to calculate how high the block slides up the ramp. Also, the ramp angle θ does not matter. A steep ramp would shorten the amount of time the block stays on the ramp, but would not change the fact that the block eventually "gets back" all the kinetic energy it starts with. The same conclusion applies to a light versus a heavy block. This example illustrates how energy considerations can simplify problem solving.

QUESTION 11-5

A car is cruising down a very steep mountain road at 20 m/s. (That is about 48 mph.) Suddenly, a deer rushes onto the road. The driver slams on the brakes and skids for 30 meters, stopping just in time.

How much do the tires heat up as a result of this skid? In other words, calculate the temperature change of the tires.

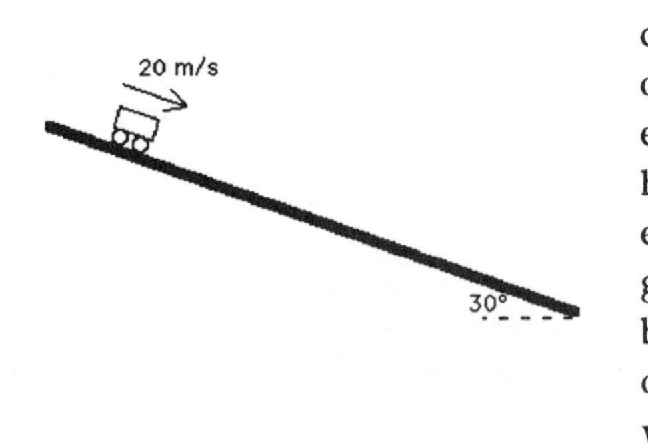

To find a number, let us assume the following: The car has mass $m = 1500$ kg, about one and a half tons. The other kinds of energy generated by the skid, such as the energy carried by sound waves, are negligible. 50% of the heat energy generated by the skid is absorbed by the tires, each of which gets an "equal share." (The remaining heat goes into the road and air.) For each joule of energy absorbed by a tire, it heats up by 4.0×10^{-5} degrees Celsius. The angle of incline of the road is 30°. (That is unusually large. 10° would be more typical.)

ANSWER 11-5

Before addressing this difficult problem, let me first talk about a simpler one. Suppose the car skidded along a flat road.

Then we would not have to worry about gravitational potential energy. It would always equal zero.

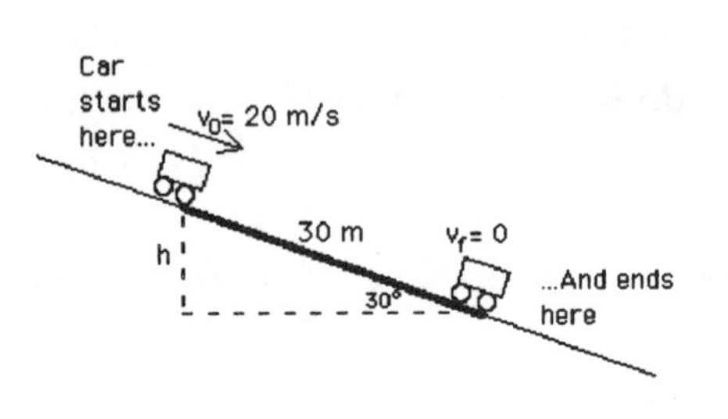

Therefore, during the car's skid, its initial kinetic energy would convert entirely into heat energy: $K_i = (\text{Heat})_f$, by conservation of energy.

But in this problem, things get more complicated. As the car skids, it loses not only kinetic energy, but also gravitational potential energy, because it ends up lower than it started. In fact, the car ends up lower by $h = 15$ meters, as you can see from trigonometry.

$$h = (30 \text{ meters}) \sin 30° = 15 \text{ meters}.$$

If we say that the car's final potential energy (at the end of the skid) is $U_f = 0$, then its initial potential energy is $U_i = mgH$, where $h = 15$ meters. Including these potential energy terms in our conservation of energy equation, we get

$$\begin{aligned} E_i &= E_f, \\ K_i + U_i &= K_f + U_f + (\text{Heat})_f \qquad (1) \\ \frac{1}{2}mv_0^2 + mgh &= \frac{1}{2}mv_f^2 + 0 + (\text{Heat})_f. \end{aligned}$$

Since the car comes to rest at the end of its skid, $v_f = 0$. So, Eq. (1) reduces to $\frac{1}{2}mv_0^2 + mgh = (\text{Heat})_f$. In words, this formula tells us that the car's initial kinetic *and potential* energy convert into heat. As a result, more heat gets generated than would have been the case if the car had skidded along a flat road.

Substituting in the numbers gives

$$\begin{aligned}\text{Heat}_f &= \frac{1}{2}mv_0^2 + mgh \\ &= \frac{1}{2}(1500 \text{ kg})(20 \text{ m/s})^2 + (1500 \text{ kg})(9.8 \text{ m/s}^2)(15 \text{ m}) \\ &= 5.2 \times 10^5 \text{ J}.\end{aligned}$$

Of this total heat, one half gets absorbed by the tires. And the heat absorbed by the tires gets "shared" among four tires. So, each individual tire gets only one fourth of the full "tire heat." So, an individual tire gets only one fourth of one half of the total heat generated by the skid: $H_{\text{one tire}} = \frac{1}{8}\text{Heat}_f = \frac{1}{8}(5.2 \times 10^5 \text{ J}) = 6.5 \times 10^4 \text{ J}$. Since a tire heats up by 4.0×10^{-5} degrees Celsius per joule of absorbed energy, the temperature change is $\Delta\text{Temp} = (6.5 \times 10^4 \text{ J})\left(4.0 \times 10^{-5} \frac{°\text{C}}{\text{joule}}\right) = 2.6°\text{C}$. You could probably notice this temperature difference; the tires would feel warmer.

QUESTION 11-6

The equilibrium length of a spring is its "natural" length, i.e., its length when neither stretched nor compressed. When compressed a distance x from its equilibrium length, a spring carries elastic potential energy $U_e = \frac{1}{2}kx^2$. (Here, "k" is the stiffness constant from the formula $F_{\text{spring}} = -kx$.)

Consider a frictionless block of mass m held against a compressed spring as shown. The spring is compressed distance d from equilibrium. When released, the block is pushed rightward. The spring is not attached to the block: So, the block keeps sliding even after losing contact with the spring. It slides along frictionlessly until reaching the semicircular ramp of radius r. The block ascends the ramp and shoots straight up into the air.

Express your answers in terms of k, d, r, m, and g.

(a) Why is the potential energy stored in a spring given by the formula $U_e = \frac{1}{2}kx^2$? Derive that formula. Hint: See my "theoretical digression" at the end of Question 11-3.

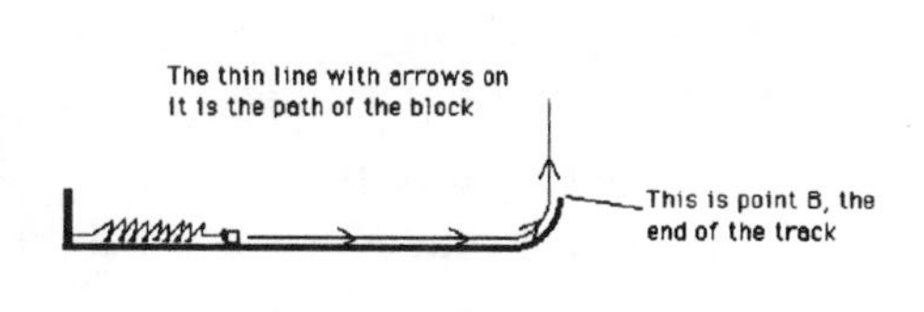

(b) What is the block's speed at point B, the "take-off" point of the semicircular ramp?

(c) What height does the block reach at its "peak" (i.e., its highest point) as it flies upward through the air?

ANSWER 11-6

(a) The potential energy stored in an object equals the amount of work it took to get the object into its present position or configuration. So, the potential energy stored in a compressed spring equals the work it takes to compress the spring that far. How can we calculate that work?

It would be easy if the spring exerted a constant force $-F$. In that case, the work needed to compress it would be Fx. But as the spring gets more and more compressed, it "fights back" with a bigger and bigger force. Indeed, the force it exerts is proportional to how far it is been compressed:

$F_{\text{spring}} = -kx$. So, it takes less work to compress the spring from $x = 0$ to $x = 1$ cm than it takes to compress it from $x = 1$ cm to $x = 2$ cm. The force is not constant.

As explained in the textbook, to calculate the work done by a nonconstant force, we must integrate that force over distance: $W = \int \mathbf{F} \cdot d\mathbf{x}$. Since the spring exerts a backwards (negative) force $F_{\text{spring}} = -kx$, the person or object compressing the spring must push with (at least) force $F = kx$. Otherwise, the person cannot "overcome" the spring's force. So, the work needed to compress the spring from equilibrium ($x = 0$) to an arbitrary compression distance x is $W = \int \mathbf{F} \cdot dx = \int_0^x (kx)dx = \frac{1}{2}kx^2 - \frac{1}{2}k0^2 = \frac{1}{2}kx^2$. As just noted, the work done to compress the spring is the potential energy now stored in the spring: $U_e = \frac{1}{2}kx^2$.

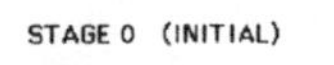

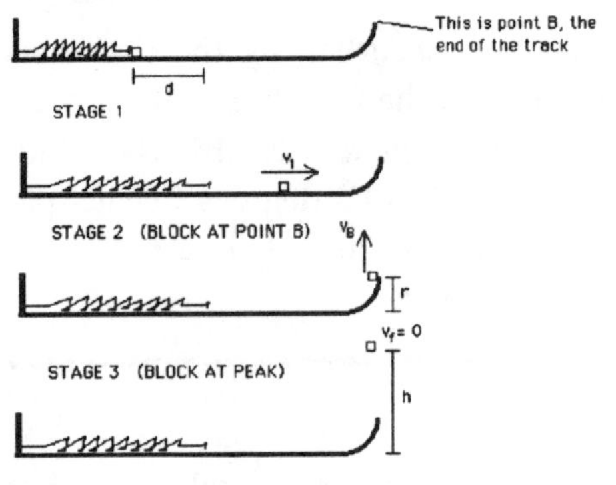

(b) Let us think through the process. It helps to make a diagram corresponding to each stage.

The mass starts at rest, with the spring compressed (stage 0). So, the system has elastic potential energy. As the spring de-compresses, pushing the block outward, the elastic potential energy converts into the kinetic energy of the block (stage 1). The block retains all this kinetic energy until reaching the ramp. As the block ascends the ramp, some (but not all) of the kinetic energy converts into gravitational potential energy (stage 2). The block then flies straight up, gradually slowing down, until reaching its peak, at which point it momentarily comes to rest (stage 3). So, the kinetic energy at the peak is $K_f = 0$.

Conservation of energy tells us that the total energy stays constant throughout:

$$E_0 = E_1 = E_2 = E_3,$$

We are trying to solve for v_B, the speed at point B. Some of you probably solved "step by step": You set the initial energy equal to the energy at stage 1, and then you set the energy at stage 1 equal to the energy at stage 2. This "sequential" problem-solving strategy is fine. Indeed, in collision problems, you have no choice but to proceed in this manner. (We will see why in Chapters 12 and 13.) But here, since $E_0 = E_1$, and $E_1 = E_2$, it is quicker to just set $E_0 = E_2$ directly, thereby saving a step.

At point B, the block is a height $h = r$ off the ground, and hence its gravitational potential energy is $U = mgr$. At that time, the spring is no longer compressed; it is returned to its natural length. So, the spring no longer carries elastic potential energy. Putting all this together gives

$$E_0 = E_2$$

$$K_0 + U_{0,\text{grav}} + U_{0,e} = K_2 + U_{2,\text{grav}} + U_{2,e}$$

$$0 + 0 + \frac{1}{2}kd^2 = \frac{1}{2}mv_B^2 + mgr + 0.$$

Solving for v_B gives $v_B = \sqrt{\frac{kd^2 - 2mgr}{m}}$. *Some of you may have used $F = ma$ and kinematics to find the block's speed when it leaves the spring.* ***This cannot*** *work if you use constant-acceleration kinematic formulas such as $v^2 = v_0^2 + 2a\Delta x$, because the spring exerts a nonconstant force. Therefore, the block's acceleration is not constant.* ***Conservation of energy gives you a way to solve problems even when the forces and accelerations are not constant.***

(c) Once in the air, the block only feels the constant force of gravity. Therefore, you *can* use constant-acceleration kinematics, if you wish. (Keep in mind that the "initial" height is $y_0 = r$, not $y_0 = 0$.) But it is just as straightforward to continue using conservation of energy.

At its peak, the block is motionless and therefore has no kinetic energy: $K_3 = 0$. Its potential energy at that point is $U_3 = mgh$, where h is the height we are seeking. Since $E_0 = E_1 = E_2 = E_3$, you can generate an equation with h in it by setting E_3 equal to E_0, or by setting E_3 equal to E_1, or by setting E_3 equal to E_2. The choice is yours. Since we want the final answer to contain "d," I will set

$$E_0 = E_3$$
$$K_0 + U_{0,\text{grav}} + U_{0,e} = K_3 + U_{3,\text{grav}} + U_{3,e}$$
$$0 + 0 + \frac{1}{2}kd^2 = 0 + mgh + 0.$$

Physically, this formula tells us that all the initial elastic potential energy stored in the spring eventually gets converted into the gravitational potential energy of the block at its peak. Solving for h, the peak height, yields $h = \frac{kd^2}{2mg}$. Let us double-check this expression for h using dimensional (units) analysis. The units of k are newtons per meter, since a spring exerts force $F = -kx$, and hence $k = \frac{-F}{x}$. So, the units of $\frac{kd^2}{2mg}$ are

$$\frac{(\frac{\text{newtons}}{\text{meter}})(\text{meters})^2}{(\text{kg})(\frac{\text{meters}}{\text{s}^2})}. \qquad (*)$$

But remember that a newton is a $(\text{kg})(\frac{\text{meter}}{\text{s}^2})$. So the denominator of (*) is newtons! Therefore, (*) simplifies to

$$\frac{(\frac{\text{newtons}}{\text{meter}})(\text{meters})^2}{\text{newtons}} = \frac{(\text{newtons})(\text{meters})}{\text{newtons}} = \text{meters},$$

the correct units for h, a distance.

QUESTION 11-7

A cannonball of mass $m = 10$ kg is dropped from rest from a height $H = 7.0$ m above the top of a large, vertically-oriented spring, which gets compressed from its relaxed (equilibrium) position when the cannonball lands on it. The spring has spring constant $k = 5000$ N/m. What is the maximum compression of the spring? In other words, how far does the cannon ball compress the spring?

ANSWER 11-7

Always start by sketching the "stages" of the physical process, with the relevant velocities and positions labeled. This can help you sort out which kinds of energy are present.

I have let s denote the maximum compression of the spring. When the spring is maximally compressed, $v = 0$; the ball temporarily comes to rest while it "turns around."

Notice that the ball falls through total height $H + s$, not H. So, if you set $mgH = \frac{1}{2}ks^2$, you got the wrong answer. Instead you should set $mg(H + s) = \frac{1}{2}ks^2$.

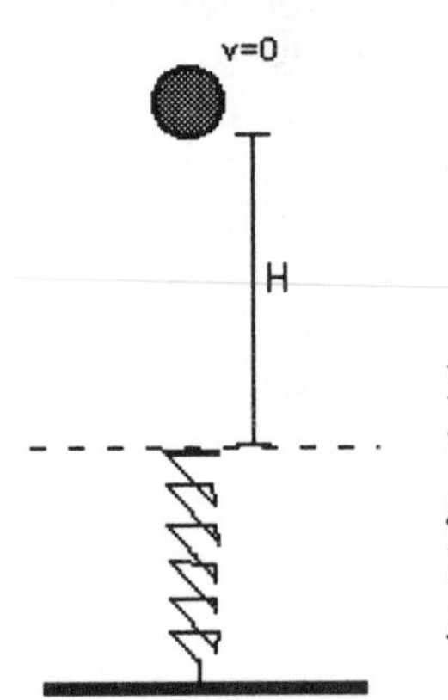

Some of you probably solved in steps as follows:

(1) Use energy conservation (or old-fashioned kinematics) to calculate the ball's speed v_1 when it reaches the spring.

(2) Use energy conservation again, this time going from stage 1 to stage 2, to figure out s.

This strategy works fine, and I will use it in a minute. Let me mention that you could go from stage 0 to stage 2 directly, setting $mgH = \frac{1}{2}ks^2$. *This shortcut works only when no heat or other non-mechanical (i.e., "waste") energy gets generated, for reasons we will see next chapter.* Instead of using the shortcut, however, I will use the two-step approach outlined above.

Stage 0 to stage 1.

We need a "reference" point from which to measure heights. Let us say the top of the relaxed spring (in stage 0) is $h = 0$. Then the ball's initial height is $H = 7.0$ m, and its stage 1 height is 0. Alternatively, if you choose the ball's final height in stage 2 as your reference point, then the ball's initial height is $H + s$, and its stage 1 height is s. Either choice of reference points works fine, because they agree that the *difference* between the ball's initial height and its stage 1 height is $\Delta y = H$. Using my original choice of reference points, I get $U_{0,\text{grav}} = mgH$, and $U_{1,\text{grav}} = 0$.

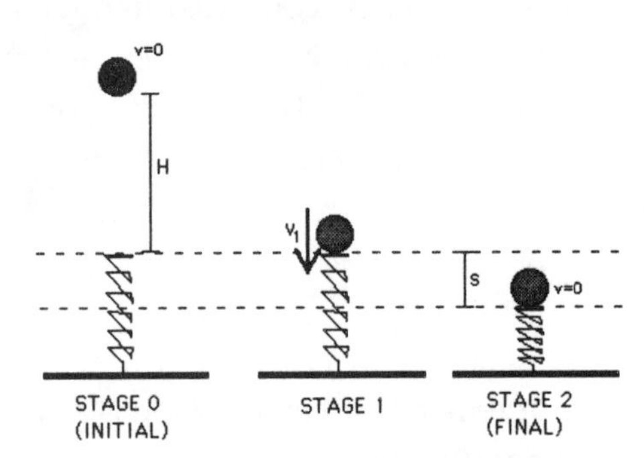

In this problem, another form of potential energy is present: The spring's elastic potential energy. But between stage 0 and stage 1, the spring is not yet compressed; no energy is stored in it. So $U_{0,\text{spring}} = U_{1,\text{spring}} = 0$.

Putting all this together, I get

$$E_0 = E_1$$

$$K_0 + U_{0,\text{grav}} + U_{0,\text{spring}} = K_1 + U_{1,\text{grav}} + U_{1,\text{spring}}$$

$$0 + mgH + 0 = \frac{1}{2}mv_1^2 + 0 + 0.$$

Solve for v_1 to get $v_1 = \sqrt{2gH}$, the same answer you obtain using the old kinematic formula $v_1^2 = v_0^2 + 2a\Delta y$.

Stage 1 to stage 2.

Since I have picked the top of the relaxed spring as my reference point, the ball's final height is $h = -s$. The minus sign indicates that the ball ends up below the reference point. So, conservation of energy gives us

$$E_1 = E_f$$

$$K_1 + U_{1,\text{grav}} + U_{1,\text{spring}} = K_f + U_{f,\text{grav}} + U_{f,\text{spring}}$$

$$\frac{1}{2}mv_1^2 + 0 + 0 = 0 + mg(-s) + \frac{1}{2}ks^2.$$

When we substitute in our result from step 1, $v_1 = \sqrt{2gH}$, this equation becomes $mgH = 0 + mg(-s) + \frac{1}{2}ks^2$. Add mgs to both sides to get $mg(H + s) = \frac{1}{2}ks^2$, the same equation obtained using the shortcut mentioned above.

Algebra starts here. To solve for s, first write the equation in standard quadratic form: $\frac{1}{2}ks^2 - mgs - mgH = 0$. Now use the quadratic formula.

$$s = \frac{mg \pm \sqrt{(mg)^2 - 4(-mgH)(\frac{1}{2}k)}}{2(\frac{1}{2}k)}$$

$$= \frac{(10\text{ kg})(9.8\text{ m/s}^2) \pm \sqrt{[(10\text{ kg})(9.8\text{ m/s}^2)]^2 - 4[-(10\text{ kg})(9.8\text{ m/s}^2)(7.0\text{ m})](\frac{1}{2}5000\ \frac{\text{N}}{\text{m}})}}{2(\frac{1}{2}5000\ \frac{\text{N}}{\text{m}})}$$

$$= 0.54 \text{ meters or } -0.50 \text{ meters.}$$

So, $s = 0.54$ m must be the answer, because s cannot be negative.

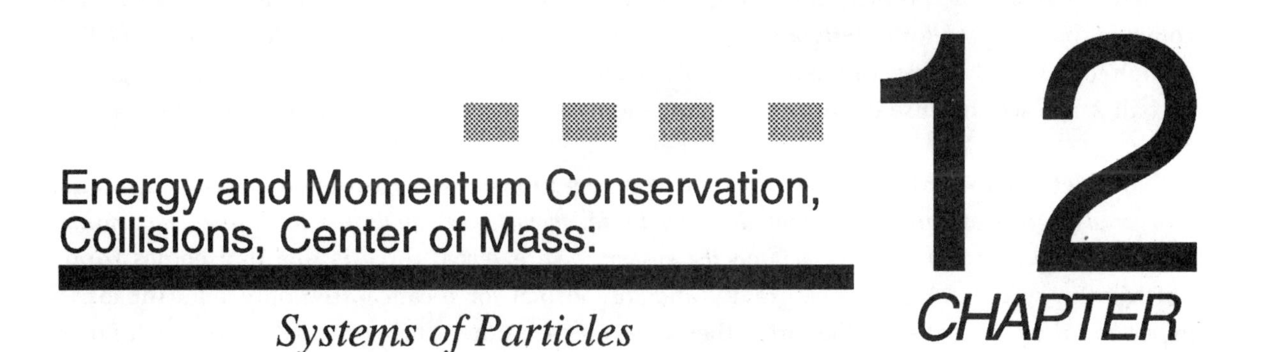

Energy and Momentum Conservation, Collisions, Center of Mass:

Systems of Particles

12 CHAPTER

QUESTION 12-1

Consider two pool balls sliding frictionlessly across a pool table. Before the collision, ball 1 slides leftward at 2.0 m/s, and ball 2 is motionless. After the "head-on" collision, ball 1 slides leftward at 0.50 m/s. Both balls have mass $m = 0.10$ kg.

(a) What is the velocity (speed and direction) of ball 2 after the collision?

(b) During the collision, the balls heat up slightly. How many joules of "dissipated" energy (e.g., heat and sound energy) are generated during the collision? (By the way, this heat will not be enough to raise the temperature of the balls noticeably.)

ANSWER 12-1

(a) In collision problems, and in other problems involving multiple "stages," begin by sketching the important stages of the process, carefully labeling the velocities, even the unknown ones. For brevity, let me leave out units.

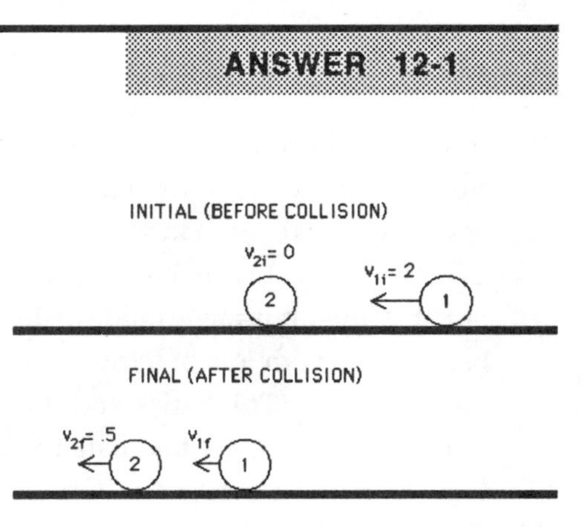

In my notation, subscripts "1" and "2" refer to balls 1 and 2; and subscripts "i" and "f" mean "initial" and "final." Notice that v_{1f}, the post-collision velocity of ball 1, is the only unknown velocity. On the diagram, I have "guessed" that v_{1f} is leftward. If I am wrong, then my numerical answer for v_{1f} will come out negative when I solve for it.

The most common mistake in this problem is to set the initial kinetic energy equal to the final kinetic energy. Energy is, of course, conserved. But during the collision, not all of it stays in the form of *kinetic* energy. Some of the initial kinetic energy gets converted into heat, sound, and other "waste" forms of energy. So, $K_f < K_i$, as we will see in part (b). In other words, the collision is *inelastic*. This does not necessarily mean the balls stick together, as happens during a "completely inelastic" collision. "Inelastic" means simply that some heat and/or sound energy gets created, and therefore the initial *kinetic* energy does not equal the final *kinetic* energy. By contrast, an "elastic" collision involves no heat or sound, because it is perfectly bouncy. So, during an elastic collision, the kinetic energy does not change.

Before solving the problem numerically, let us see what is happening intuitively. Before the collision, ball 1 has a lot of leftward "oomph." During the collision, ball 1 "loses" some of that oomph to ball 2. By conservation of momentum, the oomph lost by ball 1 equals the oomph gained by ball 2. To see this, use the computer simulation, which lets you adjust the "bounciness" of the balls.

Now let us formalize these intuitive considerations into a problem solving strategy. *Conservation of momentum says that the total initial momentum equals the total final momentum, provided no net "external" force acts on the system.* An external force is one that comes from outside the system. Here, since gravity and the normal force cancel, the only relevant forces experienced by the balls are the forces they exert on each other during the collision. Such forces are called "internal," because they result from one part of the system acting on another part of the system. (The "system" here is the two balls, not either ball individually.) Because the balls are subject only to internal forces, momentum is conserved.

As shown in the textbook, an object's momentum ("oomph") is $\mathbf{p} = m\mathbf{v}$. Since momentum is a vector, we would normally deal separately with the x-component and the y-component. I will demonstrate this later on. Here, however, everything happens along a straight line. We need not worry about vector components.

I will choose leftward as positive. By conservation of momentum, using the above diagrams, we get

$$p_i = p_f$$

$$mv_{1i} + mv_{21} = mv_{1f} + mv_{2f}$$

$$(.10 \text{ kg})(2.0 \text{ m/s}) + 0 = (.10 \text{ kg})v_{1f} + (.10 \text{ kg})(.50 \text{ m/s}).$$

In many collision problems, conservation of momentum does not give us sufficient information to solve. In those cases, we must also use conservation of energy. But here, momentum conservation generates an equation in which v_{1f} is the only unknown. So, we can solve for v_{1f}. Doing so yields $v_{1f} = 1.5 \text{ m/s}$. Let us hook up this answer with the above intuitive considerations. Ball 1 starts out at 2 m/s, but then collides with ball 2, which ends up sliding at $.5 \text{ m/s}$. Intuitively speaking, ball 1 starts with lots of leftward oomph, but then gives one quarter of that oomph to ball 2 during the collision. (This quick-and-dirty reasoning is valid because both balls have the same mass.) Therefore, we expect that ball 1 retains three-quarters of its oomph during the collision, because total "oomph" is conserved. Our numerical answer confirms this prediction.

(b) To solve this part, use energy conservation. Since the balls slide along a flat table, potential energy is neither gained nor lost, and hence we can set it equal to zero (or otherwise ignore it). Before the collision, all of the system's energy is "stored" in the kinetic energy of ball 1. During the collision, ball 1 loses some of its kinetic energy. *Part* of the kinetic energy lost by ball 1 transfers into the kinetic energy of ball 2. The rest of the kinetic energy lost by ball 1 gets converted into heat and sound. The higher the inelasticity, the more heat and sound get created. To get a feel for this, experiment with the computer simulation.

Since all the potential energy terms cancel out (or equal zero), setting the total initial energy equal to the total final energy gives

$$E_i = E_f$$

$$K_{1i} + K_{2i} = K_{1f} + K_{2f} + \text{Heat}_f,$$

and hence

$$\begin{aligned}\text{Heat}_f &= (K_{1i}+K_{2i})-(K_{1f}+K_{2f})\\ &= \left(\frac{1}{2}mv_{1i}^2+\frac{1}{2}mv_{2i}^2\right)-\left(\frac{1}{2}mv_{1f}^2+\frac{1}{2}mv_{2f}^2\right).\end{aligned}$$

I am using "Heat" to denote heat, sound, and all other dissipated energy. In words, this equation says that the heat produced during the collision "comes from" the lost kinetic energy, i.e., the initial kinetic energy minus the final kinetic energy.

Now we can substitute in the numbers, including our result from part (a) that $v_{1f} = 1.5$ m/s. This yields

$$\begin{aligned}\text{Heat}_f &= \left(\frac{1}{2}mv_{1i}^2+\frac{1}{2}mv_{2i}^2\right)-\left(\frac{1}{2}mv_{1f}^2+\frac{1}{2}mv_{2f}^2\right)\\ &= \left\{\frac{1}{2}(.1\text{ kg})(2\text{ m/s})^2+0\right\}-\left\{\frac{1}{2}(.1\text{ kg})(1.5\text{ m/s})^2+\frac{1}{2}(.1\text{ kg})(.5\text{ m/s})^2\right\}\\ &= \{0.20\text{ J}\}-\{0.125\text{ J}\}\\ &= 0.07\text{ J},\end{aligned}$$

which would heat up the balls by a few thousandths of a degree. We now see that the system started off with 0.20 joules of kinetic energy, but ended up with only 0.125 joules of kinetic energy. Those "lost" 0.07 joules converted into heat.

QUESTION 12-2

Again consider two pool balls of equal mass M. Before colliding with ball 2, ball 1 travels rightward at 2.0 m/s. Ball 2 is motionless before the collision. After the collision, ball 1 is observed to be sliding at 1.0 m/s at a 45° angle to its initial direction of motion; see the diagram below.

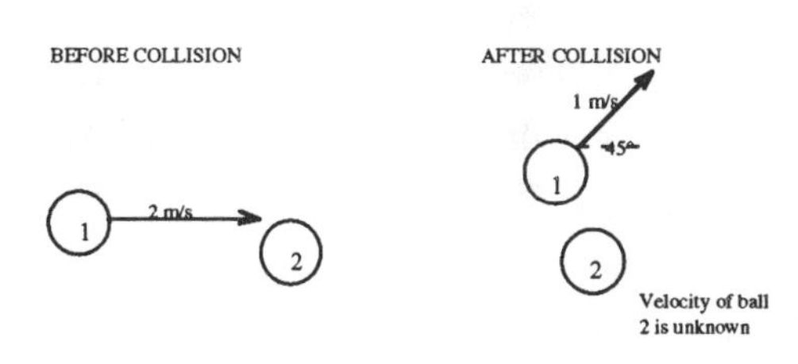

(a) After the collision, what is the magnitude and direction of the velocity of ball 2?

(b) Was the collision elastic? What fraction of the initial kinetic energy was lost in the collision?

ANSWER 12-2

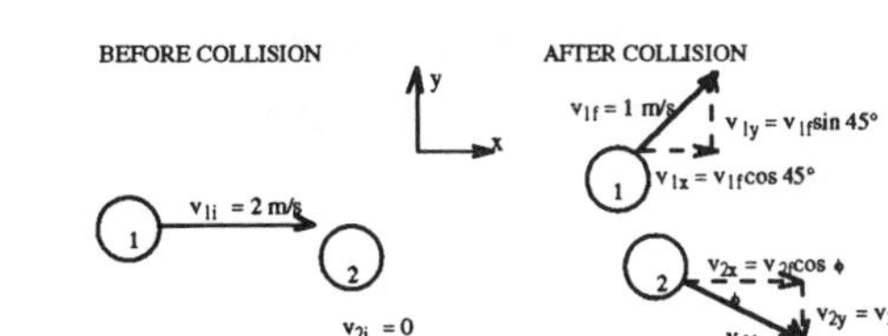

(a) As always in collision problems, start with carefully labeled diagrams of the relevant stages. Here, we must worry about directions, because the motion does not all happen along a straight line. You should divide up all velocity vectors into components. Since the initial direction of ball 1 is rightward, I will choose rightward as one of my coordinate axes.

I have guessed that ball 2 ends up moving rightward and "downward." This makes intuitive sense, as you can see by colliding actual balls or by experimenting with the computer simulation. We are solving for v_{2f} and ϕ, the speed and direction of ball 2 after the collision.

The most common error is to set the initial kinetic energy equal to the final kinetic energy. As explained in Question 12-1 above, some of the initial kinetic energy might get converted into heat, sound, and other "waste" forms of energy. Therefore, the final kinetic energy might be less than the initial. By contrast, since heat does not "count" as a kind of momentum (oomph), momentum conservation always applies during collisions.

Since momentum is a vector, and since this problem involves two-dimensional motion, we can employ our usual "trick": *Deal separately with the x- and y-components of momentum,* just as you dealt separately with x- and y-components of forces. We can separately solve for v_{2x} and v_{2y}, the post-collision x- and y-velocity of ball 2. Then we can vectorially add those components to obtain ball 2's total final velocity.

Let us start with the x-motion. Only ball 1 has initial x-directed momentum. Both balls have x-directed momentum after the collision. From the above diagram, I get

$$(x\text{-direction}) \qquad p_{ix} = p_{fx}$$
$$Mv_{1i} + 0 = Mv_{1f}\cos 45° + Mv_{2x}$$

The M's cancel, and we know $v_{1i} = 2$ m/s and $v_{1f} = 1$ m/s. So, we can immediately solve for v_{2x} to get $v_{2x} = v_{1i} - v_{1f}\cos 45° = 2\text{ m/s} - (1\text{ m/s})\cos 45° = 1.29\text{ m/s}$.

Now I will consider the y-momenta. Initially, neither ball has any y-ward motion. After the collision, ball 1 has some positive y-directed motion, and ball 2 has negative y-directed motion (in my coordinate system). So,

$$(y\text{-direction}) \qquad p_{iy} = p_{fy}$$
$$0 + 0 = Mv_{1f}\sin 45° - Mv_{2y}.$$

So, $v_{2y} = v_{1f}\sin 45° = (1\text{ m/s})\sin 45° = 0.71\text{ m/s}$.

Now that we have the final x- and y-components of ball 2's velocity, we can "Pythagoras" them together: $v_{2f} = \sqrt{v_{2x}^2 + v_{2y}^2} = \sqrt{(1.29\text{ m/s})^2 + (0.71\text{ m/s})^2} = 1.5\text{ m/s}$. We can also find the angle ϕ using trig. Since $\tan\phi = \frac{v_{2y}}{v_{2x}}$, we get $\phi = \tan^{-1}\frac{v_{2y}}{v_{2x}} = \tan^{-1}\left(\frac{0.71\text{ m/s}}{1.29\text{ m/s}}\right) = 29°$.

To get a feel for how the final velocity of ball 2 depends on how "head-on" or "glancing" the collision is, experiment with the computer simulation. Or better yet, play pool.

(b) We are trying to find the fraction of the initial kinetic energy lost during the collision. What formula should we use? Let K_i and K_f denote the total initial and final kinetic energies. The amount of kinetic energy lost is $K_{\text{lost}} = K_i - K_f$. For instance, if we begin with 10 joules and end with 8 joules of kinetic energy, then we lost 2 joules. But we need a formula for the *fraction* of the initial kinetic energy lost. Well, in the numerical example just given, 2 joules of the original 10 joules of kinetic energy are lost. So, the fraction of kinetic energy lost is 2 out of 10, which corresponds to $\frac{2}{10} = 0.2$. Generalizing from this example, you can see that the fraction of kinetic energy lost is fraction of initial K lost $= \frac{K_{\text{lost}}}{K_i} = \frac{K_i - K_f}{K_i}$. So, using $K = \frac{1}{2}Mv^2$, we get

$$\text{fraction of initial } K \text{ lost} = \frac{\left(\frac{1}{2}Mv_{1i}^2 + \frac{1}{2}Mv_{2i}^2\right) - \left(\frac{1}{2}Mv_{1f}^2 + \frac{1}{2}Mv_{2f}^2\right)}{\frac{1}{2}Mv_{1i}^2 + \frac{1}{2}Mv_{2i}^2}$$

$$= \frac{\left(v_{1i}^2 + v_{2i}^2\right) - \left(v_{1f}^2 + v_{2f}^2\right)}{v_{1i}^2 + v_{2i}^2} \quad \left[\text{canceling out the } \frac{1}{2}M \text{ factors}\right]$$

$$= \frac{\{(2 \text{ m/s})^2 + 0\} - \{(1 \text{ m/s})^2 + (1.47 \text{ m/s})^2\}}{(2 \text{ m/s})^2 + 0}$$

$$= 0.21.$$

So, about a fifth of the initial kinetic energy was "lost." That energy did not just disappear. It converted into heat, sound, etc. The collision was inelastic. In an elastic collision, none of the initial kinetic energy gets lost.

Remember, since energy is a "number," not a vector, you do not have to worry about components. That is why I could add and subtract kinetic energies "pointing" in different directions.

QUESTION 12-3

Block 1, which has mass m_1, slides down a frictionless ramp of height h_1 and angle θ, as shown in the diagram. Simultaneously, block 2, of mass m_2, slides down another frictionless ramp of angle θ and height h_2. The two blocks started from rest. Somewhere on the frictionless floor between the two ramps, the two blocks collide and stick together.

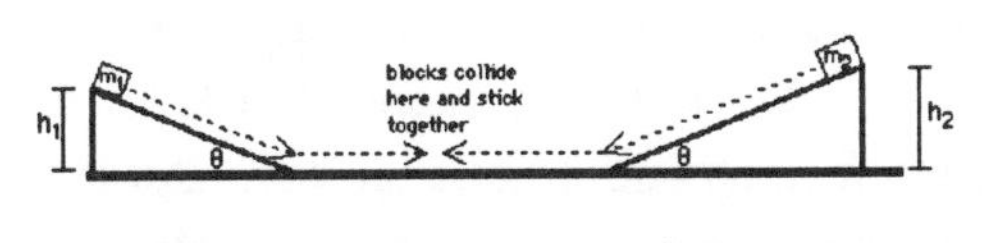

After the collision, what is the magnitude and direction of the blocks' velocity? Call that velocity v_f. Express v_f in terms of g and the given constants, i.e., m_1, h_1, etc. Assume that the bottoms of the ramps are rounded so that the blocks exit the ramps smoothly, without "bouncing" or slowing down.

ANSWER 12-3

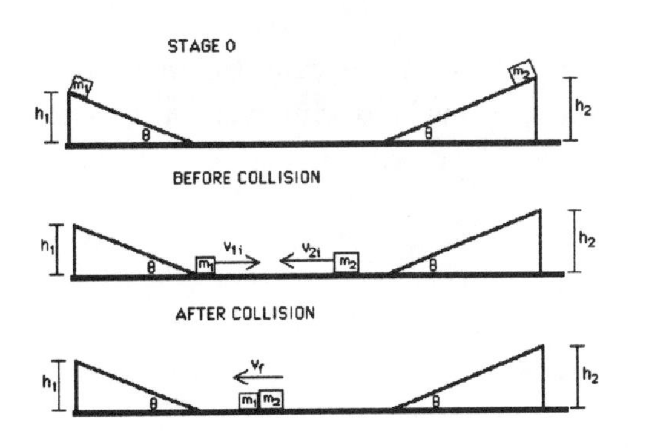

This one is conceptually more complicated than the previous two problems. We would better formulate a strategy before proceeding. To organize your thoughts and to avoid making careless errors, diagram the different stages of the process.

The intermediate stage is after the blocks have slid off their ramps, but before they collide.

The most common error is to set the initial potential energy (at stage 0) equal to the post-collision kinetic energy (when the blocks are stuck together). But during the collision, some of the blocks' kinetic energy might convert into heat. Therefore, the post-collision kinetic energy might be less than the pre-collision kinetic energy. In fact, we *know* heat gets generated, because the collision is (completely) inelastic. So, we cannot use energy conservation to deal with the collision. Instead, we must use momentum conservation to calculate v_f.

Does this mean we should set the initial momentum in stage 0 equal to the final post-collision momentum? No! During the collision, no external forces act on the blocks, and hence momentum is conserved. (The only forces the blocks feel during the collision are the "internal" forces they exert on each other.) But while the blocks slide down their ramps, the external force of gravity acts on them. So, momentum is *not* conserved while the blocks slide down he ramps. Therefore, the stage 0 momentum does not equal the momentum either before or after the collision.

Let me summarize what is conserved when:

Stage 0 → Before collision (blocks slide down ramps)

–Energy conserved, with no heat generated.

–Momentum not conserved, due to the external force of gravity.

Before collision → After collision (blocks collide and stick)

–Heat produced, so energy conservation not useful.

–Momentum conserved, as always during a collision.

This step-by-step description immediately suggests a problem-solving strategy.

(A) Use energy conservation to calculate v_{1i} and v_{2i}, the blocks' pre-collision speeds on the floor.

(B) Use momentum conservation to figure out the post-collision speed of the blocks.

Here we go...

Step A: Stage 0 → Before collision

During this part of the physical process, the blocks do not interact. They separately slide down their ramps. So, we can consider each block separately, as if the other did not even exist.

Let is first consider block 1. At the top of its ramp (stage 0), the block sits motionless, and therefore has no kinetic energy. But it has gravitational potential energy $U_{top} = m_1gh_1$. At the bottom of the ramp, all that potential energy has converted into kinetic energy. So, setting $E_{\text{block 1, top of ramp}} = E_{\text{block 1, bottom of ramp}}$, we get

$$\text{BLOCK 1} \quad K_{top} + U_{top} = K_{bottom} + U_{bottom}$$
$$0 + m_1gh_1 = \frac{1}{2}m_1v_{1i}^2 + 0$$

Solving for v_{1i} gives $v_{1i} = \sqrt{2gh_1}$. By equivalent reasoning, the velocity of block 2 as it slides along the floor, before it collides with block 1, is $v_{2i} = -\sqrt{2gh_2}$. The minus sign indicates that block 2 travels "backwards" compared to block 1. I have implicitly chosen rightward as positive.

Step B: Before collision → After collision

Now that we know the pre-collision velocities of the blocks, we can use momentum conservation. After the blocks collide, they stick together, traveling as one blob with mass $m_1 + m_2$ and speed v_f.

$$p_{\text{before collision}} = p_{\text{after collision}}$$
$$m_1v_{1i} + m_2v_{2i} = (m_1 + m_2)v_f.$$

Solve for v_f, and set $v_{1i} = \sqrt{2gh_1}$ and $v_{2i} = -\sqrt{2gh_2}$ from above, to get

$$v_f = \frac{m_1v_{1i} + m_2v_{2i}}{m_1 + m_2}$$
$$= \frac{m_1\sqrt{2gh_1} - m_2\sqrt{2gh_2}}{m_1 + m_2}.$$

If this velocity is positive, then the stuck-together blocks slide rightward, i.e., block 1 overpowers block 2 during the collision.

This problem illustrates a general strategy for solving complicated multi-step problems. With the help of drawings, break the physical process up into stages, and "solve" each stage separately. Specifically, you must think about what is conserved and what is not conserved during each stage.

QUESTION 12-4

Supplemental topic. Skip this unless rocket motion is covered.

A certain kind of rocket propels itself as follows: Inside the engine, liquid hydrogen and liquid oxygen explosively combine to form steam, which shoots out the back of the rocket. When the steam shoots backward, the rocket gets propelled forward. In outer space, the only non-negligible forces experienced by the rocket and the steam are the forces they exert on each other.

When the engine is operational, the mass of steam shot out the back of the rocket per unit time, in kilograms per second, is $C = \frac{dm}{dt}$. Those steam molecules have an extremely high velocity v_s relative to the rocket.

Suppose the rocket is sitting motionless in space when the engine gets turned on. The mass of the rocket (including its fuel) is M. What is the acceleration of the rocket immediately after the engine is switched on? Express your answer in terms of M, C, and v_s.

ANSWER 12-4

Your first impulse might be to use Newton's 2nd law, $a = \frac{F}{M}$, where F is the force exerted by the engine. But we do not know F, and cannot find it using the standard methods. We will have to reason differently. (We *will* eventually find F, though by a new method.)

The trick is to use conservation of momentum, considering a very small time interval dt. We will figure out dv, the rocket's change in velocity during that infinitesimal time interval. Then, to obtain the acceleration, we will use the definition of acceleration, $a = \frac{dv}{dt}$. Momentum is conserved because no external forces act on the rocket or steam.

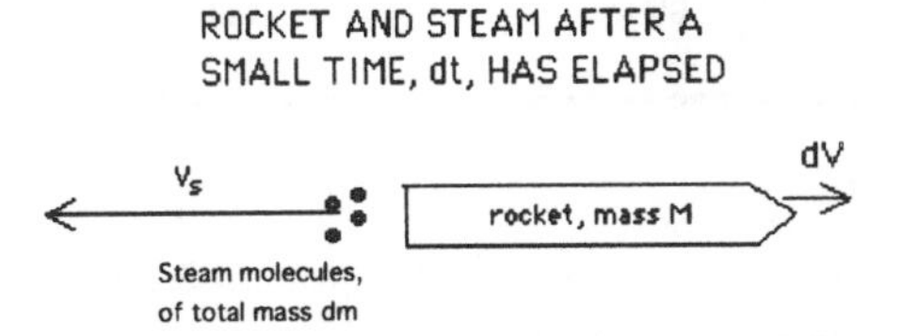

As usual, I will start by drawing pictures. Initially, the rocket and its fuel are sitting still. I will not bother drawing that. The final configuration is after a small amount of time, dt, has elapsed. At this point, a small mass of steam dm has shot out the back of the rocket, which now moves forward with tiny speed dV. I have chosen a very small time interval dt partly because we are interested in the rocket's *initial* acceleration, i.e., how quickly it is speeding up during the first moments of its trip. Strictly speaking, in the above diagram, the mass of the rocket should be $M - dm$, because the rocket has "lost" mass dm of fuel. But since dm is tiny compared to M, the rocket's mass is still very nearly M.

Because the rocket and steam start off motionless, they have initial momentum $p_i = 0$. When steam gets shot out backwards with a certain momentum, the rocket shoots forward with an equal momentum, so that the *total* momentum of the rocket and steam remains zero. In symbols,

$$p_i = p_f$$
$$0 = M(dV) - (dm)v_2$$

Solving for dV yields $dV = \frac{(dm)v_s}{M}$. That is the rocket's speed after accelerating for time dt. So, the rate at which the rocket speeds up is

$$
\begin{aligned}
a &= \frac{dV}{dt} \\
&= \frac{(dm)v_s}{(dt)M} \quad \left[\text{since } dV = \frac{(dm)v_s}{M}\right] \\
&= Cv_s \quad \left[\text{since } C = \frac{dm}{dt}\right].
\end{aligned}
$$

Some interesting physics is lurking in this answer. Notice that the acceleration takes the form $a = \frac{F}{M}$, where $F = Cv_s$. So, the force on the rocket is proportional to the velocity with which steam gets shot out the back, and also proportional to how much steam gets released per second. That makes physical sense. We also see that the force on the rocket is constant, because both C and v_s are constant.

Does this mean the rocket's acceleration is constant? No, because the rocket's mass decreases as it uses up fuel. Since a constant force acts on a mass that gets lighter and lighter, the acceleration increases. We can quantify this effect. The rocket loses mass at rate $C = \frac{\Delta m}{\Delta t}$. Therefore, at arbitrary time t, it has lost mass $\Delta m = Ct$. So, the rocket's mass at time t is its initial mass minus its "lost" mass:

$$
\begin{aligned}
m(t) &= M_{\text{initial}} - \Delta m. \\
&= M - Ct.
\end{aligned}
$$

Therefore, the rocket's acceleration at time t is $a(t) = \frac{F}{m(t)} = \frac{Cv_s}{M - Ct}$. By integrating this nonconstant acceleration over time, you could derive the rocket's speed as a function of time. I will leave that as an exercise for the reader. (I have always wanted to say that!)

QUESTION 12-5

Mass 1, which is 30 kg, is 10 meters east and 20 meters north of a reference point. Mass 2, which is 40 kg, is 5.0 meters east and 8.0 meters south of the reference point. And mass 3, which is 50 kg, is 20 meters due west of the reference point. With respect to the reference point, what is the center-of-mass position of the three masses?

ANSWER 12-5

Before I do the math, let me intuitively explain what "center of mass" *means*, and where the formulas come from.

The basic concept. On a physics exam, ten people get a 70, and twenty people get an 80. What is the class average?

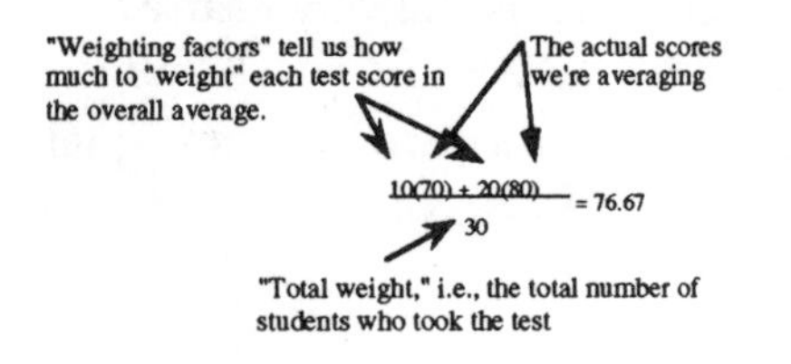

Without doing any calculations, you can almost guess the answer. First of all, the average will be between 70 and 80. Second of all, it will be closer to 80 than to 70, because more people scored 80. Indeed, since twice as many people got 80 as got 70, we expect the class average to be twice as close to 80 as it is to 70.

Now let us calculate the average score. This is indeed twice as close to 80 as it is to 70.

The center-of-mass position is nothing more than the weighted average of position. But the weighting factors are not the number of people. Rather, the weighting factors are the masses of the relevant objects. Let me clarify this with an example.

One-dimensional example. Let us say a system consists of two masses, $m_1 = 10$ kg and $m_2 = 20$ kg. Mass 1 is sitting at position $x_1 = 70$ meters, while mass 2 is sitting at $x_2 = 80$ meters. Where is the center-of-mass position, x_{cm}?

Without doing any math, we can nearly guess the answer. First of all, x_{cm} will be between $x_1 = 70$ meters and $x_2 = 80$ meters. Second of all, it will be closer to $x_2 = 80$ meters than to $x_1 = 70$ meters, because more mass is sitting at $x_2 = 80$ meters. Indeed, since twice as much mass is at $x = 80$ m, we expect the average position to be twice as close to $x = 80$ m as it is to $x = 70$ m.

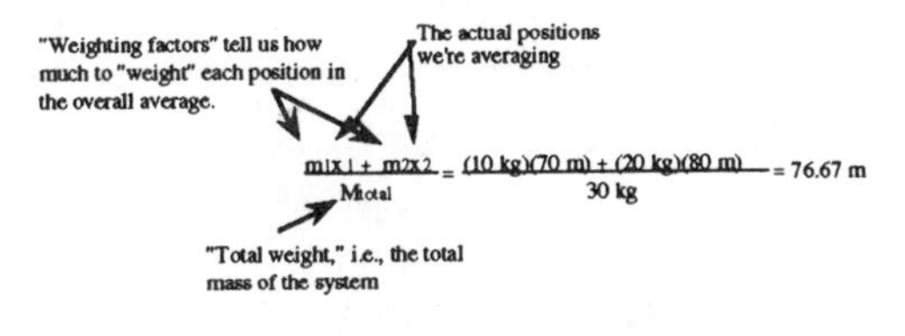

Calculating the weighted average gives us the following figure.

Now that I have shown where the center of mass formula comes from, let me generalize to the two-dimensional case. The basic idea is the same. But now, we must find the system's average x-position and also its average y-position. Should we try to do this all in one big step? No! As you have seen numerous times before (with projectiles, forces, etc.), we can treat the x-stuff and the y-stuff *separately*. That is, we can calculate x_{cm}, and then we can separately calculate y_{cm}. Afterwards, we can express the center of mass as a vector "built" from x_{cm} and y_{cm}.

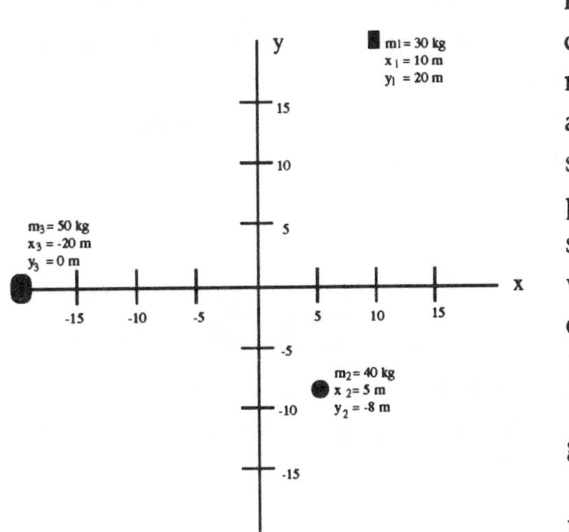

In the problem at hand, we had better organize the given information with a diagram:

Using the formula for weighted average, we can now *separately* calculate x_{cm} and y_{cm}.

$$
\begin{aligned}
x_{cm} &= \frac{m_1 x_1 + m_2 x_2 + m_3 x_3}{M_{total}} \\
&= \frac{(30 \text{ kg})(10 \text{ m}) + (40 \text{ kg})(5 \text{ m}) + (50 \text{ kg})(-20 \text{ m})}{(30 + 40 + 50) \text{ kg}} \\
&= -4.2 \text{ meters.}
\end{aligned}
$$

When calculating x_{cm}, I did not care about the y-positions of the masses. Similarly, I will now calculate y_{cm}, and my calculation will ignore the x-positions of the masses.

$$
\begin{aligned}
x_{cm} &= \frac{m_1 y_1 + m_2 y_2 + m_3 y_3}{M_{total}} \\
&= \frac{(30 \text{ kg})(20 \text{ m}) + (40 \text{ kg})(-8 \text{ m}) + (50 \text{ kg})(0 \text{ m})}{(30 + 40 + 50) \text{ kg}} \\
&= 2.3 \text{ meters.}
\end{aligned}
$$

So, the center-of-mass position is 4.2 meters west and 2.3 meters north of the reference point. Of course, you could draw a triangle and use Pythagorean theorem to find the total distance from the reference point to the center of mass. You could also express x_{cm} using $\hat{i}$, $\hat{j}$, notation:

$$\begin{aligned}\mathbf{r}_{cm} &= x_{cm}\hat{i} + y_{cm}\hat{j}\\ &= (-4.2 \text{ m})\hat{i} + (2.3 \text{ m})\hat{j}.\end{aligned}$$

I do not care what notation you use. The point is, to find the center of mass, you should *separately* solve for x_{cm} and y_{cm} (and if necessary, z_{cm}).

QUESTION 12-6

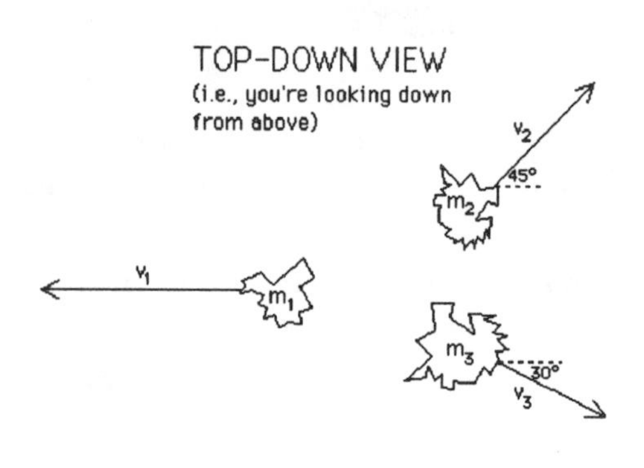

A bomb explodes into three fragments. Immediately after the explosion, the first fragment, of mass $m_1 = 2.0$ kg, travels leftward at $v_1 = 100$ m/s. The second fragment, of mass $m_2 = 3.0$ kg, moves at a 45° angle as shown, at $v_2 = 80$ m/s. The third fragment, of mass $m_3 = 4.0$ kg, moves at the angle shown, with velocity $v_3 = 50$ m/s.

Was the bomb moving before it exploded? If so, what was its speed and direction?

ANSWER 12-6

During an explosion, the fragments "push off" of each other. Indeed, the fragments are sent flying not by some external force, but entirely by the "push off" forces they exert on each other. So, momentum is conserved.

From the given information, you can calculate the final momentum of the bomb fragments. This final total momentum equals the initial momentum of the bomb before it blew up.

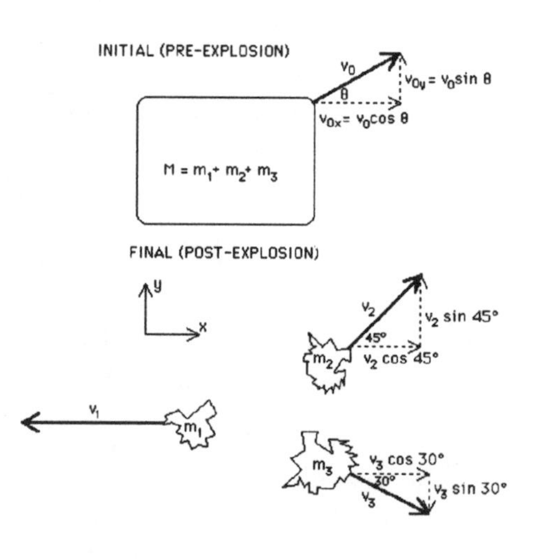

As always, you can deal with whole vectors, or you can break them up into vector components. I will use components to separately deal with the x-directed and y-directed momentum. So, in my drawing of the initial (pre-explosion) and final (post-explosion) situation, I will break up the velocity vectors in the usual way. Notice that I am forced to take a complete guess about the unexploded bomb's direction of motion. Perhaps it is not moving at all. If so, I will get $v_0 = 0$ when I solve numerically.

Looking first at the horizontal momentum, and picking the rightward direction as positive, we get

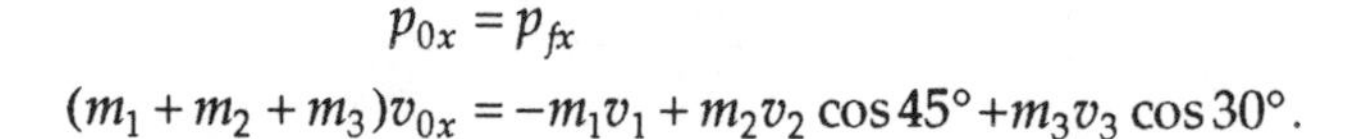

$$\begin{aligned}p_{0x} &= p_{fx}\\ (m_1 + m_2 + m_3)v_{0x} &= -m_1 v_1 + m_2 v_2 \cos 45° + m_3 v_3 \cos 30°.\end{aligned} \quad (1)$$

Solve for v_{0x} to get

$$
\begin{aligned}
v_{0x} &= \frac{-m_1 v_1 + m_2 v_2 \cos 45° + m_3 v_3 \cos 30°}{m_1 + m_2 + m_3} \\
&= \frac{(-2 \text{ kg})(100 \text{ m/s}) + (3 \text{ kg})(80 \text{ m/s})\cos 45° + (4 \text{ kg})(50 \text{ m/s})\cos 30°}{2 \text{ kg} + 3 \text{ kg} + 4 \text{ kg}} \\
&= 15.9 \text{ m/s}.
\end{aligned}
$$

By similar reasoning, we can find the bomb's initial y-velocity. Given my choice of coordinate axes, the y-momentum of fragment 3 is negative.

$$
\begin{aligned}
p_{0y} &= p_{fy} \\
(m_1 + m_2 + m_3) v_{0y} &= 0 + m_2 v_2 \sin 45° - m_3 v_3 \sin 30°.
\end{aligned}
\tag{2}
$$

Solve for v_{0y} to get

$$
\begin{aligned}
v_{0y} &= \frac{0 + m_2 v_2 \sin 45° + m_3 v_3 \sin 30°}{m_1 + m_2 + m_3} \\
&= \frac{0 + (3 \text{ kg})(80 \text{ m/s})\sin 45° - (4 \text{ kg})(50 \text{ m/s})\sin 30°}{2 \text{ kg} + 3 \text{ kg} + 4 \text{ kg}} \\
&= 7.7 \text{ m/s}.
\end{aligned}
$$

Both v_{0x} and v_{0y} turned out positive. Given my choice of coordinates, this means the unexploded bomb traveled northward and eastward, as drawn in my "initial" picture. Had I obtained a negative value for v_{0x}, then I would know the unexploded bomb traveled northward and westward (instead of eastward).

Now that we know v_{0x} and v_{0y}, we can obtain the total pre-explosion speed with Pythagoras' theorem: $v_0 = \sqrt{v_{ox}^2 + v_{oy}^2} = \sqrt{(15.9 \text{ m/s})^2 + (7.7 \text{ m/s})^2} = 18 \text{ m/s}$. Using trig, we can calculate the angle θ in my "initial" picture. Since $\tan\theta = \frac{v_{0y}}{v_{0x}}$, $\theta = \tan^{-1}\left(\frac{v_{0x}}{v_{0y}}\right) = \tan^{-1}\left(\frac{7.7 \text{ m/s}}{15.9 \text{ m/s}}\right) = 26°$.

QUESTION 12-7

Deb has purchased a batch of defective baseballs. For some unknown reason, they tend to explode. Deb tosses one of these baseballs straight up. While in mid-air, it explodes into two fragments, one of which has mass $m_1 = 0.20$ kg, the other of which has mass $m_2 = 0.10$ kg. The two fragments land on the ground at the same time.

The heavier fragment lands 3.0 meters to the left of Deb. The lighter fragment lands in some thick grass to the right of Deb, thick enough to hide the fragment.

Where should Deb look for the lighter fragment? First, see if you can answer using "raw intuition." Then, try to justify your answer.

ANSWER 12-7

Intuitively speaking, the explosion pushes the heavier fragment (m_1) to Deb's left and the lighter fragment (m_2) to Deb's right. At first glance, we might guess that the lighter fragment lands 3 meters to Deb's right. But the lighter fragment is only half as heavy as the other one. So, we might expect it to fly farther away from Deb than the heavier fragment does. Since it is twice as light, maybe it lands twice as far away from Deb–6 meters instead of 3 meters.

That guess turns out to be correct. We can arrive at that answer using two different methods: conservation of momentum, and center-of-mass considerations. In this answer, I hope to show you that those two "different" problem-solving techniques are closely related. You can "derive" one technique from the other. They are really two different ways of saying the same thing.

In this case, center-of-mass considerations get you to the answer faster. But I will solve first using momentum conservation, because it is probably more familiar to you at this point.

Method 1: Momentum conservation

For simplicity, let us assume the ball explodes at its peak. (You get the same final answer if the explosion occurs at some other point, though the reasoning becomes more complex.) Since both fragments land at the same time, the explosion must have sent each fragment traveling sideways. To see why, think about momentum conservation. At its peak, immediately before the explosion, the baseball is motionless. So, its "initial" momentum is $\mathbf{p}_{\text{before}} = 0$. Therefore, immediately after the explosion, the total momentum of the two fragments must still be zero. In other words, the post-collision momentum of fragment 1 must cancel the post-collision momentum of fragment 2. Formally,

$$\begin{aligned} \mathbf{p}_{\text{before}} &= \mathbf{p}_{\text{after}} \\ 0 &= \mathbf{p}_1 + \mathbf{p}_2 \\ &= m_1 v_1 + m_2 v_2, \end{aligned}$$

and hence $m_1\mathbf{v}_1 = -m_2\mathbf{v}_2$. Here, $\mathbf{v}_1$ and $\mathbf{v}_2$ denote the velocities of fragments 1 and 2 immediately after the explosion. Remember, momentum and velocity are vectors; they have directions. Since $m_1\mathbf{v}_1 = -m_2\mathbf{v}_2$, it follows that the two fragments fly off in opposite directions.

So far, so good. But why does this imply that both fragments fly off sideways? Why could not it be the case that fragment 1 flies off at an upward angle? Well, if that were the case, then fragment 2 would fly off at a downward angle, because the two fragments fly off in opposite directions. But if fragment 2 explodes off at a downward angle while fragment 1 explodes off at an upward angle, then fragment 2 would land first. *We are told, however, that both fragments land at the same time.* So, it cannot be the case that one fragment shoots upwards while the other shoots downwards. They must both shoot off sideways.

This argument was hard to follow. Let me summarize it with pictures.

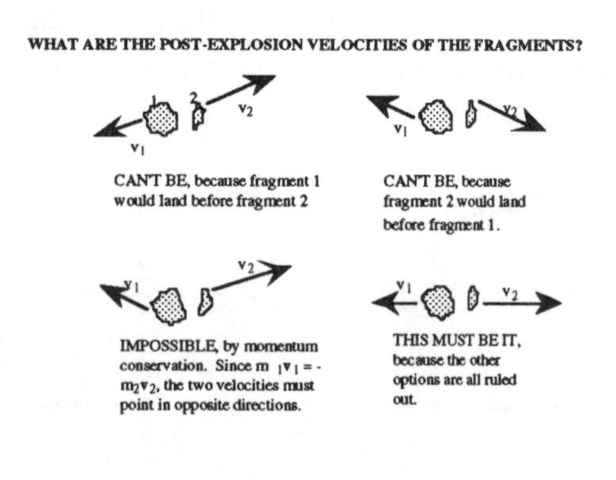

By momentum conservation, $m_1\mathbf{v}_1 = -m_2\mathbf{v}_2$, and hence $\mathbf{v}_2 = -\frac{m_1}{m_2}\mathbf{v}_1 = \frac{0.20\text{ kg}}{0.10\text{ kg}}\mathbf{v}_1 = 2\mathbf{v}_1$. So, fragment 2 leaves the explosion traveling twice as quickly as fragment 1, and in the opposite direction. But I am trying to find x_2, the distance away from Deb that fragment 2 lands. We know that fragment 1 lands a distance $x_1 = 3.0$ meters left of Deb. How do these distances relate to $\mathbf{v}_1$ and $\mathbf{v}_2$, the post-explosion velocities? That is a projectile motion problem.

As always with projectiles, you can *separately* consider the x-direction and y-directed motion. Since the only force acting on the fragments after the explosion is gravity, the fragments experience no *horizontal* forces. So, they both have horizontal velocity $a_x = 0$. Therefore, we can immediately relate their post-explosion velocities to the relevant distances. Remember, $\mathbf{v}_1$ and $\mathbf{v}_2$ are the *initial* horizontal velocities of the fragments.

$$x_1 = x_0 + v_1 t + \frac{1}{2} a_x t^2$$
$$= 0 + v_1 t$$

$$x_2 = x_0 + v_2 t + \frac{1}{2} a_x t^2$$
$$= 0 + v_2 t.$$

We do not know the "fall time" t of either fragment. But we do know that both fragments land at the same time. So, "t" for fragment 1 equals "t" for fragment 2. Therefore, these equations tell us that, since $v_2 = 2v_1$, x_2 is twice x_1. In other words, these equations tell us that, for each fragment, the distance it lands from Deb is proportional to its initial (post-explosion) speed. Since fragment 2 has twice as much initial speed as fragment 1, it covers twice as much horizontal distance before landing, and therefore lands twice as far from Deb. Since $x_1 = 3.0$ m, we conclude that $x_2 = 6.0$ m.

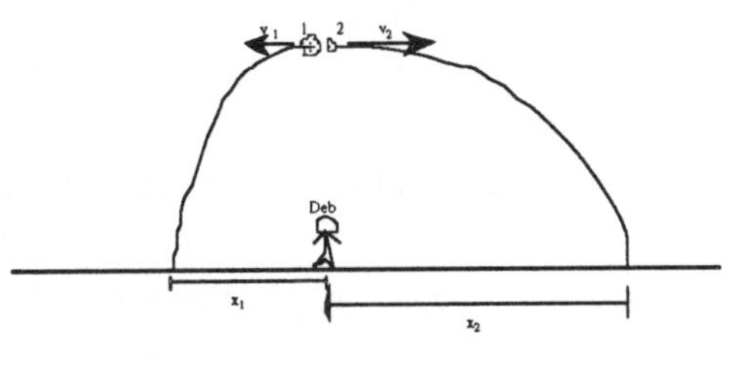

Notice that we were able to complete the problem without knowing $\mathbf{v}_1$, $\mathbf{v}_2$, or t. We just had to know the ratio of the post-explosion velocities.

Before re-solving this problem using center-of-mass considerations, let me summarize in a single equation all the reasoning used so far. We started with momentum conservation in the x-direction:

$$p_{x\,\text{before}} = p_{x\,\text{after}}$$
$$0 = p_1 + p_2$$
$$= m_1 v_1 + m_2 v_2$$
$$= m_1 \frac{x_1}{t} + m_2 \frac{x_2}{t},$$

where in the last step I used the above projectile motion formulas. Multiply this momentum conservation equation through by t to get

$$m_1 x_1 + m_2 x_2 = 0. \qquad (*)$$

Soon, I will explain why I just bothered to obtain that expression.

Method 2: Center-of-mass considerations

As shown in the textbook, it is often useful to consider the motion of a system's center of mass. This is because the center of mass usually undergoes "simpler" motion than any other part of the system. Remember, when you use Newton's 2nd law to find a system's acceleration, you are really finding the acceleration of the center of mass: $\sum \mathbf{F}_{\text{ext}} = m\mathbf{a}_{\text{cm}}$, where $\mathbf{F}_{\text{ext}}$ are the *external* forces, by which I mean the forces acting from outside the system. In this example, once the baseball leaves Deb's hand, the only external force is gravity. When the ball explodes, *internal* forces push the two fragments apart. (An internal force is a force exerted by one part of the system on another part.) The whole point of considering $\sum \mathbf{F}_{\text{ext}} = m\mathbf{a}_{\text{cm}}$ is this; to calculate the motion of the baseball's center of mass, we do not need to consider those complicated internal forces. The external forces, and nothing else, determine the trajectory of the ball's center of mass. And this is true whether the ball explodes into two fragments, or explodes into two hundred fragments, or stays whole.

Let us apply this reasoning to the problem at hand. Since only gravity acts on the ball, the center-of-mass acceleration is downward. Therefore the ball's center of mass goes straight up and then falls straight back down, "landing" right next to Deb. This is true whether or not the baseball explodes. If it does not explode, then the center of mass is the center of the baseball itself. In that case, the center of the baseball lands right next to Deb. By contrast, if the ball explodes into two fragments, then the center of mass is an imaginary point in space somewhere between the two fragments. But that imaginary point does exactly what you would expect. It falls straight down (accelerating at $9.8\ \text{m/s}^2$) and lands right next to Deb. Here is my point: Even if the center of mass is an imaginary point in space, it follows the same trajectory that the baseball *would* have followed had it remained whole.

This conclusion applies in more complicated situations, too. For instance, suppose I throw a baseball at an angle. As we have seen before, if the baseball stays whole, it traces a parabolic path. What if the ball explodes? Then the center of mass traces the *same* parabolic path that the whole ball would have followed if it had not exploded. So, using projectile motion reasoning, you could easily find where the center of mass lands, even if the individual fragments spread out all over the place.

Now we can apply these insights to the problem at hand. If the baseball had not exploded, it would have landed right next to Deb. Therefore, even when it *does* explode, its center of mass (an imaginary point) still lands right next to Deb. Let us call Deb's position $x = 0$, and let us call leftward the positive direction. So, we know that the center of mass lands at $x = 0$, and fragment 1 lands at $x_1 = 3.0\ \text{m}$. From this information, we can figure out where fragment 2 lands. As shown in Question 12-5 above, the formula for a system's center of mass (i.e., its "average" position) is

$$x_{\text{cm}} = \frac{m_1 x_1 + m_2 x_2 + \ldots}{M_{\text{total}}}. \tag{1}$$

Normally, we would know x_1 and x_2, and we would solve for x_{cm}. Here, we know $x_{\text{cm}} = 0$ and $x_1 = 3.0\ \text{m}$, and we are solving for x_2. Multiply Eq. (1) through by M_{total}, and use the known information to get

$$\begin{aligned} M_{\text{total}} x_{\text{cm}} &= m_1 x_1 + m_2 x_2 \\ 0 &= m_1 x_1 + m_2 x_2 \\ &= (0.20\ \text{kg})(0.30\ \text{m}) + (0.10\ \text{kg}) x_2. \end{aligned} \tag{2}$$

Solve for x_2 to get $x_2 = -0.60\ \text{m}$. The minus sign indicates rightward. This is the same answer obtained above using momentum conservation.

Indeed, we did not just get the same answer. We got the exact same equation. Recall that, in method 1 above, we started with momentum conservation and then used projectile motion reasoning. The "final" formula we got, by combining all this reasoning, was Eq. (*). But Eq. (*) is the same as Eq. (2)! So, momentum conservation and center-of-mass considerations are, in a sense, two different ways of saying the same thing. The same intuitions apply to both methods. In problems such as this, center-of-mass considerations get you to the answer more quickly.

QUESTION 12-8

An artist has created a "mobile" consisting of two glued-together pieces: A plastic ring of mass 0.50 kg and radius 0.50 meters, and a thin metal rod of mass 1.50 kg and length 0.50 meters. Point *B* is in the middle of the arc subtended by the rod.

The artist wants to attach a string to the middle of the rod and hang the whole thing from the ceiling. And when he does so, he wants the ring to be parallel to the ground. In other words, he wants the plane of the ring to be horizontal, not vertical. Therefore, the artist decides to stick some clay onto point B. What mass of clay must be piled onto point B to ensure that the ring will balance properly when hung from the string attached to the middle of the rod?

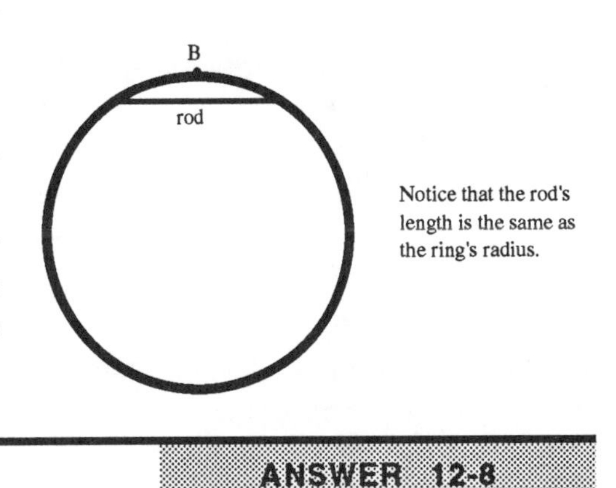

ANSWER 12-8

From the textbook, we know that objects "balance" when pivoted at the center of mass. (To see this, balance a ruler or pencil on the tip of your finger.) So, the artist must pile enough clay onto point B so that the center of mass of the whole system is located at the center of the rod. If you did not realize that this is a center of mass problem, go back and give it another try, before reading on.

At first, this looks scary; we might need to take integrals around the ring and along the rod. But we can completely avoid integrals. Think of the overall system as consisting of three parts: A ring of mass $m_{\text{ring}} = 0.50$ kg, a rod mass mass $m_{\text{rod}} = 1.50$ kg, and a piece of clay of unknown mass m_{clay}. The rote formula for the center-of-mass position of the system is $x_{\text{cm}} = \frac{m_{\text{ring}}x_{\text{ring}} + m_{\text{rod}}x_{\text{rod}} + m_{\text{clay}}x_{\text{clay}}}{M_{\text{total}}}$. But what do I mean by x_{ring}, the "position of the ring?" Do I mean the position of point B, or the position where the rod is attached to the rod, or what? Well, intuitively speaking, "x_{ring}" should be the weighted average of the positions of all the different parts of the ring. In other words, when we say the "position of the ring," we mean the center of mass of the ring. And by the "position of the rod," we mean the position of the center of mass of the rod. In general: When calculating the overall center of mass of a multi-part system, use the centers of mass of the parts.

So, in the above center-of-mass formula, x_{ring} is the center of mass of the ring, and x_{rod} is the center of mass of the rod.

Intuitively, the center of mass of the ring is the imaginary point right at the center of the ring. The center of mass of the rod is simply the center of the rod.

In the diagram below, I have marked with heavy dots the centers of mass of the ring, rod, and clay. I have also drawn in a triangle that allows us to calculate the coordinates of the rod's center. Let me arbitrarily say that the center of the ring is where $x = 0$. (You will get the same final answer no matter where you place the origin.) So, $x_{\text{ring}} = 0$. Since point B is "one radius" away from the center of the ring, $x_{\text{clay}} = r = 0.50$ m. Notice that the x-direction is "up the page."

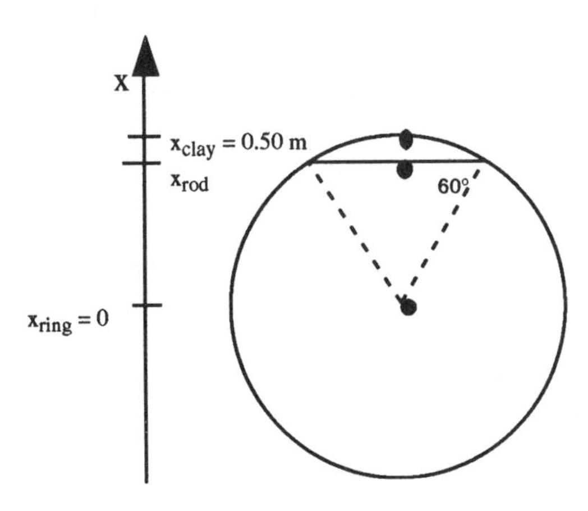

Let us use geometry and trig to find x_{rod}, the distance from the center of the ring to the center of the rod. The rod, like both dotted lines, is "one radius" long. Therefore, the triangle is equilateral. Hence, as this mini-diagram shows, the distance from the center of the ring to the center of the rod is 0.43 m.

x_{rod} = rsin 60° = (0.50 m)sin 60° = 0.43 m.

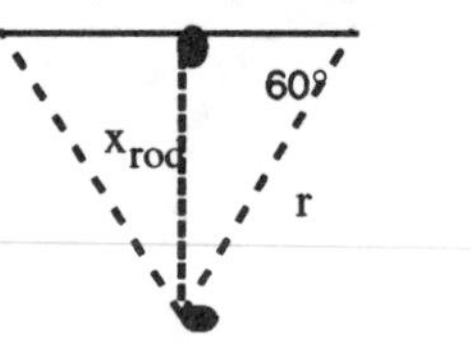

We are trying to find the mass of clay (m_{clay}) needed so that the overall system's center of mass is located at the center of the rod. In other words, we want the overall center of mass to be x_{rod}. So, we can invoke the above center-of-mass formula, and set $x_{cm} = x_{rod} = 0.43$ m. $x_{cm} = \frac{m_{ring}x_{ring} + m_{rod}x_{rod} + m_{clay}x_{clay}}{m_{ring} + m_{rod} + m_{clay}}$.
Algebra starts here. To solve this equation for m_{clay}, multiply through by the denominator of the right-hand side to get $(m_{ring} + m_{clay} + m_{rod})x_{cm} = m_{ring}x_{ring} + m_{clay}x_{clay} + m_{rod}x_{rod}$. Bring all the m_{clay} terms to the left-hand side, and all the other terms to the right-hand side, to get $m_{clay}(x_{cm} - x_{clay}) = m_{ring}(x_{ring} - x_{cm}) + m_{rod}(x_{rod} - x_{cm})$. Now divide through by $x_{cm} - x_{clay}$, and substitute in all the known information, remembering that the artist wants the overall center of mass to be located at the rod's center: $x_{cm} = x_{rod} = 0.43$ m.

$$
\begin{aligned}
m_{clay} &= \frac{m_{ring}(x_{ring} - x_{cm}) + m_{rod}(x_{rod} - x_{cm})}{x_{cm} - x_{clay}} \\
&= \frac{m_{ring}(x_{ring} - x_{cm})}{x_{cm} - x_{clay}} \qquad [(x_{rod} - x_{cm}) = 0, \text{ since } x_{rod} = x_{cm}] \\
&= \frac{(0.50 \text{ kg})(0 - 0.43 \text{ m})}{0.43 \text{ m} - 0.50 \text{ m}} \\
&= 3.1 \text{ kg}.
\end{aligned}
$$

End of algebra.

That is a lot of clay, probably more than the artist can pile onto point *B*. Intuitively speaking, a big mass of clay is needed to "pull" the center of mass of the overall system all the way to the rod.

QUESTION 12-9

Daredevil Dana plans to go over a waterfall in a canoe. Do not try this trick yourself; Dana is a trained professional with little regard for her bodily well-being. Her mass is $m_D = 50$ kg. The canoe's mass is only $m_c = 10$ kg, and its length is $L = 4.0$ m. The canoe's mass is evenly distributed along the whole length of the canoe.

(a) Dana starts out by "training" in a placid lake. She stands at the front of the canoe. Realizing that the canoe could tip over, Dana walks to the center of the canoe. Due to Dana's walking, how far does the canoe move relative to the water? Before plugging in numbers, solve in terms of symbols (m_D, m_c, and L).

(b) When Dana stops walking at the center of the canoe, does the canoe keep drifting or come to rest? Justify your answer.

(c) *Hard.* Now Dana is riding her canoe in a river whose current flows at speed $v_r = 2.0$ m/s. How fast would Dana have to run in the canoe so that the canoe "stands still" in the water (i.e., so that the canoe does not move with respect to the river bank)?

(d) *Very hard.* Now Dana stands in the front of the canoe, in that same river. She is 10 meters from the waterfall, and the river flows at $v_r = 2.0$ m/s. So, the front of the boat will take 5 seconds to reach the waterfall. But Dana is anxious to finish her stunt. She wants the front of the boat to reach the waterfall in less than 5 seconds. What can she do, and how much time can she "save" (i.e., how many seconds *before* time $t = 5$ s will the front of the boat reach the waterfall)?

ANSWER 12-9

(a) Intuitively, when Dana walks backwards in the canoe, the canoe "recoils" forward. You can experience this by walking on a skateboard. We can find the size of this recoil using either center-of-mass considerations or conservation of momentum. I will demonstrate both ways, and try to show that they are really two different ways of saying the same thing.

Method 1: Center-of-mass considerations

As explained in Question 12-7 above, the center of mass of the overall system (consisting of Dana and her canoe) "responds" only to external forces. In this case, no (horizontal) external force acts on any part of the system. The only sideways forces felt by Dana and the canoe are the internal forces they exert on each during Dana's walk. So, the overall center of mass stays motionless, exactly the same as if Dana had not walked. You saw a similar phenomenon in Question 12-7 above. There, the baseball's center of mass lands in the same place whether or not the baseball explodes. Here, the system's center of mass undergoes the same motion–or more specifically, the same *lack* of motion–whether or not Dana walks. Because the center of mass would have stayed still if Dana had not walked, the center of mass continues to stay still even when Dana *does* walk.

Since the overall center of mass does not move, but Dana travels backwards, the boat must move forward to "compensate." We need to figure out how far forward.

In this kind of problem, you should organize your thoughts by drawing "before" and "after" pictures, exactly as if you were using conservation laws. For convenience, I will call the initial position of the back of the canoe $x = 0$. (But you could equally well pick any other point as your origin, provided your pictures reflect the fact that the canoe moves forward.) Since the canoe's mass is evenly distributed, its center of mass sits right in the middle of the canoe. I will mark the canoe's center of mass with a heavy dot. Let me call "s" the distance by which the canoe moves forward. We are solving for s.

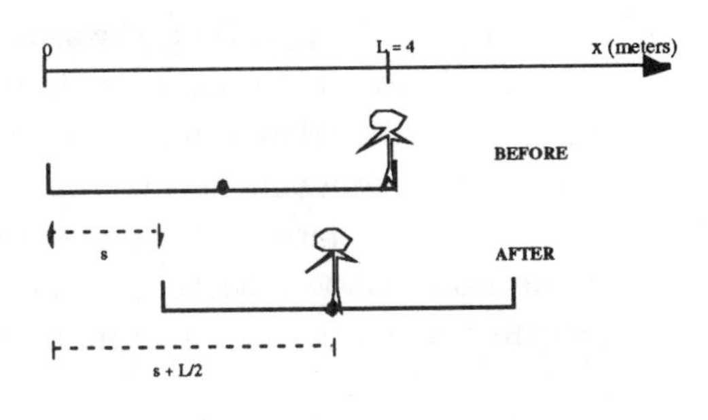

The diagram reveals a crucial fact: Dana's final position, which is also the center of the canoe's final position, is $s + \frac{L}{2}$. This is because Dana ends up "half a canoe" away from the back of the canoe. Notice also that Dana does not walk a distance $\frac{L}{2}$ with respect to the water. Sure, she walks half a canoe-length *with respect to the canoe*. But because the canoe moves forward "under her feet," Dana does not end up half a canoe-length behind her starting point *with respect to the water*. The above diagram shows this clearly.

To solve for s, we can use the central physical insight discussed above: The overall system's center of mass does not move. So, we can set $x_{\text{cm before}} = x_{\text{cm after}}$. As mentioned in Question 12-7, when plugging "x_{Dana}" and "x_{canoe}" into the center-of-mass formula, you must use Dana's and the canoe's center-of-mass positions. So, the canoe's initial position is $x_{\text{canoe before}} = \frac{L}{2} = 2.0$ meters. And the canoe's final position is the same as Dana's final position: $x_{\text{canoe after}} = x_{\text{Dana after}} = s + \frac{L}{2}$.

Given all this, I get

$$x_{\text{cm before}} = x_{\text{cm after}}$$
$$\frac{m_D x_{\text{Dana before}} + m_c x_{\text{canoe before}}}{M_{\text{total}}} = \frac{m_D x_{\text{Dana after}} + m_c x_{\text{canoe after}}}{M_{\text{total}}}$$
$$\frac{m_D L + m_c \frac{L}{2}}{m_D + m_c} = \frac{m_D(s+\frac{L}{2}) + m_c(s+\frac{L}{2})}{m_D + m_c}.$$

Since the only unknown in this equation is s, we are done with the physics.

Algebra starts here. Multiply through by $m_D + m_c$ to get $m_D L + m_c \frac{L}{2} = m_D(s+\frac{L}{2}) + m_c(s+\frac{L}{2})$. Cancel the $m_c \frac{L}{2}$ terms, subtract $m_D \frac{L}{2}$ from both sides, and then divide through by $m_D + m_c$ to get

$$\begin{aligned} s &= \frac{m_D}{m_D + m_C}\left(\frac{L}{2}\right) \\ &= \frac{50 \text{ kg}}{50 \text{ kg} + 10 \text{ kg}}\left(\frac{4.0 \text{ m}}{2}\right) \\ &= 1.7 \text{ m}. \end{aligned}$$

End of algebra.

So, when Dana walks backwards 2.0 meters with respect to the canoe, the canoe moves forward under her feet by 1.7 meters. Therefore, Dana ends up only 0.3 meters behind where she started (with respect to the water).

Let me now re-solve this problem using

Method 2: Momentum conservation

Let us say Dana walks backwards at speed v_{rel} relative to the canoe. Let v_D and v_c denote Dana's velocity and the canoe's velocity with respect to the water. Using momentum conservation, we can find the relationship between v_D and v_c. With that information, we will then invoke kinematics to figure out how far the canoe moves.

Before Dana starts walking, the total momentum of the system is zero, because nothing moves. While Dana walks, the boat moves forward (positive velocity) and Dana moves backwards (negative velocity). From momentum conservation,

$$\begin{aligned} p_0 &= p_f \\ 0 &= m_C v_C + m_D v_D. \end{aligned} \tag{1}$$

Since v_c is the canoe's speed with respect to the water, v_D must be Dana's speed *with respect to the water*, not respect to the canoe. To see the difference, let us just pretend that Dana walks with respect to the canoe at $v_{\text{rel}} = -2.0$ m/s (the minus sign indicating "backwards"), but that the canoe moves forward under her feet at $v_c = 1.7$ m/s. Then Dana's velocity with respect to the water is only $v_D = -0.3$ m/s. The formula I just used is $v_{\text{rel}} = v_D - v_c$. For reasons that will become clear in a minute, let me solve this relative-velocity formula for v_D to get $v_D = v_c + v_{\text{rel}}$, and substitute that expression into Eq. (1) to get

$$0 = m_c v_c + m_D v_c + m_D v_{\text{rel}}. \tag{2}$$

At this point, I have related the canoe's speed through the water to Dana's speed relative to the boat. But we want the *distance* traveled by the canoe. If we assume the canoe and Dana travel at constant velocity (no acceleration), then the distances they travel are given by $x = vt$. Hence, the boat travels unknown distance $s = v_c t$ relative to the water; and Dana walks a distance $x_{rel} = v_{rel} t$ relative to the canoe. So, the "trick" here is to multiply Eq. (2) through by t to get

$$\begin{aligned} 0 &= m_c v_c t + m_D v_c t + m_D v_{rel} t \\ &= m_c s + m_D s + m_D \frac{L}{2}, \end{aligned}$$

where I invoked the fact that Dana travels a distance $x_{rel} = \frac{L}{2}$ relative to the boat. Solve this equation for s to get $s = \frac{m_D}{m_D + m_c}(\frac{L}{2})$, the same answer obtained above using center-of-mass considerations. Once again, we see that center-of-mass considerations and conservation of momentum are, in a deep sense, two different ways of saying the same thing. In this particular problem, however, center-of-mass considerations were easier to use.

(b) I have conflicting intuitions on this one. Part of me wants to focus on the fact that the overall system's center of mass does not move. So, when Dana stops, the canoe must stop as well, or else the overall center of mass would "drift." This sounds reasonable. But another part of me wants to say that once the canoe is set in motion, it will not just come to rest for no reason; an object in motion tends to stay in motion. What makes the canoe stop?

The center-of-mass reasoning is correct: The canoe must stop when Dana stops, or else the overall center of mass would keep moving through the water, which is impossible since no external force sets the overall system into motion. But this does not explain, intuitively, what makes the boat stop. Fortunately, an explanation lies waiting at Dana's feet. To see what I mean, imagine yourself running across a rug, when suddenly you "slam on the brakes" (i.e., you abruptly stop running). When you stop, the rug slides forward in the direction you were running. This goes to prove that when you stop walking or running, your feet exert a force on the rug or floor–or canoe. And that force points in the direction you were walking. Now, when Dana starts walking backwards, the boat moves forward. But when Dana abruptly stops walking, she exerts a backwards force on the canoe. This backwards force is exactly big enough to stop the canoe.

By the way, the force Dana exerts on the canoe is "internal," by which I mean one part of the overall system acting on another part. So, that force does not screw up the center-of-mass considerations or momentum conservation considerations just discussed.

(c) Once again, either momentum conservation or center-of-mass reasoning works. Let me use center of mass. When Dana stands still in the canoe, the overall system gets carried along by the river at speed $v_r = 2.0$ m/s. As discussed in problem 12-7, the overall system's center of mass "behaves" the same way no matter whether the system stays whole or explodes into pieces or whatever. Since the center of mass gets carried along at speed v_r when Dana stands still in the canoe, the center of mass *continues* to drift along at speed v_r when Dana starts walking in the canoe. No matter how fast Dana runs, the overall center of mass continues to get carried along at that speed.

Dana must run so as to make the canoe stop drifting forward. Therefore, Dana must run forwards in the canoe, to make it "recoil" backwards.

When Dana runs forward at the "proper" speed v_{Dana}, the canoe will be motionless (with respect to the river bank): $v_{canoe} = 0$. Therefore, Dana's speed with respect to the canoe will also be her speed with respect to the river bank. And remember, by the above center-of-mass considerations, the overall system's center-of-mass velocity will be $v_{cm} = v_r = 2.0$ m/s, no matter what Dana does.

Start with the usual formula for center of mass: $x_{\text{cm}} = \frac{m_D x_{\text{Dana}} + m_c x_{\text{canoe}}}{M_{\text{total}}} = \frac{m_D x_{\text{Dana}} + m_c x_{\text{canoe}}}{m_D + m_c}$. This gives us a relationship between Dana's position and the overall center-of-mass position. But we are solving for Dana's *velocity*, given that the overall center-of-mass velocity is $v_{\text{cm}} = v_r = 2.0$ m/s. How can we get from a formula about center-of-mass *position* to a formula about center-of-mass *velocity*? Simply take the derivative with respect to time of the above equation. Since $\frac{dx_{\text{cm}}}{dt} = v_{\text{cm}}$, $\frac{dx_{\text{Dana}}}{dt} = v_{\text{Dana}}$, and $\frac{dx_{\text{canoe}}}{dt} = v_{\text{canoe}}$, we get $v_{\text{cm}} = \frac{m_D x_{\text{Dana}} + m_c v_{\text{canoe}}}{m_D + m_c}$. At this point, we can immediately solve for v_{Dana}, and set $v_{\text{canoe}} = 0$ and $v_{\text{cm}} = 2.0$ m/s, to get

$$\begin{aligned} v_{\text{Dana}} &= \frac{(m_D + m_c)v_{\text{cm}} - m_c v_{\text{canoe}}}{m_D} \\ &= \frac{(m_D + m_c)v_{\text{cm}}}{m_D} \qquad \text{[since } v_{\text{canoe}} = 0] \\ &= \frac{(50 \text{ kg} + 10 \text{ kg})(2.0 \text{ m/s})}{50 \text{ kg}} \\ &= 2.4 \text{ m/s}. \end{aligned}$$

The only center-of-mass formula you need to remember (or write on your cheat sheet) is the one for x_{cm}. You can derive the center-of-mass velocity and acceleration formulas by differentiating with respect to t.

(d) As seen in part (a), Dana can make the canoe shoot forwards by running backwards. So, ironically, Dana can make the canoe reach the waterfall more quickly by running *away* from the waterfall (i.e., by running backwards in the canoe).

To fully exploit this effect, Dana can run all the way to the back of the canoe. This makes the canoe shoot forward (compared to where it otherwise would have been) by an unknown distance I will call D. We can solve for D using center-of-mass considerations. Then, we can calculate how much time gets "saved" as a result of the canoe's shooting a distance D ahead of where it otherwise would have been.

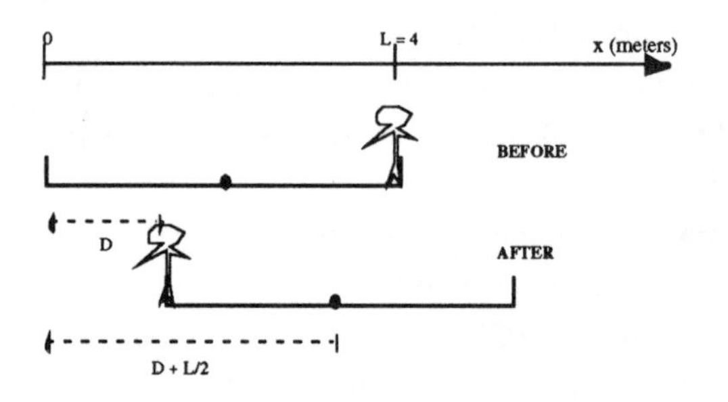

To solve for D, we can repeat the reasoning of part (a), except now Dana ends up at the back of the canoe. So, her "final" position is $x_{\text{Dana after}} = D$, while the center of the canoe's final position is $x_{\text{canoe after}} = D + \frac{L}{2}$. Set the "before" overall center of mass equal to the "after" overall center of mass to get

$$\begin{aligned} x_{\text{cm before}} &= x_{\text{cm after}} \\ \frac{m_D x_{\text{Dana before}} + m_c x_{\text{canoe before}}}{M_{\text{total}}} &= \frac{m_D x_{\text{Dana after}} + m_c x_{\text{canoe after}}}{M_{\text{total}}} \\ \frac{m_D L + m_c \frac{L}{2}}{m_D + m_c} &= \frac{m_D D + m_c (D + \frac{L}{2})}{m_D + m_c}. \end{aligned}$$

Solving for D yields, after a few steps of algebra, $D = \frac{m_D}{m_D + m_c} L = \frac{50 \text{ kg}}{50 \text{ kg} + 10 \text{ kg}}(4.0 \text{ m}) = 3.3$ m. Because the canoe is light compared to Dana, it shoots forward a large distance, almost a full canoe-length.

You might be worried about the validity of the above technique, where I set $x_{\text{cm before}} = x_{\text{cm after}}$. In this problem, unlike in part (a), the overall center of mass does not stay still. Rather, it drifts down the river at speed v_r. So, why did I set $x_{\text{cm before}} = x_{\text{cm after}}$? Well, I was working in a frame of reference that floats down the river at speed v_r. In other words, I was solving for D *as seen by an observer who also floats down the river.* According to that observer, the Dana-and-canoe system's center of mass is indeed motionless. Therefore, the above problem-solving technique is valid. D is how far the canoe shoots forward, *over and above the distance it drifts.* Put another way, it is how far Dana's walk makes the canoe shoot forward, compared to where the canoe would have been had Dana stood still.

So, Dana's walk causes the canoe to get $D = 3.3$ meters closer to the waterfall than it otherwise would have been. For this reason, the canoe will reach the waterfall sooner than it otherwise would have. But how much sooner? Well, after Dana stops walking, the canoe drifts toward the waterfall at speed $v_r = 2.0$ m/s. Since the velocity is constant, the relationship between distance traveled and velocity is $v = \frac{x}{t}$. Solve for t to find the amount of time the canoe would take to travel distance $D = 3.3$ meters: $t = \frac{D}{v_r} = \frac{3.3 \text{ m}}{2.0 \text{ m/s}} = 1.7$ s. If Dana had not made the canoe shoot forward, then it would had needed to travel an "extra" distance $D = 3.3$ meters before reaching the waterfall. I just calculated the time the canoe would have taken to travel that extra distance. By making the canoe shoot forward, Dana "saves" herself exactly this much time. She reaches the waterfall 1.7 seconds sooner than she otherwise would have.

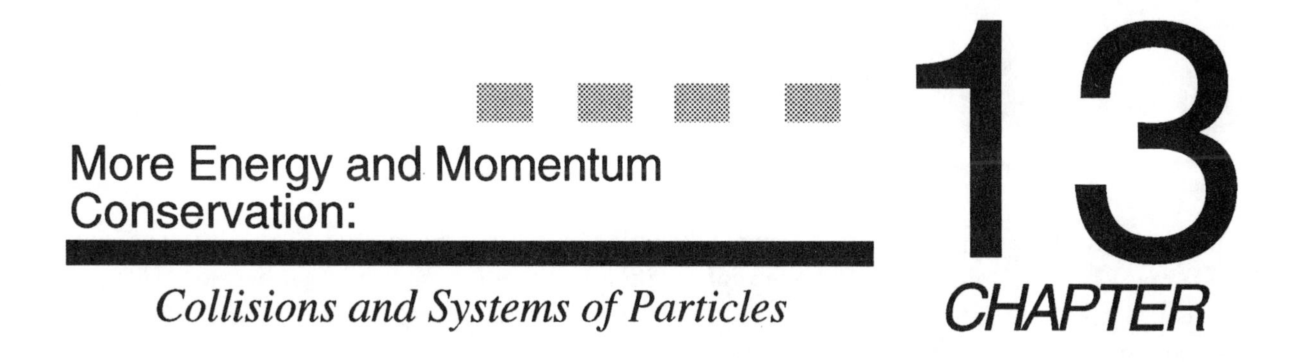

13 CHAPTER More Energy and Momentum Conservation: *Collisions and Systems of Particles*

QUESTION 13-1

Two pool balls, both of mass M, collide "head on" on top of a frictionless pool table. Before the collision, ball 1 moves rightward at speed 5.0 m/s, and ball 2 is motionless. The collision is not elastic. Instead, during the collision, 20% of the initial kinetic energy carried by ball 1 gets converted into heat, sound and other "dissipative" forms of energy.

After the collision, what is the speed of ball 2?

ANSWER 13-1

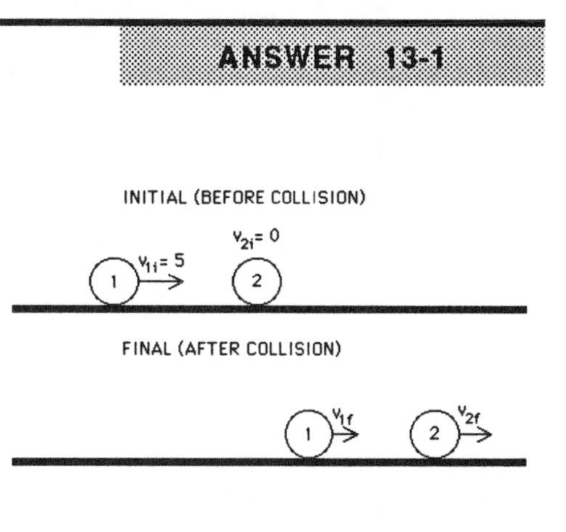

As always in collision problems, make "before" and "after" diagrams.

Since momentum is *always* conserved during a collision, no matter whether it is elastic or not, you should start by considering momentum conservation. Here, we do not have to worry about vectors, because all the motion happens along one dimension. So, from the diagram,

$$
\begin{aligned}
p_i &= p_f \\
Mv_{1i} + 0 &= Mv_{1f} + Mv_{2f} \qquad (1) \\
M(5 \text{ m/s}) + 0 &= Mv_{1f} + Mv_{2f}.
\end{aligned}
$$

The M's cancel. We are solving for v_{2f}. But Eq. (1) contains a second unknown, namely v_{1f}. We need more information. Since we have already used momentum conservation, let us now try energy conservation.

The potential energy of the balls does not change, because they slide along a flat pool table (instead of a hill). But the collision is inelastic; some heat gets produced. So, the conservation of energy equation becomes

$$E_i = E_f$$
$$K_{1i} + K_{2i} = K_{1f} + K_{2f} + \text{Heat}_f$$
$$\frac{1}{2}Mv_{1i}^2 + \frac{1}{2}Mv_{2i}^2 = \frac{1}{2}Mv_{2f}^2 + \frac{1}{2}Mv_{2f}^2 + \text{Heat}_f \quad (2)$$
$$\frac{1}{2}M(5.0\text{ m/s})^2 + 0 = \frac{1}{2}Mv_{1f}^2 + \frac{1}{2}Mv_{2f}^2 + \text{Heat}_f.$$

By "Heat," I mean all the heat, sound, and other dissipative energy created during the collision. Eqs. (1) and (2) still do not give us enough information to solve for v_{2f} and v_{1f}, because Heat_f is unknown. Or is it? According to the problem, Heat_f is 20% of the initial kinetic energy. The initial kinetic energy, you can see from Eq. (2), is $\frac{1}{2}M(5.0\text{ m/s})^2$. So, $\text{Heat}_f = .2[\frac{1}{2}M(5.0\text{ m/s})^2]$. Using this fact to modify Eq. (2), we get

$$\frac{1}{2}M(5.0\text{ m/s})^2 + 0 = \frac{1}{2}Mv_{1f}^2 + \frac{1}{2}Mv_{2f}^2 + .2\left[\frac{1}{2}M(5.0\text{ m/s})^2\right]. \quad (2')$$

Now we have enough information to solve for v_{2f}, because Eqs. (1) and (2') are two equations in two unknowns (v_{1f} and v_{2f}). It is just a matter of algebra to finish things off.

Algebra starts here. Starting with Eq. (2'), subtract $.2[\frac{1}{2}M(5.0\text{ m/s})^2]$ from both sides, and then cancel the $\frac{1}{2}M$'s to get $(5\text{ m/s})^2(1-.2) = v_{1f}^2 + v_{2f}^2$, which simplifies to

$$20\text{ m}^2/\text{s}^2 = v_{1f}^2 + v_{2f}^2. \quad (*)$$

Now take Eq. (1) and cancel out the M's to get $5\text{ m/s} = v_{1f} + v_{2f}$, and hence $v_{1f} = 5\text{ m/s} - v_{2f}$. Substitute this expression for v_{1f} into (*), and leave out the units, to get $20 = (5 - v_{2f})^2 + v_{2f}^2$, which eventually simplifies to $2v_{2f}^2 - 10v_{2f} + 5 = 0$. The quadratic formula immediately gives

$$v_{2f} = \frac{10 \pm \sqrt{100 - 4(2)(5)}}{4} = \frac{10 \pm \sqrt{60}}{4} = \frac{10 \pm 7.746}{4} = 4.4\text{ m/s or } 0.6\text{ m/s}.$$

End of algebra.

Which of these two "answers" is correct? That is not easy to see. It is $v_{2f} = 4.4\ m/s$. Here is how I know. Look at Eq. (1) above. It tells us that if $v_{2f} = 4.4\ m/s$, then the final speed of ball 1 is $v_{1f} = 0.6\ m/s$. This makes sense, because it means ball 2 stays "ahead" of ball 1.

But suppose I "guessed" that the correct answer for ball 2 is $v_{2f} = 0.6\ m/s$. Then, from Eq. (1), the final speed of ball 1 would be $v_{1f} = 4.4\ m/s$. Ball 1 would be going in the same direction as ball 2, but faster. So, ball 1 would have to "pass" ball 2. Since the balls collide head-on, the only way ball 1 could pass ball 2 would be to tunnel right through ball 2. That cannot happen. Since balls 1 and 2 end up traveling in the same direction, ball 1 cannot end up faster than ball 2. For this reason, the physically sensible answer is $v_{2f} = 4.4\ m/s$ and $v_{1f} = 0.6\ m/s$, instead of vice versa.

QUESTION 13-2

Block 1, initially motionless at the top of a ramp of height H, slides down the frictionless ramp and then along the frictionless floor. Block 1 collides *elastically* with block 2, which was initially sitting motionless on the floor. Block 2 is heavier than block 1. Consequently, after the collision, block 1 slides back towards the ramp, and then slides part way up the ramp. Block 1 has mass m, while block 2 has mass $5m$.

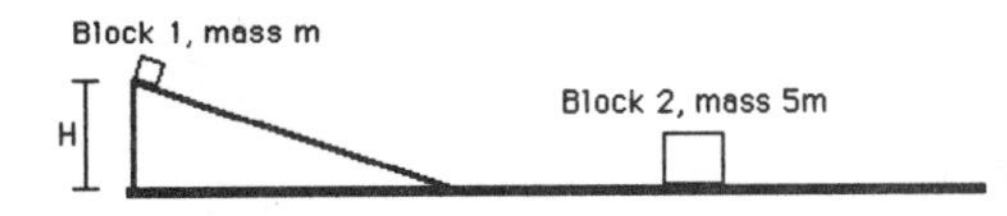

What peak height h does block 1reach as it slides back up the ramp? (As usual, by "height," I mean a vertical distance off the ground, not the distance traveled along the ramp.) Express your answer for h in terms of H. *If you do not have time to complete the algebra, clearly indicate your steps and show how you could use the results from your uncompleted algebra to reach the final answer.*

ANSWER 13-2

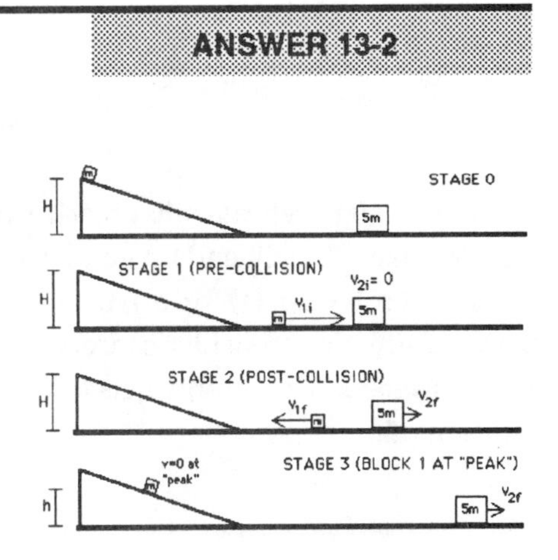

Before writing any formulas, let us think through what happens, using drawings. Block 1 starts off with a certain amount of gravitational potential energy. As it slides down the ramp, its potential energy converts into kinetic energy; the block speeds up. When block 1 collides with block 2, block 1 "gives" some of its kinetic energy to block 2, and "keeps" the rest of its kinetic energy. (Because the collision is elastic, none of the kinetic energy converts into heat.) So, after the collision, block 1 slides backwards with a certain percentage of its pre-collision kinetic energy, while block 2 slides slowly rightward. As block 1 slides back up the ramp, all of its kinetic energy eventually converts into potential energy, at which point block 1 momentarily comes to rest.

To see how the final height of block 1 depends on the mass of block 2, play with the computer simulation. You will see that, the heavier block 2 is, the faster block 1 rebounds during the collision, and therefore the higher block 1 slides up the ramp.

Notice that you cannot solve the problem in one step by setting the initial (stage 0) potential energy equal to the final (stage 3) total energy, because that equation would contain two unknowns, h and v_{2f}. *Do not forget about the kinetic energy of block 2.* As usual in complex problems, you must work step by step through the different stages of the physical process. The drawings above suggest a three-part strategy:

(A) Stage 0 to stage 1: Use conservation of energy to find the speed of block 1 after it slides down the ramp, but before it collides with block 2. Call that speed v_{1i}. Conservation of momentum does not work in this step, because the external force of gravity increases the momentum of block 1 as it slides down the ramp.

(B) Stage 1 to stage 2: To deal with the collision, start with conservation of momentum, and also use conservation of energy if necessary, to find the post-collision velocity of block 1. Call this velocity v_{1f}, the speed with which block 1 slides backwards towards the ramp.

(C) Stage 2 to stage 3: Again use conservation of energy to figure out how high up the ramp block 1 slides. Conservation of momentum does not apply here, for the same reason as in step A above.

Each of these steps is not impossibly difficult. The difficult part is formulating the overall strategy. Anyway, let us implement (A) through (C).

Step A: Stage 0 → Stage 1. Energy conserved as block slides down ramp.

As block 1 slides down the ramp, all of its initial potential energy converts into kinetic energy. No heat gets generated, because the block slides frictionlessly. Setting the energy at stage 0 equal to the energy at stage 1, we get

$$E_0 = E_1$$
$$K_0 + U_0 = K_1 + U_1$$
$$0 + mgH = \frac{1}{2}mv_{1i}^2 + 0.$$

Solve for v_{1i} to get $v_{1i} = \sqrt{2gH}$.

Step B: Stage 1→ Stage 2. Collision reasoning.

To find v_{1f}, let us start with conservation of momentum. Picking rightward as positive, I get

$$mv_{1i} + 0 = -mv_{1f} + 5mv_{2f}$$
$$m\sqrt{2gH} + 0 = -mv_{1f} + 5mv_{2f}. \tag{1}$$

The minus sign indicates that block 1 moves leftward after the collision. (If I do not put in that minus sign "by hand," I will end up getting a negative value for v_{1f} when I solve for it mathematically.) We cannot yet solve Eq. (1) for v_{1f}, because that equation contains a second unknown, namely v_{2f}. To gather more information, consider conservation of energy. Since the collision is elastic, no heat gets generated, and hence no kinetic energy gets lost. Also, the potential energy of the blocks does not change during the collision. So,

$$E_1 = E_2$$
$$\frac{1}{2}mv_{1i}^2 + 0 = \frac{1}{2}mv_{1f}^2 + \frac{1}{2}(5\text{ m})v_{2f}^2 \tag{2}$$
$$\frac{1}{2}m(2gH) + 0 = \frac{1}{2}mv_{1f}^2 + \frac{1}{2}(5\text{ m})v_{2f}^2,$$

since $v_{1i} = \sqrt{2gH}$ from step A. Equations (1) and (2) are two equations in two unknowns, v_{1f} and v_{2f}. So, we can solve for v_{1f}. The algebra takes longer than usual. Some instructors would not penalize you more than a point or two for simply writing, "since (1) and (2) contain only two unknowns, I can solve for v_{1f}, and use it in the next step." Check your instructor's policy on this. For the record, the algebra eventually yields $v_{1f} = \frac{2}{3}v_{1i} = \frac{2}{3}\sqrt{2gH}$. That is the speed of block 1 as it slides back towards the ramp. It "rebounds" with two thirds of its pre-collision speed.

Step C: Stage 2→ Stage 3. Energy conserved as block 1 slides up ramp.

Some students wonder whether to include block 2 in their energy conservation equation in this part of the problem. You can if you want, but you do not have to. Here is why. After the collision, block 2 no longer interacts with block 1. Block 1 behaves as if block 2 did not exist. So, you could write an energy conservation equation for block 1 alone. However, if you include a kinetic energy term for block 2, it does not matter. After the collision, block 2 slides forever rightward with constant velocity v_{2f}. So, to account for block 2, you would include a $\frac{1}{2}(5m)v_{2f}^2$ term on *both* sides of the energy conservation equation written below. Those terms would cancel.

As block 1 ascends the ramp, all of its kinetic energy converts into potential energy. At its "peak," block 1 momentarily comes to rest. So, setting the stage 2 energy equal to the stage 3 energy, looking only at block 1, gives

$$E_2 = E_3$$
$$K_2 + U_2 = K_3 + U_3$$
$$\frac{1}{2} = mv_{1f}^2 + 0 = 0 + mgh,$$

where h is the block's "peak height" on the ramp, i.e., the height at which it momentarily stops before sliding back down. Solving for h yields $h = \frac{v_{1f}^2}{2g}$. If you did not complete the algebra in step B, then this is your final answer, provided that you correctly write down the equations needed to find v_{1f}, and provided you clearly describe (in words) how to solve for v_{1f} using those equations. Since I completed the algebra in step B, I can set $v_{1f} = \frac{2}{3}\sqrt{2gH}$ to get

$$h = \frac{v_{1f}^2}{2g} = \frac{\left(\frac{2}{3}\sqrt{2gH}\right)^2}{2g} = \frac{4}{9}H.$$

So, when block 2 is five times heavier than block 1, block 1 rebounds fast enough to slide about half way up the ramp. The computer simulation confirms this, and lets you change the mass of block 2, to see how the "rebound height" changes. What do you think happens if block 2 is hundreds of times heavier than block 1? If it is only a little heavier?

QUESTION 13-3

A bomb has lots of chemical energy stored inside of it. When a certain chemical reaction occurs, the bomb explodes, releasing the chemical energy. During the explosion, about 25% of the initial chemical energy gets converted into heat and sound energy. The rest of the initial chemical energy goes into the kinetic energy of the bomb fragments.

Suppose a very small bomb, of mass $M = 3.0$ kg and with 1000 joules of chemical energy stored inside it, sits motionless. When the bomb explodes, it breaks into two fragments. The first fragment, which has mass $m_1 = 2.0$ kg, flies northward.

What is the speed and direction of the second fragment?

(Yes, you have enough information. Trust me.)

ANSWER 13-3

Let us first think intuitively. Imagine two skaters, Ann and Bob, on slippery ice. They stand motionless right next to each other. If Ann pushes Bob northward, what happens to Ann? She "recoils" southward. Analogously, if one fragment flies north, then the other fragment, we expect, flies south. Let us write down the conservation of momentum equation to see if this intuition holds. The initial momentum, i.e., the momentum of the unexploded bomb, is $\mathbf{p}_0 = 0$. So, the sum of the momenta of the two fragments must be zero. In vector notation,

$$\mathbf{p}_0 = 0 = \mathbf{p}_{1f} + \mathbf{p}_{2f}. \tag{1}$$

From Eq. (1), we see that $\mathbf{p}_{2f}$ must point southward. Why? Since $\mathbf{p}_{1f}$ and $\mathbf{p}_{2f}$ add up to 0, $\mathbf{p}_{2f}$ must be a vector the same length as $\mathbf{p}_{1f}$, but pointing in the opposite direction. In rough terms, the two vectors must "cancel."

Given this insight, we can now make an "after" diagram. (The "before" diagram is not worth drawing, because the bomb just sits motionless.) Fragment 1 has mass $m_1 = 2.0$ kg. The two fragments taken together must weigh about the same as the unexploded bomb. Therefore, since the unexploded bomb had mass 3.0 kg, fragment 2 must have mass $m_2 = 1.0$ kg.

AFTER EXPLOSION

v_{1f}

2 kg

1 kg

v_{2f}

We are looking for v_{2f}, the final velocity of fragment 2. Let us invoke the usual collision/explosion reasoning. Start with conservation of momentum. Letting north be the "positive" direction, and substituting our information into momentum conservation Eq. (1), gives

$$0 = (2\text{ kg})v_{1f} - (1\text{ kg})v_{2f}. \tag{1'}$$

Because we do not know v_{1f}, we cannot immediately solve Eq. (1′) for v_{2f}. So, let us use energy information. We are told that 25% of the bomb's initial chemical energy transforms into heat and sound. The remaining 75% of the bomb's chemical energy converts into the kinetic energy of the two fragments. Since the bomb initially packed 1000 joules of chemical energy, we know that .75(1000 joules) = 750 joules of kinetic energy end up in the two fragments. So

$$\begin{aligned} 750\text{ J} &= K_{1f} + K_{2f} \\ &= \frac{1}{2}(2\text{ kg})v_{1f}^2 + \frac{1}{2}(1\text{ kg})v_{2f}^2. \end{aligned} \tag{2}$$

Equations (1′) and (2) are two equations in two unknowns, v_{1f} and v_{2f}. We are done, except for algebra.

Algebra starts here. Solving Eq. (1′) for v_{1f} gives $v_{1f} = \frac{v_{2f}}{2}$. Substitute that into Eq. (2) to get

$$\begin{aligned} 750\text{ J} &= \frac{1}{2}(2\text{ kg})\left(\frac{v_{2f}}{2}\right)^2 + \frac{1}{2}(1\text{ kg})v_{2f}^2 \\ &= \left(\frac{3}{4}\text{ kg}\right)v_{2f}^2, \end{aligned}$$

and hence

$$v_{2f} = \sqrt{\frac{750\text{ J}}{\frac{3}{4}\text{ kg}}} = 32\text{ m/s}.$$

End of algebra.

You could substitute that answer for v_{2f} back into Eq. (1), and solve for v_{1f}, to get $v_{1f} = 16$ m/s. In words, since fragment 1 is twice as heavy, it ends up going only half as fast, because both fragments carry the same "oomph" (momentum). What would happen if both fragments were equally massive?

QUESTION 13-4

A block of mass m sits at the top of a ramp of height H and ramp angle θ. I push the block, giving it an initial speed v_0 down the ramp. Due to friction, the block does not speed up as it slides down the ramp; instead, it slides down the ramp at constant velocity. After sliding off the ramp, the block skids along the floor, quickly coming to rest due to friction.

The coefficient of friction between the block and the ramp equals the coefficient of friction between the block and the floor, because the ramp and floor are coated with the same material.

(a) What is the coefficient of kinetic friction?

(b) As the block slides along the ramp and then along the floor, how much heat gets produced due to the frictional rubbing? (By "heat," I actually mean all forms of dissipative energy. In this case, most of that energy really does take the form of heat.)

ANSWER 13-4

(a) We are trying to find μ, the coefficient of kinetic friction. How can we find it? Given the focus of this chapter, you might try conservation of energy. But energy conservation can only tell us how much heat gets produced. Here, we must revert to the work-energy theorem, or old-fashioned force considerations.

Consider the work energy theorem, $W_{net} = \Delta K$. While sliding down the ramp, the block neither gains nor loses speed. So, its kinetic energy does not change: $\Delta K = 0$. Therefore, the net work down on it is $W_{net} = 0$. Since $W_{net} = F_{\parallel net}\Delta x$, it follows that the net force on the block as it slides down the ramp is $F_{\parallel net} = 0$.

Straightforward force-and-acceleration reasoning leads to the same conclusion. Because the block neither speeds up nor slows down, its acceleration on the ramp is $a = 0$. Therefore, the net force on it is $F_{net} = 0$.

Apparently, no matter how we approach this problem, we must consider forces. So, let us start with a free-body diagram.

We are using old-fashioned force considerations instead of energy considerations. In some problems, forces get you to the answer quicker. This is one of them.

As just noted, since the block stays at constant velocity in the x-direction, $a_x = 0$. It has no y-acceleration either, because it neither jumps off nor burrows into the ramp: $a_y = 0$. So, Newton's 2nd law gives us

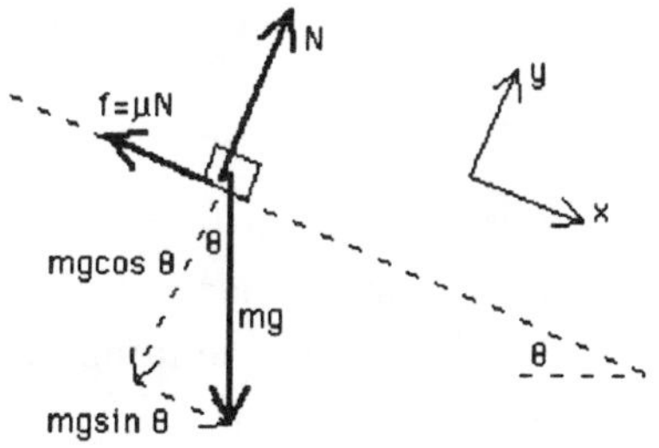

$$\begin{aligned}\sum F_x &= ma_x \\ mg\sin\theta - \mu N &= 0\end{aligned} \qquad (1)$$

$$\begin{aligned}\sum F_y &= ma_y \\ N - mg\cos\theta &= 0.\end{aligned} \qquad (2)$$

These two equations contain only two unknowns, μ and N. Therefore, we can find μ. Solve Eq. (2) for N to get $N = mg\cos\theta$. Then solve Eq. (1) for μ, and plug in that normal force, to obtain

$$\mu = \frac{mg\sin\theta}{N} = \frac{mg\sin\theta}{mg\cos\theta} = \tan\theta.$$

(b) To solve this part, you could use force reasoning to calculate how far the block travels before coming to rest. You could then use the definition of work to figure out how much work friction performs on the block while it slides down the ramp, and how much work friction performs while the block slides across the floor. The total work done by friction equals the heat generated by friction. (Intuitively, the frictional work "saps" energy from the block; and all this sapped energy converts into heat.) So, you would set Heat $= W_{\text{by fric}} = W_{\text{by fric, ramp}} + W_{\text{by fric, floor}}$.

Although this strategy works fine, energy conservation provides a shortcut.

The block starts off with both kinetic and potential energy. As it slides down the ramp, the block loses all of its potential energy. But unlike usual, the potential energy does not convert into kinetic energy; the block does not speed up. Instead, the block's potential energy gets converted into heat. This heat gets generated as the block rubs against the ramp during its slide. Then, while sliding across the floor, the block loses all of its kinetic energy. That kinetic energy gets converted to heat. You should draw pictures. I will just draw the "initial" and "final" configurations, leaving out the intermediate one (with the block at the bottom of the ramp).

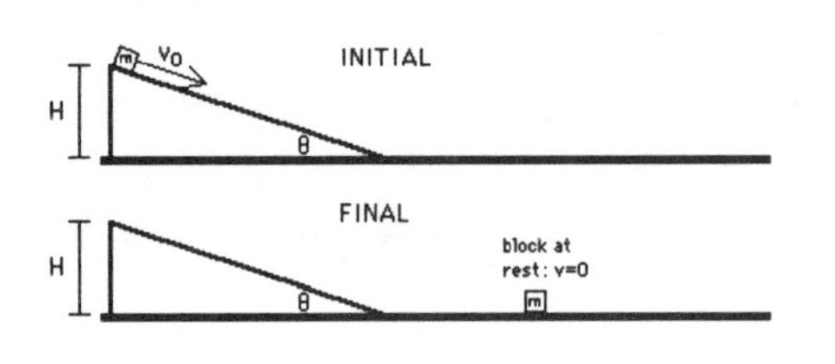

In summary, all of the block's initial energy gets converted into heat, since the block has no kinetic or potential energy after it comes to rest on the floor.

Formally,

$$E_i = E_f$$
$$K_i + U_i = K_f + U_f + \text{Heat}_f$$
$$\frac{1}{2}mv_0^2 + mgH = 0 + 0 + \text{Heat}_f.$$

So, the heat generated is $\text{Heat}_f = \frac{1}{2}mv_0^2 + mgH$.

Notice that the ramp angle did not matter, nor did the distance traveled by the block, nor did μ. Intuitively, if the block slides farther along the floor, that is because the floor is slippery, i.e., μ is small. A small μ implies that the block rubs less "frictionfully" against the floor, and hence, heat gets generated at a small rate. But of course, when μ is small, the block slides longer. In summary, when μ is small, heat gets generated at a low rate, but for a long time (since the block slides far). By contrast, if μ is big, then heat gets generated at a high rate, but only for a short time, because the block quickly comes to rest. In both cases, the same amount of heat gets produced.

QUESTION 13-5

A small frictionless block, of mass m, slides rightward at initial speed v_0 towards a ramp of mass M. The ramp, unlike most of the ramps in this course, is not attached to the floor. Instead, the ramp is free to slide frictionlessly along the floor. So, as the block slides up the ramp, the ramp starts sliding rightward. The block's initial speed is small enough that the block does not slide all the way to the top of the ramp. Instead, the block slides part way up, and then slides back down.

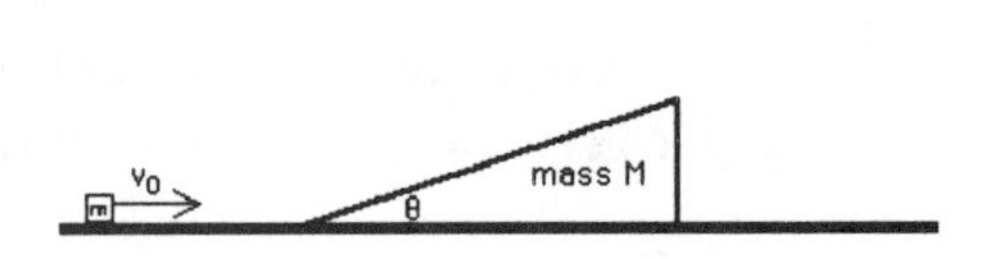

Express your answers in terms of g and the given constants–m, M, and v_0.

(a) What is the peak height reached by the block? In other words, when the block slides up the ramp, what is the highest point it reaches before turning around and sliding back down? I am looking for the vertical height off the floor.

(b) What is the final speed of the ramp, i.e., the ramp's speed after the block has slid "backwards" off the ramp? *Set up, but do not solve,* the relevant equation or equations.

ANSWER 13-5

To solve this, we must carefully picture what is going on. As the block starts to ascend the ramp, the ramp starts sliding rightward. Eventually, the block reaches its peak. At this point, the block is neither sliding up nor sliding down the ramp; it is just sitting still with respect to the ramp, which now slides rightward. Then the block slides down the ramp, which continues to speed up. Finally, the block leaves the ramp and slides leftward along the floor, while the ramp continues sliding rightward. See the computer simulation.

(a) To solve this part, you will invoke momentum conservation and also energy conservation. Usually, these conservation laws provide enough information to solve completely. But here, you need *additional* information, information you can obtain only by having a physical insight about the motion. Specifically, you must figure out a crucial fact about the block's speed at its peak (on the ramp).

You might think the block's speed at its peak is $v=0$. That is not quite right. At its peak, the block is motionless *with respect to the ramp.* But the ramp itself is sliding rightward with some speed I will call v_1. So, at its peak, the block gets "carried along" at that same speed, v_1. (Similarly, if you sit still inside a car cruising at 50 mph, then you get "carried along" at 50 mph with respect to the road.) In summary: When the block reaches its peak, its velocity is the same as the ramp's. See the computer simulation.

Given this insight, we can now proceed in the usual way. Sketch the different stages of the physical process. I will let "u_f" denote the ramp's final velocity.

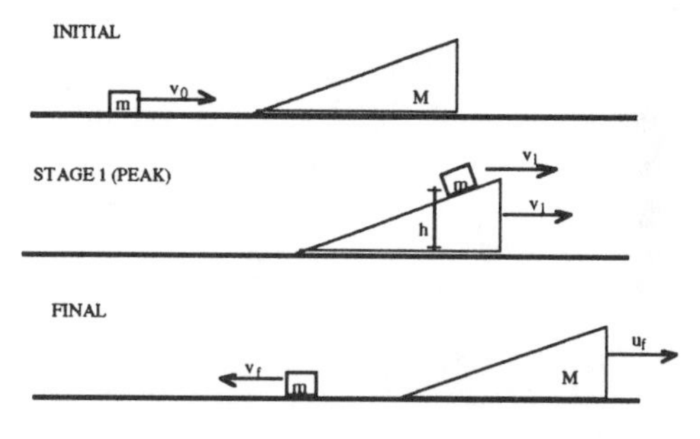

We are solving for h. So, let us start with energy conservation, going from the initial stage to stage 1. Since everything slides frictionlessly, no heat gets generated. As just discussed, in stage 1, the block and ramp "share" the same (unknown) velocity v_1. So,

$$E_0 = E_1$$
$$K_{0,\text{ block}} + K_{0,\text{ ramp}} + U_{0,\text{ block}} = K_{1,\text{ block}} + K_{1,\text{ ramp}} + U_{1,\text{ block}} \tag{1}$$
$$\frac{1}{2}mv_0^2 + 0 + 0 = \frac{1}{2}mv_1^2 + \frac{1}{2}Mv_1^2 + mgh.$$

Crucially, at stage 1, both the block *and* the ramp have kinetic energy. Since this equation contains two unknowns, h and v_1, we need more information. So, we can use momentum conservation. Or can we?

You might think that momentum is not conserved as the block slides up the ramp, because an external force, gravity, acts on the block. This is partly correct: Gravity does indeed muck up the system's momentum. But remember, momentum is a vector; it has both x- and y-components. In this problem, we are concerned only with *horizontal* (x-directed) momentum. That is because, at all three relevant stages, the block and ramp move entirely horizontally. Gravity is a y-directed force. It changes the system's y-directed momentum as the block slides up the ramp. But a

y-directed force does not affect the system's x-momentum. The only x-directed force component felt by the ramp comes from the normal exerted by the block. And the only x-directed force component felt by the block comes from the normal exerted by the ramp. These are *internal* forces, by which I mean forces exerted by one part of the system on another. So, x-momentum is conserved.

From the diagrams, we immediately get

$$\begin{aligned} p_0 &= p_1 \\ mv_0 + 0 &= mv_1 + Mv_1. \end{aligned} \tag{2}$$

Since we now have two equations in the two unknowns, h and v_1, we can solve for h.

Algebra starts here. Solve Eq. (2) for v_1 to get $v_1 = \frac{mv_0}{(M+m)}$. Substitute that into Eq. (1) to get

$$\begin{aligned} \frac{1}{2}mv_0^2 &= \frac{1}{2}m\left(\frac{mv_0}{M+m}\right)^2 + \frac{1}{2}M\left(\frac{mv_0}{M+m}\right)^2 + mgh \\ &= \frac{1}{2}(m+M)\left(\frac{mv_0}{M+m}\right)^2 + mgh \\ &= \frac{1}{2}\left(\frac{m^2}{M+m}\right)v_0^2 + mgh. \end{aligned}$$

Now isolate h to get

$$\begin{aligned} h &= \frac{\frac{1}{2}mv_0^2 - \frac{1}{2}\left(\frac{m^2}{M+m}\right)^2 v_0^2}{mg} \\ &= \frac{\frac{1}{2}mv_0^2\left(1-\frac{m}{M+m}\right)}{mg} \\ &= \frac{\frac{1}{2}v_0^2\left(1-\frac{m}{M+m}\right)}{g} \\ &= \frac{\frac{1}{2}v_0^2\left(\frac{M+m}{M+m}-\frac{m}{M+m}\right)}{g} \qquad \left[\text{since } 1 = \frac{M+m}{M+m}\right] \\ &= \frac{v_0^2}{2g}\left(\frac{M}{m+M}\right). \end{aligned}$$

End of algebra.

Notice that, in the limit as M gets *much* bigger than m, the fraction $\frac{M}{m+M}$ approaches 1, and hence h approaches $\frac{v_0^2}{2g}$. That is the same h we would get if the ramp were nailed down, as you can confirm using energy conservation. Physically, a very heavy ramp stay almost motionless, and therefore *might as well* be nailed down. On the other hand, what happens if M is very small? Play around with the computer simulation.

(b) There is a clever shortcut, using some of your reasoning from part (a). Before showing you the trick, however, I will address this problem the long but straightforward way.

This is a disguised collision problem. The block and ramp stay in contact for a while, and then separate. In a pool ball collision, the "contact" lasts only a few hundredths of a second. Here, the contact lasts several seconds, perhaps. But the contact time does not matter: Conservation of momentum and energy still apply. Indeed, since the block slides frictionlessly along the ramp, no heat gets generated; the "collision" is elastic! Therefore, the answer to this problem is the same as if a pool ball of mass m and speed v_0 elastically collided head-on with a heavier pool ball of mass M. Your textbook derives these elastic collision formulas.

"Before" the collision corresponds to my "initial" drawing, while "after" the collision corresponds to my final drawing. So, I will ignore the intermediate stage 1, and set $E_0 = E_f$ and $p_0 = p_f$. (Of course, I would get the same answers setting $E_1 = E_f$ and $p_1 = p_f$.)

Let us start with momentum. Notice that I am calling the final velocity of the ramp "u_f." From the diagrams above, choosing rightward as positive,

$$\begin{aligned} p_0 &= p_f \\ mv_0 + 0 &= -mv_f + Mu_f. \end{aligned} \tag{3}$$

The minus sign indicates my "guess" that the block ends up sliding backwards. (If I am wrong, the math will yield a negative value for v_f.) This equation contains two unknowns, the block's final velocity and the ramp's final velocity. To obtain more information, invoke energy conservation. Since no heat gets generated, and since the masses end up with the same potential energy they started with $(U_0 = U_f = 0)$, we get

$$\begin{aligned} E_0 &= E_1 \\ K_{0,\text{ block}} + K_{0,\text{ ramp}} &= K_{f,\text{ block}} + K_{f,\text{ ramp}} \\ \frac{1}{2}mv_0^2 + 0 &= \frac{1}{2}mv_f^2 + \frac{1}{2}Mu_f^2. \end{aligned} \tag{4}$$

Eqs. (3) and (4) are two equations in the two unknowns, v_f and u_f. So, we could solve for those final velocities. To conserve paper, I will not do so.

Before closing, let me describe the tricky shortcut you could have used. How do you think the ramp's final speed relates to its speed at stage 1, when the block reaches the peak? In symbols, how does u_f relate to v_1? By visualizing the process, perhaps with the help of the computer simulation, you can see that $u_f = 2v_1$. Here is why:

As the block slides up the ramp, the block exerts a normal force on the ramp for time t_{up}. A component of this normal force points rightward. As a result, the ramp speeds up by v_1. When the block slides down the ramp, it exerts the *same* normal force on the ramp, as you can confirm with a force diagram. Also, the block takes the same time to slide down as it took to slide up: $t_{\text{down}} = t_{\text{up}}$. So, during the descent, the ramp feels the same normal force for the same amount of time as when the block ascended. Therefore, the ramp gains as much speed during the descent as it gained during the ascent.

In summary: The ramp gains speed v_1 as the block slides up; and then the ramp gains an additional amount of speed v_1 as the block slides back down. So, the ramp ends up with speed $u_f = 2v_1$.

Using this fact, you could have quickly solved part (b) by using your part (a) equations to obtain v_1, and then doubling that value to get u_f.

QUESTION 13-6

In deep outer space, a tiny space probe of mass $M = 50$ kg travels at speed $v_0 = 2.0$ m/s. (The space probe contains no astronauts.) At a certain instant, an internal explosion occurs, splitting the space probe into two pieces of mass $m_1 = m_2 = 25$ kg. During the explosion, about 100 joules of kinetic energy are given to the system. In other words, after the explosion, the system has 100 more joules of kinetic energy than it started with. Neither piece leaves the original line of motion of the unexploded space probe.

Find the post-explosion velocity of each piece.

ANSWER 13-6

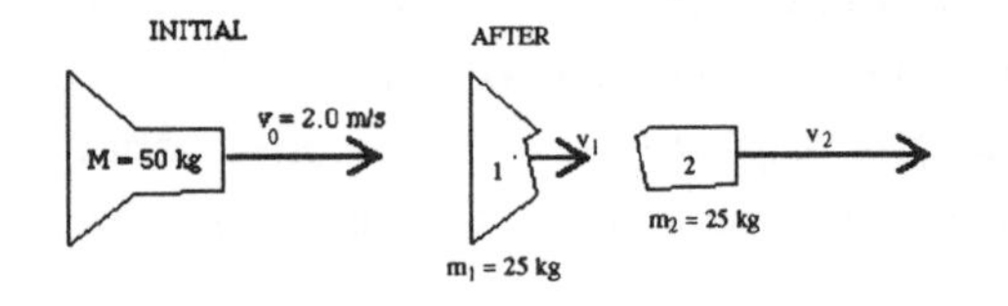

As always in this kind of collision/explosion problem, sketch "before" and "after" diagrams.

I have chosen rightward as the positive direction. In the diagram, I have assumed that v_1, the final speed of piece 1, is forward (i.e., positive). If this assumption is wrong, I will end up getting a negative value for v_1.

As always in problems where you cannot use forces, the "tools" we have at our disposal are conservation of energy and conservation of momentum. Let us start with momentum conservation. Because all the motion happens along a straight line, we do not need to use vector components. Since both pieces have the same mass, let me call that mass m.

$$p_i = p_f \qquad (1)$$
$$Mv_0 = mv_1 + mv_2.$$

Because this equation contains two unknowns, v_1 and v_2, we must also consider energy conservation. The problem tells us that 100 joules of kinetic energy get imparted to the pieces. In other words, 100 joules of the space probe's internal energy (probably chemical energy) get converted into kinetic energy during the explosion. This does not mean the system ends up with 100 J of kinetic energy. It means the system ends up with 100 more joules of kinetic energy than it started with. In summary,

$$E_i = 100 \text{ J of chemical energy} + \text{space probe's initial kinetic energy},$$
$$E_f = \text{pieces' final kinetic energy}.$$

So,

$$E_i = E_f \qquad (2)$$
$$100\text{ J} + \frac{1}{2}Mv_0^2 = \frac{1}{2}mv_1^2 + \frac{1}{2}mv_2^2.$$

A common mistake is to write $\frac{1}{2}Mv_0^2 = \frac{1}{2}mv_1^2 + \frac{1}{2}mv_2^2 + 100$ J. You can see why that is wrong by substituting in the actual initial kinetic energy, $\frac{1}{2}Mv_0^2 = \frac{1}{2}(50\text{ kg})(2\text{ m/s})^2 = 100$ J. Solving the incorrect equation for the final kinetic energy would give $\frac{1}{2}mv_1^2 + \frac{1}{2}mv_2^2 = 0$, instead of the correct value given by Eq. (2), $\frac{1}{2}mv_1^2 + \frac{1}{2}mv_2^2 = 200$ J. Remember, the pieces end up with *more* kinetic energy than the system started with.

Now we have two equations in the two unknowns, v_1 and v_2. We are done with the physics thinking, at least until we complete some algebra.

Algebra starts here. Isolating v_2 in Eq. (1) gives

$$v_2 = \frac{Mv_0 - mv_1}{m} = \frac{(50\text{ kg})(2.0\text{ m/s}) - (25\text{ kg})v_1}{25\text{ kg}} = 4\text{ m/s} - v_1.$$

Substitute that expression for v_2 into Eq. (2), and plug in the numbers, to get

$$100\text{ J} + \frac{1}{2}(50\text{ kg})(2.0\text{ m/s})^2 = \frac{1}{2}(25\text{ kg})v_1^2 + \frac{1}{2}(25\text{ kg})(4\text{ m/s} - v_1)^2$$
$$100\text{ J} + 100\text{ J} = (12.5\text{ kg})v_1^2 + (12.5\text{ kg})[16\text{ m}^2/s^2 - (8\text{ m/s})v_1 + v_1^2)$$
$$200\text{ J} = (25\text{ kg})v_1^2 - (100\text{ kg}\cdot\text{m/s})v_1 + 200\text{ J}.$$

Subtract 200 J from both sides, and divide through by 25 kg, to get

$$0 = v_1^2 - (4\text{ m/s})v_1$$
$$= v_1(v_1 - 4\text{ m/s}),$$

from which we get $v_1 = 0$ or 4.0 m/s.

End of algebra.

Both of those answers are "right." If we say $v_1 = 0$, then solving for v_2 gives $v_2 = 4.0\text{ m/s}$. And if we say $v_1 = 4.0\text{ m/s}$, then solving for v_2 gives $v_2 = 0$. Our conclusion is that, after the explosion, one piece moves forward at 4.0 m/s, while the other piece sits motionless in space. In my drawing, piece 1 is "ahead" of piece 2, and hence, it is the one that ends up moving forward.

QUESTION 13-7

On a frictionless track, a block of mass $m = 0.10\text{ kg}$ slides rightward at $v_0 = 1.0\text{ m/s}$. Just ahead of it is a larger block of mass $M = 9.0\text{ kg}$, moving in the same direction with the same speed. At the right end of the track is a rubber wall. The large mass bounces off the wall perfectly elastically, and then slams into the small block, again perfectly elastically.

(a) Set up, but do not solve, the equation or equations needed to find the speed v of the small block after the large mass slams into it.

(b) Without doing the math, estimate the value of v, and justify your answer.

ANSWER 13-7

(a) A wall is effectively an infinite mass, because it is "nailed down" to the Earth. When something collides with a wall, the wall does not budge; it acquires no kinetic energy. And during an elastic collision, no kinetic energy is lost. Therefore, during its collision with the wall, the large mass keeps *all* of its kinetic energy. So, the large mass rebounds off the wall at 1.0 m/s, the same speed with which it hit the wall. This makes sense, because elastic collisions are "perfectly bouncy." By contrast, if the large mass had collided inelastically with the wall, it would have rebounded with less than its initial speed.

So, immediately before colliding with the small block, the large mass moves leftward at speed $v_0 = 1.0\text{ m/s}$, the same speed with which the small block travels rightward. Intuitively speaking, we expect the large mass to "win" the collision. We expect that after the collision, the large mass continues to travel leftward, because the small block has insufficient "oomph" to make the large mass turn around. As always, to keep track of all this information, sketch the relevant stages of the process. Remember, the large mass is moving leftward because it rebounded off the (undrawn) wall at the right end of the track.

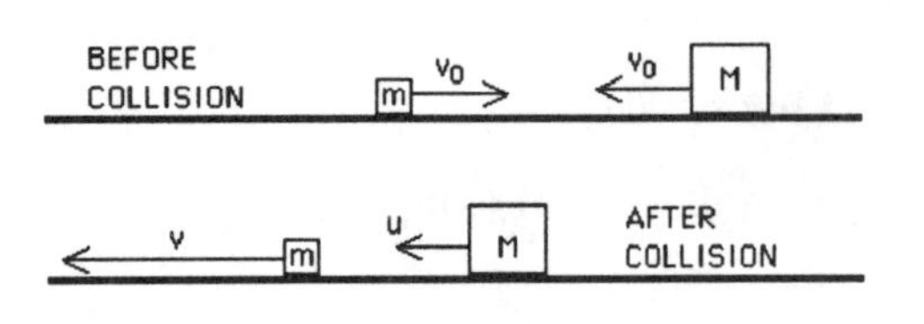

I have let "u" denote the post-collision speed of the large mass, so as not to confuse it with "v," the post-collision speed of the small block. We are solving for v, given $m = 0.10\text{ kg}$, $M = 9.0\text{ kg}$, and $v_0 = 1.0\text{ m/s}$.

I will use standard collision/explosion/interaction reasoning: Start with conservation of momentum, and then use conservation of energy if more information is needed.

Remember that momenta, unlike energies, have directions. I will pick rightward as positive. So, by momentum conservation,

$$\begin{aligned} p_{\text{before}} &= p_{\text{after}} \\ mv_0 - Mv_0 &= -mv - Mu. \end{aligned} \tag{1}$$

We cannot yet solve Eq. (1) for v, because that equation contains a second unknown, namely u.

To obtain more information, consider conservation of energy. During most collisions, heat gets generated. But since this collision is elastic, no heat (or sound, etc.) gets produced. Therefore, no kinetic energy is lost. Notice that energy has no direction; all kinetic energies are positive.

$$\begin{aligned} E_{\text{before}} &= E_{\text{after}} \\ K_0 + U_0 &= K_f + U_f + \text{No heat} \\ \left(\frac{1}{2}mv_0^2 + \frac{1}{2}Mv_0^2\right) + 0 &= \left(\frac{1}{2}mv^2 + \frac{1}{2}Mu^2\right) + 0 + 0. \end{aligned} \tag{2}$$

At this point, we have two equations in the two unknowns, v and u. Therefore, it is just a matter of algebra to solve for v, the small block's post-collision speed. You were not required to complete the algebra.

(b) The large mass is 90 times heavier than the small mass. Therefore, the collision should not slow down the large mass very much: it continues moving leftward at nearly 1.0 m/s. To see what this implies about the little block, it helps to switch your frame of reference. Imagine an observer riding leftward at 1.0 m/s. This observer "keeps up" with the big block. So, according to this observer, the large mass is (nearly) motionless the whole time. Before the collision, says this observer, the little block approaches the big block at speed $2v_0 = 2\text{ m/s}$. Why?

Imagine yourself driving northward at 50 mph, while your friend drives southward toward you, also at 50 mph. If the cars start 100 miles apart, they meet after just one hour. In other words, you "see" your friend as approaching you at 100 mph, the relative speed of the two cars. Similarly for the blocks.

What about after the collision? According to the observer, since the collision is elastic, the little block loses no kinetic energy during the collision. So, it rebounds leftward at speed $2v_0$. In other words, the rate at which the little block speeds away from the big mass after the collision is 2.0 m/s, the same relative speed with which it approaches the big mass.

Given this conclusion, we can switch back to our "original" observer, who sits motionless next to the air track. He agrees that, after the collision, m moves leftward at speed $2v_0 = 2.0\ \text{m/s}$ *with respect to* M. But block M is itself moving leftward at speed $v_0 \approx 1.0\ \text{m/s}$. So, block m's speed with respect to the air track is $2v_0 + v_0 = 3v_0 = 3.0\ \text{m/s}$.

This estimate is very close: the exact numerical answer works out to be $v = 2.96\ \text{m/s}$.

See your textbook for a deeper discussion of reference frames in collisions.

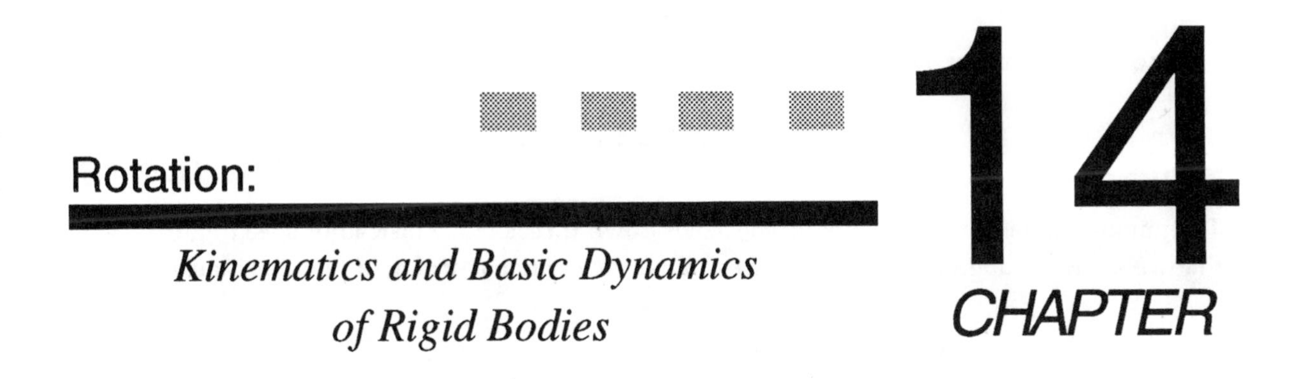

Rotation: Kinematics and Basic Dynamics of Rigid Bodies

CHAPTER 14

QUESTION 14-1

A huge turbine is rotating at 30 rad/s. When the power is turned off at time $t = 0$, the turbine decelerates due to air resistance and internal friction. It stops spinning at $t = 10$ seconds. Assuming the turbine decelerates at a steady rate, find

(a) the angle (in radians) through which the turbine rotates between $t = 0$ and $t = 10$ s, and

(b) the number of revolutions made by the turbine between $t = 0$ and $t = 10$ s.

ANSWER 14-1

(a) The problem tells us that the turbine's initial angular speed is $\omega_0 = 30$ rad/s, and its final angular speed is $\omega = 0$ rad/s. This process of coming to rest takes time $t = 10$ s. We need to calculate the angle ϕ through which it spins during those 10 seconds. (Some books use θ to denote the angle.)

I find rotational motion harder to conceptualize than linear motion. For this reason, I like to solve rotation problems *by thinking in terms of an analogy with a linear problem.* Let me demonstrate this technique, and you can decide if you like it.

First, we must think of an analogous linear problem. Here, a turbine's rotational speed slows down at a constant rate. So, by analogy, consider a car slowing down with constant deceleration. We know the car's initial and final speed (analogous to initial and final angular speed), as well as the time t it takes to slow down. We want to know how much distance x the car travels in that time, since distance (in linear motion) is analogous to angle (in rotational motion). How could we find x? Well, all of our old constant-acceleration formulas for x contain the acceleration. So we need to know the car's acceleration. Specifically, we could solve for acceleration using

$$v = v_0 + at. \tag{1}$$

Then, we could take that acceleration and substitute it into

$$x = x_0 + v_0 t + \frac{1}{2}at^2, \tag{2}$$

to solve for x. (We equally well could use $v^2 = v_0^2 + 2a\Delta x$.)

To deal with the turbine, we can use the rotational analogues of these linear kinematic formulas. First, we must find its angular acceleration using the rotational analog of Eq. (1), namely

$$\omega = \omega_0 + \alpha t. \tag{1*}$$

Solving for α immediately gives

$$\alpha = \frac{\omega - \omega_0}{t} = \frac{0 - 30 \text{ rad/s}}{10 \text{ s}} = -3.0 \text{ rad/s}^2.$$

The minus sign indicates that the turbine slows down (decelerates) instead of speeding up. Now that we know the angular acceleration, we can use the rotational analog of Eq. (2), namely

$$\begin{aligned} \phi &= \phi_0 + \omega_0 t + \frac{1}{2}\alpha t^2 \\ &= 0 + (30 \text{ rad/s})(10 \text{ s}) + \frac{1}{2}(-3.0 \text{ rad/s}^2)(10 \text{ s})^2 \qquad (2^*) \\ &= 150 \text{ rad.} \end{aligned}$$

This problem illustrates a general rotational problem-solving strategy: Think in terms of linear analogies. Here, we used the fact that angle is analogous to distance; angular speed is analogous to linear speed; and angular acceleration is analogous to linear acceleration. See your textbook for a detailed discussion of these analogies, complete with an "analogy table."

(b) One full revolution, i.e., one full turn around a circle, corresponds to an angle of 2π radians. So, if ϕ is expressed in radians, the corresponding number of revolutions is $\frac{\phi}{2\pi}$. Equivalently, you can think in terms of unit conversions: $150 \text{ rad} \times \frac{1 \text{ revolution}}{2\pi \text{ rad}} = 24$ revolutions.

QUESTION 14-2

Inside a small cyclotron, electromagnetic fields make an ion move in a circle of radius $R = 2.0$ m. The particle's initial speed, at $t = 0$, is $v_0 = 10$ m/s That is the regular (linear) speed, not the angular speed. The electromagnetic fields make the ion move faster and faster. In fact, the cyclotron is tuned in such a way that the ion's angular acceleration is known to be $a_{cyc} = 15 \text{ rad/s}^2$.

(a) What is the ion's angular speed at later time $t_1 = 5.0$ s?

(b) Between $t = 0$ and $t_1 = 5.0$ s, how much distance does the ion cover? This problem asks for a linear distance, not an angle.

ANSWER 14-2

(a) To solve this part, you must use a combination of rotational and linear kinematic reasoning. As shown in Question 14-1, you can simplify the rotational reasoning by thinking in terms of an analogy with a similar linear situation. But in this problem, you must also invoke formulas that are neither purely linear nor purely rotation. I am referring to the formulas relating linear speed to angular speed, and linear acceleration to angular acceleration.

The problem gives us an initial linear speed v_0, and asks for the *angular* speed at a later time. So, we must relate a linear speed v to an angular speed ω (or alternatively, a linear acceleration to an angular acceleration.)

To build up the relevant intuitions, consider two points on a spinning wheel. Point A sits on the edge of the wheel, while point B sits halfway between the center and the edge.

Suppose we spin the wheel. How does the *angular* speed ω_A of point A relate to the *angular* speed ω_B of point B? You might be tempted to say that ω_A is bigger, because A travels faster. Well, it is true that point A moves faster than point B. To see this, focus on the fact that A and B both trace a full circle in the same amount of time; but point A traces a *bigger* circle. Therefore, point A covers more distance per time than point B covers. This all goes to show that A's *linear* speed, v_A, is bigger than B's linear speed, v_B. In other words, A covers more *distance* per time. But angular speed is not distance per time; it is *angle* per time. And as just noted, A and B both trace out a full circle (2π radians) in the *same* time. So, both points share the same angular speed: $\omega_A = \omega_B$. In summary: Although all points on the wheel "spin together" and therefore share the same angular speed, the "outermost" points are moving faster, which means they have greater linear speed. To experience this, ride on a merry-go-round, first near the center, then near the edge.

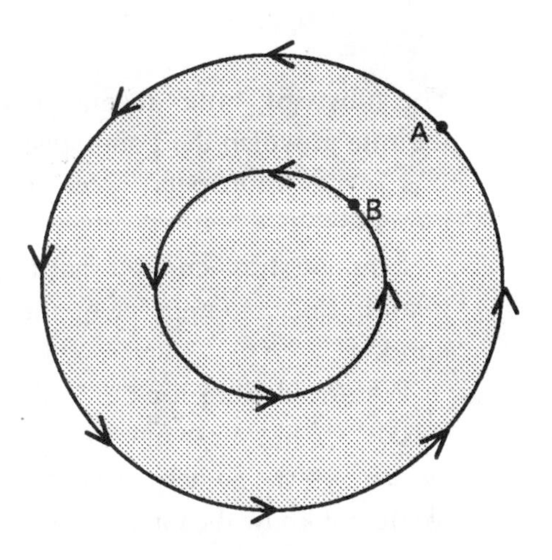

This still does not tell us how angular speed relates to linear speed. It just tells us that two points can have the same angular speed, *without* also having the same linear speed. Let us use intuitive reasoning to find the relationship between v and ω.

Well, intuitively, the faster the wheel spins, the faster any given point on the wheel moves. For instance, what happens to v_A when we double the angular speed of the wheel? Point A traces twice as many circles per second as it did previously. In other words, point A moves twice as fast. This goes to show that the linear and angular speeds are proportional: $v \sim \omega$.

Now let us see how v relates to r, the distance from the center of the circle to the point of interest. Suppose the wheel spins with angular speed ω. As shown above, point A covers more distance per time than point B, because it traces a bigger circle (more distance) in the same time. But how much more distance per time? Let us say point A is twice as far from the center as point B: $r_A = 2r_B$. It follows that the circle traced out by point A has exactly twice the circumference of the circle made by point B, since $C = 2\pi r$, and $r_A = 2r_B$. So, in the time it takes point B to cover distance $2\pi r_B$, point A covers *twice* that distance. Since speed is distance per time, this all goes to show that the speed of a point on the wheel is proportional to r, its distance from the center of the circle: $v \sim r$.

In summary, the linear speed of a point moving in a circle is proportional to its angular speed, and is also proportional to the point's distance from the center. So, it should come as no surprise that the relation between v and w is

$$v = \omega r. \tag{1}$$

This relationship holds only if the angular speed is expressed in radians per time. If you are given w in degrees per time or revolutions per time, convert to radians per time before using Eq. (1).

By similar reasoning, or simply by differentiating both sides of Eq. (1) with respect to t, you can obtain the relationship between linear acceleration and angular acceleration. Since $\frac{dv}{dt} = a_{\text{tang}}$, and $\frac{d\omega}{dt} = \alpha$, we get

$$a_{tang} = \alpha r. \tag{2}$$

The subscript "tang" means "tangential," the component of the acceleration *along the circle,* i.e., the rate at which the point speeds up or slows down. You must carefully distinguish a_{tang} from a_{radial}, the "centripetal" acceleration. I will come back to this distinction later, and belabor it mercilessly.

I cannot stress enough that Eqs. (1) and (2) are not mere analogies. They are actual equations relating linear to angular quantities. Using one of these relations, we can solve the problem at hand.

We are looking for the ion's "final" angular speed, given its angular acceleration. The ion speeds up, like a car merging onto a highway. To calculate the car's final linear speed, we would use $v = v_0 + at$. So here, we can use the rotational analog of that formula, namely $\omega = \omega_0 + \alpha t$. But we do not know the ion's initial angular speed, ω_0. Fortunately, we can get ω_0 from Eq. (1):

$$\begin{aligned}
\omega &= \omega_0 + \alpha_{cyc} t_1 \\
&= \frac{v_0}{R} + \alpha_{cyc} t_1 \qquad \text{[since } v = \omega r\text{, from Eq. (1)]} \\
&= \frac{10 \text{ m/s}}{2.0 \text{ m}} + (15 \text{ rad/s}^2)(5 \text{ s}) \\
&= 5 \text{ s}^{-1} + 75 \text{ rad/s} \\
&= 80 \text{ rad/s}.
\end{aligned}$$

At first glance, I appear to have "fudged" the units, by adding 5 s^{-1} to 75 rad/s. Apparently, I arbitrarily inserted a factor of "radians" into the 5. With most units, you are not "allowed" to do this. But radians are a so-called "dimensionless" unit, because an angle is just a *number* instead of a mass or length or time. (Ask your instructor for more details about this, if you are interested.) Since dimensionless units are just numbers, you can insert them or remove them at will from your answers. So for instance, 5 rad/s and 5 s^{-1} are two different ways of saying the same thing, as are 15 rad/s^2 and 15 s^{-2}.

(b) I see two equally good ways to solve this. Since we are looking for a (linear) distance covered, we could use an old linear kinematic formula such as $x = x_0 + v_0 t + \frac{1}{2}at^2$. Alternatively, you could solve for ϕ, the angle traversed by the ion. If ϕ is expressed in radians, then the relation between x and ϕ is

$$x = r\phi. \tag{3}$$

From trigonometry, you should recognize Eq. (3) as the *definition* of angle (in radians). Remember, "angle" is defined to be the ratio of arc length to radius: $\phi = \frac{x}{r}$, which might look more familiar if I wrote "s" instead of x for arc length. In any case, you should know Eq. (3) by heart. Interestingly, we can derive Eq. (1) above ($v = r\omega$) by differentiating both sides of Eq. (3) with respect to t. But I preferred to give a more intuitive justification of Eq. (1), in order to emphasize the difference between linear and angular speed.

Let me solve for x using both of the methods mentioned above.

Method 1: Solve for x "directly" using linear kinematics

Start with $x = x_0 + v_0 t + \frac{1}{2}at^2$. We are given $v_0 = 10$ m/s and $t_1 = 5$ s. But we do not know the linear acceleration a. The "a" in this equation refers to the rate at which the particle speeds up or

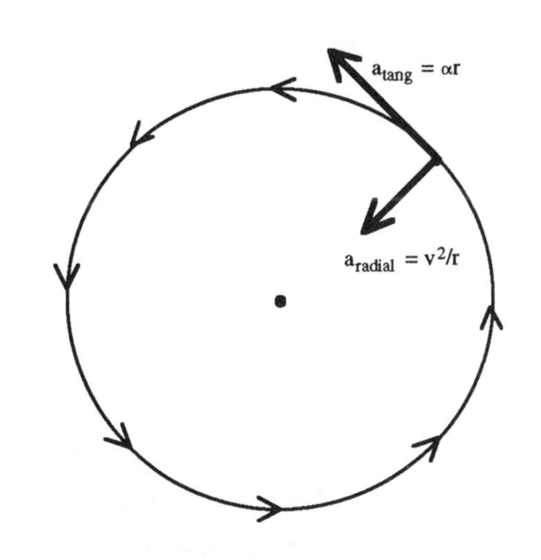

slows down *along its direction of motion*. If the particle stays at constant speed, then the "a" in that kinematic equation is 0. Therefore, we do not want to plug in the centripetal acceleration $a_{radial} = \frac{v^2}{r}$. Remember, a particle undergoing circular motion has a radial acceleration *even if it is moving at constant speed*. Roughly speaking, the forces that cause a_{radial} are what keep the particle moving in a circle.

To reiterate: When something travels in a circle, it always has a "radial" acceleration, the acceleration that keeps the object moving in a circle (as opposed to flying off in straight line). This radial acceleration exists whether or not the particle stays at constant speed. By contrast, the "tangential" acceleration along the direction of motion, if it exists, encodes how fast the particle speeds up or slows down. From Eq. (2) above, $a_{tang} = \alpha r$. *This* is the acceleration we must use in linear kinematic formulas such as $x = x_0 + v_0 t + \frac{1}{2}at^2$, because a_{tang} is the acceleration along the direction of motion, the acceleration that tells us how quickly the particle speeds up or slows down.

Using this insight, we get

$$\begin{aligned} x &= x_0 + v_0 t_1 + \frac{1}{2} a_{tang} t_1^2 \\ &= 0 + v_0 t_1 + \frac{1}{2}(\alpha_{cyc} R) t_1^2 \qquad [\text{since } a_{tang} = \alpha r] \\ &= 0 + (10\ \text{m/s})(5.0\ \text{s}) + \frac{1}{2}(15\ \text{s}^{-1})(2.0\ \text{m})(5.0\ \text{s})^2 \\ &= 420 \text{ meters.} \end{aligned}$$

Now I will re-solve this problem "indirectly," by first obtaining the angle covered by the ion, and then "converting" that angle into a distance.

Method 2: Find the angle ϕ, then use $x = \phi r$ to get the distance.

Start with the rotational analog of $x = x_0 + v_0 t + \frac{1}{2}at^2$, namely

$$\begin{aligned} \phi &= \phi_0 + \omega_0 t + \frac{1}{2}\alpha t^2. \\ &= 0 + \frac{v_0}{R} t_1 + \frac{1}{2}\alpha_{cyc} t_1^2 \qquad [\text{since } v = \omega r, \text{ from part (a)}] \\ &= 0 + \frac{10\ \text{m/s}}{2.0\ \text{m}}(5.0\ \text{s}) + \frac{1}{2}(15\ \text{rad/s}^2)(5.0\ \text{s})^2 \\ &= 210 \text{ rad.} \end{aligned}$$

Now use the relationship between arc length and angle to get

$$\begin{aligned} x &= R\phi \\ &= (2.0\ \text{m})(210\ \text{rad}) \\ &= 420\ \text{m}, \end{aligned}$$

in agreement with the answer obtained above by a different method. Notice that I exercised my "right" to insert and delete factors of radians at will.

This problem illustrates a general problem-solving technique. When you have to deal with both linear and rotational quantities in the same problem, you can use linear kinetics and/or rotational kinematics, and you can invoke the formulas

$$x = r\phi$$
$$v = r\omega$$
$$a = r\alpha$$

to travel back and forth between "rotation land" and "linear land."

QUESTION 14-3

A thread is wrapped around a cylindrical spool, of radius $R = 2.0$ cm, whose central axis is fixed on a support. Somebody's hand pulls the thread off the spool. This causes the spool to rotate. The acceleration of the person's hand is constant. Starting from rest, it takes 10 seconds for 5.0 meters of thread to be pulled off.

What is the angular speed of the spool at time $t = 2.0$ s?

ANSWER 14-3

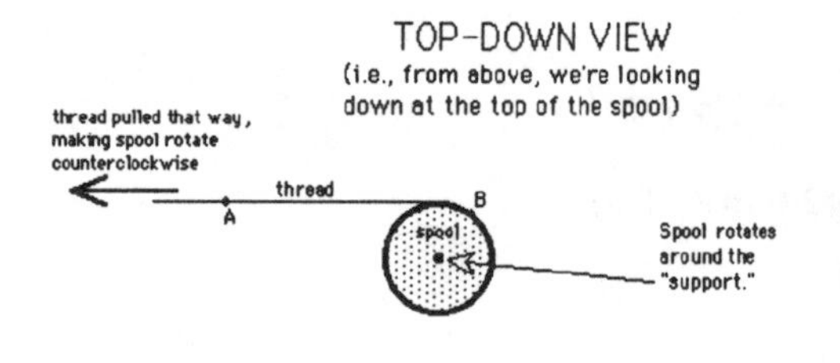

To solve this completely, we must picture what is happening. On the diagram, I label some points for future reference.

The hand, and therefore the thread, accelerates at a constant rate. We are not given the acceleration, but we can probably figure it out from the information supplied. Unfortunately, the problem does not ask us for the thread's acceleration or velocity. It asks for the spool's *angular* speed. How does that relate to the thread's motion? Intuitively, as the thread moves faster and faster, the spool spins faster and faster, because the thread makes the spool rotate. If we can figure out the relationship between the thread's speed and the spool's angular speed, it should help us solve the problem.

Well, since the whole thread moves together (without stretching), the speed of point A on the thread must equal the speed of point B on the thread. But point B stays in contact with the edge of the spool. In other words, point B on the thread moves together with the edge of the spool. Therefore, the thread has the same speed as the edge of the spool does. In symbols,

$$v_{\text{thread}} = v_{\text{edge of spool}}. \tag{1}$$

By the same kind of reasoning, or by differentiating both sides of Eq. (1) with respect to t, you can see that

$$a_{\text{thread}} = a_{\text{edge of spool, tang}} \tag{1'}$$

Equation (1) or (1′) is the physical insight needed to solve this problem. ***You will use this insight extensively in the next chapter, when we solve problems with ropes slung over rotating pulleys.***

Here is a suggested strategy: First, figure out the thread's velocity v at time $t = 2$ s. That v also "belongs" to the edge of the spool, by Eq. (1). But as we saw in Question 14-2 above, the linear speed v of a point on a rotating object relates in a simple way to the object's angular speed: $v = r\omega$. So, once we know the linear speed of the edge of the spool, we will immediately be able to find the spool's angular speed.

Step A: Find the thread's linear speed.

Since the thread covers distance $x = 5$ meters in time $t = 10$ s, you might be tempted to say that its velocity is $v = \frac{x}{t} = \frac{5\text{ m}}{10\text{ s}} = 0.5$ m/s. This is indeed the thread's average velocity. But remember, it is speeding up (accelerating). It starts out at $v_0 = 0$ m/s and ends up moving somewhat faster than 0.5 m/s. Right at time $t = 2$ s, it might be moving slower or faster than its average velocity of $\bar{v} = 0.5$ m/s. To figure out v at $t = 2$ s, we must calculate the thread's acceleration, and then use a constant-acceleration kinematic formula.

Since we know the thread covers $x = 5$ meters in time $t = 10$ s, the best formula to invoke is $x = x_0 + v_0 t + \frac{1}{2}at^2$. We can set $x_0 = 0$, and we know the thread starts at $v_0 = 0$. So, solving for acceleration gives $a = \frac{2x}{t^2} = \frac{2(5.0\text{ m})}{(10\text{ s})^2} = 0.10\text{ m/s}^2$. Now that we know the acceleration, we can substitute it into

$$\begin{aligned} v &= v_0 + at \\ &= 0 + (0.10\text{ m/s}^2)(2.0\text{ s}) \\ &= 0.20\text{ m/s}. \end{aligned}$$

I set $t = 2.0$ s, because we are interested in the thread's velocity at that time, not at $t = 10$ s.

Step B: Relate the thread's linear speed to the spool's angular speed.

As argued above, and encoded in Eq. (1), the thread's speed equals the (linear) speed of the edge of the spool. And the speed of the edge of the spool relates to the angular speed of the spool by $v = r\omega$. So, solving for ω at time $t = 2$ seconds gives

$$\begin{aligned} \omega &= \frac{v}{R} \\ &= \frac{0.20\text{ m/s}}{0.020\text{ m}} \\ &= 10\text{ rad/s}. \end{aligned}$$

This is the spool's angular speed right at time $t = 2.0$ s. As time increases, the spool spins faster and faster, i.e., ω increases.

By the way, you also could have solved this problem by finding the thread's acceleration, using $a_{\text{thread}} = a_{\text{edge of spool, tang}} = \alpha R$ to calculate the spool's angular acceleration, and then substituting that α into $\omega = \omega_0 + \alpha t$ to find the spool's angular speed at $t = 2$ s.

QUESTION 14-4

(This problem involves rotational dynamics as well as kinematics. If you are not learning dynamics until next week, skip this for now.)

A merry-go-round is a horizontal disk, pivoted frictionlessly at its center, and free to spin around. (Kids and physicists like to ride on them.) By pushing the edge of the merry-go-round, Monique makes it spin. Her push is a constant force directed at a 30° angle to an imaginary line tangent to the merry-go-round; see the diagram below. As the merry-go-round speeds up, Monique runs along next to it, and continues pushing at a 30° angle. The merry-go-round was initially at rest.

Monique's push is represented by the dark arrow.

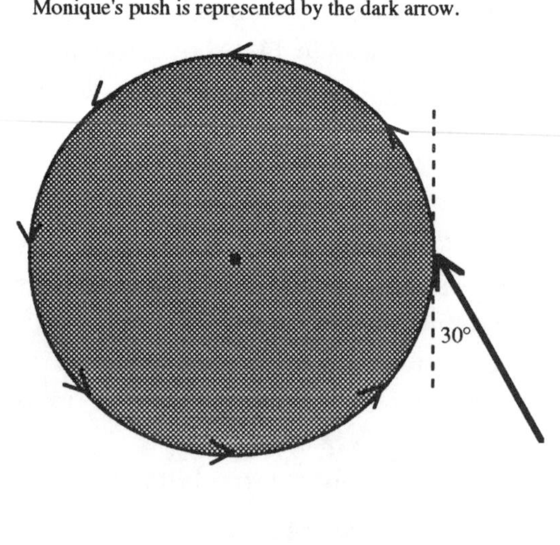

The merry-go-round has mass M, radius R, and rotational inertia $I = \frac{1}{2}MR^2$. Observation reveals that, due to Monique's pushing, the merry-go-round completes 1 revolution in time T.

Express all answers in terms of M, R, T, and $30°$.

(a) What is the merry-go-round's angular speed at time $\frac{T}{3}$?

(b) With what force is Monique pushing?

ANSWER 14-4

This looks hopelessly confusing. But it is not, *if* you think in terms of an analogy with a linear problem. Let us picture what is going on. Monique's push creates a *torque* that makes the merry-go-round spin. Just as a constant force causes an object to speed up with constant acceleration, a constant torque causes an object to rotate faster and faster, with constant angular acceleration. So, the merry-go-round spins faster and faster. We are told how much time the merry-go-round takes to cover a certain angle, and we are asked how fast it is spinning at a given time, and what force was required to make it spin that fast. To see what this looks like, and to see how the strength of the push affects the motion, experiment with the computer simulation.

What is the analogous linear problem? Well, suppose Monique pushes a frictionless desk in a *straight line* across the floor. And suppose we are given the amount of time T it takes the desk to cover a certain distance x. Could we figure out how fast the desk is going at time $\frac{T}{3}$? Sure. You could not use $v = \frac{x}{T}$, because the velocity is not constant. But we could figure out the desk's acceleration using

$$x = x_0 + v_0 t + \frac{1}{2}at^2, \tag{1}$$

with $t = T$. Then we could calculate the desk's velocity at time $t = \frac{T}{3}$ by substituting that acceleration into

$$v = v_0 + at, \tag{2}$$

with $t = \frac{T}{3}$ instead of $t = T$. We could also calculate the force on the desk by substituting that acceleration into Newton's 2nd law, $F_{\text{net}} = ma$. So, the linear version of the merry-go-round problem is not so bad. To solve merry-go-round problem itself, we just need to use the rotational analog of the above formulas and reasoning.

(a) "Analogizing" the above strategy, we can first use the rotational analog of Eq. (1) to find the merry-go-round's angular acceleration, α. Then, we can substitute that angular acceleration into the rotational analog of Eq. (2), to find the merry-go-round's angular speed. Remember, ϕ is analogous to x, ω is analogous to v, and α is analogous to a.

So, the analog of Eq. (1) is $\phi = \phi_0 + \omega_0 t + \frac{1}{2}\alpha t^2$. The merry-go-round starts at rest ($\omega_0 = 0$), and we can set $\phi_0 = 0$. At time T, the merry-go-round has spun through one revolution, which corresponds to $\phi = 2\pi$ radians. So, solving for α gives $\alpha = \frac{2\phi}{t^2} = \frac{2(2\pi)}{T^2} = \frac{4\pi}{T^2}$. Notice that I used $t = T$ in that equation, instead of $t = \frac{T}{3}$, because in time T (*not* $\frac{T}{3}$), the merry-go-round completes one revolution.

Now substitute that angular acceleration into the rotational analog of Eq. (2) to get

$$\begin{aligned}\omega &= \omega_0 + \alpha t \\ &= 0 + \left(\frac{4\pi}{T^2}\right)\frac{T}{3} \\ &= \frac{4\pi}{3T}.\end{aligned}$$

I plugged in $t = \frac{T}{3}$, because that is the time at which we are finding the angular speed.

(b) In a linear version of this problem, we would find the force on the desk by substituting the already-known acceleration into the linear version of Newton's 2nd law, $F_{net} = ma$. So here, we can invoke the rotational version of Newton's 2nd law, using the angular acceleration found above:

$$\begin{aligned}\tau &= I\alpha \\ &= \left(\frac{1}{2}MR^2\right)\left(\frac{4\pi}{T^2}\right) \\ &= \frac{2\pi MR^2}{T^2}.\end{aligned}$$

I just used the fact that torque is a "rotational push" (i.e., the rotational analog of force); and the rotational inertia I is an object's resistance to getting spun, just like mass m is an object's resistance to getting moved. So, we know Monique's torque. Unfortunately, we are asked for her *force*. How does force relate to torque? I could just write down the answer. But let me motivate the formula intuitively.

Roughly speaking, the size of the torque encodes the effectiveness of a push at making an object rotate. Imagine pushing a door open. Needless to say, the harder you push, the more quickly it swings open. So, the torque τ is proportional to the force F with which you push. But *where* you push also matters. Suppose you push near the hinge, and then you push near the door knob. The door responds more when you push near the door knob. Indeed, you can confirm that, other things being equal, the farther away from the hinge you push, the more the door responds. (Try it!) The hinge is the "pivot point" around which the door rotates. So, this experiment suggests that torque is proportional to r, the distance from the pivot point to the "contact point" where the force acts.

By the intuitive arguments given so far, τ is proportional to the force F and also proportional to r. So, you might guess that $\tau = Fr$. But one small complication remains. To understand this complication, consider the following diagram showing different angles at which a given force could be applied to the door. The arrows represent forces.

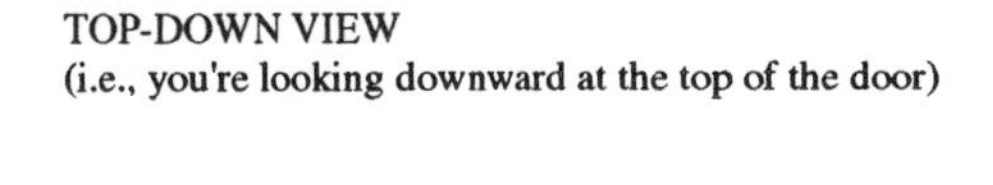

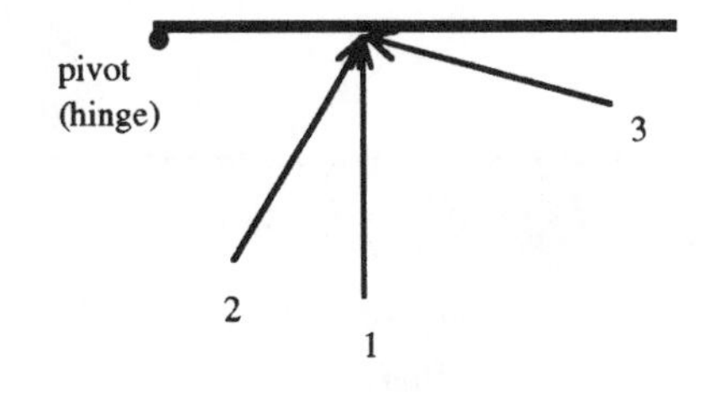

Forces 1, 2, and 3 all act the same distance r away from the pivot. And all three forces have the same magnitude. Nonetheless, the three forces are not all equally effective at making the door rotate. Intuitively, force 3 hardly makes the door rotate at all, while force 1 makes it rotate the most; it is "best" to push perpendicular to the door. If this does not make immediate intuitive sense, go to the door and *try* it. So, force 1 creates a big torque, while force 3 creates a small torque. Only the force–or part of a force–perpendicular to the door "contributes" to the force's ability to cause rotation, i.e., contributes to the torque.

How can we formalize this intuitive realization? First, draw a vector from the pivot point to the contact point where the force acts. Call that vector **r**. On the door, this vector points rightward. We just saw that only force components perpendicular to **r** contribute to the torque. So, let $F_\perp$ denote the component of the force perpendicular to **r**. Then, the torque is given by $\tau = F_\perp r$. Your book probably writes this formula as $\tau = \mathbf{r} \times \mathbf{F}$, a cross product. The magnitude of the cross product is given by my formula. You can also check that the book's formula $\tau = |\mathbf{r} \times \mathbf{F}| = Fr\sin\theta$, where θ is the angle between **F** and **r**, "agrees" with my version of the formula, $\tau = F_\perp r$. I like my version because it encourages you to draw pictures and break vectors into components. Let me show you what I mean.

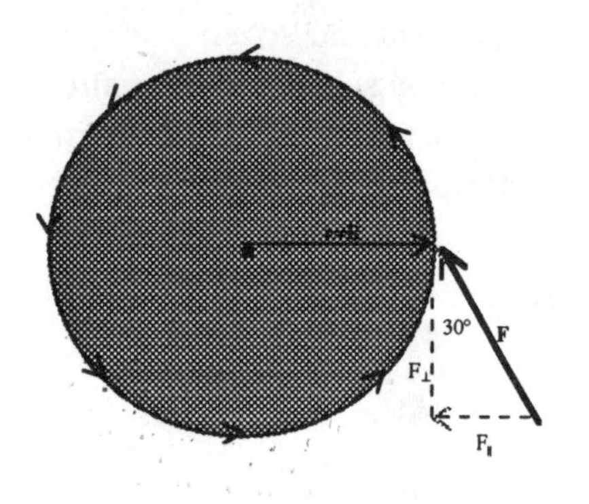

For Monique's push, the radius vector **r** from the pivot point to the contact point extends straight rightward. Only the component of her force perpendicular to **r**, namely $F_\perp$, helps to make the merry-go-round spin; only that component contributes to the torque. So,

$$\begin{aligned}\tau &= F_\perp r \\ &= (F\cos 30°)R.\end{aligned} \tag{3}$$

(This agrees with the equation you would get using $\tau = Fr\sin\theta$. By drawing **F** and **r** tail to tail, you can see that the angle between them is $120°$. Since $\sin 120° = \cos 30°$, the different formulas "agree.")

Given Eq. (3), we can finish this problem. Above, using $\tau = I\alpha$, we figured out that Monique's torque is $\frac{2\pi MR^2}{T^2}$. Solve Eq. (3) for F, and substitute in this torque to get

$$\begin{aligned}F &= \frac{\tau}{R\cos 30°} \\ &= \frac{\left(\frac{2\pi MR^2}{T^2}\right)}{R\cos 30°} \\ &= \frac{2\pi MR}{T^2\cos 30°}.\end{aligned}$$

QUESTION 14-5

(Supplemental topic. Skip this unless your instructor says you need to be able to derive rotational inertias.)

In Question 14-4, I told you that the rotational inertia of a uniform solid disk, pivoted at its center, is $I = \frac{1}{2}MR^2$, where M is the total mass and R is the radius. In this problem, I want you to *derive* that result. In other words, start with the definition of rotational inertia, and show that for a uniform solid disk, $I = \frac{1}{2}MR^2$.

ANSWER 14-5

Here is a strategy you can use to find the rotational inertia of any uniform solid object. (I will also note how to modify the strategy if the object has non-uniform density ρ; but I will not work out an example of this.) The strategy looks a bit confusing in the abstract, but makes sense once you see it applied. The idea is to break the object into tiny "pieces," and figure out the infinitesimal rotational inertia dI of each such piece. Then, add together all those pieces to obtain the total rotational inertia of the object. Here is the details:

(1) Make a careful drawing of the object, and "shade in" all the mass that is an arbitrary distance r from the pivot point. More precisely, shade in all the mass that is between r and $r + dr$ from the pivot point. (Pick r to be less than the radius of the whole object.) Call that shaded-in mass element "dm." So, dm is the amount of mass inside the infinitesimal volume element between r and $r + dr$.

(2) Write an expression for dm. If the object is uniform, the ratio of the "shaded" volume to the total volume equals the ratio of the shaded mass to the total mass. In symbols, $\frac{dm}{M} = \frac{dV}{V}$, and hence $dm = \frac{m}{V}dV$, where dV is the shaded volume and V is the total volume. Notice that $\frac{m}{V}$ is the density ρ. If the density is not constant, you must still use $dm = \rho dV$, provided you keep in mind that ρ will not equal $\frac{m}{V}$ at every point. Only when the density is constant does it equal the total mass per total volume.

If the object is two-dimensional, then use $dm = \frac{m}{A}dA$, where dA is the shaded area and A is the total area.

(3) By definition, the tiny rotational inertia "contributed" by mass element dm is $dI = r^2 dm$. To find the total rotational inertia, we must add up all these infinitesimal contributions. In other words, we must integrate: $I = \int dI = \int r^2 dm$. The limits of this integral are the smallest possible r to the largest possible r.

Now I will implement the 3-part strategy.

Step 1: Careful drawing with "shaded-in" region.

Here, dm is the infinitesimal mass contained in the dark ring between r and $r + dr$. Notice that I picked r to be less than R, the radius of the disk.

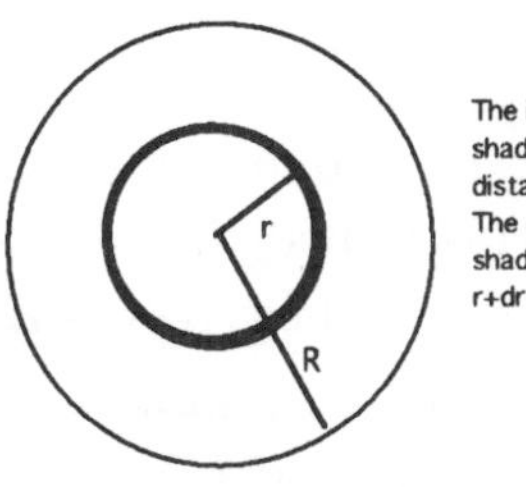

The inner edge of the shaded (blackened) area is a distance r from the center. The outer edge of the shaded area is a distance r+dr from the center

Step 2: Write expression for dm

Here, the object is two-dimensional, so I will use areas instead of volumes. As noted above, density is mass per volume; but in this case, we will use the "area mass density," by which I mean the mass per area: $\rho = \frac{m}{A}$. So, we get $dm = \rho dA = \frac{m}{A}dA$. We are given M. We need expressions for A and dA.

The area of the whole disk is $A = \pi R^2$. But the area of the shaded-in region, dA, is the length times the width of that ring, because the ring is very thin. The "length" of the ring, i.e., its circumference, is $2\pi r$. And its width is dr. So, $dA = (\text{length})(\text{width}) = 2\pi r dr$. Therefore,

$$\begin{aligned} dm &= \frac{M}{A} dA \\ &= \frac{M}{\pi R^2} 2\pi r dr \\ &= \frac{2Mr}{R^2} dr. \end{aligned}$$

Step 3: Integrate over $dI = r^2 dm$ to get the total rotational inertia, I.

By definition, the rotational inertia of the shaded-in ring is $dI = r^2 dm$, which in this particular problem works out to be

$$\begin{aligned} dI &= r^2 dm \\ &= r^2 \frac{2Mr}{R^2} dr \\ &= \frac{2Mr}{R^2} r^3 dr. \end{aligned}$$

Think of the disk as consisting of a bunch of concentric rings. Now we are going to "add up" the rotational inertia of all those rings to find the total rotational inertia of the disk. The smallest ring has a tiny radius; r is very close to 0. The largest ring has radius $r = R$. So, when we integrate over dI, the limits are $r = 0$ to $r = R$:

$$\begin{aligned} I &= \int_{r=0}^{r=R} dI = \int_{r=0}^{r=R} r^2 dm \\ &= \int_0^R \frac{2M}{R^2} r^3 dr \\ &= \frac{2M}{R^2} \int_0^R r^3 dr \\ &= \frac{2M}{R^2} \left(\frac{1}{4} r^4 \right) \Big|_0^R \\ &= \frac{1}{2} MR^2. \end{aligned}$$

If you need to be able to derive rotational inertia in this manner, I recommend working some of the sample problems provided in the textbook chapter, *without looking at the textbook's solution* until you have applied this strategy yourself.

QUESTION 14-6

(This problem contains rotational dynamics. Save it until next week if you have not reached this point.)

A huge uniform solid disk, of mass M and radius R, is mounted on a frictionless wall, pivoted at its center. An essentially massless rope is wrapped around an essentially massless circular ridge located a distance $\frac{R}{3}$ from the center of the disk. (It is kind of like a giant yo-yo.) In order to make the disk spin, someone pulls on the rope horizontally, as drawn below, with constant force F_p. Let me emphasize that the disk spins in place; it does not roll or otherwise move to a different place. (In physicists' jargon, "the disk's center of mass does not move.")

Unfortunately, the disk was mounted too low on the wall. As a result, the bottom of the disk scrapes against the floor. The floor exerts a constant frictional force f on the wheel.

Suppose the disk starts from rest at $t = 0$, and the rope is pulled for time t_1, at which point the person stops pulling.

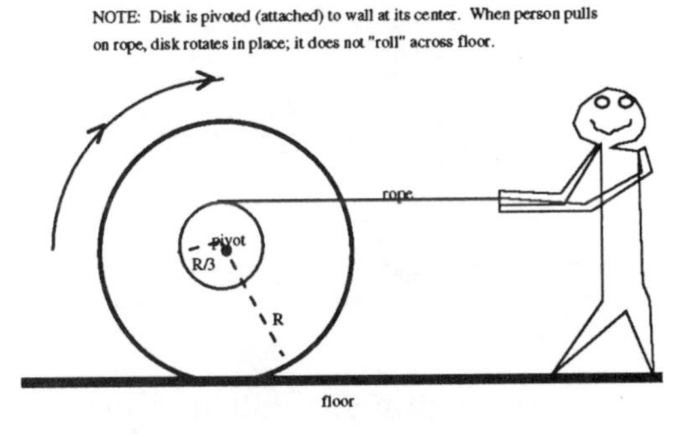

Express all answers in terms of F_p, f, M, R, and t_1.

(a) What is the angular speed of the disk at time t_1?

(b) What is the *radial* acceleration of a point on the edge of the disk at time t_1?

(c) How long, after the person lets go of the rope, will it take for the disk to come to rest?

ANSWER 14-6

Let us first picture what is going on physically. While the person pulls the rope, the disk gradually spins faster and faster; but friction keeps it from gaining as much angular speed as it otherwise would have gained. When the person lets go of the rope, the disk keeps spinning, but starts to slow down due to the frictional scraping. In other words, the angular speed decreases. Eventually, the wheel stops spinning.

This whole process is analogous to the following car trip: First you floor the accelerator, causing the car to speed up. Then you take your foot off the pedal. Consequently, the car starts to slow down, due to friction and air resistance. The car eventually comes to rest.

Conceptually, this question resembles Question 14-4, the merry-go-round problem. The one complication is that multiple forces, and hence multiple torques, act on the disk. To figure out how to deal with this situation, consider the linear analogy. What did you do, in linear motion problems, when multiple forces acted on an object? You drew a careful free-body diagram, and used it to write an expression for the net force. Next, you invoked Newton's 2nd law, $F_{\text{net}} = ma$. After finding the acceleration, you substituted it into old kinematic formulas to solve for the quantities of interest.

In rotational problems, we can use the analogous strategy:

(1) Draw a careful force diagram, breaking up each force vector into its components parallel ($F_{\parallel}$) and perpendicular ($F_{\perp}$) to the relevant radius vector. (It helps to draw in the radius vectors. In general, each separate force "has" a separate radius vector **r**.)

(2) From the diagram, "read off" the torques, and use them in the rotational version of Newton's 2nd law, $\sum \tau = I\alpha$. If you are given all the forces, you can solve for the angular acceleration α. Or, if you are given the angular acceleration, you can solve for an unknown torque.

(3) If you obtained a in step (2), substitute it into rotational kinematic formulas to solve for the quantities of interest.

(a) I will now implement this strategy.

Step 1: Free-body diagram.

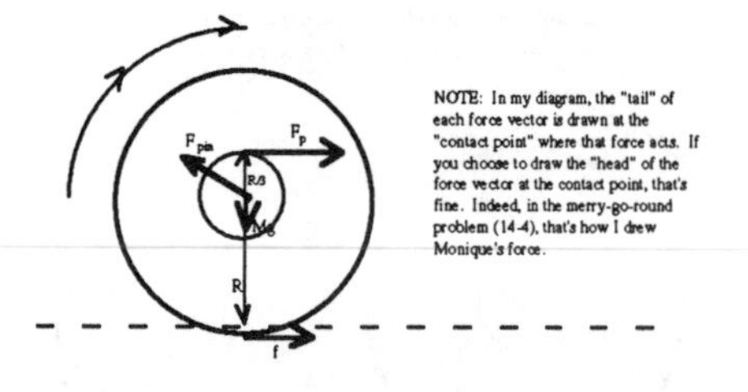

F_{pin} is the force exerted by the pin joint that holds the wheel to the wall. We do not know exactly in which direction it points. But in this problem, F_{pin} does not matter, because it acts right on the pivot point. A force acting directly on the pivot point exerts no torque. To see this intuitively, push on the hinge of your door. The door does not rotate. To see it mathematically, recall that the "r" in $\tau = F_{\perp} r$ denotes the distance from the pivot point to the "contact point" where the force acts. If the force acts right on the pivot point, then $r = 0$, and hence $\tau = 0$. So, F_{pin} generates no torque. For the same reason, neither does Mg. **(Crucial point: The force due to gravity always acts on an object's center of mass.)**

Friction points rightward, not leftward. Here is why. The rope's pull makes the wheel spin clockwise. Friction tries to resist that clockwise motion. Since friction acts on the bottom of the disk, it pushes rightward, in order to create a counterclockwise torque.

I did not need to break up $\mathbf{F}_p$ or $\mathbf{f}$ into components, because both forces point entirely perpendicular to the relevant radius vector. For $\mathbf{f}$, the radius vector points straight down and has length R. For $\mathbf{F}_p$, the radius vector points straight up, with length $\frac{R}{3}$.

Step 2: Newton's 2nd law in rotational form

Using the force diagram, we can write down the torque due to each force, add up those torques, and set $\sum \tau = I\alpha$. Before doing so, we must pick which direction counts as "positive." In class, you may learn about the "right-hand rule," which says (among other things) that a torque points perpendicular to the plane of rotation. For present purposes, however, we can intuitively think of a torque as being clockwise or counterclockwise. So, the torque due to F_p is positive (clockwise), while the torque due to friction is negative (counterclockwise). Using $\tau = F_{\perp} r$, we immediately get

$$\tau_{\text{rope}} = F_p \frac{R}{3},$$
$$\tau_{\text{fric}} = -fR.$$

Recall from above that gravity and the pin joint contribute no torque, because those forces act on the pivot point and hence $r = 0$. Therefore, from Newton's 2nd law

$$\sum \tau = I\alpha$$
$$F_p \frac{R}{3} - fR = \frac{1}{2} M R^2 \alpha,$$

where I have used $I = \frac{1}{2} M R^2$ for a uniform solid disk, a result derived in the previous problem. (For future reference, your textbook contains a table of rotational inertias. A disk is a solid cylinder that is not very long.) Solve this equation for the angular acceleration to get

$$\alpha = \frac{F_p \frac{R}{3} - fR}{\frac{1}{2} M R^2}.$$

Step 3: Rotational kinematics

We are trying to find the disk's angular speed at time t_1. We know its angular acceleration. In linear motion problems, if I know the acceleration and want to find v at a given time t, I use $v = v_0 + at$. So here, by analogy, we can use

$$\begin{aligned}\omega &= \omega_0 + \alpha t \\ &= 0 + \frac{F_p \frac{R}{3} - fR}{\frac{1}{2}MR^2} t_1.\end{aligned}$$

In summary, this question resembles a linear problem with multiple forces. The overall problem-solving strategy is analogous.

(b) You might want to plug in $a = \alpha R$. But as discussed in Question 14-2 above, we must carefully distinguish between the radial and the tangential acceleration.

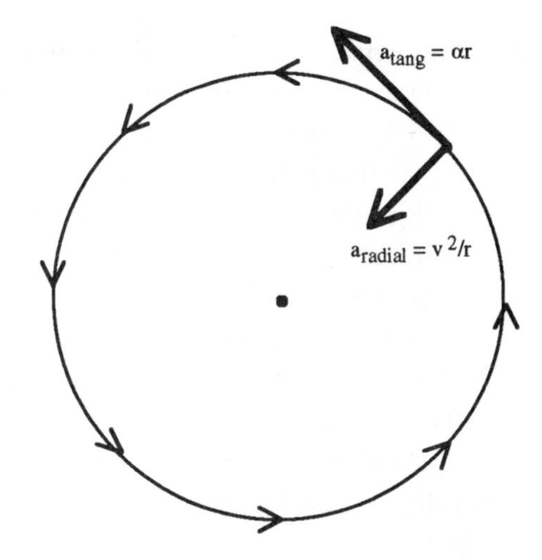

The radial (centripetal) acceleration is "responsible" for keeping the point moving in a circle. By contrast, the tangential acceleration points along the direction of motion, and specifies the rate at which the point speeds up or slows down. The formula "$a = \alpha R$" refers only to the *tangential* acceleration. By contrast, as we saw in Chapter 10 of this study guide, the radial acceleration of a point moving in a circle of radius r is $a_{radial} = \frac{v^2}{r}$. A point on the edge of the disk traces a circle of radius R. So, to solve for a_{radial}, we just need to know v, the point's speed at time t_1.

Fortunately, we can make use of our part (a) answer. The relationship between an object's angular speed ω, and the linear speed v of a point on that object, is $v = \omega r$, where r is the distance of that point from the pivot. At the edge of the disk, $r = R$. So, using our result from part (a), and setting $r = R$, we get

$$\begin{aligned}a_{radial} &= \frac{v^2}{r} \\ &= \frac{(\omega R)^2}{R} \\ &= \omega^2 R \\ &= \left(\frac{F_p \frac{R}{3} - fR}{\frac{1}{2}MR^2} t_1\right)^2 R,\end{aligned}$$

which you can simplify if you want.

(c) The disk slows down due to friction, because the rope no longer pulls on it. We can recycle the strategy of part (a). First, draw a modified force diagram reflecting the fact that the rope no longer acts. Next, use $\sum \tau = I\alpha$ to find the disk's angular acceleration. Here, we expect a negative angular acceleration, because the disk slows down. Finally, use that deceleration in rotational kinematic formulas to find out how much time the disk takes to stop. Intuitively, the bigger the deceleration, the more quickly the disk comes to rest.

Step 1: Free-body diagram.

Everything is the same as in part (a), except F_p is gone. I will not bother to draw this. Just erase F_p from your previous force diagram.

Step 2: Newton's 2nd law in rotational form.

So now, the only torque acting on the disk is frictional. Since the disk spins clockwise, I will choose clockwise as positive. Friction generates a negative (counterclockwise) torque, $\tau_{\text{fric}} = -fR$. So,

$$\sum \tau = I\alpha$$
$$-fR = \frac{1}{2}MR^2\alpha,$$

and hence $\alpha = -\frac{2f}{MR}$, which is negative as expected. This does not mean the disk starts spinning backwards. The disk continues spinning clockwise, but slower and slower.

Step 3: Kinematics.

At this point, we know the angular deceleration and want to calculate the time it takes the disk to stop completely. To do so, consider a linear (car) analogy. The disk resembles a car that suddenly slams on the brakes. Given the car's deceleration, how much time does it take to stop? Well, clearly we would need to know the car's initial speed immediately before the brakes get slammed, because the faster the car was moving, the longer it takes to stop. So, let us say we are given that initial speed, v_0. We know the car's final speed is $v = 0$. Therefore, we could solve for the "braking time" by starting with $v = v_0 + at$ and solving for t.

Analogously, in this problem, we are finding the time it takes the disk to reach final angular velocity $\omega = 0$. From part (a), we know the disk's "initial" angular speed right when "the brakes get slammed," i.e., right when the person stops pulling the rope. So, the "initial" angular speed in this part of the problem is the final angular speed from part (a). "Analogizing" $v = v_0 + at$ gives us $\omega = \omega_0 + \alpha t$. Solve for t, invoke our part (a) answer for ω_0, and use the deceleration just obtained in step 2 above, to get

$$\begin{aligned} t &= \frac{\omega - \omega_0}{\alpha} \\ &= \frac{0 - \frac{F_p \frac{R}{3} - fR}{\frac{1}{2}MR^2} t_1}{-\frac{2f}{MR}} \\ &= t_1\left(\frac{F_p}{3f} - 1\right) \end{aligned}$$

This answer has the right units. Let us check to see if it makes intuitive sense. According to this formula, the harder the person pulled the rope (F_p) before letting go, the more time the disk takes to stop thereafter. That us sensible. And according to this formula, a bigger frictional force leads to a smaller "stop time." Again, that makes sense. Finally, this formula says that the stopping time is proportional to t_1, the time for which the disk was accelerated before the person released the rope. That is reasonable, because the longer the person pulled the rope, the more the disk speeds up, and hence the longer it takes to slow back down.

Rotational Dynamics:

Multi-block Problems, Rolling, and More

15 CHAPTER

QUESTION 15-1

Hard. Ask your instructor if you need to be able to do this kind of problem.

Consider the system drawn below. The blocks have mass m and $2m$, and surface of the "plateau" is frictionless. The pulley is a uniform solid disk of mass m_p and radius r_p. The other object is a uniform solid cylinder mounted so that it can rotate around its vertical axis. The solid cylinder has mass M and radius R. An essentially massless cord is wrapped around the cylinder. So, when the block falls, the cylinder and disk rotate.

Suppose block m is released from rest.

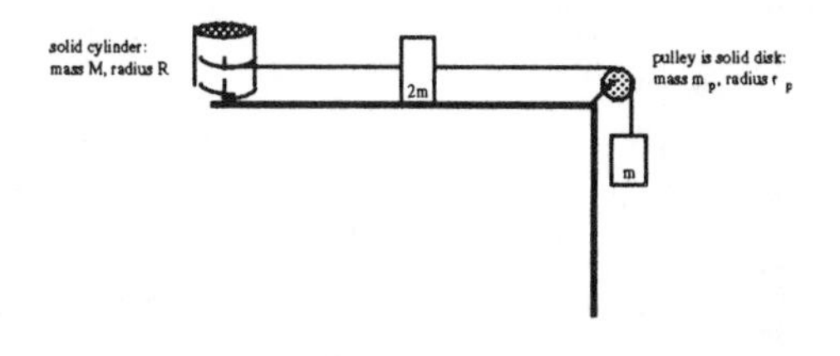

(a) How fast is block m moving after it has fallen through vertical distance s? Solve using the "multi-block" problem-solving strategy introduced in Chapter 8, making appropriate modifications to account for the fact that some of the masses undergo rotational instead of linear motion.

(b) Again solve for the block's speed after it is fallen through distance s, using conservation of energy.

ANSWER 15-1

(a) In this problem, the multi-block strategy turns out to be harder than conservation of energy. But in some problems, you *must* use the multi-block method. For instance, if the plateau were not frictionless, block 2 would generate heat during its slide. Since you would not know how much heat, conservation of energy would be difficult to use, more difficult than the multi-block strategy I am about to implement.

My plan is to find the blocks' acceleration, a, and then plug it into an old kinematic formula to find v, the blocks' speed after block m falls through distance s. Here is the complete multi-block strategy from Chapter 8.

(1) Draw a free-body diagram for each individual object,
(2) Write Newton's 2nd law (along the direction of motion) for each object separately,
(3) Algebraically solve for the acceleration, and then
(4) Use that acceleration in a kinematic equation to figure out something about the motion, or in a Newton's 2nd law equation to calculate a force.

Of course, we must modify this strategy in places, to account for the rotational motion. Instead of listing the modifications ahead of time, I will introduce them as I am going along, and summarize them at the end.

Step 1: Free-body diagrams.

Let "rope 1" denote the rope that runs from block m to block $2m$, via the pulley. If the pulley were massless, then the tension would be the same at both ends of rope 1. But since the pulley has mass, the tension in the vertical segment of rope 1 differs from the tension in the horizontal segment of rope 1. Why? Well, if the tension in both segments of rope 1 were the same, then the net torque on the pulley would be 0. If the net torque were 0, the pulley would not rotate. But the pulley *does* rotate as block m falls. Therefore, the tension in the vertical segment of rope 1, which I will call T_1, must exceed the tension in the horizontal segment, T_2. Only because $T_1 > T_2$ does the pulley rotate. The free-body diagrams reflect the fact that $T_1 \neq T_2$. But as always, both "ends" of a single segment of rope exert the same tension. So for instance, the vertical segment of rope 1 pulls down on the pulley exactly as hard as it pulls up on block 1.

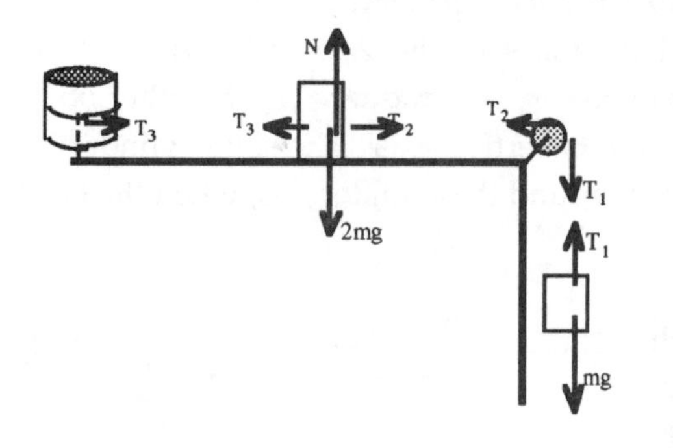

Since the rope connecting block $2m$ to the cylinder passes over no pulley–i.e., since it is a single segment of rope–the tension is the same at both ends. But as we saw in Chapter 8, *different* ropes carry different tensions, even in the absence of massive pulleys and other complications. So, I will call the tension in that rope T_3.

Actually, I have left out some of the forces acting on the cylinder and pulley. Specifically, I have omitted the gravitational forces ($m_p g$ and Mg), as well as the forces due to the "supports" that hold the pulley and cylinder in place. Here is why I am ignoring those forces: In this problem, the pulley and cylinder do not move through space; they merely rotate in place. This goes to show that the net *force* on each of those two objects is 0. Because the pulley and cylinder rotate but do not move "translationally," we care only about the torques. And the forces I have omitted exert no torque. For instance, gravity acts on the center of mass of the pulley. But the center of mass is the pivot point. So, the radius vector (also called the moment arm) **r** associated with gravity is $r = 0$; that force exerts no torque. The same argument applies to gravity acting on the cylinder, and the "support" forces holding those objects in place. My force diagrams for the rotating objects include only the "torque-generating" forces.

Step 2: Newton's 2nd law for each separate object

Given these diagrams, you can now write Newton's 2nd law for each of the four objects, using the linear form of the law ($\sum F = ma$) for moving objects and the rotational form ($\sum \tau = I\alpha$) for spinning objects.

Let us start with block m. I will choose the downward direction as positive. This choice commits me to calling clockwise positive for the pulley, rightward positive for block $2m$, and counterclockwise positive for the cylinder. That is because, when block 1 falls down, the pulley spins clockwise, block $2m$ gets dragged rightward, and the cylinder spins counterclockwise. (Remember from the multi-block strategy that your choice of positive directions must reflect the fact that the objects all "move together.") For block m, I get

$$\sum F_y = ma$$
$$\text{(BLOCK } m\text{)} \quad mg - T_1 = ma,$$

which contains two unknowns, a and T_1. As always in multi-block problems, we must consider the other objects.

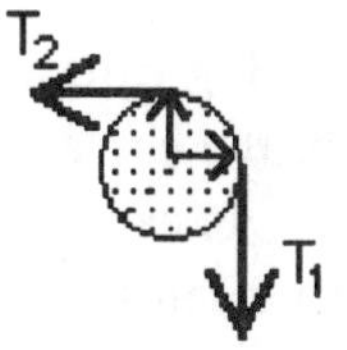

The pulley has two torques acting on it, due to the two tensions. Recall that $\tau = F_{\perp} r$. Since both forces act on the edge of the pulley, both of the relevant radius vectors have length r. Also, notice that each force acts entirely perpendicular to the relevant radius vector.
So, we do not need to break those tension vectors into components. Finally, notice that T_1 pulls the pulley clockwise, while T_2 pulls it counterclockwise. Given my choice of positive direction, the torque due to T_2 is negative. Put all this together, and use $I_{\text{solid disk}} = \frac{1}{2}mr^2$:

$$\sum \tau_{\text{pulley}} = I\alpha_{\text{pulley}}$$
$$\text{(PULLEY)} \quad T_1 r_p - T_2 r_p = \left(\frac{1}{2} m_p r_p^2\right)\alpha_{\text{pulley}}.$$

Now let us write Newton's 2nd law for block $2m$. Gravity and the normal force cancel out. Besides, we care only about the forces along the direction of motion:

$$\sum F_x = (2m)a$$
$$\text{(BLOCK } 2m\text{)} \quad T_2 - T_3 = 2ma.$$

Remember, since the blocks "move together," the downward acceleration of block m equals the rightward acceleration of block $2m$.

On to the cylinder. The tension acts on the edge of the cylinder, a distance R from the pivot point (at the center of the cylinder). That force is perpendicular to the relevant radius vector. The table of rotational inertias in your textbook says that $I = \frac{1}{2}MR^2$ for a solid cylinder. So,

$$\sum \tau_{\text{cylinder}} = I\alpha_{\text{cylinder}}$$
$$\text{(CYLINDER)} \qquad T_3 R = \left(\frac{1}{2}MR^2\right)\alpha_{\text{cylinder}}.$$

At this point, we have four equations. But there is six unknowns: a, α_{pulley}, α_{cylinder}, T_1, T_2, and T_3. In a standard linear multi-block problem, all four objects would share the same acceleration. But here, we have got three different accelerations, two of them angular and one of them linear. And we cannot even assume that $\alpha_{\text{pulley}} = \alpha_{\text{cylinder}}$. To see why, pretend the pulley is much smaller than the cylinder. By visualizing, you can see that the cylinder would rotate much slower than the pulley. We need more information.

Here is the key physical insight that applies to all problems of this sort. Since the ropes and blocks move together, the blocks' acceleration equals the ropes' acceleration. In other words, the rate at which the blocks speed up equals the rate at which the ropes speed up, because the blocks and ropes drag each other along, so to speak. But rope 1 stays in contact with the rim of the pulley. In other words, rope 1 and the rim of the pulley "move together." Therefore, the rope's acceleration equals the tangential acceleration (i.e., the acceleration along the direction of motion) of the rim of the pulley. And by similar reasoning, the ropes' acceleration equals the tangential acceleration of the edge of the cylinder. Putting all this together, we conclude that the blocks, the rim of the pulley, and the edge of the cylinder all move together; they share the same linear acceleration:

$$a_{\text{blocks}} = a_{\text{tang, rim of pulley}} = a_{\text{tang, edge of cylinder}}. \tag{5}$$

These are *tangential* accelerations directed along the direction of motion of the spinning rims and edges. They are *not* centripetal (radial) accelerations directed toward the centers of the circles.

Given Eq. (5), we can relate the linear and angular accelerations in this problem. As explained in Question 14-2 above, the tangential acceleration of a point on a rotating object relates to the object's angular acceleration by $a_{\text{tang}} = r\alpha$. So,

$$\begin{aligned} a_{\text{tang, rim of pulley}} &= r_p \alpha_{\text{pulley}} \\ a_{\text{tang, edge of cylinder}} &= R\alpha_{\text{cylinder}}. \end{aligned} \tag{6}$$

Combining Eqs. (5) and (6), and using "a" to denote the blocks' acceleration, gives

$$\begin{aligned} \alpha_{\text{pulley}} &= \frac{a}{r_p} \\ \alpha_{\text{cylinder}} &= \frac{a}{R}. \end{aligned}$$

So, we can express the angular accelerations in terms of the linear acceleration, or vice versa.

Substitute those expressions for a_{pulley} and a_{cylinder} into the above PULLEY and CYLINDER equations. For easy reference, I will also rewrite the BLOCK m and BLOCK $2m$ equations:

$$\text{Block } m \quad mg - T_1 = ma \tag{1}$$

$$\text{Pulley} \quad T_1 r_p - T_2 r_p = \left(\frac{1}{2} m_p r_p^2\right)\frac{a}{r_p} \tag{2}$$

$$\text{Block } 2m \quad T_2 - T_3 = 2ma \tag{3}$$

$$\text{Cylinder} \quad T_3 R = \left(\frac{1}{2} MR^2\right)\frac{a}{R}. \tag{4}$$

At this point, we have four equations in four unknowns (a, T_1, T_2, and T_3). We can therefore solve for the blocks' acceleration, a.

Step 3: Algebraically solve for the acceleration

To do so efficiently, divide both sides of Eq. (2) by r_p. All the r_p factors cancel out. Similarly, when you divide Eq. (4) through by R, all R factors cancel out. After performing those cancellations, add together the four equations. The tensions all cancel, leaving us with $mg = (m + \frac{1}{2}m_p + 2m + \frac{1}{2}M)a$, and hence

$$a = \frac{mg}{3m + \frac{1}{2}m_p + \frac{1}{2}M}.$$

Step 4: Kinematics.

We now know the downward acceleration of block 1. We want its velocity v after accelerating downward through a vertical distance $\Delta y = s$. So, using one of our standard constant-acceleration kinematic formulas, looking at the vertical component, I get

$$v_y^2 = v_{0y}^2 + 2a_y\Delta y$$
$$= 0 + 2\left(\frac{mg}{3m + \frac{1}{2}m_p + \frac{1}{2}M}\right)s,$$

and hence

$$v_y = \sqrt{\frac{2mgs}{3m + \frac{1}{2}m_p + \frac{1}{2}M}}.$$

Before proceeding to part (b), let me summarize the "rotational modifications" to the linear multi-block problem solving strategy:

Step 1: Free-body.

–Diagrams must show contact point of forces, so that you can calculate torques.

–Tensions in different segments of cord are different.

Step 2: Newton's 2nd law.

–Use rotational form of Newton's 2nd law for rotating objects.

–Use the relation between angular and linear acceleration, $a = \alpha r$. The contact point between the rope and each object moves at the same acceleration.

(b) Now I will re-solve the problem using conservation of energy. As always, diagram the different stages.

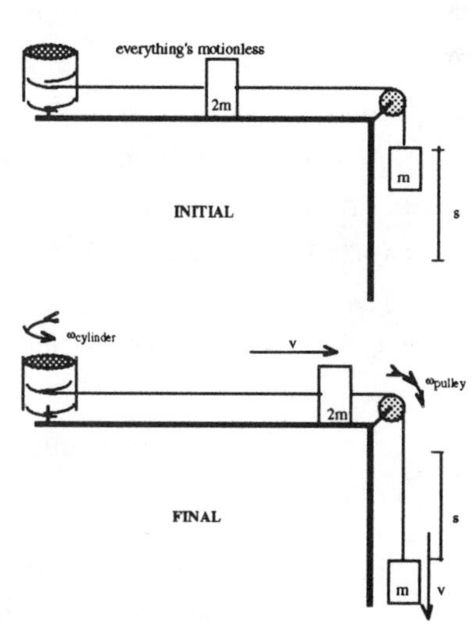

The two blocks travel at the same speed v; but the pulley and cylinder might have different angular speeds, for reasons I will discuss shortly.

The initial kinetic energy is clearly $K_i = 0$. What about the final kinetic energy? The blocks' kinetic energies are purely translational; they are moving but not spinning. But the pulley's and cylinder's kinetic energies are entirely rotational; they are spinning but not moving. The formula for rotational kinetic energy is what you would expect by analogy with $K_{\text{trans}} = \frac{1}{2}mv^2$, namely $K_{\text{rot}} = \frac{1}{2}I\omega^2$.

Let us now consider potential energy. Only block m undergoes a change in U; the other objects end up at the same height they started. If we say the final potential energy of block m is $U_f = 0$, then its initial potential energy is $U_i = mgs$, since it falls through height $h = s$.

Negligible heat gets generated as the objects move and spin.

Putting all this together, we get

$$\begin{aligned} E_i &= E_f \\ K_i + U_i &= K_f + U_f \\ 0 + mgs &= \frac{1}{2}mv^2 + \frac{1}{2}\left(\frac{1}{2}m_p r_p^2\right)\omega_{\text{pulley}}^2 + \frac{1}{2}(2m)v^2 + \frac{1}{2}\left(\frac{1}{2}MR^2\right)\omega_{\text{cylinder}}^2 + 0. \end{aligned} \tag{7}$$

Before solving for v, we must reexpress the other two unknowns in this equation, ω_{pulley} and ω_{cylinder}, in terms of v. The essential physical insight is the same one we used in part (a): The whole rope moves together. Specifically, the blocks' speed v is also the ropes' speed. But the rope "drags along" the rim of the pulley and the edge of the cylinder. In other words, the rim of the pulley and the edge of the cylinder have the same speed v as the rope, which has the same speed v as the blocks:

$$v_{\text{block}} = v_{\text{rim of pulley}} = v_{\text{edge of cylinder}}. \tag{8}$$

But as shown in Question 14-2 above, the linear speed of a point on a rotating object relates to the object's angular speed by $v = \omega r$. So,

$$\begin{aligned} v_{\text{rim of pulley}} &= r_p \omega_{\text{pulley}} \\ v_{\text{edge of cylinder}} &= R\omega_{\text{cylinder}}. \end{aligned} \tag{9}$$

Letting "v" denote the block's speed, and combining Eqs. (8) and (9), gives us

$$\begin{aligned} \omega_{\text{pulley}} &= \frac{v}{r_p} \\ \omega_{\text{cylinder}} &= \frac{v}{R}. \end{aligned}$$

So for instance, if the cylinder is smaller than the pulley, it spins faster. (You see an equivalent phenomenon with gears, for essentially the same reason.) Substituting these angular velocity expressions into Eq. (7), the conservation of energy equation, yields

$$\begin{aligned} mgs &= \frac{1}{2}mv^2 + \frac{1}{2}\left(\frac{1}{2}m_p r_p^2\right)\omega_{\text{pulley}}^2 + \frac{1}{2}(2m)v^2 + \frac{1}{2}\left(\frac{1}{2}MR^2\right)\omega_{\text{cylinder}}^2 \\ &= \frac{1}{2}mv^2 + \frac{1}{2}\left(\frac{1}{2}m_p r_p^2\right)\left(\frac{v}{r_p}\right)^2 + \frac{1}{2}(2m)v^2 + \frac{1}{2}\left(\frac{1}{2}MR^2\right)\left(\frac{v}{R}\right)^2. \end{aligned}$$

Since v is the only unknown, we are done with the physics. Only algebra remains.

Algebra starts here. Notice that the r_p's and R's cancel. Multiply through by 2, and factor v^2 out front, to get $2mgs = v^2\left(m + \frac{1}{2}m_p + 2m + \frac{1}{2}M\right)$. Now just isolate v to get

$$v = \sqrt{\frac{2mgs}{3m + \frac{1}{2}m_p + \frac{1}{2}M}},$$

the same answer obtained above using the multi-block strategy. Cool, huh?

This problem demonstrates that conservation of energy sometimes provides a shortcut in multi-block problems. But not always. For instance, if you would been asked for the tension in one or more of the rope segments, you would have had no choice but to consider forces and accelerations. Similarly, if the problem involves friction, then heat gets produced, and there is no easy way to calculate how much. In such cases, you need the multi-block strategy of part (a).

QUESTION 15-2

Consider a prototype weed wacker, consisting of a thin inflexible string attached to a motor at its end, as drawn below. The string rotates around its end. The string has length $L = 0.20$ m and mass $M = 0.015$ kg.

(a) What (constant) torque must the motor exert on the string in order to bring it from rest to full speed in time $t = 0.50$ s? "Full speed" is 1200 revolutions per minute. Neglect wind resistance.

motor makes this "attachment point" spin

(b) (*Do this part only if you need to be able to derive rotational inertias.*) Derive the rotational inertia of this inflexible string.

ANSWER 15-2

(a) The torque τ exerted by the motor on the string must be big enough so that the string speeds up from initial angular speed $\omega_0 = 0$ to final angular speed $\omega = 1200$ rev/min in time $t = 0.50$ s. How does torque relate to the rate at which the string speeds up? If the answer does not immediately pop into your head, think of a linear analogy. Instead of a torque making a string spin faster and faster, think of a force making a car move faster and faster. How would we relate the force to the rate at which the car speeds up? Well, the rate at which the car speeds up is its acceleration a, which we could relate to the forces using $\sum F = ma$.

The same reasoning, "analogized" into rotational language, applies here. After figuring out the rate at which the string speeds up–i.e., its angular acceleration α–we can substitute that value into the rotational version of Newton's 2nd law, $\sum \tau = I\alpha$, and solve for the torque.

So, I will begin by calculating α. To do so, I must first express the final angular speed in units of radians per time. Otherwise, some of our rotational formulas will not hold. (It is not worth the time right now to give a theoretical digression explaining why this is so.) $\omega = 1200 \frac{\text{rev}}{\text{min}} \times \frac{1\text{ min}}{60\text{ s}} \times \frac{2\pi\text{ rad}}{1\text{ rev}} = 126\,\text{rad/s}$. So, the string speeds up from 0 rad/s to 126 rad/s in 0.50 seconds. You might be able to see "in your head" that the angular acceleration is therefore 252 rad/s^2. More formally, we can start with the rotational analog of $v = v_0 + at$, namely $\omega = \omega_0 + \alpha t$, to get $\alpha = \frac{\omega - \omega_0}{t} = \frac{126\text{rad/s} - 0}{0.50\text{ s}} = 252\,\text{rad/s}^2$. Now we can substitute that angular acceleration into $\sum \tau = I\alpha$ to find the torque. The motor exerts the only torque felt by the string. So,

$$\begin{aligned}\tau &= I\alpha \\ &= \left(\frac{1}{3}ML^2\right)\alpha \\ &= \frac{1}{3}(0.015\text{ kg})(0.20\text{ m})^2(252\text{ rad/s}^2) \\ &= 0.050\text{ N}\cdot\text{m}\end{aligned}$$

I used the textbook's rotational inertia formula for a long thin rod pivoted at its *end*, not at its center. (The string here rotates around an attachment point located at the end of the string.)

(b) I laid out the strategy in Question 14-5. Let me just dive right in.

Step 1: Shade in dm.

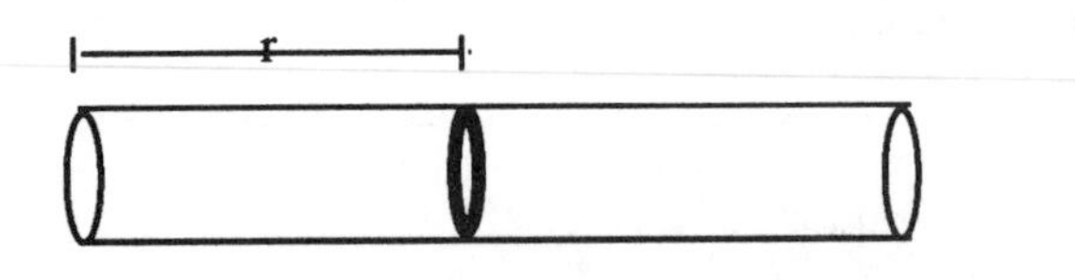

I will treat the string as a long thin cylinder. The pivot point is the left end, and the axis of rotation goes into the page (*not* through the middle of the cylinder). Start by "shading in" all the mass that is between r and $r+dr$ from the pivot point.

I have exaggerated the string's width. Let A denote the string's cross-sectional area (i.e., the area of the "circles" at the ends of the string). Do not worry: A will cancel out of the final answer. The shaded-in region is a thin disk of area A and length ("thickness") dr, a distance r from the pivot point. So, dm is shaped like a flattened penny.

Step 2: Write $dm = \rho dV$ or $dm = \rho dA$, where ρ is the density.

For a uniform mass distribution, the density is constant, $\rho = \frac{\text{mass}}{\text{volume}} = \frac{M}{V}$. Recall that dV is the infinitesimal volume of the shaded-in region (i.e., the region containing dm), while V is the volume of the whole string. Since the string's volume is $V = (\text{length})(\text{area}) = LA$, its density is $\rho = \frac{M}{V} = \frac{M}{LA}$. Since the shaded-in region has "length" dr and area A, its infinitesimal volume is $dV = Adr$. Putting all this together, I get

$$\begin{aligned} dm &= \rho dV \\ &= \left(\frac{M}{LA}\right)(Adr) \\ &= \frac{M}{L}dr. \end{aligned}$$

Step 3: "Add up" the rotational inertias of all the little mass elements, by integrating.

By definition, the infinitesimal rotational inertia of a mass element dm is $dI = r^2 dm$. Here, that is the rotational inertia of a thin "slice" of the string. Now we just need to sum up the rotational inertias of all those slices, to obtain the total rotational inertia of the string. What are the limits of the integral? Well, the slices start right on the pivot point (the attachment point), where $r = 0$. The slice farthest from the pivot point is a distance $r = L$ away. So, r ranges from 0 to L:

$$\begin{aligned} I &= \int_{r=0}^{r=L} dI = \int_{r=0}^{r=L} r^2 dm \\ &= \int_0^L r^2 \frac{M}{L} dr \\ &= \frac{M}{L}\int_0^L r^2 dr \\ &= \frac{M}{L}\left(\frac{1}{3}r^3\right)\Big|_0^L \\ &= \frac{1}{3}ML^2. \end{aligned}$$

QUESTION 15-3

Hard. Ask your instructor if you need to be able to do this kind of problem.

A bicycle wheel consists of a rim (radius $R = 0.40$ m) and essentially massless spokes. Suppose we mount the bicycle wheel on the wall, so that the wheel is parallel to the wall, and so that it is free to spin around its center, but not free to move translationally. We wrap a very light string around the rim, and attach a block of mass $m = 0.80$ kg to the end of the string.

In order to figure out the mass M of the rim, a physicist releases the block from rest and times how long it takes the mass to fall through a set distance $L = 1.5$ m. This time turns out to be $t = 0.78$ s. From this data, calculate the mass M of the rim.

ANSWER 15-3

I have deliberately made this look tricky, to illustrate that even the most formidable problem involving multiple objects can be addressed using the multi-block strategy laid out in Question 15-1. Usually, we would know all the masses, and we would figure out something like a "fall time." Here, we know the fall time and want to calculate a mass. So, the same reasoning applies as in the "standard" case. We will use the same equations. The only difference is what is the "given" and what is the unknown. So, I will start gathering information by working through the usual strategy. Please review Question 15-1 if you are not somewhat comfortable with the multi-block strategy:

(1) Draw separate force diagrams for each moving or spinning object,

(2) Write Newton's 2nd law (along the direction of motion) for each object separately, and then

(3) Use algebra to solve for the relevant acceleration, force or mass.

Step 1: Force diagrams.

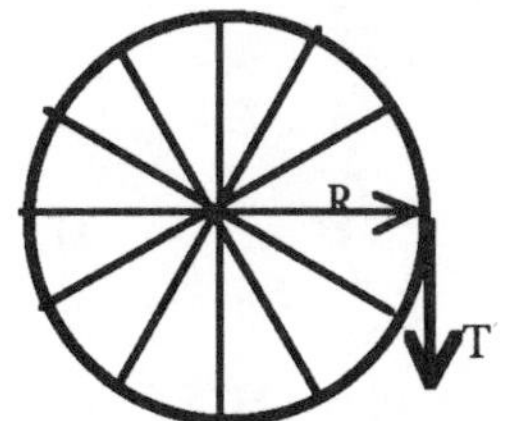

Since the bicycle wheel rotates but does not move translationally, we care only about the torques, not the forces. So, our free-body diagram can omit any force that does not cause the wheel to rotate, i.e., that does not generate a torque. Since gravity acts on the center of the wheel–right on the pivot point–it creates no torque, because the "r" in $\tau = F_{\perp} r$ is zero. Similarly, the "support" force holding the wheel in place also acts on the pivot point, and therefore exerts no torque. The only force that "torques" the wheel is the tension in the rope. Because there is only one segment of rope, the tension is the same at both ends.

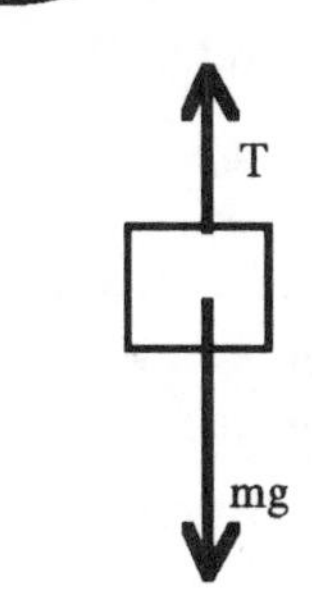

The most common error is to set $T = mg$. But if T equals mg, then the block experiences no net force; it just hangs there. Since the block accelerates downward, we know that mg must "overcome" T.

Step 2: Newton's 2nd law for each separate object

Use the linear form ($\sum F = ma$) for translationally-moving objects, and the rotational form ($\sum \tau = I\alpha$) for rotating objects.

I will pick downward as the positive direction for the block. This choice commits me to calling clockwise positive for the wheel. Why? Because when the block falls, the wheel turns clockwise; and our choice of positive directions must reflect the fact that the objects move together.

Let us start with the wheel. The only torque acting on it is $\tau = T_{\perp}R$, where $T_{\perp}$ is the component of the tension vector perpendicular to **r**. By "**r**," I mean the radius vector connecting the pivot point to the contact point. Here, the radius vector points rightward and has length R, as the above diagram shows. So, the tension **T** points entirely perpendicular to **r**: $T_{\perp} = T$. Since we can neglect the spokes, the wheel is approximately a "hoop," which is really a very thin cylindrical shell (of small "length"). So, from the textbook, $I = MR^2$. In summary,

$$\sum \tau = I\alpha \qquad (1)$$
$$TR = MR^2\alpha$$

Now consider the block. Since it moves translationally, use the linear form of Newton's 2nd law. Remember, I have chosen downward as the positive direction:

$$\sum F = ma \qquad (2)$$
$$mg - T = ma$$

Eqs. (1) and (2) contain four unknowns (M, T, a, and α). Can we gather more information? Well, intuitively, there is a connection between the block's acceleration a and the wheel's angular acceleration α. After all, the faster the block falls, the faster the wheel spins. Let me briefly review the relationship between a and α that applies in this kind of problem.

Since the rope and block move together, $a_{\text{block}} = a_{\text{rope}}$. But the rope stays in contact with the rim of the wheel; the rope and the rim of the wheel move together. Therefore, the rope's acceleration equals the tangential acceleration of the wheel's rim: $a_{\text{rope}} = a_{\text{tang, rim of wheel}}$. Putting all this together, we conclude that the block moves together with the rim of the wheel: $a_{\text{block}} = a_{\text{tang, rim of wheel}}$. But as we have seen before, the tangential acceleration of a point on a rotating object is related to the object's angular acceleration by $a_{\text{tang}} = r\alpha$. So, $a_{\text{block}} = a_{\text{tang, rim of wheel}} = R\alpha_{\text{wheel}}$. In summary, letting "$a$" denote the block's acceleration and α denote the wheel's angular acceleration, we get $a = R\alpha$. We can substitute this result into Eq. (1) to eliminate one of the four unknowns. Let do so:

$$TR = MR^2\frac{a}{R}. \qquad (1')$$

Do we now have sufficient information to solve for M? Well, we know $m = 0.80$ kg and $g = 9.8\ \text{m/s}^2$. Unfortunately, Eqs. (1') and (2) still contain three unknowns: M, a, and T. We need one more piece of information.

To get it, consider the experimental data. The block takes time $t = 0.78$ s to fall through distance $y = L = 1.5$ m. The block accelerates, beginning from rest ($v_{0y} = 0$). So, starting with the constant-acceleration kinematic equation $y = y_0 + v_{0y}t + \frac{1}{2}a_y t^2$, we can solve for the acceleration. Since downward is positive, the final position is bigger than the initial position: $y_0 = 0$ and $y = 1.5$ m. Solve for a_y to get $a_y = \frac{2(y - y_0)}{t^2} = \frac{2(1.5\ \text{m} - 0)}{(0.78\ \text{s})^2} = 4.9\ \text{m/s}^2$. Now that we know the acceleration, Eqs. (1') and (2) contain only two unknowns, M and T. So, we can finally solve for M.

Step 3: Algebra

In Eq. (1′), notice that all the R's cancel. Remarkably, that equation simplifies to $T = Ma$. Substitute this expression for T into Eq. (2) to get $mg - Ma = ma$. Now add Ma to both sides, subtract ma from both sides, and divide through by a. This gives

$$\begin{aligned} M &= m\left(\frac{g}{a} - 1\right)a \\ &= (0.80\ \text{kg})\left(\frac{9.8\ \text{m/s}^2}{4.9\ \text{m/s}^2} - 1\right) \\ &= 0.80\ \text{kg}. \end{aligned}$$

By coincidence, the wheel's mass turns out to equal the block's mass. Do not draw general conclusions from this: I happened to pick the numbers so that things would work out in this way.

QUESTION 15-4

An object of mass M, radius R, and rotational inertia kMR^2 rolls without slipping down an inclined plane that makes an angle θ with the horizontal. Here, k might equal $\frac{1}{2}$ (for a solid cylinder), or $\frac{2}{3}$ (for a spherical shell), or some other value between 0 and 1. Let us just call that number "k" instead of plugging in a specific value. It is a "gyration constant," not a spring constant.

After the object has rolled a distance x along the ramp, what is the speed v of its center of mass? In other words, how fast is the object as a whole moving down the ramp after it has rolled a distance x? Your answer may include M, R, k, θ, and x.

ANSWER 15-4

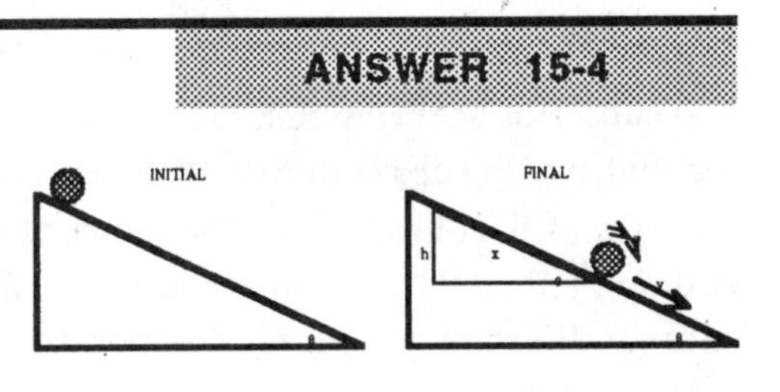

You can address this problem using forces and torques. Your textbook demonstrates this in detail, and some instructors emphasize this technique. But a quicker way is to use–you guessed it–conservation of energy. As always when conserving energy, sketch the relevant stages of the process.

At its "final" point, the object moves (translates) down the ramp *and also* rotates. It has both a linear speed v and a rotational speed ω.

Intuitively, as the object rolls down the ramp, its gravitational potential energy converts into energy of motion, i.e., kinetic energy. The kinetic energy comes in two forms. Because the object as a whole moves down the ramp, it has "translational" kinetic energy, given by the old formula $K_{\text{trans}} = \frac{1}{2}Mv^2$. But an object can also possess "extra" energy of motion because it is rotating. By analogy with $K_{\text{trans}} = \frac{1}{2}Mv^2$, the relevant formula is $K_{\text{rot}} = \frac{1}{2}I\omega^2$.

Let us say the object's final potential energy is $U_f = 0$. Since it starts off a height $h = x\sin\theta$ above its final position (as the diagram shows), its initial potential energy is $U_i = Mgh = Mgx\sin\theta$.

What about heat? It is reasonable to expect that some heat gets generated as the object rolls, by the following argument: The object rolls because friction makes it roll; if we "turn off" friction, the object just slides down the ramp without rotating. But frictional rubbing tends to make an object heat up.

Fortunately, **no heat gets generated**. It is true that friction is "on." But when an object rolls, it does not "rub" against the surface. At any given moment, the object's point of contact with the ramp is *motionless*. (See your textbook for more on this.) Put another way, *static* friction (instead of kinetic "sliding" friction) causes the object to roll. And static friction involves no rubbing or scraping. Consequently, no heat gets produced. Which is great, because otherwise we could not use energy conservation so easily.

Putting all this together yields

$$\begin{aligned} E_0 &= E_f \\ U_0 + K_0 &= U_f + K_f \\ Mgh + 0 &= 0 + K_{\text{trans}} + K_{\text{rot}} \\ Mgx\sin\theta + 0 &= 0 + \frac{1}{2}Mv^2 + \frac{1}{2}I\omega^2 \\ Mgx\sin\theta + 0 &= 0 + \frac{1}{2}Mv^2 + \frac{1}{2}(kMR^2)\omega^2 \end{aligned} \qquad (1)$$

Look what this equation tells us physically. The initial "reservoir" of potential energy, instead of all getting converted to translational movement, gets "split up" between translational and rotational motion. Due to this splitting, the rolling object ends up moving slower than an object that slides (instead of rolls) down the ramp..

Unfortunately, we cannot yet solve for v, because Eq. (1) contains a second unknown, ω. We need to express ω in terms of v. Usually, an object's linear and angular speeds are completely independent; you cannot write one in terms of the other. But as your textbook explains in gory detail, when an object rolls without slipping, its linear and angular speeds are related by

$$v_{\text{cm}} = R\omega \quad \text{ROLLING WITHOUT SLIPPING CONDITION.}$$

The subscript "cm" emphasizes that v is the center-of-mass speed of the object as a whole, i.e., how fast the center of the object moves. You must carefully distinguish v_{cm} from the speed with which a point on the edge of the object circles the center of the object. We saw earlier that such a point's speed is $v = r\omega$. Although this formula looks just like the rolling without slipping condition, it means something entirely different, and applies under different circumstances. For any rotating object, $v = r\omega$ *always* tells you a point's speed *with respect to the center of mass*. By contrast, $v_{\text{cm}} = R\omega$ applies ***only*** when an object rolls without slipping, and tells us how fast the center of mass itself travels.

The center-of-mass velocity is what appears in the translational kinetic energy formula $K_{\text{trans}} = \frac{1}{2}Mv_{\text{cm}}^2$ and in *all other linear motion formulas*. But we physicists are too lazy to write out the "cm" subscript in all those formulas.

Let us return to the problem at hand. Using the rolling without slipping condition, we can write the object's angular speed in terms of its linear center-of-mass speed: $\omega = \frac{v}{R}$. Put that into energy conservation Eq. (1) above to get

$$\begin{aligned} Mgx\sin\theta &= \frac{1}{2}Mv^2 + \frac{1}{2}(kMR^2)\left(\frac{v}{R}\right)^2 \\ &= \frac{1}{2}Mv^2 + \frac{1}{2}kMv^2. \end{aligned}$$

Notice that the R's and M's cancel! Solve for v to get $v = \sqrt{\frac{2gx\sin\theta}{1+k}}$. It is physically counterintuitive that the object's speed after rolling through a given distance does not depend on how heavy or how big it is; M and R do not matter! All that matters is "shape," as encoded by the "k" constant. (Remember, the k "belonging to" a cylinder differs from the k belonging to a spherical shell, which differs from the k belonging to a solid cylinder...) For instance, if you roll two solid cylinders down a ramp, they will reach the bottom at the same time and the same speed, even if one cylinder is much bigger and/or heavier than the other. But a solid cylinder will "beat" a cylindrical shell, no matter what size objects you choose to "race." Confirm this using the formula we just got for v. In addition, go into the physics lab and try it!

QUESTION 15-5

A thin rod of length 1.0 meter is pivoted at its center and mounted on a wall, so that it is free to rotate parallel to the plane of the wall. Attached to its ends are small blocks of mass m and $2m$, respectively. Suppose this "dumbbell" is released from rest from a horizontal position.

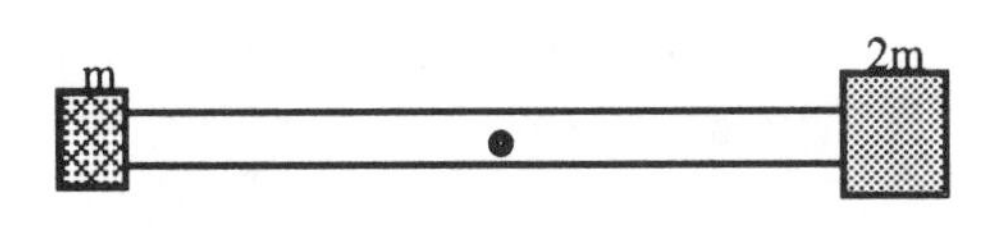

The $2m$ block swings down. When the $2m$ block reaches its bottom-most point, what is its speed v? Answer this question

(a) When the rod has mass 0,

(b) When the rod has mass $6m$.

Hint: Do not use torque-and-angular-acceleration reasoning. It will not work, for reasons I will explain below.

ANSWER 15-5

Let me first explain why you cannot just calculate the net torque, use $\sum \tau = I\alpha$ to obtain the angular acceleration, and then use kinematics to find what you want. Those old kinematic formulas apply only when the acceleration (or angular acceleration) is *constant*. The angular acceleration is constant only if the net torque is constant. But here, as the blocks swing, the torque keeps changing. To see why, consider the torque on block $2m$, at two different times.

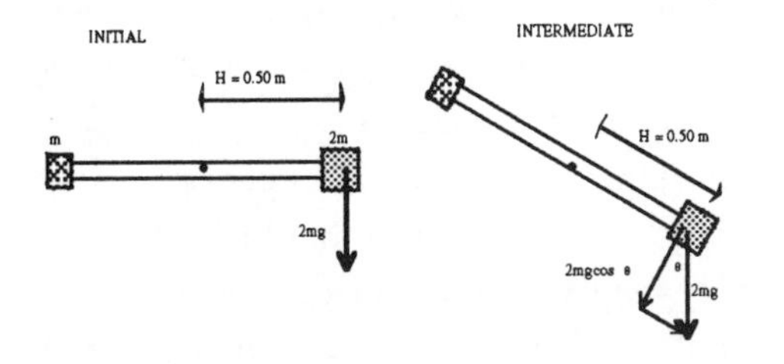

Initially, the torque on that block is $\tau = F_{\perp}r = (2mg)H$. But remember, only the part of the force perpendicular to the relevant radius vector "contributes" to the torque. So, in the intermediate stage, $\tau = F_{\perp}r = (2mg\cos\theta)H$, which *differs* from the initial torque. Indeed, this expression tells us that the torque is a function of the angle through which the rod has swung. Since the angle keeps changing, so does the torque. (The same reasoning applies to block m.) Because the net torque is not constant, the angular acceleration is not constant, and hence our old constant-acceleration kinematic formulas are not valid. We need another problem-solving strategy.

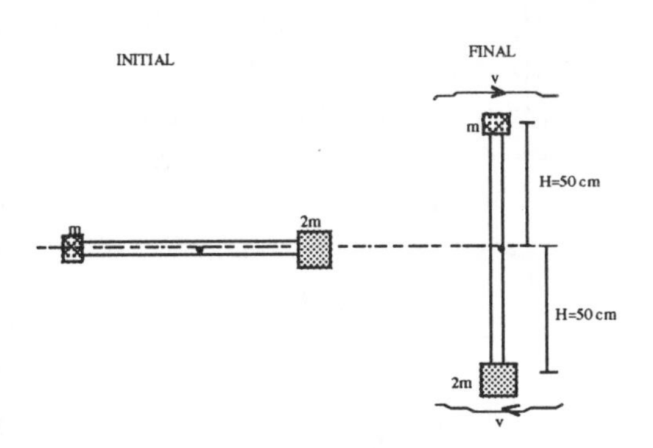

Well, what did we do in linear problems when the forces were not constant, e.g., when a spring pushed a block? We used conservation laws instead of forces and accelerations. The same techniques work in rotation problems. Here, we will use energy conservation.

(a) As always when using energy conservation, start by sketching the relevant stages of the process.

Both blocks end up with the same speed v, because the whole system "spins together" at the same angular speed ω, and both blocks are the same distance $r = H$ from the pivot. (Since block $2m$ travels in a circle, its speed around the center is $v = \omega r$. The same goes for block m. Therefore, since ω and r are the same for both blocks, they share the same speed v.)

Let us first figure out what kinds of energy are involved. There is gravitational potential energy, because masses rise and fall. There is also kinetic energy. But is it translational or rotational? In other words, do the blocks "rotate" with the rod, or do they translationally move through space? Either perspective is valid, because the blocks are small compared to the length of the rod, and therefore they can be treated as point masses. Let me sum this up in a hint: **Kinetic energy hint #1: When writing the kinetic energy of a point mass, you can use either $K_{\text{trans}} = \frac{1}{2}mv^2$ or $K_{\text{rot}} = \frac{1}{2}I\omega^2$, but not both.** To see why you have this choice, you must remember that the rotational inertia of a point mass is $I_{\text{point mass}} = mr^2$, and that the linear speed of a point on a rotating object relates to the angular speed by $v = \omega r$. Using those facts, you can see that for a point mass,

$$\text{POINT MASS} \quad K_{\text{rot}} = \frac{1}{2}I\omega^2 - \frac{1}{2}(mr^2)\left(\frac{v}{r}\right)^2 = \frac{1}{2}mv^2 = K_{\text{trans}}.$$

So, for a point mass, "translational" and "rotational" kinetic energy are two different ways of saying the same thing. Is this also true for non-point masses? NO!!! **Kinetic energy hint #2: For a non-point mass, $K_{\text{trans}} = \frac{1}{2}mv^2$ and $K_{\text{rot}} = \frac{1}{2}I\omega^2$ are different. You must use one or the other, or both, depending on the physical situation. If an object is pivoted, then it has only rotational kinetic energy, not translational.**

In part (a), the rod is massless, and hence, only the blocks matter. Since they are point masses, kinetic energy hint #1 applies. In order to pave the way for part (b), I will treat the blocks as rotating objects, and use K_{rot} instead of K_{trans}. But that is just my arbitrary choice.

What about potential energy? Intuitively, since block $2m$ falls, it loses potential energy. And block m rises, gaining potential energy. Indeed, from the diagrams, we can see that block m rises through height H, and therefore gains potential energy $\Delta U_1 = mgH$, while block $2m$ loses twice that potential energy: $\Delta U_2 = -(2m)gH$. So, we can say that block m starts off with potential energy $U_i = 0$ and ends up with $U_f = mgH$. Similarly, we can say block $2m$ starts off with $U_i = 2mgH$ and ends up with $U_f = 0$. Or, we can say that block $2m$ starts off with $U_i = 0$ and ends up with $U_f = -2mgH$. The choice is yours, provided block $2m$ ends up with $2mgH$ joules less potential energy than it starts with. I will make this second choice. In other words, I will choose the dashed line on the sketches as my "zero point" where $h = 0$.

No heat gets produced as the blocks swing. Kinetic and potential cover all bases. So, we are ready to use conservation of energy:

$$E_0 = E_f$$

$$U_0 + K_0 = U_f + K_f$$

$$U_{\text{block } m,\, i} + U_{\text{block } 2m,\, i} + K_0 = U_{\text{block } m,\, f} + U_{\text{block } 2m,\, f} + K_f$$

$$0 + 0 + 0 = mgH + (-2gmH) + \frac{1}{2}(I_{\text{block } m})\omega^2 + \frac{1}{2}(I_{\text{block } 2m})\omega^2$$

$$0 = mgH - 2mgH + \frac{1}{2}(mH^2)\omega^2 + \frac{1}{2}(2mH^2)\omega^2,$$

where in the last step I used the rotational inertia of a point mass, $I = Mr^2 = MH^2$. (For both blocks, $r = H$.)

The only unknown in this equation is the angular speed ω. So, we can solve for ω, and then set $v = \omega r$ to find the linear speed v of the blocks. (Or you can plug $\omega = \frac{v}{r} = \frac{v}{H}$ directly into the above energy-conservation equation, and solve for v directly.)

Algebra starts here. Divide through by mH. This leaves

$$\begin{aligned} 0 &= g - 2g + \frac{1}{2}H\omega^2 + H\omega^2 \\ &= -g + \frac{3}{2}H\omega^2. \end{aligned}$$

So, $\omega = \sqrt{\frac{2g}{3H}}$. Therefore,

$$\begin{aligned} v &= r\omega \\ &= H\omega \qquad \text{[since } r = H \text{ for these blocks]} \\ &= H\sqrt{\frac{2g}{3H}} \\ &= \sqrt{\frac{2gH}{3}} \\ &= \sqrt{\frac{2(9.8 \text{ m/s}^2)(0.50 \text{ m})}{3}} \\ &= 1.8 \text{ m/s}. \end{aligned}$$

End of algebra.

(b) The reasoning here duplicates the reasoning of part (a). The only modification is to include the potential and kinetic energy of the rod.

Let us start with potential energy. Again, I will choose the dotted line (in my drawing) as the zero point. Which point on the rod should we consider when calculating its potential energy? After all, the right end of the rod falls through distance H, while the left end rises by that same height. Indeed, half of the rod rises, while the other half falls. So, "on average," the rod neither rises nor falls. Let me give a shortcut for finding the "average" potential energy. **Potential energy hint #1: When writing the gravitational potential energy of an object, consider its center of mass. Measure your heights to the center of mass, not to the "ends."** According to this hint, when calculating the rod's potential energy, we should focus attention on the center. But the center neither rises nor falls. So, the initial height of the rod is $h = 0$, and the final height is also $h = 0$, because the midpoint of the rod stays put. Therefore, the rod's potential energy does not change as it rotates. Its initial potential energy equals its final potential energy, and hence, those terms cancel out of the energy conservation equation. I like this conclusion, because it agrees with the above intuitive reasoning about what the rod is doing "on average."

What about the rod's final kinetic energy? Since it is rotating but not translating, it has only K_{rot}. Look up the rotational inertia of a rod around its center to get $I_{\text{rod}} = \frac{ML^2}{12}$ The rod's length and mass are $L = 2H$ and $M = 6$ m. Of course, the rod and the blocks "share" the same angular speed, because they rotate together. So, the modified conservation of energy equation looks exactly like our part (a) equation, except with a new term for the rod's kinetic energy. I will just copy my equation from part (a), indicating the new term with boldface:

$$0 = mgH - 2mgH + \frac{1}{2}(mH^2)\omega^2 + \frac{1}{2}(2mH^2)\omega^2 + \frac{1}{2}\left[\frac{1}{12}(\mathbf{6\ m})(\mathbf{2}H)^2\right]\omega^2.$$

Once again, we can solve for ω, and then use $v = \omega r$ to obtain the blocks' speed.

Algebra starts here. Divide through by mH. This leaves

$$\begin{aligned} 0 &= g - 2g + \frac{1}{2}H\omega^2 + H\omega^2 + H\omega^2 \\ &= -g + \frac{5}{2}H\omega^2. \end{aligned}$$

So, $\omega = \sqrt{\frac{2g}{5H}}$, and hence

$$\begin{aligned} v &= r\omega \\ &= H\omega \qquad \text{[since } r = H \text{ for these blocks]} \\ &= H\sqrt{\frac{2g}{5H}} \\ &= \sqrt{\frac{2gH}{5}} \\ &= \sqrt{\frac{2(9.8\ \text{m/s}^2)(0.50\ \text{m})}{5}} \\ &= 1.4\ \text{m/s}. \end{aligned}$$

less than our part (a) answer. That is because, when the rod has mass, it tends to "slow the system down" with its rotational inertia.

QUESTION 15-6

In physics lab, Carrie experiments with various rolling objects. Specifically, she rolls objects down an inclined track of length $L = 1.5$ meters, that makes a $\theta = 30°$ angle with the horizontal. Unfortunately, while Carrie is not looking, her evil lab partner Winthorp removes the barrier from the bottom of the track, and places the track so that its bottom is even with the edge of the lab bench, as drawn below.

Carrie releases a solid ball (radius $R = 0.050$ m, mass $M = 0.30$ kg) from rest from the top of the track. It rolls without slipping down the track, off the bottom, and then slams into the wall of the room. The top of the lab bench is a height $H = 1.2$ m above the floor, and its edge is a distance $s = 0.60$ meters from the wall.

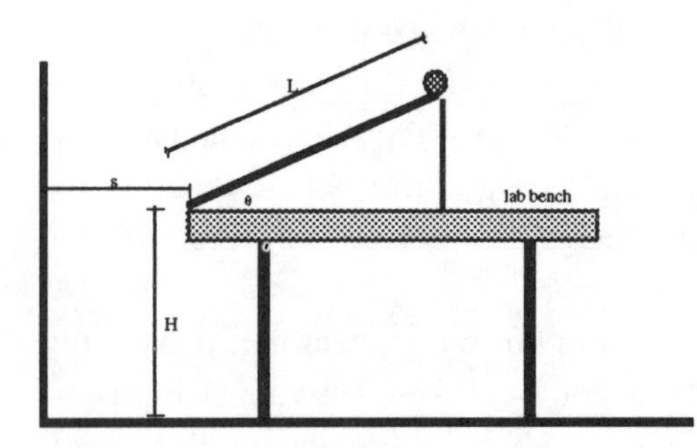

Express your answers symbolically before plugging in numbers.

(a) What is the ball's angular speed just as it leaves the track?

(b) What is the ball's angular speed right before it hits the wall?

(c) How far from the bottom of the wall does the ball hit the wall?

ANSWER 15-6

(a) This is nothing more than a "ball rolls down a ramp" problem. Although you can use forces and torques, it is quicker to use conservation of energy. As the ball rolls down, its potential energy converts into translational and rotational kinetic energy, because the ball moves translationally and also spins.

Before solving this problem, I will review the rolling without slipping condition. My hidden agenda for this review manifests itself in part (b). Intuitively, the center-of-mass speed v_{cm} is how fast the ball as a whole travels. In most situations, v_{cm} bears no relation to ω, the angular speed. For instance, if we mounted the ball on an axle and made it spin, then the angular speed would be large, but the translational speed would be $v_{cm} = 0$. By contrast, if we took the ball and threw it without much spin, then v_{cm} would be large, and ω would be small. In most cases, v_{cm} and w are not related in any simple way.

The major exception is when an object rolls without slipping. In that special case, the ball's translational speed is proportional to its angular speed; the faster it moves, the faster it spins. Confirm this by rolling something down an incline. As shown in the textbook, the precise relationship between v_{cm} and ω is

$$v_{cm} = R\omega \quad \text{ROLLING WITHOUT SLIPPING CONDITION.}$$

I cannot stress enough that this relation holds *only* when an object rolls without slipping. This caveat becomes crucial in part (b).

But enough about rolling. Let us invoke energy conservation. As always, start by sketching the relevant stages of the physical process.

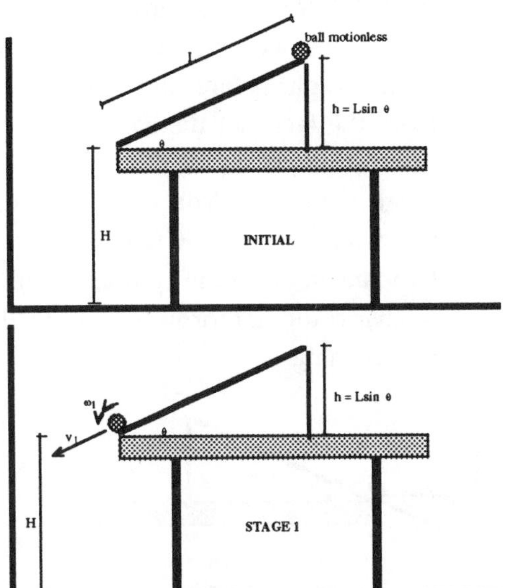

The ball starts off a height $h = L\sin\theta$ above its position at stage 1. So, we can say the stage 1 potential energy is $U_1 = 0$, while the initial potential energy is $U_0 = Mgh = MgL\sin\theta$. Or, if we measure heights from the floor up, we could say the stage 1 potential energy is $U_1 = MgH$ while the initial potential energy is $U_0 = Mg(H + L\sin\theta)$. No matter where you place your "reference point," you should agree that the ball starts off with $MgL\sin\theta$ *more* potential energy than it ends up with in stage 1. It is this *difference* in potential energy that matters when we use conservation of energy.

What about kinetic energy? Initially, the ball has none: $K_0 = 0$. But when the ball reaches the bottom of the track, it has both translational *and* rotational kinetic energy. Fortunately, when an object rolls without slipping, it does not "rub" against the surface; *static* (instead of kinetic) friction keeps it rolling So, no frictional heat gets generated. Putting all this together, we get

$$E_0 = E_1$$
$$K_0 + U_0 = K_1 + U_1$$
$$0 + Mgh = \frac{1}{2}Mv_1^2 + \frac{1}{2}I\omega_1^2 + 0$$
$$MgL\sin\theta = \frac{1}{2}M(R\omega_1)^2 + \frac{1}{2}\left(\frac{2}{5}MR^2\right)\omega_1^2,$$

where in the last step I used the rolling without slipping condition ($v = R\omega$), and the "book value" for a solid sphere's rotational inertia, $I = \frac{2}{5}MR^2$. Cancel the M's, and solve for ω_1 to get $\omega_1 = \sqrt{\frac{10gL\sin\theta}{7R^2}} = \sqrt{\frac{10(9.8\text{m/s}^2)(1.5\text{m})\sin 30°}{7(0.050\text{m})^2}} = 65\text{s}^{-1}$. (Remember, since "rad" is a dimensionless unit, you can insert factors of radians at will. So, this answer is in radians per second.) That is about 10 revolutions per second.

(b) The ball leaves the track with a "spin rate" of 65 rad/s. As it flies through the air, does this rate of rotation speed up, slow down, or stay the same? You can decide this issue experimentally by throwing a ball or pen into the air with "spin." Try it! Once in the air, the object's rate of rotation stays the same, neither speeding up nor slowing down. Why is this?

Well, just as a force is required to speed up or slow down an object's linear velocity, a torque is needed to change an object's angular velocity. But as the ball flies through the air, the only force acting on it is gravity. Gravity, as we have seen, acts on the *center* of mass, which happens to be the "pivot point" around which the ball rotates. Since gravity acts directly on the pivot point, the "r" in $\tau = F_\perp r$ is zero. So, gravity exerts no torque on the ball. Therefore, the ball in mid-air experiences no angular acceleration; its angular velocity stays constant the whole time it is in flight. Right before hitting the wall, the ball's angular speed is still $\omega_1 = 65\ \text{s}^{-1}$.

(c) Does the fact that the ball keeps spinning in mid-air alter its projectile motion? If not, then we can use standard projectile reasoning. Again, you can decide this issue experimentally by throwing a ball with spin. You will see that it traces out the same parabola it would have traced had you imparted no spin. (I am neglecting the effects of air. In softball or baseball, a "curve ball" happens because the spinning ball interacts with the air around it. In a vacuum, even a talented pitcher could not throw a curve ball!) So, we can indeed use projectile reasoning, as if the ball were not spinning.

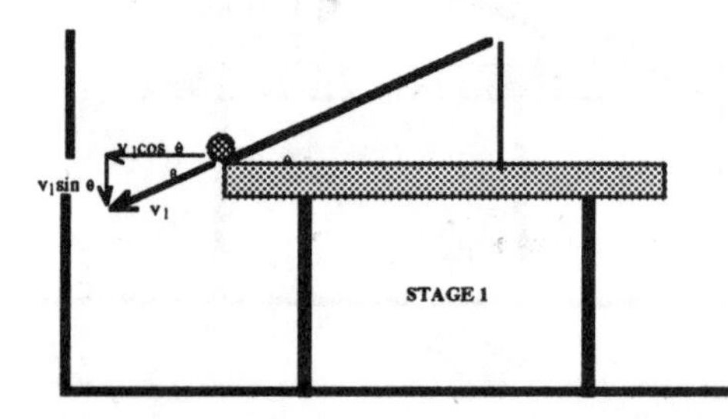

The projectile's "initial" velocity is v_1, its speed when it leaves the track. So, the "final" velocity from part (a) is the "initial" velocity here. As the above sketch shows, $\mathbf{v}_1$ points at a downward angle, because it is parallel to the track. Break up this initial velocity into its sideways and vertical components.

Crucially, the θ in this velocity vector diagram is the track's angle, $\theta = 30°$. Do we know v_1 itself? Well, from part (a), we know the ball's angular speed when it leaves the roof, $\omega_1 = \sqrt{\frac{10gL\sin\theta}{7R^2}}$. From the rolling without slipping condition, we can immediately find the ball's linear speed at that point: $v_1 = R\omega_1 = R\sqrt{\frac{10gL\sin\theta}{7R^2}} = \sqrt{\frac{10gL\sin\theta}{7}} = \sqrt{\frac{10(9.8\text{m/s}^2)(1.5\text{m})\sin 30°}{7}} = 3.24\,\text{m/s}$. That is the projectile's "initial" speed. We are solving for y, the vertical distance above the floor at which the ball strikes the wall.

So, a reasonable place to start is the vertical displacement equation, $y = y_0 + v_{0y}t + \frac{1}{2}a_y t^2$. I will choose the upward direction as positive. Given this choice, the ball's acceleration *and* its initial vertical velocity are negative (downward). Its initial vertical position is $y_0 = H$, while its initial vertical velocity is $v_{0y} = -v_1 \sin\theta$. So,

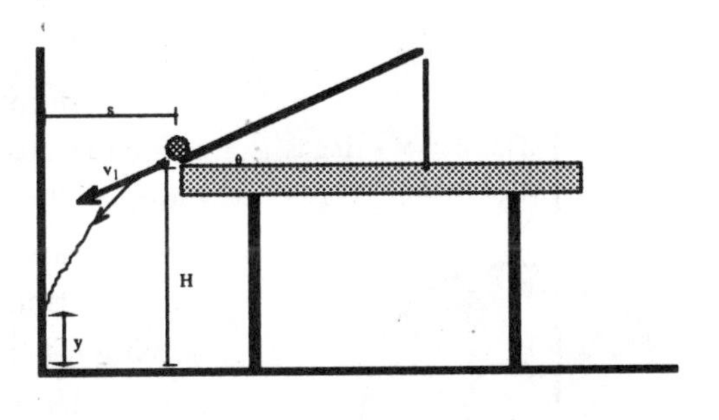

$$\begin{aligned} y &= y_0 + v_{0y}t + \frac{1}{2}a_y t^2 \\ &= H + (-v_1 \sin\theta)t - \frac{1}{2}gt^2. \end{aligned} \qquad (1)$$

The only unknown besides y is t. To find it, consider the horizontal aspects of the ball's motion. The projectile covers horizontal distance $x = s$ in time t. While in the air, the ball experiences no horizontally-directed force. So, $a_x = 0$, and hence

$$\begin{aligned} x &= x_0 + v_{0x}t + \frac{1}{2}a_x t^2 \\ s &= 0 + (v_1 \cos\theta)t + 0. \end{aligned} \qquad (2)$$

Solve for t to get $t = \frac{s}{v_1 \cos\theta}$. Substitute that expression back into Eq. (1), and recall from above that $v_1 = 3.24$ m/s, to obtain

$$\begin{aligned} y &= H + (-v_1 \sin\theta)t - \frac{1}{2}gt^2 \\ &= H + (-v_1 \sin\theta)\frac{s}{v_1 \cos\theta} - \frac{1}{2}g\left(\frac{s}{v_1 \cos\theta}\right)^2 \\ &= H - s(\tan\theta) - \frac{gs^2}{2v_1^2 \cos^2\theta} \\ &= 1.2 \text{ m} - (0.6 \text{ m})\tan 30° - \frac{(9.8 \text{ m/s}^2)(0.60 \text{ m})^2}{2(3.24 \text{ m/s})^2 \cos^2 30°} \\ &= 0.63 \text{ meters.} \end{aligned}$$

(Since the ball does not fly through the air in a straight line, you cannot figure out its "fall distance" y_{fall} using $\tan 30° = \frac{y_{\text{fall}}}{s}$.)

QUESTION 15-7

Consider a yo-yo (mass M, radius R) with a very light string wrapped around an "inner ridge" of radius $\frac{R}{3}$. The yo-yo is approximately a solid disk. Somebody holds the end of the string motionless and releases the yo-yo from rest. As a result, the yo-yo falls and rotates.

What is the tension in the string? Express your answer in terms of M, g, and R.

Hint: You can think of the yo-yo as "rolling" down the string.

ANSWER 15-7

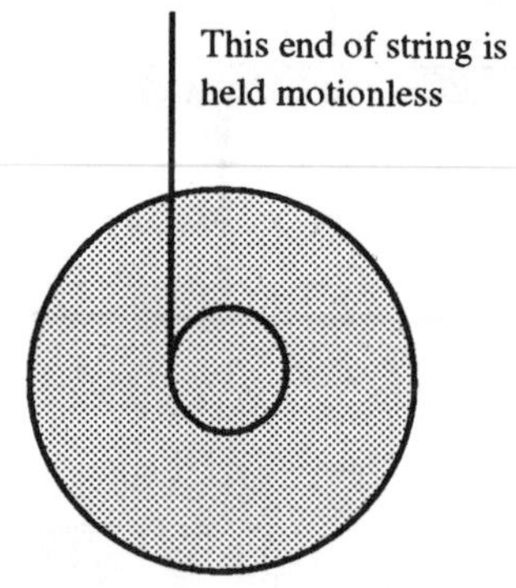

Because the yo-yo "rolls without slipping," you might think to use conservation of energy, with both rotational and translational kinetic energy. That technique would help you find, for instance, how fast the yo-yo moves after falling through a given distance. But energy conservation cannot tell you the *forces* acting on the yo-yo. Since we are solving for the tension force, we should revert to old-style force reasoning.

Unfortunately, things are complicated, because the yo-yo both moves *and* rotates. The forces *and* torques both play a role. Which shall we use?

In answer to this, let me give a general hint. **Rotational/linear motion hint:** An object's linear and rotational motion are in a sense *independent*. Even if the object is rotating, you can use linear force reasoning ($\sum F = ma$) to find the center-of-mass acceleration. And even if the object as a whole is accelerating, you can use torque reasoning ($\sum \tau = I\alpha$) to find the angular acceleration. In other words, you should *separately* deal with the linear and rotational aspects of the motion, instead of mixing them all together.

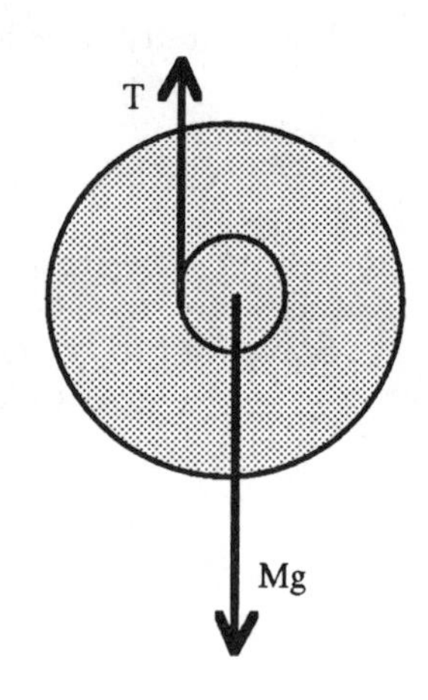

Let us see how this hint applies here. As the yo-yo falls, it gets faster and faster, accelerating downward. According to the hint, we can relate this downward acceleration to the forces acting on the yo-yo, *without* taking into account the fact that the yo-yo spins as well. In other words, we can treat this as a linear force problem, completely ignoring the yo-yo's rotational motion. So, I will use the usual strategy, starting with a force diagram.

If I pick downward as positive, Newton's 2nd law gives me

$$\begin{aligned} \sum F &= ma \\ Mg - T &= Ma. \end{aligned} \tag{1}$$

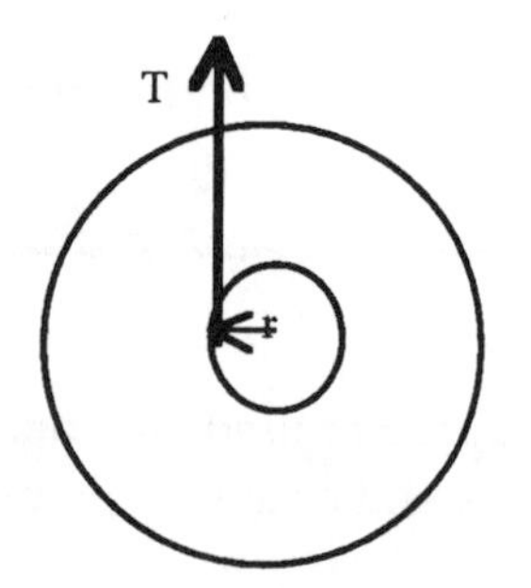

Notice that $T < Mg$; the gravitational force "beats" the tension, and hence the yo-yo accelerates downward. You might be tempted to plug in $a = g$. But the yo-yo does not free-fall. Intuitively, the string keeps it from falling at "full speed." So, the acceleration is less than g. Unfortunately, we do not know its value. So, we cannot yet solve Eq. (1) for T. We need more information. How can we get it?

According to the rotational/linear motion hint, just as we addressed the linear motion (while ignoring the rotational motion), we can also deal with the rotational motion (while ignoring the linear motion). That is, we can use torque reasoning to address the rotational aspects of the yo-yo's motion, without worrying about the fact that the yo-yo also accelerates linearly downward. Since the rotational and linear motion are in a sense independent, we can treat them separately. So, let us use torques and angular acceleration, to see if they yield useful information. From the force diagram, the torque exerted by the string is $\tau = F_{\perp}r = \frac{TR}{3}$, because the "radius vector" **r** extends from the pivot point (the center) to the "contact point" where the string pulls. Notice also that the force is perpendicular to the radius vector.

By contrast, gravity acts directly on the pivot point around which the yo-yo rotates, namely the center. So, the radius vector for gravity has length $r = 0$. Therefore, gravity exerts no torque. This makes physical sense; if we cut the string and let the yo-yo fall, it does not rotate. The tension, not gravity, is what makes the yo-yo spin. Writing Newton's 2nd law in rotational form, and remembering that a uniform solid disk has rotational inertia $I = \frac{1}{2}MR^2$, I get

$$\sum \tau = I\alpha$$
$$T\frac{R}{3} = \frac{1}{2}MR^2\alpha. \qquad (2)$$

Do we now have sufficient information to solve for the tension T? Well, by separately considering the linear and rotational motion of the yo-yo, we generated force Eq. (1) and torque Eq. (2). But these two equations contain three unknowns, T, a, and α. To solve, we need one more piece of information.

That is where the hint comes in. The yo-yo rolls without slipping down the string. So, we can use the rolling without slipping condition, usually written in the form $v_{cm} = r\omega$. But here, we need a relationship between the linear and angular *accelerations*, not velocities. Fortunately, you can differentiate both sides of $v_{cm} = r\omega$ with respect to t. Since $\frac{dv_{cm}}{dt} = a_{cm}$ and $\frac{d\omega}{dt} = \alpha$, this differentiation yields

$$a_{cm} = r\alpha, \quad \text{ROLLING WITHOUT SLIPPING CONDITION}$$

Remember, the "a" in $F = ma$ is the center-of-mass acceleration, which is the acceleration of the yo-yo as a whole. But what is "r" here? Is it R or $\frac{R}{3}$? The yo-yo rolls along the string, which is a distance $r = \frac{R}{3}$ from the center. So, the "rolling radius" is indeed $r = \frac{R}{3}$. Therefore, from the rolling without slipping condition, we can re-express the angular acceleration as $\alpha = \frac{a}{r} = \frac{a}{\frac{R}{3}} = \frac{3a}{R}$. Substitute this into Eq. (2) to get

$$T\frac{R}{3} = \frac{1}{2}MR^2\frac{3a}{R}. \qquad (2')$$

Eqs. (1) and (2') are two equations in two unknowns, T and a. So, it is just a matter of algebra to solve for T.

Algebra starts here. Notice first that all the factors of R in Eq. (2') cancel, leaving us with $\frac{T}{3} = \frac{3}{2}Ma$. Solve for a to get $a = \frac{2T}{9M}$. Substitute that expression for a into Eq. (1) to get

$$Mg - T = M\frac{2T}{9M}$$
$$= \frac{2}{9}T.$$

Now add T to both sides, and then divide through by $1 + \frac{2}{9}$, to get

$$T = \frac{Mg}{1+\frac{2}{9}} = \frac{9}{11}Mg.$$

End of algebra.

So, the upward tension is almost (but not quite) as big as the downward gravitational force. This implies that the yo-yo accelerates downward very slowly. Indeed, the acceleration works out to be $a = \frac{2}{11}g$, only about 18% of its free-fall acceleration.

I cannot overemphasize the general problem-solving strategy illustrated by this problem: When an object moves both translationally and rotationally, and you cannot use conservation laws, then you can *separately* deal with the linear motion (by writing $\sum F = ma$) and the rotational motion (by writing $\sum \tau = I\alpha$).

QUESTION 15-8

Very hard. Do not worry if you cannot get it. You will learn something by giving it a go.

A yo-yo is essentially a uniform solid disk of mass M and radius R, with a small "groove" cut out so that the string can be wrapped around an "inner ridge." Jim puts his yo-yo on the floor as pictured below, and gives the string a brief horizontal tug. As a result, a horizontal tension force briefly acts on the top of the inner ridge. Jim notices that his tug causes the yo-yo to *immediately start rolling without slipping across the floor*, though the floor is nearly frictionless.

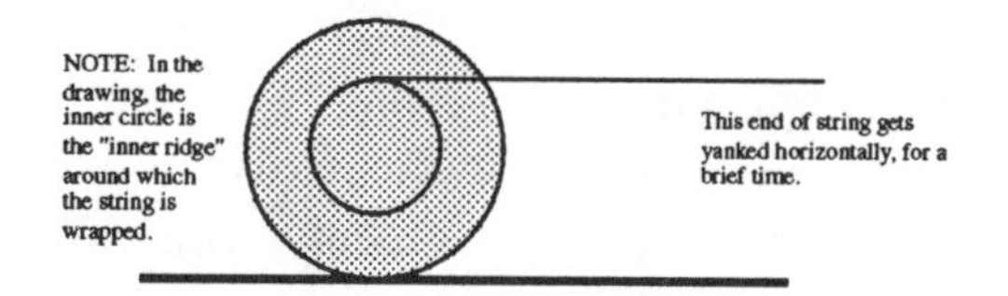

What is the radius of the inner ridge of Jim's yo-yo? Express your answer as a fraction of R.

Hint: At first, this looks hopeless, because I have not told you the tension force or the time over which that force acts. But you may freely use F and t in your reasoning, because these quantities cancel out of your final answer.

ANSWER 15-8

Let us start by thinking physically about what happens when Jim yanks the string. The yo-yo as a whole acquires some rightward velocity, v. The yo-yo also acquires some angular speed ω. Intuitively, if the inner ridge is very small, then Jim's pull gives the yo-yo little spin. By contrast, if the inner ridge is large, then the yo-yo acquires lots of spin. So, the size of the ridge "controls" the angular speed ω acquired by the yo-yo.

Given this insight, how can we find the radius r of the inner ridge? Well, as just discussed, the tug gives the yo-yo both angular and linear speed. These speeds are "properly tuned" so that the yo-yo rolls without slipping. As you have seen before, when an object rolls without slipping, its center-of-mass velocity v_{cm} (i.e., how fast the object-as-a-whole moves) and its angular velocity ω are related by

$$v_{cm} = R\omega \quad \text{ROLLING WITHOUT SLIPPING CONDITION.}$$

(Since the yo-yo rolls along the floor, not along the string, its "rolling radius" is R.) The above intuitive discussion suggests that a bigger inner ridge leads to a larger ω. We need to find the ridge radius r that produces the "right" ω, by which I mean the angular speed satisfying the rolling without slipping condition.

This suggests a strategy. Somehow, we must write an expression for v, the center-of-mass velocity after Jim's yank. (Let me omit the "cm" subscript). Then, we must write an expression for ω, the angular speed after Jim's yank. Finally, we can set $v = R\omega$. With luck, we will be able to solve that equation for r, the radius of the inner ridge.

But how can we write expressions for v and ω? That is where the hint from last problem comes in handy. Recall from above that when an object moves translationally and rotationally, we can deal *separately* with the linear and rotational aspects of its motion. Therefore, we can express v the old-fashioned way by setting $\sum F = ma$, solving for the acceleration, and then using kinematics to obtain v. This strategy safely ignores the fact that the yo-yo also rotates.

Similarly, we can express w using the rotational analogs: Set $\sum \tau = I\alpha$, solve for the angular acceleration, and then use rotational kinematics to express ω.

Let me summarize and then implement the overall strategy.

(1) Using linear motion reasoning ($\sum F = ma$ and linear kinematics), write an expression for v, the yo-yo's center-of-mass speed immediately after it is yanked. In this step, you can ignore the rotational motion.

(2) Using rotational motion reasoning ($\sum \tau = I\alpha$ and rotational kinematics), write an expression for ω, the yo-yo's angular speed immediately after it is yanked. In this step, you can ignore the linear motion.

(3) Using your expressions from steps 1 and 2, set $v = R\omega$. With luck, you can solve this equation for r, the inner ridge's radius. Remember, $v = R\omega$ is the rolling without slipping condition; I am omitting the "cm" subscript.

<u>*Step 1: Linear motion reasoning to express v.*</u>

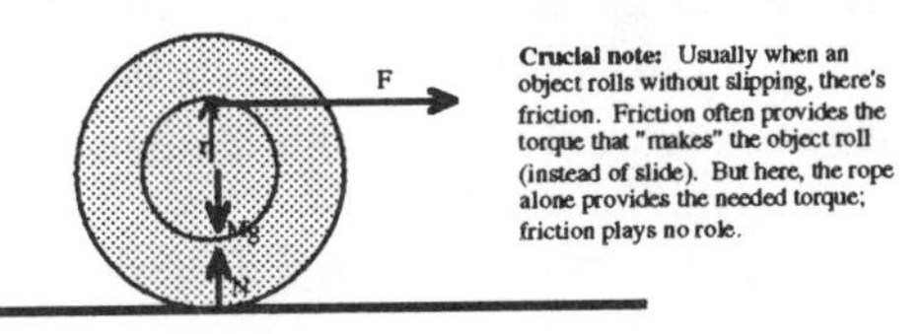

As always, start with a force diagram. This diagram applies *while Jim is pulling the string*, because we are interested in the yo-yo's motion immediately after he stops pulling. For future reference, I have included the "radius vector" r associated with Jim's tug force.

When Jim yanks the yo-yo, he applies an (unknown) horizontal force F for (unknown) time t. The normal force cancels gravity. Besides, we are interested in the yo-yo's *horizontal* motion. So, we need to find its horizontal acceleration. Newton's 2nd law gives

$$\sum F_x = Ma_x$$
$$F = Ma.$$

So, the x-directed acceleration is $a = \frac{F}{M}$. That acceleration acts for time t, the interval during which Jim pulls the string. Since the yo-yo starts at rest, the its "final" velocity (when Jim stops pulling) is

$$v = v_0 + at$$
$$= 0 + \frac{F}{M}t. \qquad (1)$$

(By the way, the momentum-impulse formula gets you to that same expression.)

So far, I have written an expression for the yo-yo's linear velocity v.

<u>*Step 2: Rotational motion reasoning to express ω.*</u>

Now let us use analogous reasoning to calculate the angular speed the yo-yo acquires from Jim's yank. Gravity exerts no torque, because it acts on the "pivot point " around which the yo-yo rotates, namely its center. So, for gravity, the "r" in $\tau = F_{\perp}r$ is 0. By contrast, the normal force acts on the bottom of the yo-yo, a distance $r = R$ from the pivot point. The radius vector associated with N points straight down, from the center to the bottom of the yo-yo. (My diagram does not include this radius vector.) But **N** points *parallel* to its radius vector. So, $N_{\perp} = 0$; the normal force exerts no torque, either. Only the yank force F exerts a torque.

Since that horizontal force points entirely perpendicular to the associated (vertical) radius vector, $\tau = F_{\perp}r = Fr$. So,

$$\sum \tau = I\alpha$$
$$Fr = \frac{1}{2}MR^2\alpha,$$

where I have used the rotational inertia of a solid disk (cylinder). Solve for angular acceleration to get a $\alpha = \frac{2Fr}{MR^2}$, and use it in the rotational analog of $v = v_0 + at$ to get

$$\omega = \omega_0 + \alpha t$$
$$= 0 + \frac{2Fr}{MR^2}t. \quad (2)$$

At this point, we have expressions for the linear speed v and angular speed ω of the yo-yo immediately after it is yanked. We obtained these expressions by *separately* considering the linear and rotational aspects of the yo-yo's motion.

Step 3: Apply the rolling without slipping condition.

We want the yo-yo to be rolling without slipping immediately after it is yanked. In other words, we want our v and ω to obey

$$v = R\omega \quad \text{ROLLING WITHOUT SLIPPING CONDITION}.$$

So, let us substitute our expressions for v and ω from Eqs. (1) and (2) into this condition, to see if we can solve for the inner ridge's radius, r:

$$v = R\omega$$
$$\frac{F}{M}t = R\left(\frac{2Fr}{MR^2}t\right).$$

We can indeed solve for r, because the F's, M's, and t's cancel! Divide both sides by $\frac{F}{M}t$ to get $1 = R\frac{2r}{R^2}$. Isolating r yields $r = \frac{1}{2}R$. So, since Jim's yo-yo rolls without slipping immediately after getting yanked, the inner ridge's radius must be half the overall radius.

This result does not change if the floor has friction. Why? Because, during the brief time Jim pulls the string, the pull force is *huge* compared to the force of friction. Therefore, we can safely neglect the forces and torques due to friction, if we are just trying to address the yo-yo's motion *immediately* after getting yanked.

QUESTION 15-9

(Extra hard. Ask your teacher if you need to be able to do problems of this complexity.)

Consider a soccer ball, which is approximately a hollow spherical shell of mass M and radius R. A player kicks the ball horizontally so as to give it initial speed v_0; but the kick does not give the ball any spin. So, the initial angular speed is 0. In other words, the ball initially slides. But the soccer field is not frictionless. In fact, the coefficient of kinetic friction between the field and the ball is μ. Consequently, as the ball slides across the field, it gradually starts rotating. Eventually, it starts rolling without slipping.

How fast is the ball moving when it first starts to roll without slipping? Express your answer in terms of v_0, M, R, g, and μ.

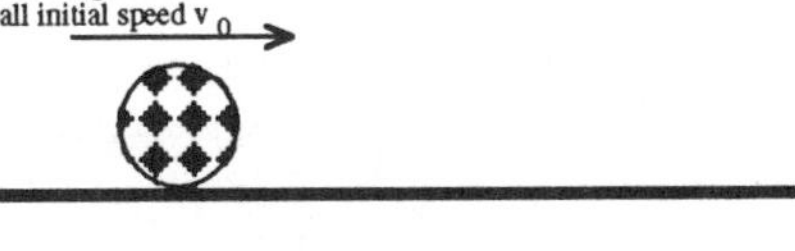

Hint: The "kick" force gives the ball speed v_0, but then "turns off." So, as the ball slides across the field, the kick force no longer acts.

ANSWER 15-9

Let us first intuitively picture what is happening. Unlike in the previous problem, the object does not roll immediately after getting "kicked." The soccer ball initially slides, with no rotation. As the ball slides across the field, friction tends to resist its motion. So, the ball slows down, at least until it starts rolling. As the ball slows down, it also starts to rotate, slowly at first, but getting faster and faster. To see this, slide a ball across a waxed floor, and play with the computer simulation. In summary, as the ball slides, its linear velocity v decreases, while its angular velocity ω increases. We are looking for the ball's speed when it starts to roll without slipping. But as we have seen before, when the ball starts rolling, its linear and angular speeds are related by

$$v_{cm} = R\omega \quad \text{ROLLING WITHOUT SLIPPING CONDITION.}$$

So, in this problem, we are really finding how fast the ball is moving when its velocity has decreased enough, and its angular speed has increased enough, that $v = R\omega$. (I will omit the "cm" subscript for the rest of this answer.) How can we find when the velocities reach this point?

Well, as I have emphasized in the previous two problems, *an object's linear and rotational motion can be treated separately*. Using linear force-and-acceleration reasoning, we can write an expression for the ball's velocity v as a function of time. Using rotational torque-and-angular-acceleration reasoning, we can write the angular velocity ω as a function of time. Then, we can set $v = R\omega$. Solving that equation should allow us to find the time at which the ball starts rolling. But once we have that time, kinematics should enable us to find the corresponding velocity, distance, or whatever we want.

In brief, this problem conceptually resembles Question 15-8 above. We can invoke a similar strategy:

(1) Using linear motion reasoning ($\sum F = ma$ and linear kinematics), write an expression for v, the ball's center-of-mass speed as a function of time. In this step, you can ignore the rotational motion.

(2) Using rotational motion reasoning ($\sum \tau = I\alpha$ and rotational kinematics), write an expression for ω, the ball's angular speed as a function of time. In this step, you can ignore the linear motion.

(3) Using your expressions from steps 1 and 2, set $v = R\omega$. Solve for t, the time at which the ball starts to roll without slipping. Remember, $v = R\omega$ is the rolling without slipping condition; I am omitting the "cm" subscripts.

(4) Once you know this time, linear kinematics should tell you how fast the ball is moving at this point.

Now I will implement the strategy.

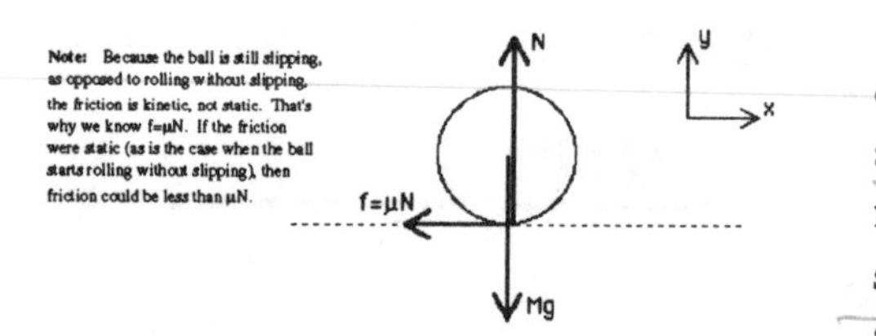

Step 1: Linear motion reasoning to express v.

As always, start with a free-body diagram. This diagram applies *after* the kick is over, but before the ball starts rolling. For this reason, the force due to the kick does not appear. Also, the friction is kinetic, because the ball *slides* (and rotates) all the way up to the time at which it starts rolling.

Since the ball does not move vertically, the vertical forces cancel: $N = Mg$. Therefore, the frictional force is $f = \mu N = \mu Mg$, leftward. Since the ball moves rightward, friction slows it down. To reflect this fact, I will choose rightward as positive, thereby making friction negative. Newton's 2nd law (in the horizontal direction) gives

$$\sum F_x = Ma_x$$
$$-\mu Mg = Ma.$$

Solve for the acceleration to get $a = -\mu g$.

Now that we know the ball's acceleration, we can easily express the velocity as a function of time, using an old kinematic formula,

$$\begin{aligned} v &= v_0 + at \\ &= v_0 - \mu g t. \end{aligned} \tag{1}$$

So far, I have expressed the ball's linear velocity as a function of time. As expected, v decreases with time. Because the linear and rotational motion are independent, nowhere in this purely linear reasoning did I take into account the fact that the ball rotates as it slides.

Step 2: Rotational motion reasoning to express ω.

Instead of using forces, acceleration, and linear kinematics, we must use torques, angular acceleration, and rotational kinematics–an exact rotational analogy to step 1.

From the force diagram, we see that gravity exerts no torque, because it acts on the center of mass, and hence does not make the ball rotate. (Mathematically, the "r" in $\tau = F_\perp r$ is zero.) The normal force also exerts no torque, because that force points radially towards the center of the ball. Let me explain. Since the normal force acts on the bottom of the ball, its radius vector points straight down, from the center to the bottom of the ball. So, the radius vector for N is vertical. But N is also vertical. No component of N points perpendicular to its radius vector. Hence, $N_\perp = 0$; that force generates no torque. Only friction creates a torque. This makes sense; if the field were frictionless, the ball would slide forever instead of starting to rotate. Intuitively, friction is responsible for "torquing" the ball, making it spin.

As we saw in step 1, the frictional force is $f = \mu Mg$. (In a minute, I will discuss whether the corresponding torque gets a minus sign.) Since friction acts on the bottom of the ball, a distance $r = R$ from the center, the radius vector has length R. Also, friction points perpendicular to its radius vector. So, $\tau_{\text{fric}} = F_\perp r = \mu MgR$.

At first glance, you might want to throw a minus sign into $\tau_{\text{fric}} = \mu MgR$, because the friction vector points backwards. But with torques, "plus" and "minus" correspond to clockwise and counterclockwise (or vice versa), not to right and left. As the ball slides rightward, it starts to rotate clockwise. Therefore, since we chose rightward as positive for our forces, we can keep things simple (when we use $v = R\omega$ later on) by calling clockwise positive for our torques. Because friction makes the ball spin clockwise, the torque it creates is positive.

Since no torques other than τ_{fric} act on the ball, Newton's 2nd law (in rotational form) says

$$\sum \tau = I\alpha$$
$$\mu MgR = \frac{2}{3}MR^2\alpha,$$

where I have used the textbook's expression for the rotational inertia of a spherical shell, $I = \frac{2}{3}MR^2$. Solve for α to get $\alpha = \frac{3\mu g}{2R}$.

Now that we know the angular acceleration, we can express the ball's angular speed as a function of time by using the rotational analog of $v = v_0 + at$, namely

$$\omega = \omega_0 + \alpha t$$
$$= 0 + \frac{3\mu g}{2R}t. \tag{2}$$

So far, by *separately* considering the linear and rotational motion of the ball, we have obtained expressions for its linear speed v and angular speed ω, as functions of time.

Step 3: Use the rolling without slipping condition to solve for t.

We are looking for the time (and eventually, the velocity) at which the ball starts to roll without slipping. From the rolling without slipping condition, this occurs when $v = R\omega$. So, using Eqs. (1) and (2) from steps 1 and 2 above, let us set $v = R\omega$ and see what pops out:

$$v = R\omega$$
$$v_0 - \mu g t = R\left(\frac{3\mu g}{2R}t\right).$$

We can immediately solve for t, the time at which the ball starts to roll. To complete the algebra, cancel the R's on the right-hand side, then isolate t to get $t = \frac{2v_0}{5\mu g}$. So, the time it takes the ball to start rolling without slipping is proportional to its initial speed, but inversely proportional to the coefficient of friction. Does the computer simulation confirm this?

Step 4: Kinematics.

We want the ball's speed v at time $t = \frac{2v_0}{5\mu g}$. Again using Eq. (1), $v = v_0 - \mu g t$, we get

$$v = v_0 - \mu g t$$
$$= v_0 - \mu g \frac{2v_0}{5\mu g}$$
$$= v_0 - \frac{2}{5}v_0$$
$$= \frac{3}{5}v_0.$$

Interestingly, this answer does not depend on the slipperiness of the field (as encoded by μ). No matter how hard you kick the ball (provided you give it no spin), and no matter how rough or smooth the field, the ball starts rolling without slipping when it has slowed down to three fifths (60%) of its initial speed. Of course, if the field is slipperier, the ball takes longer to slow down this much. Does the simulation confirm this?

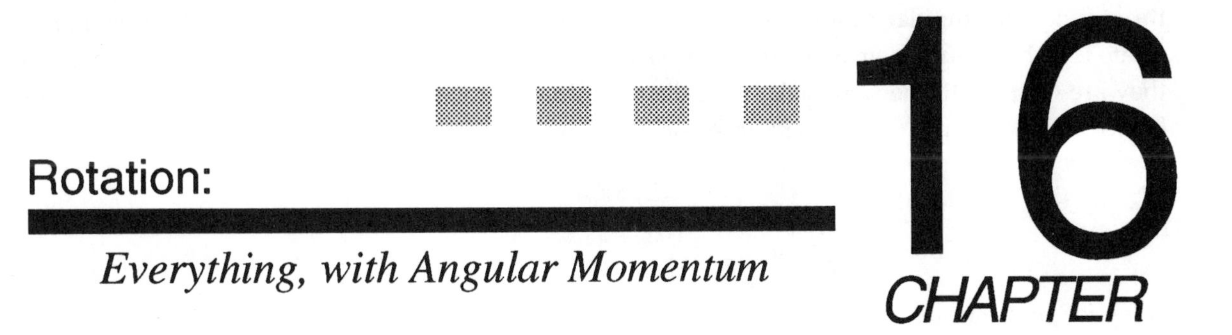

Chapter 16

Rotation:

Everything, with Angular Momentum

QUESTION 16-1

Three identical helicopter blades are mounted on a vertical axle. The axle and blades rotate around the vertical axis with angular speed ω_0. The axle's rotational inertia with respect to this axis is negligible. A fourth helicopter blade, identical to the other three, is suddenly mounted on the axle, so that all four blades now rotate with the axle. (Assume that the fourth blade does not "unbalance" the axle.)

(a) What is the angular speed of the helicopter blades after the fourth blade is mounted? Yes, you have enough information.

(b) What fraction of the initial kinetic energy is lost as a result of the fourth blade's becoming coupled to the axle? Where does this "lost" energy go?

ANSWER 16-1

(a) The hardest part of this problem is figuring out where to begin. We are not given enough information to use torques. Perhaps considering a linear analog will help us to think up a strategy.

Here, a blade at rest is suddenly "stuck" to an already rotating system, at which point the "old" and "new" blades rotate as a single system. So, a linear analog would involve (say) a block, initially at rest, that suddenly becomes stuck to an already-moving second block. This sounds exactly like what happens when one clay block collides with another. How would you solve for the post-collision velocity of the clay blocks? Conservation of momentum! (Energy conservation would not work, because heat gets created during the inelastic collision.)

By analogy, in the helicopter blade problem, we can try conservation of angular momentum. Angular momentum is conserved, because no external torque acts on the system; the only torques experienced by the blades and the axle are the "internal" torques they exert on each other.

As always in conservation problems, sketch "before" and "after" diagrams. If your diagram looks different, that is fine, provided the four blades are identical and are all attached to the axle in the same way.

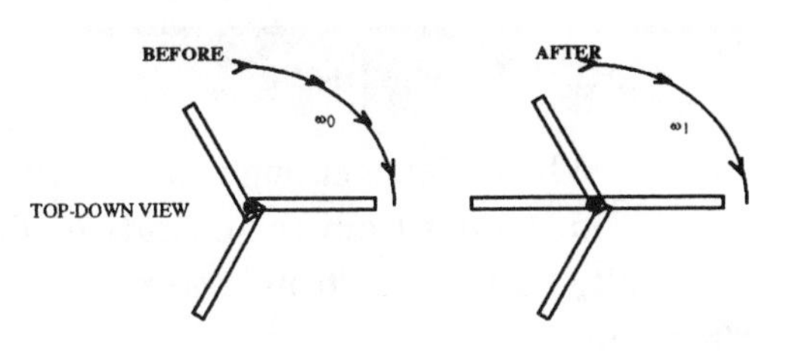

By analogy with $p = mv$, the formula for angular momentum is $L = I\omega$. I do not know the rotational inertias of the blades; but I *do* know that they all have the *same* rotational inertia. Call it "I." Initially, three blades are spinning: $L_0 = 3I\omega_0$, three times the angular momentum of a single blade. But after the "collision," the four blades rotate together at unknown angular speed, ω_f. (They share the same final angular speed, because they are coupled to the same axle.) So, the total final angular momentum is the sum of the four blades' angular momenta. By angular momentum conservation,

$$L_0 = L_f$$
$$3I\omega_0 = 4I\omega_f$$

Solve for ω_f to get $\omega_f = \frac{3}{4}\omega_0$.

As a result of the collision, the blades slow down to 75% of their initial angular speed. It is just like when a sliding block of mass 3m collides with and sticks to a stationary block of mass m. The stuck-together blocks travel at 75% of the big block's initial speed.

(b) You have answered this question before, in the context of linear inelastic collisions. The reasoning here is equivalent. To find what fraction of kinetic energy is lost, we must calculate the initial kinetic energy K_0 and the final kinetic energy K_f. The amount of kinetic energy lost is $K_{\text{lost}} = K_0 - K_f$. Therefore, fraction of kinetic energy lost $= \frac{K_{\text{lost}}}{K_0} = \frac{K_0 - K_f}{K_0}$. In analogy with $K_{\text{trans}} = \frac{1}{2}mv^2$, an object's rotational kinetic energy is $K_{\text{rot}} = \frac{1}{2}I\omega^2$. So, the initial kinetic energy, taking all three blades into account, is $K_0 = 3(\frac{1}{2}I\omega_0^2)$. The final kinetic energy is

$$\begin{aligned} K_f &= 4\left[\tfrac{1}{2}I\omega_f^2\right] \\ &= 4\left[\tfrac{1}{2}I(\tfrac{3}{4}\omega_0)^2\right] \qquad \text{(since } \omega_f = \tfrac{3}{4}\omega_0\text{, from above.)} \\ &= \frac{9}{8}I\omega_0^2. \end{aligned}$$

Therefore,

$$\begin{aligned} \text{fraction of kinetic energy lost} &= \frac{K_{\text{lost}}}{K_0} = \frac{K_0 - K_f}{K_0} \\ &= \frac{3(\frac{1}{2}I\omega_0^2) - \frac{9}{8}I\omega_0^2}{3(\frac{1}{2}I\omega_0^2)} \\ &= \frac{1}{4}. \end{aligned}$$

25% of the initial kinetic energy gets lost during the inelastic collision. Most of this energy converts into heat. Specifically, heat gets created at the contact point where blade 4 couples to the axle.

QUESTION 16-2

When a large enough star "burns up" its nuclear fuel, it "goes supernova": A huge explosion spews out matter from the outside of the star, while the inside of the star collapses in on itself so forcefully that the electrons and protons fuse into neutrons. The result is a tiny but incredibly dense "neutron star."

Consider a star three times the Sun's mass of $M_s = 1.99 \times 10^{30}$ kg, and $\sqrt{2}$ times the Sun's radius of $R_s = 6.96 \times 10^8$ m. The star happens to have the Sun's period of rotation, $T = 25$ days. In other words, the star takes 25 days to complete one rotation around its axis. For simplicity, assume that when this star goes supernova, essentially all its mass ends up in the resulting neutron star, whose radius is only $r = 14,000$ meters. What will be the period of rotation of the neutron star? Answer symbolically before plugging in numbers.

ANSWER 16-2

Intuitively, the faster an object spins, the less time it takes to complete a rotation. So, period is inversely proportion to the angular speed. To find the neutron star's period, we need to know by how much the star's ω increases or decreases during the collapse.

This is a rare example in which we cannot figure out a problem-solving strategy based on a linear analogy. Let us figure out a technique by "process of elimination."

We know absolutely nothing about the relevant forces or torques. So, conservation laws seem like our best bet. We could try energy conservation; but as the star smooshes in on itself, extra heat probably gets generated. (Indeed, astronomers know that lots of electromagnetic energy also gets produced, including prodigious quantities of visible light.) Since the star's center of mass does not move during the collapse process, conservation of linear momentum will not give us useful information. The only remaining conserved quantity we have learned is angular momentum.

Is angular momentum conserved during the collapse? In general, it is conserved when no external torque acts on the system. And indeed, as the star collapses, no outside force comes in and torques it. The collapsing star is just like a spinning ice skater who initially holds her arms out, and then pulls her arms in. As she yanks in her arms, she spins faster and faster. Here is why:

The angular momentum of a rotating object is $L = I\omega$. As the skater pulls in her arms, no external torque acts on her. So, her angular momentum stays constant. But as her arms get closer to her axis of rotation (a vertical line through the middle of her body), her rotational inertia ("resistance to rotation") decreases. Remember, an object spins more readily when most of its mass is concentrated near the center. Rotational inertia is higher when more of the mass lies farther from the pivot point. That is why a cylindrical shell has higher rotational inertia than a solid cylinder of the same mass: The shell has *all* of its mass concentrated far from the axis of rotation, while the solid cylinder has some of its mass near the axis and some of its mass far from the axis. In any case, since the skater's rotational inertia I decreases as she pulls in her arms, and since her angular momentum $L = I\omega$ stays constant, her angular speed ω must increase, to "compensate" for the decreasing I. That is why she spins faster and faster.

Similarly, when the star collapses into a smaller sphere, its rotational inertia decreases. Therefore, in order for L to stay constant, its angular speed must increase. Using this reasoning, we can figure out by how much the angular speed ω increases upon collapse. Then, using the fact that period is inversely proportional to angular speed, we can calculate by how much the period decreases from its pre-collapse value of $T = 25$ days.

Let I_0 and ω_0 refer to the uncollapsed star, while I_f and ω_f refer to the star after collapse (i.e., the neutron star). Initially, the star is approximately a solid sphere of radius $\sqrt{2}R_s$ and mass $3M_s$. So, its initial rotational inertia is $I_0 = \frac{2}{5}(3M_s)(\sqrt{2}R_s)^2$. But it collapses into a sphere of radius $r = 14,000$ meters: $I_f = \frac{2}{5}(3M_s)r^2$. By conservation of angular momentum, since no external torque acts,

$$L_0 = L_f$$
$$I_0\omega_0 = I_f\omega_f, \quad (1)$$
$$\frac{2}{5}(3M_s)(\sqrt{2}R_s)^2\,\omega_0 = \frac{2}{5}(3M_s)r^2\omega_f.$$

Since the factors of $\frac{2}{5}(3M_s)$ cancel, we can immediately solve for the final angular speed in terms of the initial angular speed:

$$\omega_f = \frac{2R_s^2}{r^2}\omega_0 = \frac{2(6.96\times10^8\text{ m})^2}{(14{,}000\text{ m})^2}\omega_0 = (4.94\times10^9)\omega_0.$$

So, when it collapses, the star's angular speed increases by a factor of about 5 billion! Therefore, since angular speed and period are inversely proportional, the star's period *decreases* by that same factor. Its new period will be about a five-billionth of 25 days: $T_f = \frac{T_0}{4.94\times10^9} = \frac{25\text{ days}}{4.94\times10^9} = 5.1\times10^{-9}$ days, which is about 0.0004 seconds.

In actuality, a typical period of rotation for a neutron star is more like 0.03 seconds. Our calculation does not take into account the angular momentum "carried off" by the material spewed out during the supernova explosion. Since lots of angular momentum gets carried away by this material, less angular momentum is left behind in the neutron star.

Notice that I was able to solve this problem without knowing the precise relationship between angular speed and period. I just had to know that they are inversely proportional. But for future reference, let me derive the precise relationship.

By definition, angular speed is how much angle per time an object covers: $\omega = \frac{\Delta\phi}{\Delta t}$, if the angular speed is constant. Now, consider a full rotation. An object spins through one rotation (2π radians) in time T, one period. So, $\omega = \frac{\Delta\phi}{\Delta t} = \frac{2\pi}{T}$. That is the relationship we are looking for. You can substitute this expression into conservation of angular momentum Eq. (1) above to get

$$\frac{2}{5}(3M_s)(\sqrt{2}R_s)^2\,\omega_0 = \frac{2}{5}(3M_s)r^2\omega_f$$
$$\frac{2}{5}(3M_s)(\sqrt{2}R_s)^2\frac{2\pi}{T_0} = \frac{2}{5}(3M_s)r^2\frac{2\pi}{T_f}. \quad (1)$$

Cancel the $\frac{2}{5}(3M_s)$ factors, and also the 2π factors. Then solve for T_f to get $T_f = \frac{r^2}{2R_s^2}T_0 = \frac{T_0}{4.94\times10^9}$, the same answer obtained above using proportionalities.

QUESTION 16-3

A long thin rod (mass M, length L) has a very small block of mass m glued to its right end. The left end is attached to a wall with a pin joint, which acts as a pivot around which the rod can rotate.

The rod is released from rest from a horizontal configuration. It swings down and crashes into the wall. What is its angular speed immediately before crashing?

ANSWER 16-3

Given the previous problems in this chapter, you might dive right into angular momentum conservation. But angular momentum is not conserved here. It is conserved only when no external torque acts on the system. Here, gravity "torques" the rod and mass, making them swing down.

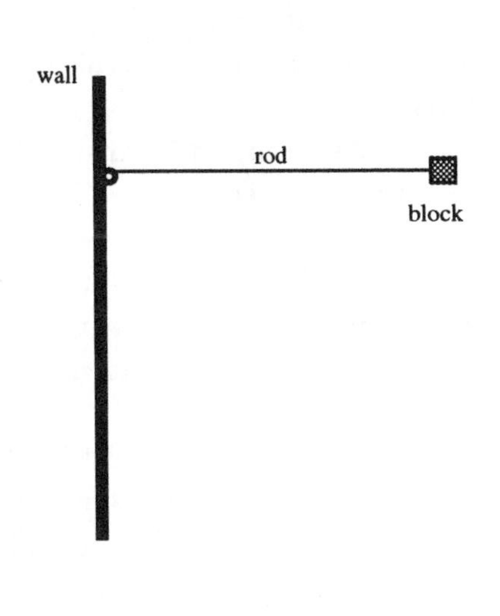

Given this insight, you might think to use $\sum \tau = I\alpha$ to find the angular acceleration, and then use rotational kinematics to find the angular speed. Unfortunately, your old kinematic formulas apply only when the acceleration (or angular acceleration) is *constant*. But here, as the rod swings down, the torque on it keeps changing. Roughly speaking, this is because the radius vectors **r** run along the rod, while the force vectors (due to gravity) point straight down. As the rod swings, its angle keeps changing, and hence, the angle that **r** makes with **F** keeps changing. Please re-read Answer 15-5 if you are the slightest bit confused; it is essential to understand why the torques here are not constant.

Since the angular acceleration is not constant, we should use conservation laws. But as noted above, angular momentum is not conserved. And linear momentum is not conserved, because external forces (such as gravity) act on the system. This leaves conservation of energy. Let us try it.

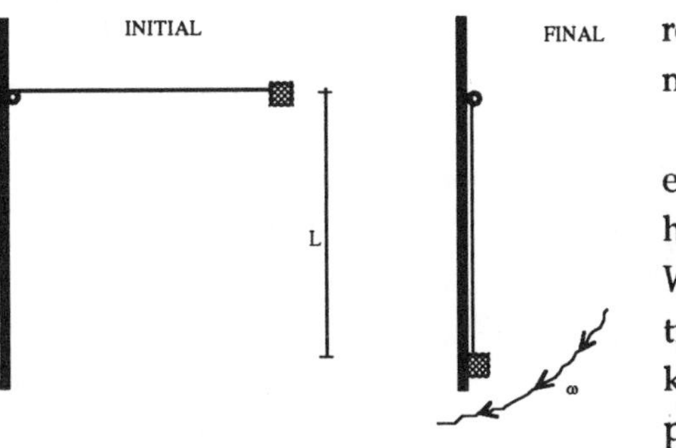

As always when using energy conservation, draw the relevant stages of the process. The "final" configuration is a millionth of a second *before* the block hits the wall.

Though the rod and block start off motionless, they end up with lots of kinetic energy (immediately before hitting the wall). Is it translational, rotational, or both? Well, since the block is a point mass, we can say that it has translational *or* rotational kinetic energy, but not both. (See kinetic energy hint #1 in Answer 15-5.) Since the rod is a non-point mass that is pivoted, its kinetic energy is entirely rotational. (See kinetic energy hint #2 in Answer 15-5.) Because I must use rotational kinetic energy to describe the rod, I might as well use K_{rot} to describe the block, too. So, the total final kinetic energy is $K_f = \frac{1}{2} I_{\text{total}} \omega^2$.

A common mistake is to write $I_{\text{total}} = \frac{1}{3}(M+m)L^2$. But since the rod and block are different shapes, you have to be more careful when calculating the total rotational inertia. A long thin rod has $I_{\text{rod}} = \frac{1}{3}ML^2$. A point mass has $I_{\text{block}} = mr^2$; and since the block is glued a distance $r = L$ from the pivot point, $I_{\text{block}} = mL^2$. So, the total rotational inertia is

$$\begin{aligned} I_{\text{total}} &= I_{\text{rod}} + I_{\text{block}} \\ &= \frac{1}{3}ML^2 + mL^2 \\ &= \left(\frac{1}{3}M + m\right)L^2, \end{aligned}$$

which differs from $\frac{1}{3}(M+m)L^2$. The rod's mass M "gets" a $\frac{1}{3}$, while the block's mass m does not. In any case, since the rod and block swing together, they share the same angular speed ω. We are solving for that ω.

Now let us think about potential energies. A common error is to write $U_0 = (M+m)gL$, since the block falls through height $h = L$. Well, it is true that the block falls through height L. But what about the rod? Sure, the tip of the rod falls through height L. But a point near the pivot falls though only a few centimeters. Which point on the rod should we consider when deciding how far "the rod" falls?

As discussed in Answer 15-5, and encoded in my "potential energy hint #1," you must always measure potential energies from the center of mass. So, the relevant question is, how far does the *middle* of the rod fall? The answer, you can see from the above diagrams, is $h = \frac{L}{2}$, not $h = L$. In summary: Since the block falls through height $h = L$, we can say its final potential energy is $U_f = 0$ and its initial potential energy is $U_{\text{block, 0}} = mgL$. But the rod's center of mass falls through height $h = \frac{L}{2}$. So, if we say its final potential energy is $U_f = 0$, then its initial potential energy is $U_{\text{rod, 0}} = \frac{MgL}{2}$.

(You could also calculate the potential energy of the whole system by finding its overall center of mass, and seeing how far that point falls. But it is generally easier to write potential energies for each "part" of the system, and add them up. You get the same expression either way, of course.)

So, energy conservation gives us

$$
\begin{aligned}
E_0 &= E_f \\
U_0 + K_0 &= U_f + K_f \\
\left(mgL + Mg\frac{L}{2}\right) + 0 &= 0 + \frac{1}{2} I_{\text{total}} \omega^2 \\
&= 0 + \frac{1}{2}(I_{\text{block}} + I_{\text{rod}})\omega^2 \\
&= 0 + \frac{1}{2}\left(mL^2 + \frac{1}{3}ML^2\right)\omega^2 \\
&= \frac{1}{2}\left(\frac{1}{3}M + m\right)L^2\omega^2.
\end{aligned}
$$

Solve for ω to get

$$
\omega = \sqrt{\frac{mgL + Mg\frac{L}{2}}{\frac{1}{2}(\frac{1}{3}M + m)L^2}} = \sqrt{\frac{g(m + \frac{1}{2}M)}{L(\frac{1}{6}M + \frac{1}{2}m)}} = \sqrt{\frac{g(6m + 3M)}{L(M + 3m)}}.
$$

QUESTION 16-4

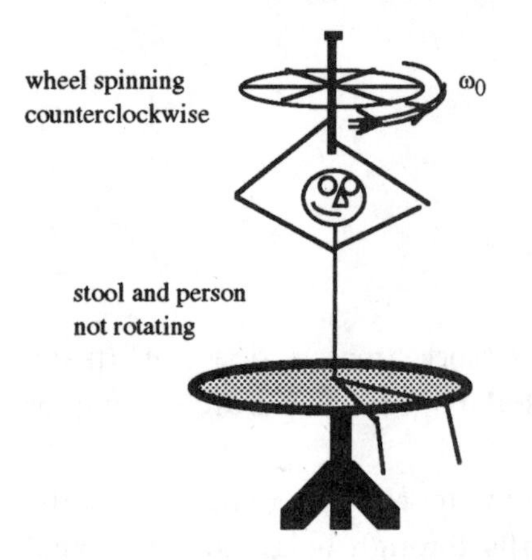

A person sits on a stool that is free to rotate. The person and stool together have rotational inertia $I_s = 12 \text{ kg} \cdot \text{m}^2$, where "s" stands for stool. Above her head, the person holds a spinning bicycle wheel of rotational inertia $I_w = 0.60 \text{ kg} \cdot \text{m}^2$. The plane of the wheel is parallel to the floor, and the center of the wheel is directly above the center of the stool. Initially, the bicycle wheel spins counterclockwise (as seen from above) at $\omega_0 = 6\pi$ rad/s. Suddenly, the person flips the wheel over, so that it is still parallel to the ground, and still directly over the center of the stool, but now spinning clockwise, still at 6π rad/s.

(a) After the wheel flip, what is the angular speed ω_s of the stool and person?

(b) Assuming the wheel ends up at the same height it started, how much work does the person perform while flipping the wheel?

ANSWER 16-4

(a) You might try to use energy conservation. But when the person flips the wheel, she does *work* on the wheel. You can confirm this by flipping a spinning bicycle wheel; it takes a lot of effort! In other words, she coverts some of her internal chemical energy into kinetic energy. So, the system ends up with *more* kinetic energy than it started with. Since we do not know how *much* more, we cannot use energy conservation.

What about angular momentum? A torque is applied to the wheel, in order to flip it. But that torque is supplied by the *person*, who is part of the system. So, it is an internal torque, by which I mean a torque exerted by one part of the system on another. Angular momentum conservation fails only when an external torque acts on the system. So here, conservation of angular momentum applies.

As always, start off by sketching the different stages of the process. To keep the drawings less cluttered, I will not include the person. But she is there.

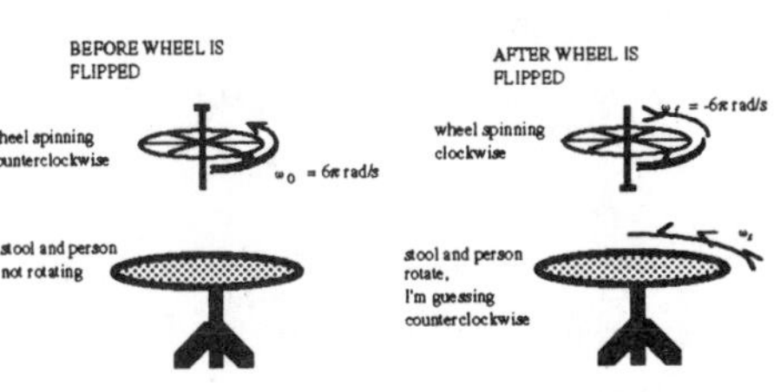

I have picked counterclockwise as my positive direction of rotation. So, the wheel's post-flip angular velocity is negative, i.e., clockwise. My conservation intuitions tell me that the stool and person end up rotating counterclockwise, to "compensate" for the flipped wheel.

Given these drawings, I am ready to apply conservation of angular momentum. Remember, the stool and person initially have no angular momentum.

$$L_0 = L_f$$

$$I_w \omega_0 = I_w(-\omega_0) + I_s \omega_s$$

$$(0.60 \text{ kg}\cdot\text{m}^2)(6\pi \text{ s}^{-1}) = (0.60 \text{ kg}\cdot\text{m}^2)(-6\pi \text{ s}^{-1}) + (12 \text{ kg}\cdot\text{m}^2)\omega_s$$

Solve for ω_s to get

$$\begin{aligned}\omega_s &= \frac{2I_w\omega_0}{I_s} \\ &= \frac{2(0.60 \text{ kg}\cdot\text{m}^2)(6\pi \text{ s}^{-1})}{12 \text{ kg}\cdot\text{m}^2} \\ &= (0.60)\pi \text{ s}^{-1} \\ &= 1.9 \text{ s}^{-1}.\end{aligned}$$

So, the stool and person end up spinning 10 times slower than the wheel initially spun. That is because the stool and person are much more "rotationally heavy," i.e., have much higher rotational inertia.

(b) As explained above, the person does work to flip the wheel; she "burns calories." Intuitively speaking, if she does (say) 20 joules of work, then the system should end up with 20 more joules of kinetic energy than it started with. This intuition is encoded in the work-energy theorem: $W_{\text{net}} = \Delta K = K_f - K_0$.

Here, the wheel neither gains nor loses kinetic energy; flipping it does not change its rate of rotation. Remember, energy has no direction; the wheel's kinetic energy is the same whether it is spinning 6π rad/s counterclockwise or 6π rad/s clockwise. But the person and stool *do* gain kinetic energy, because they started at rest but end up spinning. So, the work done by the person all gets "transferred" into the rotational kinetic energy of the stool and person.

The work-energy theorem gets us to that same conclusion. The final kinetic energy contains contributions from the stool and also the wheel, while the initial kinetic energy belongs entirely to the wheel.

$$\begin{aligned} W_{\text{net}} &= K_f - K_0. \\ &= \left\{\frac{1}{2}I_s\omega_s^2 + \frac{1}{2}I_w(-\omega_0)^2\right\} - \frac{1}{2}I_w\omega_0^2 \\ &= \frac{1}{2}I_s\omega_s^2 \\ &= \frac{1}{2}(12\text{ kg}\cdot\text{m}^2)(1.9\text{ s}^{-1})^2 \\ &= 22\text{ J}. \end{aligned}$$

Since nothing other than the person does work on the system, her work is the net work.

QUESTION 16-5

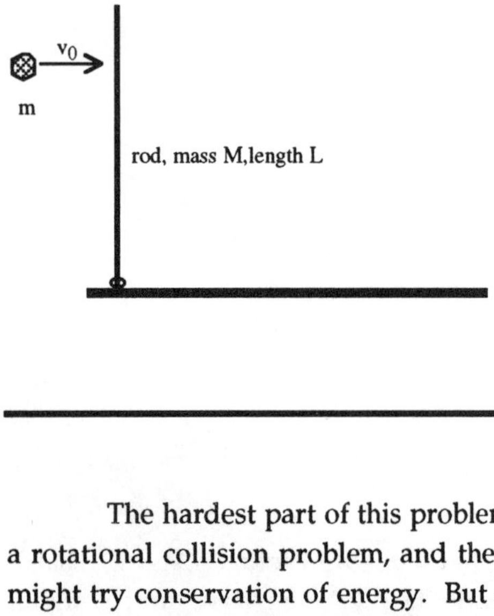

Very hard problem.

A rod of mass M and length L is attached to the floor with a pin joint, around which the rod is free to rotate. Initially, the rod is vertical and motionless. But then, a very small piece of clay (mass m) flies in horizontally at speed v_0 and hits the rod a distance $\frac{3L}{4}$ from the floor. The clay sticks to the rod.

Just before the rod and clay slam into the floor, what is the speed v_f of the clay? You need not complete all the algebra, provided you set up the right equation or equations and clearly indicate how you would solve.

ANSWER 16-5

The hardest part of this problem is figuring out how to get started. It is pretty clear that this is a rotational collision problem, and therefore we should use conservation laws. But which ones? You might try conservation of energy. But during the completely inelastic (non-bouncy) collision, heat gets generated, and we do not know how much. We cannot set the initial kinetic and potential energy equal to the final kinetic energy, because that equation would not take heat into account.

Angular momentum conservation does not apply for the whole process, either. As the rod and clay swing down, gravity exerts an external torque that speeds up the system. So, during the swing-down, the system's angular momentum increases; it is not conserved. Similar considerations apply to linear momentum; during the swing-down, gravity supplies an external force that "mucks up" momentum conservation. Apparently, none of our conservation laws apply for this whole physical process.

You have dealt with this dilemma before, back in Chapter 13 of this study guide. Remember those complicated problems in which blocks slid down hills, then collided, and so on? The trick was to divide the overall physical process into stages. During some stages, energy was conserved (with no heat). During other stages, momentum was conserved. By treating each stage separately, you could work your way, step by step, to the answer.

Similar reasoning applies here. We have got a collision, followed by a "swing-down." Let us draw the relevant stages. Remember, whenever the process includes a collision, draw a stage corresponding to immediately before the collision, and another stage corresponding to immediately after the collision.

INITIAL (Right before collision)

STAGE 1 (Right after collision)

3L/4

FINAL

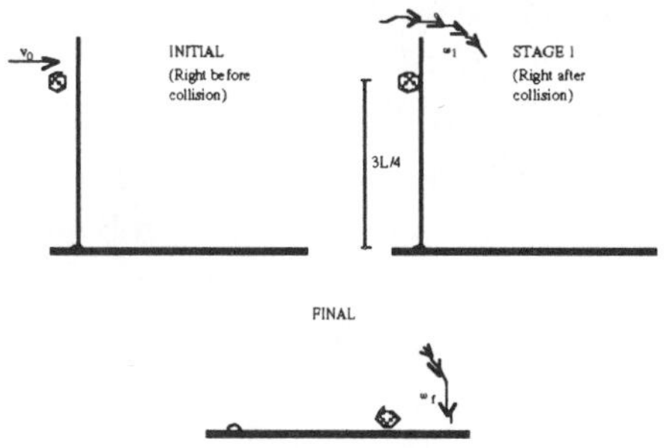

Notice that, *immediately* after the collision, the rod has not yet moved more than a millionth of a centimeter from its vertical configuration. But it *has* acquired angular speed. Now we can work step by step.

Initial → Stage 1.

During the collision, we cannot use energy conservation, because of the heat generated. So, we must use momentum. A common mistake is to combine linear and angular momentum into the same equation, by writing $mv_0 = (I_{clay} + I_{rod})\omega_1$, or something like that. You cannot do this, because linear and angular momentum are different physical quantities *with different units*. It is like equating apples to oranges. This mistake is seductive, because you *are* allowed to combine translational and rotational kinetic energies in the same equation. That is different, because K_{rot} and K_{trans} are both the same "kind" of quantity, by which I mean they both have the same units (joules). With momentum, however, you must write separate equations for linear momentum vs. angular momentum.

Is linear momentum conserved? At first glance, it appears to be, because during the collision, we are considering the horizontal momentum; and the only external force is the vertical pull of gravity. But the pin joint also exerts a force during the collision. To see why, pretend the pin joint gets disconnected, so that the rod just stands on the floor. Then, during the collision, the rod would get knocked rightward, and would end up bouncing or sliding along the floor. The pin joint is what keeps the bottom of the rod from flying rightward during the collision. Therefore, the pin joint must exert a leftward force on the rod during the collision. This external force "prevents" linear momentum from being conserved.

Does this pin joint force also prevent angular momentum from being conserved? No! The pin joint force acts directly on the pivot point around which the rod rotates. As we have seen before, any force acting on the pivot point exerts no torque, because the "r" in $\tau = F_{\perp}r$ is zero. And because the rod does not swing appreciably during the collision, gravity does not get a chance to exert a substantial torque. So, angular momentum is indeed conserved during the collision.

At this point, we must figure out how to express the angular momentum of the clay before the collision. Let us start with our general definition, $L = I\omega$. For a point mass, $I = mr^2$, where r is the point mass' distance from the pivot point. But what is the clay's "angular speed?" It does not seem to have one. But remember the relationship between linear and angular speed, $v = r\omega$. The "v" in that formula refers to the "tangential" part of the object's velocity, i.e., the component of the velocity perpendicular to the radius vector **r**. (By definition, **r** extends from the pivot point to the point mass.)

Let me call that velocity component $v_\perp$. So, $v_\perp = r\omega$. Hence, for a point mass,

$$
\begin{aligned}
L_{\text{point mass}} &= (I_{\text{point mass}})\omega \\
&= (mr^2)\frac{v_\perp}{r} \\
&= mv_\perp r.
\end{aligned}
$$

I have just derived the "special" formula for the angular momentum of a point mass. Your textbook probably writes it $\mathbf{L}_{\text{point mass}} = m\mathbf{v} \times \mathbf{r}$, a cross product. But remember, the cross product "picks off" the components of vectors that are perpendicular to each other. And "$v_\perp$" is the component of $\mathbf{v}$ perpendicular to $\mathbf{r}$. So, my formula is just another way of writing the magnitude of the cross product.

This formula *does not* apply to a non-point mass. But with a point mass, you can use either $L = I\omega$ or $L = mv_\perp r$. In this problem, I will use $L = mv_\perp r$ to describe the clay before the collision, and $L = I\omega$ to describe it after the collision. See your textbook for a more detailed discussion of the angular momentum of a point mass.

Notice that *immediately* before the collision, when the clay is a millionth of a centimeter from the rod, the clay's velocity v_0 is entirely perpendicular to the radius vector from the pivot point to the clay. That radius vector has length $r = \frac{3L}{4}$. Also, after the collision, the clay and rod "share" the same angular speed, because they swing together. So, by conservation of angular momentum,

$$
\begin{aligned}
L_0 &= L_1 \\
mv_\perp r &= \left[I_{\text{clay}} + I_{\text{rod}}\right]\omega_1, \\
mv_0\frac{3}{4}L &= \left[mr^2 + \frac{1}{3}ML^2\right]\omega_1 \\
&= \left[m\left(\frac{3}{4}L\right)^2 + \frac{1}{3}ML^2\right]\omega_1,
\end{aligned}
$$

which we can solve for ω_1 to get

$$
\omega_1 = \frac{\frac{3}{4}mv_0}{L(\frac{9}{16}m + \frac{1}{3}M)}.
$$

Now that we know the system's angular speed right after the collision, we can deal with the "swing-down."

<u>*Stage 1 → Final.*</u>

As mentioned above, neither linear nor angular momentum is conserved during the swing-down, because gravity supplies an external force and torque that speeds up the system. But energy is conserved, and no heat gets generated. (Remember, "final" is right *before* the system crashes into the floor.) Let us think about what kinds of energy are involved.

At stage 1, the clay and rod have rotational kinetic energy. Remember from Question 15-5 that we *must* treat the rod's kinetic energy as rotational, because it is pivoted. (Even though the rod' center of mass moves through space, all that motion is already "built into" the rotational kinetic energy.) By contrast, since the clay is a point mass, we can treat its kinetic energy as either translational or rotational, but not both. Let me treat it as rotational, because the clay and rod share the same angular speed.

With potential energies, we must be careful to get the heights right. The clay falls through distance $h = \frac{3}{4}L$. So, we can say its final potential energy is $U_f = 0$ and its stage 1 potential energy is $U_1 = mgh = mg\frac{3}{4}L$. You might be tempted to say the rod's stage 1 potential energy is mgL. But remember, you must measure your heights to the center of mass. The center of the rod falls through height $h = \frac{L}{2}$. So, if we say the rod's final potential energy is $U_f = 0$, then its stage 1 potential energy is $U_1 = Mgh = \frac{MgL}{2}$.

Putting all this together yields

$$E_1 = E_f$$

$$U_1 + K_1 = U_f + K_f$$

$$\left(mg\frac{3}{4}L + Mg\frac{L}{2}\right) + \frac{1}{2}\left[I_{\text{clay}} + I_{\text{rod}}\right]\omega_1^2 = 0 + \frac{1}{2}\left[I_{\text{clay}} + I_{\text{rod}}\right]\omega_f^2$$

$$\left(mg\frac{3}{4}L + Mg\frac{L}{2}\right) + \frac{1}{2}\left[m\left(\frac{3}{4}L\right)^2 + \frac{1}{3}ML^2\right]\omega_1^2 = 0 + \frac{1}{2}\left[m\left(\frac{3}{4}L\right)^2 + \frac{1}{3}ML^2\right]\omega_f^2.$$

You can immediately solve for ω_f, since we know ω_1 from above. The algebra is messy, and the problem does not require you to complete it. So, let us just say we have solved for ω_f. Unfortunately, we are not done. The problem does not ask for the system's final angular speed ω_f. It asks for the clay's final linear speed v_f. But as we have seen before, for a point on a rotating object, $v = r\omega$. So, once you find ω_f, you can easily get v_f using $v_f = r\omega_f = \frac{3}{4}L\omega_f$.

The algebra is not the point here. I want you to realize the similarity between this problem and those complicated energy/momentum conservation examples from Chapter 13. You must divide the overall physical process into stages, and then think about what is conserved and what is not conserved during each separate transition from one stage to another.

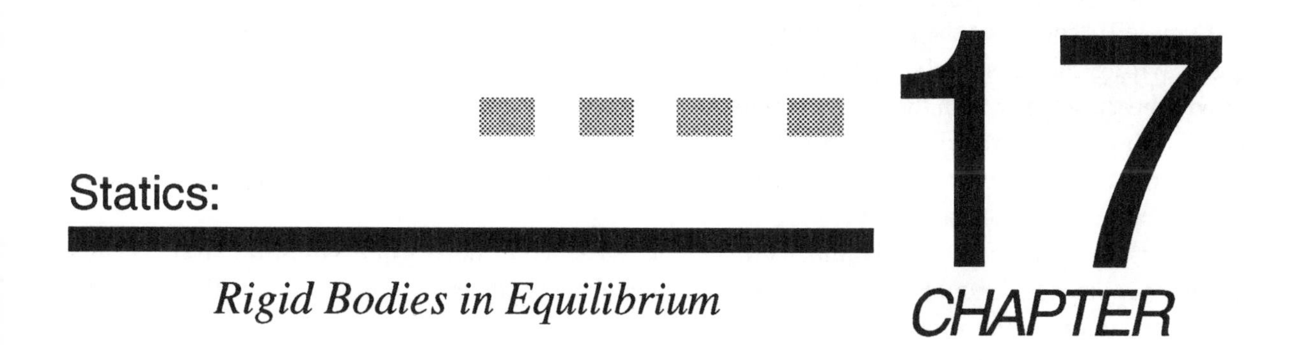

QUESTION 17-1

A long wooden plank (mass $M = 20$ kg, length $L = 8.0$ meters) is hinged to the outside wall of a building by a pin joint. A rope attached to the plank at distance $s = 5.0$ meters from the wall holds the plank in a horizontal configuration, as drawn below. The rope makes a 45° angle with the wall.

(a) What is the tension in the rope?

(b) What is the magnitude and direction of the force exerted by the pin joint on the plank?

(c) The maximum tension that the rope can withstand without breaking is $T_{\text{max}} = 1000$ newtons. How far out onto the plank can a man of mass $m = 50$ kg safely walk?

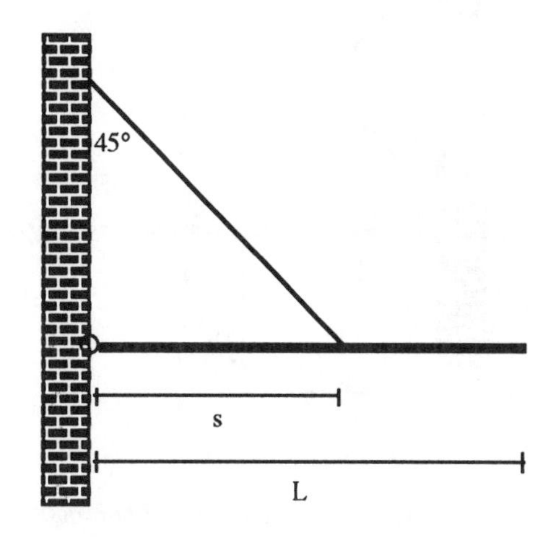

ANSWER 17-1

Conceptually, statics problems are straightforward. Since a static object neither moves nor rotates, its linear and angular acceleration are $a = \alpha = 0$. Therefore, from Newton's 2nd law, the net force and net torque on the object are both zero. So, by setting $\sum F_x = 0$, $\sum F_y = 0$, and $\sum \tau = 0$, you can usually gather enough information to solve the problem.

What makes these problems difficult is the complexity of the force diagrams. You must be extremely careful to include *all* the forces, and to calculate the torques correctly.

The strategy needed to solve a statics problem is almost always the same. I will lay it out for future reference:

Step 1: Draw a careful force diagram of the object in which you are interested. Include only the forces acting *on* the object, not the forces exerted by the object.

Step 2: Pick a reference pivot point around which the object could rotate. In other words, choose the point you will "count" as the pivot. (You will usually have a choice, as I will show below. There is no right or wrong choice; any pivot point will "work.") The reference pivot point is the point from which you will measure the "r's" in the torque equation, $\tau = F_{\perp}r$.
Step 3: Using your force diagram, set $\sum \tau = 0$, $\sum F_x = 0$, and/or $\sum F_y = 0$. This usually generates enough equations to solve for whatever you want. Remember to write your torques with respect to the reference pivot point chosen in step 2.

By the way, it is usually most efficient to write your net torque equation first, because sometimes it provides enough information to solve the whole problem. But often, you will also need to consider the y-forces, the x-forces, or both.
I will now implement this strategy.
Step 1: Force diagram

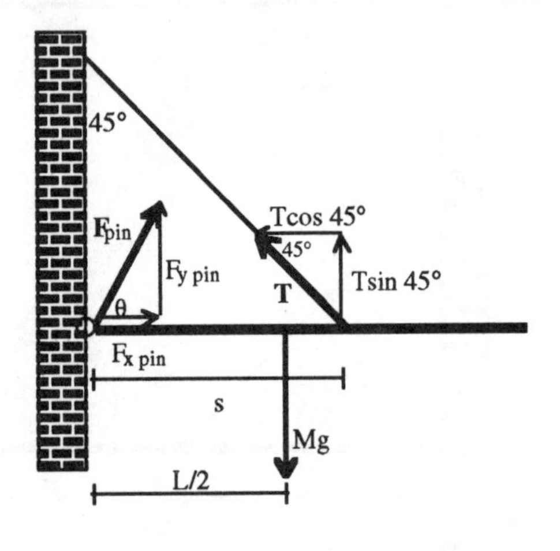

Remember that gravity acts on the center of mass of an object. So here, it acts half way down the plank. Also, do not forget that the pin joint also exerts a force. To see this intuitively, pretend the pin joint were vaporized. Then, the left end of the plank would swing down. Since the pin joint prevents this from happening, it must be helping to the hold the plank up, by exerting an upward force. In addition, the pin joint must be counteracting the leftward component of tension, or else the plank would dig into the wall. So, I am guessing that the pin joint exerts an upward and rightward force on the plank. If these guesses are wrong, I will find out by getting negative values for $F_{x\,\text{pin}}$ or $F_{y\,\text{pin}}$. The angle θ in that force diagram might differ from 45°.

Step 2: Choice of reference pivot point

Although you could choose the right end or the center of the plank, the math will work out quickest if you choose the left end, where the rod is pivoted to the wall with the pin joint.

Step 3: Torque and force equations

I will start with torques. $\mathbf{F}_{\text{pin}}$ exerts no torque, because it acts right on the reference pivot point. Therefore, the "r" in $\tau_{\text{pin}} = (F_{\perp\,\text{pin}})r$ is zero. But gravity and the tension both exert torques. For gravity, the radius vector extends from the pivot point to the middle of the rod. This radius vector is horizontal. Since gravity is entirely perpendicular to its radius vector, $\tau_{\text{grav}} = F_{\perp}r = Mg\frac{L}{2}$, clockwise.

For the tension, the relevant radius vector is horizontal, with length s. The component of T perpendicular to that radius vector is $T\sin 45°$. So, the torque due to tension is $\tau_{\text{grav}} = F_{\perp}r = (T\sin 45°)s$, counterclockwise.

Putting all this together, choosing counterclockwise as positive, yields

$$\sum \tau = (T\sin 45°)s - Mg\frac{L}{2} = 0. \tag{1}$$

(a) We can solve for T using Eq. (1) alone; no force equations are needed. That is why I advise you to write the net torque equation first; sometimes, it gets you all the way to the answer. Isolating T yields $T = \frac{MgL}{2s(\sin 45°)} = \frac{(20\text{ kg})(9.8\text{ m/s}^2)(8.0\text{ m})}{2(5.0\text{ m})(\sin 45°)} = 220\text{ N}$, well below the "maximum allowed tension" of 1000 N.

(b) We have already gathered information from the net torque equation. To solve this part, we must continue implementing the standard statics strategy, by considering the net force equations. I will choose rightward as positive for the horizontal forces, and upward as positive for the vertical forces. From the force diagram,

$$\sum F_x = F_{x\,\text{pin}} - T\cos 45° = 0. \tag{2}$$

$$\sum F_y = F_{y\,\text{pin}} + T\sin 45° - Mg = 0. \tag{3}$$

(You could equally well write $F_{\text{pin}}\cos\theta$ and $F_{\text{pin}}\sin\theta$ in place of $F_{x\,\text{pin}}$ and $F_{y\,\text{pin}}$.) We can immediately solve these equations for $F_{x\,\text{pin}}$ and $F_{y\,\text{pin}}$, because we found the tension T in part (a). From Eq. (2), $F_{x\,\text{pin}} = T\cos 45° = (220\text{ N})\cos 45° = 157\text{ N}$. From Eq. (3),

$$F_{y\,\text{pin}} = Mg - T\sin 45° = (20\text{ kg})(9.8\text{ m/s}^2) - (220\text{ N})\sin 45° = 39\text{ N}.$$

So, the magnitude of the pin joint's force is $F_{\text{pin}} = \sqrt{F_{x\,\text{pin}}^2 + F_{y\,\text{pin}}^2} = \sqrt{(157\text{ N})^2 + (39\text{ N})^2} = 160\text{ N}$. From trig, $\tan\theta = \frac{F_{y\,\text{pin}}}{F_{x\,\text{pin}}}$, and hence $\theta = \tan^{-1}\left(\frac{F_{y\,\text{pin}}}{F_{x\,\text{pin}}}\right) = \tan^{-1}\left(\frac{39\text{ N}}{157\text{ N}}\right) = 14°$.

(c) Let D denote the distance from the pivot point to where the man can stand, such that the rope *barely* does not break. So, the tension in the rope will reach its maximum allowed value, $T_{\max} = 1000\text{ N}$. The new force diagram looks like this.

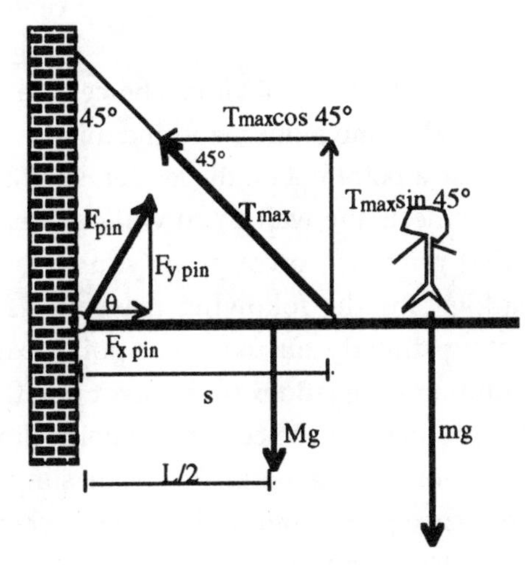

Note that F_{pin} *is different from its value in part (b), because now the pin must support extra weight.*

As usual, start with torques, letting D denote the man's distance from the pin joint. By the reasoning of part (a), we get $\sum\tau = (T_{\max}\sin 45°)s - Mg\frac{L}{2} - mgD = 0$. Be sure to use $T_{\max} = 1000\text{ N}$ instead of your T from part (a). The only unknown in this equation is D. Solving gives

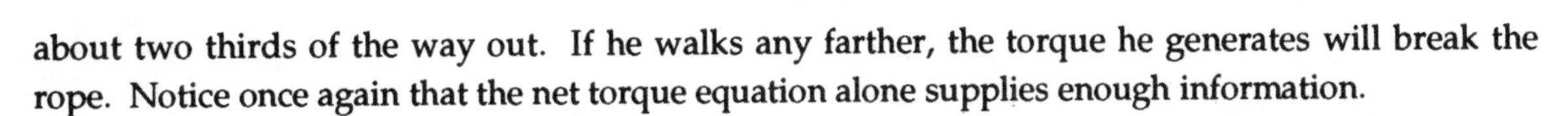

$$D = \frac{(T_{\max}\sin 45°)s - Mg\frac{L}{2}}{g}$$

$$= \frac{(1000\text{ N})(\sin 45°)(5.0\text{ m}) - (20\text{ kg})(9.8\text{ m/s}^2)\frac{(8.0\text{ m})}{2}}{(50\text{ kg})(9.8\text{ m/s}^2)}$$

$$= 5.6\text{ meters},$$

about two thirds of the way out. If he walks any farther, the torque he generates will break the rope. Notice once again that the net torque equation alone supplies enough information.

QUESTION 17-2

Consider a ladder of length $L = 4.0$ m and mass $M = 12$ kg, in the corner of a garage as drawn below. The ladder makes a 30° angle with the wall. The wall is frictionless, but the coefficient of static friction between the ladder and the floor is $\mu = 0.30$.

How high up the ladder can a man of mass $m = 60$ kg safely climb?

ANSWER 17-2

If the person climbs too high, then the frictional force exerted by the floor will not be big enough to hold the ladder in place. The biggest frictional force the floor can generate is $f = \mu N_f$, where N_f is the normal force exerted by the floor on the ladder.

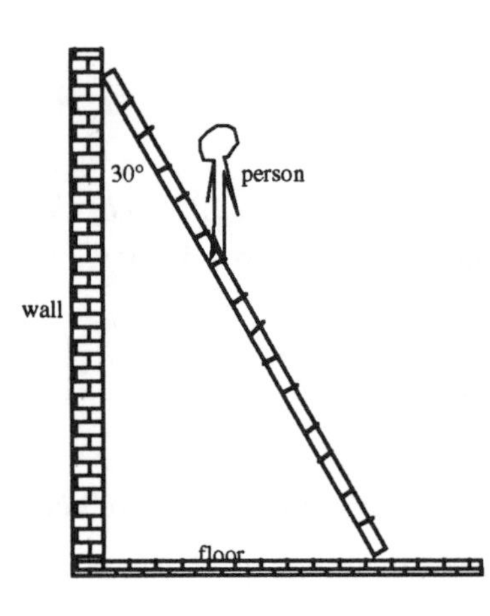

Let us invoke the statics strategy outlined in the previous problem.

Step 1: Force diagram

I have let s denote the distance up the ladder that the person can safely climb. We are solving for s. Normal forces, please recall, alway point perpendicular to the surface exerting the force. The floor's frictional force prevents the bottom end of the ladder from sliding rightward. Therefore, friction must point leftward. Remember that the gravitational force on the ladder, Mg, acts on its center of mass, a distance $\frac{L}{2}$ from either end.

Step 2: Choice of reference pivot point.

The most obvious choices are the bottom end of the ladder, the top end of the ladder, and the middle of the ladder. But you could even choose a point not on the ladder, such as the corner of the room where the floor meets the wall. You will get the same final answer for s no matter what reference pivot point you choose. I will pick the bottom of the ladder, for the following reason: As we have seen before, any force acting directly on the pivot point exerts no torque. Therefore, if the bottom of the ladder is my pivot point, then neither N_f nor f generates a torque. So, my choice of pivot point eliminates two forces from my torque equation! In general, you can simplify your math by picking your reference pivot point to be a place where many forces act. Then, none of those forces muddies up your torque equation.

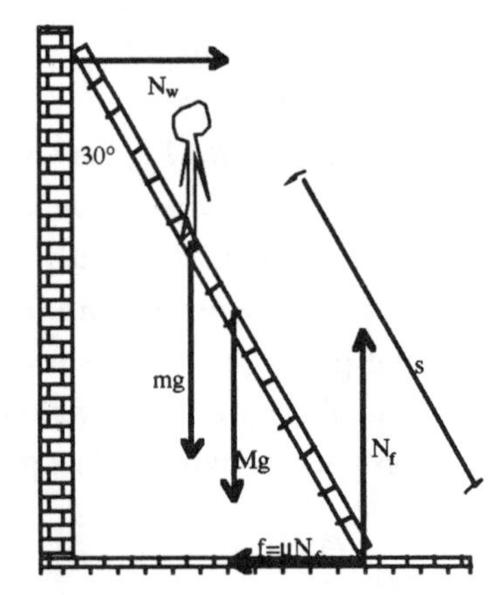

If you chose the top of the ladder, then see my force diagram and torque equation at the very end of this answer.

Step 3: Torques and forces.

I will start with torques. As just shown, my choice of pivot point ensures that only Mg, mg, and N_w exert torques. For all three of those forces, the radius vectors run along the ladder, from the pivot point (bottom of ladder) to the contact point where the force acts. So, to calculate the torques, I must break up those forces into their components parallel and perpendicular to the radius vectors, i.e., parallel and perpendicular to the ladder.

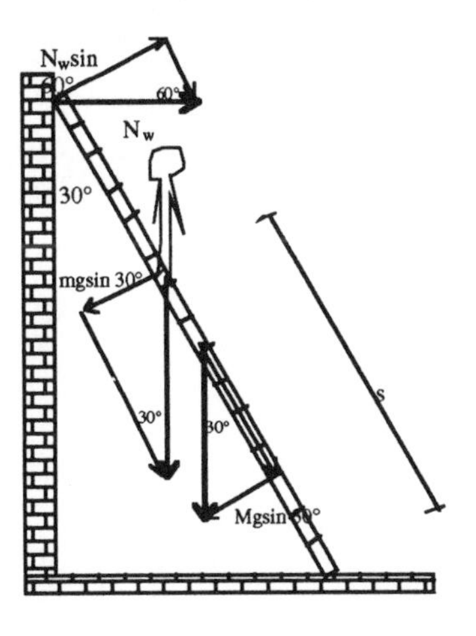

I have only bothered to label the force components perpendicular to the ladder, because only those components generate torques. I will arbitrarily pick counterclockwise as my positive direction. The two gravitational forces try to spin the ladder counterclockwise *with respect to the pivot point, the bottom of the ladder*. But the wall's normal force prevents this from happening, by torquing the ladder clockwise. Remembering that the gravitational force on the ladder acts at its center, a distance $r = \frac{L}{2}$ from the bottom, and remembering that $\tau = F_{\perp}r$, we get

$$\sum \tau = (N_w \sin 60°)L - (mg \sin 30°)s - (Mg \sin 30°)\frac{L}{2} = 0. \tag{1}$$

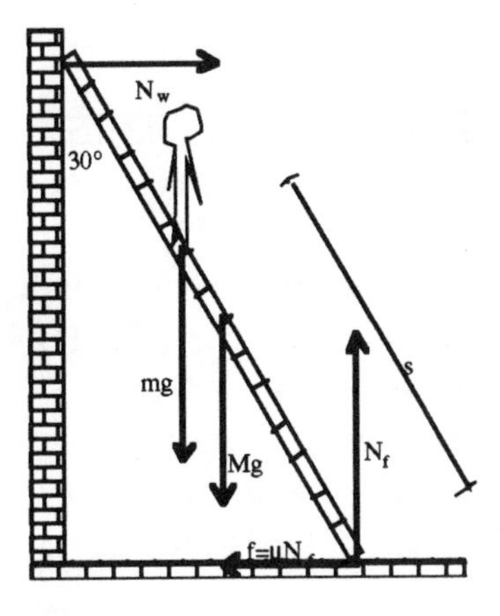

We cannot solve this equation for s, because it contains another unknown, namely N_w. As always in statics problems, you can gather more information by considering the net forces.

To calculate the torques, we had to break up force vectors along "tilted" coordinate axes, parallel and perpendicular to the ladder. You might think that we are required to continue using those tilted coordinates when we sum up forces. Fortunately, that is not true. We can switch back to "regular" coordinates if we wish. And indeed, that is worth doing, because all the forces point horizontally and vertically. So, let us pick our x-direction to be rightward and our y-direction to be upward. Then we can ignore the vector components drawn above, and work directly with the un-broken-up forces.

We immediately get

$$\sum F_x = N_w - \mu N_f = 0 \tag{2}$$

$$\sum F_y = N_f - mg - Mg = 0. \tag{3}$$

Remember, forces do not "care about" radius vectors; only torques do.

At this point, Eqs. (1) through (3) contain only three unknown (s, N_w, and N_f). So, it is just a matter of algebra to solve for s.

Algebra starts here. Solve Eq. (3) for N_f to get $N_f = (M+m)g$. Substitute that into Eq. (2), and solve for N_w, to get $N_w = \mu(M+m)g$. Finally, put that expression for N_w into torque Eq. (1) to get

$$(N_w \sin 60°)L - (mg\sin 30°)s - (Mg\sin 30°)\frac{L}{2} = 0.$$

$$[\mu(M+m)g\sin 60°]L - (mg\sin 30°)s - (Mg\sin 30°)\frac{L}{2} = 0.$$

Isolate s to get

$$\begin{aligned} s &= \frac{[\mu(M+m)g\sin 60°]L - Mg\sin 30° \frac{L}{2}}{mg\sin 30°} \\ &= \frac{[\mu(M+m)\sin 60°]L - M\sin 30° \frac{L}{2}}{m\sin 30°} \qquad \text{(cancelling the } g\text{'s)} \\ &= \frac{[(0.3)(12\text{ kg} + 60\text{ kg})\sin 60°](4.0\text{ m}) - (12\text{ kg})\sin 30° \frac{4.0\text{ m}}{2}}{(60\text{ kg})\sin 30°} \\ &= 2.1\text{ m}, \end{aligned}$$

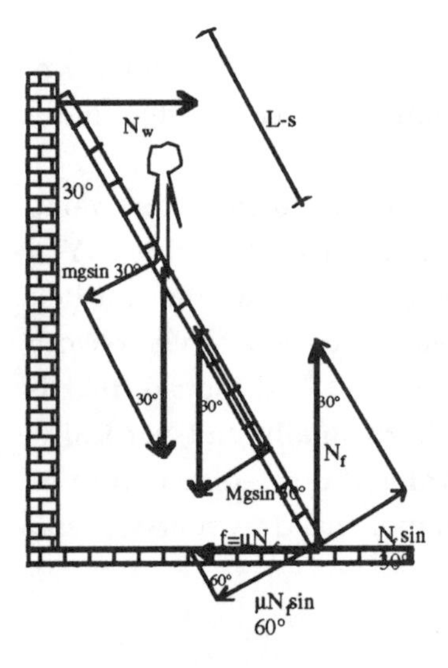

slightly more than half way up the ladder.
End of algebra.

Before moving on, let me discuss how things would have differed had I chosen the top of the ladder, instead of the bottom of the ladder, as my reference pivot point. Force Eqs. (2) and (3) would not change, because only the torques "care" about pivot points and radius vectors. But the torque equation would differ from Eq. (1). With the new choice of pivot, N_w no longer exerts a torque (because it acts directly on the pivot); but now N_f and f do create torques.

Notice that the person stands a distance $L-s$ from the new pivot. Picking clockwise (with respect to the *top* of the ladder) as my positive direction, I get

$$\sum \tau = (mg\sin 30°)(L-s) + (Mg\sin 30°)\frac{L}{2} + (\mu N_f \sin 60°)L - (N_f \sin 30°)L = 0. \qquad (1^*)$$

When I use Eq. (1*), (2), and (3) to solve for s, I get the same answer obtained above using a different reference pivot point. Indeed, I would end up with the same answer no matter what reference pivot point I choose.

QUESTION 17-3

A thick wooden board of length $L = 2.0$ meters and mass $m = 50$ kg rests on two supports, as shown below. The second support is a distance $\frac{3}{4}L = 1.5$ meters from the left end of the board. Someone wants to place an anvil of mass $M = 200$ kg as far rightward on the board as possible, so that the board does not flip over. How far to the right of the second support can the anvil safely be placed?

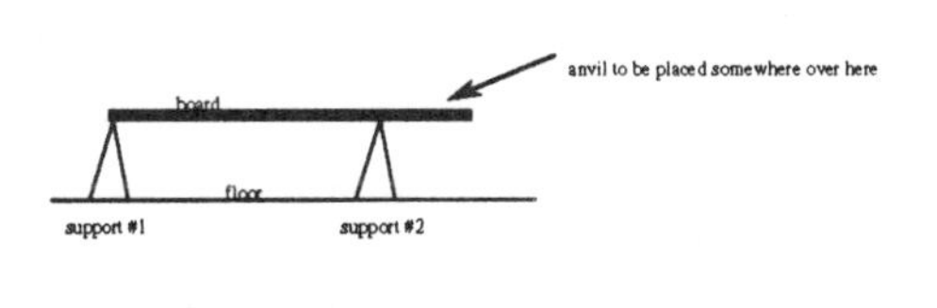

ANSWER 17-3

The standard statics problem-solving strategy will get us most of the way, but not all the way, to the answer. We will need to have an additional physical insight. But before discussing that insight, let me plow through the usual procedure, to show why it fails to take us all the way to the answer.

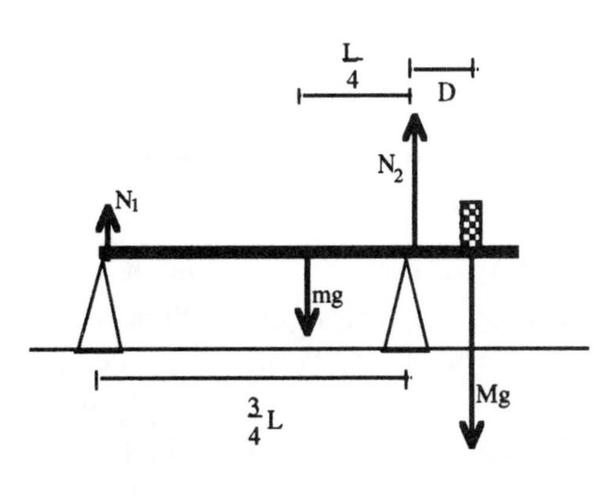

Step 1: Force diagram

Let D denote the anvil's distance from the 2nd support. That is what we are solving for. Crucially, the upward normal forces exerted by the two supports might be different. Intuitively, since support #2 is closer to the anvil, we expect it to bear more of the brunt. My diagram reflects this guess.

Notice that gravity acts on the middle of the board. You need not consider the board as consisting of two "parts." But if you did, that is o.k., provided you accurately calculated the mass of each part and draw gravity as acting on the middle of each part.

Step 2: Choice of reference pivot point

If the object were rotating instead of stationary, it would rotate around support #2. So, I will choose that as my reference point. But I equally well could have chosen support #1. Either choice of reference point leads to the same final answer.

Step 3: Torque and force equations

Since N_2 acts directly on the pivot point, it exerts no torque. Let me choose clockwise as positive. Mg and N_1 both try to make the board rotate clockwise, while mg tries to make it rotate counterclockwise with respect to support #2. So, from the diagram, remembering to measure our r's from the reference pivot point (support #2), we get

$$\sum \tau = MgD + N_1 \frac{3}{4}L - mg\frac{L}{4} = 0. \qquad (1)$$

Because this equation contains two unknowns (D and N_1), we need more information. Since no horizontal forces exist, we can consider only the vertical forces:

$$\sum F_y = N_1 + N_2 - Mg - mg = 0. \tag{2}$$

Now we have got two equations in three unknowns (D, N_1, and N_2). We lack sufficient information to solve for D. That is why we need a physical insight to complete the problem.

Imagine placing the anvil just to the right of support #2, and then nudging the anvil farther and farther rightward. As you do so, the board gets "closer and closer" to tipping over. When the board tips over, its left end comes off support #1. So, when we place the anvil as far rightward as "safely" possible, the board *barely* does not tip over. In other words, the left end of the board *barely* stays on support #1. The left of the board kind of "flutters" on top of support #1, without really resting on that support. Therefore, the normal force exerted by that support is $N_1 \approx 0$.

This insight allows us to complete the problem easily. When you set $N_1 \approx 0$ in Eq. (1), all that remains is $MgD - mg\frac{L}{4} = 0$, which we can immediately solve for D, and cancel the g's, to get

$$D = \frac{mgL}{4Mg} = \frac{mL}{4M} = \frac{(50 \text{ kg})(2.0 \text{ m})}{4(200 \text{ kg})} = 0.12 \text{ m},$$

half way between support #2 and the right end of the board.

QUESTION 17-4

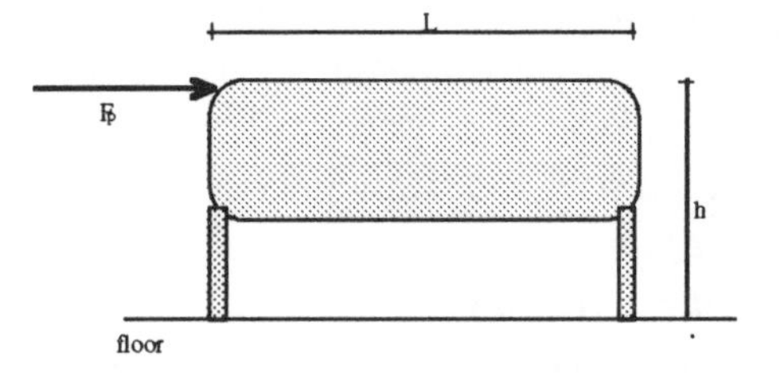

Consider a wood frame taken from the foot of a bed. The frame is $L = 2.0$ meters long and $h = 1.0$ meters high, with thin legs. Its mass is $M = 50$ kg. Someone pushes the frame across the floor *at constant speed* by applying a horizontal force F_p (of unknown magnitude) to the top left corner. I am deliberately not telling you the legs' height. The coefficient of kinetic friction between the legs and the floor is $\mu = 0.25$.

What is the frictional force exerted by the floor on the back leg (i.e., the left leg in the drawing)?

Hint: Sometimes it is convenient to write torques as $\tau = r_\perp F$ instead of $\tau = F_\perp r$.

ANSWER 17-4

A common mistake is to think that both legs support half the weight of the frame, in which case $N_{\text{back leg}} = \frac{1}{2}Mg$, and hence $f_{\text{back leg}} = \mu N_{\text{back leg}} = \frac{1}{2}\mu Mg$. But by pushing a chair or desk across the floor, you can "feel" that the front and back legs do not bear the brunt equally. The front legs support more of the weight. This becomes most noticeable if you push the chair such that it is nearly tipping over. In that case, the back legs almost come off the ground, while the front legs support almost all the weight. Even in less extreme cases, however, the front legs carry more of the load. So here, the normal force on the front leg exceeds the normal force on the back leg.

I have partly given this problem away by including it in a "statics" chapter. Although the frame is not static, it is moving at constant speed, neither speeding up nor slowing down (nor rotating). So, its acceleration is $a = 0$, and its angular acceleration is $\alpha = 0$. Therefore, by Newton's 2nd law, the forces must cancel, and the torques must cancel. For this reason, we can invoke the usual statics strategy.

Step 1: Force diagram

I have let subscripts "1" and "2" refer to the front and back leg, respectively. As just discussed, the front leg experiences a bigger normal force, and therefore a bigger frictional force. We are solving for f_2. Because the friction is kinetic, we can use $f = \mu N$. By contrast, static friction sometimes exerts a force less than its maximum allowed value of μN, and hence, you cannot always assume that $f_s = \mu_s N$.

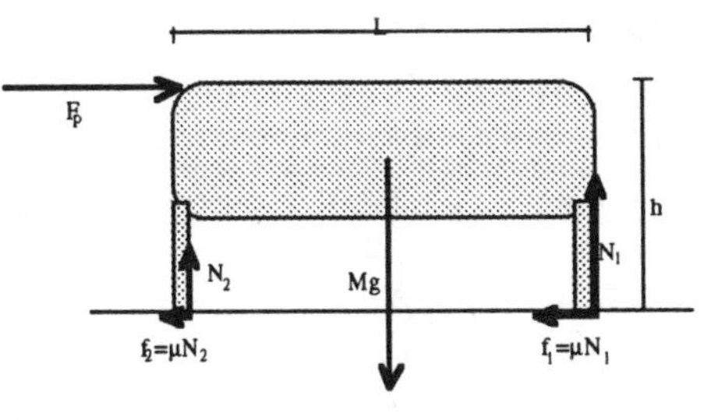

Step 2: Choose reference pivot point

If the frame were to tip over, it would tip over its front leg. So, a good reference pivot point is where the front leg touches the floor. But you would eventually reach the same final answer using any other choice of reference point.

Step 3: Torque and force equations

Let us start with torques, as usual. Neither the normal nor the frictional force on the front leg generates a torque, because those forces act on the pivot point.

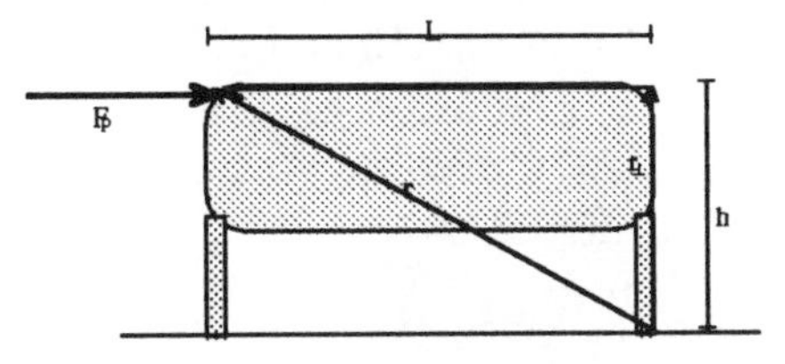

In this problem, instead of breaking force vectors into components parallel and perpendicular to radius vectors, it is easier to break radius vectors into their components parallel and perpendicular to force vectors. Let me first demonstrate this with the push force. By definition, the radius vector runs from the pivot point to the contact point where the force acts.

By definition, $r_\perp$ is the component of **r** perpendicular to the force vector. Since the push force is horizontal, $r_\perp$ is the vertical part of **r**. From the diagram, we see that $r_\perp = h$. So, the torque generated by the push force is $\tau_p = r_\perp F_p = hF_p$, clockwise. We figured this out without calculating any angles.

Let us use the same "trick" to find the torque due to gravity.

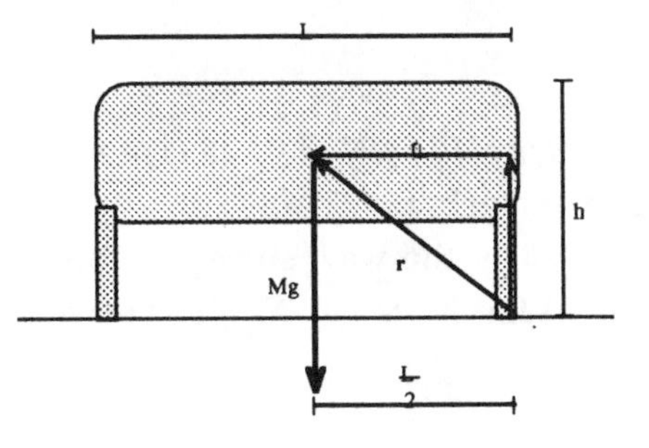

Since gravity points vertically, we want the horizontal component of **r**. Gravity acts on the center of mass, which is midway between the two legs. So, $r_\perp = \frac{L}{2}$, as the diagram shows. Therefore, the torque due to gravity is $\tau_{\text{grav}} = r_\perp F_{\text{grav}} = \frac{L}{2}Mg$, counterclockwise (with respect to the front leg.) It is cool that we could calculate τ_{grav} without knowing exactly where gravity acts, i.e., where the frame's center of mass is located. We needed to know only that it is midway between the two legs. We did not need to know its height off the floor.

Since the forces acting directly on the pivot point (the front leg) exert no torque, the only torques left to address are those acting on the back leg. I will not bother to draw in the relevant **r**; it points horizontally, from the bottom of the front leg to the bottom of the back leg, and has length $r \approx L$ (since the leg is thin). Because f_2 also points horizontally, it exerts no torque: $f_{2\perp} = 0$. But N_2 points entirely perpendicular to that radius vector. So, we do not have to break up any vectors into components: $\tau_2 = r_\perp N_2 = LN_2$, clockwise.

Adding these torques together, choosing clockwise as positive, I get

$$\begin{aligned}\sum \tau &= \tau_p + \tau_{\text{grav}} + \tau_2 \\ &= hF_p - \frac{L}{2}Mg + LN_2 = 0.\end{aligned} \tag{1}$$

Since this equation contains two unknowns (F_p and N_2), we need more information. Considering the horizontal and vertical forces, and choosing rightward and upward as my positive directions, I get

$$\sum F_x = F_p - \mu N_1 - \mu N_2 = 0. \tag{2}$$

$$\sum F_y = N_1 + N_2 - Mg = 0. \tag{3}$$

Now we have three equations in the three unknowns (F_p, N_1, and N_2). So, we can solve for N_2, and then set $f_2 = \mu N_2$. Only algebra remains.

Algebra starts here. First, solve Eq. (2) for F_p to get $F_p = \mu(N_1 + N_2)$. From Eq. (3), you can see that $(N_1 + N_2) = Mg$. So, $F_p = \mu(N_1 + N_2) = \mu Mg$. Substitute that result into Eq. (1), and solve for N_2, to get

$$\begin{aligned}N_2 &= \frac{\frac{L}{2}Mg - hF_p}{L} \\ &= \frac{\frac{L}{2}Mg - h(\mu Mg)}{L} \\ &= Mg\left[\frac{1}{2} - \frac{h\mu}{L}\right] \\ &= (50\text{ kg})(9.8\text{ m/s}^2)\left[\frac{1}{2} - \frac{(1.0\text{ m})(0.25)}{2.0\text{ m}}\right] \\ &= 184\text{ newtons,}\end{aligned}$$

and hence $f_2 = \mu N_2 = (0.25)(184\text{ newtons}) = 46\text{ newtons}$.

End of algebra.

By the way, since we now know $N_2 \approx 180$ newtons (to two significant digits), we can solve Eq. (3) for N_1 to get $N_1 \approx 310$ newtons, confirming the intuition that the front leg bears more of the burden.

QUESTION 17-5

Hard. Ask your instructor if you need to be able to address this kind of problem.

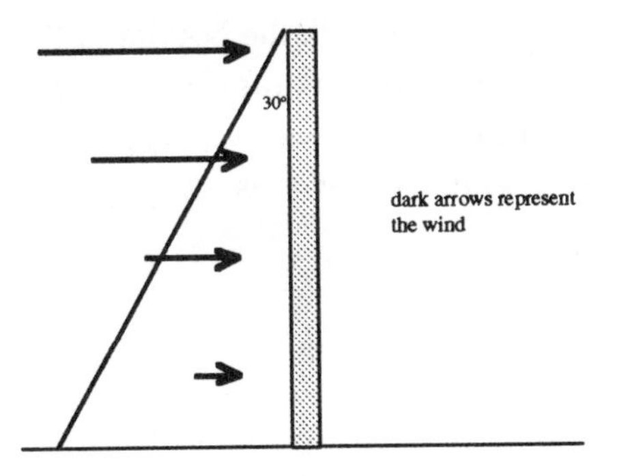

A tower in Alaska gets buffeted by strong winds. And the wind gets stronger at higher altitude. In other words, the wind blowing on the top of the tower is stronger than the wind blowing on the bottom of the tower. Specifically, the force exerted by the wind *per length of tower* is bh, where h is the height and b is a constant. This formula tells us that if you go twice as high up the tower, you will feel twice as strong wind. The total height of the tower is H.

A cable that makes a 30° angle with the tower is attached to its top. What force must the cable exert in order to prevent the tower from blowing over? Answer in terms of b and H.

ANSWER 17-5

The cable must counteract the torque due to the wind. The hardest part of the problem is calculating τ_{wind}, because the force due to the wind is not constant. We cannot just say that force is bH, because the force varies at different heights. But we know the force exerted by the wind per length of tower, i.e., the force exerted by the wind on a small segment of tower. This suggests a strategy. Starting with the force dF acting on a tiny segment of tower, we can find the torque dt on that tiny segment. Then, we can add up (i.e., integrate) the torques acting on all the segments of tower, to get the total torque on the whole tower.

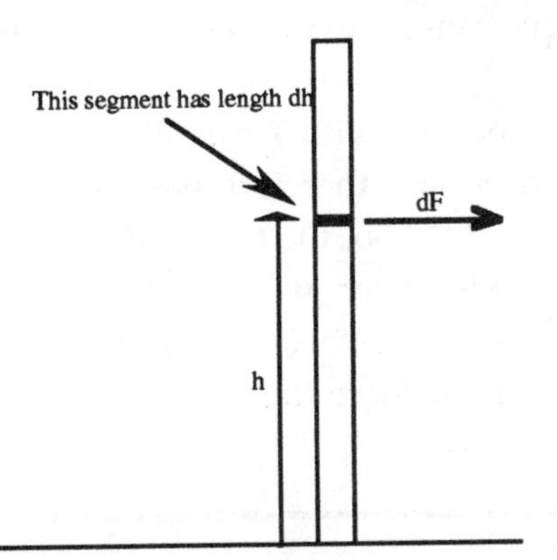

Consider the infinitesimal "slice" of the tower shaded in below. That slice is a distance h off the ground, and has length (height) dh.

The trick is to realize that the force dF on that infinitesimal segment is

$$dF = \frac{\text{force}}{\text{length}}(\text{length})$$
$$= bh(dh).$$

You can also reach this conclusion by realizing that when I told you the "force per unit length" equals bh, I was saying that $\frac{dF}{dh} = bh$, and hence $dF = bh(dh)$.

At this stage, a common error is to integrate over dF to obtain the total wind force F_{wind} on the tower, and then multiply that force by H (or $\frac{H}{2}$) to get the torque. But because the wind force is not "uniformly distributed" over the whole tower, we do not know whether (on average) it acts on the middle of the tower or elsewhere. So, we do not know what "r" to plug into $\tau_{wind} = (F_{wind\perp})r$. Instead, we need to calculate the infinitesimal torque $d\tau$ on *each* segment of tower. Then we can integrate over the $d\tau$'s. To derive the relevant expression for $d\tau$, I will continue to consider the shaded-in region of the tower drawn above.

If we pick the bottom of the tower as our reference pivot point, then the relevant radius vector extends vertically from the bottom of the tower to the shaded-in region. This radius vector has length rh, and points entirely perpendicular to the force vector $d\mathbf{F}$. So, $dF_{\perp} = dF$. Hence, the torque on that tiny section of tower is

$$\begin{aligned} d\tau &= dF_{\perp} r \\ &= (bhdh)h \\ &= bh^2 dh \end{aligned}$$

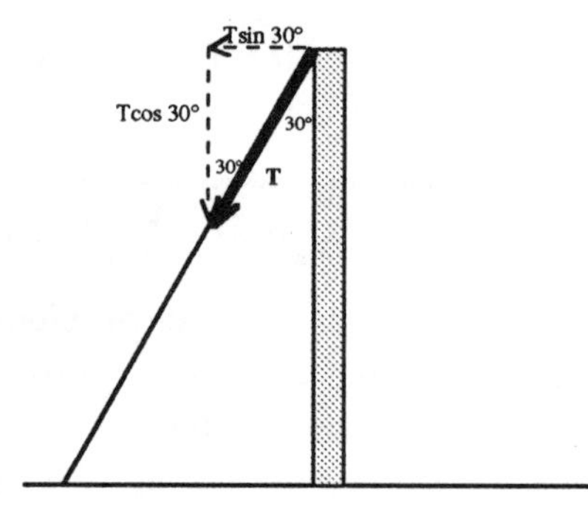

Now that we know the wind's torque on an infinitesimal section of tower, we can add up those tiny torques to obtain the total torque due to the wind. Since the tower extends from $h = 0$ to $h = H$, those are the limits of integration: $\tau_{\text{wind}} = \int d\tau = \int_0^H bh^2 dh = \frac{bh^3}{3}\Big|_0^H = \frac{bH^3}{3}$.

As noted above, the cable must counteract this torque, so that the net torque equals zero. To see how the tension in the cable relates to the torque caused by the cable, use a force diagram.

The radius vector associated with $\mathbf{T}$ extends from the bottom to the top of the tower, and therefore has length H. The component of tension perpendicular to this radius vector is $F_{\perp} = T\sin 30°$. So, the torque due to the cable is $\tau_{\text{cable}} = F_{\perp} r = (T\sin 30°)H$, counterclockwise. By contrast, the wind's torque is clockwise. So, picking clockwise as positive, I get $\sum \tau = \tau_{\text{wind}} - \tau_{\text{cable}} = \frac{bH^3}{3} - (T\sin 30°)H = 0$, and hence $\frac{bH^3}{3} = (T\sin 30°)H$. Solve for the tension to get $T = \frac{bH^2}{3\sin 30°} = \frac{2bH^2}{3}$. That is the force the cable must exert to keep the tower standing.

QUESTION 17-6

A rod of mass M and length L is attached to a wall with a pin joint, around which the rod is free to rotate. A horizontal rope attached to the end of the rod holds the rod at an angle θ_0 to the horizontal. Express your answers in terms of M, L, θ_0, and g.

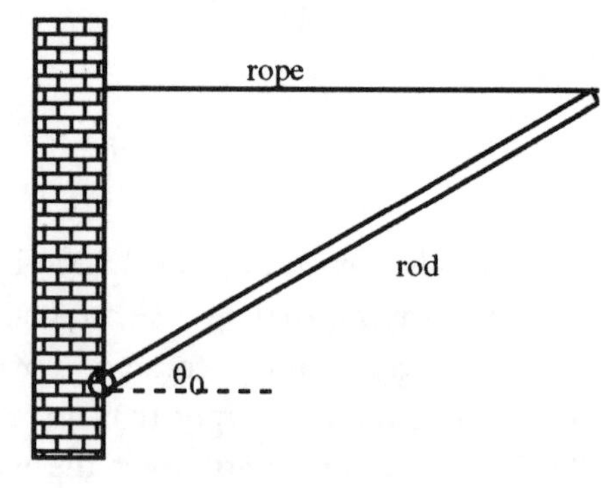

(a) What is the tension in the rope.

(b) Suddenly, the rope breaks. As a result, the rod swings down. *Immediately* after the rope breaks, what is the rod's angular acceleration?

(c) While swinging, at the moment the rod is horizontal, what is its angular speed?

ANSWER 17-6

(a) Since this is a standard statics problem, I will use the standard strategy.

Step 1: Force diagram

I know that $\mathbf{F}_{\text{pin}}$ must point upward and rightward, to cancel the downward force of gravity and the leftward force of tension.

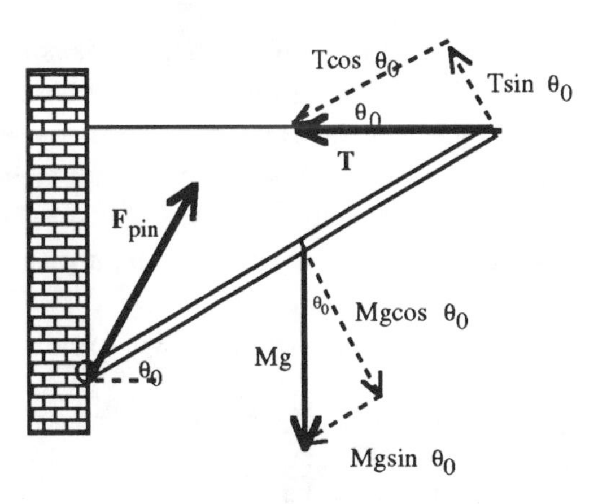

Step 2: Choice of reference pivot point

In parts (b) and (c), the rod will swing around the pin joint. Therefore, it makes sense to choose that point as our reference pivot point.

Step 3: Torque and force equations

As usual, start with torques. Since $\mathbf{F}_{\text{pin}}$ acts directly on the pivot point, it exerts no torque. Gravity acts on the center of mass, which is halfway up the rod, a distance $r = \frac{L}{2}$ from the pivot. That radius vector runs along the rod. The component of gravity perpendicular to **r** is $F_\perp = Mg\cos\theta_0$. So, the clockwise torque due to gravity is $\tau_{\text{grav}} = F_\perp r = (Mg\cos\theta_0)\frac{L}{2}$. Of course, you get this same expression by writing $\tau = r_\perp F$. Draw in the radius vector and break it into components to see that the component of **r** perpendicular to gravity, i.e., the horizontal component of **r**, is $r_\perp = \frac{L}{2}\cos\theta_0$. Therefore, $\tau_{\text{grav}} = r_\perp F = (\frac{L}{2}\cos\theta_0)Mg$, in agreement with the expression from three sentences ago.

By similar reasoning, tension exerts $\tau_{\text{rope}} = F_\perp r = (T\sin\theta_0)L$, clockwise. So, picking counterclockwise as positive, I get $\sum\tau = (Mg\cos\theta_0)\frac{L}{2} - (T\sin\theta_0)L = 0$. This equation alone provides enough information to solve for T; we need not consider the net forces. Cancel out the L's, and isolate T to get $T = \frac{Mg\cos\theta_0}{2\sin\theta_0} = \frac{Mg}{2}\cot\theta_0$.

(b) After the cable snaps, the rod starts to swing. *So, this is no longer a statics problem.* It is a rotational dynamics problem.

Here, we are looking for the angular acceleration α immediately after the cable breaks. This is the rotational analog of finding an acceleration in a linear problem. In a linear problem, given a force diagram, we would set $\sum F = ma$ to find the acceleration. Analogously, here we can set $\sum\tau = I\alpha$.

When the cable breaks, the tension vanishes, leaving gravity as the only torque on the rod. We calculated this torque in part (a): $\tau_{\text{grav}} = (Mg\cos\theta_0)\frac{L}{2}$. We also know from the textbook that a rod pivoted at its end has rotational inertia $I = \frac{ML^2}{3}$. Substitute all this into Newton's 2nd law to get

$$\sum\tau = I\alpha$$

$$(Mg\cos\theta_0)\frac{L}{2} = \frac{1}{2}ML^2\alpha.$$

Cancel the M's and solve for angular acceleration to get

$$\alpha = \frac{Mg\frac{L}{2}\cos\theta_0}{\frac{1}{3}ML^2} = \frac{3g\cos\theta_0}{2L}.$$

(c) The rod's angular acceleration is *not* constant. As the rod swings down, its angular acceleration does not stay at the value calculated in part (b). Why? As shown in part (a), the torque (and hence the angular acceleration) generated by gravity depends on the angle of the beam: $\tau_{\text{grav}} = (Mg\cos\theta)\frac{L}{2}$. As the beam swings from $\theta = \theta_0$ to $\theta = 0$ (horizontal), the torque increases from $(Mg\cos\theta_0)\frac{L}{2}$ to $(Mg\cos 0)\frac{L}{2}$. Therefore, the angular acceleration increases, too. *For this reason, we cannot use constant-acceleration kinematic equations such as* $\omega^2 = \omega_0^2 + 2\alpha\Delta\theta$. We must find another strategy.

As the rod swings down, gravity exerts an external torque that "ruins" angular momentum conservation. But energy is conserved, and no heat gets generated.

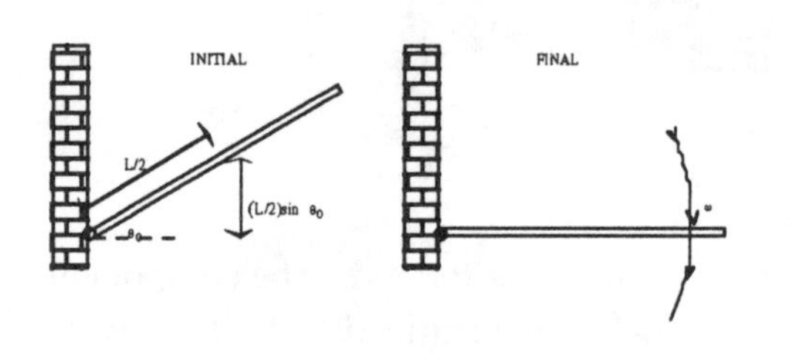

Start by sketching the relevant stages of the process. By "final," I mean when the rod reaches the horizontal. The rod starts at rest, with no kinetic energy. We are trying to find its final angular speed, ω. Since the rod is pivoted, its final kinetic energy is entirely rotational, $K_f = I\omega^2$.

Let us think about potential energy. Here, we run into a potential confusion (no pun intended): To which point on the rod should we measure "h" in the gravitational potential energy formula $U = Mgh$? As always, measure h to the center of mass, *not* to the tip. The diagram shows us that the middle of the rod falls through a distance $h = \frac{L}{2}\sin\theta_0$. So, if we say the rod's final potential energy is $U_f = 0$, then its initial potential energy is $U_0 = Mgh = Mg\frac{L}{2}\sin\theta_0$.

Putting all this together,

$$\begin{aligned} E_0 &= E_f \\ K_0 + U_0 &= K_f + U_f \\ 0 + Mg\frac{L}{2}\sin\theta_0 &= \frac{1}{2}I\omega^2 + 0 \\ &= \frac{1}{2}\left(\frac{1}{3}ML^2\right)\omega^2, \end{aligned}$$

since $I = \frac{1}{3}ML^2$ for a rod pivoted at its end. We can immediately solve for ω to get $\omega = \sqrt{\frac{3g\sin\theta_0}{L}}$.

QUESTION 17-7

A cylinder (mass M, radius R) has a thin thread wrapped around it, and sits on a ramp of ramp angle θ, as drawn below. The segment of thread between the cylinder and the nail is horizontal. What is the smallest coefficient of static friction μ_s needed to ensure that the cylinder stays in place? Express your answer in terms of M, R, θ, and g.

ANSWER 17-7

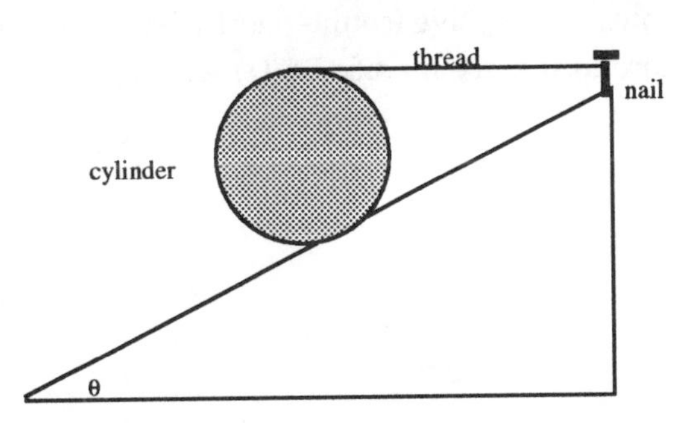

We are trying to find the coefficient of friction, μ_s. The only formula I know with μ_s in it is $f_s \le \mu_s N$. So, if we can find the normal and frictional forces, then we can easily set $\mu_s = \frac{f_s}{N}$. Let me save until the end of this answer a discussion of why sometimes $f_s < \mu_s N$, but why $f_s = \mu_s N$ in this particular problem.

Since we are trying to find forces, and since the object is motionless, let us use the standard statics strategy.

Step 1: Force diagram

For clarity, I will enlarge the cylinder. I am guessing, reasonably enough, that friction points down the ramp. We will see later that friction actually points *up* the ramp. Given my diagram, this means friction will come out negative.

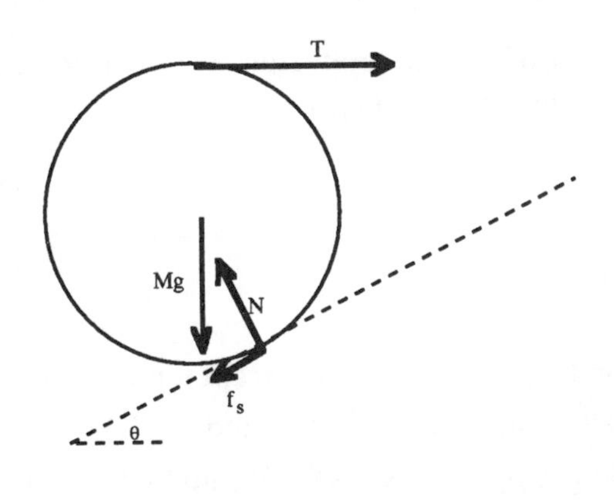

Step 2: Choice of reference pivot point

I see two reasonable choices: the center of the cylinder, or the edge of the cylinder where it touches the ramp. As we have seen before, your choice of reference point cannot affect the final answer. But here, choosing the center of the cylinder simplifies the mathematics greatly. If you choose the point where the cylinder touches the ramp, then the radius vector from that point to the top of the cylinder where tension acts has unknown (and hard to find!) length. By contrast, if we choose the center as our reference pivot point, things work out smoothly.

Step 3: Torque and force equations

With respect to the center, Mg exerts no torque, because the "r" in $\tau = F_\perp r$ is zero. The normal force exerts no torque, either, because no component of **N** points perpendicular to the relevant radius vector. (For **N**, the radius vector extends from the center of the cylinder to the "contact point" where the ramp touches the cylinder. **N** is entirely parallel to this radius vector.) So, the only torques come from tension and friction. Let me draw in the radius vectors.

The frictional force points entirely perpendicular to its radius vector, which has length R. So, $\tau_{\text{fric}} = f_\perp r = \mu_s NR$, clockwise. By similar reasoning, the torque due to tension is $\tau_{\text{thread}} = T_\perp r = TR$, clockwise. So, picking clockwise as positive and setting the net torque equal to zero gives

$$\sum \tau = \mu_s NR + TR = 0. \tag{1}$$

This equation tells me right away that I screwed up when picking the direction of friction. The sum of two positive numbers, $\mu_s NR$ and TR, cannot be zero! In order for the torques to sum to zero, one of them must be negative (counterclockwise). Friction must point *up* the ramp, so as to exert a counterclockwise torque. Let me modify Eq. (1) accordingly.

$$\sum \tau = -\mu_s NR + TR = 0. \text{ (corrected)} \tag{1}$$

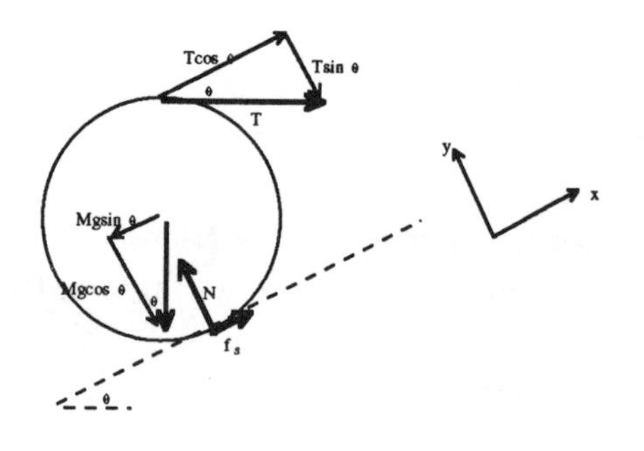

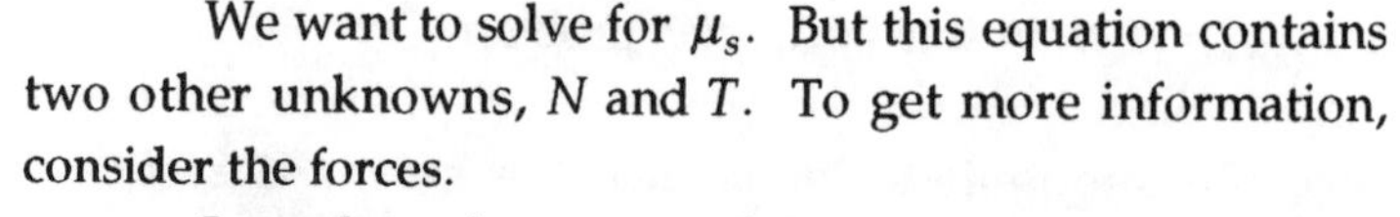

We want to solve for μ_s. But this equation contains two other unknowns, N and T. To get more information, consider the forces.

I need to choose coordinate axes along which to break up the force vectors. I could choose the standard horizontal and vertical coordinates, or the "tilted" coordinates usually used in ramp problems. (You will get the same answer either way, of course.) Let me use tilted coordinates.

I have also taken this opportunity to correct my friction vector. From the force diagram,

$$\sum F_x = \mu_s N + T\cos\theta - Mg\sin\theta = 0. \tag{2}$$

$$\sum F_y = N - T\sin\theta - Mg\cos\theta = 0. \tag{3}$$

Now we have three equations in the three unknowns (N, T, and μ_s). We are done, except for algebra.

Algebra starts here. To complete the algebra efficiently, first solve the corrected Eq. (1) for T to get $T = \mu_s N$. Substitute that expression for T into Eq. (2) to get $\mu_s N + \mu_s N\cos\theta - Mg\sin\theta = 0$. Solve this equation for N to get $N = \frac{Mg\sin\theta}{\mu_s(1+\cos\theta)}$. Therefore, since $T = \mu_s N$, we have $T = \mu_s \frac{Mg\sin\theta}{\mu_s(1+\cos\theta)} = \frac{Mg\sin\theta}{1+\cos\theta}$. Finally, substitute these expressions for N and T into Eq. (3) to get $\frac{Mg\sin\theta}{\mu_s(1+\cos\theta)} - \frac{Mg\sin\theta}{1+\cos\theta}\sin\theta - Mg\cos\theta = 0$. Divide through by Mg. Then multiply through by $\mu_s(1+\cos\theta)$. This gives

$$\sin\theta - \mu_s\sin^2\theta - \mu_s\cos\theta(1+\cos\theta) = 0.$$

Expand out the last term on the left-hand side to get $\sin\theta - \mu_s\sin^2\theta - \mu_s\cos\theta - \mu_s\cos^2\theta = 0$. Since $\sin^2\theta + \cos^2\theta = 1$, this equation reduces to $\sin\theta - \mu_s - \mu_s\cos\theta = 0$. Finally, isolate μ_s to get $\mu_s = \frac{\sin\theta}{1+\cos\theta}$. **End of algebra.**

Let me now launch on a brief...

Theoretical digression: When does $f_s = \mu_s N$?

(This digression reprises Question 9-3(b).)

Kinetic friction always equals $\mu_k N$. But the force of static friction can be less than $\mu_s N$. Here is why:

Imagine putting an eraser on your textbook, and holding the book horizontally. What is the force of friction on the eraser? Zero, because the book is not tilted, and therefore no friction is "needed" to hold the eraser in place. (If there *were* a frictional force, it would be the *only* horizontal force, and therefore the eraser would accelerate across the book. This does not happen.) So, in that case, $f_s = 0$, which is much less than $\mu_s N$.

Now you start tilting the textbook. As a result, friction must counteract the component of gravity that pulls the eraser along the book. You can confirm that friction must be $f_s = mg\sin\theta$, to hold the eraser still. So, as the tilt angle gets bigger and bigger, friction must increase, in order to hold the eraser in place. This goes to show that static friction can "adjust" itself, getting larger or smaller as necessary to hold an object still.

Of course, if you tilt the book too steeply, friction cannot hold the eraser any more. Although static friction is adjustable, its *maximum* possible value is $f_{s\,\max} = \mu_s N$.

In this problem, a certain amount of friction is needed to hold the cylinder in place. For the sake of argument, let us say f worked out to be 10 newtons, and the normal force worked out to be 20 newtons. So, if the coefficient of friction were 0.7, then the cylinder would stay in place, because $f_s = 10$ newtons is less than $f_{s\,\max} = \mu_s N = (0.7)(20\text{ newtons}) = 14\text{ newtons}$. Things also "work" if $\mu_s = 0.6$ or 0.5. But if $\mu_s = 0.4$, then the biggest frictional force the ramp can generate is $f_{s\,\max} = \mu_s N = (0.4)(20\text{ newtons}) = 8\text{ newtons}$, less than the 10 newtons needed to hold the cylinder in place. So, we see that $\mu_s = 0.5$–the μ_s for which f_s equals $\mu_s N$–is the *smallest* coefficient of static friction that keeps the cylinder in place. For this reason, whenever you are trying to make μ_s as small as possible, you can set $f_s = \mu_s N$.

Gravitation: Part I

18 CHAPTER

◆ *Feel free to use the table of astronomical constants supplied in your textbook. First, solve in terms of symbols. Then look up the relevant numbers.*

QUESTION 18-1

Suppose you weigh 580.00 newtons (that is about 130 pounds) when you are standing on a beach near San Francisco. How much will you weigh at the top of a nearby mini-mountain, which is about 500 meters high?

ANSWER 18-1

By definition, "weight" is the gravitational force on an object. Normally, we assume that the strength of gravity anywhere near the Earth's surface is the same: $F_{\text{grav}} = mg$, where "g" is the acceleration due to gravity at sea level. But the *precise* gravitational force that the Earth exerts on an object of mass m is

$$F_{\text{grav}} = \frac{GM_E m}{r^2}, \tag{1}$$

where M_E is the Earth's mass, and r is the distance *from the center of the Earth to the object*. On the mountain, you are slightly farther from the Earth's center, and therefore you weigh slightly less.

Here is one way you could solve. First, using Eq. (1), set r equal to the Earth's radius, set $F_{\text{grav}} = 580\text{ N}$, and solve for your mass, m. Then substitute that m into Eq. (1) again, this time with $r = (\text{Earth's radius} + 500\text{ m})$, to solve for your new weight at the top of the mountain. This method works fine. But let me show you an elegant shortcut, one that saves you the bother of looking up G or M_E.

Let R_E denote the Earth's radius, i.e., the distance from the center of the Earth to ground level. At sea level, your distance from the Earth's center is R_E. From Eq. (1), we have the following *ratio* of weights:

$$\frac{F_{\text{grav at top of mountain}}}{F_{\text{grav at ground level}}} = \frac{\frac{GM_E m}{(R_E+500\text{ m})^2}}{\frac{GM_E m}{R_E^2}} = \frac{R_E^2}{(R_E+500\text{ m})^2}. \tag{2}$$

The Earth's radius is $R_E = 6.37 \times 10^6$ m. Substitute that into Eq. (2) to get

$$\frac{F_{\text{grav at top of mountain}}}{F_{\text{grav at ground level}}} = \frac{(6.37 \times 10^6)^2}{(6.37 \times 10^6 \text{ m} + 500 \text{ m})^2} = 0.99984, \tag{2'}$$

which is close to one, because the mountain's height is tiny compared to the Earth's radius. So, at the top of the mountain, you weigh 99.984% of your regular weight. Remember, "F_{grav}" is just another way of saying "weight"

$$\begin{aligned}\text{Weight at top of mountain} &= (\text{Weight at ground level})\frac{F_{\text{grav at top of mountain}}}{F_{\text{grav at ground level}}} \\ &= (580.00 \text{ N})(0.99984) \\ &= 579.91 \text{ N},\end{aligned}$$

about a third of an ounce less than your regular weight. This difference is hardly noticeable. That is why we can safely assume, for most calculations, that gravity has about the same strength anywhere near the Earth's surface.

QUESTION 18-2

When the Sun uses up its "fusion" fuel, it will collapse into a "white dwarf" star, which contains almost all the mass of the Sun compressed into a sphere the size of the Earth. If an object is dropped from distance $d = 1.0$ meter above the white dwarf's surface, how much time does it take to land? Your answer may include G, the Sun's mass M_s, the Sun's radius R_s, and the Earth's radius R_E. (But you need not use all those quantities.)

Hint: As you found in Question 17-1 above, the acceleration due to gravity at all points *near the surface* of the Earth (or at all points *near the surface* of a star) is approximately constant.

ANSWER 18-2

In this problem, we imagine dropping an object (of mass m) near the surface of the white dwarf. You do not need to know m, because it cancels out of our answer. We can solve for the acceleration by finding the gravitational force and then using Newton's 2nd law, $F = ma$. Next, we can use that acceleration in a constant-acceleration kinematic formula, as noted in the hint.

In the formula $F_{\text{grav}} = \frac{GMm}{r^2}$, r is the distance from the object to the *center* of the star (or planet or whatever). Here, the distance from the center of the white dwarf to the object approximately equals the radius of the white dwarf, which is R_E. (You could use $R_E + 1$ meter; but that extra meter will not make an appreciable difference in your answer, because $R_E >> 1$ meter.) So, the force exerted by the white dwarf on the object is $F_{\text{grav}} = \frac{GM_s m}{R_E^2}$. Remember, the white dwarf contains the mass of the *Sun* in a sphere the size of the *Earth*.

Since no force other than gravity acts on the object after it is dropped, this gravitational force is the *net* force. Therefore, we can use Newton's 2nd law, $F_{\text{net}} = ma$, to find the object's acceleration:

$$a = \frac{F_{\text{grav}}}{m} = \frac{GM_s}{R_E^2}$$
$$= \frac{(6.67\times10^{-11}\ \text{N}\cdot\text{m}^2/\text{kg}^2)(1.99\times10^{30}\ \text{kg})}{(6.37\times10^6\ \text{m})^2}$$
$$= 3.27\times10^6\ \text{m/s}^2,$$

about 300,000 times the acceleration due to gravity on Earth. Notice that the gravitational acceleration does not depend on the object's mass m. This is true on Earth, too. A bowling ball and pebble, dropped at the same time from the same height, reach the ground simultaneously.

By the way, physicists usually give the name "g" to the gravitational acceleration near a planet's (or star's or moon's) surface. On Earth, $g = 9.8\ \text{m/s}^2$. On the white dwarf, we just obtained $g_{\text{white dwarf}} = 3.27\times10^6\ \text{m/s}^2$.

Now that we know the object's acceleration, we can use our old displacement formula to see how much time it takes to fall through height $y = d = 1$ meter. I will choose upward as positive. So, the acceleration is negative: $a_y = -g_{\text{white dwarf}}$. We also have $v_0 = 0$, $y_0 = d = 1$ meter, and $y = 0$. Solve $y = y_0 + v_{0y}t + \frac{1}{2}a_y t^2$ for t to get

$$t = \sqrt{\frac{2(y-y_0)}{a_y}} = \sqrt{\frac{2(0-d)}{-g_{\text{white dwarf}}}} = \sqrt{\frac{2(0-1.0\ \text{m})}{-3.27\times10^6\ \text{m/s}^2}} = 7.8\times10^{-4}\ \text{s},$$

about a thousandth a second. Gravity is *that* intense.

QUESTION 18-3

In outer space, two spherical shells are configured as drawn below. They both have radius R. Sphere 1, of mass M_1, is centered at $x = 0$. Sphere 2, of mass M_2, is centered at $x = 3R$.

Find the total gravitational force (magnitude and direction) on a particle of mass m sitting on

(a) Point A, located at $x = \frac{R}{2}$.

(b) Point B, located at $x = 1.75R$; and

(c) Point C, located at $x = 5R$.

Express your answers in terms of M, R, and any other constants you need.

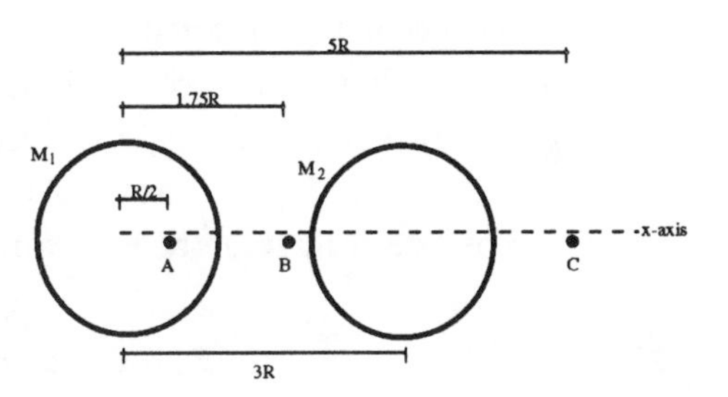

ANSWER 18-3

This problem illustrates two important principles. The first concerns the gravitational force exerted by a spherical shell. See your textbook for a good derivation of these results:

Gravity exerted by spherical shell:

- **If you are inside the shell, it exerts no force on you.**
- **If you are outside the shell, it exerts a force on you *as if* all the shell's mass were concentrated at the center of the shell.**

This tell us, for instance, that sphere 1 exerts no force on a particle at point A. But sphere 1 *does* pull on an object at points B and C.

The other important principle at work here is the

Superposition principle:

The total gravitational force exerted by a bunch of masses is simply the sum of the individual forces created by each separate mass.

Here, the superposition principle tells us that the total force felt at a point equals the force created by sphere 1, *plus* the force created by sphere 2. In other words, the total force is the sum of the individual forces. This is nothing new.

By applying these two principles, we can answer all three parts of this question.

(a) Point A is inside sphere 1. So, sphere 1 exerts no force on that point. But sphere 2 *does* pull on a particle at point A. (Being inside sphere 1 does not "shield" point A from the gravitational pull of sphere 2.)

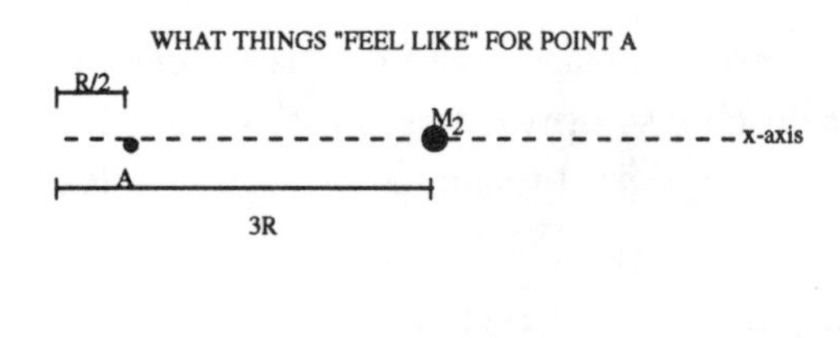

At point A, sphere 2 creates a force *as if* all the mass of sphere 2 were concentrated at the center of sphere 2. In summary, point A "feels" as if sphere 1 did not exist, and sphere 2 were a point mass at $x = 3R$.

Since point A is a distance $r = 3R - \frac{R}{2} = 2.5R$ from (the center of) M_2, the force felt by a mass m at point A is

$$F_A = \frac{GM_2m}{r^2} = \frac{GM_2m}{(2.5R)^2}, \text{ rightward.}$$

(b) Since point B is outside both spheres, it experiences a force due to both of them. Sphere 1 pulls it leftward, while sphere 2 pulls it rightward. Therefore, if I choose rightward as positive, sphere 1 exerts a negative (backwards) force.

Since point B is located $r = 1.75R$ from the center of sphere 1, a mass m at point B feels $F_1 = \frac{GM_1m}{(1.75R)^2}$, leftward. Remember, it is *as if* all of M_1 were concentrated at the center of that sphere. Similarly, as the diagram shows, point B is a distance $r = 1.25R$ from the center of sphere 2. So, sphere 2 exerts a rightward force $F_2 = \frac{GM_2m}{(1.25R)^2}$ on mass m at point B. By the superposition principle, the total force experienced by mass m at point B is

$$\mathbf{F}_B = \mathbf{F}_1 + \mathbf{F}_2 = -F_1 + F_2 = -\frac{GM_1m}{(1.75R)^2} + \frac{GM_2m}{(1.25R)^2}.$$

(c) Point C also sits outside both spheres. Furthermore, both spheres pull point C leftward. So, those two forces combine to create a bigger force. This contrasts with part (b), in which the forces partially canceled each other.

Let me choose leftward as positive in this part of the problem. Notice that point C is $r = 5R - 3R = 2R$ from the center of sphere 2, and $r = 5R$ from the center of sphere 1. So, a mass m at point C feels leftward force

$$F_C = F_1 + F_2 = \frac{GM_1m}{(5R)^2} + \frac{GM_2m}{(2R)^2}.$$

QUESTION 18-4

This one's tricky. Once you see the answer, it makes a lot of sense; but it is hard to figure out the "trick" on your own. Spend about eight minutes, and if you are not making progress, jump to the answer. No difficult math is needed.

Two spherical holes (i.e., hollowed-out sections) are dug into a formerly uniform solid sphere of radius R. The holes are spheres of diameter R, as drawn below. Notice that one "edge" of each hole touches the center of the bigger sphere.

NOTE: The white areas are the dug-out holes. The grey areas are "solid matter."

R

s

particle, mass m

The mass of the big sphere, *before* it was hollowed out, was M.

With what gravitational force does the hollowed-out sphere attract a particle of mass m located a distance s from the center of the big sphere, as drawn below? (The particle lies on an imaginary straight line connecting the center of the big sphere to the centers of the holes.)

ANSWER 18-4

The trick is to apply the superposition principle in a clever way. Let me briefly review that principle. Suppose an object is gravitationally attracted to three masses, m_1, m_2, and m_3. What is the net force on the object? Well, by summing up forces in the usual way, we get

$$F_{\text{due to } m_1} + F_{\text{due to } m_2} + F_{\text{due to } m_3} = F_{\text{due to all three masses}} \tag{1}$$

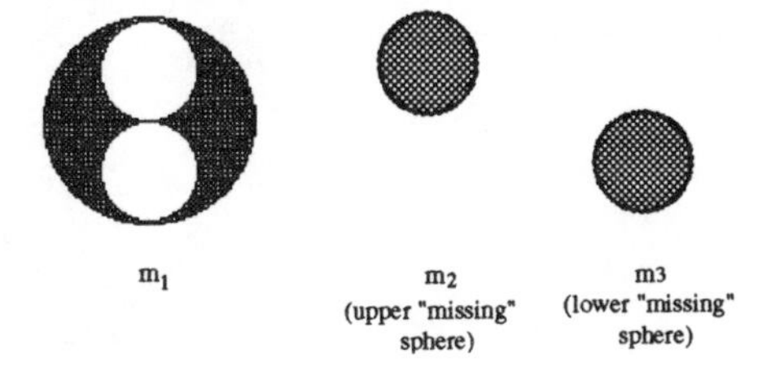

In this problem, let m_1 be the hollowed-out sphere, and let m_2 and m_3 be the smaller spheres that are "missing" from the whole big sphere.

These three masses, when "pasted" together, form the "whole" sphere. So, in this problem, Eq. (1) tells us this.

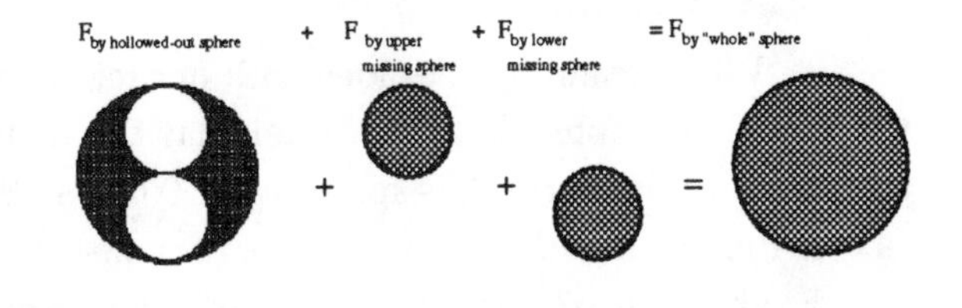

(2) $F_{\text{by hollowed-out sphere}} = F_{\text{by whole sphere}} - F_{\text{by upper missing sphere}} - F_{\text{by lower missing sphere}}$

Now here is the cool trick: Subtract $F_{\text{by upper missing sphere}}$ and $F_{\text{by lower missing sphere}}$ from both sides to get this.

So, we can find the force exerted by the hollowed-out sphere simply by calculating the force exerted by the "whole" sphere (*before* it was hollowed out), and subtracting the forces that would be generated by the small "missing" spheres.

First I will calculate $F_{\text{by "whole" sphere}}$. We are told that the whole sphere had mass M. It is center was located a distance $r = s$ from the particle of mass m. Therefore, the force exerted by the whole sphere on mass m is $F_{\text{by whole sphere}} = \frac{GMm}{s^2}$.

Now I will calculate $F_{\text{by upper missing sphere}}$. As the above diagrams show, this missing sphere has radius $\frac{R}{2}$. What is its mass? Well, the volume of the whole sphere is $V_{\text{whole}} = \frac{4\pi R^3}{3}$. Since the upper "missing" sphere has radius $\frac{R}{2}$, its volume is

$$V_{\text{upper missing sphere}} = \frac{4\pi(\frac{R}{2})^3}{3} = \frac{1}{8}\left(\frac{4\pi R^3}{3}\right) = \frac{1}{8}V_{\text{whole}}.$$

So, the upper missing sphere has one eighth the volume of the whole sphere. Therefore, it also has one eighth the mass of the whole sphere, assuming uniform density. In summary, the mass of the upper missing sphere is $\frac{M}{8}$. The same conclusion applies to the lower missing sphere.

Furthermore, as you can see from the diagram given in the problem, the center of the upper missing sphere (i.e., the center of that hole) is located a distance $s+\frac{R}{2}$ from the particle. So,

$$F_{\text{by upper missing sphere}} = \frac{G(\frac{M}{8})m}{(s+\frac{R}{2})^2}.$$

Similarly, the center of the lower missing sphere is a distance $s-\frac{R}{2}$ from the particle. So,

$$F_{\text{by lower missing sphere}} = \frac{G(\frac{M}{8})m}{(s-\frac{R}{2})^2}.$$

Putting all this together, I get

$$\begin{aligned} F_{\text{by hollowed-out sphere}} &= F_{\text{by whole sphere}} - F_{\text{by upper missing sphere}} - F_{\text{by lower missing sphere}} \\ &= \frac{GMm}{s^2} - \frac{G(\frac{M}{8})m}{(s+\frac{R}{2})^2} - \frac{G(\frac{M}{8})m}{(s-\frac{R}{2})^2}. \end{aligned}$$

That is the answer, which you can simplify if you want. Cool, huh?

QUESTION 18-5

While scanning the skies with her telescope, an astronomer discovers a meteor hurtling toward the Earth. Her observations reveal that the meteor, when it is a distance $s = 5.00 \times 10^7$ meters from the center of the Earth, has speed $v_0 = 500$ m/s. She estimates its mass to be $m = 9.00 \times 10^4$ kg.

(a) If the Earth had no atmosphere, how fast would the meteor by traveling when it crashed into the Earth? You are allowed to look up any constant you want from any table in the textbook.

(b) Assuming the meteor is not too big, what really happens to it? Where does all its energy go?

ANSWER 18-5

(a) Your first inclination might be to calculate the meteor's acceleration, and then use an old kinematic formula such as $v^2 = v_0^2 + 2a\Delta x$. But as the meteor gets closer to Earth, the gravitational force on it increases, since the "r" in $F_{\text{grav}} = \frac{GM_E m}{r^2}$ gets smaller. Therefore, its acceleration increases. Not only does the meteor speed up; it speeds up at a faster and faster rate. So, those old constant-acceleration kinematic formulas are not valid. We need another method.

When forces are not constant, we can usually address the problem with conservation laws. Here, we can use energy conservation. Intuitively, as the meteor falls toward Earth, its potential energy decreases, while its kinetic energy increases to compensate. So, it speeds up.

Unfortunately, we cannot use our old $U = mgh$ formula for gravitational potential energy. Remember, by definition, potential energy is the amount of work it takes to get the system into its current position, i.e., the amount of work done "storing up" the potential energy. When you lift an object through height $\Delta y = h$ against the gravitational force $F_{\text{grav}} = mg$, then the work you perform is $W = F\Delta y = mgh$. But $F_{\text{grav}} = mg$ applies only near the Earth's surface. Therefore, $U = mgh$ applies only near the Earth's surface.

In general, the force due to gravity is $F_{\text{grav}} = \frac{GMm}{r^2}$. To calculate the work needed to fight that force through a distance Δr, we cannot just find $F_{\text{grav}}\Delta r = \frac{GMm}{r^2}\Delta r$, because the force is not constant with respect to r. Instead, we must integrate the force over distance:

$$U = W_{\text{to fight force}} = \int \frac{GMm}{r^2}dr = -\frac{GMm}{r}.$$

That is the formula for gravitational potential energy. See the textbook for a more careful derivation that deals rigorously with the integration constant.

At first, the minus sign seems weird. But it makes physical sense. Intuitively, as you fall toward the Earth, your potential energy decreases. Let us see if our formula agrees with this intuition. As we fall toward Earth, "r" gets smaller. Therefore, $U = -\frac{GMm}{r}$ gets bigger, by which I mean a bigger *negative* number. A bigger *negative* number, however, is really a *smaller* number; -10 is less than -8. So, $U = -\frac{GMm}{r}$ indeed gets smaller when r gets smaller, exactly as we intuitively require.

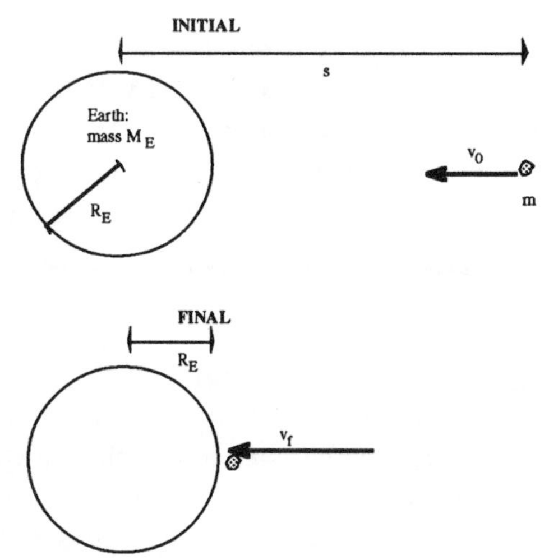

As the above deviation shows, the "r" in $U = -\frac{GMm}{r}$ is the same "r" in $F_{\text{grav}} = \frac{GMm}{r^2}$ It gets measured from the centers of the relevant (spherical) objects.

In any case, let us conserve energy.

We are solving for v_f, the meteor's speed right before crashing. As the diagram shows, the meteor at that point is a distance R_E from the center of the Earth. So, using $U = -\frac{GMm}{r}$, we get

$$E_0 = E_f$$

$$K_0 + U_0 = K_f + U_f$$

$$\frac{1}{2}mv_0^2 + \left(-\frac{GM_E m}{s}\right) = \frac{1}{2}mv_f^2 + \left(-\frac{GM_E m}{R_E}\right).$$

Notice that the m's cancel: The meteor's final speed does not depend on how heavy it is. This is not surprising; we already knew that a rock and a pen dropped from the same height hit the ground with the same velocity. In any case, solve for v_f, and set $R_E = 6.37 \times 10^6$ m and $M_E = 5.98 \times 10^{24}$ kg, to get

$$v_f = \sqrt{v_0^2 + 2GM_E\left(\frac{1}{R_E} - \frac{1}{s}\right)}$$
$$= \sqrt{(500\ \text{m/s})^2 + 2(6.67\times10^{-11}\ \text{N}\cdot\text{m}^2/\text{kg}^2)(5.98\times10^{24}\ \text{kg})\left(\frac{1}{6.37\times10^6\ \text{m}} - \frac{1}{5\times10^7\ \text{m}}\right)}$$
$$= 10{,}500\ \text{m/s}.$$

(b) Fortunately, the meteor probably would not even reach the Earth. Air resistance "takes away" some the meteor's kinetic energy. The "lost" kinetic energy converts into heat–and even light! Due to its "rubbing" against the atmosphere, the meteor heats up so much that it glows. You can see these glowing meteors sometimes; they are called "falling stars." Usually, a meteor completely burns up before reaching the ground.

QUESTION 18-6

A so-called "globular cluster" of stars is spherically symmetric, and contains thousands of stars. Let us consider a cluster containing about ten thousand stars. Its mass is 1.8×10^{34} kg.

Suppose a space rock is located outside the cluster, a distance $R = 5.0\times10^{17}$ meters from the center. (That is about 50 light-years, by which I mean the distance traveled by light in 50 years.)

What is the space rock's escape velocity? In other words, how much initial speed must the rock have in order to *barely* get *very* far from the cluster? By "very far," I mean an effectively infinite distance. If the rock's velocity is less than the value you calculate, the rock will either fall into the cluster or end up in orbit around the cluster.

Derive the formula you end up using; do not plug in a pre-derived textbook formula. Express your answer in terms of symbols before plugging in numbers.

ANSWER 18-6

Do not treat escape velocity problems as a new *kind* of problem. They are an application of energy conservation. You need not memorize any formulas.

Since the cluster is spherically symmetric (i.e., "made of" concentric spherical shells of matter), and since the rock is outside the cluster, we can "pretend" that all the matter is concentrated at the cluster's center. See Question 18-3 above. I will assume the rock's initial velocity is directed away from the cluster; but you get the same answer no matter what direction it is going. As the rock gets farther away, it slows down, due to the cluster's gravitational pull. (Similarly, when you throw your pen straight up, it gradually slows down.) If the initial speed is greater than or equal to the escape velocity, the rock never comes to a complete stop; it keeps getting farther away from the cluster, forever. But if $v_0 = v_{esc}$, the rock eventually comes to rest at its "peak" distance from the cluster, and then falls back "down" toward the cluster–just like your pen. We are solving for v_{esc}, the smallest initial speed the rock can have so that it never "falls back."

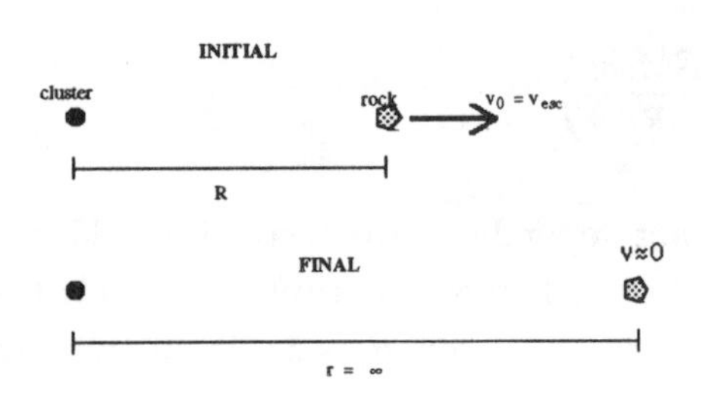

As always when using conservation laws, diagram the relevant stages. In my drawings, I "pretend" that the cluster is a point mass concentrated at the center of the cluster.

We want the initial velocity to be the escape velocity, i.e., the speed necessary to get the rock a "final" distance $r = \infty$ from the cluster. Here is another crucial point: If the rock *barely* escapes, then it has used up essentially all its initial speed by the time it gets very far away. In other words, when it reaches $r = \infty$, it is *barely* moving. So, its final speed is essentially $v = 0$.

Let m denote the rock's mass. Recall that the gravitational potential energy is $U = -\frac{GMm}{r}$. Throwing all this information into conservation of energy gives

$$\begin{aligned} E_0 &= E_f \\ U_0 + K_0 &= U_f + K_f. \\ -\frac{GMm}{R} + \frac{1}{2}mv_{\text{esc}}^2 &= -\frac{GMm}{\infty} + \frac{1}{2}mv_f^2 \\ &= 0 + 0. \end{aligned}$$

You will hear some people say that in escape velocity problems, you should set the final energy to zero. This is not something you should memorize. Rather, it is something that naturally "pops out" of the energy conservation equation in this particular situation.

Solve for v_{esc} to get

$$v_{\text{esc}} = \sqrt{\frac{2GM}{R}} = \sqrt{\frac{2(6.67 \times 10^{-11}\ \text{N} \cdot \text{m}^2/\text{kg}^2)(1.8 \times 10^{34}\ \text{kg})}{(5.0 \times 10^{17}\ \text{m})}} = 2200\ \text{m/s},$$

about 1.5 miles per second. That is not fast at all for heavenly bodies.

QUESTION 18-7

As Jupiter orbits the Sun in an approximately circular path, its average distance from the Sun is 5.19 times the Earth's average distance from the Sun. From this information alone, without looking up any astronomical constants, you are going to figure out how many years it takes Jupiter to complete one revolution around the Sun. In other words, you are going to calculate the length of a "Jovian year."

(a) What famous astronomical law can you use to solve this problem?

(b) *Derive* that law, and use it to answer the question.

ANSWER 18-7

(a) This problem involves a relationship between a planet's period of revolution T and its distance from the Sun r (i.e., the radius of its orbit). This should tip you off that Kepler's 3rd law applies.

Consider satellites in orbit around a central body. For instance, we could be talking about Saturn's moons orbiting Saturn, or weather satellites orbiting the Earth, or planets orbiting the Sun. Let r denote the radius of a satellite's orbit, and let T denote its period, the time it takes to complete one full revolution. According to Kepler's 3rd law, the ratio $\frac{r^3}{T^2}$ is the same for *any* two satellites orbiting the central body. So for instance, $\frac{r^3}{T^2}$ for Jupiter is the same as $\frac{r^3}{T^2}$ for Venus.

(b) In the problem at hand, we can use this law to compare Earth to Jupiter: $\frac{r_{\text{Earth}}^3}{T_{\text{Earth}}^2} = \frac{r_{\text{Jupiter}}^3}{T_{\text{Jupiter}}^2}$. We can immediately obtain T_{Jupiter}, because we know that $\frac{r_{\text{Jupiter}}}{r_{\text{Earth}}} = 5.19$. Solve the above equation for T_{Jupiter} to get

$$T_{\text{Jupiter}} = T_{\text{Earth}}\left(\frac{r_{\text{Jupiter}}}{r_{\text{Earth}}}\right)^{\frac{3}{2}}$$
$$= (1 \text{ year})(5.19)^{\frac{3}{2}}$$
$$= 11.8 \text{ years},$$

So, Kepler's 3rd law is clearly useful. But where does it come from? The following derivation applies to circular orbits. (With clever math, you can make a similar derivation work for elliptical orbits.)

Planets travel in approximately circular orbit around the Sun; the ellipses are not too "eccentric." As we saw in study guide Chapter 10, an object in circular motion has radial acceleration $a_{\text{radial}} = \frac{v^2}{r}$. Therefore, the net radial force on the planet must be $\sum F_{\text{radial}} = \frac{mv^2}{r}$, where m is the planet's mass. The Sun's gravitational pull provides the only radial force–indeed the only force at all. Put another way, the Sun provides the centripetal force that keeps the planet in orbit:

$$\sum F_{\text{radial}} = ma_{\text{radial}}$$
$$\frac{GM_s m}{r^2} = \frac{mv^2}{r}, \qquad (1)$$

where M_s is the Sun's mass, and r is the distance from the Sun to the planet. We could easily solve for v, the planet's velocity. But we want T, the period of revolution. How does T relate to v?

Well, we know that in time T, the planet completes one full revolution. One full revolution corresponds to one circumference, a distance $2\pi r$. So, in time T, the planet travels through distance $2\pi r$. Therefore, since the planet's speed is (approximately) constant,

$$v = \frac{\text{distance}}{\text{time}} = \frac{2\pi r}{T}. \qquad (2)$$

Eqs. (1) and (2) provide all the information needed to derive Kepler's 3rd law. Only algebra remains.

Substituting Eq. (2) into Eq. (1) gives

$$\frac{GM_s m}{r^2} = \frac{m\left(\frac{2\pi r}{T}\right)^2}{r} = \frac{m4\pi^2 r}{T^2}.$$

Cancel the m's, multiply through by r^2, and divide through by $4\pi^2$, to get

$$\frac{r^3}{T^2} = \frac{GM_s}{4\pi^2} \qquad \textbf{(Kepler's third law)}$$

I cannot stress enough that the m's cancel. Therefore, for any planet, no matter how massive, $\frac{r^3}{T^2}$ is equal to $\frac{GM_s}{4\pi^2}$, a *constant* that depends only on the Sun's mass, *not* the planet's mass. In other words, since $\frac{GM_s}{4\pi^2}$ is the same for all planets, the ratio $\frac{r^3}{T^2}$ is the same for all planets. That is why I could set $\frac{r^3_{\text{Earth}}}{T^2_{\text{Earth}}} = \frac{r^3_{\text{Jupiter}}}{T^2_{\text{Jupiter}}}$ above.

QUESTION 18-8

A rocket takes off from Earth with initial speed $v_0 = \sqrt{gR_E}$ straight up, where R_E is the Earth's radius, and g is the acceleration due to gravity near the Earth's surface. That is not fast enough to escape. Neglecting air resistance, how high above the Earth's surface does the rocket reach at its peak (highest point), before it falls back down? Express your answer symbolically as a multiple of R_E. Do the whole problem without looking up any numbers.

Hint: To express your answer as a multiple of R_E, you must get the "g" out of your answer.

ANSWER 18-8

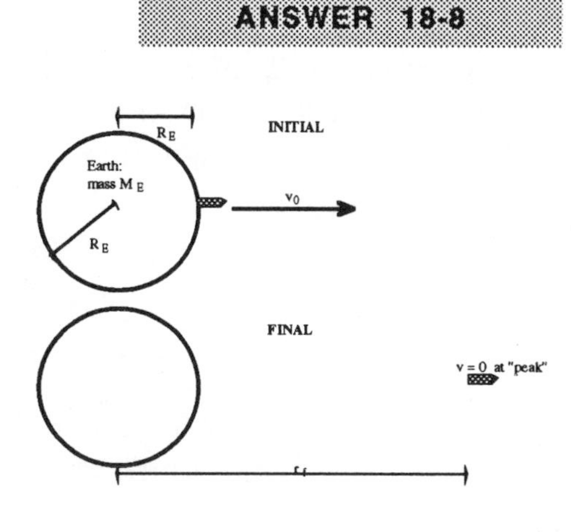

As the rocket gets farther and farther from Earth, it feels a smaller and smaller gravitational force. Therefore, its acceleration (a deceleration in this case) gets smaller. Since the acceleration is not constant, we should not use force-and-acceleration reasoning. Instead, we should use conservation of energy. As always, start with pictures.

Here, r_f denotes the rocket's peak distance from the center of the Earth. We are actually solving not for r_f, but for $r_f - R_E$, the rocket's peak distance above the *surface* of the Earth.

A common error is to set the initial potential energy equal to 0, since the rocket starts on the ground. But remember, we are not using $U_{\text{grav}} = mgh$, because the rocket does not experience a constant force mg. Instead, we must use $U_{\text{grav}} = -\frac{GMm}{r}$, which does *not* equal 0 on the Earth's surface. This new expression for U_{grav} has its "zero point" at $r = \infty$, very far away.

I will now solve for r_f. Let m denote the rocket's mass. The rocket at its peak, just like a ball at its peak, momentarily comes to rest: $v = 0$. Since the rocket begins at distance $r = R_E$ from the Earth's center, energy conservation gives us

$$E_0 = E_f$$

$$U_0 + K_0 = U_f + K_f.$$

$$-\frac{GM_E m}{R_E} + \frac{1}{2}mv_0^2 = -\frac{GM_E m}{r_f} + 0 \qquad (1)$$

$$-\frac{GM_E m}{R_E} + \frac{1}{2}m(gR_E) = -\frac{GM_E m}{r_f},$$

where in the last step I substituted in the given initial velocity, $v_0 = \sqrt{gR_E}$.

At this point, we could solve for r_f. But the answer comes out messy, not a simple multiple of R_E. To get a "pretty" expression for r_f, we must get rid of the "g" in Eq. (1). To do so, keep in mind that g is the acceleration due to gravity near the Earth's surface. From Newton's 2nd law, the acceleration due to gravity is $a = \frac{F_{\text{grav}}}{m}$. So, the gravitational acceleration near the Earth's surface (a distance R_E from the Earth's center) is

$$g = \frac{F_{\text{grav near Earth's surface}}}{m} = -\frac{\frac{GM_E m}{R_E^2}}{m} = \frac{GM_E}{R_E^2}. \tag{2}$$

This harks back to Question 18-2, in which we solved for "g" on a white dwarf in terms of G. In any case, we can substitute this expression for g into Eq. (1) above to get

$$-\frac{GM_E m}{R_E} + \frac{1}{2}m\left(\frac{GM_E}{R_E^2}R_E\right) = -\frac{GM_E m}{r_f}.$$

The left-hand side simplifies to $-\frac{GM_E m}{2R_E}$. So, we can solve for r_f to get $r_f = 2R_E$. The rocket reaches a peak height of two Earth-radiuses from the center of the Earth. Therefore, its peak is exactly *one Earth radius above the Earth's surface.* (Recall from my diagram that the rocket's peak height above the Earth's *surface* is $r_f - R_E$.)

Gravitation: Part II

19 CHAPTER

QUESTION 19-1

In this problem, pretend you are an astronaut standing on a tiny moon (belonging to another planet). Its radius is only $R = 10{,}000$ meters (about 6 miles), and its mass is $M = 1.0 \times 10^{16}$ kg.

(a) If you stood on a bathroom scale on this moon, what would it register?

(b) How fast would you need to run to put yourself into orbit? Could you do it?

ANSWER 19-1

(a) A scale registers your weight, the force gravity exerts on you. Since you are standing on the surface of the moon, your distance from its center is R. The moon's mass is M. Let us say your mass is $m = 60$ kg, which on Earth corresponds to 132 pounds. On the moon, you weigh

$$\begin{aligned} W &= F_{\text{grav}} \\ &= \frac{GMm}{R^2} \\ &= \frac{(6.67 \times 10^{-11}\ \text{N}\cdot\text{m}^2/\text{kg}^2)(1.0 \times 10^{16}\ \text{kg})(60\ \text{kg})}{(10000\ \text{m})^2} \\ &= 0.40\ \text{Newtons}, \end{aligned}$$

which is about an ounce.

(b) We must find the minimum speed needed to achieve orbit. Well, if you run *barely* fast enough to go into orbit, your orbit will be low, about a millimeter above the surface of the planet. In other words, the radius of your orbit will be about R, the moon's radius. So, the problem is *really* asking for the speed of an orbit of radius R. If you run that fast, you will achieve orbit.

To obtain this speed, employ the usual circular-motion reasoning. In a circular orbit of radius R, you are undergoing radial acceleration $a_{\text{radial}} = \frac{v^2}{R}$. Therefore, the net radial force on you is $\sum F_{\text{radial}} = \frac{mv^2}{R}$. But once you are in orbit, the only force (radial or otherwise) acting on you is $F_{\text{grav}} = \frac{GMm}{R^2}$. So,

$$\begin{aligned} \sum F_{\text{radial}} &= ma_{\text{radial}} \\ \frac{GMm}{R^2} &= m\frac{v^2}{R}. \end{aligned}$$

Solve for v to get $v = \sqrt{\frac{GM}{R}} = \sqrt{\frac{(6.67\times10^{-11}\,\text{N}\cdot\text{m}^2/\text{kg}^2)(1.0\times10^{16}\,\text{kg})}{10000\,\text{m}}} = 8.2\,\text{m/s}$, which corresponds to running the 100-meter dash in 12.2 s.

QUESTION 19-2

The Earth is about 1.50×10^{11} meters from the Sun. You know how much time the Earth takes to complete a revolution around the Sun. Finally, you know the universal gravitational constant, $G = 6.67\times10^{-11}\ \text{m}^3/\text{s}^2\cdot\text{kg}$. From this information, *and nothing else*, estimate the mass of the Sun.

ANSWER 19-2

The "trick" here is to see that Kepler's 3rd law applies. But let me solve as if I did not see the trick. In the process, I will end up rederiving Kepler's 3rd law.

We know the period of the Earth's orbit around the Sun is $T = 1.00$ year. Let us convert to seconds: $T = 1\text{ year}\times\frac{365\text{ days}}{\text{year}}\times\frac{24\text{ hours}}{\text{day}}\times\frac{60\text{ minutes}}{\text{hour}}\times\frac{60\text{ seconds}}{\text{minute}} = 3.15\times10^7$ seconds. Roughly speaking, the period tells us how quickly the Earth whips around the Sun. How does this quickness relate to the Sun's mass? Well, the Earth traces an approximately circular orbit of radius $r = 1.50\times10^{11}$ meters around the Sun. Intuitively, the bigger the Sun, the stronger the force it exerts on Earth, and hence the faster the Earth circles the Sun. Let us see if we can quantify this intuition using circular motion reasoning. Since the Earth traces a circle, its radial acceleration (i.e., its acceleration toward the Sun) is $a_{\text{radial}} = \frac{v^2}{R}$. The Sun's gravitational pull provides the radial force that causes the Earth to accelerate in this way. By Newton's 2nd law,

$$\sum F_{\text{radial}} = ma_{\text{radial}}$$
$$\frac{GM_sM_E}{r^2} = M_E\frac{v^2}{r},$$

where M_s and M_E are the masses of the Sun and Earth. The M_E's cancel, leaving us with

$$\frac{GM_s}{r^2} = \frac{v^2}{r}, \tag{1}$$

where r is the Sun-to-Earth distance. We can solve this for M_s. Unfortunately, the answer would contain v, the Earth's speed around the Sun. We are not given v. But we know Earth's *period*. Intuitively, the faster the Earth moves, the less time it takes to complete a full orbit. So, we expect an inverse relationship between v and T. Indeed, we can find the exact relationship by keeping in mind that in time T, the Earth covers one circumference, a distance $2\pi r$. Since the Earth's speed is (nearly) constant,

$$v = \frac{\text{distance}}{\text{time}} = \frac{2\pi r}{T}. \tag{2}$$

Equations (1) and (2) provide all the information we need. Substituting Eq. (2) into Eq. (1) gives

$$\frac{GM_s}{r^2} = \frac{\left(\frac{2\pi r}{T}\right)^2}{r} = \frac{4\pi^2 r}{T^2}. \tag{*}$$

By manipulating this equation, we can make Kepler's 3rd law pop out: $\frac{r^3}{T^2} = \frac{GM_s}{4\pi^2}$. But here, we want to solve Eq. (*) for M_s. Doing so yields

$$M_s = \frac{4\pi^2 r}{T^2}\left(\frac{r^2}{G}\right) = \frac{4\pi^2 r^3}{GT^2} = \frac{4\pi^2(1.50\times10^{11}\ \text{m})^3}{(6.67\times10^{-11}\ \text{m}^3/\text{s}^2\cdot\text{kg})(3.15\times10^7\ \text{s})^2} = 2.01\times10^{30}\ \text{kg},$$

which agrees with the book value, $M_s = 1.99\times10^{30}$ kg, to within rounding errors.

QUESTION 19-3

Consider a rocket, of mass m, that starts on the surface of the Earth, takes off with some initial velocity, and travels along a straight path to the Moon. After receiving its initial velocity, the rocket no longer uses its thrusters. (A realistic flight path would be curvy, but let us keep things simpler.)

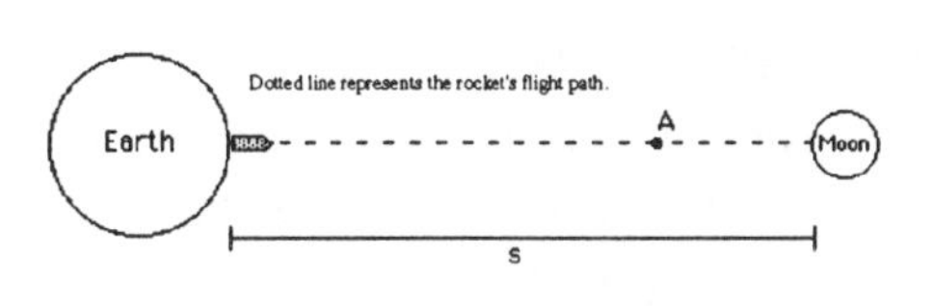

Let s denote the distance from the *surface* of the Earth to the *surface* of the Moon. Let M_E, M_m, R_E, and R_m denote the masses and radii of the Earth and Moon. Express your answers in terms of G and the constants listed above. Do not plug in numbers.

(a) Let point A refer to the spot in space where the net force on the rocket is zero. How far from the rocket's take-off point is point A? Set up, but do not solve, the relevant equation(s).

(b) What is the minimum initial speed the rocket must have in order to reach point A? Let x denote your answer to part (a), and leave your answer in terms of x.

(c) Suppose that when the rocket passes point A, it is barely moving. How fast will it be traveling right before crashing into the Moon? Again, leave your answer in terms of x.

ANSWER 19-3

(a) All along its flight path, the rocket is subject to two forces. The Moon's gravitation tries to pull the rocket forward, while the Earth's gravitation tries to pull it backwards. When the rocket is close to Earth, the Earth's gravitation "wins out," and the net force on the rocket is backward. That does not mean the rocket stops; it just means the rocket slows down. But when the rocket gets close enough to the Moon, the Moon's gravitation wins out. Therefore, the rocket starts to speed up. Here is a force diagram when the Earth is still "winning."

Right at point A, the net force is zero: $\sum F = F_m - F_E = 0$. In other words, at point A, the forces cancel: $F_E = F_m$.

Let x denote the distance from the take-off point to point A. Then, the distance from point A to the center of the Earth is $x + R_E$. The distance from point A to the center of the Moon is $(s-x)+R_m$, as the diagram shows. These distances are the "r's" you must substitute into the gravitational force formula, $F_{\text{grav}} = \frac{GMm}{r^2}$.

Let m denote the rocket's mass. Since the Earth's and Moon's force on the rocket cancel at point A,

$$F_E = F_m$$
$$\frac{GM_E m}{(x+R_E)^2} = \frac{GM_m m}{(s-x+R_m)^2}.$$

The G's and m's cancel, leaving us with $\frac{M_E}{(x+R_E)^2} = \frac{M_m}{(s-x+R_m)^2}$. Now it is just algebra to solve for x. You were not required to complete the algebra.

(b) Let v_0 denote the minimum initial speed the rocket must have to reach point A. We are trying to solve for v_0. You might think to use $v^2 = v_0^2 + 2a\Delta x$, with $v = 0$ at point A. But as the rocket flies from Earth to Moon, the Earth's gravitational pull gets weaker, while the Moon's gravitational pull gets stronger. The net force on the rocket is not constant. Therefore, the acceleration is not constant. You cannot use constant-acceleration kinematic formulas such as $v^2 = v_0^2 + 2a\Delta x$. Instead, we must use conservation laws, in this case, energy conservation.

If v_0 is the *minimum* initial speed needed to reach point A, then at point A, the rocket will hardly be moving; it will *barely* reach that point. So, we can set $v_A \approx 0$. When using conservation of energy, remember that *the potential energy of the rocket is due to both the Earth and the Moon.*

Since energy is a scalar (i.e., just a number), it does not have a direction associated with it. By contrast, if we were dealing with forces, we would have to give F_m and F_E opposite signs, because those forces point in opposite directions. But the potential energies $U_{\text{due to Earth}}$ and $U_{\text{due to Moon}}$ do not have directions. They "automatically" get the same sign. Remember that the "r" in the potential energy formula is the distance from the rocket to the *center* of the Earth or Moon. As the diagram shows, the rocket starts off a distance $r = R_E$ from the Earth's center, and a distance $r = s + R_m$ from the Moon's center. So,

$$E_0 = E_A$$
$$K_0 + U_{0,\text{ due to Earth}} + U_{0,\text{ due to Moon}} = K_A + U_{A,\text{ due to Earth}} + U_{A,\text{ due to Moon}}$$
$$\frac{1}{2}mv_0^2 + \left(-\frac{GM_E m}{R_E}\right) + \left(-\frac{GM_m m}{s+R_m}\right) = 0 + \left(-\frac{GM_E m}{x+R_E}\right) + \left(-\frac{GM_m m}{s-x+R_m}\right).$$

Cancel the m's and solve for v_0 to get

$$v_0 = \sqrt{2G\left[\frac{M_E}{R_E} + \frac{M_m}{s+R_m} - \frac{M_E}{x+R_E} - \frac{M_m}{s-x+R_m}\right]}.$$

(c) If the rocket is hardly moving when it passes point A, its initial velocity must have been the v_0 we calculated in part (b). But you never have to use that v_0, unless you want to. By conservation of energy, the rocket's final energy (right before crashing into the Moon) is the same as its energy at point A, which is the same as its initial energy: $E_0 = E_A = E_f$. Therefore, we can solve for v_f, the crash velocity, either by setting $E_f = E_0$ or by setting $E_f = E_A$. I will choose the second option. (I have been lazy about drawing pictures of the relevant phases of the process.)

In part (b), we already wrote an expression for E_A, the rocket's total energy at point A. To express the final energy, you must see that immediately before crashing, the rocket is a distance $r = R_m$ from the Moon's center and $r = s + R_E$ from Earth's center. So,

$$E_A = E_f$$

$$K_A + U_{A,\text{ due to Earth}} + U_{A,\text{ due to Moon}} = K_f + U_{f,\text{ due to Earth}} + U_{f,\text{ due to Moon}}$$

$$0 + \left(-\frac{GM_E m}{x + R_E}\right) + \left(-\frac{GM_m m}{s - x + R_m}\right) = \frac{1}{2}mv_f^2 + \left(-\frac{GM_E m}{s + R_E}\right) + \left(-\frac{GM_m m}{R_m}\right).$$

Solve for v_f to get

$$v_f = \sqrt{2G\left[\frac{M_E}{s + R_E} + \frac{M_m}{R_m} - \frac{M_E}{x + R_E} - \frac{M_m}{s - x + R_m}\right]}.$$

You would get the same answer by setting $E_f = E_0$. By the way, it is interesting to realize that if the rocket has enough initial velocity to reach point A, then it "automatically" reaches the Moon. Once the rocket passes point A, the Moon's gravity "wins out" over the Earth's gravity, and therefore the rocket gets accelerated the rest of the way to the Moon.

QUESTION 19-4

A rocket of mass m drifts in a circle at constant speed v_0 around the planet Venus, which has mass M.

(a) How far from the center of Venus is the rocket? Answer in terms of M, v_0, and G.

(b) An explosion in the engine chamber suddenly increases the rocket's speed by 25% (along its direction of motion). Make a rough sketch of the rocket's new orbit, being sure to indicate on your drawing where the rocket was when the explosion occurred.

(c) *(Very hard: Ask your instructor if you need to be able to answer questions like this.)* In its new orbit, what is the maximum distance that the rocket reaches from the center of Venus? Again, answer in terms of M, v_0, and G. Actually, just set up the relevant equation or equations; but make sure it is possible to solve for r_{max} in terms of the given constants.

ANSWER 19-4

(a) Because the rocket traces out a circle, we can exploit the fact that its radial acceleration is $a_{\text{radial}} = \frac{v^2}{R}$. Venus' gravitational pull provides the radial force that makes the rocket continue moving in a circle. So, by Newton's 2nd law,

$$\sum F_{\text{radial}} = ma_{\text{radial}}$$

$$\frac{GMm}{r_0^2} = m\frac{v^2}{r_0}.$$

I am letting r_0 denote the rocket's distance from the center of Venus. Solve for it to get $r_0 = \frac{GM}{v_0^2}$.

(b) This is NOT INTUITIVE. You just have to know that when a satellite gravitationally orbits a central body, only two "shapes" are possible: circles and ellipses. Since the rocket gets knocked out of its circular orbit, it must end up in an ellipse (assuming, of course, that it does not get knocked completely out of orbit, which happens if the rocket acquires the needed "escape velocity." This is not the case here.) Crucially, the orbit is "closed": The rocket eventually returns to where it started. Since the rocket speeds up, the ellipse is bigger than the original circle.

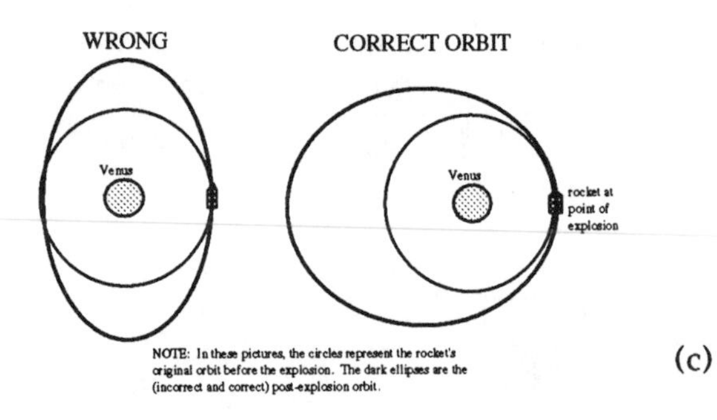

NOTE: In these pictures, the circles represent the rocket's original orbit before the explosion. The dark ellipses are the (incorrect and correct) post-explosion orbit.

What is really weird, however, is the orientation of the ellipse. You might expect it to be "elongated" along the direction in which the rocket was heading when the explosion occurred. It turns out that the elongation occurs at right angles to this. A computer simulation can help you to visualize the orbit.

Notice that Venus is the "focus," not the center of the correct ellipse.

(c) In its new elliptical orbit, the rocket starts out moving 25% faster than v_0. So, its "initial" speed is $v_1 = \frac{5v_0}{4}$. We also know that, immediately after the explosion, the rocket is still a distance $r_0 = \frac{GM}{v_0^2}$ from Venus. (In the brief time over which the explosion occurs, the rocket cannot move very far.) We are solving for its maximum distance from Venus, r_{max}. Here is a picture with all this information.

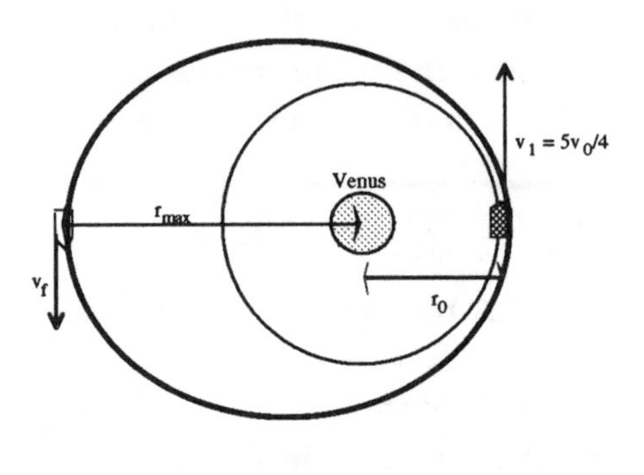

Here is a crucial point: In an elliptical orbit, the rocket's speed is not constant. It slows down as it gets farther from the planet, and speeds up again as it "falls" closer to the planet. Here is one way of understanding this effect: As the rocket gets farther from Venus, its potential energy increases. Therefore, its kinetic energy must decrease, to compensate. In other words, as the rocket gets "higher," it slows down. The same thing happens to a ball when you throw it. The upshot is that v_f, the rocket's speed when $r = r_{max}$, is *less than* v_1.

The preceding discussion suggests that we can use energy conservation to solve this problem. Let us try it. I will set the rocket's energy immediately after the explosion equal to its "final" energy at r_{max}:

$$\begin{aligned} E_1 &= E_f \\ U_1 + K_1 &= U_f + K_f \\ -\frac{GMm}{r_0} + \frac{1}{2}mv_1^2 &= -\frac{GMm}{r_{max}} + \frac{1}{2}mv_f^2. \end{aligned} \tag{1}$$

From part (a), $r_0 = \frac{GM}{v_0^2}$. We also know that $v_1 = \frac{5v_0}{4}$. But Eq. (1) still contains two unknowns, r_{max} and v_f. Therefore, we need more information before we can solve for r_{max}.

We have already used energy conservation. Is anything else conserved? Well, the rocket's linear momentum keeps changing; it is not conserved. What about angular momentum? L is conserved except when an external torque acts on the system. So, we must decide if the rocket experiences an external torque.

The rocket feels a gravitational force pulling it toward Venus. But Venus is the "pivot point" around which the rocket moves. In other words, the relevant "radius vector" points from Venus to the rocket. So, the gravitational force is *parallel* to the radius vector. In other words, $F_\perp = 0$. Since $\tau = F_\perp r$, the rocket feels no torque. This means **angular momentum is conserved.**

As explained in Answer 16-5, the angular momentum of a point mass can be written

$$L_{\text{point mass}} = mv_\perp r,$$

where $v_\perp$ is the component of the point mass' velocity perpendicular to the radius vector. Fortunately, as the above diagram shows, the rocket immediately after the explosion moves entirely perpendicular to the radius vector (from Venus to the rocket). This is true also at r_{max}. So, we need not break up any velocity vectors. Let us set the post-explosion angular momentum equal to the "final" angular momentum at r_{max}:

$$\begin{aligned} L_1 &= L_f \\ mv_1 r_0 &= mv_f r_{max} \end{aligned} \tag{2}$$

Remember, we know $r_0 = \frac{GM}{v_0^2}$ from part (a), and we know $v_1 = \frac{5v_0}{4}$. So, Eqs. (1) and (2) contain only two unknowns, r_{max} and v_f. We are done, except for algebra.

This problem illustrates a general point: **In orbital motion, energy and angular momentum are both conserved.**

QUESTION 19-5

Jupiter spins on its axis with a period of $T = 10$ hours. How high above Jupiter's surface would a satellite have to be to remain in stationary ("geosynchronous") orbit above Jupiter's equator? (By "stationary," I mean the satellite always stays directly above the *same* spot on the planet.) Use $M = 1.90 \times 10^{27}$ kg, and $R = 7.14 \times 10^7$ m, for the mass and radius of Jupiter.

ANSWER 19-5

Always start with a well-labelled diagram, to avoid confusing your variables.

We want the satellite always to stay over the same point on Jupiter. Roughly speaking, the satellite must "keep up" with Jupiter's rotation. Therefore, since Jupiter completes one rotation in time $T = 10$ hours, the satellite must circle Jupiter in that *same* amount of time:

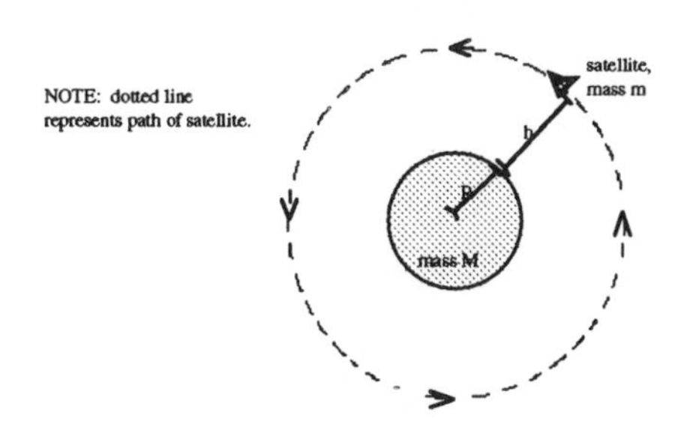

$$T_{satellite} = T_{Jupiter} = 10 \text{ hours} \times \frac{3600 \text{ seconds}}{\text{hour}} = 36000 \text{ seconds}.$$

Since we are ultimately seeking a relationship between the satellite's distance from the planet and its orbital period, a good place to start is with circular-motion reasoning. The satellite traces out a circle of radius $r = R + h$, where R is Jupiter's radius, and h is the satellite's height above Jupiter's surface. (See the diagram.) Jupiter's gravitation supplies the force that keeps the satellite moving circularly. So,

$$\begin{aligned} \sum F_{radial} &= ma_{radial} \\ \frac{GMm}{r^2} &= m\frac{v^2}{r}. \\ \frac{GMm}{(R+h)^2} &= m\frac{v^2}{R+h}, \end{aligned} \tag{1}$$

since $r = R + h$. The m's cancel. Physically, this means the satellite's mass is irrelevant.

We are trying to solve for h, the satellite's height above the planet's surface. Unfortunately, Eq. (1) contains a second unknown, namely v, the satellite's speed. How can we find v?

Well, as we deduced above, the satellite's period is $T = 10$ hours. How does period relate to velocity? During time T, the satellite travels one complete circumference, a distance $2\pi r$. Since the satellite travels at constant speed, its velocity is related to the distance covered by

$$v = \frac{\text{distance covered}}{\text{time}} = \frac{2\pi r}{T}. \tag{2}$$

Now we have enough information to solve for h. Substitute Eq. (2) into Eq. (1), and remember that $r = R + h$, to get

$$\frac{GMm}{(R+h)^2} = \frac{\left[\frac{2\pi(R+h)}{T}\right]^2}{R+h}.$$

Cancel the m's, then isolate h to get

$$\begin{aligned} h &= \sqrt[3]{\frac{GMT^2}{4\pi^2}} - R \\ &= \sqrt[3]{\frac{(6.67\times10^{-11}\ \text{m}^3/\text{kg}\cdot\text{s}^2)(1.90\times10^{27}\ \text{kg})(36000\ \text{s})^2}{4\pi^2}} - 7.14\times10^7\ \text{m} \\ &= 1.61\times10^8\ \text{m} - 7.14\times10^7\ \text{m} \\ &= 8.94\times10^7\ \text{m}. \end{aligned}$$

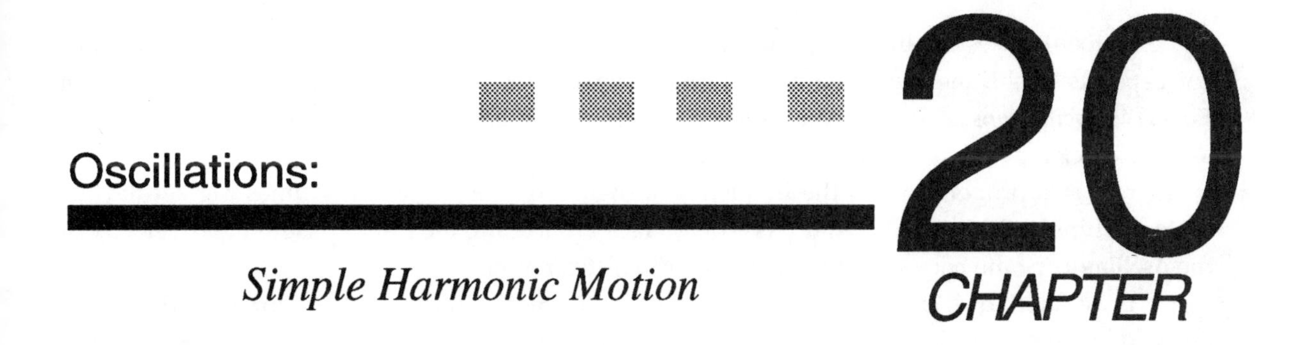

QUESTION 20-1

A block of mass m is attached to one end of a spring, the other end of which is attached to a wall. The floor is frictionless, and the spring has spring constant k. Let $x = 0$ denote the block's position when the spring is at its equilibrium length.

Suppose the block is displaced to $x = D$. Instead of releasing the block from rest, however, someone shoves the block towards $x = 0$ with speed v_0, right at time $t = 0$.

(a) What is the block's velocity at arbitrary later time t_1? *(Do not worry about the "phase constant" unless your instructor says that you need to be able to find it.)*

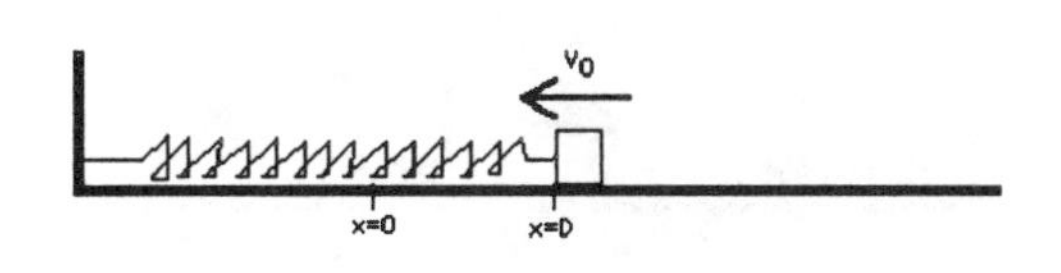

(b) During its oscillations, when the block is momentarily motionless, what is its acceleration?

(c) When the block passes through $x = 0$, what is its speed?

ANSWER 20-1

Many people approach oscillatory motion problems by scanning the textbook for applicable formulas. In this chapter, I will show how you can solve all such problems using a small set of formulas and problem-solving techniques. I will try to explain where the relevant formulas come from, without getting caught up in mathematical details.

Let us begin by picturing how this problem differs from the "standard" one in which the block is displaced and then released from rest. When released from rest from $x = D$, the block speeds up as the spring un-stretches. Then, after passing through $x = 0$, the block slows down as the spring compresses. Eventually, the block reaches $x = -D$, at which point it momentarily comes to rest before "turning around" and moving in the other direction.

But when the block is given an initial shove, the "extra" speed makes it pass through $x = 0$ so fast that the block compresses the spring more than usual; the block turns around at $x = -A$, where $A > D$. Then it oscillates back and forth between $x = -A$ and $x = +A$. To see all this, experiment with a real spring or with the computer simulation, which allows you to adjust the initial velocity.

(a) We have a standard strategy for finding $x(t)$, $v(t)$, or $a(t)$. For reasons I will explain in Question 20-2, the position of any simple harmonically oscillating object takes the following form:

$$x = A\cos(\omega t + \phi). \tag{1}$$

Some textbooks use sine instead of cosine. It does not matter, because using sine vs. cosine just corresponds to shifting the value of the "phase constant" ϕ by 90°. The key feature encoded in Eq. (1) is *oscillations*. A is the maximum position, i.e., the amplitude of oscillation. So, $x = \pm A$ are the block's positions at its turning points, when it momentarily comes to rest. The angular frequency ω is proportional to the frequency of oscillation, i.e., how many times per second the system whips back and forth. And ϕ is a phase factor, encoding the initial position and velocity of the oscillator. Some books call the phase factor δ instead of ϕ.

So, if we can find A, ω, and ϕ, then Eq. (1) gives us the block's position at any time. And once we know $x(t)$, we can easily differentiate to find $v(t)$, since $v = \frac{dx}{dt}$. Hence, the "meat" of this problem (and ones like it) is finding A, ω, and ϕ, so that we can substitute them into Eq. (1).

Finding A, ω, and ϕ is like solving three separate problems, each with its own little strategy. As we work through the details, do not lose sight of the fact that all these "subproblems" are part of the larger attempt to find $x(t)$, and ultimately $v(t)$, using Eq. (1).

Subproblem 1: Finding the amplitude, A

Sometimes, you can "guess" A intuitively by picturing the motion. For instance, if the block were released from rest from $x = D$, then it would oscillate back and forth between $x = +D$ and $x = -D$; the amplitude would be $A = D$. But here, things are more complicated, due to the initial velocity. Usually, energy conservation provides the easiest way to find A. Here is how:

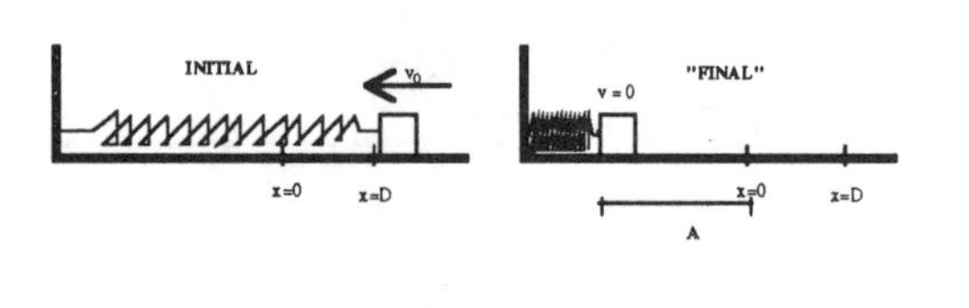

The total energy of the spring-and-block system is the kinetic energy of the block ($K = \frac{1}{2}mv^2$) plus the elastic potential energy of the spring ($U_e = \frac{1}{2}kx^2$, where x is the displacement from equilibrium.) This total energy is conserved. So, when the block moves slowly near its "turn-around" points (low kinetic energy), the spring is highly stretched or compressed (high potential energy). And when the potential energy is low (near equilibrium), that is when the block moves fastest (high kinetic energy). Using conservation of energy, we can solve for A. Let us organize our thoughts with pictures. By "final," I mean when the block reaches $x = -A$, at which point it momentarily comes to rest.

So,

$$\begin{aligned} E_0 &= E_f \\ K_0 + U_0 &= K_f + U_f \\ \frac{1}{2}mv_0^2 + \frac{1}{2}kD^2 &= 0 + \frac{1}{2}k(-A)^2. \end{aligned}$$

Solve for A to get $A = \sqrt{\frac{mv_0^2 + kD^2}{k}}$. As expected, $A > D$. Furthermore, the larger the initial speed, the larger A gets.

So far, I have used energy conservation to solve for the amplitude of oscillation. But remember, besides amplitude, we also need to find the angular frequency ω and the phase constant ϕ, so that we can substitute them into $x = A\cos(\omega t + \phi)$.

Subproblem 2: Finding the angular frequency, ω.

In Question 20-2 below, I will demonstrate a strategy for finding the angular frequency of *any* simple harmonic oscillator. For now, let me just state the specific result for a spring, $\omega = \sqrt{\frac{k}{m}}$, and discuss its physical meaning. The spring constant k tells us the spring's stiffness. Remember, k

comes from $F=-kx$, the force exerted by a spring. If we stretch two springs by the same distance x, then the stiffer spring–the one with higher k–exerts a bigger force. The formula $\omega\sqrt{\frac{k}{m}}$ tells us that a stiffer spring causes a higher ω; the block whips back and forth more times per second. This makes sense; we expect a loose spring (low k) to lead to lackadaisical oscillations (low ω). The formula $\omega\sqrt{\frac{k}{m}}$ also tells us that, other things being equal, a heavier block oscillates fewer times per second. Again, this makes sense.

Subproblem 3: Finding the phase factor, ϕ.

Here is a general strategy for finding ϕ, once you know A and ω. You will always know the block's position at one particular time, usually $t=0$. Eq. (1) gives us the position at any time, including $t=0$. So, starting with Eq. (1), $x=A\cos(\omega t+\phi)$, substitute the known initial position into the left side, and substitute $t=0$ into the right side. Then solve for ϕ.

Here, we know the initial position is $x_0=D$ at time $t=0$. So, from $x=A\cos(\omega t+\phi)$, we get

$$\begin{aligned} D &= A\cos(\omega 0+\phi). \\ &= A\cos\phi. \end{aligned}$$

To isolate ϕ, divide through by A, and then take the arccosine of both sides. This yields $\phi=\cos^{-1}(\frac{D}{A})$, where $A=\sqrt{\frac{mv_0^2+kD^2}{k}}$, as we found in subproblem 1 above.

End of subproblem 3.

By solving separate subproblems, we have found the block's amplitude A, angular frequency ω, and phase constant ϕ. So, we know the block's position at any time t:

$$\begin{aligned} x &= A\cos(\omega t+\phi) \\ &= \sqrt{\frac{mv_0^2+kD^2}{k}}\cos\left(\sqrt{\frac{k}{m}}t+\left[\cos^{-1}\left(\frac{D}{A}\right)\right]\right). \end{aligned}$$

From $x(t)$, we can now find $v(t_1)$ by differentiating $x(t)$ with respect to t, and then setting $t=t_1$. Make sure you take the derivative before, not after, substituting in the specific time value. Doing so yields

$$\begin{aligned} v(t_1) &= -A\omega\sin(\omega t_1+\phi) \\ &= -\sqrt{\frac{mv_0^2+kD^2}{k}}\sqrt{\frac{k}{m}}\sin\left(\sqrt{\frac{k}{m}}t_1+\left[\cos^{-1}\left(\frac{D}{A}\right)\right]\right). \end{aligned}$$

(b) You might be tempted to say that, when the block is not moving, $a=0$. But just because $v=0$ does not mean $a=0$. Remember, acceleration is the rate at which velocity changes. When the block reaches its turning point, it momentarily comes to rest. But a moment before reaching the turning point, the block was moving in one direction; and a moment after reaching the turning point, the block will move in the other direction. So, at its turning point, the block's velocity is *changing*, even though it is temporarily at rest. In symbols, $a\neq 0$, even though $v=0$. (You have seen this before. When you throw a ball straight up, the acceleration at the peak is $a=-g$, even though the velocity at the peak is $v=0$.)

The block, when momentarily motionless, is at $x=\pm A$, the turning point. So, this problem is really asking us to find the acceleration when $x=\pm A$.

I see two ways of proceeding. The clever shortcut is to use Newton's 2nd law, $F_{net} = ma$. The force exerted by a spring is $F = -kx$. So, Newton's 2nd law says that $-kx = ma$. Therefore, when $x = A$, Newton's law implies that $-kA = ma$. Solving for a gives us

$$\begin{aligned} a &= -\frac{k}{m}A \\ &= -\frac{k}{m}\sqrt{\frac{mv_0^2 + kD^2}{k}}, \end{aligned}$$

where I just used our expression for the amplitude obtained in part (a).

Now I will outline the second way of finding this acceleration. In part (a) above, we already found the block's velocity as a function of time. So, we can find the acceleration by differentiating:

$$\begin{aligned} a &= \frac{dv}{dt} = \frac{d}{dt}[-A\omega\sin(\omega t + \phi)] \\ &= -A\omega^2\cos(\omega t + \phi) \\ &= -\sqrt{\frac{mv_0^2 + kD^2}{k}}\frac{k}{m}\cos\left(\sqrt{\frac{k}{m}}t + \phi\right). \end{aligned}$$

That is the acceleration at arbitrary time t. But we want the acceleration right when the block reaches its turn-around point, $x = +A$ or $x = -A$. We could solve this the long way by calculating at what time x equals A, and then substituting this time into the acceleration formula just derived. But it is quicker to realize that when the block reaches its maximum displacement, the force on it–and therefore the acceleration–is as big as it ever gets.

Can you tell what is the biggest possible value of $a = -\sqrt{\frac{mv_0^2 + kD^2}{k}}\frac{k}{m}\cos\left(\sqrt{\frac{k}{m}}t + \phi\right)$, just by looking? Yes! The cosine function oscillates between +1 and −1. Therefore, a is biggest when $\cos(\omega t + \phi)$ reaches its peak value of ± 1. So, set $\cos(\omega t + \phi) = \pm 1$ to get

$$a_{max} = \pm\sqrt{\frac{mv_0^2 + kD^2}{k}}\frac{k}{m},$$

the same value obtained above using $F = ma$.

Enough math. Physically, the crucial insight is that the restoring force–and therefore the acceleration–is proportional to x, the displacement. The farther the system gets from equilibrium, the bigger its acceleration (back toward equilibrium). This may seem paradoxical, because the system moves slowest when it is farthest from equilibrium. See the computer simulation. The paradox disappears, however, if you remember that acceleration is the *change* in velocity per time. While the block turns around, its velocity changes rapidly, because the block must slow down, turn around, and then speed up in the other direction. This is true even though the velocity itself is small near the turn-around points.

(c) I see two ways to solve: conservation of energy, or "oscillation" kinematics. By the way, a common mistake is to write something like $v^2 = v_0^2 + 2a\Delta x$. But remember, the spring's force is not constant, and therefore the acceleration is not constant. For this reason, you cannot use your old constant-acceleration kinematic formulas. Instead, you must use the kinematic formulas that apply to simple harmonic motion. Before showing you the kinematic approach, however, let me conserve energy:

Method 1: Conservation of energy

Since $U_e = \frac{1}{2}kx^2$, the block has no potential energy when it passes through $x = 0$. Intuitively speaking, when the spring reaches its equilibrium length ($x = 0$), it is neither stretched nor compressed, and therefore has no energy stored inside it. At $x = 0$, all the potential energy has converted into kinetic energy. The block at that point has the most kinetic energy, and therefore the most speed, of its entire oscillation. If we let v_{max} denote the block's speed at $x = 0$, then conservation of energy tells us

$$E_i = E_{\text{at } x=0}$$
$$K_i + U_i = K_{\text{at } x=0} + U_{\text{at } x=0}$$
$$\frac{1}{2}mv_0^2 + \frac{1}{2}kD^2 = \frac{1}{2}mv_{max}^2 + 0.$$

Solve for v_{max} to get $v_{max} = \sqrt{v_0^2 + \frac{k}{m}D^2}$.

Notice that energy conservation supplied us with a shortcut in parts (a) and (c).

Method 2: Kinematics

Starting from our "fundamental" oscillation kinematic formula, $x = A\cos(\omega t + \phi)$, we can differentiate to get $v = \frac{dx}{dt} = -A\omega\sin(\omega t + \phi)$. We know A, ω, and ϕ. We want v when $x = 0$. So, we could work the long way by solving for the time at which $x = 0$, and then substituting that time into $v = -A\omega\sin(\omega + \phi)$. Fortunately, you can sidestep this math by formalizing the following insight: At $x = 0$, the block reaches its maximum velocity. Since $\sin(\omega t + \phi)$ oscillates between -1 and $+1$, $v = -A\omega\sin(\omega t + \phi)$ reaches it maximum value when $\sin(\omega t + \phi) = \pm 1$. So, plug $\sin(\omega t + \phi) = \pm 1$ into $v = -A\omega\sin(\omega t + \phi)$ to get

$$v_{max} = \pm A\omega = \pm\sqrt{\frac{mv_0^2 + kD^2}{k}}\sqrt{\frac{k}{m}} = \pm\sqrt{v_0^2 + \frac{k}{m}D^2},$$

the same answer obtained above using energy conservation. The plus and minus signs indicate that the oscillating block passes through equilibrium "in both directions," with equal (and oppositely directed) speeds.

QUESTION 20-2

Hard, but you probably need to be able to solve problems like this. Ask your instructor.

Consider a small block (mass m) inside a frictionless hemisphere-shaped bowl of radius R. The block is attached to a spring, the other end of which is nailed to a point inside the bowl. When the block is at the bottom of the bowl, the spring is at its equilibrium length. Notice that the spring stretches and compresses along the "arc" of the bowl.

Suppose the block is displaced through a small angle θ_0 and released from rest. The block oscillates.

In this problem, you may assume that θ_0 is small enough that $\sin\theta_0 \approx \theta_0$.

(a) Let T denote the period of oscillation when the block is released from θ_0. Now suppose the block is grabbed and released from rest from initial angle $2\theta_0$. Is the new period of oscillation bigger, smaller, or the same as T? Do not use formulas; answer qualitatively.

(b) Let us return to the case where the block gets released from rest from angle θ_0. How much time does the block take to reach equilibrium (i.e., the bottom of the bowl)?

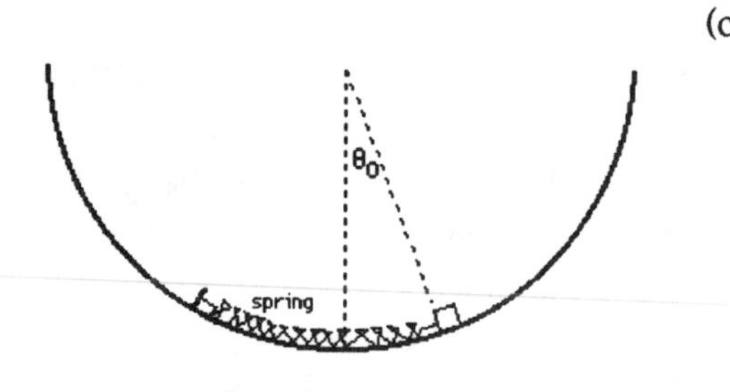

(c) How fast is the block moving when it reaches equilibrium? (Hint for energy lovers: When the block is displaced from equilibrium by angle θ, its vertical distance above the bottom of the bowl is given approximately by $\frac{R\theta^2}{2}$. The answer explains why.)

ANSWER 20-2

(a) This part asks whether changing the *size* of the oscillations (the amplitude) affects the *period* (how much time an oscillation takes). Intuitively, you might think that bigger oscillations take more time, because the object has more distance to cover. But when the oscillations are bigger, the object also moves faster, on average. It turns out that when you double the amplitude, thereby doubling the distance covered during an oscillation, the object's average speed also doubles. Therefore, the object takes the same time to complete the oscillation; the period stays the same. This counterintuitive conclusion applies to *any* oscillator, provided the oscillations are small enough that the motion is approximately "simple harmonic." Frankly, you have got to see it to believe it. Using a shoe-hanging-by-shoelace, or any other "pendulum," first make small oscillations. Then do bigger oscillations. You will see that no matter what amplitude (below 30° or so) you choose, the period–the amount of time it takes to swing back and forth once–is the *same*. Here is the pithy summary: In simple harmonic motion, period and frequency are independent of amplitude.

(b) Before solving, I will clarify some notation. Let x denote the block's distance from equilibrium, as measured by the "arc length" along which the block travels. So, x is an arc length along the inside of the bowl. From the definition of angle, $x = R\theta$.

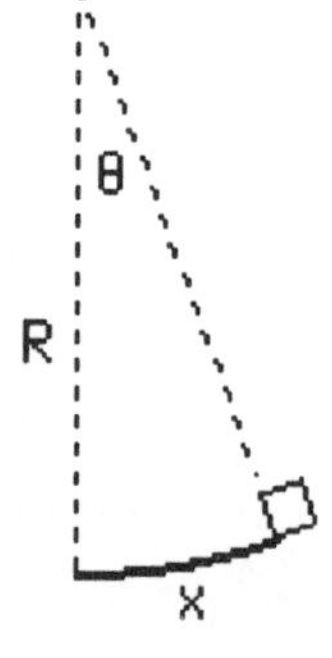

Using this relationship between linear and angular displacement, I will solve this problem in terms of x's instead of θ's.

I see two related ways to proceed. The longer way is kinematic. Start with $x = A\cos(\omega t + \phi)$. Solve for A, ω, and ϕ, using the subproblem strategies demonstrated in Question 16-1. Then solve for the time at which the block reaches equilibrium ($x = 0$) by setting $x = A\cos(\omega t + \phi) = 0$.

Before working through this method, let me show you a shortcut, which relies on the following physical insight: In going from its initial position ($x_0 = R\theta_0$) to equilibrium ($x = 0$), the block completes exactly one quarter of a full cycle. How do I know? Well, during a full oscillation, the block goes from $x = R\theta_0 \to x = 0 \to x = -R\theta_0 \to x = 0 \to x = R\theta_0$. You can see this by picturing the oscillation in your head, or by watching a pendulum. So, $x = R\theta_0 \to x = 0$ corresponds to the first quarter of the overall oscillation. Since a full oscillation takes time T, the block takes time $t = \frac{T}{4}$ to reach equilibrium, i.e., to complete the first quarter of the oscillation. Hence, you can solve for t by calculating T and dividing by 4.

How can we find T? As shown in the book, since a complete oscillation corresponds to a "phase increase" of 2π, the relationship between a harmonic oscillator's angular frequency and its period is $\omega T = 2\pi$, from which we get $T = \frac{2\pi}{\omega}$.

At this point, we are out of shortcuts. Whether we use $x = A\cos(\omega t + \phi)$ or $T = \frac{2\pi}{\omega}$, we desperately need to know the angular frequency ω.

My strategy for finding ω will not make any sense unless I review some of the underlying physics. Why do pendulums, masses on springs, blocks in bowls, and hundreds of other physical systems all undergo simple harmonic (oscillatory) motion? Because, for *all* such systems, the net force (or net torque) on the object is a "restoring" force, by which I mean it takes the following form:

$$F_{\text{restoring}} = -(\text{stuff})x = ma, \tag{1}$$

where x is the distance by which the system is displaced from equilibrium. In words, this formula says that the force always tries to push or pull the system back toward equilibrium (hence, the minus sign); and furthermore, the force gets bigger as the object gets farther from equilibrium. "Stuff," the proportionality constant, depends on what specific system we are talking about. For a spring, stuff $= k$. For a pendulum, stuff equals something else. My point is that, for *any* harmonic oscillator, the net force takes the form $F = -(\text{stuff})x$ for *some* stuff.

Since $a = \frac{d^2x}{dt^2}$, Eq. (1) is a second-order differential equation. As your textbook proves, the solution to this differential equation is

$$x = A\cos(\omega t + \phi), \tag{2}$$

where

$$\omega = \sqrt{\frac{\text{stuff}}{m}}. \tag{3}$$

So, now you see where the fundamental oscillation kinematic equation $x = A\cos(\omega t + \phi)$ comes from. It mathematically follows from Eq. (1), which applies whenever the net force on a system is "restoring." We can also see where $\omega_{\text{spring}} = \sqrt{\frac{k}{m}}$ comes from; it is just a special case of the more general Eq. (3), because stuff $= k$ for a spring.

In order to find a system's angular frequency ω using Eq. (3), we need to know "stuff" from Eq. (1). In other words, we need to know the net force on the object when it is displaced distance x from equilibrium. To find "stuff," therefore, you must use old-fashioned force reasoning:

<u>**Stuff-finding strategy:**</u>

(1) Draw a force diagram of the system when it is displaced an arbitrary distance x from equilibrium. (Do not draw the system at equilibrium, because then the net force on it will be 0)

(2) Write down the net force on the system along the direction of motion.

(3) By rewriting that net force in the form $F_{\text{net}} = -(\text{stuff})x$, figure out stuff.

Once you find stuff, just substitute it into Eq. (3) to calculate the system's angular frequency. And once you know ω, you can use it in $x = A\cos(\omega t + \phi)$ or in $T = \frac{2\pi}{\omega}$ to find practically anything you want. My point is that the stuff-finding strategy is an essential part of many oscillation problems, including this one. Let me now implement this strategy.

Stuff-finding strategy step 1: Force diagram

I have broken up gravity into its components parallel and perpendicular to the direction of motion, which is along the surface of the bowl. Notice also that the spring, since it is stretched along the surface of the bowl, pulls the block along the surface of the bowl.

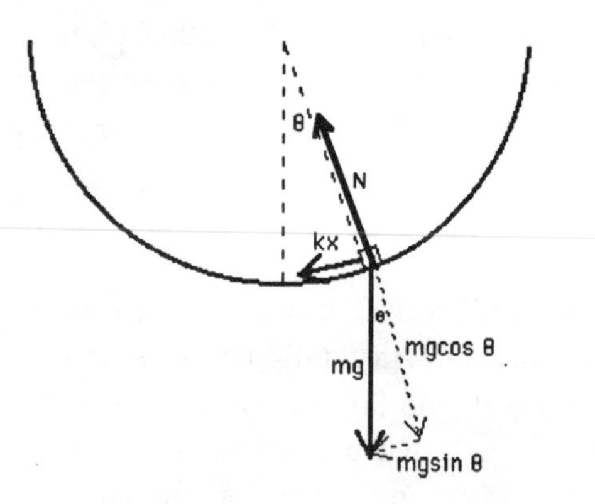

Stuff-finding strategy step 2: Write the net force along the direction of motion

The normal force is completely perpendicular to the direction of motion, i.e., perpendicular to the surface of the bowl. From the diagram, you can see that the net force along the direction of motion is $F_{\text{net}} = -kx - mg\sin\theta$. The minus signs reflect the fact that both forces point "backward" toward equilibrium.

Stuff-finding strategy step 3: Rewrite F_{net} in the form $F_{\text{net}} = -(\text{stuff})x$, and read off "stuff"

We need the right-hand side of the F_{net} equation to contain only constants and x's, not θ's. To get rid of the θ, use the hint given in the problem, along with the relationship between x and θ. From the hint, θ is small enough that $\sin\theta \approx \theta$. So, the F_{net} equation reduces to $F_{\text{net}} \approx -kx - mg\theta$. But by the definition of angle, $\theta = \frac{\text{arclength}}{\text{radius}} = \frac{x}{R}$. So, the F_{net} equation becomes

$$\begin{aligned} F_{\text{net}} &\approx -kx - mg\frac{x}{R} \\ &= -\left(k + \frac{mg}{R}\right)x. \end{aligned}$$

Since in general $F_{\text{net}} = -(\text{stuff})x$ for a simple harmonic oscillator, we immediately see that in this problem, $\text{stuff} = k + \frac{mg}{R}$. In this expression for stuff, k is due to the spring, and $\frac{mg}{R}$ is due to the curvature of the bowl.

End of stuff-finding strategy.

Now we are home free. Remember, the whole point of finding "stuff" was so that we could find the angular frequency using Eq. (3) above, $\omega = \sqrt{\frac{\text{stuff}}{m}}$. Let us do it:

$$\begin{aligned} \omega &= \sqrt{\frac{\text{stuff}}{m}} \\ &= \sqrt{\frac{k + \frac{mg}{R}}{m}} \\ &= \sqrt{\frac{k}{m} + \frac{g}{R}}. \end{aligned}$$

So, the period of oscillation is $T = \frac{2\pi}{\omega} = \frac{2\pi}{\sqrt{\frac{k}{m} + \frac{g}{R}}}$. As shown above, the time it takes the block to reach equilibrium is one quarter of a full oscillation:

$$t = \frac{1}{4}T = \frac{\pi}{2\sqrt{\frac{k}{m} + \frac{g}{r}}}.$$

Alternatively, after finding ω, you could have found t by substituting $\omega = \sqrt{\frac{k}{m} + \frac{g}{R}}$ into $x = A\cos(\omega t + \phi)$, setting $x = 0$, and solving for t. To do this, you would first need to calculate the phase constant ϕ using the strategy I laid out in Answer 16-1(a), "subproblem 3." It turns out here that $\phi = 0$.

This was a very complicated problem. Please skim over my answer again to see how the parts of my solution all fit together. Once you understand the overall structure of the reasoning, you can solve almost any oscillation problem.

(c) As often happens, you can use either energy conservation or kinematics to solve for the velocity. Usually, conservation of energy is quicker. But since we have already completed most of the kinematics in parts (a) and (b), I will show you that method first.

Kinematic method:

Differentiate $x = A\cos(\omega t + \phi)$ to get $v = -\omega A\sin(\omega t + \phi)$. Since the object was released from rest from $x = R\theta_0$, it oscillates between $x = +R\theta_0$ and $x = -R\theta_0$. We also know its angular frequency from part (b): $\omega = \sqrt{\frac{\text{stuff}}{m}} = \sqrt{\frac{k}{m} + \frac{g}{R}}$. And as you can check, the phase constant is $\phi = 0$. So, our general expression for the velocity of a simple harmonic oscillator becomes

$$\begin{aligned} v &= -\omega A\sin(\omega t + \phi) \\ &= -\sqrt{\frac{k}{m} + \frac{g}{R}}R\theta_0 \sin\left(\sqrt{\frac{k}{m} + \frac{g}{R}}t\right). \end{aligned} \qquad (*)$$

We want to know the velocity when the block reaches $x = 0$. Fortunately, in part (b), we found the time at which this happens. Just substitute that t into Eq. (*) to get

$$\begin{aligned} v &= -\sqrt{\frac{k}{m} + \frac{g}{R}}R\theta_0 \sin\left(\sqrt{\frac{k}{m} + \frac{g}{R}}\frac{\pi}{2\sqrt{\frac{k}{m} + \frac{g}{R}}}\right) \\ &= -\sqrt{\frac{k}{m} + \frac{g}{R}}R\theta_0 \sin\left(\frac{\pi}{2}\right) \\ &= -\sqrt{\frac{k}{m} + \frac{g}{R}}R\theta_0, \end{aligned}$$

since $\sin(\frac{\pi}{2}) = 1$. You could have "guessed" that the sine factor would equal 1, by realizing that the block reaches its maximum speed at equilibrium. And indeed, the v given by Eq. (*) is maximized when the sine factor reaches its largest possible value, 1.

Now I will re-solve this problem using

Conservation of energy:

Diagram the relevant stages. I will omit the "final" stage, the block at equilibrium.

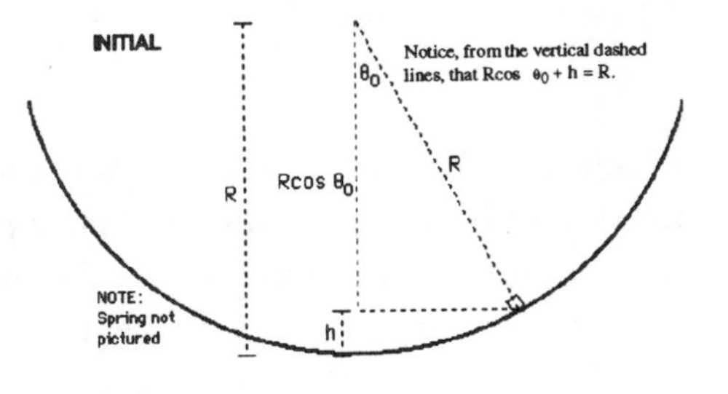

When the block reaches the bottom of the bowl, the spring is at equilibrium, and therefore the elastic potential energy is $U_{f,e} = 0$. Let us say that the bottom of the bowl is where $U_{\text{grav}} = 0$. So, $U_{f,\text{grav}} = 0$. But from the diagram, we see that the initial height is

$$\begin{aligned} h &= R - R\cos\theta_0 \\ &= R(1 - \cos\theta_0) \\ &= R\left(1 - \left[1 - \frac{\theta_0^2}{2}\right]\right) \\ &= \frac{R\theta_0^2}{2}. \end{aligned}$$

In the second-to-last step, I used a small-angle approximation, $\cos\theta \approx 1-\frac{\theta^2}{2}$.

Since the block is a point mass, I can treat it is kinetic energy as either translational or rotational. Because we are solving for the linear speed v, I will go with translational.

$$E_i = E_f$$
$$K_i + U_{i,e} + U_{i,\text{ grav}} = K_f + U_{f,e} + U_{f,\text{ grav}}$$
$$0 + \frac{1}{2}kx_0^2 + mgh = \frac{1}{2}mv^2 + 0 + 0$$
$$\frac{1}{2}k(R\theta_0)^2 + mg\frac{R\theta_0^2}{2} = \frac{1}{2}mv^2.$$

Solve for v to get $v = \left(\sqrt{\frac{k}{m}+\frac{g}{R}}\right)R\theta_0$, the same answer obtained above using kinematics.

QUESTION 20-3

A spring, with spring constant $k = 0.50$ N/m, has an $m = 0.20$ kg mass attached to its end. During its (horizontal) oscillations, the maximum speed achieved by the mass is $v_{max} = 2.0$ m/s.

(a) What is the period of the system?

(b) What is the amplitude of the motion?

ANSWER 20-3

We are trying to extract information about the motion of a simple harmonic oscillator. By starting with

$$F_{\text{restoring force}} = -(\text{stuff})x = ma, \tag{1}$$

you can use calculus to derive how the oscillator's position changes with time:

$$x = A\cos(\omega t + \phi), \text{ where} \tag{2}$$

$$\omega = \sqrt{\frac{\text{stuff}}{m}}. \tag{3}$$

Recall that A is the amplitude; ω is the angular frequency; and ϕ is a phase constant encoding the initial position of the oscillator. In this problem, we care only about A and ω, not ϕ.

(a) Intuitively, the more quickly the mass oscillates back and forth, the less time it takes to complete one oscillation. So, we expect the angular frequency w to be inversely proportional to the period T. Given this insight, I will solve the problem by first calculating ω, and then using it to find T.

For a spring, $F = -kx$. So, from Eq. (1), stuff $= k$. Substitute this result into Eq. (3) to get

$$\omega = \sqrt{\frac{\text{stuff}}{m}} = \sqrt{\frac{k}{m}} = \sqrt{\frac{0.50 \text{ N/m}}{0.20 \text{ kg}}} = 1.6 \text{ s}^{-1}.$$

How can we relate ω to the period of oscillation, T? Well, in time T, the system completes one full oscillation. One full oscillation corresponds to $\sin(\omega t + \phi)$ completing a full cycle. This happens every time $(\omega t + \phi)$ increases by 2π. So, in time T, $(\omega t + \phi)$ increases by 2π. Therefore,

$$\omega T = 2\pi, \tag{4}$$

which we can immediately solve for T to get $T = \frac{2\pi}{\omega} = \frac{2\pi}{1.6\ \mathrm{s}^{-1}} = 4.0$ s.

(b) Picture a block oscillating on a spring. It moves fastest when it is in the middle, i.e., at equilibrium ($x = 0$). Intuitively, the faster the block moves when it passes through equilibrium, the more it stretches or compresses the spring. In other words, a big v_{max} should correspond to a big amplitude A.

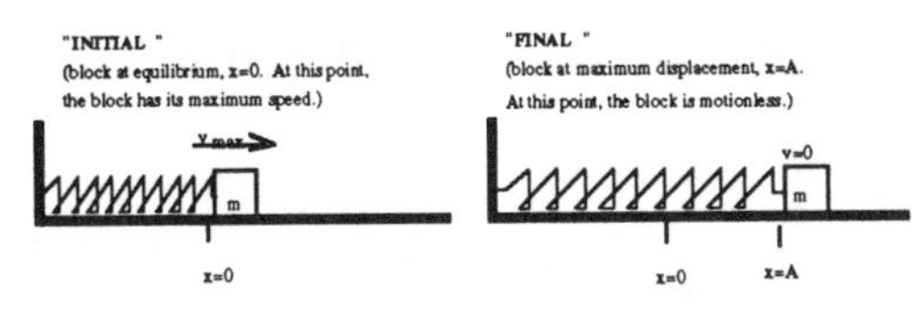

To formalize this intuition, we can use energy conservation. As always, draw pictures.

The potential energy stored in a spring is $U = \frac{1}{2}kx^2$. So, the "initial" potential energy is $U_0 = 0$, while the "final" potential energy is $U_f = \frac{1}{2}kx^2 = \frac{1}{2}kA^2$. Therefore, energy conservation gives us

$$E_0 = E_f$$
$$K_0 + U_0 = K_f + U_f$$
$$\frac{1}{2}mv_{max}^2 + 0 = 0 + \frac{1}{2}kA^2.$$

Solve for A to get

$$A = \sqrt{\frac{mv_{max}^2}{k}} = \sqrt{\frac{(0.20\ \mathrm{kg})(2.0\ \mathrm{m/s})^2}{0.50\ \mathrm{N/m}}} = 1.3\ \text{meters}.$$

It was also possible to solve for A "kinematically" as follows: Differentiate Eq. (2) above to get $v = -A\omega\sin(\omega t + \phi)$. This expression reaches its biggest value when $\sin(\omega t + \phi) = \pm 1$. So, $v_{max} = A\omega(\pm 1) = \pm A\sqrt{\frac{k}{m}}$. Solve for A to get $A = v_{max}\sqrt{\frac{m}{k}}$, which is the same as $A = \sqrt{\frac{mv_{max}^2}{k}}$, the answer we just obtained using energy conservation.

QUESTION 20-4

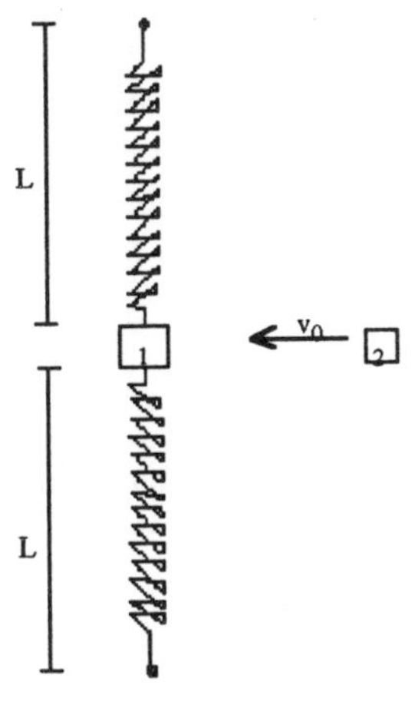

This one is extremely hard, especially part (c). But give it a try. Diagrams make it possible.

Consider block 1, of mass M, attached to two identical springs, both of spring constant k and *equilibrium length 0*. The block is attached to the springs as shown below. The ends of the springs not attached to the block are nailed to the tabletop. The distance from the block to the nails is L. The tabletop is frictionless.

Block 2, of mass $\frac{M}{2}$, slides towards block 1 as shown, with speed v_0. After the collision, block 2 is observed to be motionless.

Express your answers in terms of M, L, k, and v_0.

(a) Was mechanical energy (i.e., the sum of kinetic and potential) conserved, gained, or lost during the collision? If mechanical energy was gained or lost, how much?

(b) After the collision, how far leftward from its initial position will block 1 slide before momentarily coming to rest?

(c) Block 2 crashed into block 1 at $t = 0$. As a result, block 1 slides leftward, then turns around and slides rightward. At what time will block 1 crash back into block 2?

ANSWER 20-4

(a) Let us start off, as always, by making a before and after diagram for the collision. The before diagram is given in the problem.

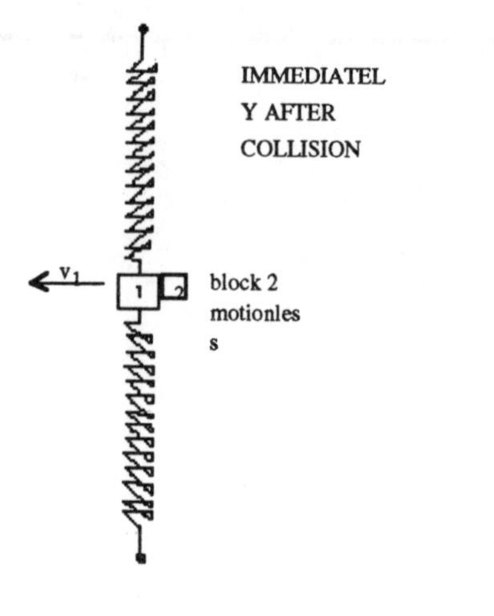

Notice that, *immediately* after the collision, block 1 has not had time to move very far, and therefore the springs are no more stretched than they were initially. So, the system has no more or no less potential energy than it started with. But block 1 has acquired a speed, call it v_1.

If we just *assume* the collision was elastic, then no heat gets generated, and hence no mechanical energy gets lost. But we cannot assume this. Maybe heat *is* created during the collision. To see whether this is the case, we must find v_1, and then use it to check whether the system lost kinetic energy (and hence mechanical energy) during the collision. If the system lost kinetic energy, we know it converted into heat, sound, and other "dissipative" kinds of energy.

As always, to solve collision problems, invoke momentum conservation:

$$p_0 = p_1$$

$$0 + \frac{M}{2}v_0 = Mv_1 + 0.$$

Solving for v_1 yields $v_1 = \frac{v_0}{2}$.

Given this result, we can compare the "before" and "after" mechanical energies. Since the potential energy of the system does not change during the collision, we can check whether mechanical energy gets lost simply by seeing if kinetic energy is lost. In other words, we must see if K_0 is bigger than K_1, the kinetic energy right after the collision. Well,

$$\begin{aligned} K_{\text{lost}} &= K_0 - K_1 \\ &= \frac{1}{2}\left(\frac{M}{2}\right)v_0^2 - \frac{1}{2}Mv_1^2 \\ &= \frac{1}{2}\left(\frac{M}{2}\right)v_0^2 - \frac{1}{2}M\left(\frac{v_0}{2}\right)^2 \qquad \left[\text{since } v_1 = \frac{v_0}{2}, \text{ from above}\right] \\ &= \frac{1}{4}Mv_0^2 - \frac{1}{8}Mv_0^2, \end{aligned}$$

which equals $\frac{1}{8}Mv_0^2$. That is how much mechanical energy is lost. It equals half the initial mechanical energy.

(b) You can complete this part without finding the angular frequency ω. To do so, use conservation of energy. Because heat gets generated during the collision, you cannot set the initial (pre-collision) energy equal to the final energy. But no more heat gets created as block 1 slides leftward and eventually reaches its "turn-around" point, where it momentarily stops. So, you *can* set the post-collision energy equal to the final energy. As always when using energy conservation, sketch the relevant stages.

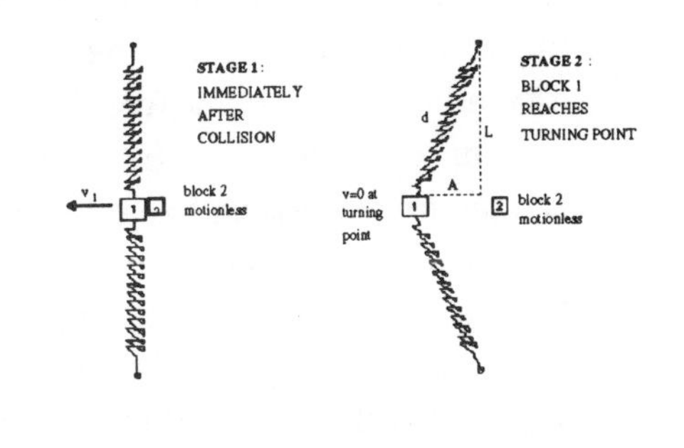

The turning point, remember, is where block 1 reaches its maximum leftward displacement, $x = A$. This "A" would be the amplitude of block 1's oscillations, if block 2 were not sitting there ready to interfere with block 1's motion. In other words, if block 2 vanished, block 1 would oscillate leftward and rightward between $x = +A$ and $x = -A$, where $x = 0$ is its equilibrium position.

Let us think about the potential energies. The energy stored in a stretched spring is $U = \frac{1}{2}ks^2$, where s is the distance by which the spring is stretched *from its equilibrium length*. These springs have equilibrium length 0. So, in stage 1, both springs are stretched by length $s = L$ from equilibrium, and hence, each spring has potential energy $\frac{1}{2}kL^2$. But in stage 2, the springs are stretched even farther, a distance $d = \sqrt{L^2 + A^2}$, as the diagram shows. Furthermore, we must take into account the fact that there is two springs.

I will solve for A by setting $E_1 = E_2$. No new heat gets generated after the collision is complete.

$$E_1 = E_2$$

$$K_1 + U_{1,\text{ top spring}} + U_{1,\text{ bottom spring}} = K_2 + U_{2,\text{ top spring}} + U_{2,\text{ bottom spring}}$$

$$\frac{1}{2}Mv_1^2 + \frac{1}{2}kL^2 + \frac{1}{2}kL^2 = 0 + \frac{1}{2}kd^2 + \frac{1}{2}kd^2$$

$$\frac{1}{2}M\left(\frac{v_0}{2}\right)^2 + 2\left(\frac{1}{2}kL^2\right) = 2\left(\frac{1}{2}k[L^2 + A^2]\right),$$

where in the last step I used $v_1 = \frac{v_0}{2}$ from part (a), and $d = \sqrt{L^2 + A^2}$ from the above diagram. Despite its messiness, this equation contains no unknown other than A. To solve quickly, notice that a factor of $2(\frac{1}{2}kL^2)$ lives on both sides of the equation, and therefore cancels out. Isolate A to get

$$A + \sqrt{\frac{Mv_0^2}{8k}}.$$

(c) Let us picture what happens after the first collision. Block 2 comes to rest, and just sits there at $x = 0$, where block 1 started. Meanwhile, block 1 slides leftward until reaching $x = A$, at which point it turns around and speeds back towards $x = 0$. When it again reaches $x = 0$, block 1 crashes into block 2.

So really, the problem asks how much time t it takes an oscillating object (block 1), knocked from its equilibrium position, to return to equilibrium. Let T denote the period of oscillation. You might be tempted to say $t = T$. But the motion of block 1 described above is *half* a complete cycle.

A whole cycle would involve block 1 going from $x=0 \to x=+A \to x=0 \to x=-A \to x=0$. Here, block 1 completes only the first half of that trip. So, $t=\frac{T}{2}$. If we can find the period of oscillation, we are home free. To find T, you have no choice but to use $T=\frac{2\pi}{\omega}$.

You could also solve by starting with $x=A\cos(\omega t+\phi)$, and finding the time t at which $x=0$, i.e., the time at which block 1 crashes into block 2.

But whether you use $T=\frac{2\pi}{\omega}$ or $x=A\cos(\omega t+\phi)$, you *need* to know the angular frequency ω. And because this is not just a simple block on a spring, you cannot assume $\omega=\sqrt{\frac{k}{m}}$. Recall from Question 16-2 that, in general, $\omega=\sqrt{\frac{\text{stuff}}{\text{m}}}$, where "stuff" is the proportionality constant in $F_{\text{net}}=-(\text{stuff})x$. To find ω, we must find "stuff," using the stuff-finding strategy I showcased in Question 16-2:

Stuff-finding strategy:

(1) Draw a force diagram of the system when it is displaced an arbitrary distance x from equilibrium. (Do not draw the system at equilibrium, because then the net force on it will be 0.)

(2) Write down the net force on the system *along the direction of motion.*

(3) By rewriting that net force in the form $F_{\text{net}}=-(\text{stuff})x$, figure out stuff.

Once we have stuff, we can quickly finish the problem by finding $\omega=\sqrt{\frac{\text{stuff}}{m}}$, then calculating the period $T=\frac{2\pi}{\omega}$, then using $t=\frac{T}{2}$ to find the travel time of block 1 between crashes.

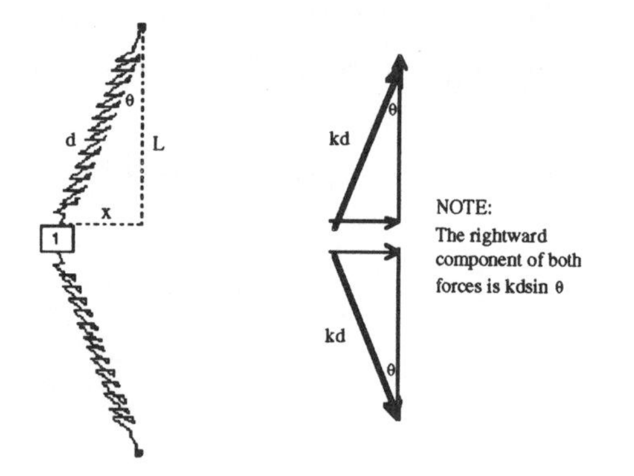

Stuff-finding strategy step 1: Force diagram

I will draw the system displaced an arbitrary distance x from equilibrium. The x-direction is leftward. You could set $x=A$, but you do not have to. For easy reference, let me draw the system itself next to the corresponding force diagram.

Since each spring is stretch a distance d from equilibrium, each spring exerts a force $F=kd$, in the directions shown.

Stuff-finding strategy step 2: Write net force along direction of motion

Since the block only moves leftward and rightward, we are interested only in the horizontal components of these forces. From the diagram, we see that the net force along the direction of motion, due to both springs, is $F_{x\,\text{net}}=-2kd\sin\theta$.

Stuff-finding strategy step 3: Express F_{net} in the form $F_{\text{net}}=-(\text{stuff})x$, and find stuff.

We need to rewrite $F_{x\,\text{net}}=-2kd\sin\theta$ so that "x" appears on the right-hand side. To do so, you must look at the above diagram and realize that, by definition, $\sin\theta=\frac{\text{opposite}}{\text{hypotenuse}}=\frac{x}{d}$. So,

$$\begin{aligned}F_{x\,\text{net}}&=-2kd\sin\theta\\&=-2kd\frac{x}{d}\\&=-(2k)x.\end{aligned}$$

This is a major coincidence! We see that stuff $=2k$, the same "stuff" we would get if the mass were oscillating in the y-direction. I cannot overemphasize that could not have safely guessed that stuff $=2k$. Things just worked out that way.

End of stuff-finding strategy.

We can now write $\omega = \sqrt{\frac{\text{stuff}}{m}} = \sqrt{\frac{2k}{m}}$, and substitute that into the period formula to get $T = \frac{2\pi}{\omega} = \frac{2\pi}{\sqrt{\frac{2k}{m}}}$. So, the time it takes block 1 to complete half an oscillation (i.e., to slide from $x = 0$ to $x = +A$ to $x = 0$) is

$$t = \frac{T}{2} = \frac{\pi}{\sqrt{\frac{2k}{m}}} = \pi\sqrt{\frac{m}{2k}}.$$

By the way, once we found ω, we could have solved for the time at which $x = 0$ by setting $x = A\cos(\omega t + \phi) = 0$. It turns out that $\phi = 90°$, and hence we can eliminate the phase factor by writing $x = A\sin(\omega t)$. This is because a sine curve is out of phase with a cosine curve by 90°. So, $x = 0$ when $A\sin(\omega t) = 0$, which occurs when $\omega t = \pi$. Hence, $t = \frac{\pi}{\omega} = \frac{\pi}{\sqrt{\frac{2k}{m}}}$, the same answer just obtained.

QUESTION 20-5

Hard. Ask your instructor if you need to be able to handle difficult rotational oscillatory motion problems such as this.

A uniform solid disk, of mass M and radius R, is mounted on an axle through its center. In the diagram below, the axle extends into (and out of) the page. The disk is free to spin about its axle, but not to move or roll anywhere. A spring of spring constant k is attached to an essentially massless peg that sticks out of the disk a distance $\frac{R}{2}$ from the axle. The other end of the spring is attached to a wall. In the diagram below, the spring is at its equilibrium length.

When the disk is rotated a small angle from equilibrium, the distance by which the spring stretches is approximately equal to the arc length through which the spring is pulled (i.e., the arc length traced out by the peg). Also, if the angle is small, the direction of the force exerted by the stretched or compressed spring is almost exactly perpendicular to an imaginary line connecting the peg to the axle.

(a) Someone displaces the disk through 1°, lets it go, and times the period of oscillation with a stopwatch. Then she displaces the disk through 0.5°, and again times the period. In which case will she measure the period to be bigger?

(b) When it is displaced by 1° and released, what is the disk's period of oscillation? *Hint: Instead of forces, use torques. Develop a problem-solving technique by "analogizing" the strategies and formulas you used above in linear oscillation problems.*

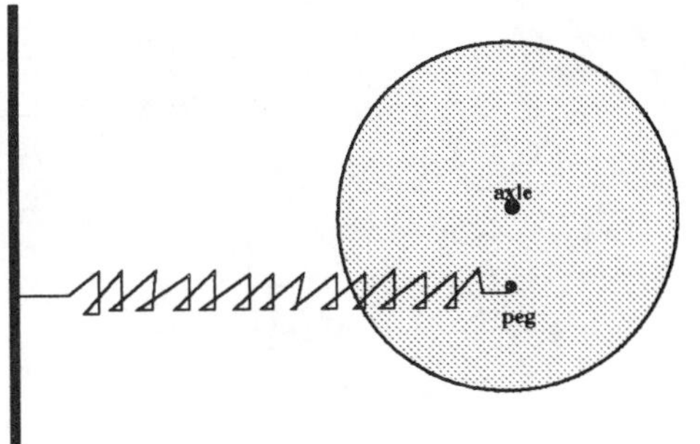

ANSWER 20-5

(a) You might be tempted to say the period is bigger when the disk gets released from 1°, because it has more angle to spin through. But when the amplitude is bigger, the disk spins faster, on average. When released from 1° instead of 0.5°, the disk has twice as much angle to cover; but it spins (on average) twice as quickly. So, the *period*–i.e., the time needed to complete one oscillation–is the same either way. In general, the period of an oscillator's motion is completely independent of its *amplitude*, no matter whether the oscillations are linear or rotational, unless the oscillations are so big that the motion is no longer "simple harmonic." (Simple harmonic motion is when the restoring force or restoring torque is approximately proportional to the object's displacement from equilibrium.) To see this, play with various oscillating systems.

(b) How would we solve for the period T in a linear oscillation problem? Well, since $T = \frac{2\pi}{\omega}$, we would need to know the angular frequency, ω. In general, $\omega = \sqrt{\frac{\text{stuff}}{m}}$, where "stuff" is the proportionality constant in $F_{\text{net}} = -(\text{stuff})x$. To find stuff, we would use the stuff-finding strategy implemented in Questions 16-2 and 16-4.

Here, we must use the rotational analog of this problem-solving technique. The period is still $T = \frac{2\pi}{\omega}$, where ω is the angular frequency, *not* the angular velocity. In general, for rotational oscillations, the angular frequency is $\omega = \sqrt{\frac{\text{stuff}}{I}}$, since the rotational analog of mass is rotational inertia. And now, "stuff" is the proportionality constant in the net *torque* equation $\tau_{\text{net}} = -(\text{stuff})\theta$, which is analogous to $F_{\text{net}} = -(\text{stuff})x$. So you see, there is nothing fundamentally "new" about this problem. It is just a matter of "rotationalizing" our old strategy.

The meat of the problem is finding "stuff." Once we do so, it is easy to find $\omega = \sqrt{\frac{\text{stuff}}{I}}$, and then to calculate the period, $T = \frac{2\pi}{\omega}$. I will use my stuff-finding strategy, rotationalized:

Stuff-finding strategy (rotationalized):

(1) Draw a force diagram of the system when it is displaced an arbitrary angle θ from equilibrium. (Do not draw the system at equilibrium, because then the net torque on it will be 0.)

(2) Write down the net torque on the system.

(3) By writing net torque in the form $\tau_{\text{net}} = -(\text{stuff})\theta$, figure out stuff.

Now I will implement this strategy.

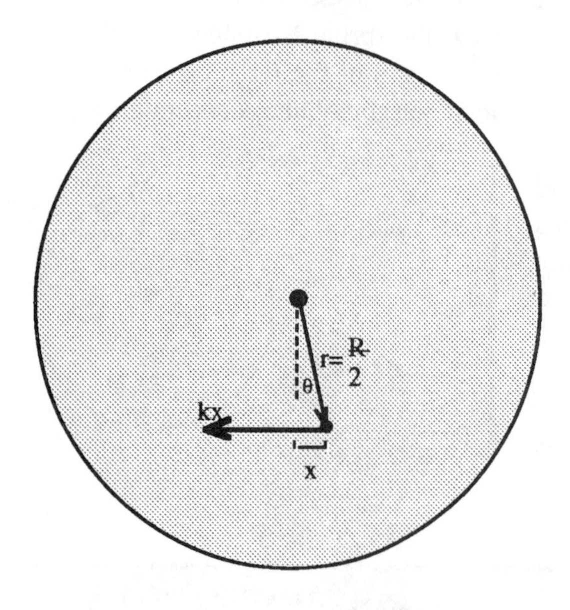

Rotationalized stuff-finding strategy step 1: Force diagram

The force acts a distance $r = \frac{R}{2}$ from the center. I have omitted the forces exerted by gravity and the axle on the disk, because those forces both act directly on the pivot point (the center), and therefore exert no torque. The problem tells us that the distance x by which the spring is stretched from equilibrium approximately equals the arc length through which the spring has swung. I have tried to approximate that "x" on the diagram.

Rotationalized stuff-finding strategy step 2: Write net torque

The question tells us that to good approximation, the spring's force points perpendicular to the radius vector connecting the axle to the peg. (The diagram shows this to be a good approximation–and better still when $\theta = 1°$.) So, the component of the force vector perpendicular to $\mathbf{r}$ is $F_{\perp} \approx kx$. Since we have displaced the disk counterclockwise, we should call counterclockwise positive. So, the torque is negative, because it tends to spin the disk "backwards" (clockwise) towards equilibrium:

$$\tau_{net} = -F_{\perp}r$$
$$= -(kx)\frac{R}{2}.$$

Rotationalized stuff-finding strategy step 3: Express torque in form $\tau_{net} = -(\text{stuff})\theta$*, and find stuff*

We need the right-hand side of the net torque equation to contain θ instead of x. Fortunately, the problem tells us that x is approximately the arc length through which the spring has rotated. From the definition of angle, $\theta = \frac{\text{arc length}}{\text{radius}} = \frac{x}{r}$. So here, $x = r\theta$, where $r = \frac{R}{2}$. Setting $x = \frac{R}{2}\theta$ in the net torque equation gives

$$\tau_{net} = -(kx)\frac{R}{2}.$$
$$= -\left(k\frac{R}{2}\theta\right)\frac{R}{2}$$
$$= -\left(k\frac{R^2}{4}\right)\theta.$$

So, we see that $\text{stuff} = \frac{kR^4}{4}$.

End of stuff-finding strategy

Now that we know stuff, we can obtain the angular frequency, and from that, the period. By analogy with $\omega = \sqrt{\frac{\text{stuff}}{m}}$, the angular frequency of a rotationally oscillating system is $\omega = \sqrt{\frac{\text{stuff}}{I}}$. For a uniform solid disk, $I = \frac{1}{2}MR^2$. So,

$$\omega = \sqrt{\frac{\text{stuff}}{I}} = \sqrt{\frac{\frac{kR^4}{4}}{\frac{1}{2}MR^2}} = \sqrt{\frac{k}{2m}}.$$

So, the period is

$$T = \frac{2\pi}{\omega} = \frac{2\pi}{\sqrt{\frac{k}{2M}}} = 2\pi\sqrt{\frac{2M}{k}}.$$

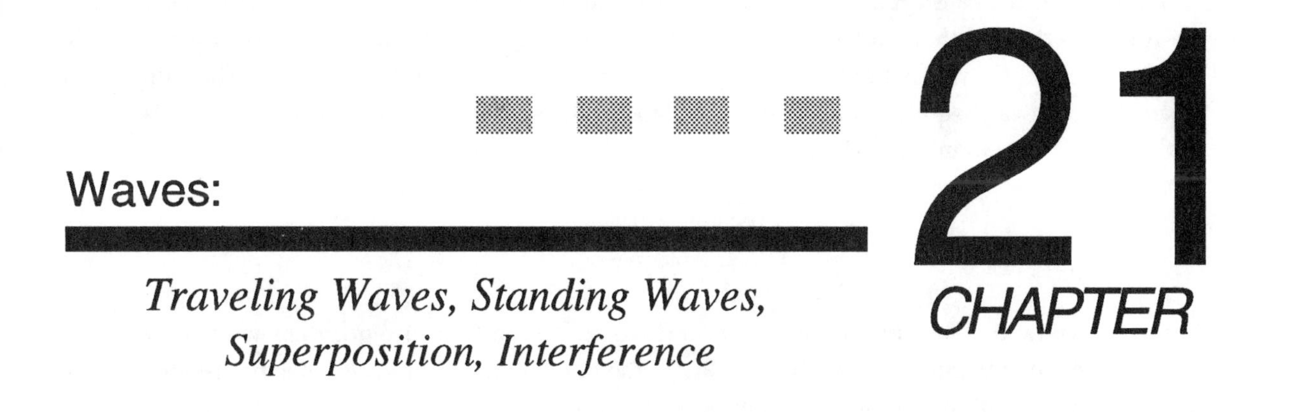

Waves: Traveling Waves, Standing Waves, Superposition, Interference

CHAPTER 21

QUESTION 21-1

Consider a spring, of equilibrium length zero, mass M, and spring constant k. One end is attached to a wall. The other end is held by a person. Suppose the person stretches the spring a distance L, and shakes it back and forth N times over time interval t_1.

(a) What is the velocity of the waves produced?

(b) What is the wavelength of the waves produced?

(c) Now suppose the person walks farther from the wall, so that the spring is stretched a length $2L$. The person again shakes it N times in time t_1. What are the velocity and wavelength of the waves produced? Are your answers the same or different from your answers in parts (a) and (b)?

ANSWER 21-1

(a) You might think that the wave velocity depends on how quickly the person shakes the spring. But as you can confirm by playing with a slinky, the speed of the waves is independent of how many shakes per second you apply. Let me formalize this in a "wave hint." By "medium," I mean the substance through which the waves travel. **Wave hint #1: The speed with which waves propagate through a medium does NOT depend on the frequency with which they were produced, or on their size or shape. The speed depends *only* on properties of the medium, specifically, its "tautness" and "heaviness."** For a one-dimensional medium such as a spring, "tautness" is encoded by the tension, and "heaviness" corresponds to the *linear mass density* μ, which is the mass per length. This makes intuitive sense. The "tighter" the spring is stretched, the faster we expect waves to travel. By contrast, if the spring is heavy, we expect the waves to propagate more lackadaisically. As your textbook derives, the exact relationship is

$$v = \sqrt{\frac{\text{Tension}}{\mu}}. \tag{1}$$

(You can tell from context whether μ means linear mass density or friction coefficient.) Here, a spring of mass M extends over a distance L. So, $\mu = \frac{M}{L}$.

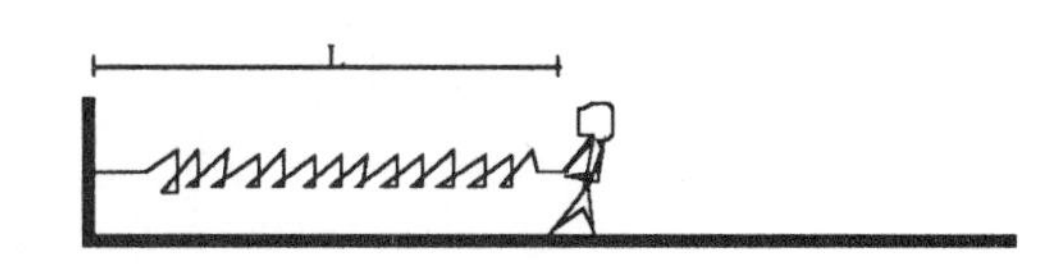

What about the tension in the spring? Recall that tension in a rope or spring (or whatever) is the force exerted by that rope or spring. Since the spring is stretched a distance $x = L$ from its equilibrium length, it pulls on the person and on the wall with force $F = -kL$. (The minus sign means "back toward equilibrium.") So, the tension in the spring is Tension $= kL$.

Substitute all this into Eq. (1) to get

$$v = \sqrt{\frac{\text{Tension}}{\mu}} = \sqrt{\frac{kL}{\frac{M}{L}}} = L\sqrt{\frac{k}{M}}.$$

(b) So far, we know the velocity of the waves, and we are trying to figure out their wavelength. It is not clear where to begin. You just have to know the basic relation between velocity, frequency, and wavelength, a relation that applies to *all* kinds of waves:

$$v = \lambda f \tag{2}$$

where λ is wavelength, and f is frequency, the number of times per second that a point in the medium oscillates back and forth. (Some books denote frequency with the Greek letter ν.) Where does Eq. (2) come from? Let us derive it using a picture.

I will let capital T denote the period. As we have seen in other contexts, the period (i.e., the time per oscillation) is the inverse of frequency (i.e., the number of oscillations per unit time): $f = \frac{1}{T}$. For instance, if a point oscillates at $f = 5$ cycles per second, then its period is $T = \frac{1}{5}$ of a second.

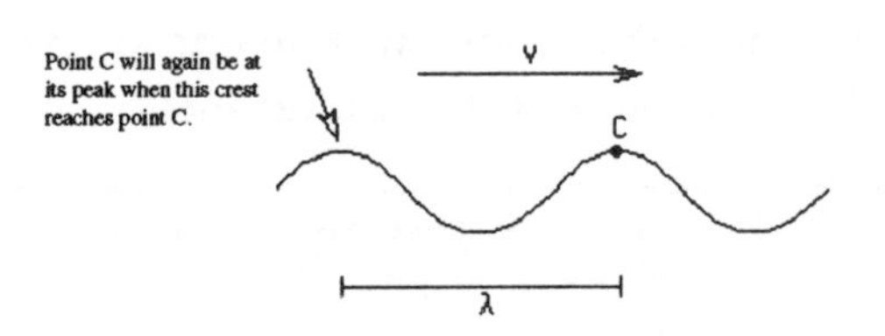

To see how T relates to the speed and wavelength, consider point C.

As the waves propagate rightward, point C moves downward and then upward. The period T is the time it takes point C to again reach its "peak." This happens when the next crest reaches point C. As the diagram shows, the next crest reaches point C when the waves have traveled one full wavelength, λ. In summary, the period T is the time needed for the waves to travel a distance λ. But since the waves travel at constant speed (assuming a uniform medium), their velocity is $v = \frac{\text{distance covered}}{\text{time}} = \frac{\lambda}{T}$. Since $f = \frac{1}{T}$, this equation for v is another way of writing Eq. (2). So, that is where Eq. (2) comes from.

Now let us use Eq. (2) to solve the problem at hand. We need to know the frequency, so that we can substitute it into Eq. (2) and solve for λ. Another universal wave hint applies: **Wave hint #2: The frequency of the waves equals the frequency of the oscillations generating the waves. Frequency does *not* depend on the medium.** Each time the person shakes the rope, she creates another wave. So, she created N waves in t_1 seconds. Therefore, the frequency of the waves is $f = \frac{N}{t_1}$. Solve Eq. (2) for λ, and substitute in this expression for f along with our part (a) answer for v:

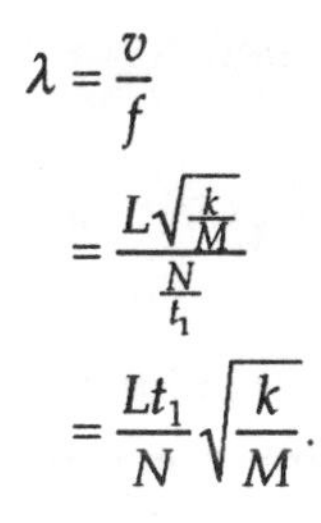

$$\lambda = \frac{v}{f} = \frac{L\sqrt{\frac{k}{M}}}{\frac{N}{t_1}} = \frac{Lt_1}{N}\sqrt{\frac{k}{M}}.$$

(c) The reasoning here is the same as above, except the spring is stretched a distance $x = 2L$ instead of $x = L$. How does this change things? Well, the spring is more taut: Tension $= kx = k(2L)$, double its previous value. The mass density decreases by 50%, because the same mass is now "spread out" over twice the distance: $\mu = \frac{M}{2L}$. Putting all this into the formula for v in part (a) yields

$$v = \sqrt{\frac{\text{Tension}}{\mu}} = \sqrt{\frac{k(2L)}{\frac{M}{2L}}} = 2L\sqrt{\frac{k}{M}},$$

So, by doubling the spring's length, we double the wave speed. This simple relation does not hold in general. It is true only for certain media, e.g., a spring of equilibrium length 0.

The new frequency is the same as before, $f = \frac{N}{t_1}$. Why? Because the wave frequency depends *only* on the frequency of the oscillations that generate the waves, not on the medium. As mentioned in part (b), this is a general fact about waves.

In summary, when we double the spring's length, the velocity doubles, and the frequency stays the same. From the relation $v = \lambda f$, it follows that the wavelength also must double.

QUESTION 21-2

A string is stretched between two walls, and jiggled so as to produce a standing wave with 5 humps. The frequency of this mode is observed to be $f_5 = 120$ Hz.

(a) For this string, what is the fundamental frequency, i.e., the frequency of the one-humped standing wave?

(b) Now suppose the string is soaked in acid, and then dried. As a result, the tension in the string decreases by a factor of 9, but everything else stays the same. What is the new fundamental frequency?

Hint: You can solve this without knowing any other relevant quantities, such as the length of the string, because all such quantities cancel out of your final answers.

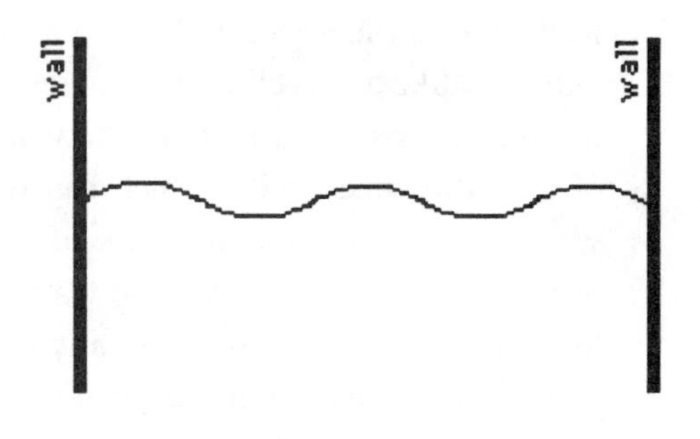

ANSWER 21-2

(a) For this problem, you need only the fundamental wave relation $v = \lambda f$, along with a formula about the wavelength of standing waves, a formula you can "derive" entirely by making pictures.
Let us start with

$$v = \lambda f, \tag{1}$$

which I just derived in Question 21-1. The new formula we need is

$$\lambda_n = \frac{2L}{n} \qquad \textbf{[standing wave equation only]} \qquad (2)$$

Here, n denotes the "mode," i.e., the number of humps on the standing wave. (Up humps and down humps both count as humps. "Humps" are sometimes called "loops.") L is the length of the string, the gas-tube, or whatever medium the standing wave inhabits. And λ_n is the wavelength of the n-th mode, i.e., the wavelength of the standing wave when there is n humps.

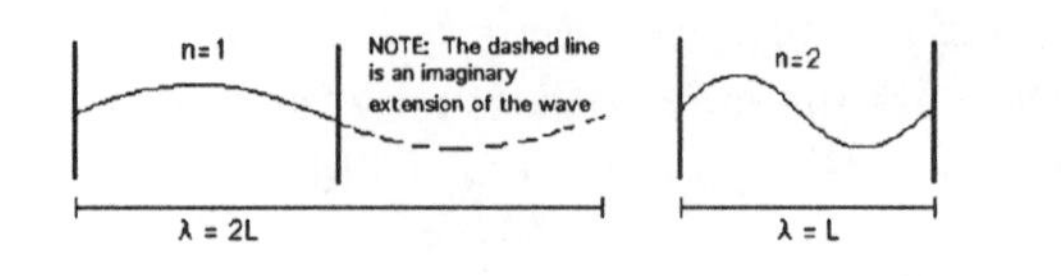

To see where Eq. (2) comes from, just draw a bunch of pictures.

In each case, $\lambda_n = \frac{2L}{n}$, as you can confirm. The same formula holds for $n = 5$ or any other n.

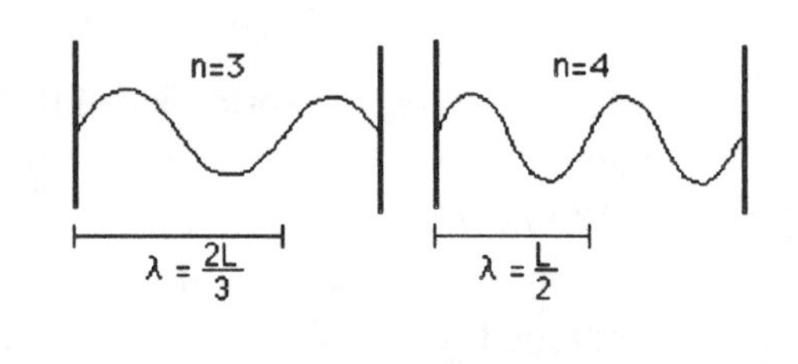

Now that we know the intuitions underlying Eqs. (1) and (2), let us solve the problem at hand. The frequency of the $n = 5$ mode is $f_5 = 120\text{ Hz}$. We want to find the frequency f_1 of the $n=1$ mode. Here is how we can reason:

From Eq. (1), the speed of the waves is $v = \lambda_5 f_5$. *A fundamental fact about waves is that their speed depends only on the medium, not on the frequency or wavelength.* The tension and mass density of the medium are all that matters. Therefore, if we made a one-humped standing wave on the string, its "velocity" $v = \lambda_1 f_1$ would be the same as the velocity of waves on the five-humped string, $v = \lambda_5 f_5$. In brief,

$$v = \lambda_5 f_5 = \lambda_1 f_1. \qquad (*)$$

I must stress that "v" is *not* the speed with which a point on the string oscillates back and forth. Rather, it is the speed with which crests and troughs propagate along the string. For a standing wave, unlike a traveling wave, crests and troughs do not propagate in this way. So, it really does not make sense to talk about the velocity v of a standing wave. For standing waves, you can think of v as the speed with which the wave would propagate if it were not "trapped" between two walls. Or, if you like to think of a standing wave as resulting from the "superposition" of two oppositely-directed traveling waves, then "v" is the speed of those traveling waves. (See your textbook for more on this.) In any case, you can still use $v = \lambda f$ when addressing standing waves, provided you keep in mind what that v does–and does not–mean.

So, we can use Eq. (*) to solve this problem. From Eq. (2), the wavelength of the five-humped pattern is $\lambda_5 = \frac{2L}{5}$, while the one-humped wave has length $\lambda_1 = 2L$. Finally, we are given $f_5 = 120\text{ Hz}$. Solve Eq. (*) for f_1 to get

$$f_1 = \frac{\lambda_5 f_5}{\lambda_1} = \frac{\frac{2L}{5}(120\text{ Hz})}{2L} = \frac{(120\text{ Hz})}{5} = 24\text{ Hz}$$

Intuitively, when we change n from 5 to 1, we increase the wavelength by a factor or 5, and therefore decrease the frequency by that same factor.

(b) Here, we must again use wave hint #1 from Question 21-1 above: The speed of a wave depends only on the tautness and heaviness of the medium. For a one-dimensional medium, this wave hint gets encapsulated by

$$v = \sqrt{\frac{\text{restoring force factor}}{\text{mass factor}}} = \sqrt{\frac{\text{Tension}}{\mu}} \tag{3}$$

where μ is the linear mass density. According to Eq. (3), if the tension decreases by a factor of 9, then the wave velocity decreases by a factor of 3.

How does this affect the frequency of standing waves? Well, according to Eq. (1) applied to the one-humped standing wave, $v = \lambda_1 f_1$. If we reduce v by a factor of 3, and if the wavelength remains unchanged, then the frequency decreases by a factor of 3. And indeed, the wavelength of the $n = 1$ mode does not change; it is determined entirely by how much "room" the wave has. Specifically, $\lambda_1 = 2L$, no matter what. So, the fundamental frequency f_1 must decrease by a factor of 3. It is now $f_{1\text{ new}} = \frac{f_{1\text{ old}}}{3} = \frac{24\text{ Hz}}{3} = 8\text{ Hz}$.

QUESTION 21-3

(Extra hard problem.) A bungee cord, of mass M, has equilibrium length 0. When stretched a distance x, the cord exerts a force $F = -Cx^2$, where C is a "stiffness" constant.

Suppose one end of the cord is attached to a wall. The cord is stretched a distance d horizontally (in the x-direction). The end of the cord that is not attached to the wall is shaken so as to produce sinusoidal traveling waves. The machine generating the waves is turned on at $t = 0$, and turned off at the exact moment the first wave reaches the midpoint of the cord. An observer notices that N complete waves were produced.

(a) What is the frequency of the waves? Express your answer in terms of M, C, d, and N.

(b) By watching a single point on the rope, an observer notices that the maximum distance it reaches from equilibrium is R. What is the maximum speed achieved by that point as it oscillates? Answer in terms of M, C, d, N, and R. Note: The textbook formula for a sinusoidal traveling wave is $y = A\sin(kx \pm \omega t)$, whatever that means.

(c) Suppose the cord is now stretched by a distance $2d$, and attached to the same wave-generating machine used in part (a). What will be the frequency of the waves?

ANSWER 21-3

(a) We are solving for the frequency f, given some information about the mass and force of the cord. So, it looks as if we can calculate the wave speed $v = \sqrt{\frac{\text{Tension}}{\mu}}$, where μ is the linear mass density. Given v, how can we find f? I see two ways. But before showing you these methods, let me first obtain v, because we are helpless without it.

The cord is stretched a distance d. The force it exerts is $F = Cd^2$ towards equilibrium. But remember, the force exerted by a cord or spring (or whatever) is the tension. So, Tension $= Cd^2$. Also, since the cord has mass M, and since its length is d, the mass per unit length is $\mu = \frac{M}{d}$. Putting all this together, we get

$$v = \sqrt{\frac{\text{Tension}}{\mu}} = \sqrt{\frac{Cd^2}{\frac{M}{d}}} = \sqrt{\frac{Cd^3}{M}}. \tag{1}$$

Now that we know v, we can find the frequency, using either of two methods.

Method 1: The direct way.

The frequency of the waves equals how many waves were generated per second. We know that N waves were produced in time t_f, where t_f denotes the time at which the machine got turned off. So, the frequency is $f = \frac{N}{t_f}$. To find f, we need to know t_f.

Well, the machine got switched off when the first wave reached the midpoint, i.e., when the first wave traveled a distance $\frac{d}{2}$. So, t_f is the time a wave takes to travel through distance $\frac{d}{2}$. Since the waves travel at constant speed, we have

$$v = \frac{\Delta x}{\Delta t} = \frac{\frac{d}{2}}{t_f}.$$

Solve this for t_f to get

$$\begin{aligned} t_f &= \frac{d}{2v} \\ &= \frac{d}{2\sqrt{\frac{Cd^3}{M}}}, \end{aligned} \tag{2}$$

where I invoked our expression for v from Eq. (1) above. So, the frequency is

$$\begin{aligned} f &= \frac{N}{t_f} \\ &= \ldots \text{algebra} \ldots \\ &= 2N\sqrt{\frac{Cd}{M}}. \end{aligned}$$

Now I will again solve for f using...

Method 2: The cute way.

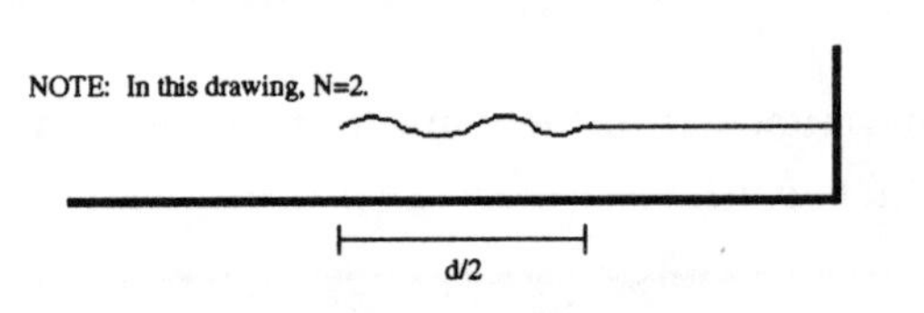

Here is a picture of the rope right when the machine gets turned off.

N waves are spread out over a distance $\frac{d}{2}$. Therefore, the distance taken up by a single wave is $\frac{\frac{d}{2}}{N}$. For instance, on the diagram, since $N = 2$, the distance taken up by a single wave is $\frac{\frac{d}{2}}{2} = \frac{d}{4}$. But the distance taken up by a single wave is the wavelength λ. So, $\lambda = \frac{\frac{d}{2}}{N}$.

Now that we know the wavelength and velocity, we can easily solve for frequency using the basic wave relation, $v = \lambda f$:

$$f = \frac{v}{\lambda} = \frac{\sqrt{\frac{Cd^3}{M}}}{\left(\frac{\frac{d}{2}}{N}\right)} = 2N\sqrt{\frac{Cd}{M}},$$

the same answer obtained above using another method.

(b) The wave velocity "v" that appears in all your wave formulas is *not* the speed with which a point on the rope oscillates back and forth. It is the speed with which crests and troughs propagate along the rope. To see this, stretch out a slinky and shake your end of it. Although crests shoot up and down the slinky, an individual coil of the slinky merely oscillates in place. For "transverse waves," these oscillations are at right angles to the direction in which the waves move. So for instance, if your slinky is stretched from north to south, then the waves propagate southward; but individual coils oscillate east and west. In this problem, the waves propagate rightward along the rope; but individual points on the rope oscillate up and down (vertically). We are looking for the maximum vertical speed achieved by the point on the rope. So, the answer is not the wave speed "v," which tells us the rightward velocity of the crests and troughs.

As discussed in your textbook, the general equation for sinusoidal waves, leaving out "phase constants," is

$$y = A\sin(kx \pm \omega t), \tag{3}$$

where A is the amplitude, i.e., the maximum displacement in the y-direction; and k and ω are "abbreviations" I will discuss shortly. In Eq. (3), k is *not* a spring constant. Here, x is the rightward distance *along* the slinky, and y is the vertical displacement from equilibrium.

In this problem, the "trick" is to realize that we are solving for the (maximum) velocity in the y-direction, the direction in which a point on the rope oscillates. Velocity is the rate at which position changes. So, the point's speed is

$$\begin{aligned} v_y &= \frac{dy}{dt} \\ &= \frac{d}{dt}[A\sin(kx \pm \omega t)] \\ &= \omega A\cos(kx \pm \omega t). \end{aligned}$$

Since $\cos(kx \pm \omega t)$ oscillates between $+1$ and -1, v_y reaches its biggest value when $\cos(kx \pm \omega t) = \pm 1$. So, setting $\cos(kx \pm \omega t) = 1$ in our expression for v_y yields

$$v_{y\,\max} = \omega A. \tag{4}$$

At this stage, to find $v_{y\,\max}$, we just need to find the amplitude A and the angular frequency ω.

Well, the problem tells us that the point's maximum displacement from equilibrium is $y_{\max} = R$. This maximum displacement is the amplitude, because the point oscillates back and forth between $y = y_{\max}$ and $y = -y_{\max}$. So, $A = R$.

What about angular frequency? Instead of just regurgitating the relevant formula, let me show where it comes from. Consider a given point on the cord, and suppose that at time $t = 0$, the point is at its maximum displacement, $y = A$. How much time passes before that point returns to the same spot, i.e., the same y? By definition, the answer is T, the period of oscillation of the waves. So, for given x, whatever value y has at time $t = 0$, y has that same value one full oscillation later, at time T:

$$y(0) = y(T) \tag{*}$$

Into Eq. (*), substitute Eq. (3) above, to get

$$A\sin(kx+\omega 0) = A\sin(kx+\omega T).$$

So, $\sin(kx+0) = \sin(kx+\omega T)$. From trig, the sine function "repeats" itself when the angle increases by 2π. In other words, $\sin(\theta) = \sin(\phi)$ only if $\phi = \theta$ or $\phi = \theta + 2\pi$, etc. So, $\sin(kx+0) = \sin(kx+\omega T)$ if $(kx+0)+2\pi = (kx+\omega T)$. Cancel the kx's to get $\omega T = 2\pi$, which I can rewrite as

$$\omega = \frac{2\pi}{T} = 2\pi f. \tag{5}$$

In the last step, I used $f = \frac{1}{T}$. In words, the "angular frequency" is 2π times the frequency. You have seen this relationship before, in simple harmonic motion. This is no coincidence; when sinusoidal waves travel down the rope, individual points on the rope oscillate back and forth in simple harmonic motion.

In part (a), we found the wave frequency, $f = 2N\sqrt{\frac{Cd}{M}}$. So, we can immediately solve Eq. (4) for $v_{y\,\max}$:

$$\begin{aligned} v_{y\,\max} &= \omega A \\ &= (2\pi f)A \\ &= \left(2\pi 2N\sqrt{\frac{Cd}{M}}\right)R. \end{aligned}$$

By the way, you can derive the formula for "k" in $y = A\sin(kx \pm \omega t)$ using reasoning similar to the above. The result turns out to be $k = \frac{2\pi}{\lambda}$.

(c) When the cord is stretched to $2d$, the wave velocity increases, because the cord now has more tension, Tension $= C(2d)^2$, and less mass per unit length, $\mu = \frac{M}{2d}$. But the Question asks how the *frequency* changes. A fundamental property of waves is that their frequency depends *only* on the frequency of the oscillations producing the waves. Since the waves are produced by the same machine used in part (a), the resulting wave frequency will be the same as in part (a).

In summary: The frequency stays the same. The velocity increases. Therefore, since $v = \lambda f$, the wavelength also increases.

QUESTION 21-4

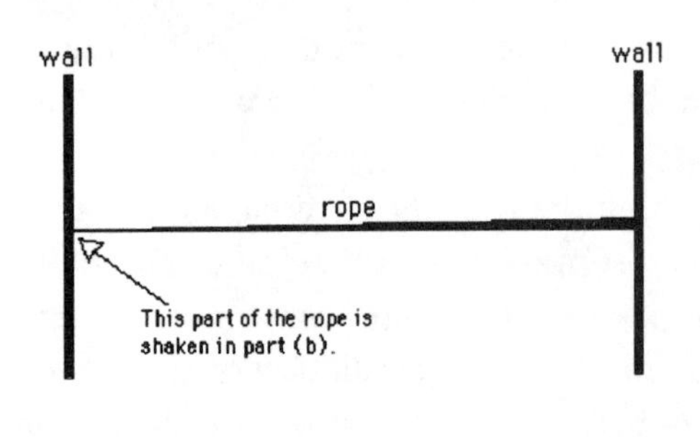

A rope, of length L, gradually thickens from one end to the other. Indeed, if we let x denote the distance from the left end of the rope, the linear mass density is $\mu(x) = Bx^2$, where B is a constant.

Suppose the rope is stretched between two walls, as shown below. The tension in the rope is F_T. (I am avoiding the symbol "T," so as not to confuse tension with period.) The tension is roughly the same throughout the rope, because it is stretched horizontally; it does not matter that the rope is not uniform.

The left end of the rope is shaken at frequency f, starting at time $t = 0$ and ending just as the very first wave reaches the right end of the rope.

In this problem, L, B, f, and F_T are the given constants.

(a) What are the units of the constant B? (There is more than one possible answer.)

(b) *(Very hard problem, but try it.)* At what time t does the first wave produced by the shaking reach the right end of the rope?

(c) Make a rough drawing of what the rope looks like when the first wave produced by the shaking reaches the right end of the rope. You can do this part even if you did not complete part (b).

ANSWER 21-4

(a) Linear mass density is a mass per distance: $\mu = \frac{\text{mass}}{\text{distance}}$. So, since $\mu = Bx^2$,

$$B = \frac{\mu}{x^2} \text{ "=" } \frac{\frac{\text{mass}}{\text{distance}}}{(\text{distance})^2} = \frac{\text{mass}}{(\text{distance})^3},$$

and therefore must have units of $\frac{\text{kg}}{\text{meter}^3}$, or $\frac{\text{grams}}{\text{cm}^3}$, or something like that. So, B has the units of density, i.e., mass per volume.

(b) If the linear mass density were constant, this would be easier. You would reason as follows: The speed of the waves is $v = \sqrt{\frac{F_T}{\mu}}$. Since the waves travel at constant speed down a rope of length L, we get $v = \frac{\Delta x}{\Delta t} = \frac{L}{t}$, where t is the time it takes a wave to travel across the whole rope. Equating these two expressions for v gives $\sqrt{\frac{F_T}{\mu}} = \frac{L}{t}$, which you could immediately solve for t.

Unfortunately, this reasoning fails here, because μ is not constant, and therefore v is not constant. Let us picture what is happening. As a wave travels from left to right, the linear mass density of the rope increases. Therefore, since $v = \sqrt{\frac{F_T}{\mu}} = \sqrt{\frac{F_T}{Bx^2}}$, the velocity decreases as x increases. Intuitively, as the wave travels rightward, it slows down, because the rope gets "heavier."

As usual, when a quantity such as velocity or acceleration is not constant, and when you cannot use conservation laws, you should write the quantity in calculus form, to see what information can be extracted. Velocity is distance per time: $v = \frac{dx}{dt}$. So,

$$v = \frac{dx}{dt} = \sqrt{\frac{F_T}{Bx^2}}. \tag{1}$$

This is a first-order differential equation in x and t. So, with luck, we can find $x(t)$, the position of the first wave as a function of time. Given that function, we can calculate at what time t the first wave reaches the right end of the rope, i.e., the time at which $x(t) = L$.

So, to make progress, we must solve Eq. (1) for x. Fortunately, it is a so-called "separable" differential equation. You can get all the "x" terms on one side and all the "t" terms on the other. To do so, multiply both sides by x and by dt to get $(x)dx = \sqrt{\frac{F_T}{B}}dt$. Now we can integrate both sides, keeping in mind that is a constant, to get

$$\begin{aligned} \int (x)dx &= \sqrt{\frac{F_T}{B}} \int dt \\ \frac{1}{2}x^2 &= \sqrt{\frac{F_T}{B}}t + (\text{integration constant}) \end{aligned} \tag{2}$$

The integration constant equals 0, because physically we know $x = 0$ at $t = 0$.

We could solve Eq. (2) for x to obtain the first wave's position as a function of time. The function comes out to be

$$x = \sqrt{2t\sqrt{\frac{F_T}{B}}}.$$

But on second thought, I am not really interested in x as a general function of time. I am interested in the time at which $x = L$. So, solve Eq. (2) for t, and set $x = L$: $t = \frac{1}{2}L^2\sqrt{\frac{B}{F_T}}$. That is how long it takes the first wave to traverse the rope. Neat, huh?

(c) At time t, the first wave produced at the left end of the rope reaches the right end. But remember, the first wave is not the only wave on the rope. As the first wave traveled across the rope, new waves kept getting produced at the left end. So, the whole rope is "full" of waves. As explained above, the waves closer to the right end move slower than the waves near the left end, because the rope gets "heavier" near the right end. In other words, v decreases as we move rightward. Does this mean the frequency decreases as the waves move rightward? No. The frequency does not change when the medium changes. Here is why:

However many waves per second are created at the left end must eventually reach the right end. Otherwise, waves would have to "appear" or "disappear" or "bunch up." Waves can speed up, slow down, and change shape, but they cannot just appear and disappear (unless they get absorbed, or "dissipate away" due to internal "friction" in the medium). For instance, if 4 waves per second pass the point $x = 10$ centimeters down the rope, then 4 waves per second must also pass the point $x = 20$ centimeters down the rope. If only 3 waves per second pass $x = 20$ cm, then one wave per second would have to "vanish" between $x = 10$ cm and $x = 20$ cm. Which does not happen.

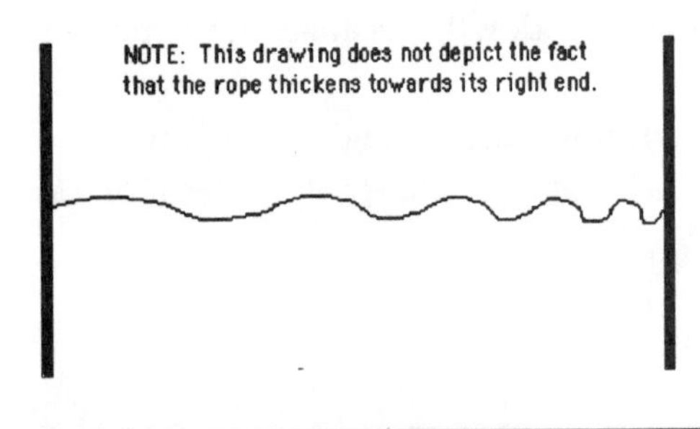

So, as the waves move rightward, their frequency f stays the same, but their velocity v decreases. It follows from the basic wave relation, $v = \lambda f$, that the wavelength decreases as the waves move rightward. Therefore, the rope looks roughly like this.

QUESTION 21-5

A horizontally-stretched string is attached to a vibrating prong, which vibrates up and down at frequency $f = 90.0$ Hz. The other end of the string runs over a massless pulley and is attached to a block. The horizontal segment of string has length $s = 100$ cm and mass $m = 20.0$ g. The vertical segment of string is short enough that you can neglect its mass.

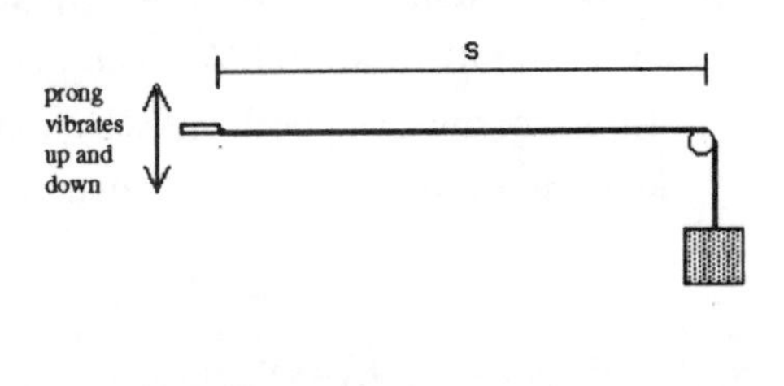

What must be the mass M of the block if we want the horizontal segment of string to resonate in a four-humped standing wave? Express your answer in terms of symbols before plugging in numbers.

ANSWER 21-5

In this problem, it is hard to decide how to begin. When that is the case, a good "trick" is to start at the end, and work backwards. Let me show you what I mean.

We want the block's mass to be such that the string looks like this.

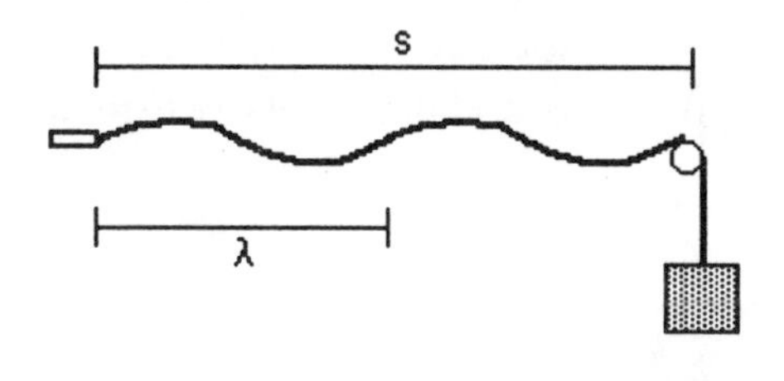

From the picture, you can see that two complete waves "fit" onto the horizontal segment of string. (Remember, a complete wave has both an upward and a downward hump.) Each wave "fills" half the string. So, the wavelength is

$$\lambda = \frac{s}{2}. \qquad (*)$$

You could also obtain Eq. (*) from the standing wave equation $\lambda_n = \frac{2s}{n}$, where n is the number of humps on the string and s is the string's length. Here, $n = 4$. Recall from Question 21-2 that you can *derive* $\lambda_n = \frac{2s}{n}$ by drawing pictures just like the one above.

In general, the frequency of waves equals the frequency of the oscillations producing the waves. So here, $f = 90.0$ Hz. In summary, we know λ and f, and we are trying to find the mass of the block. How can we do so?

You need to have the insight that the block's mass determines the tension in the string. And the tension in the string helps to determine the wave velocity. But from the basic wave relation $v = \lambda f$, we can immediately find the wave velocity, since we know $\lambda = \frac{s}{2}$:

$$\begin{aligned} v &= \lambda f \\ &= \frac{s}{2} f. \end{aligned} \qquad (1)$$

Given that velocity, we can now calculate the tension in the string. And once we have the tension, we can figure out how heavy the block must be to create that tension.

Tension and velocity are related by

$$v = \sqrt{\frac{\text{Tension}}{\mu}}, \qquad (2)$$

where μ is the linear mass density of the string. Since the horizontal segment of string has mass m and length s, its mass per unit length is $\mu = \frac{m}{s}$. So, we can solve Eq. (2) for Tension:

$$\begin{aligned} \text{Tension} &= \mu v^2 \\ &= \frac{m}{s} v^2 \\ &= \frac{m}{s}\left(\frac{s}{2} f\right)^2 \qquad \left[\text{from eq. (1), } v = \frac{s}{2} f\right] \\ &= \frac{msf^2}{4}. \end{aligned}$$

In this reasoning, m is the string's mass, *not* the block's mass.

Now that we know the tension in the string, we can calculate the block's mass, M. Intuitively, a heavier block creates more tension. To figure out the exact relation between the block's mass and the string's tension, draw a force diagram of the block.

Assuming the waves on the string do not appreciably disturb the block, the block does not move. It is acceleration is $a = 0$. So, Tension and gravity cancel: Tension $= Mg$. Solve for M, and use the above expression for Tension, to get

T

$$\begin{aligned} M &= \frac{\text{Tension}}{g} \\ &= \frac{msf^2}{4g} \\ &= \frac{(.0200\ \text{kg})(1.00\ \text{m})(90\ \text{s}^{-1})}{4(9.8\ \text{m/s}^2)} \\ &= 4.13\ \text{kg}. \end{aligned}$$

mg

I converted the string's length into meters, and the string's mass into kilograms, to mesh with my units of "g." Recall also that Hertz are cycles per second: Hz "=" s^{-1}.

QUESTION 21-6

Consider a long slinky carrying the wave pulses drawn below, in the directions shown.

(a) Which pulse travels faster?

(b) What does the slinky look like when the pulses overlap completely? Make a rough sketch.

(c) What does the slinky look like a second after the pulses have overlapped? Make a rough sketch.

(d) Suppose this experiment is repeated, except this time with a stiffer (though equally heavy) slinky. How does this affect your answers to parts (a) through (c)? Describe in words what differs and what stays the same.

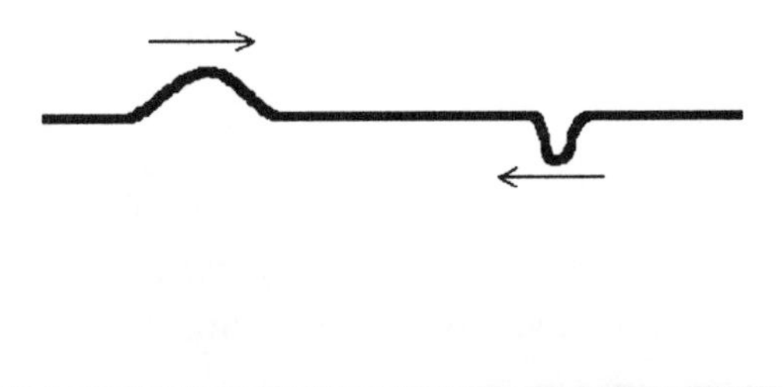

ANSWER 21-6

In this problem, you must qualitatively apply some of the core wave concepts.

(a) You might think that one wave travels faster than the other, because they have different "wavelengths." But the velocity of a wave depends *only* on the medium through which it travels. The tautness and heaviness of the medium are all that matters. (Mathematically, $v = \sqrt{\frac{\text{Tension}}{\mu}}$, where Tension refers to the medium, and μ is the linear mass density of the medium.) So, both pulses travel at the same speed.

If you are worried about $v = \lambda f$, just keep in mind that the pulse with a bigger "wavelength" has a lower "frequency." So, $v = \lambda f$ is the same for both pulses.

(b) When waves overlap, the resulting amplitude is the sum of the amplitudes of the individual waves. This "superposition" principle is a fundamental property of all kinds of waves. So, when the waves overlap completely, the "net" wave is the sum of the individual waves.

(c) After they finish overlapping, the waves continue to travel in their separate directions, *with their shapes unchanged.*

When waves "collide," they pass through one another without affecting each others' size, shape, speed, or anything else. To see this, recruit a friend to create a pulse at one end of a slinky, while you generate a pulse at the other end. You will see that the waves overlap, and then continue on their merry way as if no "collision" ever happened. A more dramatic example of this phenomenon happens at every party. The sounds waves comprising dozens of separate conversations all pass right through each other. Yet, you can hear the sound waves coming from the person with whom you are flirting, as if those waves had not ever "interfered" with other people's sound waves.

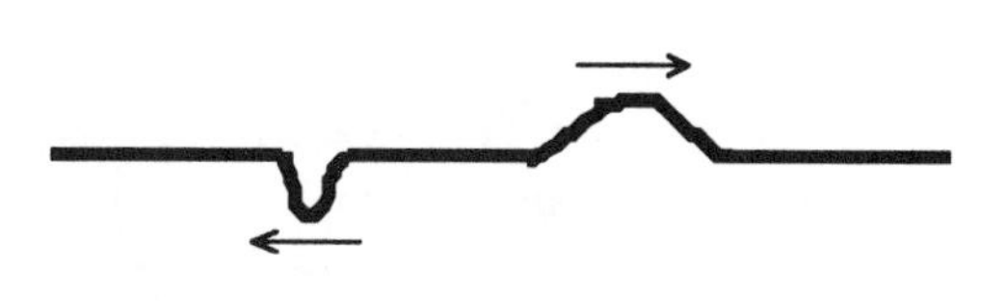

(d) A "stiffer" slinky means the tension is higher than it was before. Therefore, the pulses travel faster than before, though they still "share" the same speed. (You can see this intuitively, or from the formula $v = \sqrt{\frac{\text{Tension}}{\mu}}$.) Therefore, the two pulses reach the center of the slinky (where they overlap) quicker than they did before. The shape of the slinky during overlap, and the behavior of the pulses after they pass through each other, do not change. So, the experiment looks exactly like it did before, except everything happens faster. It is as if you took a home video of the original experiment, and played it on fast forward.

QUESTION 21-7

(Non-standard topics. Check with your instructor whether you need to understand the concepts addressed here. Many curricula do not address these topics until 2nd or 3rd semester.)

Two radio-wave sources, which you may consider to be point-like, are sitting a distance L apart on the ground, as drawn below (top down view). Source 2 is on the end of a sidewalk that is perpendicular to an imaginary line connecting the two sources. By turning a knob, the controller can turn on source 1, source 2, or both sources together. The sources always emit one single frequency. When both sources are turned on together, the waves they produce are in phase. The speed of radio waves is c, the speed of light.

(a) Suppose only source 1 is turned on. In one minute, the source produces a total of E joules of energy. (All that energy takes the form of electromagnetic wave energy.) What is the intensity of those radio waves at source 2?

(b) Now sources 1 and 2 are turned on, and emit radio waves of frequency f. Suppose you want to place your radio on the sidewalk. How far from source 2 should you position your radio so as to get the "cleanest" reception? (You cannot just place the radio right next to source 2.) Answer in terms of L, c, and f. It is o.k. if you do not complete the algebra.

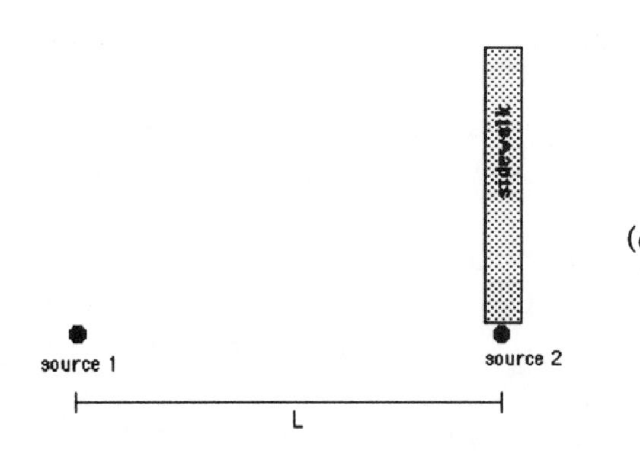

(c) Now a radio-wave detector is placed midway between the two sources. When source 2 alone is turned on, the intensity of waves at the detector is measured to be I_0. If both sources are turned on at once, what will be the intensity at that detector? Express your answer in terms of I_0.

ANSWER 21-7

(a) Intensity is the (average) rate of energy flow per area passing through an imaginary small area. For instance, in this problem, suppose you held up your hand near source 2. Let A denote the area of your hand, and make sure your palm faces source 1. If the average power of radio waves (i.e., the average rate of energy flow) passing through your hand is P_{hand}, then the intensity of radio waves at your hand (and hence at source 2) is $I = \frac{P_{\text{hand}}}{A}$. Intensity corresponds to "brightness" of visible light, and "loudness" of sound.

In this problem, we are told that E joules of energy are generated in time $t = 60$ seconds. So, the average power generated by source 1, in watts (joules/second), is $P = \frac{\text{energy}}{\text{time}} = \frac{E}{60}$. That energy radiates outward from source 1, spreading uniformly in all directions. To figure out the intensity at source 2, imagine a sphere of radius L centered at source 1. (Source 2 is on the surface of this imaginary sphere.) The imaginary sphere goes into the ground, but that does not matter. The energy created by source 1 all passes through the sphere. And because the energy radiates uniformly in all directions, no part of the sphere receives more power than any other part.

Since energy gets produced at rate $P = \frac{E}{60}$, it follows that energy passes through the imaginary sphere at rate $P = \frac{E}{60}$, because energy cannot "disappear." (For instance, if the source creates 5 joules/second of energy, then 5 joules per second of energy must pass through the sphere.) The surface area of a sphere is $4\pi r^2$. So, through a sphere of surface area $A = 4\pi L^2$, energy flows at rate $P = \frac{E}{60}$. Therefore, the intensity at all points on the imaginary sphere–including source 2–is

$$I = \frac{P}{A} = \frac{\frac{E}{60}}{4\pi L^2} = \frac{E}{240\pi L^2}.$$

(b) The radio's reception is cleanest if, when a crest from source 1 reaches the radio, a crest from source 2 reaches the radio at the same time. In this case, the waves constructively "add" together, producing a "combined" crest of bigger amplitude. Put another way, the radio's reception is best if the waves from the two sources constructively interfere at the radio, assuming we cannot just put the radio right "on top of" source 2.

As shown in the textbook, since the waves are emitted from the two sources in phase, they will still be in phase at the radio *if* the path difference is an integer number of wavelengths. By "path difference," I mean $d_1 - d_2$, where d_1 is the distance from source 1 to the radio, and d_2 is the distance from source 2 to the radio. Constructive interference occurs at the radio if

$$\text{path difference} = d_1 - d_2 = n\lambda, \tag{1}$$

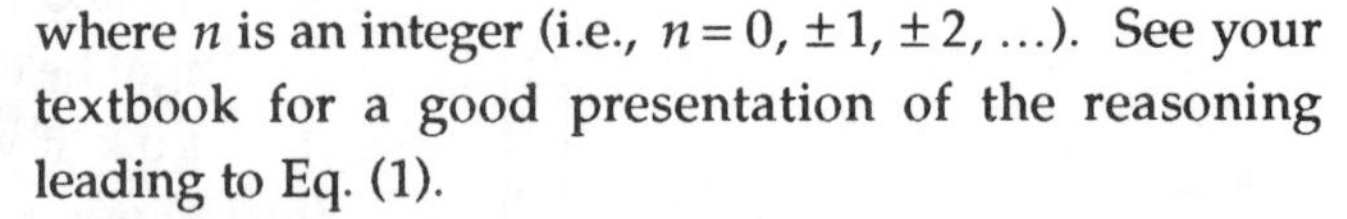

where n is an integer (i.e., $n = 0, \pm 1, \pm 2, \ldots$). See your textbook for a good presentation of the reasoning leading to Eq. (1).

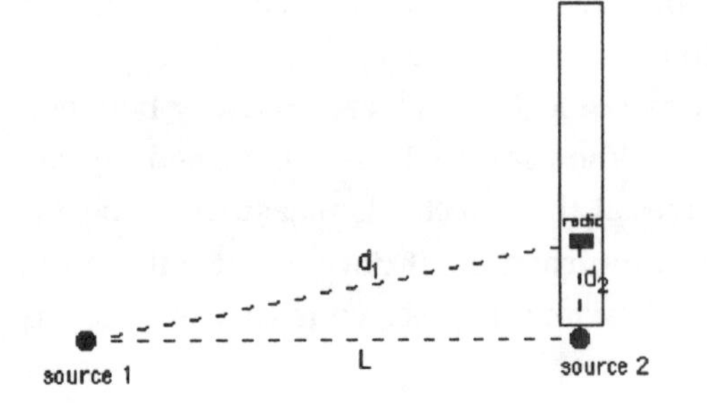

In this problem, we are solving for d_2, because that path length is the distance from source 2 to the radio. So, we need to rid Eq. (1) of the other unknowns, namely d_1 and λ.

From the diagram, you can see that $d_1 = \sqrt{L^2 + d_2^2}$. So, Eq. (1) simplifies to

$$\sqrt{L^2 + d_2^2} - d_2 = n\lambda\,. \tag{1'}$$

And from the basic wave relation $v = \lambda f$, we have $\lambda = \frac{c}{f}$. So $\sqrt{L^2 + d_2^2} - d_2 = n\frac{c}{f}$. Now it is just a matter of algebra to solve for d_2.

Algebra starts here. Add d_2 to both sides of the equation, and then square both sides to get $L^2 + d_2^2 = d_2^2 + 2n\frac{c}{f}d_2 + \left(n\frac{c}{f}\right)^2$. Cancel the d_2^2's, and subtract $\left(n\frac{c}{f}\right)^2$ from both sides to get $L^2 - \left(n\frac{c}{f}\right)^2 = 2n\frac{c}{f}d_2$. Isolate d_2 to get

$$d_2 = \frac{fL^2}{2nc} - \frac{nc}{2f}\,. \tag{2}$$

End of algebra.

We are not done. We have to decide which value of n corresponds to the best reception. Your guess might be $n = 1$. But this is not necessarily so. To get the best reception, we would like constructive interference, and we would *also* like the radio to be as close to the sources as possible. Why? Because the radio signal gets weaker, i.e., less intense, as you get farther from the sources. You saw why in part (a): The same total "signal energy" gets spread out over a bigger and bigger area as you get farther way

For this reason, we want to pick the smallest d_2 that satisfies Eq. (2). On a test, you probably would not have to do anything this complicated; it would be more obvious which n to use, or else all you would be expected to do is come up with Eq. (2), making sure you write *in words* that n must be an integer. Just for the heck of it, can you figure out a quick way (besides trial and error) to find which n corresponds to the smallest d_2 in Eq. (2)?

(c) You might guess that the intensity doubles when both sources are turned on. But that guess fails to take into account the constructive interference between the waves from the two sources. Why is the interference constructive? Well, since the detector sits midway between the two sources, the distance from source 1 to the detector is $d_1 = \frac{L}{2}$. And the distance from source 2 to the detector is also $d_2 = \frac{L}{2}$.

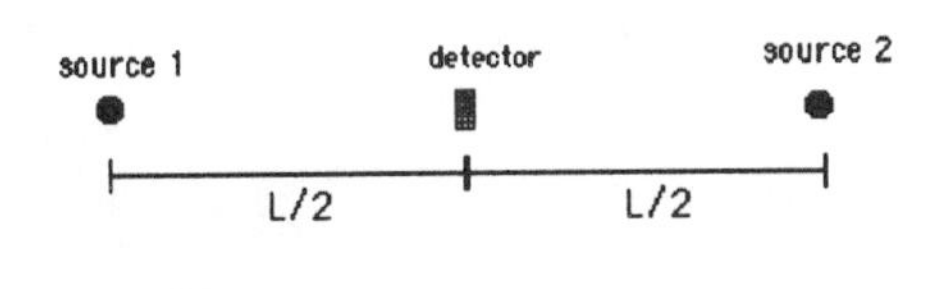

In words, the waves going from source 1 to the detector travel the same distance as the waves going from source 2 to the detector. Therefore, the path *difference* is $d_1 - d_2 = 0$. Since the waves are in phase at the sources, they are still in phase at the detector. (The "n" in the path-difference formula is 0.)

Why does this fact affect the intensity? Well, suppose the waves emitted from source 1 and source 2 both have amplitude A when they reach the detector. Then, the amplitude of the constructively-interfering waves at the detector is $2A$, because the crests "add." The textbook derives a formula telling us that the average power is proportional to the *square* of the amplitude. And intensity is proportional to the average power, as we saw in part (a). Since the amplitude doubles as a result of the constructive interference, the average power quadruples, and therefore the intensity quadruples: $I = 4I_0$.

Fluids: Statics and Dynamics

22 CHAPTER

QUESTION 22-1

A box having a square lid of side length $s = 0.050$ meters is partially evacuated. (This means that some, but not all, of the gas inside the box is pumped out.) The box, which is airtight, is then turned upside down, to see if the lid falls off. It does not. So, curious students start sticking chunks of clay to the outside of the lid, to make it heavier. They observe that the lid finally, barely falls off when its mass (including the attached clay) is $M = 0.75$ kg. The lab barometer says that air pressure is $p_0 = 1.0 \times 10^5$ Pa. (A Pascal is a newton per meter^2.)

(a) Before the lid fell off, what was the pressure inside the box?

(b) If there had been a complete vacuum inside the box, how massive a lid could the vacuum have "held up"? In other words, what mass of lid would barely have fallen off?

ANSWER 22-1

Let me first address a common preconception. Most people think that the partial vacuum inside the box "holds the lid up" when the box is upside down. In other words, "vacuums suck." This is not quite true. A vacuum is an absence of particles. An *absence* of particles cannot pull or push anything; it cannot exert a force. A vacuum *seems* to suck because something on the other side of the container wall–in this case, the air on the other side of the lid–exerts a pressure. If the box were filled with air, this upward air-pressure force would be balanced by the downward air-pressure force produced by the air inside the box. But with a vacuum inside the box, there is nothing to counterbalance the outside air pressure. And with a partial vacuum inside the box, the gas pressure in the box cannot fully counterbalance the outside air pressure.

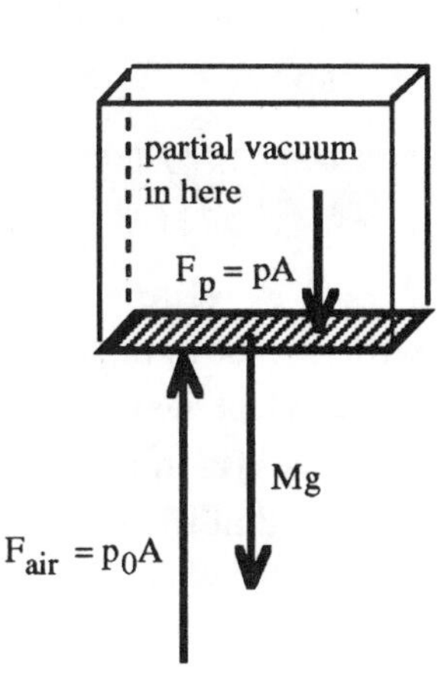

Another example: When you suck the air out of a plastic bottle, the bottle "crumples in" on itself, *not* because the vacuum sucks it in, but because the air on the outside of the bottle *pushes* the bottle walls inward.

(a) Given all this, let us draw a force diagram of the lid.

Here, A denotes the area of the lid. F_p is the force exerted by the gas inside the box, gas which is at unknown pressure p. Since the box is partially evacuated, p is less than p_0, the outside air pressure. I have use the definition of pressure, $p = \frac{\text{force}}{\text{area}}$, and hence $F = pA$. We are solving for p.

As long as the lid stays on the box, the upward forces must "beat" the downward ones. (Indeed, in that case, the top of the container walls must push down on the lid, a force I have not drawn.) But when the lid's mass exceeds $M = 0.75$ kg, the downward forces "win," and the lid falls off. Right when $M = 0.75$ kg, the lid "teeters" on the edge of barely falling off. So, the net force on the lid when $M = 0.75$ kg must be very nearly 0:

$$\sum F_y \approx 0$$
$$F_{\text{air}} - F_p - Mg = 0$$
$$p_0 A - pA - Mg = 0$$
$$p_0 s^2 - ps^2 - Mg = 0 \qquad [\text{since the area of the square lid is } A = s^2].$$

Now it is just algebra to solve for p. Doing so yields

$$p = p_0 - \frac{Mg}{s^2} = 1.0 \times 10^5 \text{ N/m}^2 - \frac{(0.75 \text{ kg})(9.8 \text{ m/s}^2)}{(0.050 \text{ m})^2} = 9.7 \times 10^4 \text{ N/m}^2,$$

97% of the outside air pressure! So, the inside of the box was "evacuated" by only 3%. This goes to show that you do not need a "complete" vacuum to cause dramatic effects.

(b) If $p = 0$ inside the box, then the F_p arrow gets erased from the above force diagram. When the mass of the lid is such that the lid barely falls off, the upward force due to air pressure exactly balances the downward pull of gravity, making the net force zero:

$$\sum F_y \approx 0$$
$$F_{\text{air}} - mg = 0$$
$$p_0 s^2 - mg = 0.$$

Solve for the lid's mass, m, to get

$$m = \frac{p_0 s^2}{g} = \frac{(1.0 \times 10^5 \text{ N/m}^2)(0.05 \text{ m})^2}{(9.8 \text{ m/s}^2)} = 26 \text{ kg!}$$

So, the lid weighs over 55 pounds. Air pressure is that strong. Usually we do not notice it, because air pressure acts on "both sides" of the container or wall, and therefore cancels itself out, so to speak.

QUESTION 22-2

A weirdly shaped tube contains mercury, as pictured below. The right arm makes a 45° angle with the horizontal. Exactly $d = 10.0$ cm of water is poured into the right arm. (By d, I mean the "length" of water in the tube, not the height of water.) How high above its initial level does the mercury rise in the left arm? The density of mercury is about 13.5 times the density of water. You may assume that the radius of the tube is the same throughout the tube, and is extremely small compared to 10 cm.

Hint: Introduce ρ_w and ρ_m to denote the density of water and mercury, respectively. These constants will cancel out of your final answer.

ANSWER 22-2

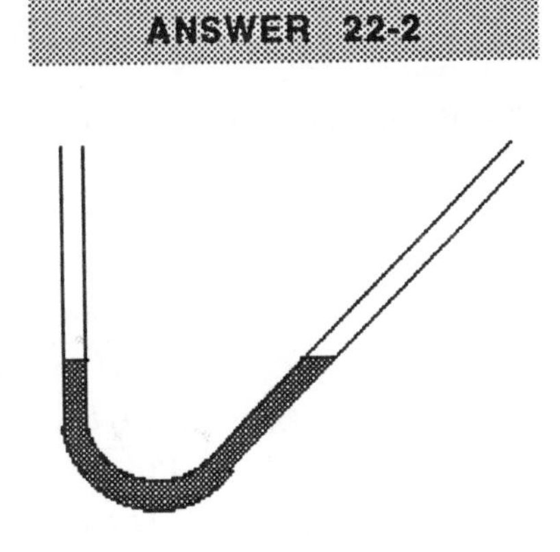

This is a funky "U-tube" problem, of which your textbook contains an example or two. Always begin such a problem by drawing BEFORE and AFTER pictures. I have let D denote the distance above its initial level (in the left arm) that the mercury gets pushed. D is also the distance by which the mercury in the right arm is pushed down, because the total "length" (amount) of mercury in the tube does not change. From the diagram, notice that the *height* by which the mercury goes down in the right tube is $D\sin 45°$, not D, because the tube is angled. For the same reason, the *height* of water in the right tube is not L, but $L\sin 45°$. I will take advantage of these heights later. Q, R, and S are conveniently-placed points, *not* distances.

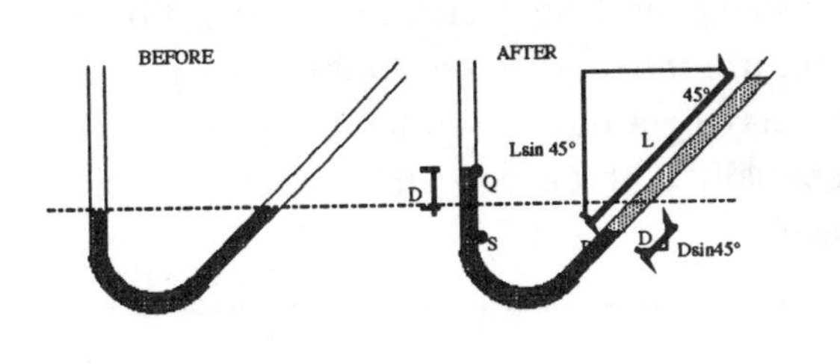

As shown in the textbook, if we consider two different points inside a static fluid, both at the same height, and both inside the same fluid, then the pressures at those two points are the same. Therefore, the pressure at point R equals the pressure at point S. (Since point R borders on mercury, this assumption is valid, even though point R is not fully "inside" mercury.) In symbols,

$$p_R = p_S. \tag{1}$$

Intuitively, the pressure at point S is caused by the column of mercury over it; and the pressure at point R is due to the column of water over it. *Something* about those two columns must be the same, in order for the pressures to be equal at R and S. We will solve the problem using a formalized version of this insight.

The pressure at a point submerged in a fluid is caused by two factors: The column of fluid over that point, and the air pressure pushing down on top of that column. For instance, the pressure at point R is due to the column of water, *plus* the air pressure pushing down on top of that column. Let p_0 denote the pressure at the top of the column. As your textbook shows, the pressure due to the fluid column is ρgh, where ρ is the fluid density and h is the *height* of the column. That is height, not length. Putting this together, the general formula for the fluid pressure a height h below the top surface of a static fluid is

$$p = p_0 + \rho gh. \tag{*}$$

See the textbook for a rigorous derivation. For now, I just want you to understand the intuitions behind this equation, as described in the previous paragraph. Here, $p_0 = p_{\text{air}}$, atmospheric pressure.

Point R is a height $h = L\sin 45°$ below the top surface of water. So, according to Eq. (*), the pressure at point R is $p_R = p_{\text{air}} + \rho_w g(L\sin 45°)$. Point S sits at the same level as point R. Notice from the diagram that it is a height $h = D\sin 45°$ below the dashed line, which is a height D below the top surface of mercury in the left arm. So, the total height of mercury above point S is $h = D + D\sin 45°$. Therefore, according to Eq. (*), the pressure at point S is $p_S = p_{\text{air}} + \rho_m g(D + D\sin 45°)$.

As discussed above, since points R and S sit at the same level inside the same fluid (mercury), those points "share" the same pressure:

$$p_R = p_S$$
$$p_{\text{air}} + \rho_w g(L \sin 45°) = p_{\text{air}} + \rho_m g(D \sin 45°).$$

The p_{air}'s cancel. So do the g's. Solving for D yields

$$\begin{aligned} D &= \frac{\rho_w L \sin 45°}{\rho_m (1 + \sin 45°)} \\ &= \frac{L \sin 45°}{13.5(1 + \sin 45°)} \qquad \left[\text{mercury is 13.5 times denser: } \frac{\rho_m}{\rho_w} = 13.5.\right] \\ &= \frac{(10.0 \text{ cm}) \sin 45°}{13.5(1 + \sin 45°)} \\ &= 0.31 \text{ cm}, \end{aligned}$$

The mercury barely budged, because it is so much denser than water.

I have just illustrated the general U-tube problem-solving strategy. First, draw careful before and after pictures. Pick relevant points inside the fluid, and use the fluid statics formula $p = p_0 + \rho g h$ to write equations about the pressures at those points. By playing around with those equations, you will usually be able to solve for the quantity of interest, especially if you keep in mind that two points at the same level in the same fluid experience the same pressure.

QUESTION 22-3

A car of mass $m = 1000$ kg sits on a huge circular movable piston of radius $R = 2$ meters, as drawn below. (The 1000 kilograms include the mass of the big piston.) A physicist, determined to show off his natural strength, wants to lift the car by pushing on the other circular piston, which has radius $r = 0.020$ meters (about an inch). The chamber is filled with water, which has density $\rho = 1000$ kg/m^3. How much force will the physicist need to apply if

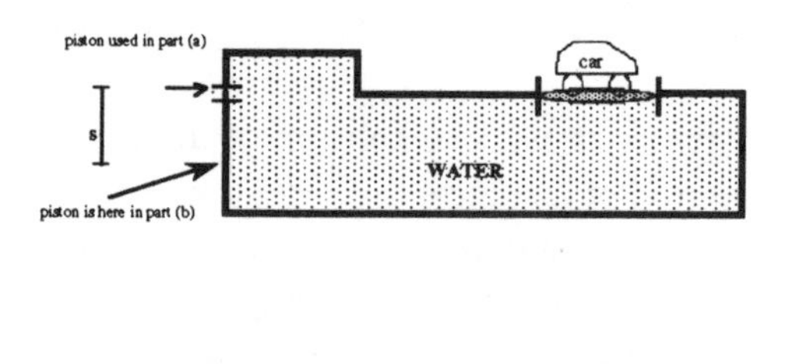

(a) The small piston is at the same height as the big piston, as drawn below?

(b) The small piston is a distance $s = 1.0$ meter below the big piston.

ANSWER 22-3

(a) Intuitively, you might think the physicist needs to apply a force equal to the car's weight. But fortunately, the fluid "helps" the physicist. Since the big and small piston are at the same height inside the same fluid, the pressure at both pistons is the same. Does this mean the force on both pistons is the same? No, because pressure is force *per area*. Since the area of the small piston is one ten-thousandth the area of the big piston, the physicist needs to apply a force only one ten-thousandth the weight of the car in order to achieve the same *pressure*, and hence, to hold the car up.

Let me work through this more slowly. The car exerts force $F_{\text{on big piston}} = Mg$. So, the pressure on the big piston is $p_{\text{big piston}} = \frac{F_{\text{on big piston}}}{A_{\text{big piston}}} = \frac{Mg}{A_{\text{big piston}}}$. Similarly, if we let F_p denote the push force exerted by the physicist, then $p_{\text{small piston}} = \frac{F_{\text{on small piston}}}{A_{\text{small piston}}} = \frac{F_p}{A_{\text{small piston}}}$.

As noted above, since the two pistons are at the same height in the same fluid, they are at the same pressure. (This qualitative fact is implicit in the formula $p = p_0 + \rho g h$; two points at the same h have the same pressure.) So,

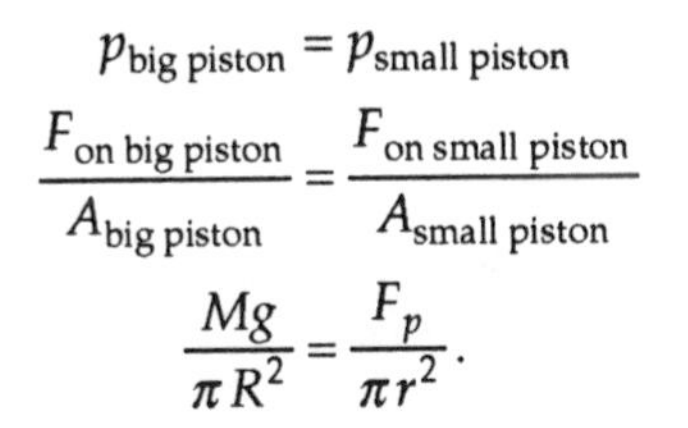

$$p_{\text{big piston}} = p_{\text{small piston}}$$

$$\frac{F_{\text{on big piston}}}{A_{\text{big piston}}} = \frac{F_{\text{on small piston}}}{A_{\text{small piston}}}$$

$$\frac{Mg}{\pi R^2} = \frac{F_p}{\pi r^2}.$$

Solve for F_p to get

$$F_p = Mg\frac{\pi r^2}{\pi R^2} = Mg\left(\frac{r}{R}\right)^2 = Mg\left(\frac{0.020 \text{ m}}{2.0 \text{ m}}\right)^2 = \frac{1}{10000}Mg.$$

Since the car's weight is $Mg = (1000 \text{ kg})(9.8 \text{ m/s}^2) = 9800 \text{ N}$, the physicist must push with only 0.98 newtons of force, which is under a quarter of a pound.

Systems that take advantage of this effect are called *hydraulics*. For instance, airplane designers build in redundant hydraulic systems to control the wing flaps, etc.

(b) Now the small piston is deeper than the big piston. So, the pressure at the two pistons is no longer the same. Intuitively, the pressure gets higher at the deeper point. Fortunately, we have a formula telling us how much higher. As discussed in Question 22-2 above, if we let p_0 denote the pressure at the top surface of a fluid, then the pressure a depth h below that surface is $p = p_0 + \rho g h$. We can arbitrarily choose what place we consider to be the "top surface." Let us pick the big piston. So, $p_0 = p_{\text{big piston}}$. Since the small piston is a depth $h = s$ below that top surface,

$$\begin{aligned} p_{\text{small piston}} &= p_0 + \rho g h \\ &= p_{\text{big piston}} + \rho g s. \end{aligned}$$

Give this equation, we can now duplicate the reasoning of part (a), in order to relate the physicist's force to the car's weight:

$$p_{\text{small piston}} = p_{\text{big piston}} + \rho g s$$

$$\frac{F_{\text{on big piston}}}{A_{\text{big piston}}} = \frac{F_{\text{on small piston}}}{A_{\text{small piston}}} + \rho g s$$

$$\frac{F_p}{\pi r^2} = \frac{Mg}{\pi R^2} + \rho g s.$$

Solve for F_p to get $F_p = Mg(\frac{r}{R})^2 + \rho g s \pi r^2$, which is our part (a) answer *plus* an additional term, $\rho g s \pi r^2$. This extra term corresponds to the extra force the physicist must exert in order to "hold up" a column of water of height s. But the radius of this column of water is only r, not R (counterintuitively enough!). Once again, hydraulics allows the physicist to push less hard than you would expect. Let us substitute in the numbers:

$$F_p = Mg\left(\frac{r}{R}\right)^2 + \rho g s \pi r^2$$
$$= (1000\text{ kg})(9.8\text{ m/s}^2)\frac{1}{10000} + (1000\text{ kg/m}^3)(9.8\text{ m/s}^2)(1.0\text{ m})\pi(0.020\text{ m})^2$$
$$= 0.98\text{ N} + 12.3\text{ N}$$
$$= 13.3\text{ N},$$

about 4 pounds. Notice that the second term, $\rho g s \pi r^2$, is responsible for most of the effort the physicist must exert. Roughly put, the fact that the piston is deep in the water contributes more to his effort than the fact that he is lifting a car.

QUESTION 22-4

Water, which has density r, is pumped (at high pressure) at rate Q into the left end of a cylindrical pipe. Q is the mass flux, i.e., the mass per time pumped into the pipe. The radius of the left end of the pipe is r_1.

The pipe gradually narrows and curves as shown below. At the other end of the pipe (i.e., the "mouth"), the radius is r_2. The mouth of the pipe is at the same height as point B.

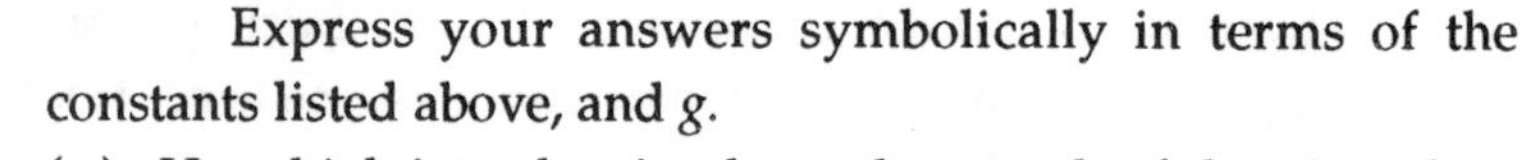

Express your answers symbolically in terms of the constants listed above, and g.

(a) How high into the air, above the mouth of the pipe, does the water shoot?

(b) If the air pressure is p_{air}, then what is the pressure exerted by the water on the wall of the pipe at point B?

ANSWER 22-4

As is often the case in steady-fluid-flow problems, you must use both "flux conservation" and Bernoulli's law to reach all the answers. Let v_2 denote the water's velocity just as it leaves the pipe.

(a) Once we find v_2, the speed at which water shoots upward out of the pipe, we can calculate how high it goes using projectile reasoning (or energy conservation, if you prefer.) The water shoots straight up with initial speed v_2. Its "final" speed at its peak is $v_f = 0$. Also, the acceleration of the water once it is in the air is $a = -g$. (I have picked upward as positive.) So, from the constant-acceleration kinematic formula $v_f^2 = v_0^2 + 2a\Delta y$, we get

$$0 = v_2^2 - 2gh. \tag{1}$$

After finding v_2, we can easily solve Eq. (1) for h, the height of the water column. As just mentioned, you could also obtain Eq. (1) using conservation of energy.

To find v_2, you might try Bernoulli's equation. But that will not get you very far, because you do not know the pressure at the left end of the pipe. Instead, you must invoke the *other* incredibly useful fluid flow equation, often called "flux conservation" or "continuity equation." Before using this equation, I will briefly review where it comes from.

Let us just say, hypothetically, that 5 gallons per second of water are pumped into the left end of the pipe. Could it be the case that only 4 gallons/second of water leave the right end of the pipe? Only if the pipe is leaky. Otherwise, where would that "extra" gallon/second of water *go*?

Intuitively, however many gallons per second enter the pipe, that is how many gallons per second should leave the pipe, because "extra" water cannot just appear or disappear. Indeed, the rate at which water passes any two points in the pipeline should be equal, for the same reason.

As the book derives, the rate at which fluid passes a point in the pipe is $R = Av$. Here, A is the cross-sectional area of the pipe, and v is the velocity of the water. You can confirm that R is a volume per time. So, "flux conservation" demands that if R_1 and R_2 denote the rate at which water passes "point 1" and "point 2" in the pipe, then

$$\begin{aligned} R_1 &= R_2 \\ A_1 v_1 &= A_2 v_2 \quad \textbf{[Flux conservation]} \end{aligned} \tag{2}$$

Eq. (2) talks about the *volume* per time of water in the pipe. Sometimes, we want to talk about *mass* per time. Since mass equals density times volume ($m = \rho V$), we can "convert" Eq. (2) into mass-per-time by multiplying through by ρ:

$$\rho A_1 v_1 = \rho A_2 v_2 \quad \textbf{[Mass conservation]} \tag{2'}$$

In this problem, we are solving for v_2, so that we can use projectile reasoning to calculate how high the water spurts. The mass flux (i.e., mass per time) pumped into the left end of the pipe is Q. In other words, $\rho A_1 v_1 = Q$. So, from mass conservation Eq. (2′), we get

$$\begin{aligned} \rho A_1 v_1 &= Q = \rho A_2 v_2. \\ &= \rho(\pi r_2^2) v_2. \end{aligned} \tag{3}$$

I have just used the fact that the cross-sectional "shape" of a cylindrical pipe is a circle, which has area πr^2. (If you look into the end of a cylinder, you see a circle.)

We can immediately solve Eq. (3) for v_2 to get $v_2 = \frac{Q}{\rho \pi r_2^2}$. That is the speed with which water shoots out of the pipe. Now just substitute that speed into kinematic Eq. (1) above, and solve for the height h, to get $h = \frac{v_2^2}{2g} = \frac{Q^2}{2g\rho^2 r_2^4}$.

(b) To solve this, we need Bernoulli's law:

$$p_1 + \frac{1}{2}\rho v_1^2 + \rho g y_1 = p_2 + \frac{1}{2}\rho v_2^2 + \rho g y_2 \quad \textbf{[BERNOULLI'S LAW]}.$$

In this formula, p is the pressure and y is the vertical position, measured from the ground up. Each side of the formula refers to a different point in the pipe. For instance, in this problem, p_2 is the pressure at the mouth of the pipe, while p_1 is the pressure at point B. Here, $y_1 = y_2$: point B and the mouth of the pipe share the same height.

I must stress that Bernoulli's law, like flux conservation, applies not only to the ends of pipes, but also to points inside the pipe.

We are trying to solve Bernoulli's equation for p_1, the water pressure at point B near the entrance of the pipe. Do we have enough information? Well, we are given ρ. The two y's are the same, and hence the $\rho g y$ terms cancel away. In part (a), we found v_2. So, the only unknowns in Bernoulli's equation, besides p_1, are v_1 and p_2.

Well, p_2 is the pressure right at the mouth of the pipe. Intuitively, just as the water leaves the pipe, its pressure "equalizes" with its surroundings. So, at the exit point, the water's pressure is p_{air}, the atmospheric pressure. **Whenever a fluid leaves a pipe (or keg or whatever) and enters the atmosphere, its pressure becomes that of the atmosphere.** So here, $p_2 = p_{air}$.

Before we can solve Bernoulli's equation for p_1, we still need to know v_1, the water's speed at point B. Well, remember that we are given the mass flux of water at the pipe's entrance, which is also the mass flux at point B (because the pipe does not widen or narrow between the entrance and point B). As mentioned above in Eq. (3), the mass flux here is just $Q = \rho A_1 v_1$. So, $v_1 = \frac{Q}{\rho A_1} = \frac{Q}{\rho \pi r_1^2}$. (You get this same expression for v_1 by using flux conservation, $A_1 v_1 = A_2 v_2$.)

At this point, we can solve Bernoulli's equation for p_1. Remember, the $\rho g y$ terms cancel, because $y_1 = y_2$. I will also use the expression for v_2 obtained above in part (a).

$$\begin{aligned} p_1 &= p_2 + \frac{1}{2}\rho v_2^2 - \frac{1}{2}\rho v_1^2 \\ &= p_{air} + \frac{1}{2}\rho\left(\frac{Q}{\rho \pi r_2^2}\right)^2 - \frac{1}{2}\rho\left(\frac{Q}{\rho \pi r_1^2}\right)^2 \\ &= p_{air} + \frac{Q^2}{2\rho \pi^2}\left[\frac{1}{r_2^4} - \frac{1}{r_1^4}\right]. \end{aligned}$$

In most fluid-flow problems, combining the flux conservation equation with Bernoulli's law gets you to the answer.

QUESTION 22-5

Consider the following oddly-shaped container.

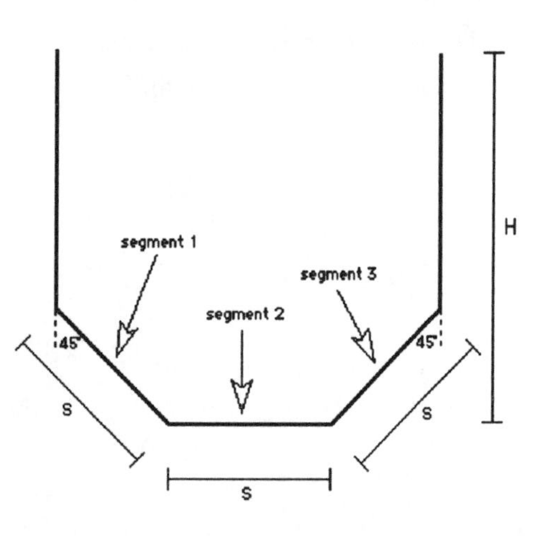

The container extends "into the page" a distance L. As the diagram shows, the bottom of the container is composed of three segments.

Suppose the container is filled to the brim with ammonia, which has density ρ. Let p_0 denote air pressure. Express your answers in terms of r, g, H, s, L, and p_0.

(a) What is the force exerted by the fluid on "segment 2" of the bottom of the container?

(b) *(Hard. Ask your instructor if you need to solve this kind of problem.)* What is the force exerted by the fluid on segment 1?

(c) What is the total force due to the fluid on the whole bottom of the container (all three segments)? If you did not get an answer to part (b), just make one up and use it here.

ANSWER 22-5

(a) This part is relatively straightforward, because segment 2 is horizontal, and therefore all points on segment 2 are at the same depth. Consequently, the fluid pressure is the same at all points of segment 2. Call that pressure p_2. Using the "column of fluid" derivation shown in the textbook, you can get $p_2 = p_0 + \rho g H$. The area of segment 2 is $A_2 = sL$, because that segment is a rectangle of width s and length L (into the page). Remembering that pressure is force per area ($p = \frac{F}{A}$), we can solve for the force F_2 on segment 2:

$$\begin{aligned} F_2 &= p_2 A_2 \\ &= (p_0 + \rho g H)sL. \end{aligned}$$

The force due to fluid pressure always points perpendicular to the surface on which the fluid presses. The direction of this force is perpendicular to segment 2, i.e., straight downward.

(b) This part is harder, because the pressure is not the same on all "parts" of segment 1. Why not? Pressure increases with depth. The pressure on the bottom-most parts of segment 1 exceeds the pressure on the top-most points. But if we consider a bunch of points on segment 1 all at the same depth, then all those points experience the same pressure.

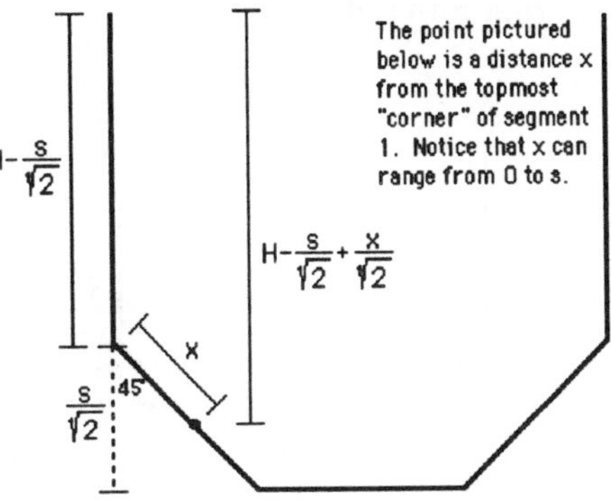

This suggests a strategy. We can break up segment 1 into tiny "slivers" consisting of points all at the same depth. Since the whole sliver experiences the same pressure, we can quickly calculate the tiny force dF acting on that tiny sliver. Then we can add up the forces acting on *all* the slivers comprising segment 1, by integrating. This gives us the total force on segment 1.

Let x denote the distance of an arbitrary point on segment 1 from the topmost point of segment 1. Consider a small "sliver" of segment 1, a sliver whose endpoint is the point drawn above. Here is a bad attempt at a 3-D view.

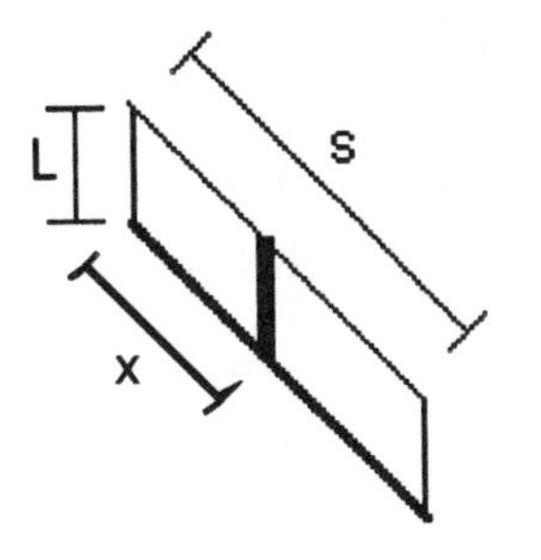

The black "shaded in" area is the sliver I am talking about. It has length L and width dx.

What is the pressure on that sliver? Well, the topmost point of segment 1 (at $x = 0$) is a distance $H - \frac{s}{\sqrt{2}}$ from the top of the liquid, as the above diagram shows. And the sliver is a vertical distance $x\cos 45° = \frac{x}{\sqrt{2}}$ below the topmost point of segment 1. So, the sliver is a distance $h = \left(H - \frac{s}{\sqrt{2}}\right) + \frac{x}{\sqrt{2}}$ below the top of the liquid. Therefore, the pressure on that sliver is $p = p_0 + \rho g\left[H - \frac{s}{\sqrt{2}} + \frac{x}{\sqrt{2}}\right]$. The (tiny) area of the sliver is length times width: $dA = L dx$. So, the force on that sliver is

$$\begin{aligned} dF &= p\,dA \\ &= \left(p_0 + \rho g\left[H - \frac{s}{\sqrt{2}} + \frac{x}{\sqrt{2}}\right]\right)L dx. \end{aligned}$$

Segment 1 is "made of" slivers running from $x = 0$ to $x = s$. To find the total force on segment 1, we must add up (integrate) the dF's acting on all those slivers.

$$\begin{aligned}F_1 &= \int_{x=0}^{x=s} dF \\ &= \int_0^s \left(p_0 + \rho g\left[H - \frac{s}{\sqrt{2}} + \frac{x}{\sqrt{2}}\right]\right) L\,dx \\ &= \left\{\left(p_0 + \rho g\left[H - \frac{s}{\sqrt{2}}\right]\right) Lx + \rho g L \frac{x^2}{2\sqrt{2}}\right\}\Bigg|_0^s \\ &= \left(p_0 + \rho g\left[H - \frac{s}{\sqrt{2}}\right]\right) Ls + \rho g L \frac{s^2}{2\sqrt{2}} \\ &= (p_0 + \rho g H) Ls - \frac{1}{2\sqrt{2}} \rho g L s^2.\end{aligned}$$

The force due to fluid pressure always points perpendicular to the surface on which the fluid pushes. So, the total force on segment 1 points at a 45° angle, downward and leftward. See the diagram below.

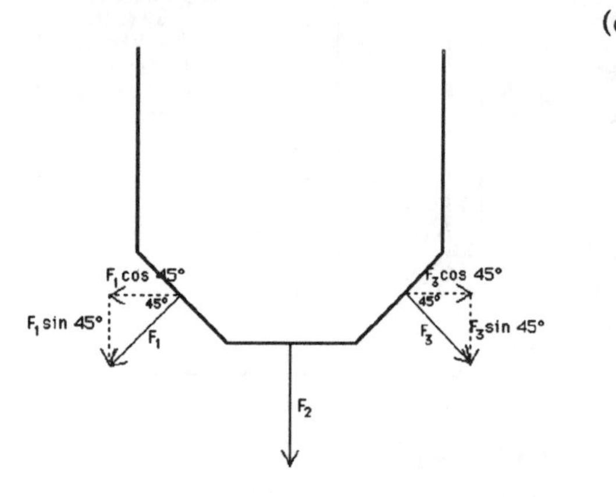

(c) By the same reasoning, we can find F_3, the force on segment 3. From symmetry, you can "guess" that F_3 has the same magnitude as F_1, but points in another direction. So, here is the force diagram for the bottom of the container. Since we are trying to find the total force *due to the fluid*, I have omitted other forces such as gravity. Also, I have broken the forces into their horizontal and vertical components.

Since these forces point in different directions, we must add them vectorially, by separately considering the horizontal and vertical components. From the force diagram,

$$\begin{aligned}\sum F_x &= F_3 \cos 45° - F_1 \cos 45° \\ &= 0 \qquad \text{[since } F_3 = F_1\text{]}.\end{aligned}$$

$$\begin{aligned}\sum F_y &= F_1 \sin 45° + F_2 + F_3 \sin 45° \\ &= 2F_1 \sin 45° + F_2, \qquad \text{[since } F_3 = F_1\text{]}.\end{aligned}$$

where F_1 and F_2 are the answers to parts (b) and (a).

When you complete the math, it turns out that the net force exerted by the ammonia on the bottom of the container exactly equals the weight of ammonia in the container. This makes intuitive sense. In fact, you could have "guessed" this result ahead of time, by the following reasoning. Consider the ammonia to be one big blob, of mass M. Since the blob is motionless, the net vertical force on it is 0. Therefore, the downward force on the blob equals the upward force on the blob. The downward force is Mg, the blob's weight. So, the upward force provided by the bottom of the container must equal Mg. By Newton's 3rd law, since the bottom of the container pushes upward on the ammonia with force Mg, the ammonia pushes downward on the bottom of the container with the same force.

You also knew ahead of time that the x-forces would cancel, since the ammonia does not tend to push the bottom of the container either rightward or leftward.

QUESTION 22-6

(Very difficult problem.) Consider water (density ρ_w) flowing through the pipe drawn below. At the exit, which has cross-sectional area A_2, the water flows into the air (pressure p_0). Water is pumped at high pressure into the entrance of the pipe, at speed v_1. The entrance has area A_1.

A small "offshoot" sticks out of the top of the fat section of pipe, as drawn below. Due to the high pressure in the fat part of the pipe, water goes up into the offshoot, and settles at a certain level h. The top of the offshoot is open to the air.

(a) Solve for h in terms of the other constants listed above (and g, if you need it). *Hint*: After the water in the offshoot settles at height h, is that water flowing or static?

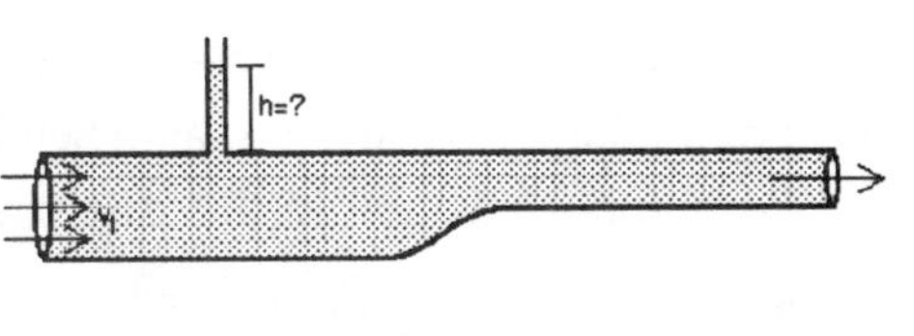

(b) Suppose we now drill a hole in the top of the thin section of pipe, and attach another offshoot. How high would water rise in that offshoot?

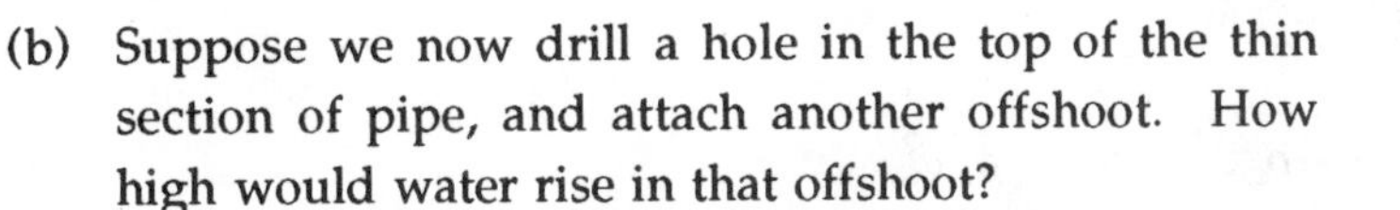

ANSWER 22-6

(a) Once it settles, the water in the offshoot is static; it is not flowing anywhere. Therefore, we can apply fluid-statics reasoning to the column of water in the offshoot. But the water in the rest of the pipe is flowing. We *cannot* apply fluid-statics reasoning to that water. Instead, we must apply Bernoulli's law and flux conservation.

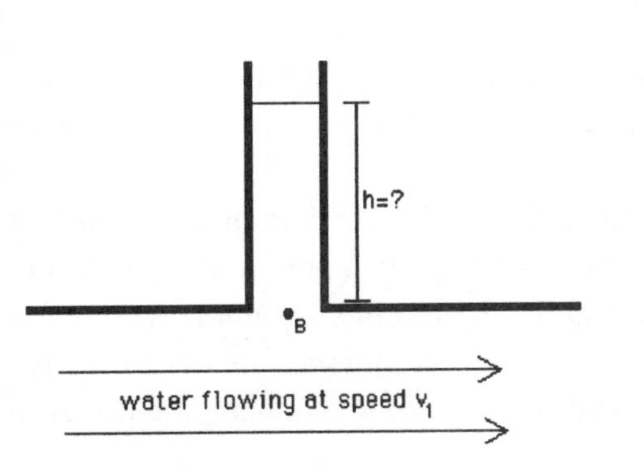

Let me "zoom in" on the offshoot.

Point B sits on the boundary between the flowing water in the pipe and the static water in the offshoot. Therefore, we can use point B in both fluid-statics *and* fluid-dynamics equations. That is the insight needed to solve the problem. Here is a general strategy for addressing complicated questions of this sort:

(1) Make a big drawing of the system, with "key" points labeled. Be sure to include a point at the boundary between flowing and static fluid (in this case, point B).

(2) Using fluid-flow or fluid-statics reasoning, write equations relating the pressures at the different points. In this manner, you will usually generate enough equations to solve for the quantity of interest.

I will now implement the strategy.

Step 1: Diagram.

It is useful to choose a point at the mouth of the pipe, if it has a mouth, because we know the pressure at the mouth. (I will show why in a minute.) I am calling that point "C." Since point E plays no role until part (b) below, ignore it for now.

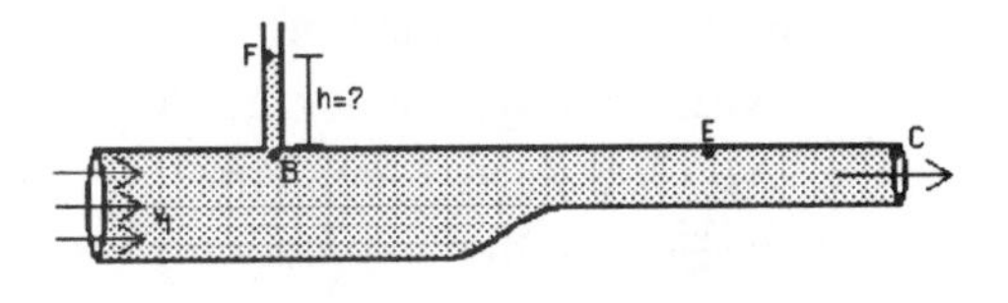

Step 2: Pressure equations

I will begin by relating point *C* to point *B*, hoping to find the pressure at point *B*. Call that pressure p_B. Later, I will relate point *B* to point *F*, to figure out how the height of water in the offshoot depends on p_B.

The speed of water at the entrance, and therefore the speed of water throughout the fat part of the pipe, is v_1. So, the speed of water at point *B* is v_1. By Bernoulli's law,

$$p_B + \frac{1}{2}\rho_w v_1^2 + \rho_w g y_B = p_C + \frac{1}{2}\rho_w v_C^2 + \rho_w g y_C.$$

Since point *C* "touches" the air, the water pressure at point *C* is the external air pressure, p_0. Also, points *B* and *C* share the same height: $y_B = y_C$. Therefore, the $\rho_w g y$ terms cancel. Consequently, Bernoulli's equation in this case simplifies to

$$p_B + \frac{1}{2}\rho_w v_1^2 = p_0 + \frac{1}{2}\rho_w v_C^2. \qquad (1)$$

We cannot yet solve Eq. (1) for p_B, because it contains another unknown, v_C; we do not know how fast water shoots out the right end of the pipe. To get more information, consider flux conservation, according to which the volume per time of water passing point *B* equals the volume per time of water passing point *C*:

$$A_1 v_1 = A_2 v_C. \qquad (2)$$

These two equations contain only two unknowns, v_C and p_B. So, we can solve for p_B. But let me save the algebra until later. For now, I will pretend we have solved for p_B. I will treat it as "known" for the rest of this problem.

Given p_B, we can figure out the height *h* of water in the offshoot, by writing an equation that relates point *B* to point *F*. The fluid in the offshoot is not moving. It is just a static column. Therefore, we must use fluid-statics equations (or equivalently, Bernoulli's equation with $v = 0$ everywhere). The pressure at the top of the column (point *F*) is the outside air pressure, p_0. Therefore, according to the "column of liquid" argument, which I will not repeat here, the pressure at point *B* is

$$p_B = p_0 + \rho_w g h, \qquad (3)$$

So, given p_B, we can solve for *h*, the height of fluid in the offshoot. I must stress that Eq. (3) works only because the fluid in the offshoot is not flowing, and because point *B* lies inside (or on the border of) that static fluid. Otherwise, Eq. (3) would be invalid.

Now, finally, I will complete the algebra.

Algebra starts here. First, I will solve Eqs. (1) and (2) for p_B. To do so, solve Eq. (2) for v_C to get $v_C = \frac{A_1 v_1}{A_2}$, put that expression back into Eq. (1), and isolate p_B to get

$$\begin{aligned} p_B &= p_0 + \frac{1}{2}\rho_w\left[\left(\frac{v_1 A_1}{A_2}\right)^2 - v_1^2\right] \\ &= p_0 + \frac{1}{2}\rho_w v_1^2\left[\left(\frac{A_1}{A_2}\right)^2 - 1\right] \end{aligned} \qquad (*)$$

Now substitute that expression for p_B into Eq. (3) to get

$$p_B = p_0 + \rho_w g h$$

$$p_0 + \frac{1}{2}\rho_w v_1^2\left[\left(\frac{A_1}{A_2}\right)^2 - 1\right] = p_0 + \rho_w g h.$$

Cancel the p_0's, and isolate h to get $h = \frac{1}{2g}v_1^2\left[\left(\frac{A_1}{A_2}\right)^2 - 1\right]$. Notice that the air pressure p_0 cancels out of the answer. Let us see why, by pretending the air pressure suddenly increases. In that case, the pressure at point C increases, and hence, the pressure inside the pipe increases. This tends to push water higher up the offshoot. But higher air pressure also means that that the top surface of water in the offshoot (at point F) gets pushed *down* harder. This increased downward push exactly cancels the increased upward push just discussed. So, increasing the air pressure leaves the height of water in the offshoot unaffected.

(b) By similar reasoning, you can derive an expression for the height of water in the new offshoot. The answer comes out to be zero. Here is why.

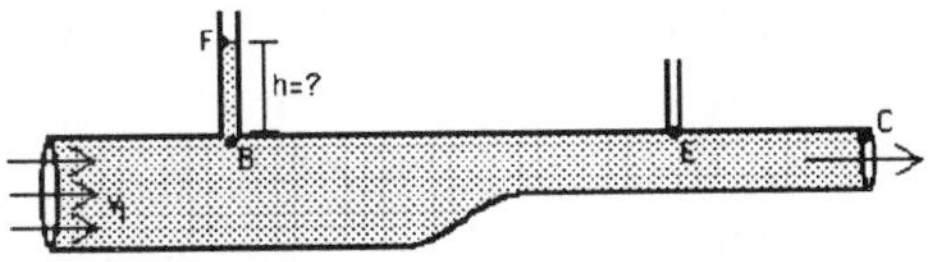

The pressure at point C, as shown earlier, is $p_C = p_0$, the outside air pressure. Now, the fluid at point E is going the same speed as the fluid at point C, because the pipe has not widened or narrowed between those two points. (You can confirm this using $A_E v_E = A_C v_C$.) And points C and E share the same height. Since $v_E = v_C$ and $y_E = y_C$, it follows from Bernoulli's law that the pressures at points C and E are the same: $p_C = p_E = p_0$.

In words, the fluid pressure at point E equals the outside air pressure. Therefore, a water molecule at point E "feels" an upward force due to the fluid pressure, but also feels a downward force due to air pressure in the offshoot; and these two forces are the same! So, the molecule does not rise up into the offshoot. (You can reach this conclusion mathematically by assuming there is a column of fluid in the offshoot, and setting $p_E = p_0 + \rho_w g h$. Since $p_E = p_0$, you get $h = 0$.) In summary, no water enters the new offshoot.

QUESTION 22-7

Consider a hollow spherical container designed to hold gas at high pressure. The container (mass $M = 0.60$ kg, radius $R = 0.10$ m) is essentially a very thin spherical shell, like a soccer ball with thin, sturdy walls. It is filled with air at high pressure, and sealed. The filled sphere is then placed into a big beaker of water. Observation reveals that the sphere floats, with exactly one third of its total volume sticking above the surface of the water. Water has density $\rho_w = 1000\ \text{kg/m}^3$.

(a) What is the mass density ρ of the pressurized air within the container?

(b) *(Skip this unless you have taken chemistry.)* Given that air at regular pressure, 1 atmosphere, has mass density $\rho_{\text{air}} = 1.20\ \text{kg/m}^3$, how many atmospheres of pressure are there inside the spherical container?

(c) The sphere of gas is now heated, and then placed back into the beaker of water. Will it float with more of its volume immersed, less of its volume immersed, or the same fraction of its volume immersed, as compared to before? (The sphere's "thermal expansion" is negligible; it does not expand appreciably when heated.)

ANSWER 22-7

(a) Archimedes' law is extremely easy to screw up, because you must worry about whether the volumes and masses appearing in the formula refer to the fluid or the material floating in the fluid. To avoid careless errors, you must see where the relevant formula comes from.

When an object floats (or sinks), part (or all) of the object is immersed, by which I mean under the surface of the fluid. The immersed part of the object has "pushed aside" some fluid. Let V_{dis} denote the volume of fluid displaced by the object. As just mentioned, V_{dis} equals the volume of the *part* of the object that is immersed. If part of the object sticks above the surface of the fluid, then V_{dis} is less than V, the whole volume of the object.

Archimedes law says that when an object is partly or fully immersed, **the weight of the object equals the weight of fluid displaced by the object**. We must "translate" that law into an equation.

Well, the weight of the object is $m_{object}g$. Since mass equals density times volume ($m = \rho V$), you can reexpress this as $m_{object}g = \rho_{object}Vg$, if the object has uniform density. (In this problem, the density is not uniform, because the "object" consists of two parts: a hollow sphere and the pressurized air inside that sphere.) Notice that the volume V in this expression is the *whole* volume of the object, because the whole volume "contributes" to its weight.

By contrast, the weight of fluid displaced is $m_{fluid\ displaced}g = \rho_{fluid}V_{dis}g$. We must use V_{dis} instead of V here; only the immersed part of the object "pushes aside" fluid, and it is the weight of this pushed-aside fluid that appears in Archimedes' law.

Given all this, let us "derive" the relevant form of Archimedes' law for this particular problem. Let V denote the volume of the sphere. The problem tells us that $\frac{1}{3}$ of the sphere sticks out of the water, and hence $\frac{2}{3}$ of the sphere is immersed. In other words, only $\frac{2}{3}$ of the sphere displaces water: $V_{dis} = \frac{2}{3}V$. Also, keep in mind that the "object" consists of the container plus the gas inside the container. So,

$$\text{Weight of water displaced} = \text{Weight of object}$$

$$m_{water\ displaced}g = m_{object}g$$

$$\rho_w V_{dis}g = (m_{container} + m_{air\ in\ container})g \tag{1}$$

$$\rho_w \frac{2}{3}Vg = (M + \rho_{gas}V)g.$$

In the last step, I used the fact that the container's mass is M, and the gas inside it has volume V and unknown density ρ_{gas}. We are solving for ρ_{gas}. Now it is just a matter of algebra. Cancel the g's and isolate ρ_{gas} to get

$$\begin{aligned}\rho_{\text{gas}} &= \frac{\frac{2}{3}\rho_w V - M}{V} \\ &= \frac{2}{3}\rho_w - \frac{M}{V} \\ &= \frac{2}{3}\rho_w - \frac{M}{\frac{4}{3}\pi R^3} \qquad \left[\text{since } V_{\text{sphere}} = \frac{4}{3}\pi r^3\right] \\ &= \frac{2}{3}(1000 \text{ kg/m}^3) - \frac{0.60 \text{ kg}}{\frac{4}{3}\pi(0.10 \text{ m})^3} \\ &= 667 \text{ kg/m}^3 - 143 \text{ kg/m}^3 \\ &= 520 \text{ kg/m}^3,\end{aligned}$$

a little more than half the density of water.

(b) Remember the ideal gas law? Although air at extremely high pressure is not a terribly ideal gas, we can use $pV = nRT$ as a reasonable approximation. Divide the ideal gas law through by V to get $p = \frac{n}{V}RT$. So, the pressure is proportional to the *molar density*, i.e., the number of moles per unit volume. But the molar density is obviously proportional to the mass density, because if you double the number of molecules, you double the mass. So, pressure is proportional to the mass density ρ, other things being equal.

How does that help us here? Well, the ratio of the density of air inside the container to "regular" air density is $\frac{\rho_{\text{air in container}}}{\rho_{\text{air at 1 atmosphere}}} = \frac{520 \text{ kg/m}^3}{1.20 \text{ kg/m}^3} = 430$. By the above ideal gas law reasoning, since the air in the container is 430 times denser than regular air, its pressure is 430 times that of regular air. So, the pressure in the container is 430 atmospheres.

(c) You might think that heating the sphere makes it lighter, because hot air tends to rise. If that is true, more of the sphere would stick out of the water. On the other hand, you might think the sphere gets heavier, because raising the temperature increases the pressure inside the container, and we just found pressure to be proportional to mass density. In that case, more of the sphere would be immersed.

Neither of these arguments is correct, for subtle and interesting reasons.

Hot air tends to rise *because* it becomes more "spread out"; the same amount of air expands to cover more volume. Therefore, it is density decreases, making it lighter. But here, the air in the container *cannot* expand. It is stuck in a fixed volume. So, its density cannot decrease. Sure, its pressure increases; the air molecules in the container bang more frequently and more violently against the inside wall of the container. But this does not increase or decrease the mass of those molecules. The sphere of gas keeps the same weight, and hence, $\frac{1}{3}$ of its volume sticks above the water, exactly as before. Remember, the ideal gas law argument showed that density is proportional to pressure *when the temperature stays the same.* If you increase the temperature, you can increase the pressure without increasing the density.

You will learn more about this *thermodynamics* stuff in a later physics (or chemistry) course.

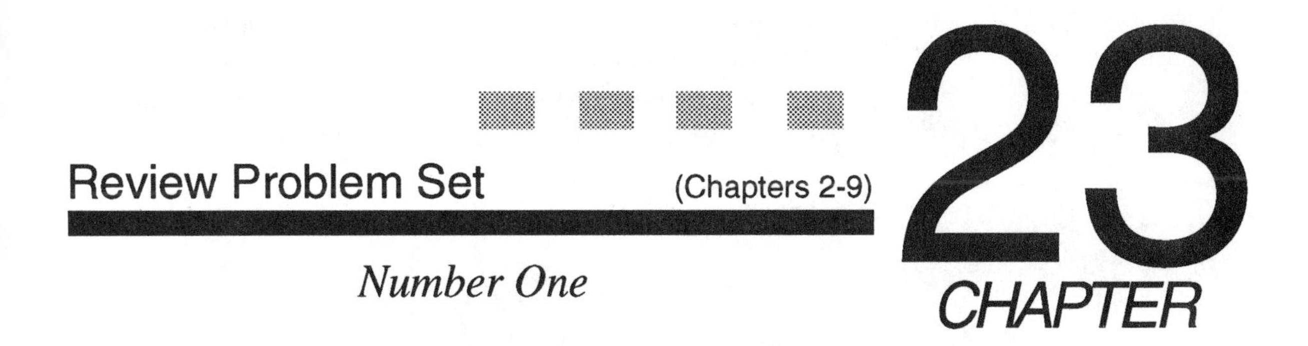

- *Warning: If your midterm covers circular motion, then work some problems from Chapter 10. Circular motion is not covered in these review problems.*

REVIEW QUESTION 1-1

You are a firefighter. Some rooms on the 6th floor of a building are burning. You want to shoot water out of the fire hose up through the 6th story windows, which are about H meters off the ground. Your most powerful fire hose is attached to the fire truck, which is parked on the street about d meters from the building.

The fire hose, which is essentially at ground level, shoots water at speed V. In order to shoot water through the 6th story window, at what angle θ to the ground should you align the fire hose? Set up, but do not solve, an equation or equations that you could solve for θ in terms of H, d, V, and any other constants you need.

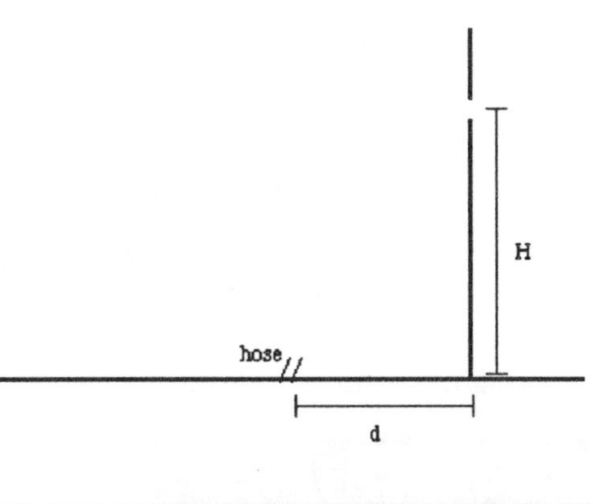

REVIEW ANSWER 1-1

We can treat the water droplets as projectiles. The fire hose must be angled such that the water travels a vertical distance $y = H$ in the same time it travels a horizontal distance $x = d$. So, as usual in projectile problems, we will need to consider both the horizontal *and* vertical aspects of the motion. Specifically, since we know both x and y, we can try breaking up the *vector* kinematic equation $\mathbf{r} = \mathbf{r}_0 + \mathbf{v}_0 t + \frac{1}{2}\mathbf{a}t^2$ into its horizontal and vertical components:

$$x = x_0 + v_{0x}t + \frac{1}{2}a_x t^2, \tag{1}$$

$$y = y_0 + v_{0y}t + \frac{1}{2}a_y t^2. \tag{2}$$

We are solving for the firing angle θ, which at first glance appears to be missing from Eqs. (1) and (2). But remember, the initial x- and y-velocity are both functions of θ.

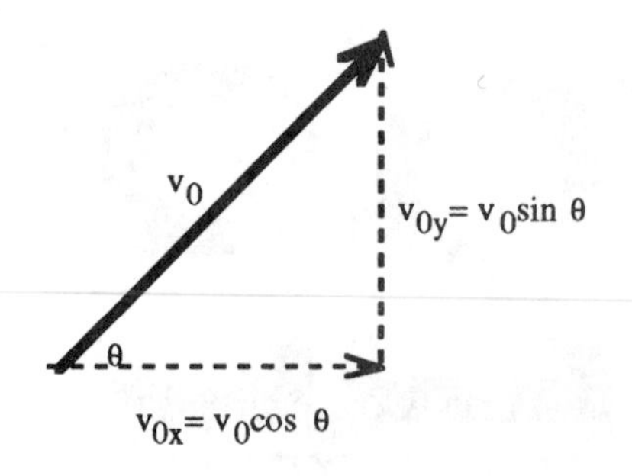

Let us start with the horizontal motion. The water travels distance $x = d$, starting from $x_0 = 0$, with initial horizontal speed $v_{0x} = V\cos\theta$. (The initial speed is given as "V.") Once the water droplets leave the hose, no horizontal force acts on them. So, as usual in projectile problems, the horizontal acceleration is $a_x = 0$; the horizontal part of the motion is uniform. Substituting all this information into Eq. (1) gives

$$\begin{aligned} x &= x_0 + v_{0x}t + \frac{1}{2}a_x t^2 \\ d &= 0 + (V\cos\theta)t + 0. \end{aligned} \tag{1*}$$

We cannot yet solve for θ, because this equation contains a second unknown, namely the flight time t of the water (as it flies from hose to window). To gather more information, consider the vertical motion. The droplets start at $y_0 = 0$ and end at $y = H$. Their initial vertical speed is $v_{oy} = V\sin\theta$. And their vertical acceleration is $a_y = -g$. (When I wrote down a positive initial vertical velocity, I implicitly picked upward as positive.) Substituting all this information into Eq. (2) yields

$$\begin{aligned} y &= y_0 + v_{0y}t + \frac{1}{2}a_y t^2 \\ H &= 0 + (V\sin\theta)t - \frac{1}{2}gt^2. \end{aligned} \tag{2*}$$

At this point, we have two equations in the two unknowns, θ and t. It is just a matter of algebra to solve for θ. You were not supposed to complete the algebra. But it was essential to set up Eqs. (1*) *and* (2*), because you need *both* to obtain θ. By the way, when you solve for θ, you get *two* answers. One of them corresponds to the water's entering the window on its way up, and the other corresponds to the water's entering the window on its way down.

REVIEW QUESTION 1-2

Consider the system of four blocks attached to each other by massless ropes, as drawn below. The blocks are placed on a frictionless ramp, of ramp angle θ. Block 4 is attached to a spring (of equilibrium length 0), the other end of which is attached to a wall at the top of the ramp. The spring is stretched parallel to the surface of the ramp. When a spring is stretched a distance x, it exerts a force $F_s = -kx$, where k is a "stiffness" constant. The minus sign indicates that the force is "backwards" compared to the direction in which the spring is stretched.

Suppose the blocks are placed close to the wall and released from rest. They slide down the ramp, stretching the spring. Eventually, they momentarily come to rest, before "bouncing" back up the ramp. At the moment the blocks come to rest, the spring is stretched a distance s. This is the moment drawn below. All of the following questions refer to this moment.

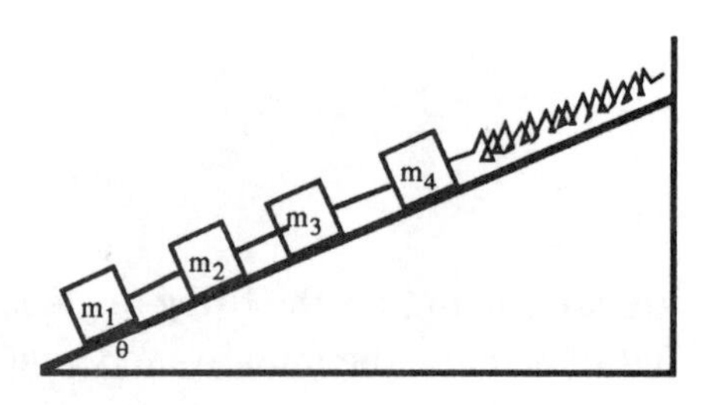

(a) Draw a free-body diagram showing all the forces acting on block 2 (of mass m_2). Label your forces.

(b) At the instant the blocks momentarily come to rest, what is block 1's acceleration? Express your answer in terms of k, s, g, and the blocks' masses. Does this acceleration equal zero?

(c) Look at your free-body diagram from part (a). Solve for each of the forces you drew, in terms of k, s, g, and the blocks' masses.

REVIEW ANSWER 1-2

(a) Block 2 "feels" a downward pull due to gravity; a "normal" force exerted by the surface of the ramp; a pull (tension) due to the rope that is also attached to block 3; and a pull (tension) due to the rope that is also attached to block 1. Although unnecessary here, for future reference I will break up gravity into its components parallel and perpendicular to the ramp. As always, the normal force points perpendicular to the surface exerting the force.

Importantly, the spring does *not* pull directly on block 2, and should not be included in the free-body diagram. The spring "acts" on block 2 *indirectly*. Here is how: By pulling on block 4, the spring increases the tension in the rope between blocks 3 and 4. Consequently, the tension in the rope between blocks 2 and 3 also gets increased. But only forces that act *directly* on the object get included in force diagrams. Your block 4 force diagram will take into account the spring's force.

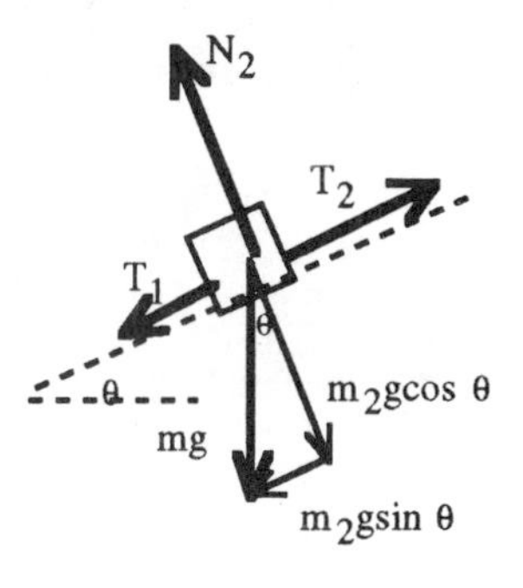

If you labeled the tensions differently, that is o.k., provided you clearly understood what is going on. Notice that the two tensions are not equal.

(b) A common mistake is to say that the acceleration is $a = 0$. (About half the students in a Berkeley class made this error.) But acceleration is not the same as velocity. Just because the velocity is 0 does not mean the acceleration is 0. *Acceleration is the rate at which velocity changes.*

Here, although the velocity is *momentarily* 0, the velocity is changing. A moment before the blocks stop, they are moving down the ramp. And a moment after the blocks come to rest, they move ("bounce") back up the ramp. The blocks begin with a down-the-ramp velocity, and end with an up-the-ramp velocity. So, the velocity is not constant; it is changing. Therefore, the acceleration is nonzero.

If this is not clear, imagine throwing a ball straight up. At its peak, the ball momentarily comes to rest: $v = 0$. But the acceleration is not 0; it is $g = 9.8\ \text{m/s}^2$, downward. That is because the ball's velocity is changing. Specifically, at its peak, the ball's velocity is changing from upward to downward. For this reason, the acceleration is nonzero. The same conclusion applies to the blocks on the ramp.

So, we will have to find the acceleration using the standard multi-block strategy.

(1) Draw a free-body diagram for each block, breaking up the force vectors into their components parallel and perpendicular to the direction of motion.

(2) Write down Newton's 2nd law (along the direction of motion) for each block separately. Since the blocks "move together" (because they are attached by ropes), they all share the same acceleration, a.

(3) Solve your equations for a, if that is what you need.

(4) Substitute that acceleration into a kinematic equation or into one of your Newton's 2nd law equations, to find what you want. You will do this in part (c).

Now I will implement this strategy:

Step 1: Free-body diagrams.

The tension is the same at both ends of a rope. For instance, the force with which the lowest rope pulls block 1 up the ramp equals the force with which that rope tries to pull block 2 down the ramp. In part (a), I called that tension force T_1. Since we already drew a force diagram for block 2, I will not repeat it here.

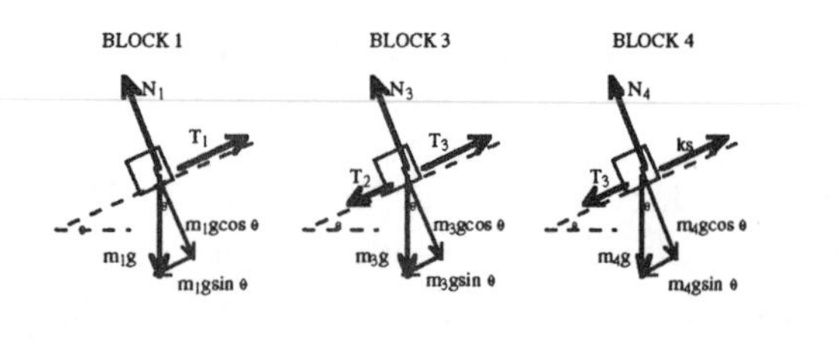

The spring pulls block 4 up the ramp. You might be tempted to throw in a negative sign, or to reverse the direction of that force. But the minus sign in $F_s = -kx$ tells you that the spring pulls in the opposite direction to which the spring is stretched. So, that minus sign is already "built into" the above force diagram; it is encoded in the direction of the *ks* arrow.

Step 2: Newton's 2nd law for each block separately.

I will choose up-the-ramp as the positive x-direction. (Down-the-ramp works equally well.) Since I care only about the direction of motion, I will not bother writing y-force equations. But if there were friction, I would need to solve for the normal forces, so that I could use $f = \mu N$.

$$\text{Block 1:} \quad \sum F_x = T_1 - m_1 g \sin\theta = m_1 a \tag{1}$$

$$\text{Block 2:} \quad \sum F_x = T_2 - T_1 - m_2 g \sin\theta = m_2 a \tag{2}$$

$$\text{Block 3:} \quad \sum F_x = T_3 - T_2 - m_3 g \sin\theta = m_3 a \tag{3}$$

$$\text{Block 4:} \quad \sum F_x = ks - T_3 - m_4 g \sin\theta = m_4 a. \tag{4}$$

We now have four equations in the four unknowns (T_1, T_2, T_3, and a). Only algebra remains.

Step 3: Algebra to solve for acceleration.

To complete the algebra quickly, add the 4 equations. The tensions cancel, leaving us with

$$ks - m_1 g \sin\theta - m_2 g \sin\theta - m_3 g \sin\theta - m_4 g \sin\theta = m_1 a + m_2 a + m_3 a + m_4 a.$$

Solving for a yields $a = \frac{ks - Mg\sin\theta}{M}$, where M is shorthand for $m_1 + m_2 + m_3 + m_4$.

You could also have obtained this acceleration by considering the 4 blocks as a single system of mass M acted upon by two external forces. The external forces are gravity and the spring. (All the tensions are internal forces, i.e., parts of the system pulling on other parts of the system.) Gravity pulls the system-as-a-whole down the ramp with force $Mg\sin\theta$. The spring pulls the system up the ramp with force *ks*. So the net force on the system-as-a-whole is $\sum F_{\text{external}} = ks - Mg\sin\theta$. That force on the system-as-a-whole accelerates the system-as-a-whole, which has mass $M = m_1 + m_2 + m_3 + m_4$. So according to Newton's 2nd law for the system-as-a-whole, $ks - Mg\sin\theta = Ma$. Solving for a gives the same answer obtained above.

By the way, if you got $a = \frac{ks - Mg\sin\theta}{m_2}$, you "mixed up" the two techniques for solving the problem. If you apply Newton's 2nd law to the system-as-a-whole to get $\sum F = ks - Mg\sin\theta$, then you must use the mass of the system-as-a-whole when you write $\sum F = Ma$. To avoid this kind of error, it is safer to write Newton's 2nd law for each block separately, and then add the equations. You can use the system-as-a-whole reasoning to double check your answer.

(c) The forces acting on block 2 are gravity $= m_2 g$, the normal force N_2, and the tensions T_1 and T_2. To solve this part correctly, you have no choice but to consider the blocks separately, using some of the 4 equations listed in part (b). This is yet another reason to use my multi-block strategy instead of considering the system-as-a-whole.

To obtain T_1, solve Eq. (1) to get

$$\begin{aligned} T_1 &= m_1(g\sin\theta + a) \\ &= m_1\left(g\sin\theta + \frac{ks - Mg\sin\theta}{M}\right) \end{aligned}$$

Substitute that expression for T_1 into Eq. (2), and solve for T_2:

$$\begin{aligned} T_2 &= T_1 + m_2(g\sin\theta + a) \\ &= m_1(g\sin\theta + a) + m_2(g\sin\theta + a) \\ &= (m_1 + m_2)\left(g\sin\theta + \frac{ks - Mg\sin\theta}{M}\right). \end{aligned}$$

So much for the tensions. We know gravity. So only the normal force remains. To find N_2, consider the forces perpendicular to the surface of the ramp. Since block 2 moves only in the x-direction (along the ramp), its acceleration in the y-direction is zero:

$$\begin{aligned} \sum F_y &= ma_y \\ N - m_2 g\cos\theta &= 0, \end{aligned}$$

and hence $N = m_2 g\cos\theta$.

REVIEW QUESTION 1-3

In outer space, an electron moves one-dimensionally under the influence of an electric field. Observations reveal that the electron's velocity is given, in meters per second, by the formula. $v = b - ct^2$, where b and c are constants, and t is the time in seconds. The electron's position at time $t = 0$ was $x_0 = 0$.

(a) Does the electric field exert a constant force on the electron? Justify your answer.

(b) What is the electron's position at time $t = 2$ seconds? Let X denote your answer. Solve for X in terms of b and c.

(c) Let us forget about the electron, and instead consider a proton, which also travels one-dimensionally in outer space. The proton is described by the following position versus time graph.

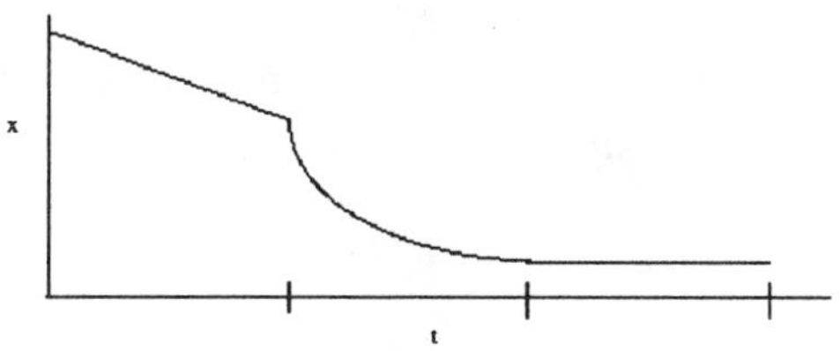

Make a rough sketch of the proton's acceleration versus time.

REVIEW ANSWER 1-3

(a) According to Newton's 2nd law, $F_{\text{net}} = ma$, if the force on the electron is constant, then the electron's acceleration is constant, and vice versa. So, we need to see if the acceleration is constant.

Acceleration is the rate at which the velocity changes. In calculus language, this translates to

$a = \frac{dv}{dt}$. Since we are given v as a function of t, we can differentiate to find $a = \frac{dv}{dt} = \frac{d(b-ct^2)}{dt} = -2ct$, since the derivative of a constant is zero. We now see that the acceleration is not constant. Instead, it changes with time. Therefore, the force acting on the electron must also change with time.

(b) You might be tempted to use a kinematic formula such as $x = x_0 + v_0t + \frac{1}{2}at^2$. But this equation, as well as $v = v_0 + at$ and $v^2 = v_0^2 + 2a\Delta x$, applies *only* when the acceleration is constant. We just saw in part (a) that the acceleration changes with time.

So, we must go back to the basic definition relating velocity to position. Velocity is the rate at which position changes. In calculus language, $v = \frac{dx}{dt}$. If we were given the position, we could differentiate to get v. But here, we are given the velocity, and we want the position. Therefore, we must do the "opposite" of differentiation, namely integration.

Multiply both sides of $v = \frac{dx}{dt}$ by dt to get $vdt = dx$. We could do a definite integral, letting time range from $t = 0$ to $t = 2$. But let me just take an indefinite integral, to obtain x as a general function of time:

$$\begin{aligned} x &= \int dx = \int vdt \\ &= \int (b - ct^2)dt \\ &= bt - \frac{ct^3}{3} + \text{(integration constant)}. \end{aligned}$$

The integration constant is the initial position, x_0. To see this, notice that at time $t = 0$, $x =$ (integration constant). Here, $x_0 = 0$.

So, the electron's position (in meters) at time $t = 2$ s is $X = b(2) - \frac{c(2)^3}{3} = 2b - \frac{8}{3}c$.

(c) We are given the position vs. time, and we want acceleration vs. time. Instead of trying to graph acceleration "directly" from the given graph, it is easier to solve in two steps. Since velocity is the rate at which position changes, the v vs. t graph is the slope of the x vs. t graph. So, we can sketch v vs. t. And since acceleration is the rate at which velocity changes, we can take the slope of the v vs. t graph to find a vs. t. These two graphical steps correspond to taking two derivatives with respect to t.

First, I will sketch v vs. t. During the first time interval, the position steadily decreases. The rate at which position decreases–i.e., the slope of the x vs. t graph–is constant. So, the velocity is constant over that time interval. And since the position is decreasing (i.e., going backwards), the velocity is negative.

During the second time interval, the position continues to decrease, but at a slower and slower rate. Physically, this means the proton continues to move backwards; but its backwards motion gets slower and slower. It is like you are cruising in reverse, and you hit the brakes. (Mathematically, the slope of the x vs. t graph is negative, but gets smaller and smaller.) Therefore, the velocity in that region is negative, but gets smaller and smaller in magnitude.

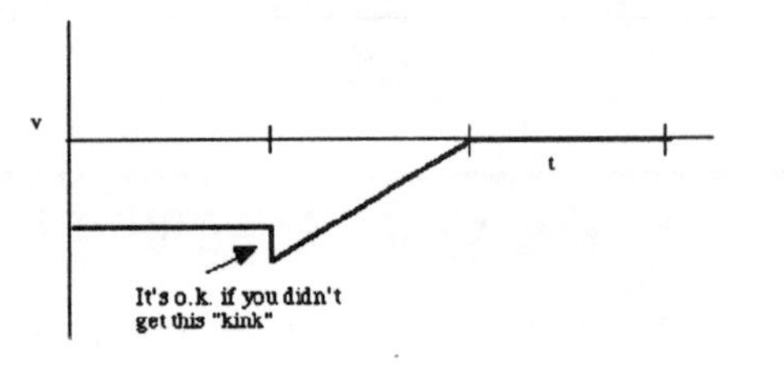

Finally, during the third time interval, the proton's position does not change. The proton sits still ($v = 0$). Mathematically, the slope of the x vs. t graph is zero in that interval.

Putting all this together, I get the following velocity vs. time graph.

Now, to graph the acceleration versus time, we just need to take the slope of the v vs. t graph. During the first time interval, the velocity is constant; the car neither speeds up nor slows down. So, $a = 0$. You can see this mathematically, too, because the slope of the v vs. t graph is 0. (At the kink, the velocity suddenly gets more negative, corresponding to a quick "burst" of negative acceleration. Do not stress about this subtlety.)

During the second time interval, the velocity gradually "increases," by which I mean the velocity gets less and less negative. The slope of the v vs. t graph is positive in that region, and is approximately constant. So, the acceleration in that region is constant (or nearly constant), and *positive*. Intuitively, something speeding up in the backwards direction has negative acceleration, and hence, something slowing down in the backwards direction must have positive acceleration.

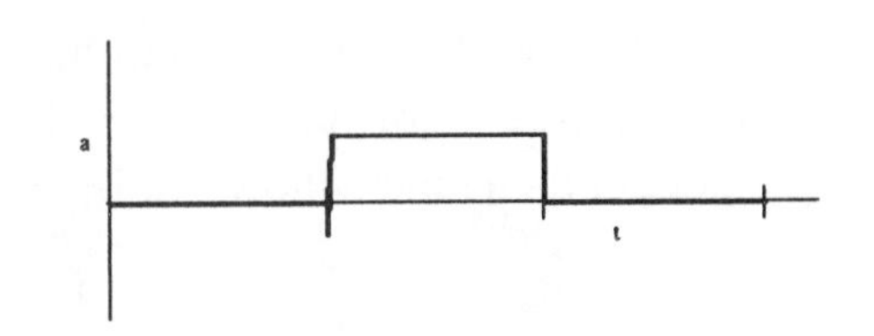

Finally, in the third time interval, the velocity is unchanging (slope = 0), and hence the acceleration is $a = 0$. Putting all this together gives the following graph.

REVIEW QUESTION 1-4

A charged particle is moving leftward at speed $V = 20$ m/s. The particle enters a box containing an electric field. The electric field exerts a *rightward*-directed force on the particle, thereby slowing the particle down. The function of the box is to slow the particle down such that, by the time the particle reaches the back of the box, the particle is motionless. In other words, the particle is supposed to *barely* reach the back of the box.

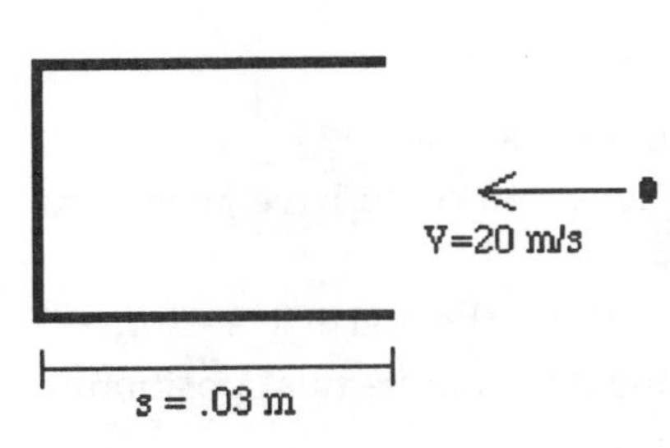

In this problem, neglect gravity. The particle's mass is $m = 1.0 \times 10^{-6}$ kg. The box is $s = .030$ meters "wide," as shown in the diagram.

By turning a knob on the box, we can strengthen or weaken the electric field, causing a bigger or smaller force to act on the particle.

(a) What force must the electric field exert on the particle in order for the box to function properly? (We want the particle to barely reach the back of the box.)

(b) When the box functions properly, the particle slows down, eventually becoming motionless at the back of the box; but then it accelerates rightward, because the force is not turned off. How fast will the particle be moving when it leaves the box?

(c) Make a rough sketch of the particle's position versus time, and also its velocity vs. time, clearly indicating the points on your graph corresponding to when the particle enters the box, reaches the back of the box, and leaves the box. Pick the open end of the box as $x = 0$, and the leftward direction as positive.

REVIEW ANSWER 1-4

(a) We want the force exerted on the particle to be such that the particle gradually slows down, coming (momentarily) to rest just as it reaches the back of the box. If the force is too weak, the particle will not slow down quickly enough, and will therefore crash into the back of the box. And if the force is too strong, the particle slows down so quickly that it "turns around" before reaching the back of the box.

This qualitative description suggests a strategy that is the reverse of our usual force problem strategy. Normally, we would know the forces on the particle, and we would want to find how far the particle travels before coming to rest, or something like that. To solve that standard problem, we would use $\sum F = ma$ to calculate the acceleration, and then we would substitute that acceleration into a kinematic formula. But here, we know how far the particle travels before coming to rest, and we are finding the force. So, we can work backwards. First, use a kinematic equation to calculate the particle's acceleration. Then, use $\sum F = ma$ to obtain the force.

The particle slows down from $v_0 = V = 20$ m/s to $v = 0$ over a distance of $\Delta x = s = .030$ meters. Given this information, the easiest way to find the (constant) acceleration is the time-independent kinematic formula, $v^2 = v_0^2 + 2a\Delta x$. Solve for the acceleration to get

$$a = \frac{v^2 - v_0^2}{2\Delta x} = \frac{0 - V^2}{2s} = \frac{0 - (20 \text{ m/s})^2}{2(.030 \text{ m})} = -6700 \text{ m/s}^2.$$

The minus sign indicates an acceleration that is backwards compared to the initial direction of motion (leftward), which I implicitly chose as positive. In other words, the particle slows down.

Since the electric force is the only force, it is the net force. Now that we have the acceleration, we can solve for F using Newton's 2nd law:

$$F_{\text{net}} = ma = m\left(\frac{-V^2}{2s}\right) = (1.0 \times 10^{-6} \text{ kg})(-6700 \text{ m/s}^2) = -.0067 \text{ N}.$$

Again, the minus sign indicates a "backwards" (rightward) push on the particle.

(b) First I will calculate this answer "the long way." Then I will show how you could have jumped to the answer with no calculations.

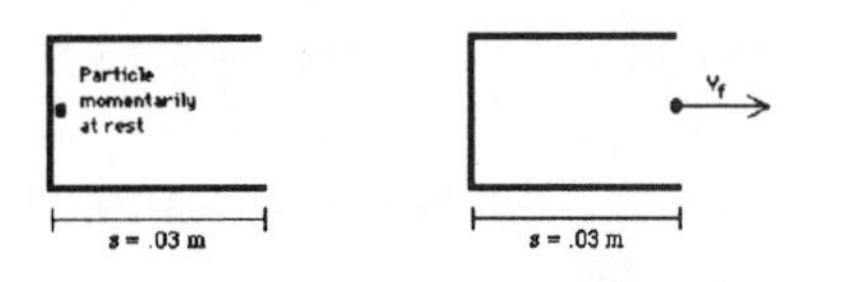

In this part of the problem, the particle's "initial" position is at the back of the box, and its final position is when it emerges from the box.

I will continue to choose leftward as positive. So, the particle's rightward displacement is negative: $\Delta x = -s$. The rightward acceleration is also negative. Since the force on the particle is the same as it was in part (a), and since the particle's mass has not changed, the acceleration is the same as above: $a = -\frac{V^2}{2s}$. Given all this information, we are trying to solve for v_f, the particle's velocity as it emerges from the box. So, we can re-use the kinematic formula from part (a). But remember, in this part of the problem, the "initial" speed is $v_0 = 0$:

$$\begin{aligned} v_f^2 &= v_0^2 + 2a\Delta x \\ &= 0 + 2\left(\frac{-V^2}{2s}\right)(-s) \\ &= V^2. \end{aligned}$$

So, the particle emerges from the box at rightward speed $V = 20$ m/s, the same leftward speed with which it entered the box.

Coincidence? Not! Picture yourself throwing a ball straight up. As you have seen, by the time it comes back down and reaches your hand, it is traveling at the same speed you threw it (but in the opposite direction, of course). All the speed it loses while going up, it regains while falling back down. Similarly, when you "throw" the particle into the box, it eventually "bounces back" at the same speed you "threw" it, but in the opposite direction. In both cases, the "thrown" object experiences a constant acceleration that slows it down and then speeds it up in the opposite direction. By seeing this analogy, you could have jumped right to the answer. Later in the course, you will learn another way to address problems such as this, using "energy conservation."

(c) You could draw the x vs. t graph and then use it to get v vs. t; or vice versa. Sketch whichever one you find easier first.

Before it enters the box, the particle moves in the positive direction at constant speed. So, its *velocity* is a horizontal (constant) straight line, while its *position* is steadily increasing. Therefore, the position graph is an upward-sloped line.

Things get messier when the particle enters the box and moves toward the back wall. The particle is decelerating, i.e., slowing down at a steady rate. So, the *velocity* decreases, eventually reaching $v = 0$ when the particle reaches the back. Does this mean the position graph goes "down"? No. Until reaching the back of the box, the particle continues to move in the positive direction. So, x is increasing. But since the particle slows down, the position increases *at a slower and slower rate.* This is indicated on the x vs. t graph by a *decreasing slope*. At the back of the box, where the particle's position momentarily stops changing, the slope of x vs. t hits zero.

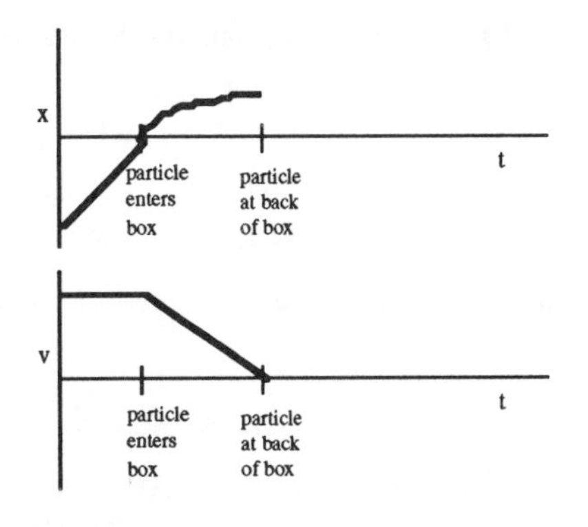

Let me draw what we have figured out so far.

Remember, I happened to choose $x = 0$ as the spot where the particle enters the box. So, before it entered the box, the particle's position must have been negative.

After the particle reaches the back, it starts to speed up in the rightward direction. So, its velocity gets more and more *negative* (rightward), at a steady rate. In fact, since the acceleration is constant, the slope of the v vs. t graph stays the same.

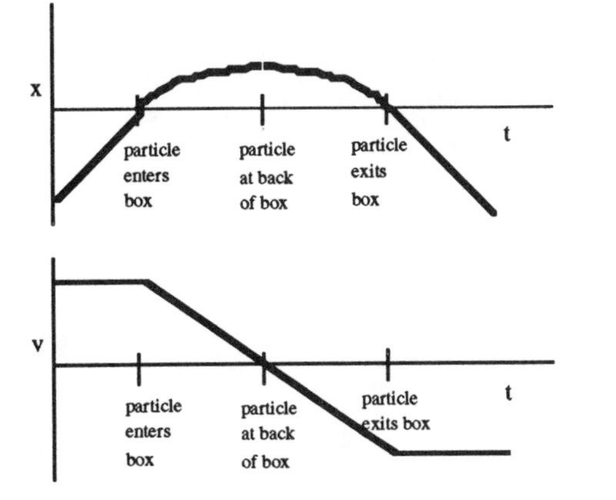

Since the particle now goes backwards, its position decreases; the x vs. t graph "comes down." And since the particle speeds up, its position decreases at a faster and faster rate. So, the slope of the x vs. t graph is negative, and gets bigger and bigger.

Finally, after the particle leaves the box (at $x = 0$), its velocity stays constant; the particle no longer speeds up or slows down. It just floats along. So, its position steadily becomes more and more negative, while its velocity graph becomes a horizontal (constant) line.

Notice the symmetry in the graphs. Also, compare these graphs to the analogous graphs describing a ball thrown straight upward. You will see that the ball-toss graphs are equivalent to the "middle part" of the above graphs, starting when the particle enters the box and ending when the particle leaves the box. That is because, in both cases, the object undergoes constant acceleration, first slowing down, then speeding up in the opposite direction.

REVIEW QUESTION 1-5

Back in the olden days, ice cubes were produced by crashing together large blocks of ice. (If you believe that, I have got a bridge to sell you.) Suppose ice block 1, of mass m_1, is released from rest from the top of a ramp of angle θ and length D. The block slides down the ramp, eventually crashing into ice block 2, which is resting next to the bottom of the ramp. This ice is not completely frictionless; the coefficient of kinetic friction between an ice block and the ramp is μ.

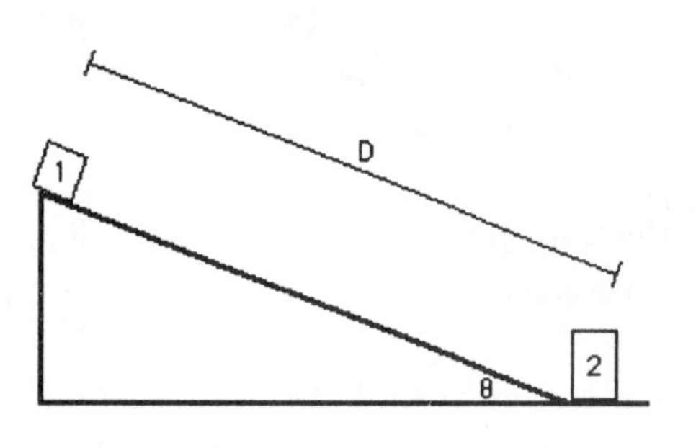

(a) How fast will ice block 1 be moving immediately before it crashes into ice block 2? Express your answer in terms of m_1, θ, D, μ, and g.

(b) *(Quite hard)* Unfortunately, the velocity you obtained in part (a) is too small to break the ice blocks into little cubes. So, the ice-breaking procedure is modified. Block 2 is placed on the bottom of the ramp. At the moment ice block 1 is released from the top of the ramp, a worker starts pushing block 2 up the ramp. Starting from rest, the worker exerts a force F_p on the block, directed parallel to the surface of the ramp. This force is big enough to make block 2 accelerate up the ramp. (The worker "runs" behind ice block 2, in order to continue applying this force.) Ice block 2 has mass m_2.

How fast is block 1 moving relative to block 2 when they crash? Express your answer in terms of $m_1, m_2, F_p, \theta, D, \mu$, and g.

REVIEW ANSWER 1-5

(a) The block starts from rest, and we are trying to calculate its velocity at the bottom of the ramp. Therefore, we need to know the rate at which it speeds up, i.e., the acceleration. To calculate the acceleration, I will invoke the usual force problem strategy:

Step 1: Force diagram.

As the block slides down, the ramp exerts a normal force, which is perpendicular to the surface. In this problem, friction resists the block's motion down the ramp, and therefore points "backwards." A component of gravity pulls the block down the ramp.

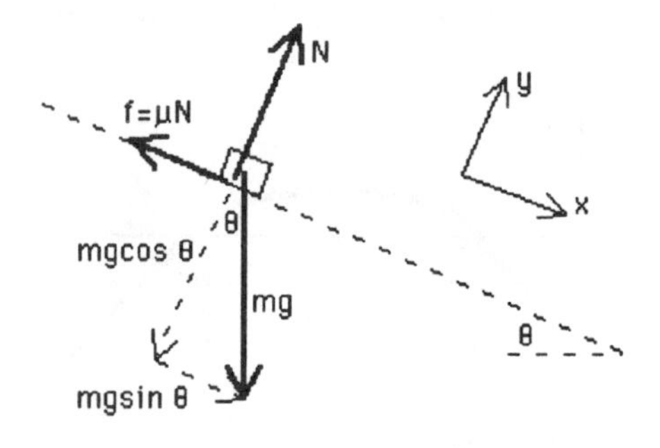

I have picked down-the-ramp as my positive x-direction.

Step 2: Newton's 2nd law, $\sum F = ma$.

We want a_x, the block's acceleration along the ramp. So, start with

$$\sum F_x = ma_x$$
$$mg\sin\theta - \mu N = ma_x. \quad (1)$$

Before we can solve Eq. (1) for a_x, we need to find N, by considering the y-directed forces. Because the block moves only in the x-direction (along the ramp), $a_y = 0$:

$$\sum F_y = ma_y$$
$$N - mg\cos\theta = 0.$$

Solve for N to get $N = mg\cos\theta$, and substitute that expression into Eq. (1). Then solve Eq. (1) for acceleration to get $a_x = g(\sin\theta - \mu\cos\theta)$.

Step 3: Kinematics

Now that we have the block's acceleration along the ramp, we can use kinematics to find its speed v at the bottom. Since we know the block travels distance $\Delta x = D$, but we do not know the "slide time" t, the most efficient formula to use is

$$v_x^2 = v_{0x}^2 + 2a_x\Delta x$$
$$= 0 + 2[g(\sin\theta - \mu\cos\theta)]D,$$

and hence $v = \sqrt{2g(\sin\theta - \mu\cos\theta)D}$.

(b) This one is tricky. You have to see that it is a meeting problem. We are looking for the blocks' speeds when they crash, i.e., *when they are at the same position*. So, we can invoke the usual "meeting strategy" of setting $x_1(t) = x_2(t)$ and solving for the time at which the blocks crash. Then we can use that meeting time to figure out the "meeting velocities" of blocks 1 and 2. Once we obtain those separate velocities, we can figure out the relative velocity.

Meeting strategy step 1: Express the two blocks' positions as functions of time.

For block 1, this is not difficult, because in part (a) we already found that block's acceleration: $a = g(\sin\theta - \mu\cos\theta)$. So, its position as a function of time is

$$\text{(BLOCK 1)}\quad x_1 = x_0 + v_{0x}t + \frac{1}{2}a_x t^2$$
$$= 0 + 0 + \frac{1}{2}g(\sin\theta - \mu\cos\theta)t^2.$$

To write a similar expression for block 2, we need to know that block's acceleration while it is getting pushed up the ramp. You can find this acceleration by drawing a force diagram and using Newton's 2nd law. Since friction resists the block's motion up the ramp, it points down the ramp.

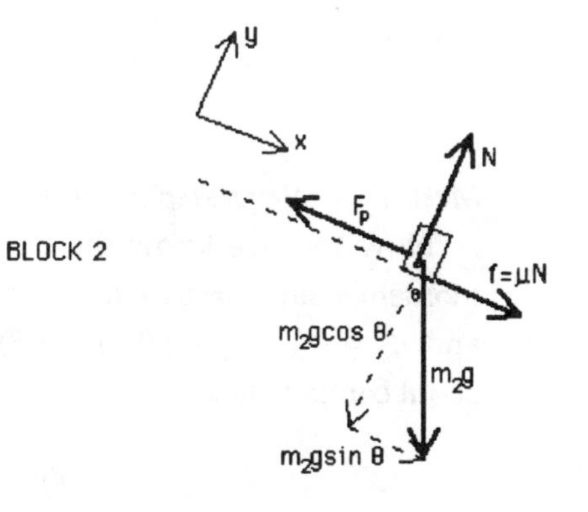

By the same reasoning used in part (a), the normal force on block 2 is $N = m_2 g\cos\theta$. To find block 2's acceleration along the ramp, I will write $\sum F_x = ma_x$, *remembering that I have chosen the down-the-ramp direction as positive. I must use the same positive direction for both blocks.*

$$\text{(BLOCK 2)} \qquad \sum F_x = m_2 a_x$$
$$m_2 g \sin\theta + \mu N - F_p = m_2 a_x$$
$$m_2 g \sin\theta + \mu m_2 g \cos\theta = m_2 a_x.$$

Solve for a_x to get

$$\text{(BLOCK 2)} \quad a_x = -\left[\frac{F_p}{m_2} - g(\sin\theta + \mu\cos\theta)\right].$$

This acceleration is negative, because it points up the ramp, while the down-the-ramp direction is positive.

In any case, now that we know the acceleration of block 2, we can express its position as a function of time. Remember, I have chosen the top of the ramp as $x = 0$, and the bottom as $x = D$. So, the initial position of block 2 is $x_0 = D$:

$$\text{(BLOCK 2)} \quad x_2 = x_0 + v_{0x}t + \frac{1}{2}a_x t^2$$
$$= D + 0 - \frac{1}{2}\left[\frac{F_p}{m_2} - g(\sin\theta + \mu\cos\theta)\right]t^2.$$

At this point, I have written the positions of both blocks as functions of time. So, I can proceed to...

Meeting strategy step 2: Set $x_1 = x_2$, and solve for the "meeting time" t.

$$x_1 = x_2$$
$$\frac{1}{2}g(\sin\theta + \mu\cos\theta)t^2 = D - \frac{1}{2}\left[\frac{F_p}{m_2} - g(\sin\theta + \mu\cos\theta)\right]t^2.$$

To complete the algebra efficiently, expand out all the terms and notice that both sides of the equation contain $\frac{1}{2}g(\sin\theta)t^2$ terms. Cancel those terms, and then move all the t terms to the left-hand side. This yields $-g\mu(\cos\theta)t^2 + \frac{F_p}{2m_2}t^2 = D$. Isolate the t to get

$$t = \sqrt{\frac{D}{\frac{F_p}{2m_2} - g\mu\cos\theta}}.$$

Meeting strategy step 3: Kinematics

Now that we know the time at which the blocks crash, we can calculate their speeds at that moment, using $v = v_0 + at$. From step 2, we know the blocks' accelerations: $a_1 = g(\sin\theta - \mu\cos\theta)$, and $a_2 = -\left[\frac{F_p}{m_2} - g(\sin\theta - \mu\cos\theta)\right]$. Since everything happens in the x-direction, I will not worry about components.

$$v_1 = v_0 + a_1 t$$
$$= 0 + g(\sin\theta - \mu\cos\theta)\sqrt{\frac{D}{\frac{F_p}{2m_2} - g\mu\cos\theta}}.$$

$$v_2 = v_0 + a_2 t$$
$$= 0 - \left[\frac{F_p}{m_2} - g(\sin\theta - \mu\cos\theta)\right]\sqrt{\frac{D}{\frac{F_p}{2m_2} - g\mu\cos\theta}}.$$

Notice that v_2 is negative, because block 2 slides *up* the ramp.

We are done with the meeting strategy. Now that we know the blocks' speeds at impact, we can calculate their relative speed. To figure out the appropriate formula, let us play around with simple numbers. Pretend that block 1 moves down the ramp at $v_1 = 10$ mph, while block 2 slides with speed $v_2 = -5$ mph, where the minus sign means "up the ramp." Then, intuitively, the blocks approach each other at relative speed 15 mph. To obtain that answer, I subtracted v_2 from v_1, since $10 - (-5) = 15$. In general, $v_{1 \text{ relative to } 2} = v_1 - v_2$. So, in this problem,

$$v_{\text{rel}} = v_1 - v_2$$
$$= g(\sin\theta - \mu\cos\theta)\sqrt{\frac{D}{\frac{F_p}{2m_2} - g\mu\cos\theta}} - \left\{-\left[\frac{F_p}{m_2} - g(\sin\theta + \mu\cos\theta)\right]\sqrt{\frac{D}{\frac{F_p}{2m_2} - g\mu\cos\theta}}\right\}$$
$$= \left(\frac{F_p}{m_2} - 2g\mu\cos\theta\right)\sqrt{\frac{D}{\frac{F_p}{2m_2} - g\mu\cos\theta}}.$$

Extra credit question for masochistic students: If the ramp is frictionless, then the crash speed v_{rel} does not depend on g. Therefore, the crash speed would be the same if we repeated this experiment on the moon. Why is this, intuitively?

REVIEW QUESTION 1-6

Crouched at the edge of a cliff, a monkey fires a gun at a hunter. The hunter is standing at the edge of another cliff 30 meters away. The two cliffs are the same height. The monkey, who has never studied physics, fires the bullet horizontally at 120 m/s.

In this problem, approximate g as 10 m/s^2. Assume the bullet was fired from cliff level. In other words, the gun is resting on the "plateau" at the edge of the cliff.

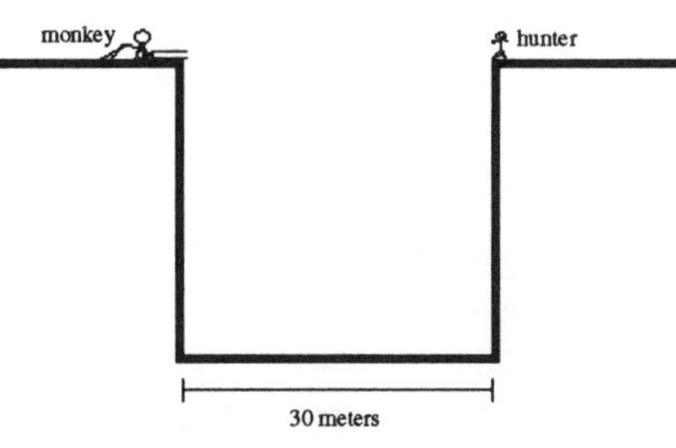

(a) How far below the hunter's feet will the bullet strike the face of the cliff directly below the hunter?

(b) What is the bullet's speed (i.e., the magnitude of its velocity) immediately before it reaches the face of the cliff directly below the hunter?

(c) Suppose the bullet is big enough to cast a shadow. The Sun is directly overhead, and hence the bullet's shadow appears on the valley floor under the bullet. What is the shadow's speed when it is half way across the valley?

REVIEW ANSWER 1-6

To organize your thoughts, sketch the bullet's trajectory.

I have let d denote the distance below the hunter's feet that the bullet strikes the face of the cliff. Notice that d is the vertical distance through which the bullet "falls."

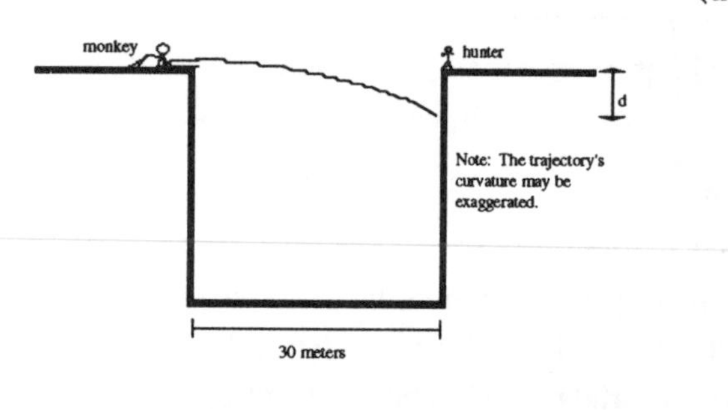

(a) Since d is a vertical distance, we can try to use the constant-acceleration vertical displacement formula, $y = y_0 + v_{0y}t + \frac{1}{2}a_y t^2$, with $y = -d$ or $+d$, depending on which direction you choose as positive. But we cannot immediately solve this equation for d, because we do not know the "flight time" t. Fortunately, we can obtain t by considering the horizontal aspect of the bullet's trip. So, as usual in projectile problems, we must separately consider the horizontal and vertical motion.

I will start with the horizontal. We are seeking the flight time t, given the bullet's horizontal displacement $x - x_0 = 30$ m. Since the monkey fires the bullet horizontally, the bullet's entire initial velocity is x-directed: $v_{0x} = v_0 = 120$ m/s. Also, the horizontal acceleration is $a_x = 0$. (Remember, once in the air, the bullet experiences no horizontal forces, and hence its *horizontal* velocity does not speed up or slow down.) So, we have

$$\begin{aligned} x &= x_0 + v_{0x}t + \frac{1}{2}a_x t^2, \\ 30\text{ m} &= 0 + (120\text{ m/s})t + 0, \end{aligned} \tag{1}$$

and hence $t = 0.25$ s. Now put that time into the vertical displacement formula. I will choose upward as positive. Therefore, if I call the top of the cliff $y_0 = 0$, then the bullet ends up *below* that, at $y = -d$. As noted above, the bullet has no initial *vertical* velocity: $v_{0y} = 0$, because it is fired horizontally. So,

$$\begin{aligned} y &= y_0 + v_{0y}t + \frac{1}{2}a_y t^2, \\ -d &= 0 + 0 - \frac{1}{2}gt^2 \\ &= -\frac{1}{2}(10\text{ m/s}^2)(0.25\text{ s})^2 \\ &= -0.31\text{ meters.} \end{aligned} \tag{2}$$

The bullet hits the cliff only $d = 0.31$ meters below the hunter's feet. That is about a foot.

(b) The most common mistake is to write a formula that "illegally" combines x-stuff with y-stuff. For instance, some students try to use $\mathbf{v} = \mathbf{v}_0 + \mathbf{a}t$ by setting $v = 120 - 10t$. But that cannot work, because 120 m/s is a *horizontal* velocity, while 10 m/s^2 is a *vertical* acceleration. You cannot just add horizontal and vertical stuff together. To see why, suppose a kid walks 3 meters east, and then 4 meters north. How far is she from her starting point? If you could just add x-stuff to y-stuff, you would say 7 meters. But instead, you must draw a triangle and use Pythagoras' theorem.

The same Pythagorean reasoning applies here. We must *separately* calculate the bullet's final horizontal velocity and final vertical velocity. From the sketch of the bullet's trajectory, we see that the final velocity will be rightward and downward.

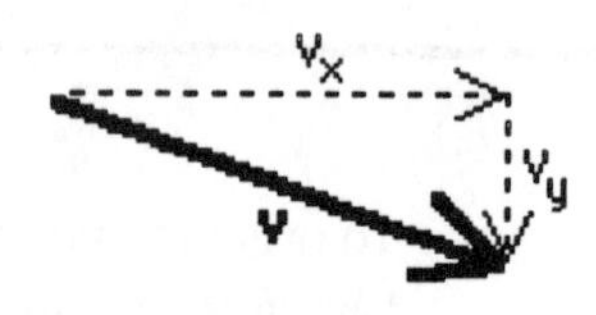

After finding v_x and v_y, we can "Pythagoras" them together to obtain the total final speed.

Let us start by finding v_x. Use the horizontal component of $\mathbf{v} = \mathbf{v}_0 + \mathbf{a}t$, namely

$$\begin{aligned} v_x &= v_{0x} + a_x t \\ &= 120 \text{ m/s} + 0, \end{aligned}$$

since a projectile has no horizontal acceleration. A projectile "keeps" all its horizontal velocity.

By contrast, the vertical velocity changes, because gravity pulls the bullet downward:

$$\begin{aligned} v_y &= v_{0y} + a_x t \\ &= 0 + (-10 \text{ m/s}^2)(0.25 \text{ s}) \\ &= -2.5 \text{m/s}. \end{aligned}$$

Here, I used the flight time calculated in part (a). The minus sign indicates that the bullet's final vertical velocity is downward. By the way, you equally well could have found v_y by substituting the "d" obtained in part (a) into $v_y^2 = v_{0y}^2 + 2a_y(y - y_0)$.

Now that we know the final horizontal and vertical velocities, we can find the total final speed: $v = \sqrt{v_x^2 + v_y^2} = \sqrt{(120 \text{ m/s})^2 + (-2.5 \text{ m/s})^2} = 120.03 \text{ m/s}$. The bullet has barely sped up.

(c) To gain insight into this problem, pretend the bullet was dropped straight down. Would its shadow move? No, because the shadow sits directly under the bullet. Therefore, if the bullet falls straight down, the point directly under the bullet does not move, and hence the shadow does not budge. Only if the bullet moves sideways does the shadow move. This goes to show that the shadow's motion "mirrors" the horizontal component of the bullet's velocity, and does not depend at all on the bullet's vertical velocity. So, this problem really asks you for v_x, the bullet's horizontal speed when it is half way across the valley.

As we just saw in part (b), however, the bullet's horizontal speed is constant, at 120 m/s, because no horizontal forces act on the bullet to speed up or slow down its sideways motion. Therefore, the shadow's speed at any point, including the half way point, is 120 m/s.

REVIEW QUESTION 1-7

Hard, but solvable if you visualize.

A spy needs to fire a special bullet across some train tracks. (The bullet contains a microfilm.) Unfortunately, just when she wants to fire her gun, a train comes by, moving leftward at constant speed $v_{\text{train}} = 20.0 \text{ m/s}$. Since time is of the essence, the spy decides to fire the bullet through a tiny gap between the first two cars of the train. The gap is barely wider than the bullet. And if the bullet so much as nicks one of the train cars, the microfilm could get destroyed.

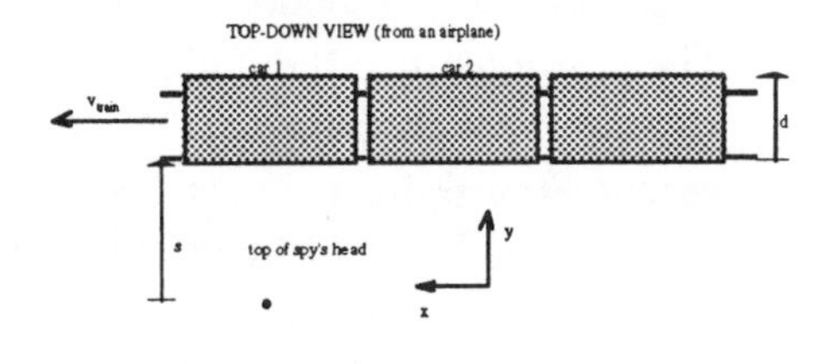

The spy is a distance $s = 20.0 \text{ m}$ from the train. The train cars are $d = 5.00$ meters wide. For convenience, let us say the x-direction is leftward, and the spy is at $x = 0$. The spy's gun fires the bullet at speed $v_{\text{bullet}} = 100 \text{ m/s}$. In order to ensure that the bullet passes cleanly through the gap,

(a) In what direction should the spy fire the bullet, and
(b) What must be the x-position of the gap at the moment the bullet is fired?
Hint: You might not use all the information I have provided.

REVIEW ANSWER 1-7

(a) This is one of those problems where you need to have a crucial physical insight, or else you cannot make any progress. You have to realize that *relative motion* is the key.

Think about the bullet as it passes through the gap. It had better not move forward (leftward) or backward (rightward) *relative to the train*, or else it will scrape the back of car 1 or the front of car 2. So, the bullet must be fired at an angle such that its x-component of velocity *relative to the train* is zero. In other words, the leftward component of the bullet's velocity must equal the train's velocity, so that the bullet neither "gains on" nor "loses ground to" the train:

$$v_{x\text{ bullet}} = v_{\text{train}}. \tag{1}$$

That is the physical insight we need. The spy should not fire the bullet in the y-direction, because if she does, the train moves forward relative to the bullet, and therefore the bullet in the gap would end up scraping the front of car 2. (Put another way, the front of car 2 would run into the sideways-moving bullet.)

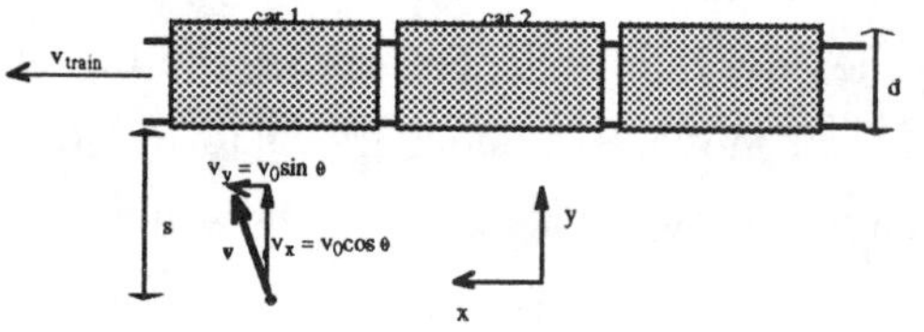

By breaking up the bullet's initial velocity into its x- and y-components, we can calculate the "firing angle" needed to ensure that Eq. (1) holds.

From Eq. (1) and this diagram, we immediately get

$$\begin{aligned} v_{\text{train}} &= v_{x\text{ bullet}} \\ &= v_{\text{bullet}} \sin\theta. \end{aligned}$$

To solve for θ, divide through by v_{bullet}, then take the arcsine of both sides:

$$\begin{aligned} \theta &= \sin^{-1}\left(\frac{v_{\text{train}}}{v_{\text{bullet}}}\right) \\ &= \sin^{-1}\left(\frac{20\text{ m/s}}{100\text{ m/s}}\right) \\ &= 11.5^\circ. \end{aligned}$$

The computer simulation can help you see why that angle works.

(b) The spy must fire the bullet at the forward angle just calculated, or else the bullet cannot pass cleanly through the gap. But the spy still needs to time her shot so that the bullet, when it reaches the train, is "even" with the gap. At first glance, you might think that you need to calculate how much time the bullet takes to reach the train, and exactly where the bullet will be when it reaches the train, so that you can figure out where the gap must be when the bullet was fired. That technique will work. But if you go through all that effort, you are not taking advantage of the crucial physical insight from part (a): The bullet, when "properly" fired, has the same leftward velocity as the train has, and therefore neither "gains on" nor "loses ground to" the train. It immediately follows that the spy should fire the bullet when the gap is "even" with her, at $x = 0$!

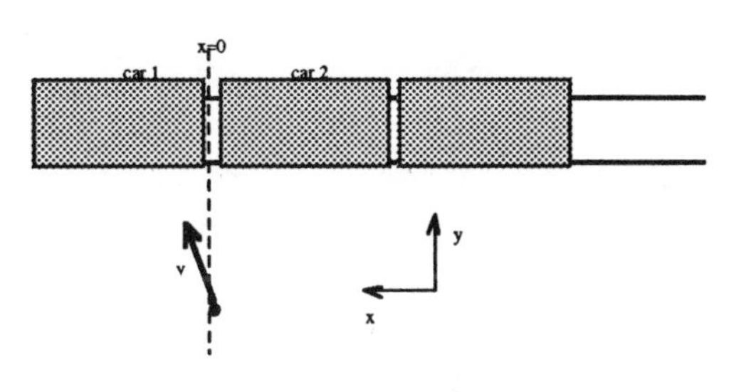

The bullet, as it rushes toward the train, does not move rightward or leftward with respect to the train. Therefore, the bullet does not "gain on" or "lose ground to" the gap. For this reason, if the bullet starts even with the gap, the bullet *stays* even with the gap for its entire trip, including the part of the trip when the bullet reaches the gap and passes through it. That is why the spy should fire the bullet when she is even with the gap, as drawn above.

Notice something interesting: It simply does not matter how much time the bullet takes to reach the train; the bullet stays even with the gap for as long as it needs to–indeed, it stays even with the gap after passing through the gap. So, the spy could be 20 meters away or 50 meters away–it just does not matter. That is kind of counterintuitive. You may need to see it to believe it. Play around with the computer simulation.

24 CHAPTER

Review Problem Set (Chapters 10-16)

Number Two

REVIEW QUESTION 2-1

Suppose I push a block of ice (mass $m = 5.0$ kg) along the floor. I let it go when it has initial speed $v_0 = 5.0$ m/s. The block comes to rest exactly $s = 20$ meters from where I let go of it. I should mention that the floor had been waxed in some places but not in others.

(a) After I let go of the block, how much work does the floor's friction do on the block?

(b) After I let go of the block, how many joules of heat (and sound and other forms of "dissipative" energy) are generated as the block slides across the floor?

REVIEW ANSWER 2-1

(a) Many students get the right answer by the following incorrect reasoning: Using kinematics, calculate the block's acceleration (deceleration) due to friction. Set $F = ma$ to find the frictional force. Then, use $W = F\Delta x$ to calculate the work. Though this reasoning yields the right answer (for reasons I will discuss below), it is *not valid*, because some parts of the floor are waxier than others. The floor is not uniformly slippery. In other words, its coefficient of friction μ is not constant. Therefore, the force of friction ($f = \mu N$) is not constant. But the work formula $W = F\Delta x$ applies only to constant forces. (Otherwise, you must use $W = \int F dx$.) Furthermore, since the force is not constant, neither is the acceleration. Therefore, you cannot use constant-acceleration kinematic formulas such as $v^2 = v_0^2 + 2a\Delta x$.

Fortunately, the textbook's work/energy chapter gives you problem-solving tools that apply even when the forces are not constant. Here, for instance, we can use the work-energy theorem:

$$W_{\text{net}} = \Delta K \quad \textbf{Work-energy theorem}.$$

After I release the block, the only forces acting on it are friction, gravity, and the normal force. Gravity and the normal force cancel, and besides, they do not point along the direction of motion. Friction is the only force parallel to the direction of motion, and hence, the only force that can perform work. So, $W_{\text{net}} = W_{\text{done by floor's friction}}$.

When released, the block starts off with kinetic energy $K_0 = \frac{1}{2}mv_0^2$, and ends up with $K_f = 0$, since it comes to rest. So, by the work-energy theorem,

$$
\begin{aligned}
W_{\text{done by floor's friction}} &= W_{\text{net}} \\
&= \Delta K \\
&= K_f - K_0 \\
&= 0 - \frac{1}{2}mv_0^2 \\
&= -\frac{1}{2}(5.0 \text{ kg})(5.0 \text{ m/s})^2 \\
&= -62 \text{ J}.
\end{aligned}
$$

The minus sign indicates that the floor "takes energy from" the ice cube, instead of giving it energy. In other words, friction slows the block down instead of speeding it up.

This answer depends *only* on the block's initial speed, not on how much distance it covers. The $s = 20$ m was unneeded information. So, a rough floor (corresponding to a short "slide distance") and a slippery floor (corresponding to a long slide distance) do the *same* work on the block. Let me explain why.

On the rougher floor, the ice cube experiences a bigger frictional force, and therefore slides for a shorter distance. So, in $W = F\Delta x$, the "F" is big and the "Δx" is small. By contrast, on the slipperier floor, the frictional force is weaker, and hence the block slides farther. So, "F" is small, but "Δx" is big. In both cases, the product $F\Delta x$ is the same.

By extending this argument, you can show that even if the floor exerts a non-constant force, it still expends the same work stopping a $v_0 = 5.0$ m/s ice cube. So, if you mistakenly assume the frictional force to be constant, you get the same (correct) answer.

(b) Since heat is a form of energy, conservation of energy is the best, and perhaps the only way to proceed. The block starts out with kinetic energy. As the block slides along, rubbing against the floor, its kinetic energy transforms into heat; the block slows down while heating up the floor and itself. By the time the block comes to rest, *all* of its initial kinetic energy has turned into heat (and sound, etc.). Gravitational potential energy plays no role, because the block stays at the same height throughout. So, by energy conservation,

$$
\begin{aligned}
E_0 &= E_f \\
K_0 &= K_f + \text{Heat} \\
\frac{1}{2}mv_0^2 &= 0 + \text{Heat},
\end{aligned}
$$

and hence $\text{Heat} = \frac{1}{2}mv_0^2 = 62$ J, exactly the (negative) amount of work done on the block by the floor. Coincidence? No! When we say the floor does negative work on the block, we are really saying that the floor "takes away" some of the block's kinetic energy and converts it into another form. (The energy cannot just disappear; it is conserved.) All of that "lost" kinetic energy converts into heat. So, the work done by friction must equal the heat.

REVIEW QUESTION 2-2

A 30° ramp of height $h = 0.80$ meters sits at the top of a cliff, as shown. This ramp is attached to the ground. Ball 1, of mass $m = 1.0$ kg and initially motionless at the top of the ramp, slides down the ramp and along the top of the cliff, eventually colliding head-on with ball 2. (The balls in this problem slide frictionlessly; *they do not roll.*) Ball 2 has mass $2m = 2.0$ kg, and was motionless before the collision. After the collision, ball 1 is motionless. The collision knocks ball 2 off the cliff. Ball 2 flies through the air, eventually landing an unknown distance d from the bottom of the cliff, which is $H = 5.0$ meters high.

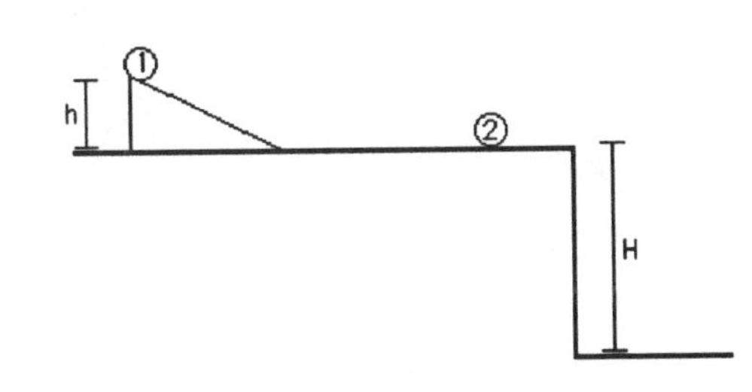

(a) Was the collision elastic? Explain your answer.

(b) Solve for d, in meters.

REVIEW ANSWER 2-2

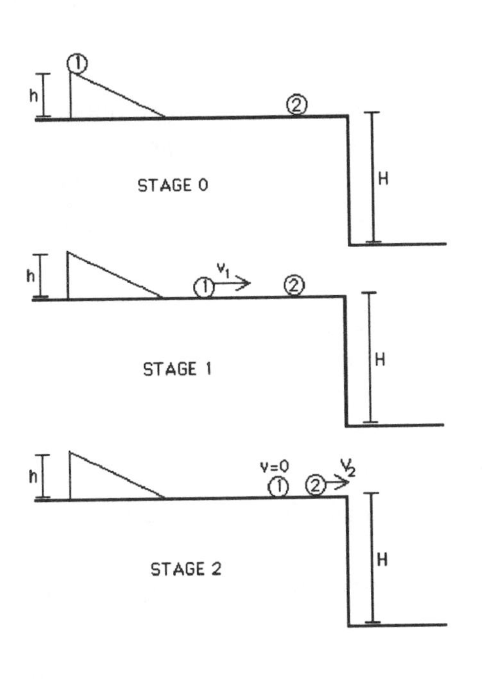

In complicated, multistep problems such as this, it is essential to visualize the different stages of the process, in order to formulate a problem-solving strategy. While ball 1 slides down the ramp, it speeds up. Then it slides (at constant speed) across the plateau until colliding with ball 2. We are told that ball 1 is motionless after the collision. So, ball 2 acquires all of ball 1's rightward "oomph." As a result, ball 2 slides off the cliff, becoming a projectile.

Here are the different stages of the process, with the velocities labeled. Small v refers to ball 1, while large V refers to ball 2. You should always organize your thoughts with pictures.

"Final" is when ball 2 lands a horizontal distance d from the bottom of the cliff.

With the help of these pictures, we can formulate a problem-solving strategy. Where ball 2 lands depends on V_2, the speed with which it slides off the cliff. V_2 depends on v_1, the speed with which ball 1 crashes into ball 2. And v_1 depends on h, the height of the ramp.

So, working forward, we can go from stage 0 to stage 1, using conservation of energy (or old-fashioned forces and kinematics) to figure out the speed of ball 1 at the bottom of the ramp. This v_1 is the speed with which ball 1 strikes ball 2. Using this v_1, we can invoke collision reasoning (conservation of momentum and/or conservation of energy) to calculate V_2, the post-collision speed of ball 2. That takes us from stage 1 to stage 2. V_2 is the "initial" speed with which ball 2 shoots off the cliff. So finally, we can use projectile reasoning, with V_2 as the "initial" velocity, to calculate the horizontal distance d traveled by ball 2 while in the air.

This strategy allows us to solve both part (a) and part (b), as we will see below. I will now implement it step by step.

Stage 0 → Stage 1: Conservation of energy.

At the top of the ramp, ball 1 has gravitational potential energy $U_0 = \text{m}gh$, or $U_0 = \text{m}g(H+h)$ if you measure heights from the bottom of the cliff. At the bottom of the ramp, the ball's potential energy is $U_1 = 0$, or $U_1 = \text{m}gH$ if you measure heights from the bottom of the cliff. Crucially, the *difference* between U_0 and U_1 is the same no matter where you choose as your "zero point." Since the ball slides frictionlessly, no heats gets generated. So, energy conservation gives us

$$E_0 = E_1$$
$$K_0 + U_0 = K_1 + U_1$$
$$0 + \text{mg}h = \frac{1}{2}mv_1^2 = 0.$$

Solve for v_1 to get $v_1 = \sqrt{2gh}$.

Stage 1 → Stage 2: Collision reasoning.

If the collision is not elastic, then some heat gets generated, i.e., some kinetic energy gets lost. In that case, you could not use $\frac{1}{2}mv_1^2 = \frac{1}{2}(2m)V_2^2$, because the right-hand side of the equation needs a "heat" term. Since we do not know ahead of time whether the collision is elastic, we should start with momentum conservation, which applies unproblematically *no matter whether heat gets produced or not*. (Remember, a collision is inelastic until proven otherwise.) In this step, we are finding V_2, the post-collision speed of ball 2, because that is the speed with which it shoots off the cliff.

Since all the motion happens along a straight line, we need not worry about vector components. After the collision, ball 1 is motionless $(p = 0)$. Remembering that ball 2 has mass $2m$, we get

$$p_{\text{stage 1}} = p_{\text{stage 2}}$$
$$mv_1 = 2mV_2.$$

Solve for V_2, and then substitute in our above expression for v_1, to get $V_2 = \frac{v_1}{2} = \frac{\sqrt{2gh}}{2}$. Intuitively, this result makes sense. Since ball 1 comes to rest, it gives all of its "oomph" to ball 2. And since ball 2 is twice as heavy, it needs to move only half as fast to carry the same oomph.

(a) Before going on to the next step, let us answer part (a). A collision is elastic if no heat gets produced, i.e., if no kinetic energy gets "lost." Therefore, to determine if the collision was elastic, we just need to check whether the pre-collision kinetic energy, K_1, equals the post-collision kinetic energy, K_2.

Well, the pre-collision kinetic energy is $K_1 = \frac{1}{2}mv_1^2$. And using our result that $V_2 = \frac{v_1}{2}$, we get

$$K_2 = \frac{1}{2}(2m)V_2^2 = \frac{1}{2}(2m)\left(\frac{v_1}{2}\right)^2 = \frac{1}{4}mv_1^2,$$

exactly half the initial kinetic energy. So, half the initial kinetic energy was converted into heat. The collision is inelastic. A collision can be "inelastic" even though the objects do not stick together. When objects stick, that is called "completely inelastic."

So much for part (a). I will now continue my step-by-step approach to part (b).

Stage 2 → Stage 3: Projectile reasoning.

Ball 2 leaves the cliff horizontally with initial speed V_2. So, in projectile "language," the components of the initial velocity are $v_{0x} = V_2$ and $v_{0y} = 0$. We are finding d, the horizontal distance covered by ball 2 while in the air. So, a good place to start is the x-component of the vector kinematic relation $\mathbf{r} = \mathbf{r}_0 + \mathbf{v}_0 t + \frac{1}{2}\mathbf{a}t^2$. Since no horizontal forces act on ball 2 while it is airborne, $a_x = 0$. Set $x_0 = 0$ and $v_{0x} = V_2$ to get $x = x_0 + v_{0x}t + \frac{1}{2}a_x t^2$.

$$d = 0 + V_2 t + 0. \qquad (1)$$

We know V_2 from above. But to solve for d, we still need t, the time of flight. As usual in projectile problems, you can extract more information by considering the vertical (as well as the horizontal) aspect of the motion. In time t, ball 2 "falls" from $y_0 = H$ to $y = 0$. We also know the initial vertical velocity is $v_{0y} = 0$, and the vertical acceleration is $a_y = -g$. (I have picked upward as positive.) So, we can use the vertical version of Eq. (1), namely $y = y_0 + v_{0y}t + \frac{1}{2}a_y t^2$.

$$0 = H + 0 - \tfrac{1}{2}gt^2. \tag{2}$$

Solve Eq. (2) for t to get $t = \sqrt{\frac{2H}{g}}$. Substitute that t into Eq. (1), along with our above expression for V_2, to get

$$\begin{aligned} d &= V_2 t \\ &= \left(\frac{\sqrt{2gh}}{2}\right)\left(\frac{\sqrt{2H}}{g}\right) \\ &= \sqrt{Hh} \\ &= \sqrt{(5.0\text{ m})(.80\text{ m})} = 2.0\text{ meters.} \end{aligned}$$

You could not solve this projectile subproblem using energy conservation. Conservation would allow you to find the ball's speed as it hits the ground, but not the horizontal distance it covered.

REVIEW QUESTION 2-3

A pendulum bob of mass m is attached to the end of a long (essentially massless) string of length L, making a pendulum. The bob is displaced as shown below. Before it is released, the bob is a height H above point X, the "bottom" of the bob's path.

When the bob reaches point X, what is the tension in the string? Express your answer in terms of m, L, H, and g.

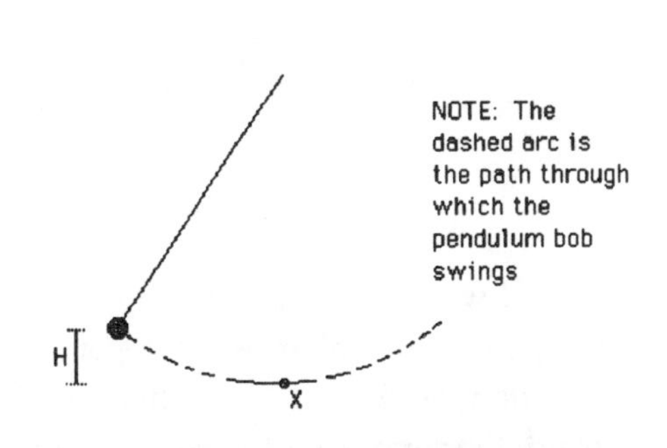

REVIEW ANSWER 2-3

Since we are trying to find a force, we cannot go wrong starting with a free-body diagram. We are interested in the forces acting on the bob when it is at point X.

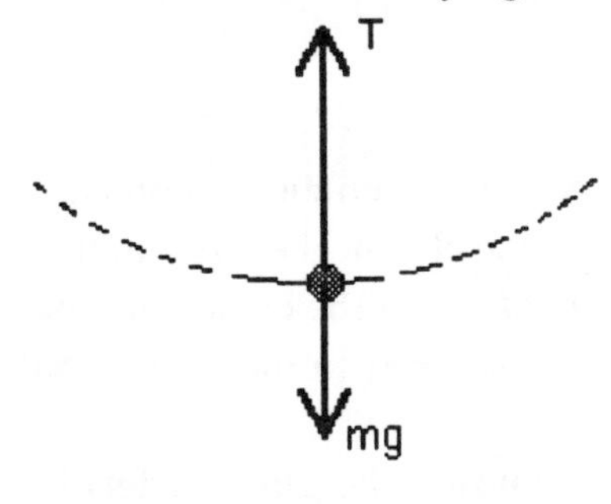

A common mistake is to set $T = mg$. But if those forces cancelled, then the net force would be zero, and hence the bob would not be accelerating. But the bob *is* accelerating, because it is tracing out a circle. As discussed *ad nauseam* in Chapter 10, when an object travels in a circle, it undergoes a radial ("centripetal") acceleration $a_{\text{radial}} = \frac{v_2}{r}$. By definition, the radial direction is towards the center of the circle, which in this case is the end of the string attached to the ceiling. So, when the bob is at point X, the radial direction points straight up; and the radius of the circle is $r = L$, the string's length. Newton's 2nd law immediately gives us

$$\sum F_{\text{radial}} = ma_{\text{radial}}$$
$$T - mg = m\frac{v^2}{r} \qquad (1)$$
$$= m\frac{v^2}{L}.$$

We cannot yet solve Eq. (1) for T, because we do not know the bob's speed v at point X. Luckily, we can find v using energy conservation. (Forces and kinematics do not work, because the net force on the bob is not constant.) Let us measure our heights from point X. The "initial" situation is when the bob gets released (from height H), and the "final" configuration is when it reaches point X. (The pictures are not worth drawing.) No heat gets generated. So,

$$E_0 = E_f$$
$$K_0 + U_0 = K_f + U_f$$
$$0 + mgH = \frac{1}{2}mv^2 + 0.$$

Solving for v yields $v = \sqrt{2gH}$. Now solve Eq. (1) for T, and use this expression for v to get

$$T = m\left(\frac{v^2}{L} + g\right)$$
$$= m\left(\frac{2gH}{L} + g\right)$$
$$= mg\left(\frac{2H}{L} + 1\right).$$

I would like you to notice something intuitive about this answer. When the bob reaches point X, the rope is performing two independent functions. First, it keeps the bob moving along a circular path. The rope would need to do this even in outer space, assuming the bob was made to swing in a circle. Second, the rope "counteracts" gravity, in order to keep the bob from falling. The rope would need to do this even if the bob hanging instead of moving. The force needed to keep the bob moving circularly is $\frac{mv^2}{L}$. The force needed to counteract gravity is mg. That is why the tension in the rope is $T = \frac{mv^2}{L} + mg$.

REVIEW QUESTION 2-4

Hard, but you can work it out step by step.

A cannon works as follows: Gunpowder is packed inside the cannon, and then the cannon ball is placed on top of the gunpowder. The gunpowder is lit. When it explodes, the chemical energy initially stored in the gunpowder is released. Approximately 25% of the energy released in the explosion takes the form of heat, sound, light, etc. The rest of the released energy gets transferred to the cannon ball and to the cannon in the form of kinetic energy.

Suppose the cannon ball has mass $m = 10$ kg, and the cannon (including the gunpowder) has mass $M = 100$ kg. The gunpowder in the cannon contains $E_c = 100{,}000$ joules of chemical energy that can be released by the explosion. When this gunpowder explodes, the cannon ball shoots off at a 30° angle to the ground, and the cannon recoils backwards (also at a 30° angle to the ground).

What is the peak height h reached by the cannon ball during its flight? h is measured from the cannon. The cannon nozzle is very short compared to h.

REVIEW ANSWER 2-4

To find how high the ball goes, we need to know v_0, the ball's initial speed when it leaves the cannon. In many problems, we are told v_0. But here, we have to extract v_0 from the given explosion information. Once we find v_0, we can invoke projectile motion reasoning to solve for the ball's peak height.

How can we find v_0? Since we do not know the relevant forces, and cannot figure them out, we have no choice but to use conservation laws. Let us break this problem into two "subproblems":

Subproblem 1: Explosion reasoning to find the ball's speed when it leaves the cannon.

Subproblem 2: Projectile reasoning to find its peak height.

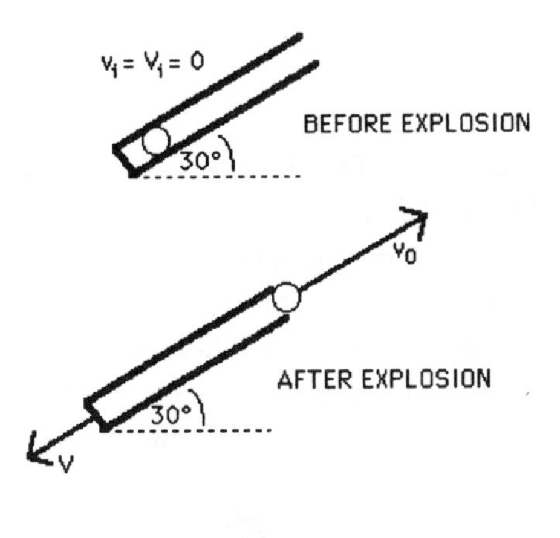

I will start with subproblem 1. As always, diagram the "before" and "after" configurations. I will let "v_0" denote the ball's post-explosion velocity, because that is the ball's "initial" speed when it becomes a projectile. Also, I will let big V denote the cannon's "recoil" velocity.

The cannon recoils in the opposite direction to the cannon ball's motion. You can see this intuitively, or by using conservation of momentum.

Because the ball and cannon initially travel along the same straight line (in opposite directions), we need not worry about vectors–at least, not until we consider the subsequent projectile motion of the ball in subproblem 2. For now, let us just talk about the explosion.

We are trying to find v_0. As always in collisions/explosions, start with conservation of momentum. Before the explosion, nothing was moving, and hence $p_{\text{before}} = 0$. After the explosion, the ball has momentum mv_0, while the cannon has momentum $-MV$. The minus sign indicates that the cannon's motion is backwards compared to the cannon ball's motion. So,

$$\begin{aligned} p_{\text{before}} &= p_{\text{after}} \\ 0 &= mv_0 - MV. \end{aligned} \tag{1}$$

We cannot yet solve Eq. (1) for v_0, because it contains a second unknown, V. To get more information, consider energy. Of the $E_c = 100{,}000$ joules of chemical energy "stored" in the gunpowder, 25% gets "wasted" as heat, etc. The remaining 75% of that initial energy, 75000 joules, gets transferred into the kinetic energy of the cannon ball and the recoiling cannon. I am neglecting the small height by which the ball rises inside the cannon nozzle. So,

$$\begin{aligned} \text{"}E_{\text{before}}\text{"} &= E_{\text{after}} \\ \frac{3}{4}E_c = 75{,}000 \text{ joules} &= \frac{1}{2}mv_0^2 + \frac{1}{2}MV^2. \end{aligned} \tag{2}$$

We now have two equations in the two unknowns, v_0 and V. To complete the algebra efficiently, first solve Eq. (1) for V to get $V = \frac{mv_0}{M}$. Put that expression for V into Eq. (2) to get

$$\frac{3}{4}E_c = \frac{1}{2}mv_0^2 + \frac{1}{2}M\left(\frac{m}{M}v_0\right)^2$$
$$= \frac{1}{2}\left(m + \frac{m^2}{M}\right)v_0^2$$
$$= \frac{1}{2}m\left(1 + \frac{m^2}{M}\right)v_0^2.$$

Isolate v_0 to get

$$v_0 = \sqrt{\frac{3E_c}{2m(1+\frac{m}{M})}} = \sqrt{\frac{3(100000\text{ J})}{2(10\text{ kg})\left(1+\frac{10\text{ kg}}{100\text{ kg}}\right)}} = 117\text{ m/s}.$$

That is how fast the cannon ball shoots out of the cannon.

Subproblem 2: Projectile reasoning or energy conservation to find peak height.

A common error is to set the ball's velocity equal to zero at the peak. But the ball does not stop at its peak. It is still moving. By throwing a pencil at an angle, you can see that a projectile at its peak moves *sideways*. So, its *vertical* velocity is $v_y = 0$; the object's upward/downward motion momentarily stops. But it retains its horizontal velocity v_x. Therefore, if you solve this subproblem using energy conservation, you must include a $\frac{1}{2}mv_x^2$ term at the peak.

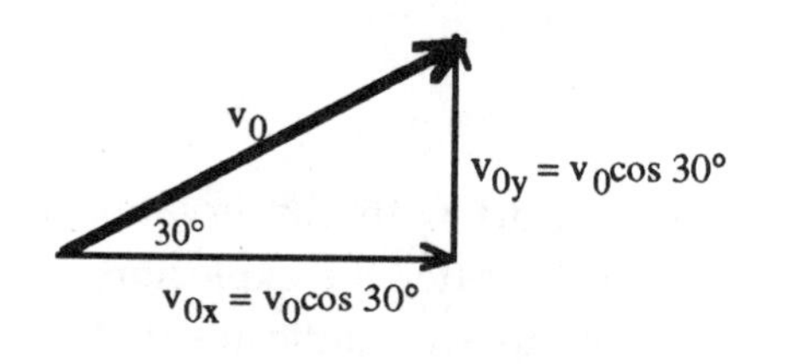

While in mid-air, the ball's *sideways* motion neither speeds up nor slows down: $a_x = 0$. So, its horizontal velocity at the peak–or at any other point in the trajectory–is $v_x = v_{0x} + a_x t = v_{0x}$, the initial x-velocity. We can easily find v_{0x}.

So, at the peak, $v_x = v_{0x} = v_0 \cos\theta$, and $v_y = 0$. Using those insights, we can now use either old-fashioned projectile reasoning or new-fashioned energy conservation to solve for h, the peak height. I will demonstrate both methods.

Projectile reasoning. We are trying to find $y = h$, the ball's peak vertical height. We know the initial and "final" vertical velocities: $v_{0y} = v_0 \sin\theta$ initially, and $v_y = 0$ at the peak. Since we do not know anything about the flight time, it is most efficient to use $v_y^2 = v_{0y}^2 + 2a_y\Delta y$, Solve for Δy to get

$$h = \Delta y = \frac{v_y^2 - v_{0y}^2}{2a_y} = \frac{0-(v_0 \sin 30°)^2}{2(-g)} = \frac{(117\text{ m/s})^2(\sin 30°)^2}{2(9.8\text{ m/s}^2)} = 170\text{ meters}.$$

Now I will again solve for h using...

Energy conservation: Since kinetic energy has no "direction," we need not worry about vector components. No more heat gets generated as the ball flies through the air. Therefore, if we relate the energy immediately after the explosion to the final energy (when the ball reaches the peak), then we need not worry about heat. Also, since the ball and cannon no longer interact, I can leave the cannon out of the equation. (If I leave the cannon in, it contributes energy terms to both sides of the equation, terms that cancel out.)

Conservation of energy demands

$$E_{\text{ball, after explosion}} = E_{\text{ball, at peak}}$$
$$K_{\text{after}} + U_{\text{after}} = K_{\text{peak}} + U_{\text{peak}}$$
$$\frac{1}{2}mv_0^2 + 0 = \frac{1}{2}mv_{\text{peak}}^2 + mgh$$
$$= \frac{1}{2}mv_{0x}^2 + mgh,$$

where in the last step I invoked the fact that the ball's velocity is purely horizontal at the peak; and this sideways velocity neither speeds up nor slows down from its initial value.

Cancel the m's and solve for h to get $h = \frac{(v_0^2 - v_{0x}^2)}{2g}$. But by the Pythagorean theorem, $v_0^2 = v_{0x}^2 + v_{0y}^2$. So, $v_0^2 - v_{0x}^2 = v_{0y}^2$. Substitute this into our expression for h to get

$$h = \frac{v_0^2 - v_{0x}^2}{2g} = \frac{v_{0y}^2}{2g} = \frac{v_0^2(\sin 30°)^2}{2g},$$

exactly the same expression obtained above using projectile reasoning.

REVIEW QUESTION 2-5

A solid cylinder of mass M and radius R is rolling without slipping towards a semicircular track, as shown below. The cylinder's rotational velocity around its central axis is ω_0. The track is attached to the floor, and the radius of the semicircle is s, where $s >> R$. The cylinder rolls up the track, without slipping, until it loses contact with the track at point P.

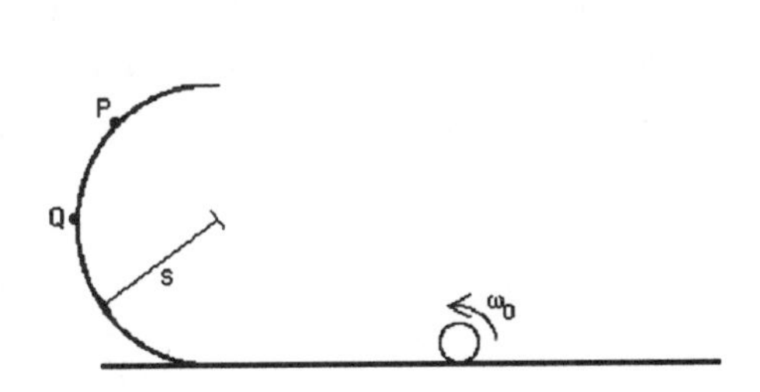

(a) *(Skip this part unless you need to be able to derive rotational inertias.)*
Derive the rotational inertia of a cylinder around its central axis.

(b) What is the cylinder's center-of-mass speed when it reaches point Q? (Point Q is where the track is vertical.) Express your answer in terms of g, R, s, and ω_0.

(c) *(Very hard.)* How high is point P off the ground? Set up, but do not solve, the equation or equations needed to express this height in terms of g, R, s, and ω_0.

REVIEW ANSWER 2-5

(a) This is nearly equivalent to deriving the rotational inertia of a disk. Please review Question 14-5 before reading on. Here is the only difference: The "shaded in" region is now a volume instead of an area, namely a cylindrical "shell." The volume of this shell is $dV = (2\pi r dr)L$, where L is the cylinder's length. But since the volume of the whole cylinder is $V = \pi R2L$, we get

$$dm = \rho dV = \frac{M}{V}dV = \frac{M}{\pi R^2 L}(2\pi r dr)L = \frac{2Mr}{R^2}dr,$$

the same as for a disk. So, the rest of the derivation in Answer 14-5 goes through unchanged.

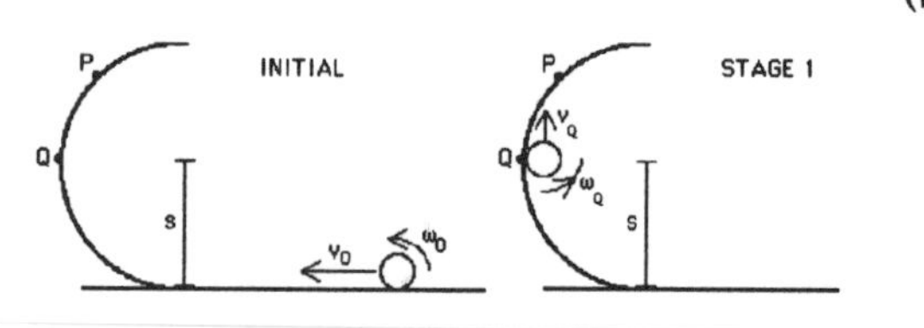

(b) Because the slope of the track is not constant, the forces and torques exerted on the cylinder are not constant. For this reason, we cannot invoke constant-acceleration kinematic equations. Instead, we must use conservation, in this case energy conservation.

As the cylinder rolls up the track, it gradually slows down, losing kinetic energy. This "lost" kinetic energy converts into gravitational potential energy. Let us diagram the relevant stages.

Initially, the cylinder rolls along the ground. So, it has lots of translational *and* rotational kinetic energy, but no potential energy. By analogy with $K_{\text{trans}} = \frac{1}{2}mv^2$, the formula for rotational kinetic energy is $K_{\text{rot}} = \frac{1}{2}I\omega^2$. For a solid cylinder, $I = \frac{1}{2}mR^2$.

Usually when an object undergoes both translational and rotational motion, v and ω bear no simple relationship to each other. But when an object rolls without slipping, we have

$$v_{\text{cm}} = R\omega, \quad \textbf{Rolling without slipping condition,}$$

where "cm" denotes center of mass. So, v_{cm} is the speed of the center of the cylinder, i.e., the speed of the cylinder as a whole. That is the "v" appearing in all our linear (translational) formulas, such as $K_{\text{trans}} = \frac{1}{2}mv^2$. From now on, I will omit the "cm" subscripts.

At point Q, the cylinder is "one radius" off the ground, at $h = s$. (Since $R << s$, we can neglect the fact that the cylinder's center starts not a ground level, but at height R off the ground.) Given all this, conservation of energy yields

$$E_0 = E_1$$
$$U_0 + K_0 = U_1 + K_1$$
$$0 + \frac{1}{2}Mv_0^2 + \frac{1}{2}I\omega_0^2 = Mgs + \frac{1}{2}Mv_Q^2 + \frac{1}{2}I\omega_Q^2.$$

Set $I = \frac{1}{2}MR^2$. We are trying to solve for v_Q in terms of ω_0 and the other givens. So, on the left-hand side of the equation, we can invoke the rolling without slipping condition to replace v_0 with $\omega_0 R$. Similarly, on the right-hand side, we can use $\omega_Q = \frac{v_Q}{R}$ to get rid of the ω_Q. Substituting all this into the energy conservation equation yields

$$\frac{1}{2}M(\omega_0 R)^2 + \frac{1}{2}\left(\frac{1}{2}MR^2\right)\omega_0^2 = Mgs + \frac{1}{2}Mv_Q^2 + \frac{1}{2}\left(\frac{1}{2}MR^2\right)\left(\frac{v_Q}{R}\right)^2.$$

Now it is just algebra to solve for v_Q. Cancel the M's and isolate v_Q to get, after a few more steps of math, $v_Q = \sqrt{R^2\omega_0^2 - \frac{4}{3}gs}$.

(c) To solve this, you need a physical insight about what is going on when the cylinder leaves the track. While still on the track, the cylinder presses against the track, and vice versa. As the cylinder approaches point P, this normal force decreases. Right when the cylinder loses contact with the track, the normal force goes to $N = 0$.

How does that help us? Well, right at point P, the cylinder is still moving along a circular path of radius $r = s$. So, maybe we can extract useful information from circular-motion reasoning. Here is the idea: Since the cylinder's center of mass traces out a circle of radius s, it must undergo a radial (centripetal) acceleration $a_{\text{radial}} = \frac{v^2}{s}$. This radial acceleration is caused by the radial forces acting on the cylinder, which I will draw below. But as the cylinder gets higher and higher, it slows down. Eventually, it gets so slow that $\frac{v^2}{s}$ is *less than* the radial acceleration generated by the radial forces. That is when the cylinder falls off.

Notice something crucial here. The cylinder falls off the track *before* it slows down to $v = 0$. That is because v is still nonzero when $\frac{v^2}{s}$ gets "too small." More precisely, the cylinder falls off when $\frac{v^2}{s}$ is a billionth of a percent less than a_{radial}. So, we can just set $a_{\text{radial}} \approx \frac{v^2}{s}$. From above, we also know that the "fall off" point is where $N \approx 0$. Let me draw a force diagram at point P.

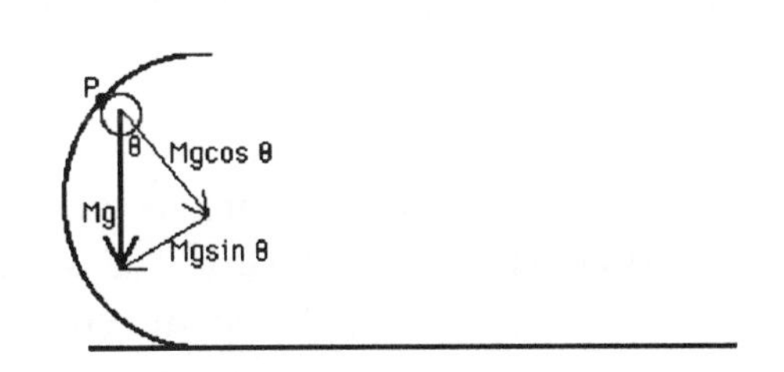

I have broken up gravity into its tangential and radial components. (The radial component is toward the center of the circle.) So, from Newton's 2nd law,

$$\sum F_{\text{radial}} = Ma_{\text{radial}}$$
$$Mg\cos\theta = M\frac{v_P^2}{s}.$$

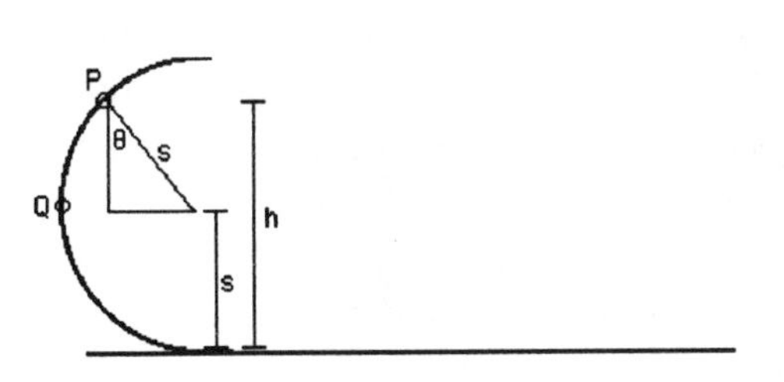

At first glance, Eq. (1) seems useless, because it does not contain the height h. But Eq. (1) *does* contain h, at least implicitly. To see why, consider the following diagram. The θ here is the same angle appearing in the force diagram. The hypotenuse goes from point P to the center of the circle, and therefore has length s.

The vertical leg of the triangle has length $s\cos\theta$. Therefore, as the diagram shows,

$$\begin{aligned} h &= s + s\cos\theta \\ &= s(1+\cos\theta). \end{aligned} \tag{2}$$

Equations (1) and (2) still do not provide enough information to solve for h, because those equations contain three unknowns–h, θ, and v_p, the cylinder's speed at point P. We need another equation, preferably one with v_p in it.

Well, in part (b), we found the cylinder's speed at point Q, using conservation of energy. The same reasoning works here. I will not repeat all the steps. The only change from part (b) is that the cylinder's height off the ground is now h instead of s. So, our part (b) energy conservation equations becomes

$$E_0 = E_p$$
$$U_0 + K_0 = U_p + K_p$$
$$0 + \frac{1}{2}Mv_0^2 + \frac{1}{2}I\omega_0^2 = Mgh + \frac{1}{2}Mv_p^2 + \frac{1}{2}I\omega_0^2.$$

$$\frac{1}{2}M(\omega_o R)^2 + \frac{1}{2}\left(\frac{1}{2}MR^2\right)\omega_0^2 = Mgh + \frac{1}{2}Mv_p^2 + \frac{1}{2}\left(\frac{1}{2}MR^2\right)\left(\frac{v_p}{R}\right)^2, \tag{3}$$

which you can solve for v_P if you want to get $v_P = \sqrt{R^2\omega_0^2 - \frac{4}{3}gh}$. But you did not have to solve Eq. (3). Equations (1) through (3) are three equations in the three unknowns (h, θ, and v_P). So, we are done.

REVIEW QUESTION 2-6

Three boats are involved in a triple collision. Before the collision, boat 1 has mass 200 kg and speed 5.0 m/s eastward; boat 2 has mass 300 kg and speed 4.0 m/s westward; and boat 3 has mass 400 kg and speed 2.0 m/s northward. The boats all crash into each other at once, and they stick together.

(a) What is the speed and direction of the boats immediately after the collision? Express the "direction" as the angle by which the direction of motion is north of due east.

(b) What fraction of the initial kinetic energy was lost during the collision?

REVIEW ANSWER 2-6

(a) Especially because the problem supplies no picture, you had better draw one. In collision problems, it is helpful to take "before" and "after" snapshots.

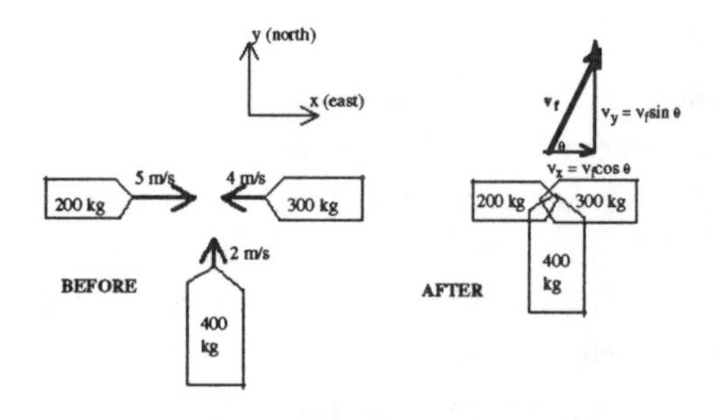

Since the collision is inelastic, heat and sound (and other dissipative forms of energy) get produced. So, we cannot set the pre-collision kinetic energy equal to the post-collision kinetic energy. Instead, we must use momentum. Since things do not all happen along a straight line, we must worry about vectors. As always, it is easiest to *separately* consider the x- and y-directed motion.

Let us start with x-momentum. Picking eastward as positive, we see that initially, the 200 kg boat has positive x-momentum, while the 300 kg boat has negative x-momentum. The heaviest boat moves purely in the y-direction, and therefore carries *no* x-momentum. So,

$$p_{x \text{ before}} = p_{x \text{ after}}$$
$$(200 \text{ kg})(5 \text{ m/s})(300 \text{ kg})(4 \text{ m/s}) = (200 \text{ kg} + 300 \text{ kg} + 400 \text{ kg})v_x.$$

You can immediately solve for v_x to get $v_x = -0.22$ m/s. The minus sign indicates westward. So, my "after" sketch is misleading. The v_f arrow should point upward and leftward, not upward and rightward. The "θ" we will get later should be greater than 90°.

Now I will consider the y-motion. Only the 400 kg boat starts off with any y-momentum. So,

$$p_{y \text{ before}} = p_{y \text{ after}}$$
$$(400 \text{ kg})(2 \text{ m/s}) = (200 \text{ kg} + 300 \text{ kg} + 400 \text{ kg})v_y,$$

and hence $v_y = 0.89$ m/s.

Now that we know both components of the final velocity, we can Pythagoras them together to obtain the total final speed: $v_f = \sqrt{v_x^2 + v_y^2} = \sqrt{(-0.22 \text{ m/s})^2 + (0.89 \text{ m/s})^2} = 0.92$ m/s. We can also solve for the angle. The above diagram, though slightly misleading, shows us that $\tan\theta = \frac{v_y}{v_x}$, and hence $\theta = \tan^{-1}\left(\frac{v_y}{v_x}\right) = \tan^{-1}\left(\frac{0.89 \text{ m/s}}{-0.22 \text{ m/s}}\right) = -76°$ or $104°$. If your calculator spat out

$-76°$, remember that for any angle ϕ, $\tan\phi = \tan(\phi + 180°)$, as you can confirm by playing around with triangles. It follows that if $\tan^{-1}(z)$ equals ϕ, then $\tan^{-1}(z)$ also equals $\phi + 180°$. As mentioned above, the physically correct answer should be greater than 90°, but less than 180°, since the boats go westward and northward. So, $\theta = 140°$.

(b) The amount of kinetic energy lost in the collision is $K_{\text{lost}} = K_i K_f$. So, the *fraction* of kinetic energy lost is fraction lost $= \frac{K_{\text{lost}}}{K_i} = \frac{K_i - K_f}{K_i}$. We need to calculate the initial and final kinetic energies:

$$K_i = \frac{1}{2}(200\text{ kg})(5\text{ m/s})^2 + \frac{1}{2}(300\text{ kg})(4\text{ m/s})^2 + \frac{1}{2}(400\text{ kg})(2\text{ m/s})^2 = 5700\text{ J}.$$

$$K_f = \frac{1}{2}(200 + 300 + 400\text{ kg})v_f^2 = \frac{1}{2}(900\text{ kg})(0.92\text{ m/s})^2 = 381\text{ J}.$$

So, fraction lost $= \frac{K_i - K_f}{K_i} = \frac{5700\text{ J} - 381\text{ J}}{5700\text{ J}} = 0.93$. 93% of the initial kinetic energy converts into heat, sound, etc.

REVIEW QUESTION 2-7

A solid cylinder of mass M and radius R rolls without slipping along the ground towards a ramp of mass m, as drawn below. The cylinder's linear speed is v_0. When it reaches the ramp, the cylinder rolls without slipping up the ramp, but it does not reach the top of the ramp. The part of the floor on which the ramp rests is frictionless; so, the ramp is free to slide.

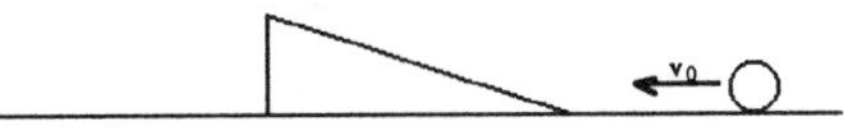

Express your answers in terms of M, R, m, v_0, and g.

(a) A red dot, which I will call point A, is painted on the outside of the cylinder. In this part of the problem, consider the cylinder while it is still rolling along the floor. At time $t = 0$, point A is touching the floor. What is the earliest time at which point A is a height R off the ground?

(b) Let H denote the cylinder's "peak" vertical height off the ground (i.e., the highest point on the ramp that the cylinder reaches). Solve for H.

REVIEW ANSWER 2-7

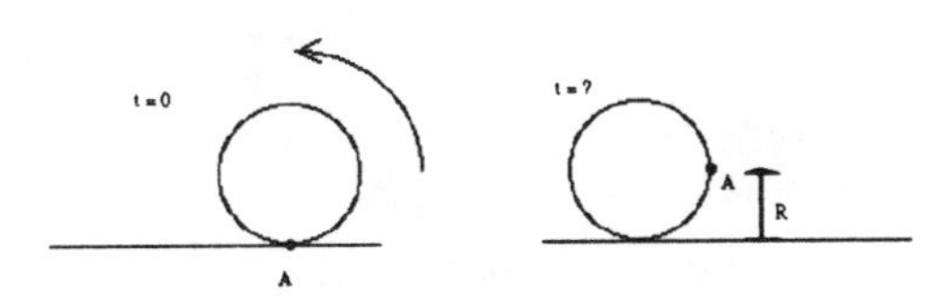

(a) Here is the cylinder at time $t = 0$, and then at the time when point A reaches a distance R off the ground. The cylinder is rolling leftward.

From these pictures, we see that the problem asks how long the cylinder takes to spin through 90°, one quarter of a full rotation. Well, since the cylinder rolls without slipping, its linear and angular speed are related by $v_{\text{cm}} = R\omega$ **Rolling without slipping condition.** So, while rolling along the floor, its angular speed is constant at $\omega_0 = \frac{v_0}{R}$. By definition, angular speed is the rate at which the cylinder spins, i.e., the angle per time. Since ω_0 is constant, $\omega_0 = \frac{\Delta\theta}{\Delta t}$. We are looking for the Δt corresponding to $\Delta\theta = \frac{\pi}{2}$ radians. Solve for Δt to get

$$\Delta t = \frac{\Delta\theta}{\omega_0} = \frac{\frac{\pi}{2}}{\frac{v_0}{R}} = \frac{\pi R}{2v_0}.$$

You get the same answer by calculating the period of rotation, and then dividing by 4: Since $T = \frac{2\pi}{\omega_0}$, point A completes a quarter circle in time $\Delta t = \frac{T}{4} = \frac{2\pi}{4\omega_0} = \frac{\pi R}{2v_0}$.

(b) If the ramp were nailed down, we could use energy conservation to calculate the peak height. But here, the ramp is free to slide. How does that change things?

Well, energy is still conserved, with no heat generated. But as the cylinder rolls up the ramp, the ramp starts to slide leftward, pushed by the normal force exerted by the cylinder. So, our energy expressions must include the ramp's kinetic energy.

To use energy conservation successfully, you also must avoid a common error. If the ramp were nailed down, then the cylinder would be motionless ($v = 0$) at its "peak" (highest point). Here, it is still true that the cylinder at its peak is motionless *with respect to the ramp*. The peak is still the "turn-around" point where the cylinder is (momentarily) neither rolling up nor rolling down the ramp. But remember, the ramp itself slides leftward. So, even when the cylinder is motionless with respect to the ramp, it is still getting carried along by the ramp, and therefore "shares" whatever speed the ramp has (relative to the floor). Analogously, when you sit "motionless" inside a car cruising at 50 mph, you are also moving at 50 mph (relative to the street). So, if we let v_f denote the ramp's speed when the cylinder reaches the peak, then v_f is also the speed with which the cylinder gets carried along at that point.

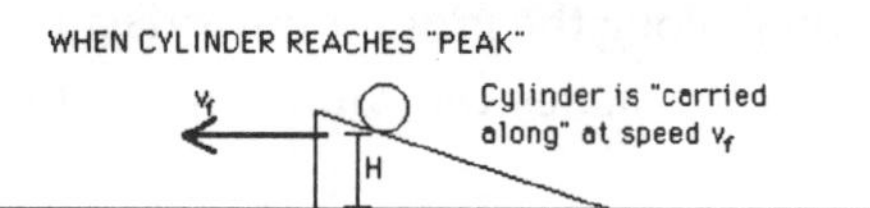

In summary: At its peak, the cylinder is no longer rolling, and hence $\omega_f = 0$. But it still has translational kinetic energy, because it is carried along at the ramp's speed.

Given this physical insight, we can try energy conservation. Remember, when it is rolling, the cylinder possesses both translational and rotational kinetic energy:

$$E_0 = E_f$$

$$U_0 + K_0 = U_f + K_f$$

$$0 + \frac{1}{2}Mv_0^2 + \frac{1}{2}I\omega_0^2 = MgH + \frac{1}{2}(M+m)v_f^2.$$

$$0 + \frac{1}{2}Mv_0^2 + \frac{1}{2}\left(\frac{1}{2}MR^2\right)\left(\frac{v_0}{R}\right)^2 = MgH + \frac{1}{2}(M+m)v_f^2, \tag{1}$$

where in the last step I set $I_{\text{solid cylinder}} = \frac{1}{2}MR^2$, and used the rolling without slipping condition ($v = \omega R$) to get rid of the ω_0 on the left-hand side. Unfortunately, we cannot yet solve Eq. (1) for H, because it contains a second unknown, v_f. How can we gather more information?

Since the cylinder "collides" with the ramp, momentum might come in handy. But we should check whether it is conserved. At first glance, momentum appears *not* to be conserved, for this reason: As the cylinder ascends the ramp, the external force of gravity slows it down. Momentum is not conserved when an external force acts on the system.

This argument, though correct, applies only to the vertical component of the momentum. By the "system," I mean the cylinder *and* ramp. The only external forces acting on the system, namely gravity and the normal forces due to the floor, point *vertically*. Vertical forces cannot "muck up" horizontal momentum conservation. The only *sideways* forces experienced by the cylinder and ramp are the normal and frictional forces they exert on each other. So, x-momentum is conserved.

I am talking about *linear momentum*, not angular momentum. Angular momentum is not conserved here. And you cannot combine linear and angular momenta into the same equation, because they have different units. I am going to write a linear momentum conservation equation, which will ignore the rotational aspects of the cylinder's motion:

$$p_{ix} = p_{fx}$$
$$Mv_0 + 0 = (M+m)v_f. \tag{2}$$

We now have two equations in two unknowns (H and v_f). Only algebra remains.

Algebra starts here. Solve Eq. (2) for v_f to get $v_f = \frac{Mv_0}{(M+m)}$. Substitute that into Eq. (1) to get

$$\frac{1}{2}Mv_0^2 + \frac{1}{2}\left(\frac{1}{2}MR^2\right)\left(\frac{v_0}{R}\right)^2 = MgH + \frac{1}{2}(M+m)\left(\frac{M}{M+m}v_0\right)^2$$
$$= MgH + \frac{1}{2}\left(\frac{M^2}{M+m}\right)v_0^2.$$

Cancel out a factor of M, and then isolate H to get $H = \frac{v_0^2}{g}\left[\frac{3}{4} - \frac{M}{2(M+m)}\right]$.

End of algebra.

How can we check this answer for plausibility? A good trick is to examine "limiting cases." For instance, what if the ramp were infinitely heavy ($m = \infty$). Then, physically speaking, the ramp is effectively nailed down. In that case, the problem reduces to an easier conservation of energy question. You can confirm that the answer to that easier question is the *same* answer you get by setting $m = \infty$ in the above formula for H, namely $H = \frac{3v_0^2}{4g}$.

REVIEW QUESTION 2-8

A hollow spherical ball (mass m, radius R) sits at the bottom of a ramp of ramp angle θ. At time $t = 0$, the ball is hit with a bat, so that it slides up the ramp. Immediately after the ball is hit, it has speed v_0 (along the ramp), but it is not rotating. Since the ramp is not frictionless, however, the ball gradually starts rotating faster and faster as it slides up the ramp. Eventually, it starts to roll without slipping. The coefficient of kinetic friction between the ball and ramp is μ.

(a) Draw a force diagram that describes the ball while it is sliding up the ramp (*after* the bat is no longer in contact with the ball), but before it starts rolling.

(b) (*Very hard. Ask your instructor if you need to be able to do this.*) How far from the bottom of the ramp is the ball when it first starts to roll without slipping? Express your answer in terms of R, v_0, θ, μ, and g.

REVIEW ANSWER 2-8

(a) *After the ball is hit*, the only forces acting on it are gravity, the normal force, and friction. Since friction resists the ball's motion up the ramp, it points down the ramp. While the ball is still sliding (as opposed to rolling without slipping), the friction is kinetic, and hence $f = \mu N$.

(b) This problem looks overwhelming at first. But it is not so bad if you break it into steps. Let us carefully picture what is happening. As the ball slides up the ramp, it gradually slows down, due to gravity and friction. But while its slows down, the ball rotates (spins) faster and faster. So, its angular speed ω increases. Eventually, v gets small enough and ω gets big enough that the rolling without slipping condition holds:

$v = \omega R$ **Rolling without slipping condition**

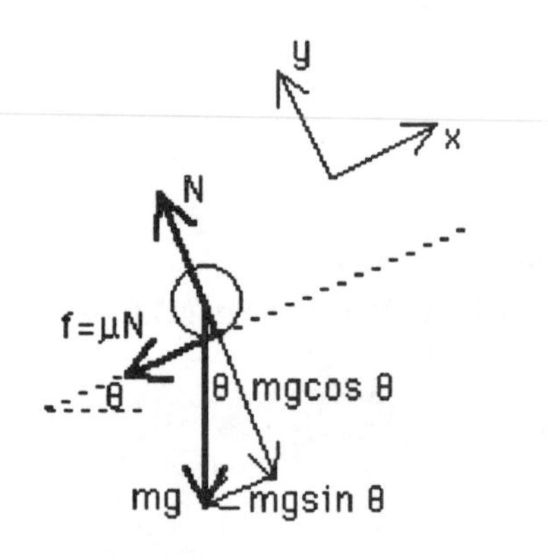

When this happens, the ball stops slipping and starts rolling. So, we must find how far the ball has traveled when it reaches the point that v equals ωR.

How can we do this? The key is to keep in mind that the ball's rotational and translational motion are *independent*. We can treat them separately. For instance, using purely linear reasoning, such as forces and acceleration, we can express the ball's velocity as a function of time. The reasoning, and the resulting equation for v, will completely ignore the fact that the ball also happens to rotate. Similarly, using purely rotational reasoning (torques and angular accelerations), we can express the ball's angular speed as a function of time. The reasoning, and the equation for ω, will be *as if* the ball were only rotating, not moving. Once we have $v(t)$ and $\omega(t)$, we can figure out when the ball starts to roll without slipping by setting $v(t) = R\omega(t)$. Finally, given that time, we can use kinematics to calculate how far it is traveled.

Let me formalize the strategy just laid out. You have seen it before, in Question 15-9.

Sliding-until-it-rolls strategy:

(1) Using linear motion reasoning ($\sum F = ma$ and linear kinematics), write an expression for v, the ball's center-of-mass velocity as a function of time. In this step, you can ignore the rotational motion.

(2) Using rotational motion reasoning ($\sum \tau = I\alpha$ and rotational kinematics), write an expression for ω, the ball's angular speed as a function of time. In this step, you can ignore the linear motion.

(3) Using your expressions from steps 1 and 2, set $v = R\omega$. Solve for t, the time at which the ball starts to roll without slipping. Remember, $v = R\omega$ is the rolling-without-slipping condition; I am omitting the "cm" subscripts.

(4) Given this time, kinematics can tell you the ball's distance from its starting point.

Here we go...

Step 1: Linear motion reasoning to express v

From part (a), we know the forces acting on the ball. Therefore, we can implement the usual strategy: Use Newton's 2nd law to find the acceleration, and then substitute that acceleration into an old kinematic formula to find the velocity.

We are interested in a_x, the acceleration along the ramp. The only x-directed forces are friction and a component of gravity, both of which point backwards (down-the-ramp). So, if I choose the up-the-ramp direction as positive, those forces are negative:

$$\sum F_x = ma_x$$
$$-mg\sin\theta\mu N = ma_x. \tag{*}$$

Before we can solve for a_x, we need to know the normal force. To find it, consider the y-forces. Since the ball slides entirely in the x-direction, $a_y = 0$:

$$\sum F_y = ma_y$$
$$N - mg\cos\theta = 0,$$

and hence $N = mg\cos\theta$. Substitute that into Eq. (*), and solve for a_x to get $a_x = -(g\sin\theta + \mu g\cos\theta)$. Given that expression for acceleration, we can now write the ball's center-of-mass (linear) velocity as a function of time, using the old kinematic formula

$$\begin{aligned} v_x &= v_{0x} + a_x t \\ &= v_0 - (g\sin\theta + \mu g\cos\theta)t. \end{aligned} \tag{1}$$

Eq. (1) confirms our intuition that the ball slows down as it ascends the ramp.

<u>Step 2: Rotational motion reasoning to express ω</u>

Use the rotational analog of the reasoning in step 1. First, invoke the rotational version of Newton's 2nd law to find the angular acceleration α. Then use a kinematic formula to find ω.

Since the ball rotates around its center, choose the center as the "pivot" point. From the force diagram, the only torque is due to friction. This is because gravity always "pulls" on the center of the ball, and therefore the "r" in $\tau = F\perp r$ is zero. Also, the normal force points parallel to the radius vector connecting the center of the ball to the "contact point" where the ball touches the ramp. No component of N points perpendicular to this radius vector. Since $N\perp = 0$, that force generates no torque. Only friction does.

The torque due to friction makes the ball spin clockwise, the direction the ball will rotate when it starts to roll. Since the ball rotates clockwise when it rolls up the ramp, and since I have chosen up-the-ramp as positive, I must therefore choose clockwise as positive. (Otherwise, things go haywire when I set $v = R\omega$ in step 3 below.) So, the frictional torque is positive, even though friction points backward.

The frictional force points entirely perpendicular to the radius vector **R** connecting the center of the sphere to the "contact point" where friction acts. So, the torque due to friction is

$$\begin{aligned} \tau_{\text{fric}} &= f\perp R \\ &= \mu N R \\ &= \mu(mg\cos\theta)R, \end{aligned}$$

where I invoked our expression for N from step 1 above. We also know from the textbook table that $I_{\text{spherical shell}} = \frac{2}{3}mR^2$. So, Newton's 2nd law in rotational form gives us

$$\sum \tau = I\alpha$$
$$\mu(mg\cos\theta)R = \frac{2}{3}mR^2\alpha,$$

and hence the angular acceleration is $\alpha = \frac{3\mu g\cos\theta}{2R}$.

Now we can easily find the angular speed as a function of time, using the rotational analog of $v = v_0 + at$, namely

$$\begin{aligned} \omega &= \omega_0 + \alpha t \\ &= 0 + \frac{3\mu g\cos\theta}{2R}t. \end{aligned} \tag{2}$$

At this point, we have used purely linear reasoning to express the ball's velocity as a function of time, and purely rotational reasoning to express its angular speed as a function of time. So, we are ready to find when the ball starts to roll without slipping.

Step 3: Set $v = \omega R$, and solve for t, the time at which the ball starts to roll

When the ball starts rolling without slipping, its linear (center-of-mass) speed and angular speed are related by $v = R\omega$, the rolling without slipping condition. From Eq. (1) in step 1, we see that v starts "too big" but gradually decreases. And from Eq. (2) in step 2, we see that ω starts small but gradually increases. Eventually, ω gets big enough, and v gets small enough, that the rolling without slipping condition is satisfied.

To find out when this occurs, use Eq. (1) and Eq. (2) to set $v = R\omega$,

$$v_0 - (g\sin\theta + \mu g\cos\theta)t = R\left(\frac{3\mu g\cos\theta}{2R}t\right).$$

Solving for t gives us

$$t = \frac{v_0}{g(\sin\theta + \frac{5}{2}\mu\cos\theta)}.$$

That is the time at which the ball starts to roll without slipping. Notice that it does not depend on the ball's mass or radius.

Step 4: Kinematics to find the quantity of interest

We want the ball's position at time $t = \frac{v_0}{g(\sin\theta + \frac{5}{2}\mu\cos\theta)}$. So, we can try $x = x_0 + v_{0x}t + \frac{1}{2}a_x t^2$. In step 1, we found the ball's acceleration along the ramp to be $a_x = -(g\sin\theta + \mu g\cos\theta)$. And its initial velocity along the ramp is $v_{0x} = v_0$. So,

$$\begin{aligned} x &= x_0 + v_{0x}t + \frac{1}{2}a_x t^2 \\ &= 0 + v_0\frac{v_0}{g(\sin\theta + \frac{5}{2}\mu\cos\theta)} - \frac{1}{2}(g\sin\theta + \mu g\cos\theta)\left[\frac{v_0}{g(\sin\theta + \frac{5}{2}\mu\cos\theta)}\right]^2 \\ &= \ldots\text{(lots of algebra)}\ldots \\ &= \frac{v_0^2(\sin\theta + 4\mu\cos\theta)}{2g[\sin\theta + \frac{5}{2}\mu\cos\theta]^2}. \end{aligned}$$

By setting $\theta = 0$, you can find the ball's position when it starts rolling, if the experiment takes place on a flat floor instead of on a ramp.

REVIEW QUESTION 2-9

A spool of thread is approximately a uniform solid cylinder of radius $R = 0.010$ meters and mass $M = 0.030$ kg. Someone drops the spool, but pulls upward on the end of the thread exactly hard enough so that the spool "spins" in place, neither rising nor falling. When the spool was first dropped, it was not spinning. It starts spinning because the person pulls on the thread.

How much time does it take for the spool to complete 10 revolutions?

REVIEW ANSWER 2-9

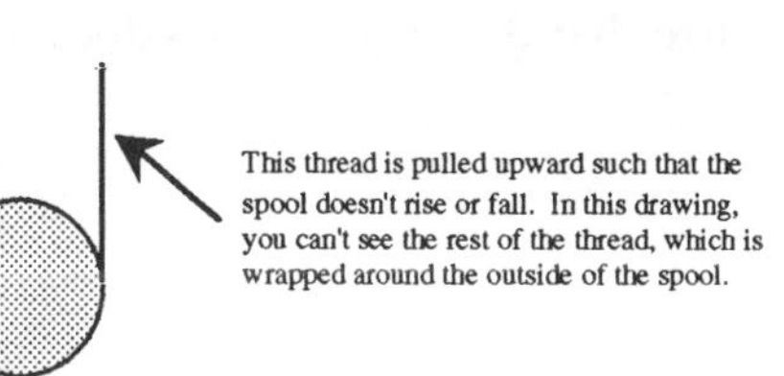

We want to know the time t it takes the spool to spin through 10 revolutions, which is

$$\theta = (10 \text{ rev})(\tfrac{2\pi \text{ rad}}{\text{rev}}) = 20\pi \text{ radians}.$$

So, we will need to use rotational kinematics. We know the initial angular speed, $\omega_0 = 0$. So, maybe we can use the rotational analog of $x = x_0 + v_0 t + \frac{1}{2}at^2$, namely $\theta = \theta_0 + \omega_0 t + \frac{1}{2}\alpha t^2$. But before we can solve that equation (or any other kinematic equation) for t, we need to know the angular acceleration, α.

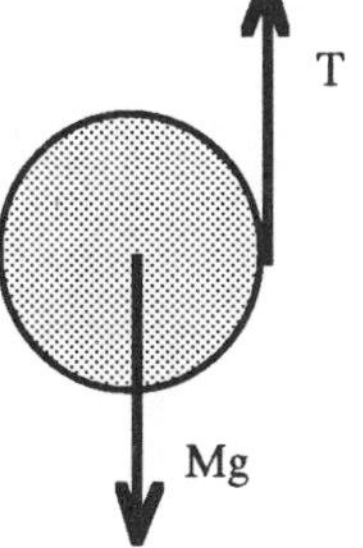

Well, just as we use forces to find accelerations, we can use torques to find the angular acceleration. Let us start with a force diagram. Gravity exerts no torque, because it acts directly on the "pivot point" around which the spool rotates. But the tension acts a distance $r = R$ from the pivot. Furthermore, T points entirely perpendicular to the (horizontal) radius vector connecting the pivot point to the contact point where the thread "touches" the spool. So, $\tau_{\text{tension}} = T\perp r = TR$. Since $I_{\text{solid cylinder}} = \frac{1}{2}MR^2$, Newton's 2nd law in rotational form gives us

$$\begin{aligned}\sum \tau &= I\alpha \\ TR &= \frac{1}{2}MR^2\alpha.\end{aligned} \qquad (1)$$

Unfortunately, we cannot yet solve for α, because we do not know the tension T in the rope.

To find T, you must keep in mind that the rotational and translational aspects of the spool's motion are independent. We can treat them separately. We just dealt with the rotational motion. So, let us look at the linear motion. The spool does not have any; it spins "in place." Therefore, its linear acceleration is $a_y = 0$. In other words, the *forces* cancel out, even though the torques do not:

$$\begin{aligned}\sum F_y &= Ma_y \\ T - Mg &= 0,\end{aligned} \qquad (2)$$

and hence $T = Mg$. *If the spool were accelerating up or down, then tension would not cancel gravity.* To maintain this constant tension, the person pulling on the thread must accelerate his hand upward. Try it!

Now that we know T, put it into Eq. (1) above and solve for the angular acceleration to get

$$\alpha = \frac{TR}{\frac{1}{2}MR^2} = \frac{(Mg)R}{\frac{1}{2}MR^2} = \frac{2g}{R}.$$

Given this angular acceleration, rotational kinematics gets us to the answer. Start with

$$\theta = \theta_0 + \omega_0 t + \frac{1}{2}\alpha t^2. \qquad (3)$$

We know $\omega_0 = 0$, and we can set $\theta_0 = 0$. We are seeking the time it takes the spool to spin through $\theta = 10 \text{ rev} = 20\pi$ radians. Throw all this into Eq. (3), and solve for t to get only a quarter of a second. Notice that the spool's mass does not matter; but its size does. Why is this, intuitively?

Review Problem Set (Chapters 1-22)

Number Three

25 CHAPTER

◆ *This collection of final exam review problems is particularly hard. The practice final exams later in this study guide are a bit easier, on average. Do not get psyched out if you cannot solve all these review problems. Use them to identify your weak spots and to practice breaking up long problems into manageable pieces.*

REVIEW QUESTION 3-1

If a planet spins too quickly on its axis, loose material on the equator will "fly off" the planet. Explain why, without using the words "centrifugal force." Then derive an expression for ω, the *biggest* angular speed a planet can have such that loose material on the equator (barely) stays on the planet. Express your answer for ω as a function of G, the planet's mass M, and the planet's radius R.

REVIEW ANSWER 3-1

First, let us answer the "why?" Suppose a person of mass m stands on the equator. Because the rotating planet "carries along" the person, the person travels in a circle around the center of the planet. Therefore, the net radial force on her must be $\frac{mv^2}{R}$; otherwise, the person would not move in a circle. For instance, if no forces were acting on her, the person would move in a straight line instead of a circle, and would therefore "fly off" the planet.

Here, the radial force is provided by gravity. If the planet spins too quickly, then v is so high that $\frac{mv^2}{R}$ exceeds the gravitational force on the person. In this case, gravity is not strong enough to keep the person moving in a circle; she "flies off" the planet. My point here is that no force tries to make the person fly off the planet. What makes the person fly off is her natural inertial tendency, in the absence of forces, to travel in a straight line at constant speed. The gravitational force must be strong enough to "bend" that natural straight-line motion into a circle.

Using this reasoning, I will now derive an expression for ω, the biggest planetary angular velocity such that the person (or pebble or whatever) stays in contact with the planet. You may think we need to know m, the person's mass. As always, if you are missing what seems like an essential quantity, introduce it and hope it cancels out of your final answer.

Let us start by considering the forces acting on the person at the equator. There is gravity, obviously. There is also an upward normal force exerted by the ground on the person. But if ω is so big that the person *barely* stays in contact with the ground, then her shoes barely press against the ground, and vice versa. In other words, the normal force is barely greater than 0. So, we can set $N \approx 0$. This leaves gravity as the only appreciable force. Gravity pulls her radially toward the center of the planet.

That is good to know, because she travels in a circle of radius R. Therefore, she must undergo a radial acceleration $a_{\text{radial}} = \frac{v^2}{R}$. So,

$$\sum F_{\text{radial}} = ma_{\text{radial}}$$
$$\frac{GMm}{R^2} = m\frac{v^2}{R}, \tag{1}$$

where I have used the formula for the gravitational attraction between two bodies. We can immediately solve for v, the person's linear speed. But we want her angular speed. Well, since the person traces out a circle, her linear and angular speed are related by

$$v_{\text{tang}} = r\omega, \tag{2}$$

where v_{tang} is the velocity *along* the circle. Here, all of the person's velocity is tangential. And $r = R$, the planet's radius. Substitute this into Eq. (1) to get

$$\frac{GMm}{R^2} = m\frac{v^2}{R} = m\frac{(R\omega)^2}{R} = mR\omega.$$

Solve for ω to get $\omega = \sqrt{\frac{GM}{R^3}}$.

It is interesting to plug in the Earth's mass and radius. You get $\omega = 0.00124$ rad/s, which corresponds to a period of $T = \frac{2\pi}{\omega} = 5060 \text{ seconds} = 1.4 \text{ hours}$. So, if the Earth's "day" were shorter than 1.4 hours, Brazilians would have a serious problem.

REVIEW QUESTION 3-2

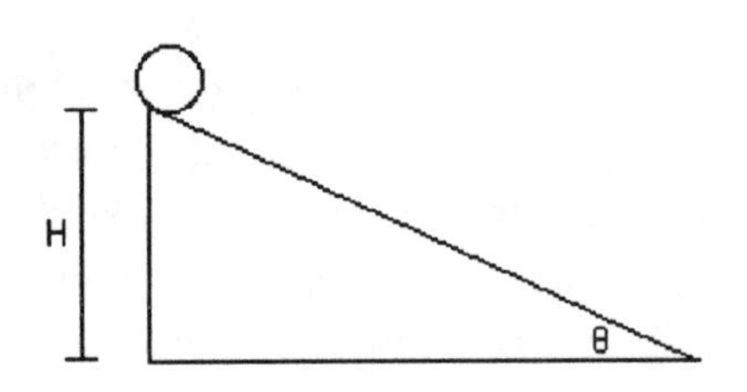

An ice sphere of mass M and radius R is released from rest from the top of a ramp of vertical height H and ramp angle θ. The coefficient of friction, μ, is small. As a result, the sphere does not roll without slipping. Instead, it slides down the ramp, gradually acquiring some rotational motion as it slides. Crucially, it never acquires enough rotational motion to roll without slipping.

What is the sphere's angular momentum when it reaches the bottom of the ramp? Answer in terms of M, R, H, θ, μ, and g.

REVIEW ANSWER 3-2

It is not clear how to approach this problem. Since we are solving for angular momentum, you might try angular momentum conservation. But angular momentum is not conserved, because an external torque (due to friction) acts on the ice as it slides down the ramp. Consequently, the ice sphere ends up with some angular momentum, even though it started off with none.

Since the ice slides down a ramp, you might think to use energy conservation. But as the ice frictionally rubs against the ramp, heat gets generated, and we do not know how much. If the ice were *rolling*, then the frictional force on it would be *static* friction, which involves no "rubbing." That is why you can treat rolling objects with energy conservation. But here, the friction is *kinetic*; the ice rubs against the ramp, generating heat. So, energy conservation gets us nowhere.

Since conservation laws do not work here, we have no choice but to use old-fashioned forces and/or torques. Let us figure out what is happening. As the ice sphere slides down the ramp, it spins faster and faster. If we can somehow figure out its angular speed ω at the bottom of the ramp, then we can easily calculate the angular momentum using $L = I\omega$. How can we find ω? Well, in the linear analog of this problem, we would be asked for the linear speed v at the bottom of the ramp. To find it, we would use Newton's 2nd law to calculate the acceleration, and then we would use kinematics. Analogous reasoning works here. We can use $\sum \tau = I\alpha$ to calculate the angular acceleration. Then we can use rotational kinematics to find ω.

As always when using forces or torques, start with a force diagram.

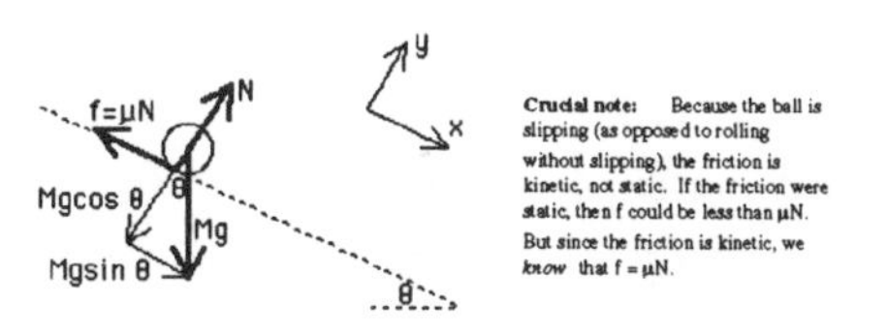

We want the net torque. Well, gravity exerts no torque, because it acts directly on the center of the sphere, which is the point around which the sphere rotates. So, the "r" in $\tau = F_\perp r$ is 0. The normal force exerts no torque, either. Here is why: For N, the relevant "radius vector" **r** goes from the center of the sphere to the contact point where the sphere touches the ramp. **N** points radially along that radius vector. No component of **N** is perpendicular to **r**: $N_\perp = 0$. Therefore, the normal force generates no torque. Only friction torques the sphere.

Because the friction is kinetic, we can set $f = \mu N$. *(For static friction, f can be less than μN. Indeed, for most rolling objects, $f < \mu N$.)* To figure out N, we must consider the y-directed forces. Since the sphere moves entirely in the x-direction (along the ramp), $a_y = 0$:

$$\sum F_y = ma_y$$
$$N - Mg\cos\theta = 0,$$

and hence $N = Mg\cos\theta$. So, the frictional force is $f = \mu N = \mu Mg\cos\theta$.

Now we can calculate the torque and use Newton's 2nd law in rotational form. Friction acts a distance $r = R$ from the pivot point (the center), and points perpendicular to the relevant radius vector. So, $\tau_{\text{fric}} = f_\perp r = (\mu Mg\cos\theta)R$. Therefore

$$\sum \tau = I\alpha$$
$$(\mu Mg\cos\theta)R = \frac{2}{5}MR^2\alpha,$$

where I invoked the textbook value for $I_{\text{solid sphere}}$. Solve for the angular acceleration to get $\alpha = \frac{5}{2}\mu g\cos\theta$.

The whole point of calculating a is to enable us to find the sphere's angular velocity at the bottom of the ramp, using rotational kinematics. For instance, we could try the rotational analog of $v = v_0 + at$, namely $\omega = \omega_0 + \alpha t$. But we do not know how much time t the sphere takes to reach the bottom. (Similarly, if we try to use $\omega^2 = \omega_0^2 + 2\alpha\Delta\theta$, we do not know the angle $\Delta\theta$ through which the sphere rotates while sliding down.) So, somehow or other, we must find the "slide time" t.

To escape this impasse, you must realize that when an object undergoes both translational (linear) and rotational motion, those two kinds of motion are *independent*. We can treat them separately. For instance, while solving for the *angular* acceleration, we completely ignored the fact that the ice also moves linearly. Similarly, we can address its linear motion with separate equations that ignore the sphere's rotation. Here is the upshot. We can use good old-fashioned forces and acceleration to find how much time the sphere takes to reach bottom. This reasoning will not take into account the sphere's rotational motion.

To calculate the sphere's "slide time" t, we need to know its acceleration down the ramp, a_x. So, we must consider the x-directed forces. Above, we obtained $f = \mu N = \mu Mg\cos\theta$. So, from the force diagram,

$$\sum F_x = Ma_x$$
$$Mg\sin\theta - \mu Mg\cos\theta = Ma_x,$$

and hence $a_x = g(\sin\theta - \mu\cos\theta)$. From trig, we know that the height and length of the ramp are related by $H = x\sin\theta$, and hence $x = \frac{H}{\sin\theta}$. So, let us use our favorite kinematic formula,

$$x = x_0 + v_{0x}t + \frac{1}{2}a_x t^2$$
$$\frac{H}{\sin\theta} = 0 + 0 + \frac{1}{2}g(\sin\theta - \mu\cos\theta)t^2,$$

which we can immediately solve for t to get $t = \sqrt{\frac{2H}{g(\sin^2\theta - \mu\sin\theta\cos\theta)}}$.

Now that we know how much time the sphere takes to reach the bottom, we can solve for its angular speed at that point. Remember, we found above (using rotational reasoning) that $\alpha = \frac{5}{2}\mu g\cos\theta$.

$$\omega = \omega_0 + \alpha t$$
$$= 0 + \frac{5}{2}\mu g\cos\theta\sqrt{\frac{2H}{g(\sin^2\theta - \mu\sin\theta\cos\theta)}}$$

So, the sphere's angular momentum at the bottom of the ramp is

$$L = I\omega = \frac{2}{5}MR^2\left[\frac{5}{2}\mu g\cos\theta\sqrt{\frac{2H}{g(\sin^2\theta - \mu\sin\theta\cos\theta)}}\right].$$

This may have become confusing. Let me reiterate the overall strategy. Using purely linear-motion reasoning (forces and acceleration and linear kinematics), we calculated how much time the ice takes to reach the bottom of the ramp. Then, using purely rotational reasoning (torques and angular acceleration and rotational kinematics), we calculated the ice sphere's angular speed at that time. (Actually, I carried out these steps in reverse order; but the logic is the same.) So, this example illustrates the general problem-solving hint that you can *separately* deal with the linear and rotational aspects of an object's motion.

REVIEW QUESTION 3-3

A small cube of light plastic, of density $.8\rho_w$ (where ρ_w is the density of water) and side length s, is placed at the bottom of a tall barrel of water, and released. The barrel has height H, where $H >> s$. The top of the barrel is open to the air. Sitting on the surface of the water at the top of the barrel is a beach ball of mass M. The cube floats upward and eventually *elastically* collides head-on with the beach ball, knocking the beach ball straight up into the air. The drag forces due to water and air are negligible. Express your answers in terms of ρ_w, s, H, M, and g.

(a) How high above the surface of the water does the beach ball reach (at its peak)? *Do not complete the algebra*; but set up all the necessary equations, and clearly indicate how you would arrive at the final answer.

(b) During the collision, how much work is done by the cube on the beach ball? Again, you need not complete the algebra.

REVIEW ANSWER 3-3

(a) This is a multistep problem. To formulate a strategy, let us picture what is going on. The cube floats upward toward the surface, accelerating due to the buoyant force. It then collides with the beach ball. During the collision, the cube probably "bounces" back downward into the water, while the beach ball gets knocked straight up into the air. Once in the air, the beach ball becomes a projectile, slowing down as it rises, eventually coming to rest at its peak.

To organize your thoughts, diagram the different stages. I will let little v denote the cube's speed, while big V denotes the beach ball's speed. Stage 1 is right before the collision, and stage 2 is right after the collision.

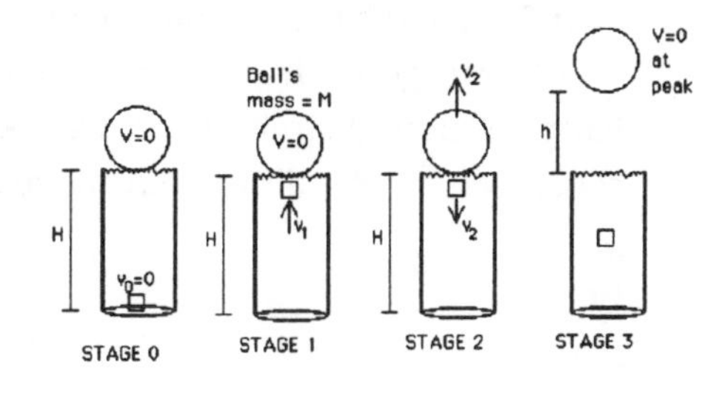

We are solving for h. These pictures suggest a three-step strategy.

(1) Using Archimedes' law, calculate the net force on the cube while it floats upwards. From this net force, obtain its upward acceleration. Then use the acceleration to find v_1, the cube's speed just before crashing into the beach ball.

(2) With collision reasoning (conservation of momentum and, if necessary, conservation of energy), calculate the post-collision speed of the beach ball. Call that speed V_2.

(3) V_2 is the beach ball's initial speed as it shoots upward. So, we can use projectile reasoning, or conservation of energy if you prefer, to figure out the ball's peak height.

As I work through each step, keep the overall strategy in mind.

Step 1: Forces and acceleration and kinematics to find v_1

First, I must find the net force on the cube while it is immersed. As always, start with a force diagram.

F_b denotes the buoyant ("floating") force, and m denotes the cube's mass.

Since density is mass per volume ($\rho = \frac{m}{V}$), the mass of the cube is $m = \rho_{cube} V_{cube} = .8\rho_w s^3$ because s^3 is the cube's volume.

The buoyant force equals the weight of fluid displaced, according to Archimedes' law. Well, the volume of water displaced by the fully-immersed cube is the whole cube's volume: $V_{\text{dis}} = s^3$. Therefore, the mass of water displaced is density times volume, $\rho_w V_{\text{dis}} = \rho_w s^3$. So, the weight of water displaced equals that mass times g, $\rho_w s^3 g$. That is the buoyant force.

Putting all this together, and using the force diagram, we get

$$\begin{aligned}\sum F_y &= ma_y \\ F_b - mg &= ma_y \\ \rho_w s^3 g - (.8\rho_w s^3)g &= (.8\rho_w s^3)a_y.\end{aligned}$$

Solve for a_y to get $a_y = \frac{g}{4}$.

Given this upward acceleration, we can immediately find the cube's speed v_1 after traveling through distance $\Delta y = H$, by using an old kinematic formula:

$$\begin{aligned}v_1^2 &= v_0^2 + 2a_y \Delta y \\ &= 0 + 2\left(\frac{1}{4}g\right)H \\ &= \frac{gH}{2},\end{aligned}$$

and hence the cube's speed immediately before striking the beach ball is $v_1 = \sqrt{\frac{gH}{2}}$.

Step 2: Collision reasoning (conservation of momentum and/or energy) to find V_2

We are going from stage 1 (before collision) to stage 2 (after collision). Since everything happens along a straight line, we need not worry about vector components when using momentum conservation. Recalling from above that the cube's mass is $m = .8\rho_w s^3$, we get

$$\begin{aligned}p_1 &= p_2 \\ mv_1 + 0 &= -mv_2 + MV_2 \qquad (1) \\ .8\rho_w s^3 v_1 &= -.8\rho_w s^3 v_2 + MV_2.\end{aligned}$$

(I have guessed that the cube bounces backwards off the ball. Hence, the minus sign.)

Even though we know v_1, we lack sufficient information to solve for V_2, because the post-collision speed of the cube (v_2) is also unknown. Fortunately, we can generate another equation using energy conservation. Since the collision is elastic, no heat gets produced. Also, during the collision, the ball doesnot have time to rise (more than a millimeter or so) above its initial height. Therefore, the pre-collision potential energy equals the post-collision potential energy. Only kinetic energy gets exchanged. So,

$$\begin{aligned}E_1 &= E_2 \\ \frac{1}{2}mv_1^2 + 0 &= \frac{1}{2}mv_2^2 + \frac{1}{2}MV_2^2. \qquad (2) \\ \frac{1}{2}(.8\rho_w s^3)v_1^2 &= \frac{1}{2}(.8\rho_w s^3)v_2^2 + \frac{1}{2}MV_2^2.\end{aligned}$$

Since we found v_1 in step 1 above, equations (1) and (2) contain just two unknowns, v_2 and V_2. Therefore, we can solve for V_2, the ball's speed right after the collision. The algebra takes a long time. On a test, some instructors would award partial credit for writing, "Use these 2 equations in 2 unknowns to solve for V_2," and then continuing to step 3. (Check your instructor's policy on this.)

Step 3: Find the ball's peak height using projectile reasoning or energy conservation

V_2 is the ball's "initial" speed (what we usually call "v_0") as it shoots up into the air We are solving for h, the maximum height it reaches above the water's surface. At its peak, the ball momentarily comes to rest. I see two good ways to find h given V_2.

Method 1: Energy conservation. Since the ball no longer interacts with the cube, we can treat the ball as a "lone" system. Immediately after getting struck, it has kinetic but no potential energy (assuming we measure our heights from the water's surface). At its peak, the ball has potential but no kinetic energy:

$$E_{2,\text{ ball}} = E_{\text{peak, ball}}$$
$$K_2 + U_2 = K_{\text{peak}} + U_{\text{peak}}$$
$$\frac{1}{2}MV_2^2 + 0 = 0 + Mgh.$$

Solve for h to get $h = \frac{V_2^2}{2g}$.

Method 2: Kinematics. Starting with the kinematic formula $V^2 = V_0^2 + 2a\Delta y$, we can substitute in $V = 0$ (at the peak), "V_0" $= V_2$, $a = -g$ (downward acceleration due to gravity), and $\Delta y = h$, to get $0 = V_2^2 + 2(-g)h$. Solve for h to get $h = \frac{V_2^2}{2g}$, the same answer obtained using energy conservation.

(b) You might try starting with $W = \mathbf{F} \cdot \Delta \mathbf{x}$. But that gets us nowhere, because we do not know the average force exerted by the cube on the ball.

To solve this problem, you must remember the work-energy theorem, according to which the net work done on an object equals the object's change in kinetic energy:

$$W_{\text{net}} = \Delta K \qquad \textbf{Work-energy theorem}$$

Intuitively, this formula says that when work gets done on an object, the object speeds up or slows down. During the brief collision, the only force doing work on the beach ball is the "contact" force of the cube. (Gravity and a buoyant force also act on the ball, but they cancel each other.) So, in this case, $W_{\text{net}} = W_{\text{by cube}}$. Therefore, to calculate the work done by the cube, we just need to find how much the ball's kinetic energy changes during the collision.

Well, its pre-collision kinetic energy is $K_1 = 0$. And its post-collision kinetic energy (*immediately* after the collision) is $K_2 = \frac{1}{2}MV_2^2$. Putting all this together yields

$$W_{\text{by cube}} = W_{\text{net}} = \Delta K$$
$$= \frac{1}{2}MV_2^2 - 0.$$

Remember you could obtain V_2 by completing the algebra in step 2 above. But this problem did not require you to do so.

REVIEW QUESTION 3-4

Consider two ropes, both of length $D = 9.0$ meters. The first rope, made of yarn, has mass $M = 1.0$ kg; while the second rope, made of hemp, has mass $4M = 4.0$ kg. A long rope of length $2D = 18$ meters is created by tying together the two separate ropes.

One end of the tied-together rope is attached to a wall, while the other end is held by a person, who creates a tension $F_T = 100$ Newtons in the rope. The person also shakes her end of the rope, making it oscillate up and down once every 0.20 seconds. Although the rope "sags" in the middle, let us say it is approximately horizontal.

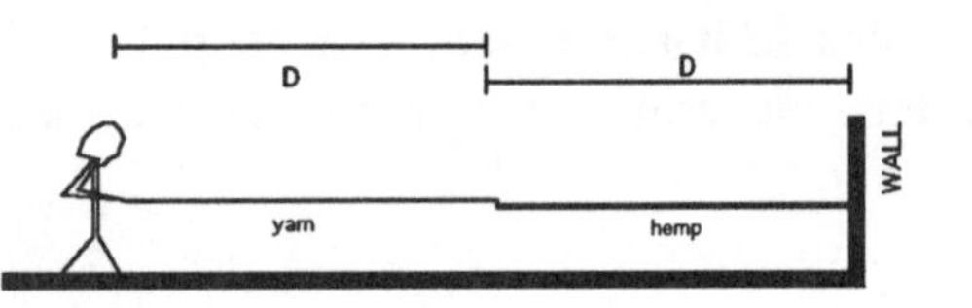

(a) How much time does a wave take to travel from the person's hand to the wall?

(b) What is the wavelength of waves on the hemp segment of rope?

REVIEW ANSWER 3-4

(a) A common mistake is to calculate the "average" mass density as $\mu_{av} = \frac{\text{mass}}{\text{distance}} = \frac{M+4M}{2D} = \frac{5\text{ kg}}{18\text{ m}}$, and then use μ_{av} to calculate the velocity. But the v you calculate in this way is *not* the correct average velocity, partly because the waves do not spend equal time on the two segments of rope.

You need to *separately* calculate v_{yarn} and v_{hemp}, the wave speed on the yarn and hemp segments. Using those speeds, you can separately calculate t_{yarn} and t_{hemp}, the time it takes a wave to cross the yarn segment and hemp segment. The total travel time from the person's hand to the wall is $t_{\text{total}} = t_{\text{yarn}} + t_{\text{hemp}}$.

As discussed *ad nauseam* in Chapter 21, *the speed of waves depends only on properties of the medium*, not on the wavelength or frequency. So, in this part of the problem, we will ignore the information about how frequently the rope gets shaken. If it were shaken faster or slower, the waves would still travel at the same speed.

Wave velocity increases with the "tautness" of the rope, and decreases with "heaviness": $v = \sqrt{\frac{\text{Tension}}{\mu}}$. On both segments of rope, the tension is $F_T = 100$ N. (Tension "equalizes" over the whole rope, if it is approximately horizontal.) So,

$$v_{\text{yarn}} = \sqrt{\frac{F_T}{\mu_{\text{yarn segment}}}} = \sqrt{\frac{F_T}{\frac{M}{D}}} = \sqrt{\frac{100\text{ N}}{\frac{1\text{ kg}}{9\text{ m}}}} = 30\text{ m/s}.$$

$$v_{\text{hemp}} = \sqrt{\frac{F_T}{\mu_{\text{hemp segment}}}} = \sqrt{\frac{F_T}{\frac{4M}{D}}} = \sqrt{\frac{100\text{ N}}{\frac{4\text{ kg}}{9\text{ m}}}} = 15\text{ m/s}.$$

Waves travel twice as fast on the lighter (yarn) segment.

Since the wave speed is constant on a given segment of rope, we can use $v = \frac{\Delta x}{\Delta t}$. Solve for t to get

$$t_{\text{yarn}} = \frac{\Delta x_{\text{yarn}}}{v_{\text{yarn}}} = \frac{D}{\sqrt{\frac{F_T}{\frac{M}{D}}}} = \frac{9\text{ m}}{30\text{ m/s}} = 0.30\text{ s}.$$

$$t_{\text{yarn}} = \frac{\Delta x_{\text{hemp}}}{v_{\text{hemp}}} = \frac{D}{\sqrt{\frac{F_T}{\frac{4M}{D}}}} = \frac{9\text{ m}}{15\text{ m/s}} = 0.60\text{ s}.$$

So, the total "travel time" is $t_{total} = t_{yarn} + t_{hemp} = 0.30\text{ s} + 0.60\text{ s} = 0.90\text{ s}$.

(b) Now that we know the speed of waves on the hemp segment, we can use

$$v_{hemp} = \lambda_{hemp} f_{hemp} \quad (1)$$

to solve for the wavelength on the hemp segment.

This raises a crucial conceptual issue: When the waves go from yarn to hemp, and therefore slow down, does their wavelength stay the same, and their frequency decrease? Or does their frequency stay the same, and their wavelength decrease? And does the answer depend on the fact that "part" of the wave reflects off the yarn/hemp interface instead of getting transmitted to the hemp?

To answer this, you must recall a fundamental fact about waves: *The frequency of waves depends on the "driving" frequency creating the waves, not on the medium.* Therefore, whatever frequency the waves have on the yarn segment, they keep on the hemp segment. To see why, let us hypothetically pretend that the frequency *does* change when the waves cross from yarn to hemp. Suppose that we have $f = 4$ waves/second on yarn, but only $f = 3$ waves/second on hemp. (Those 4 waves/second include the effects of reflected waves.) If this were the case, however, then 1 wave per second must "disappear" as the waves cross from yarn to hemp. Waves can speed up or slow down, but they *cannot* abruptly disappear. This goes to show that the frequency of waves on the hemp cannot be lower or higher than the frequency of waves on yarn. The frequency must be the same on both segments. If four waves per second are created on the yarn segment, then 4 waves per second must reach the other end of the rope, assuming the waves do not "reflect completely" at the yarn/hemp interface. So, when the medium changes, the frequency stays the same, while the velocity and wavelength "adjust." This happens when light goes from air into glass, or when sound goes from helium into air.

We are told that the person generates one full wave every 0.2 seconds. So, the period is $T = 0.2$ seconds. If you thought 0.2 was the frequency, remember that period is the amount of time per cycle, while frequency is the number of cycles per time. Since a cycle (oscillation) takes $T = 0.2$ seconds, the frequency is $f = \frac{1}{T} = \frac{1}{0.20\text{ s}} = 5.0\text{ s}^{-1}$. Intuitively speaking, since the person generates a wave every 0.2 seconds, she makes 5 waves per second.

In this problem, since $f_{yarn} = f_{hemp} = 5.0\text{ s}^{-1}$, we can solve Eq. (1) above to get

$$\lambda_{hemp} = \frac{v_{hemp}}{f_{hemp}} = \frac{15\text{ m/s}}{5.0\text{ s}^{-1}} = 3.0\text{ meters}.$$

REVIEW QUESTION 3-5

Every hundred million years or so, a huge meteor crashes into the Earth, disrupting the climate. A crackpot scientist came up with the following scheme to protect the Earth: Figure out where the meteor is going to land, and place a huge spring under it, so that the meteor "bounces" back into outer space.

Suppose a meteor of mass m has speed v_0 when it is *very* far from Earth. The meteor approaches the Earth and lands on top of a stiff spring of equilibrium length L, as drawn below. (I have drawn the spring and meteor much bigger than they really are in relation to Earth.) The mass and radius of the Earth are M and R, where $R >> L$. You may assume that the Earth is so much more massive than the meteor that the Earth stays essentially "fixed in place."

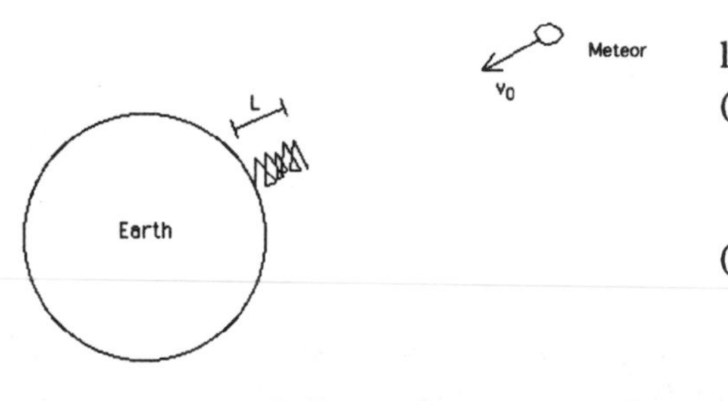

Express your answers in terms of the quantities listed above, and universal constants.

(a) What is the meteor's speed when it reaches the top of the spring? (You may neglect air resistance, since we are just making a rough calculation.)

(b) What would the spring constant have to be to prevent the meteor from reaching the Earth?

REVIEW ANSWER 3-5

The meteor speeds up as it approaches the Earth, due to the gravitational attraction.

(a) You might be tempted to use kinematics. But as the meteor gets closer to Earth, the gravitational force on it increases. Therefore, its acceleration increases. Because the acceleration is not constant, you cannot use our old constant-acceleration kinematic formulas. Instead, we must use conservation, in this case, energy conservation.

As always, make diagrams. The "initial" situation is drawn in the problem, though the meteor should be an effectively infinite distance from Earth. By subscript "t," I mean the situation when the meteor reaches the top of the spring.

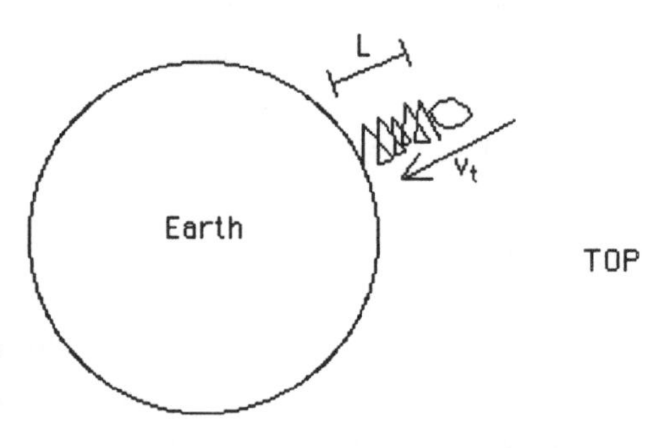

The meteor's initial distance from the center of the Earth is $r_i \approx \infty$. When it reaches the top of the spring, its distance from the Earth's center is $r_t = R + L$. I will set $R + L \approx R$, because L is tiny compared to R. Since the potential energy due to gravity is $U = -\frac{GMm}{r}$, we get

$$E_0 = E_t$$

$$K_0 + U_0 = K_t + U_t$$

$$\frac{1}{2}mv_0^2 + \left(-\frac{GMm}{\infty}\right) = \frac{1}{2}mv_t^2 + \left(-\frac{GMm}{R}\right),$$

where you could use $U_t = -\frac{GMm}{R+L}$ if you prefer. Notice that $U_0 = -\frac{GMm}{\infty} = 0$. Solve for v_t to get

$$v_t = \sqrt{v_0^2 + \frac{2GM}{R}}.$$

(b) If the spring barely prevents the meteor from reaching the Earth, the meteor stops just before reaching the surface of the Earth, i.e., a billionth of a centimeter above the surface. So, we can say the meteor momentarily comes to rest right at the surface. In other words, the spring must slow down the meteor quickly enough so that the meteor comes to rest ($v_f = 0$) just as the spring is fully compressed (through distance $x = L$). Here is the "final" situation just described.

In the final configuration, the spring has energy $U_{spring} = \frac{1}{2}kL^2$, because it is compressed a distance $x = L$ from equilibrium. When writing the gravitational potential energy, I will take advantage of the fact that $L << R$, and hence, the meteor's gravitational energy changes negligibly as it compresses the spring.

Conservation of energy tells us that $E_0 = E_t = E_f$. Therefore, we proceed either by setting E_f equal to E_t, or by setting E_f equal to E_0. I will choose the first option

$$E_t = E_f$$
$$K_t + U_t = K_f + U_f$$
$$\frac{1}{2}mv_t^2 + \left(-\frac{GMm}{R}\right) = 0 + \left(-\frac{GMm}{R}\right) + \frac{1}{2}kL^2$$
$$\frac{1}{2}m\left(v_0^2 + \frac{2GM}{R}\right) + \left(-\frac{GMm}{R}\right) = 0 + \left(-\frac{GMm}{R}\right) + \frac{1}{2}kL^2,$$

where in the last step I used our part (a) answer for v_t. The left-hand side of the equation simplifies to $\frac{1}{2}mv_0^2$. (That is because $E_t = E_0$.) So, we can easily solve for k to get

$$k = \frac{mv_0^2 + \frac{2GMm}{R}}{L^2}.$$

Although you were not required to do so, let me plug in some realistic numbers to see if Earth can be saved. I will consider a best-case scenario. The meteor starts out very slow ($v_0 \approx 0$), and is small ($m = 10000$ kg, the mass of about 160 people). Let us also say the spring is huge, about $L = 100$ meters long (the length of a football field!). Plugging in these numbers yields $k \approx 10^8$ N/m, as stiff as a rock. A spring this stiff would shatter, not compress, when the meteor crashes into it.

REVIEW QUESTION 3-6

Consider a solid cylinder (mass m, radius R) with an essentially massless axle through its center. The cylinder is free to rotate around its axle. A spring (of spring constant k) is attached to the axle, and the other end is attached to a wall, as drawn below. The cylinder is placed such that the spring is stretched from its equilibrium length. As a result, the cylinder *rolls without slipping* toward the wall, until the spring becomes compressed, at which point the cylinder rolls the other way. In other words, the cylinder oscillates back and forth, rolling without slipping the whole time.

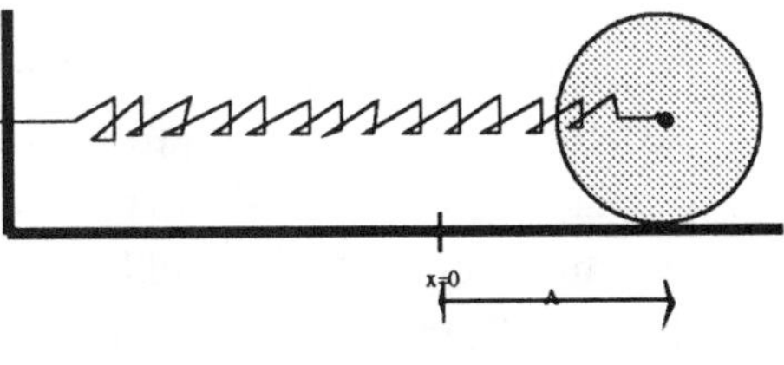

(a) The cylinder is released from rest at $t = 0$ from $x = A$. Draw a *rough* sketch of the cylinder's center-of-mass position vs. time, with $x = 0$ corresponding to the spring's equilibrium position.

(b) *(Very hard!)* What is the frequency of oscillation (i.e., how many oscillations per second does the cylinder undergo)? *Hint:* The friction here is static, not kinetic. So, the frictional force might be less than μN.

REVIEW ANSWER 3-6

(a) Although this is an exotic example of oscillatory (simple harmonic) motion, it is still simple harmonic motion. The cylinder oscillates back and forth, just like a block on a spring. So, as we have seen before, the position oscillates sinusoidally.

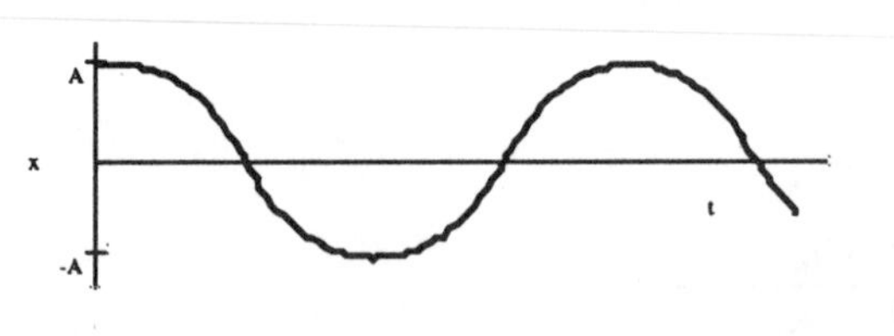

Here, the cylinder oscillates between $x = +A$ and $x = -A$. At $t = 0$, it starts at $x = +A$. So, we need to draw a sinusoidal-shaped curve that "starts" at $x = +A$ and reaches $x = -A$ at its bottom-most point.

It is a cosine.

(b) Although it is tricky, you can use the standard oscillation problem-solving strategies introduced in Chapter 20. Let me briefly review the central concepts.

A system undergoes oscillatory motion when the net force tends to push the system back toward equilibrium. For instance, in this problem, when the cylinder is to the right of $x = 0$, the stretched spring pulls it leftward. And when the cylinder is to the left of $x = 0$, the compressed spring pushes it rightward. To produce simple harmonic motion, the "restoring" force must be proportional to the system's displacement from equilibrium:

$$F_{\text{net}} = -(\text{stuff})x = ma, \tag{1}$$

where x is the displacement from equilibrium. "Stuff" denotes the relevant proportionality constant. For a *frictionless* mass on a spring, $F_{\text{net}} = -kx$, and hence stuff $= k$. For other oscillating systems, *including this one*, stuff equals something else. (As we will see below, friction plays a role, and enters into "stuff." Intuitively, friction slows everything down.) For *any* harmonic oscillator, however, the net force takes the form of Eq. (1).

Since $a = \frac{d^2x}{dt^2}$, Eq. (1) is a second-order differential equation. Using calculus, we can solve for x as a function of time. Doing so yields

$$x = A\cos(\omega t + \phi), \tag{2}$$

where

$$\omega = \sqrt{\frac{\text{stuff}}{m}}. \tag{3}$$

(Some books use sine instead of cosine. It does not matter, because "shifting" the phase factor ϕ by 90° turns a sine curve into a cosine curve.) In Eq. (2), A is the amplitude, i.e., the maximum displacement the system ever reaches from equilibrium. The phase constant ϕ, which encodes the "starting point" of the oscillation, is not worth discussing here. From our part (a) graph, we see that $\phi = 0$; the curve is an "unshifted" cosine. The angular frequency ω encodes how quickly the system oscillates. In fact, ω is related to the oscillator's frequency and period by

$$\omega = 2\pi\nu = \frac{2\pi}{T}. \tag{4}$$

where v is the number of oscillations per second, and T is the time needed to complete one oscillation. I am using "v" instead of "f" to denote frequency, so as not to confuse frequency with friction.

Do not confuse angular frequency with angular speed. For instance, the "ω" in Eq. (3) does *not* equal $\frac{v}{R}$, or $\omega_0 + \alpha t$, or any other expression for angular velocity.

Here, we are solving for v. To find it using Eq. (4), we need to know the angular frequency ω. And to find the angular frequency using Eq. (3), need to know "stuff," the proportionality constant in Eq. (1). That is, we need to calculate the net force with which the system gets yanked back toward equilibrium when displaced a distance x. Here is the technique.

Stuff-finding strategy:

(1) Draw a force diagram of the system when it is displaced an arbitrary distance x from equilibrium. (Do not draw the system at equilibrium, because then the net force on it will be 0.)

(2) Write down the net force along the direction of motion.

(3) By writing that net force in the form $F_{net} = -(\text{stuff})x$, figure out stuff.

Once you find stuff, just substitute it into Eq. (3) to calculate the system's angular frequency. And once you know ω, you can substitute it into $x = A\cos(\omega t + \phi)$ or into $v = \frac{\omega}{2\pi}$ to find practically anything you want. I will now implement this strategy.

Stuff-finding step 1: Force diagram

When stretched through distance x, the spring exerts a force $F_{spring} = -kx$ on the center of the cylinder. The minus sign means the force points "backwards" toward equilibrium. So here, the "backwards" (negative) direction is leftward. Gravity pulls downward on the center; while the normal force pushes upward on the bottom. Less apparent, though crucial, is friction. *If there were no friction, the cylinder would slide across the floor like a block, instead of rolling.*

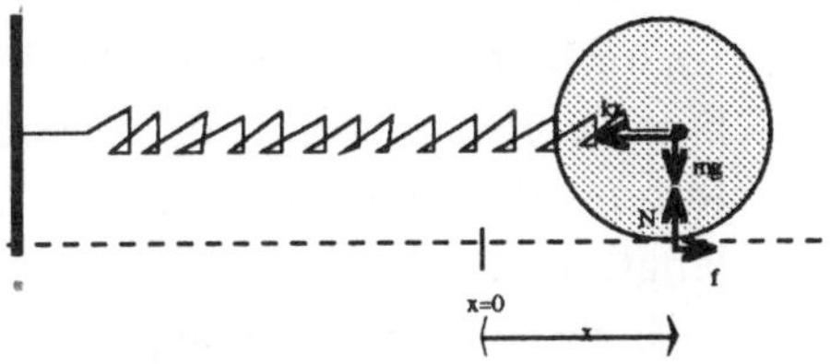

Remember, "f" denotes friction instead of frequency. When the cylinder rolls leftward from the position shown above, friction tends to make it rotate faster and faster counterclockwise. Therefore, since friction acts on the bottom, it must point rightward, in order to "torque" the cylinder counterclockwise.

Stuff-finding step 2: Write the net force along direction of motion

As just noted, the leftward direction is negative, and hence rightward counts as positive. So, the net force along the direction of motion is clearly $F_{net} = f - kx$. At this point, a common–indeed, almost universal–error is to set $f = \mu N$. Although it is true that *kinetic* friction always equals μN, static friction can be less than or equal to μN. I explain why this is so in Chapter 9, Question 9-3.

In this problem, static friction continually adjusts itself so that is it is exactly big enough to ensure that the cylinder rolls without slipping. Usually, if not always, it will be less than μN. So we must solve a separate "subproblem" to find the friction needed to make the cylinder roll. As you work through this subproblem, keep in mind how it fits into the overall strategy. We want to know f so that we can write the net force on the cylinder, and thereby find "stuff." Once we have stuff, we will use it to calculate the angular speed ω, which will immediately allow us to find the frequency of oscillation.

Subproblem: Find the friction f needed to make the cylinder roll.

When an object rolls without slipping, its linear and angular speed are related by $v = R\omega$, where ω is the angular velocity of the rotating cylinder, not the angular frequency of its oscillations. (It is confusing that the same symbol gets used to denote two different, though closely related, quantities.) To solve this problem, you must rewrite the rolling without slipping condition in terms of accelerations instead of velocities. To do so, differentiate both sides of $v = R\omega$ with respect to t to get

$$a = R\alpha \quad \textbf{Rolling without slipping condition} \tag{5}$$

Friction must be exactly the right size to ensure that the linear and angular acceleration obey this relationship. Given this insight, we can proceed as follows. By considering the forces on the cylinder, we can write an expression for its linear acceleration a. Then, by considering the torques, we can write an expression for the angular acceleration a. Finally, we can set $a = R\alpha$. With luck, we will be able to solve that equation for f, the friction.

I will start with forces. Remembering that rightward is positive, we get

$$\begin{aligned} F_{x\,\text{net}} &= ma \\ f - kx &= ma, \end{aligned} \tag{6}$$

and hence $a = \frac{f-kx}{m}$.

Now I will use torques to find the angular acceleration. Neither gravity nor the spring exerts a torque, because those forces act on the pivot point (the center) around which the cylinder rotates. So, for both those forces, the "r" in $\tau = F_{\perp}r$ is 0. The normal force does not generate a torque either, because it points radially toward the center of the cylinder, and hence $N_{\perp} = 0$. Only friction torques the cylinder. It points entirely perpendicular to the relevant radius vector, which has length R.

Is the frictional torque positive or negative? Well, for the forces, rightward is positive. When the cylinder rolls rightward, it rotates clockwise. Therefore, I am committing to calling clockwise my positive rotation direction. But friction tends to make the cylinder rotate counterclockwise. So, the frictional torque is negative:

$$\begin{aligned} \sum \tau &= I\alpha \\ -fR &= \frac{1}{2}mR^2\alpha, \end{aligned} \tag{7}$$

and hence $\alpha = -\frac{2f}{mR}$.

We have now expressed the linear and angular accelerations in terms of friction. So, by setting $a = R\alpha$ (the rolling without slipping condition), we can solve for f, the friction needed to make the cylinder roll without slipping:

$$\begin{aligned} a &= R\alpha \\ \frac{f - kx}{m} &= R\left(-\frac{2f}{mR}\right). \end{aligned}$$

Cancel the m's and R's. Then isolate f to get $f = \frac{1}{3}kx$.

End of friction subproblem.

Where are we? By solving a difficult forces-and-torques subproblem, we figured out the frictional force on the cylinder. Now we can return to our stuff-finding strategy, step 2. (If you are lost, reread the stuff-finding strategy from two pages ago, to reorient yourself.) We want the net force on the cylinder along the direction of motion. Well, from Eq. (6) above, we immediately get

$$\begin{aligned} F_{\text{net}} &= f - kx \\ &= \frac{1}{3}kx - kx. \\ &= -\frac{2}{3}kx. \end{aligned}$$

Stuff-finding strategy step 3: Write F_{net} in the form $F_{\text{net}} = -(\text{stuff})x$, and read off stuff.

We do not have to manipulate our equation for F_{net}; it is already in the proper form. We can immediately read off $\text{stuff} = \frac{2}{3}k$. So, the cylinder oscillates back and forth as if it were a frictionless block attached to a spring of spring constant $\frac{2}{3}k$ instead of k.

End of stuff-finding strategy.

Now that we know stuff, we can breeze through the rest of the problem. The whole point of finding stuff was to enable us to calculate the angular frequency: $\omega = \sqrt{\frac{\text{stuff}}{m}} = \sqrt{\frac{2k}{m}}$. And given the angular frequency, we can immediately find the frequency using $\text{frequency} = \frac{\omega}{2\pi} = \frac{1}{2\pi} = \sqrt{\frac{2k}{3m}}$. As expected, since friction slows down the oscillations, this frequency is less than the frequency of a frictionless block on a spring, $\frac{1}{2\pi}\sqrt{\frac{k}{m}}$.

Digression: Alternative problem-solving technique. Before closing, I should note that you also could have solved this problem using the rotational analog of the stuff-finding strategy. Let me race through it. Instead of using $F_{\text{net}} = -(\text{stuff})x$, you could use $\tau_{\text{net}} = -(\text{stuff})\theta$. Since the object rolls, the distance through which it moves is related to the angle through which it spins by $x = R\theta$. Above, we found the net torque to be $\tau_{\text{net}} = -fR = -(\frac{1}{3}kx)R$, where I have used our above "answer" for friction. Since $x = R\theta$, we can rewrite this expression for the net torque as $\tau_{\text{net}} = -\frac{1}{3}kxR = -\frac{1}{3}kR^2\theta$. From this equation, we can immediately read off $\text{stuff} = \frac{1}{3}kR^2$.

Since we are reasoning rotationally, we must calculate the angular frequency using the rotational analog of $\omega = \sqrt{\frac{\text{stuff}}{m}}$, namely $\omega = \sqrt{\frac{\text{stuff}}{I}}$. So, we get

$$\omega = \sqrt{\frac{\text{stuff}}{I}} = \sqrt{\frac{\frac{1}{3}kR^2}{\frac{1}{2}mR^2}} = \sqrt{\frac{2k}{3m}},$$

exactly what we obtained above using the linear version of the stuff-finding strategy. So, the frequency and period work out the same either way.

REVIEW QUESTION 3-7

Consider the system drawn below. The blocks' masses are $M = 2.0$ kg and $m = 0.50$ kg. The floor is frictionless. The two blocks slide rightward toward a horizontal spring of spring constant $k = 100$ N/m, at its equilibrium length. Block m "rides on" block M.

(a) If the little block were frictionless, what would happen when the blocks reach the spring? Answer qualitatively, without numbers or formulas.

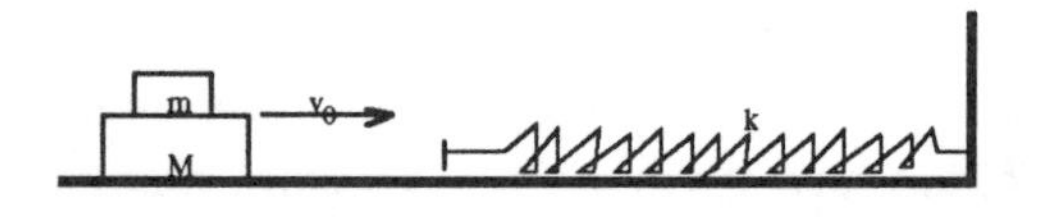

(b) *Very hard.* Now suppose the coefficient of static friction between the two blocks is $\mu_s = 0.30$. What is the biggest initial rightward velocity that the blocks can have such that block m never falls off block M?

REVIEW ANSWER 3-7

(a) When it starts compressing the spring, block M slows down. But block m does *not* slow down; it continues rightward at its original speed. Here is why: According to Newton's 2nd law, an object speeds up or slows down only if a force acts on it. But if block m is frictionless, it experiences no rightward or leftward forces. There is nothing to make it slow down.

Since block M slows down while block m keeps moving forward at full speed, block m slides off the front of block M. It is like riding in a car that suddenly slows down. Unless you are wearing a seat belt, you keep flying forward–not because a force *throws* you forward, but simply because you keep drifting forward at your original speed, while the car slows down. So, you are "thrown forward" with respect to the car. Similarly, block m feels "thrown forward" relative to block M, not because a force throws it forward, but simply because it keeps its original speed while block M slows down.

(b) As just hinted, the spring does not pull or push directly on block m. If block m stays on top of M during the spring compression, it is because static friction acts as a "seat belt" holding block m in place. While the spring compresses, block M slows down (decelerates). Therefore, it experiences a backwards (leftward) force. For this reason, whatever force functions as block m's "seat belt" must point leftward, to make block m slow down, too. The following force diagram depicts only the forces acting *on* the little block, while the spring compresses.

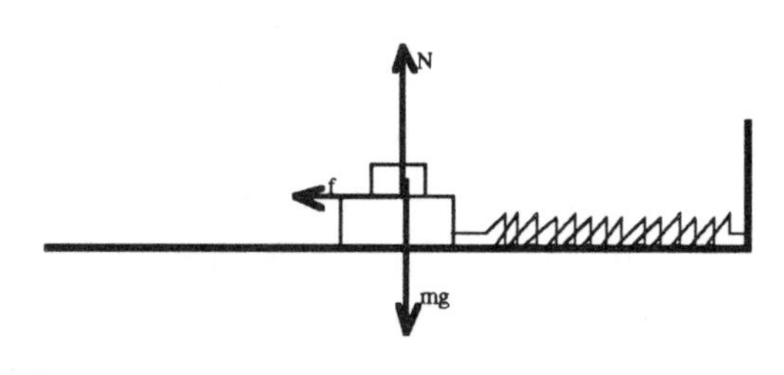

Since the friction is static, not kinetic, we cannot automatically assume $f = \mu_s N$. But here, that assumption is valid. We are looking for the *biggest* initial speed v_0 such that static friction can hold block m in place during the deceleration. But a bigger v_0 leads to a bigger spring compression. A bigger spring compression means the spring exerts a bigger force on M. And a bigger force on M leads to a bigger deceleration, by Newton's 2nd law. So, by making v_0 as big as possible, we are making block m undergo the biggest deceleration it can, without its "seat belt" breaking. So, the "seat belt"–namely static friction–must reach its maximum possible strength, $f_{\text{max}} = \mu_s N$:

$$\begin{aligned} f_{\text{max}} &= \mu_s N \\ &= \mu_s mg. \end{aligned}$$

(Since the normal force cancels gravity on the little block, $N = mg$.)

How can we calculate the biggest "safe" initial speed v_0? In other words, how can we find the biggest v_0 such that f_{max} can hold block m on top of M, even during the "worst" of the deceleration? Well, if block M decelerates too quickly, friction will not be strong enough to keep block m from sliding off block M. Since we know f_{max}, we can immediately calculate the maximum deceleration the little block can undergo. From $\sum F_x = ma$, we get $a_{\text{max allowed}} = \frac{f_{\text{max}}}{m} = -\mu_s g$. (The minus sign indicates leftward.) If the big block decelerates faster than this maximum allowed value, then the little block cannot "stay in place," because the frictional force on it is too small. Therefore, v_0 must be small enough that, when the spring compresses, it never forces the blocks to decelerate at a higher rate than $a_{\text{max allowed}} = -\mu_s g$.

But how does v_0 relate to the deceleration caused by the spring? I will address this question in two stages. Let A denote the distance by which the spring gets compressed from equilibrium. So, the blocks reach their "turn-around" point at $x = A$. We can figure out how this amplitude A relates to the blocks' deceleration. Indeed, we can solve for the "maximum allowed" amplitude, the amplitude at which the blocks reach their maximum deceleration, $a_{\text{max allowed}} = -\mu_s g$. Given this maximum allowed A, we can then calculate what initial speed v_0 causes the spring to compress to that amplitude, but no farther.

Subproblem: Find the maximum "safe" amplitude A

I see two ways to proceed. The shortcut is to remember that the force exerted by a spring is $F_s = -kx$. This is the net external force acting on the two-block system, which has mass $M+m$. So, Newton's 2nd law applied to the two-block system says

$$F_{\text{net external}} = -kx = (M+m)a. \qquad (*)$$

If you wanted to write an equation about block M only, it would have to include the frictional force exerted by m on M. (Remember Newton's 3rd law!) So, it is easier to write Eq. (*), which considers the two-block system as one big blob.

In any case, we can solve Eq. (*) for a to get $a = -\frac{kx}{m+M}$. From this equation, we see that the deceleration is largest when x is largest. But x is largest when the blocks reach their "turn-around" point, $x = A$. So, substitute $x = A$ into this expression for acceleration to get $a_{\text{max actual}} = -\frac{kA}{M+m}$. This is the biggest acceleration the system actually undergoes when the spring compresses to amplitude A. This actual acceleration had better be no bigger than $a_{\text{max allowed}}$, or else the little block falls off. So, to see how big we can "safely" make A, set

$$a_{\text{max actual}} = a_{\text{max allowed}}$$
$$-\frac{kA}{M+m} = -\mu_s g.$$

Solve for A to get $A = \frac{\mu_s g(M+m)}{k}$. Below, I will demonstrate another way you could have found $a_{\text{max actual}}$, and hence A. But for now, let me press on.

End of subproblem.

Now that we know the biggest "safe" amplitude, we can use energy conservation to figure out which initial speed v_0 makes the spring compress to exactly that amplitude. At maximum compression $(x = A)$, the blocks momentary come to rest $(v = 0)$ before shooting leftward.

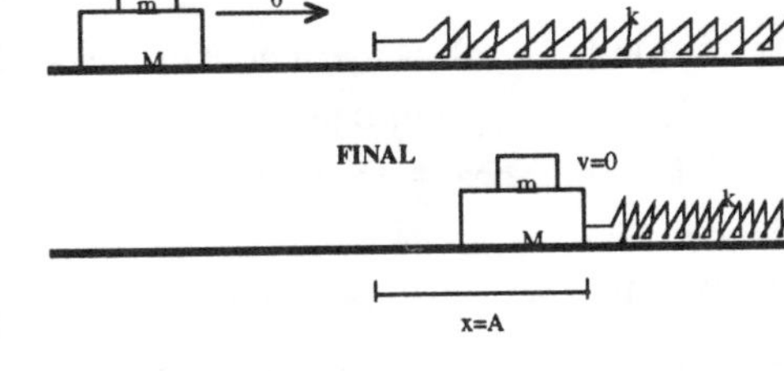

Remembering that a stretched or compressed spring carries elastic potential energy $U = \frac{1}{2}kx^2$, we get

$$E_0 = E_f$$
$$K_0 + U_0 = K_f + U_f$$
$$\frac{1}{2}(M+m)v_0^2 + 0 = 0 + \frac{1}{2}kA^2.$$

Solve for v_0, and then substitute in the maximum "safe" amplitude found in the subproblem above:

$$\begin{aligned} v_0 &= A\sqrt{\frac{k}{M+m}} \\ &= \frac{\mu_s g(M+m)}{k}\sqrt{\frac{k}{M+m}} \\ &= \mu_s g\sqrt{\frac{M+m}{k}} \\ &= (0.30)(9.8 \text{ m/s}^2)\sqrt{\frac{2.0 \text{ kg}+0.50 \text{ kg}}{100 \text{ N/m}}} \\ &= 0.46 \text{ m/s}. \end{aligned}$$

Digression. Before closing, I will show you another way to obtain $a_{\text{max actual}}$ in the above subproblem, using oscillation kinematics. Skip it if you are not interested.

Start with the general expression for the position of an oscillating system, $x = A\cos(\omega t + \phi)$. To get from position to acceleration, take derivatives:

$$v = \frac{dx}{dt} = -\omega A\sin(\omega t + \phi).$$

$$a = \frac{dv}{dt} = \frac{d^2x}{dt^2} = -\omega^2 A\cos(\omega t + \phi).$$

The acceleration oscillates with time. Since cos(...) oscillates between +1 and −1, the largest value of the acceleration occurs when $\cos(\omega t + \phi) = \pm 1$, at which point $a_{\text{max}} = \pm\omega^2 A$. In words, the maximum acceleration is proportional to the maximum displacement, as we saw above using Newton's 2nd law. Since the angular frequency of this mass-spring system is $\omega = \sqrt{\frac{\text{stuff}}{\text{mass}}} = \sqrt{\frac{k}{M+m}}$, our expression for a_{max} becomes

$$\begin{aligned} a_{\text{max}} &= -\omega^2 A \\ &= -\frac{k}{M+m}A, \end{aligned}$$

the same expression for $a_{\text{max actual}}$ obtained above. *End of digression.*

REVIEW QUESTION 3-8

On frictionless ice, a hockey puck of mass m slides at speed v_0 toward a uniform stick of mass $3m$ and length s, as drawn below. The puck's initial velocity is perpendicular to the stick. The puck strikes the stick a distance $\frac{s}{6}$ from the center of the stick. It is observed that, after hitting the stick, the puck comes to rest.

(a) Qualitatively describe the stick's motion after the collision.

(b) *(Hard.)* Consider the stick after the collision. How far from its starting point is the center of the stick, when the stick has completed one full rotation?

REVIEW ANSWER 3-8

TOP-DOWN VIEW

v_0

(a) The stick moves rightward, and also rotates clockwise around its center. Specifically, its center of mass travels in a straight line at constant speed; and the rest of the rod rotates around that point. You might think the rod would rotate around its "bottom" end. But an unpivoted object always rotates around its center of mass. To see this, throw your pen into the air. No matter how you spin it, the pen still rotates around its center.

(b) We want to know how far the stick moves in the time it takes to complete one rotation. So, we need to know the period T (i.e., the time of one rotation). Well, intuitively, the faster the stick rotates, the smaller T is. So, to calculate T, we could find the post-collision angular speed ω, and use $T = \frac{2\pi}{\omega}$.

Then, to find the distance x covered by the (center of the) stick in time T, we can use an old kinematic formula such as $x = x_0 + v_0 t + \frac{1}{2}at^2$, where $t = T$, and where v_0 denotes the stick's center-of-mass velocity immediately after the collision. On frictionless ice, the stick neither speeds up nor slows down, and hence its acceleration after the collision is $a = 0$. But we still need to know the stick's post-collision velocity, before we can use $x = x_0 + v_0 t + \frac{1}{2}at^2$.

In summary: Once we figure out the stick's post-collision linear speed v and angular speed ω, we can use kinematics to obtain the answer. But we are stuck without v and ω. How can we find them?

Since the puck and stick *collide*, conservation laws probably apply. But which ones? Energy conservation looks attractive; but during the collision, an unknown amount of heat might get generated. In other words, the collision might be inelastic. The unknown "heat" term in your energy conservation equation renders it less useful.

So, we must use momentum. A common mistake is to combine linear and angular momentum into one big "total momentum conservation" equation. But you cannot do this. Linear and angular momentum have different units. Throwing them both into the same equation is like adding apples to oranges. (By contrast, you can combine translational and rotational kinetic energy into the same equation, because K_{trans} and K_{rot} have the same units, joules.)

Since we cannot deal with all "momentum" in one step, we need to decide whether linear momentum is conserved, angular momentum is conserved, or both. Let us start with linear momentum. It is conserved if no external forces act on the system during the collision. Well, that is certainly the case; the only external forces are gravity and the normal force, which act perpendicular to the direction of motion. (Besides, those forces cancel.) The forces exerted by the puck and stick on each other are *internal* forces, i.e., forces exerted by one part of the system on another. Crucially, if the stick had been pivoted to the ice, linear momentum would not be conserved. But since the stick is free to slide along the ice, linear momentum is conserved.

Is angular momentum also conserved? Yes, because no external torque acts on the system. The only torque experienced by the puck and stick are the torques they exert on each other.

Although the rod clearly has angular momentum after the collision, it is hard to see what "carries" the initial angular momentum. Here is the deal: The puck's initial velocity is not directed radially towards the "pivot point," the center of the rod. So, part of the puck's velocity is tangential, and therefore "carries" some angular momentum. Indeed, as shown in Question 16-5, the angular momentum of a point mass can be written as $I\omega$ or as the "special" formula

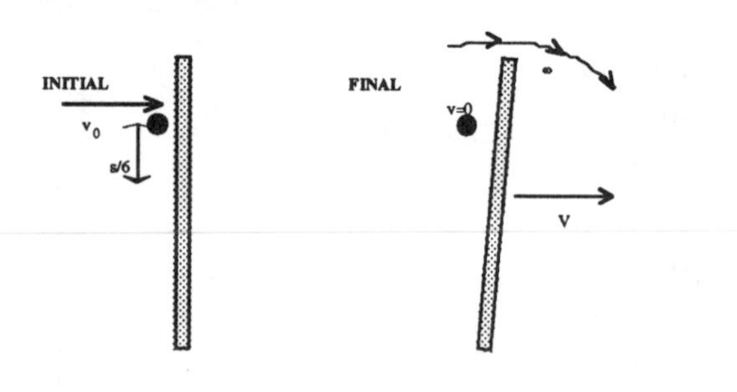

$L_{\text{point mass}} = mv_{\perp}r$. Here, $v_{\perp}$ denotes the component of **v** perpendicular to **r**, the radius vector connecting the pivot to the point mass. Immediately before the collision, the puck is a distance $r = \frac{s}{6}$ from the pivot point; and its velocity is entirely perpendicular to the relevant radius vector, which runs along the rod.

The puck's initial angular momentum is $L_0 = mv_{\perp}r = \frac{mv_0 s}{6}$. So, angular momentum conservation gives us

$$\begin{aligned} L_0 &= L_f \\ L_{0\,\text{puck}} + L_{0\,\text{stick}} &= L_{f\,\text{puck}} + L_{f\,\text{stick}} \\ mv_0 \frac{s}{6} + 0 &= 0 + I_{\text{stick}}\,\omega \\ &= \frac{1}{12}(3m)s^2\omega, \end{aligned} \tag{1}$$

where I have used the rotational inertia of a uniform rod pivoted at its center (*not* at its end). Solve for ω to get $\omega = \frac{2v_0}{3s}$.

Remember, the whole point of invoking conservation laws is to find the rod's post-collision linear and angular speeds, so that we can use kinematics to calculate how far the rod travels in the time it takes to complete one rotation. We have ω, but need v. We cannot use the rolling without slipping condition, $v_{\text{cm}} = R\omega$, because the stick does not roll. And we cannot use $v = r\omega$, because that formula refers to the velocity with which a point on the stick circles the center; but here, we want the velocity of the center itself.

Fortunately, as discussed above, linear momentum is also conserved.

$$\begin{aligned} p_0 &= p_f \\ mv_0 &= (3m)V, \end{aligned} \tag{2}$$

which we can immediately solve for V, the rod's post-collision center-of-mass speed, to get

$$V = \frac{1}{3}v_0.$$

Now we are home free. Since the stick spins with constant angular speed, we can use $\omega = \frac{\Delta\theta}{\Delta t}$. We want the time $\Delta t = T$ it takes the stick to spin through one rotation, which is 2π radians. So, $\omega = \frac{\Delta\theta}{\Delta t} = \frac{2\pi}{T}$. Solve for T, and substitute in the above expression for ω, to get

$$T = \frac{2\pi}{\omega} = \frac{2\pi}{\frac{2v_0}{3s}} = \frac{3\pi s}{v_0}.$$

The stick's center travels at constant speed $V = \frac{v_0}{3}$. So, in time T, it travels through

$$\Delta x = VT + (\text{no acceleration term})$$
$$= \left(\frac{1}{3}v_0\right)\left(\frac{3\pi s}{v_0}\right)$$
$$= \pi s,$$

about 3.1 times the length of the stick.

REVIEW QUESTION 3-9

A block of mass m slides frictionlessly with speed v_0 towards a large block (mass M) that has a semicircular "hole" cut into it, as drawn below. The semicircle has radius r.

(a) Suppose that the large block is free to slide frictionlessly, and that v_0 is big enough that the little block slides along the whole semicircle, shooting off the semicircle at point A. (The little block is still snugly in contact with the large block at point A.) What is the final speed of the large block? *Set up, but do not solve, the relevant equation or equations needed to find this speed in terms of m, M, v_0, r and g.*

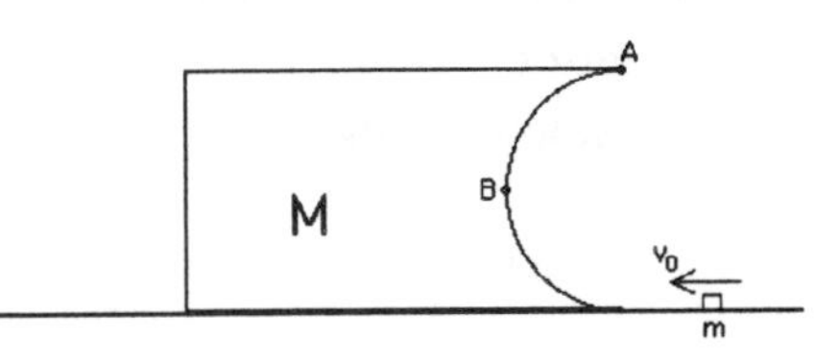

(b) What is the vertical acceleration of the small block at point B?

REVIEW ANSWER 3-9

(a) You should see that this is a collision problem, and that you will need to use conservation laws–along perhaps with some physical insights–in order to solve. This is not an "object falls off a circle" problem, because the little block does not lose contact with the "track" until it shoots by point A.

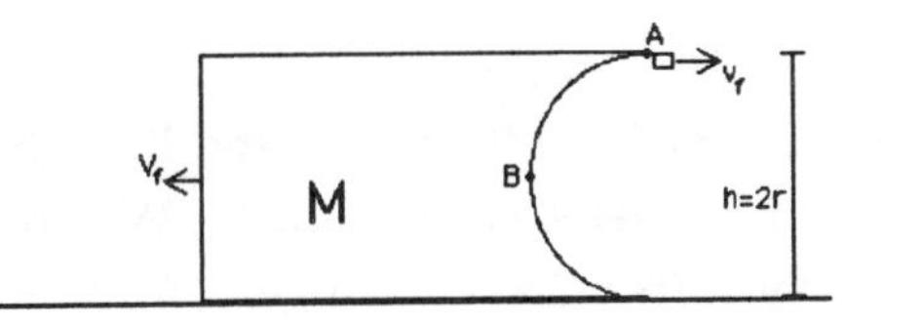

As always, accurate diagrams will help you solve. An "initial" (pre-collision) diagram is given in the problem. Here is the system immediately after the collision (i.e., immediately after the little block loses contact with the big block).

I have let little v_f and big V_f denote the final speeds of the little and big block, respectively.

To solve this, you must have the following physical insight: *Immediately* after the little block shoots off the semicircle, it is traveling straight rightward, because it shot off a horizontal segment of track (at point A).

(If you choose another point as your "final" point, you should still get the right answer, provided you are careful about vector components. Once the small block becomes a projectile, it "keeps" all its horizontal velocity. This horizontal velocity is what we will use below in our momentum conservation equation.)

Given this insight, let us apply momentum and energy conservation. (Although gravity exerts an external force on the little block, this force is vertical, and therefore does not screw up *horizontal* momentum conservation.) Looking at the horizontal momenta, and picking leftward as positive, I get

$$p_0 = p_f$$
$$mv_0 = -mv_f + MV_f. \tag{1}$$

Because both final velocities are unknown, we cannot yet solve. To gather more information, consider energy conservation. At point A, the little block is height $h = 2r$ off the ground. Because everything slides frictionlessly, no heat gets generated. So,

$$E_0 = E_f$$
$$K_0 + U_0 = K_f + U_f \tag{2}$$
$$\frac{1}{2}mv_0^2 + 0 = \frac{1}{2}mv_f^2 + \frac{1}{2}MV_f^2 + mg(2r).$$

We now have two equations in the two unknowns, v_f and V_f. You were not supposed to finish the algebra.

(b) To calculate the vertical acceleration, we could use the definition of acceleration as change in velocity per change in time, $a_y = \frac{dv_y}{dt}$. But we do not have enough information. So instead, we will have to consider the forces acting on the little block at point B, and then use $\sum F_y = ma_y$.

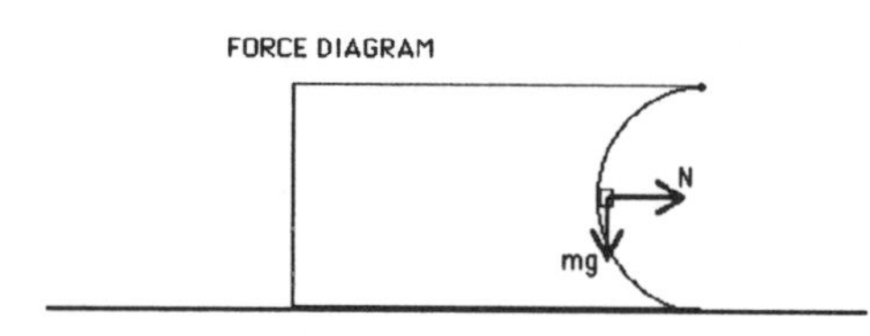

A normal force always point perpendicular to the surface exerting that force. So, the only *vertical* force is gravity:

$$\sum F_y = ma_y.$$
$$-mg = ma_y,$$

and hence $a_y = -g$. Right at point B, the little block's vertical motion is slowing down at a rate of 9.8 m/s per second.

This was easier than expected for two reasons. First, since we are calculating the vertical acceleration, we can simply ignore whatever complicated *horizontal* motion the little block experiences. Second, we do not need to know the little block's velocity at point B. We are not interested in its velocity. We are interested in the rate at which its velocity changes. And to find that acceleration, you just need to know the forces, nothing else.

REVIEW QUESTION 3-10

A solid disk (mass M, radius R) is free to rotate around an axle through its center. The axle is mounted on a wall. So, the disk is free to spin, and the plane of the disk is parallel to the wall.

A small ball of clay (mass m) is thrown toward the bottom of the disk at an angle ϕ to the vertical, as drawn below. At time $t = 0$, the clay strikes and sticks to the bottom of the wheel. As a result, the disk starts to rotate.

(a) *Hard*. What is the smallest speed v_0 with which the clay must strike the disk to ensure that the wheel rotates through 180°, i.e., to ensure that the clay rotates to the top of the wheel?

(b) *(Super hard. See how far you can go.)* Suppose the clay's actual speed v_0 is much smaller than your part (a) answer. What is the disk's angular speed at arbitrary time t? In this problem, v_0 is given.

REVIEW ANSWER 3-10

(a) The clay collides with the disk. Then, the disk and clay rotate together, causing the clay to rise up. So, this is analogous to a linear problem in which, say, a clay block crashes into a second block; and then the stuck-together blocks slide up a hill. How would you solve that linear problem? You could not use energy conservation to handle the whole problem, because the collision is inelastic. Instead, you would use momentum conservation to find the post-collision speed of the stuck-together blocks. Then, you would use energy conservation to calculate how high up the hill the blocks slide.

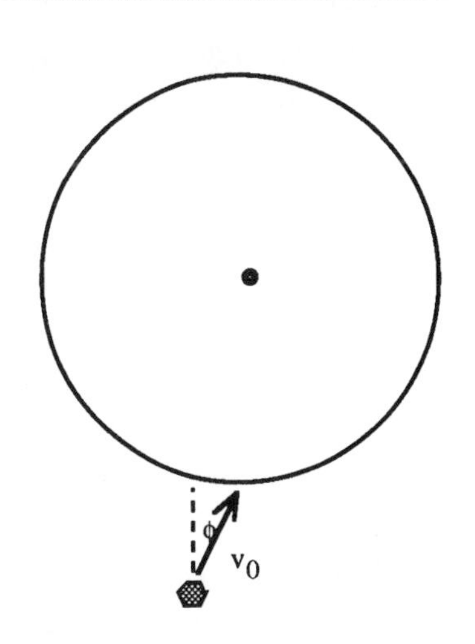

Analogous reasoning applies here. The collision is completely inelastic. Therefore, lots of heat gets generated. Since we do not know how much heat, we cannot use energy conservation for this whole problem. For instance, you cannot just set the system's initial kinetic energy equal to its final potential energy, because that equation would fail to account for the heat. Instead, we should divide this problem into steps, using energy conservation for only some of the steps. The "initial" configuration is drawn in the problem.

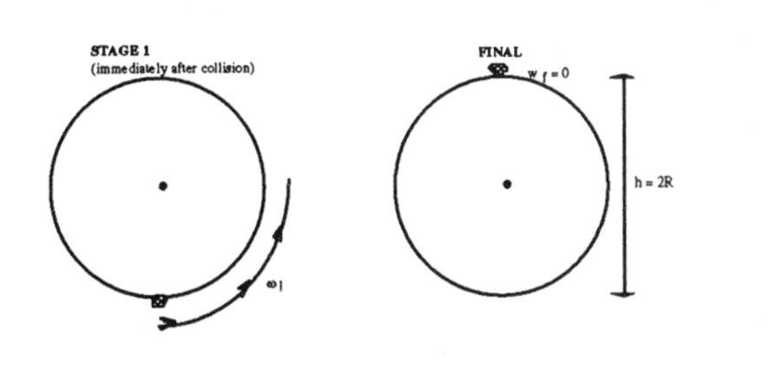

Immediately after the collision, the disk and clay rotate together with some angular speed ω_1. As the clay rises up, the angular speed slows down. If the clay *barely* reaches the top, then the disk and clay are barely moving when the clay reaches the top. So, the "final" angular speed of the disk and clay is $\omega_f = 0$. In this problem, we are trying to find the clay's initial speed v_0 such that the clay's angular speed at the top is $\omega_f = 0$. Usually in a problem of this sort, we would solve for ω_f, given v_0. Here, it is the other way around. But the same physical reasoning works.

During the collision, heat gets generated, and hence, we cannot productively use energy conservation. But angular momentum is conserved; the only torques experienced by the clay and disk during the collision are the torques they exert on each other.

Is angular momentum also conserved as the clay swings up (stage 1 to final)? No, because gravity torques the clay. That is why the clay and disk gradually slow down. But energy is conserved during this stage, with no more heat generated. The stage 1 kinetic energy converts into potential energy as the clay rises.

Given all this, we can use the following two-part strategy:

(1) Use angular momentum conservation to relate the clay's initial speed v_0 to the post-collision angular speed of the system, ω_1.

(2) Use energy conservation to relate ω_1 to the "final" angular speed, ω_f. Set $\omega_f = 0$ to solve for ω_1. Then, using your step 1 equation, solve for v_0.

It makes sense to work "backwards," doing step 2 first. Intuitively, in step 2, we can find what angular speed ω_1 the system must have after the collision, to ensure that the clay barely reaches the top. Then we can complete step 1, solving for the clay's initial speed needed to produce ω_1. Although I will do step 2 first, you get the same answer no matter in what order you complete the steps.

Step 2: Stage 1 → Final, using energy conservation

Since the clay is a point mass, you have the option of treating its kinetic energy as translational *or* rotational (but *not* both). Here, it is easier to use K_{rot} for the clay, because the clay and disk "share" the same angular speed ω_1. So, immediately after the collision, the clay and disk have rotational kinetic energy.

What about potential energy? Remember, there is no such thing as "rotational gravitational potential energy." Potential energy is always $U = mgh$, where h is measured to the center of mass of the object. So here, to calculate U, you could find the center of mass of the disk-and-clay system, and figure out through what height that center of mass rises. But it is easier to write separate potential energy terms for the disk and clay. The disk's center of mass neither rises nor falls, and hence, its stage 1 potential energy cancels its final potential energy. But the clay rises through height $h = 2R$.

Putting all this together, and recalling that $I_{\text{solid disk}} = \frac{1}{2}MR^2$ and $I_{\text{point mass}} = mr^2$, where r is the point mass' distance from the pivot point, we get

$$E_1 = E_f$$
$$U_1 + K_1 = U_f + K_f$$
$$0 + \frac{1}{2}(I_{\text{disk}} + I_{\text{clay}})\omega_1^2 = mgh + 0$$
$$\frac{1}{2}\left(\frac{1}{2}MR^2 + mR^2\right)\omega_1^2 = mg(2R)$$

Solve for ω_1 to get

$$\omega_1 = \sqrt{\frac{4mg}{(\frac{1}{2}M + m)R}}.$$

That is how fast the system must be spinning immediately after the collision, to ensure that the clay barely reaches the top.

Step 1: Initial → Stage 1, using angular momentum conservation

Now we can figure out the clay's initial speed v_0 needed to give the system angular speed ω_1 after the collision. As shown in Question 16-5, starting with $L_{\text{point mass}} = (I_{\text{point mass}})\omega$, you can derive a "special" formula for the angular momentum of a point mass: $L_{\text{point mass}} = mv_{\perp}r$, where **r** is the radius vector drawn from the pivot point to the point mass. So here, if we look at the clay a millionth of a second before it hits the bottom of the disk, the radius vector points straight down, from the center to the bottom of the disk. Crucially, $v_{\perp}$ is the component of **v** perpendicular to this radius vector.

Intuitively, a point mass' angular momentum depends on $v_{\perp}$ instead of the entire velocity, because only the perpendicular part of the velocity helps to make the system turn. For instance, if we throw clay straight upward at the bottom of the disk, the disk will not spin upon impact. Only the sideways (tangential) part of the clay's motion gets imparted to the disk as rotational motion. You have seen this intuition before, with torques. Only the "tangential" part of the force, $F_{\perp}$, contributes to the torque.

In any case, we see from the diagram that $r = R$ and $v_\perp = v_0 \sin\phi$. So, angular momentum conservation gives us

$$L_0 = L_1$$
$$mv_\perp r = (I_{\text{disk}} + I_{\text{clay}})\omega_1$$
$$m(v_0 \sin\phi)R = \left(\frac{1}{2}MR^2 + mR^2\right)\omega_1.$$

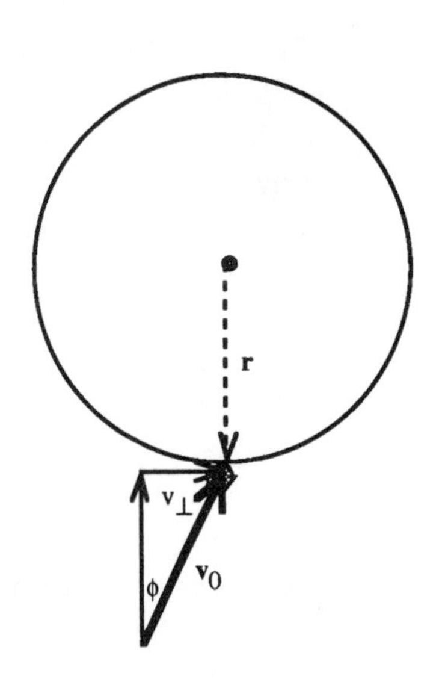

We are done with the physics. Only algebra remains. Solve this equation for v_0, and then invoke the expression for ω_1 obtained in step 2 above:

$$v_0 = \frac{(\frac{1}{2}M + m)R}{m\sin\phi}\omega_1$$
$$= \frac{(\frac{1}{2}M + m)R}{m\sin\phi}\sqrt{\frac{4mg}{(\frac{1}{2}M + m)R}}$$
$$= \sqrt{\frac{4g(\frac{1}{2}M + m)R}{m\sin^2\phi}}.$$

(b) If the clay strikes the disk softly, the clay swings up a few degrees, then swings back down, then swings up the other way. If these oscillations are small enough, the system undergoes simple harmonic motion, just like a pendulum. Indeed, this *is* a so-called "physical pendulum." More on that later.

Since the system undergoes rotational oscillatory motion instead of linear oscillatory motion, we must use the rotational analogs of all our favorite oscillation formulas and problem-solving strategies. Here, we want to know the angular speed as a function of time. Well, if we wanted to know the linear speed of a linear harmonic oscillator, we would start with our general kinematic expression for the position of any harmonic oscillator, $x = A\cos(\omega t + \delta)$, where ω is the angular frequency and δ is the phase constant. (I am using δ instead of ϕ so as not to confuse the phase constant with the angle from part (a).) Then we would differentiate to find $v = \frac{dx}{dt}$.

Here, we can proceed by analogy. Since angle is analogous to position, we get $\theta = A\cos(\omega t + \delta)$, where ω is still the angular frequency of the oscillations, *not* the angular speed. Indeed, since the same symbol gets used to denote two different quantities, I had better clarify my notation. Let ω_{rot} denote the angular velocity, and ω_{osc} denote the angular frequency of oscillation.

So, we have $\theta = A\cos(\omega_{\text{osc}}t + \delta)$. In this case, however, it is better to use sine, because at $t = 0$, the clay is at equilibrium ($\theta = 0$). Therefore, if we use $\theta = A\cos(\omega_{\text{osc}}t + \delta)$, then δ works out to be –90°. But if we use $\theta = A\sin(\omega_{\text{osc}}t + \delta)$, then $\delta = 0$. To see this, set $t = 0$ and $\theta = 0$, and solve for δ. So, to avoid needless phase factors, I will use $\theta = A\sin(\omega_{\text{osc}}t)$. Differentiate to get

$$\omega_{\text{rot}} = \frac{d\theta}{dt} = \frac{d}{dt}[A\sin(\omega_{\text{osc}}t)] = A\omega_{\text{osc}}\cos(\omega_{\text{osc}}t). \tag{1}$$

I cannot stress enough that the angular speed ω_{rot} differs from the angular frequency ω_{osc}. The angular speed tells you how many radians per second the disk spins through. The angular frequency is proportional to how many oscillations per second the system undergoes.

To obtain the system's angular speed at arbitrary time t using Eq. (1), we need to know the amplitude A and angular frequency ω_{osc}. So, we have to solve two major subproblems. I will start with the angular frequency. This will take several pages. As you slog through these details, keep in mind how finding ω_{osc} fits into the overall problem-solving strategy.

Subproblem 1: Find the angular frequency ω_{osc}

The shortcut is to plug in the formula for a "physical pendulum," as I will demonstrate below. But first, let me review the more general technique for finding angular frequencies, a technique that *always* works, not just for physical pendulums. As shown in Review Question 3-6, you can obtain angular frequencies by finding the "stuff" in $F_{net} = -(\text{stuff})x$, and then substituting it into $\omega_{osc} = \sqrt{\frac{\text{stuff}}{m}}$. We can utilize the stuff-finding strategy here, too. But we must use the rotational analog. And when we find the "rotational" stuff, we will invoke the rotational analog of $\omega_{osc} = \sqrt{\frac{\text{stuff}}{m}}$, namely $\omega_{osc} = \sqrt{\frac{\text{stuff}}{I}}$.

So, let us begin by finding stuff. Here is the strategy, with linear quantities crossed out and replaced by their rotational analogs. If this all looks mysterious, please review Question 20-5.

Stuff-finding strategy (rotationalized):

(1) Draw a force diagram of the system when it is displaced an arbitrary angle θ from equilibrium. (Do not draw the system at equilibrium, because then the net torque on it will be 0.)

(2) Write down the net torque on the system.

(3) By writing the net torque in the form $\tau_{net} = -(\text{stuff})\theta$, figure out stuff.

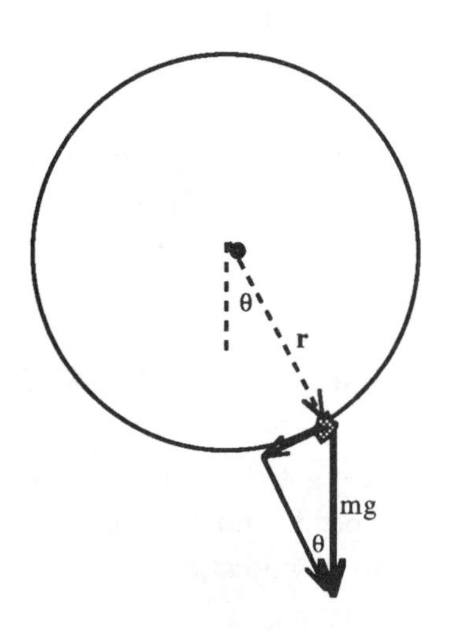

Stuff-finding step 1: Force diagram

Since we are ultimately interested in finding the torque on the system, we can leave out the gravitational force acting on the disk. That force pulls on the center of the disk, which is the pivot point around which the disk turns. So, that force cannot generate a torque. By contrast, gravity on the clay *can* torque the system.

Stuff-finding step 2: Write the net torque

Only the component of gravity perpendicular to $\mathbf{r}$, namely $F_{\perp} = mg\sin\theta$, produces a torque. The torque is clockwise when the angle θ is counterclockwise. So, the torque is negative; it tends to push the system back toward equilibrium.

$$\tau_{net} = F_{\perp}r = -(mg\sin\theta)R.$$

Stuff-finding step 3: Write τ_{net} in the form $\tau_{net} = -(\text{stuff})\theta$, and read off stuff.

Since the clay hit the disk gently, the oscillations are small. The angle θ never gets too big. Therefore, we can use the small angle approximation, $\sin\theta \approx \theta$. This gives

$$\begin{aligned}\tau_{net} &= -(mg\sin\theta)R \\ &\approx -mg\theta R \\ &= -(mgR)\theta.\end{aligned}$$

So, stuff $= mgR$.

End of stuff-finding strategy.

Now that we know stuff, we can immediately find the angular frequency using the rotational version of our usual formula for ω_{osc}. We must use the rotational inertia of the whole system, not just the clay, because the whole system gets turned by the torque we just calculated.

$$\omega_{osc} = \sqrt{\frac{\text{stuff}}{I}} = \sqrt{\frac{\text{stuff}}{I_{disk} + I_{clay}}} = \sqrt{\frac{mgR}{\frac{1}{2}MR^2 + mR^2}} = \sqrt{\frac{mg}{(\frac{1}{2}M + m)R}}.$$

Remember, the whole point of finding ω_{osc} is so that we can substitute ω_{osc} and the amplitude into Eq. (1) above, $\omega_{rot} = A\omega_{osc}\cos(\omega_{osc}t)$, to find the angular *speed* as a function of time. So, in the next subproblem, I will find the amplitude A. But first, let me show you a quick (though plug-n-chuggy) way to find ω_{osc}, treating the system as a physical pendulum.

Using reasoning similar to the above, your textbook derives the formula

$$\omega_{osc} = \sqrt{\frac{M_{total} g d_{cm}}{I}} \qquad \textbf{Physical pendulum ONLY}$$

where d_{cm} is the distance from the pivot point to the overall center of mass. (Some books actually derive the period instead of the angular frequency. If that is so, just use $\omega_{osc} = \frac{2\pi}{T}$.) To use this "special" formula, you must find the overall center of mass. Well, if we set $x = 0$ at the center of the disk, then $x = R$ at the clay. So, the center of mass position is

$$d_{cm} = \frac{m_{disk}x_{disk} + m_{clay}x_{clay}}{m_{disk} + m_{clay}} = \frac{M(0) + mR}{M + m} = \frac{m}{M + m}R.$$

Substitute that into the above formula to get

$$\omega_{osc} = \sqrt{\frac{M_{total} g d_{cm}}{I}} = \sqrt{\frac{(M + m)g\frac{m}{M+m}R}{\frac{1}{2}MR^2 + mR^2}} = \sqrt{\frac{mg}{(\frac{1}{2}M + m)R}},$$

the same expression obtained above using the stuff-finding strategy. On a test, if you are presented with a physical pendulum and you are comfortable finding centers of mass, then go ahead and use the physical pendulum formula. But you should also understand the stuff-finding strategy, so that you can solve *all* oscillation problems.

At this point, we know the system's angular frequency, and we need its amplitude.

Subproblem 2: Find the amplitude

Usually, you can find the amplitude most efficiently with energy conservation. Intuitively, all of the system's rotational kinetic energy (immediately after the collision) converts into potential energy when the system momentarily comes to rest at its maximum angle, $\theta_{max} = A$. But here, it is hard (though possible) to relate that angle to the height h of the clay at that point. Instead of energy conservation, let me use a kinematic trick. Look at Eq. (1):

$$\omega_{rot} = A\omega_{osc}\cos(\omega_{osc}t). \qquad (1)$$

We are trying to find A so that we can obtain ω_{rot} at arbitrary time t. So for now, ω_{rot} is unknown. But there is one time at which we do know ω_{rot}. At $t=0$, right when the clay hits the disk, we know the system's angular speed–we found an expression for ω_{rot} in terms of v_0 in part (a) above! Using angular momentum conservation, we obtained the equation

$$m(v_0 \sin\phi)R = \left(\frac{1}{2}MR^2 + mR^2\right)\omega_1.$$

Since v_0 is given in this part of the problem, we have $\omega_1 = \frac{m\sin\phi}{(\frac{1}{2}M+m)R}v_0$. That is the system's angular (rotational) speed ω_{rot} at $t=0$. So, set $\omega_{\text{rot}} = \omega_1$ and $t=0$ in Eq. (1), to get

$$\omega_1 = A\omega_{\text{osc}}\cos(0).$$

$$\frac{m\sin\phi}{(\frac{1}{2}M+m)R}v_0 = A\sqrt{\frac{mg}{(\frac{1}{2}M+m)R}}, \qquad \text{[since } \cos(0)=1\text{]}$$

where I have used our answer for ω_{osc} found in subproblem 1 above. Solve for the amplitude to get $A = \sqrt{\frac{m}{g(\frac{1}{2}M+m)R}}v_0\sin\phi$.

End of subproblem 2.

At this point, we know both the angular frequency and amplitude of oscillation. Substitute them into Eq. (1) to get the angular speed of the system at arbitrary time t:

$$\begin{aligned}\omega_{\text{rot}} &= A\omega_{\text{osc}}\cos(\omega_{\text{osc}}t).\\ &= \left(\sqrt{\frac{m}{g(\frac{1}{2}M+m)R}}v_0\sin\phi\right)\left(\sqrt{\frac{mg}{(\frac{1}{2}M+m)R}}\right)\cos\left(\sqrt{\frac{mg}{(\frac{1}{2}M+m)R}}t\right)\\ &= \frac{m}{(\frac{1}{2}M+m)R}v_0\sin\phi\cos\left(\sqrt{\frac{mg}{(\frac{1}{2}M+m)R}}t\right).\end{aligned}$$

O.K., that was a long, hard mess. At this point, I recommend skimming over the whole answer, skipping the fine details, to get a feel for how it all fits together. First, I found the angular frequency using the rotationalized stuff-finding strategy. Then I found the amplitude of oscillation using kinematics (though I equally well could have used energy conservation).

REVIEW QUESTION 3-11

Consider an equilateral triangular prism of side-length s and mass M, sitting on the floor. In the picture below, the prism extends into the page. The coefficient of static friction between the prism and the floor is high enough that someone can tip the prism over by pushing on the top.

(a) What is the smallest *horizontal* (rightward) force with which someone must push the top of the prism to tip it over its bottom right vertex? (Note: If the force is strong enough to "start" the prism tipping over, it will be strong enough to tip it all the way over.)

F = ?

(b) Suppose I am not strong enough to apply the force you found in part (a). How might I still be able to tip the prism over by pushing on its top? Give me as detailed advice as possible about how I should push.

REVIEW ANSWER 3-11

(a) If someone pushes with the minimum force needed to tip the prism over, then it *barely* tips over. In other words, initially it is *barely* rotating. Therefore, its angular acceleration is *very* small, just a shade larger than 0. So, we can set $\alpha \approx 0$, in which case the net torque on the prism must be $\sum \tau \approx 0$. In a sense, this is a statics problem! That is the physical insight needed to address "tipping over" problems, "rotate-the-object-over-a-ledge" problems, and other such situations.

Since this is a "statics" problem, I will invoke our old strategy.

Step 1: Force diagram

When the prism starts tipping over, the front vertex has to bear its entire weight, because the back vertex comes off the ground. For this reason, the floor's normal and frictional forces act on that front vertex. Gravity acts on the center of mass, which turns out to be one third of the way up from the bottom. But you *do not* need to know this fact in order to solve the problem, as I will show below.

Step 2: Choice of reference pivot point

Since the prism actually pivots around its bottom front vertex, we might as well choose that point. And look: Since N and f act directly on the pivot point, those forces cannot exert torques, because the "r" in $\tau = F_{\perp} r$ is 0. So, our net torque equation will be less complicated than expected.

Step 3: Set net torque and/or net force equal to 0.

As usual, I will start with torques. Since the normal and frictional forces do not exert torques, we only have to worry about F and Mg.

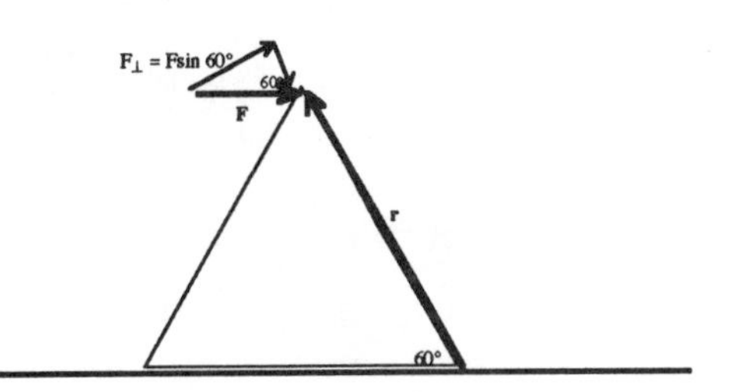

For any given force, the associated radius vector **r** runs from the pivot point to the contact point where the force acts. Let us start with F. Its radius vector runs along the side of the triangle.

The radius vector has length s, the side length. I have broken the force vector into its components parallel and perpendicular to **r**. The torque is $\tau_{\text{push}} = F_{\perp} r = (F \sin 60°)s$.

Now let us find the gravitational torque. To do so relatively painlessly, we must use the $\tau = r_{\perp} F$ form of the torque equation. For this reason, I will break up the relevant **r** vector into its components parallel and perpendicular to gravity. Without even knowing how high off the ground gravity acts, we can nevertheless see that the component of **r** perpendicular to gravity is a half side-length long, since it starts at the edge and ends at the middle of the triangle: $r_{\perp} = \frac{s}{2}$. So, the torque due to gravity is $\tau_{\text{grav}} = r_{\perp} F_{\text{grav}} = \frac{s}{2}(Mg)$.

Now we are ready to write the net torque, by adding τ_{push} to τ_{grav}. I will pick clockwise as positive. So, the gravitational torque gets a minus sign. Remember, since the prism barely tips over, the torques sum to approximately zero:

$$\sum \tau \approx 0$$

$$(F \sin 60°)s - \frac{s}{2}(Mg) = 0,$$

Solve for the push force F to get

$$F = \frac{Mg}{2\sin 60°} = \frac{Mg}{2\left(\frac{\sqrt{3}}{2}\right)} = \frac{Mg}{\sqrt{3}}.$$

Here, we got all the way to the answer with the net torque equation. We did not need to consider the net x-force or net y-force.

(b) Only the component of a force perpendicular to the relevant radius vector helps to "torque" the object. From the above force diagrams, you can see that when the person pushes horizontally, part of her force is "wasted," by which I mean part of her force points parallel to the radius vector. Since I am weak, I should push at a 30° angle to the horizontal. That way, my force points entirely perpendicular to the radius vector, and hence, *all* of my push gets "devoted" to torquing the prism. This "trick" really works; try it on a heavy desk or counter.

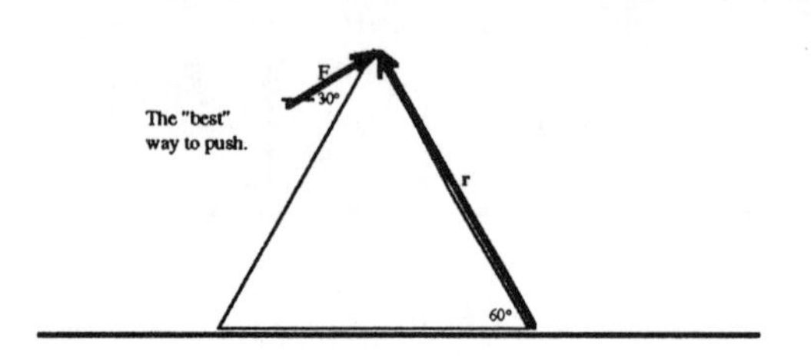

Exam Preparation

Study Tips, Cheat Sheet Tips, and Test-taking Tips

26 CHAPTER

EFFICIENT STUDY STRATEGY

◆ *Start studying several days, preferably a week, before the test. In physics, cramming does not work.*

I recommend that for each separate topic (e.g., each textbook chapter), you do the following:

(1) Review your notes from class, focusing on problem-solving strategies and the sample problems. If you need extra help with a topic, review the corresponding textbook sections.

(2) Rework the relevant chapter or chapters in this study guide, skipping algebra to save time. Just set things up.

(3) Rework almost every problem in the relevant homework assignment, working *by yourself*. Check your answers against your instructor's solutions. Some of the homework problems, however, are not worth reviewing, either because they cover a topic that will not appear on the exam or because they are too hard. Ask your instructor or teaching assistant for a list of which homework problems you need not study.

(4) If you find yourself stuck on any topic, look for appropriate practice problems, and work through them. Sources of practice problems include this reader and the worked-out sample problems in the textbook. Then, again try the homework problems that gave you trouble.

(5) After you have reviewed all the material in this way, do the appropriate *review problem set* in this study guide. See the table of contents.

(6) Ask your instructor for old midterms used in the course. This study guide also contains sample midterms and final exams.

- Do not stay up all night studying. In physics, that extra bit of knowledge will *not* compensate for the fact that you will be less clear-headed. You can score well on many problems by thinking clearly, even if you do not know every last detail.
- Make sure you know exactly what you are allowed to bring into the test (e.g., cheat sheet, calculator...).
- If you are allowed a cheat sheet, spend the time needed to make a good one. See the hints below. Otherwise, skip right to the "test-taking tips" two pages down.

CHEAT SHEET-MAKING HINTS

In many courses, the instructor allows you to bring in a page or an index card of notes. You may be tempted to cram every last formula onto your cheat sheet. My experience shows this to be counterproductive, if taken too far. Why? Two reasons:

(1) Your cheat sheet gets so cluttered that you have trouble finding what you want.

(2) With so many formulas, it gets difficult to remember what they all mean.

I recommend a leaner cheat sheet. It should include all the basic formulas, but only a few "special-situation" formulas. Let me explain what I mean.

A basic formula is an equation that you use to derive the more complicated formulas. It is a "starting point" in your problem solving. Basic formulas from this course include: Definition of instantaneous and average velocity and acceleration, the three basic constant-acceleration kinematic equations, Newton's 2nd law, "centripetal" force or acceleration, force due to spring, force due to friction, force due to gravity (near the Earth's surface, and also planetary), work and work-energy theorem, power, momentum and energy formulas (including kinetic, gravitational potential near the Earth's surface, gravitational potential in other situations, and spring potential), torque, rotational inertia, angular momentum, formulas relating linear velocity to angular velocity and linear acceleration to angular acceleration of a rotating point, basic oscillation relations (including the "stuff" formulas), the "column-of-fluid" fluid-statics formula, Archimedes law, fluid-flow formulas (Bernoulli's law and flux conservation), basic wave formulas (relating velocity to wavelength and frequency, and relating velocity to tension and linear mass density), and the standing wave wavelength formula.

You may notice that lots of rotation formulas are conspicuously absent from my list. That is because those formulas are analogous to basic linear equations. Therefore, if you include an "analogy table" on your cheat sheet, and if you have all the basic linear formulas, you automatically have many basic rotational formulas.

Let me now explain what I mean by a special-situation formula. It is a formula that applies only in one particular physical context. For instance, in the textbook, you will find a formula telling you the final velocities of two blocks that collide elastically in one dimension. Should you include this formula on your cheat sheet? Maybe, but not necessarily. First of all, you might be expected to *derive* that formula from more basic ones (conservation of momentum and conservation of energy, with $\text{Heat} = 0$). Second, when you include a special-situation formula on your cheat sheet, you might accidentally use it when it does not apply (for instance, when two blocks collide inelastically).

Some special situations show up so commonly that you should probably include them on your cheat sheet. Such formulas include: the horizontal range of a projectile that lands at the same height from which it was thrown; the rolling-without-slipping condition; sinusoidal traveling waves and standing waves; and a few others that you repeatedly encounter as you work through old homework and practice problems. When you include a special-case formula on your cheat sheet, *be sure to clearly label in which situation the formula does and does not apply*. For instance, if you include the above-mentioned elastic collision formula, be sure to indicate to yourself that it applies *only* during elastic collisions.

On your cheat sheet, you may also want to include some problem-solving strategies. That is fine, although I think you will find that after working through enough practice problems, those strategies will etch themselves permanently onto your brain.

Let me illustrate these hints by showing you two sample cheat sheets. They are designed for a midterm exam that covers kinematics and forces. *Your cheat sheet should differ from mine, because no two courses and no two instructors emphasize exactly the same material.*

BAD CHEAT SHEET

$x = x_0 + v_0t + (1/2)at^2 = (v_0 + v)t/2$
$y = y_0 + v_{0y}t + (1/2)a_yt^2 = (v_{0y} + v_y)t/2$
$v = v_0 + at = \Delta x/\Delta t = \sqrt{v_0^2 + 2a\Delta x}$
$v_y = v_{0y} + a_yt = \Delta y/\Delta t$
$a = \Delta v/\Delta t = v^2/r$
$y = (\tan\theta)x - gx^2/2v_0^2\cos^2\theta$
$R = (v_0^2\sin 2\theta)/g$
$R_{max} = v_0^2/g$
$T = (2v_0\sin\theta)/g$
$T = 2\pi/\omega$
$v_B = v_A - u$

$F = ma$
$F_x = ma_x$
$F_y = ma_y$
$F_{AB} = -F_{BA}$
$F_g = mg$
$f = \mu N$, $\mu_s = \tan\theta_c$
$N = mg\cos\theta$
$F_D = (1/2)\rho AC_Dv^2$
$F_c = mv^2/r$.

BETTER CHEAT SHEET

Kinematics

General motion: $v = dx/dt$, $a = dv/dt$
$v_{average} = \Delta x/\Delta t$, and $a_{average} = \Delta v/\Delta t$
Constant acceleration:
$x = x_0 + v_{0x}t + (1/2)a_xt^2$
$v_x = v_{0x} + a_xt$
$v_x^2 = v_{0x}^2 + 2a_x\Delta x$
} SAME FOR "y"
Projectile lands at initial height:
Find peak height using $v_{y\ peak} = 0$

v_0, θ
flight time $T = (2v_0\sin\theta)/g$
range $R = (v_0^2\sin 2\theta)/g$
(maximized at θ=45°)

Forces

$\Sigma F_x = ma_x$ (same for "y")
$F_{by\ 1\ on\ 2} = -F_{by\ 2\ on\ 1}$
(Note: Be careful about what force acts on which object.)
Particular forces:
$F_{grav} = mg$
friction $f_k = \mu_kN$, $f_s \le \mu_sN$
drag $F_D = (1/2)\rho AC_Dv^2$
where A=cross-section area
Circular motion (uniform)
$a_{radial} = v^2/r$, so
$\Sigma F_{radial} = mv^2/r$.
Period of rotation $T = 2\pi/\omega$

The bad cheat sheet does not label which formulas are general and which apply only to specific situations. This lack of labeling could cause you to misapply a formula. Also notice the potential for confusion, i.e., the two conflicting formulas for *T*. On the better cheat sheet, it is clear that those two "*T*'s" mean entirely different things. The bad cheat sheet contains many formulas that apply only to specific problems. For instance, $\mu_s = \tan\theta_c$ tells us the maximum ("critical") angle that you can tilt a ramp without making a block slide down the ramp, given the coefficient of static friction. That formula does not apply to other problems. The cheat sheet should make this clear.

TEST-TAKING TIPS

- *Do not do the problems in order.* Do the problems that are easiest for you (and worth the most points) first. Go back later and try to earn partial credit on the harder ones.
- Keep in mind the problem-solving skills you have practiced all semester, such as making sketches, formulating a strategy, etc. It is easy to panic, become sloppy, and revert to formula-plugging. To reduce your anxiety level, take practice exams ahead of time.
- Do not waste precious time completing algebra at the expense of getting to every problem. You can earn lots of partial credit by setting up the relevant equations, and describing in words your problem-solving strategy. If you are running out of time, this is an excellent way to rack up partial credit. Check with your instructor to determine how best to "weight" your time between setting things up and completing the algebra.
- If you get completely stuck on a problem, go on to the next one.
- Sometimes, test problems suffer from ambiguous phrasing. If you are the slightest bit confused, ask the proctor to clarify what the question means. Students routinely misinterpret problems because they were unwilling to bother the proctor. This is no time to be shy.

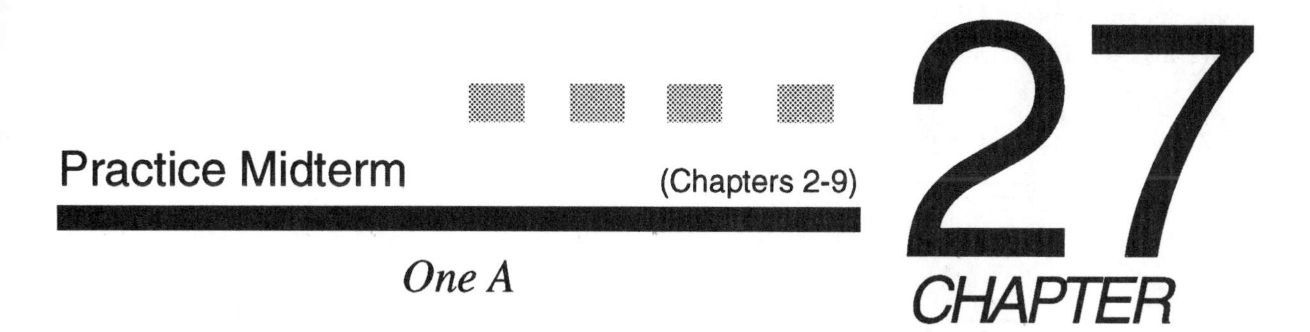

- *Do not forget about "Review Problem Set Number One," Chapter 23 of this study guide. It contains test-level problems on this material.*
- *Three questions, worth 100 points total. You have 70 minutes.*

1. *(30 points)* A turbo-boosted car is traveling along a straight highway. Its velocity as a function of time is given by $v = t^2 - 3t$, where t is in seconds and v is in meters per second.

a) *(7 points)* What is the car's instantaneous acceleration at the moment its velocity is 10 m/s?

b) *(7 points)* What is the car's average acceleration between $t = 0$ and $t = 3$ seconds? Briefly explain the physical interpretation of your perhaps-unexpected answer.

c) *(8 points)* What is the car's average velocity between $t = 0$ and $t = 3$ seconds?

d) *(8 points)* At time $t = 0$, the car is at rest at its initial position. How far is the car from its initial position when it next comes (momentarily) to rest?

2. *(35 points)* A medieval army with a cannon attacks a castle. The cannon is entrenched a distance L from the foot of the castle, and fires balls from a height d above the ground. The castle's one major weakness is a window at height H above the ground.

a) *(15 points)* Suppose the cannon is set to fire at a 60° angle to the ground. At what speed should the cannon fire the ball in order to make it enter the window? Express your answer in terms of L, d, H, g, and 60°.

b) *(15 points)* How fast is the cannon ball moving just as it enters the window? Let "v_0" denote your answer to part **a)**, and answer in terms of v_0 and the given constants.

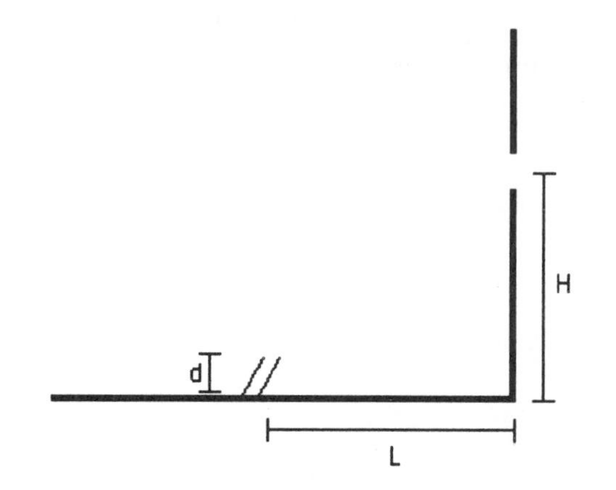

c) *(5 points extra credit)* Now the artilleryman decides to change his strategy. Instead of firing balls through the window, he will try to fire *over* the castle wall. So, he sets the cannon tofire at its maximum possible speed, v_m. Now he has to choose the new firing angle, θ. He wants the ball to clear the wall by the greatest possible margin. In other words, he wants the ball, when it is directly over the wall, to be as high off the ground as possible. What angle should he choose? Answer in terms of v_m and the other given constants.

3. *(35 points)* Consider the following system of blocks and ropes. All the ropes and pulleys are massless and frictionless, and the surfaces are frictionless. The ramp angle is $\theta = 30°$.

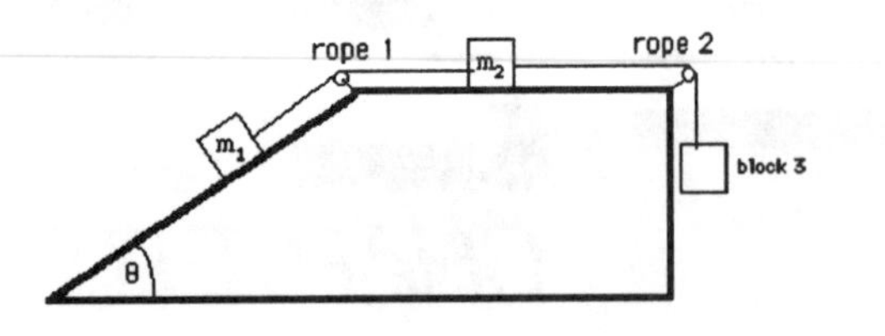

a) *(15 points)* Supposeblocks 1 and 2 have mass $m_1 = 2.0$ kg and $m_2 = 1.0$ kg, respectively. What must be the mass of block 3 to ensure that none of the blocks move?

b) *(20 points)* Suppose block 3 has the mass you calculated in part **a)**. Suddenly, rope 2 breaks at time $t = 0$. How far across the plateau does block 2 slide between $t = 0$ and $t = 1.0$ s, assuming it does not crash into the pulley during that time interval?

ANSWER 1

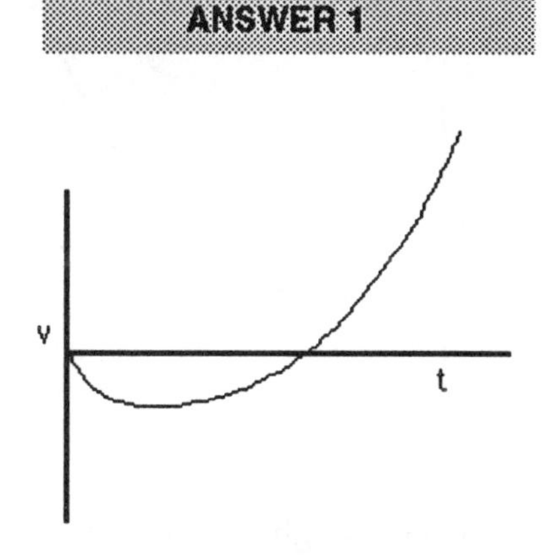

Sketch a quick graph of the given information, for future reference. Since $v = 0$ at $t = 0$, the graph starts at the origin. It initially curves down because of the negative coefficient in front of the "t" term, but eventually curves up because of the positive coefficient (+1) in front of the t^2 term. You can check this by plugging in $t = 1$ s, $t = 2$ s, etc.

a) Acceleration is the rate of change of velocity, which corresponds to the *slope* of the velocity vs. time graph. In calculus language, this means acceleration is the derivative of v with respect to t: $a = \frac{dv}{dt} = 2t - 3$. We are looking for the acceleration when the car's velocity is $v = 10$ m/s. So, we need to find at what time t the car goes 10 m/s. Then, substitute that time into the above formula for acceleration.

To calculate when the car hits $v = 10$ m/s, just set $v = 10$ in the expression for velocity: $v = 10 = t^2 - 3t$. This particular quadratic equation, $t^2 - 3t - 10 = 0$, can be factored: $(t - 5)(t + 2) = 0$. Throwing out the "unphysical" answer $t = -2$, we are left with $t = 5$ s. That is the time at which the velocity reaches $v = 10$ m/s. Substitute $t = 5$ into the above expression for acceleration to get

$$\begin{aligned} a &= 2t - 3 \\ &= 2(5) - 3 \\ &= 7 \text{ m/s}^2. \end{aligned}$$

b) Acceleration is the rate at which the car speeds up (or slows down). To calculate an *average* acceleration, we just need to know how much the car's velocity changes during the relevant time interval; we do not care about the moment-by-moment details of the motion. For instance, if the car speeds up by 20 m/s over a 5 second interval, then its *average* acceleration is 4 m/s^2. The formula I just used is $\bar{a} = \frac{\Delta v}{\Delta t} = \frac{v_f - v_0}{\Delta t}$. From $v = t^2 - 3t$, the "initial" speed at $t = 0$ is $v(0) = 0$. The "final" speed, at $t = 3$ s, is $v(3) = 3^2 - 3(3) = 0$. So,

$$\begin{aligned} \bar{a} &= \frac{\Delta v}{\Delta t} \\ &= \frac{v(3) - v(0)}{3} \\ &= \frac{0 - 0}{3} \\ &= 0. \end{aligned}$$

This seems like a mistake–unless you sketched v vs. t. The car initially backs up before "turning around" and going forward. Time $t = 3$ is when the v vs. t curve crosses the t axis. So, the car at $t = 0$ was stopped; and the car at $t = 3$ s is again stopped.

Physically, between $t = 0$ and $t = 3$ s, the car's velocity is negative, i.e., it is moving backward. But acceleration is the rate of *change* of velocity. Look at the *slope* of the v vs. t graph. At first, the car's negative velocity gets bigger and bigger; the car speeds up in the backward direction. On

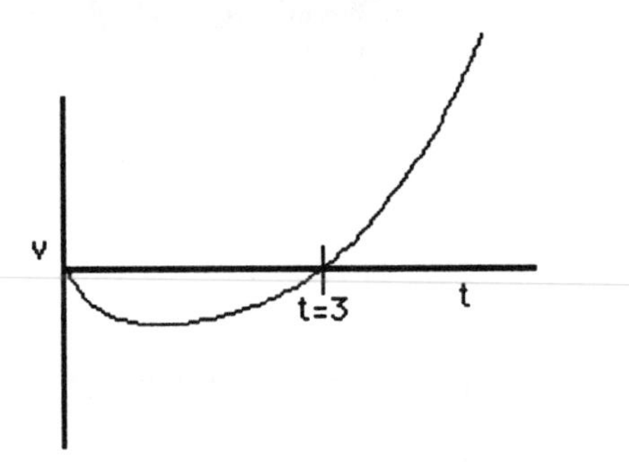

the graph, this corresponds to a negative slope. But then, the negative velocity gets smaller and smaller; the car slows down. When the car speeds up, its acceleration has the opposite sign compared to when the car slows down. In other words, the car's acceleration between $t=0$ and $t=3$ s is negative part of the time and positive part of the time. *On average*, the acceleration is 0.

Let me summarize: At $t=0$, the car is motionless. At $t=3$ s, it is again motionless. Overall, it neither sped up nor slowed down. So, its average acceleration–i.e., the average rate at which it sped up or slowed down–was zero, even though its instantaneous acceleration was sometimes negative and sometimes positive.

c) A common mistake is to say the average velocity is $\frac{v_f-v_0}{t}$. But that cannot work. To see why, suppose a car cruises at 50 mph for the whole trip. Then its initial and final velocity are both 50 mph, and hence $\frac{v_f-v_0}{t}=\frac{50\text{ mph}-50\text{ mph}}{t}=0$. But the average velocity is obviously 50 mph, not 0. The formula $\frac{v_f-v_0}{t}$ tells us something about how the car's velocity *changes*. Specifically, it gives us the car's average acceleration, *not* its average velocity.

Another common mistake is to use $\frac{v_f+v_0}{t}$. This formula works *only* if the acceleration is constant. Otherwise, it might give the wrong answer. To see why, suppose a car starts at rest ($v_0=0$), merges onto a highway, and then cruises at 50 mph for the next three hours. Intuitively, its average velocity is just under 50 mph. But $\frac{v_f+v_0}{t}=\frac{50\text{ mph}+0}{2}=25\text{ mph}$, much too low. To figure out the correct formula for average velocity, we must think physically...

Velocity is the rate at which position changes. For instance, if the car travels in a straight line a total of 200 meters over 40 seconds, then its average velocity must be 5 m/s. My point is this: To calculate an *average* velocity, we just need to know how much the car's *position* changes during the relevant time interval; we do not care about the moment-by-moment details of the motion. In symbols, $\bar{v}=\frac{\Delta x}{\Delta t}=\frac{x_f-x_0}{\Delta t}=\frac{x(3)-x(0)}{\Delta t}$. So, we need to know the car's position at $t=3$ s and at $t=0$. How can we find it? Instantaneous velocity is the rate of change of position. In formulas, this means $v=\frac{dx}{dt}$. So, to derive an expression for x, multiply both sides of the equation by dt and integrate:

$$\begin{aligned} x &= \int dx = \int v\,dt \\ &= \int (t^2-3t)dt \\ &= \frac{1}{3}t^3-\frac{3}{2}t^2+C, \end{aligned}$$

where C is an integration constant. Physically, C is x_0, the car's initial position. Now that we have an expression for x, we can substitute it into the above expression for the average velocity over the three-second interval:

$$\begin{aligned} \bar{v} &= \frac{x(3)-x(0)}{\Delta t} \\ &= \frac{\left[\frac{1}{3}(3)^3-\frac{3}{2}(3)^2+C\right]-[0-0+C]}{3} \\ &= -1.5\text{ m/s}. \end{aligned}$$

Notice that the C's canceled. The minus sign confirms what my graph told me; during the first three seconds, the car moves backwards.

d) Here, we are asked for the car's displacement between $t = 0$ and the time at which the car again comes to rest. So, we need to find the time at which $v = 0$. To find this time, substitute $v = 0$ into the expression for velocity given in the problem: $v = 0 = t^2 - 3t$. Solve by factorization: $0 = t(t-3)$, yielding $t = 0$ and $t = 3$ s as the times at which the car is momentarily stopped. Indeed, we already anticipated this result in part **b)**; see the graph. So, we are finding the car's displacement between $t = 0$ and $t = 3$ s.

Let us use the expression for x derived in part **c)**, $x = \frac{1}{3}t^3 - \frac{3}{2}t^2 + C$, to get $x(3) = -4.5 \text{ meters} + C$, and $x(0) = C$. So,

$$\begin{aligned}\Delta x &= x(3) - x(0) \\ &= -4.5 \text{ meters} + C - C \\ &= -4.5 \text{ meters.}\end{aligned}$$

In words, the car at $t = 3$ seconds is 4.5 meters behind where it started.

We also could have obtained this result using the *average* velocity found in part **c)**. The car's average velocity between $t = 0$ and $t = 3$ s is $\bar{v} = -1.5$ m/s. So, its displacement during those three seconds is $\Delta x = \bar{v}\Delta t = (-1.5 \text{ m/s})(3 \text{ s}) = -4.5$ meters.

ANSWER 2

a) We are trying to find the initial velocity v_0 such that the ball, when fired at a 60° angle, travels a horizontal distance L in the same time it reaches height H. So, we will probably be able to extract information by separately considering the horizontal and vertical aspects of the ball's motion. This is our usual projectile motion strategy.

Let us start with the vertical part of the motion. We know the ball's initial height is $y_0 = d$, and its "final" height when it enters the window is $y = H$. We can also express the vertical component of its initial velocity in the usual way.

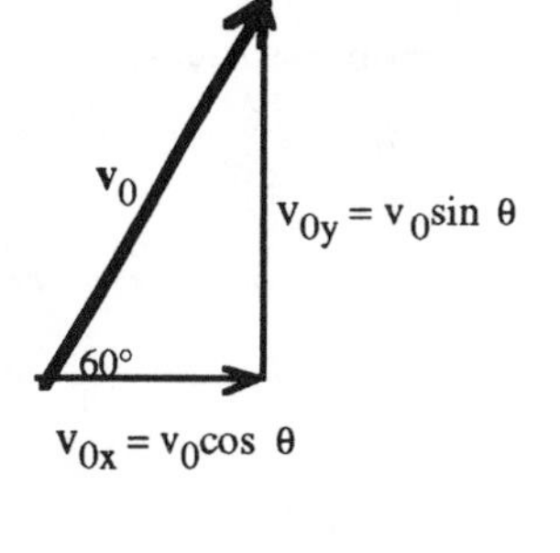

Give all this, I will start with the vertical displacement formula, picking upward as positive:

$$\begin{aligned}y &= y_0 + v_{0y}t + \frac{1}{2}a_y t^2 \\ H &= d + (v_0 \sin 60°)t - \frac{1}{2}gt^2.\end{aligned} \quad (1)$$

We are trying to solve for v_0. But Eq. (1) contains a second unknown, the "flight time" t. We need another equation. To get additional information, consider the horizontal aspect of the motion. We can set $x_0 = 0$, in which case the final x-position is $x = L$. Also, since no horizontal push or pull acts on the ball once it is in mid-air, the ball's *horizontal* velocity neither speeds up nor slows down: $a_x = 0$. Given all this, let us try the horizontal displacement equation:

$$\begin{aligned}x &= x_0 + v_{0x}t + \frac{1}{2}a_x t^2 \\ L &= 0 + (v_0 \cos 60°)t - 0.\end{aligned} \quad (2)$$

We now have two equations in the two unknowns, v_0 and t. Therefore, it is just a matter of algebra to solve for v_0. Some instructors would grant almost full credit for generating these equations and explaining in words how you would solve. Other instructors would insist that you complete the algebra.

Algebra starts here. Solve Eq. (2) for t to get $t = \frac{L}{(v_0 \cos 60°)}$, and substitute it into Eq. (1) to get

$$\begin{aligned} H &= d + v_0 \sin 60° \left(\frac{L}{v_0 \cos 60°} \right) - \frac{1}{2} g \left(\frac{L}{v_0 \cos 60°} \right)^2 \\ &= d + L \tan 60° - \frac{gL^2}{2v_0^2 \cos^2 60°}. \end{aligned}$$

Now multiply through by v_0^2, and isolate v_0 to get $v_0 = \sqrt{\frac{gL^2}{(2\cos 60°)(d + L\tan 60° - H)}}$.
End of algebra.

b) To find the ball's speed when it reaches the window, we must use the standard strategy of *separately* finding v_x and v_y, the ball's horizontal and vertical velocity. Then we can vectorially add those velocity components to obtain the total velocity. (You cannot use the formula $\mathbf{v} = \mathbf{v}_0 + \mathbf{a}t$ directly, because the motion does not happen along a straight line; $\mathbf{v}_0$ and $\mathbf{a}$ point in different directions.

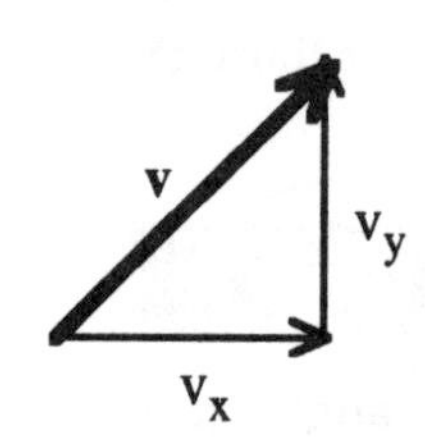

We can use the x- and y-components of either $\mathbf{v} = \mathbf{v}_0 + \mathbf{a}t$ or $\mathbf{v}^2 = \mathbf{v}_0^2 + 2\mathbf{a}\Delta\mathbf{x}$. Since we found the "flight time" in part **a)**, $t = \frac{L}{(v_0 \cos 60°)}$ I will use $\mathbf{v} = \mathbf{v}_0 + \mathbf{a}t$.

$$\begin{aligned} v_x &= v_{0x} + a_x t \\ &= v_0 \cos 60° + 0. \end{aligned}$$

The horizontal part of the velocity stays constant for the whole flight. By contrast, the vertical motion slows down:

$$\begin{aligned} v_y &= v_{0y} + a_y t \\ &= v_0 \sin 60° - gt \\ &= v_0 \sin 60° - g \frac{L}{v_0 \cos 60°}. \end{aligned}$$

To find the magnitude of the total velocity, just "Pythagoras together" the x- and y-components:

$$\begin{aligned} v &= \sqrt{v_x^2 + v_y^2} \\ &= \sqrt{(v_0 \cos 60°)^2 + \left(v_0 \sin 60° - \frac{gL}{v_0 \cos 60°} \right)^2}, \end{aligned}$$

where v_0 is your part **a)** answer. On a test, your instructor will probably allow you to express your answer to part **b)** in terms of your answer to part **a)**, provided you write in words something like "...where v_0 is the answer from above." Ask your instructor about the "rules" concerning this.

c) When I first encountered this problem, I thought the "best" angle θ corresponded to the ball's reaching its peak over the wall. But the ball clears the wall with even more leeway if it is falling when it is directly over the wall. To see this, look at the diagram below, in which I show two trajectories corresponding to two different firing angles. The lower-angled trajectory reaches its peak directly above the wall, while the higher-angled trajectory reaches its peak in front of the wall. But even so, the higher-angled trajectory clears the wall by a greater margin. So, the artilleryman should fire the ball at an angle such that it peaks a little before reaching the wall, instead of peaking directly over the wall. To see this more clearly, experiment with the computer simulation.

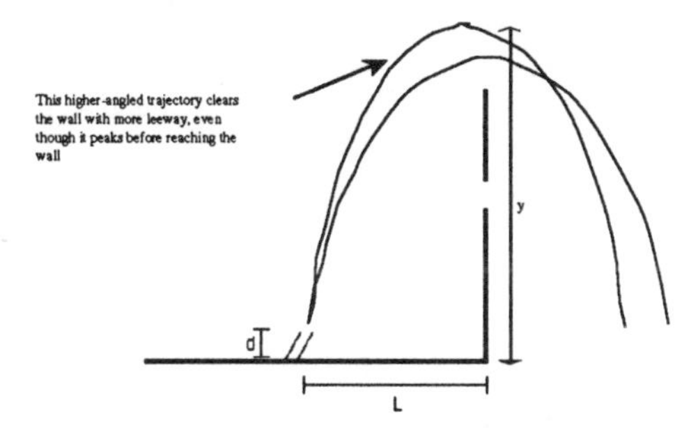

This is a maximization problem. We are trying to find the cannon angle θ that maximizes y, the ball's height when it is directly over the wall.

So, we can use the usual maximization/minimization strategy:

(1) Figure out which quantity we are trying to maximize (or minimize), and which quantity is the free variable, i.e., the variable we can "alter" in order to achieve maximization.

(2) Write the quantity we are maximizing as a function of the free variable. *The equation may contain given constants and other quantities that are "independent" of the free variable, but must contain no other variables besides the quantity you are maximizing and the free variable.*

(3) Differentiate the quantity you are maximizing with respect to the free variable, and set it equal to 0. This will generate an equation that you can solve for the free variable.

Now I will implement this strategy, step by step.

Step 1: Figure out what is getting maximized and what is the free variable

Here, we are maximizing y, the ball's height *when it is directly over the wall*. So, y is the ball's height *when it has traveled a horizontal distance L*. The variable we can "control" is θ, the firing angle. So, θ is the free variable.

Step 2: Express the quantity being maximized as a function of the free variable

We must write y as a function of θ. So, a good place to start is $y = y_0 + v_{0y}t + \frac{1}{2}a_y t^2$. The "$t$" in this equation now differs from the "t" of parts **a)** and **b)**, because the firing angle is no longer $60°$. Since the initial velocity is $v_0 = v_m$, we have $v_{0y} = v_m \sin\theta$. We also know the initial height, $y_0 = d$. Substitute all this into $y = y_0 + v_{0y}t + \frac{1}{2}a_y t^2$ to get

$$y = d + (v_m \sin\theta)t - \frac{1}{2}gt^2. \tag{3}$$

The only unknowns "allowed" in this equation are y and θ, the variable we are maximizing and the free variable. (Actually, the equation may also contain unknowns that do not depend on the free variable.) But Eq. (3) contains an "extra" unknown, namely t. And this extra unknown depends on θ; by changing the firing angle, you automatically affect the flight time. So, we must rid Eq. (3) of t.

To do so, consider the horizontal aspect of the projectile's motion. From the horizontal displacement equation $x = x_0 + v_{0x}t + \frac{1}{2}a_x t^2$, with initial position $x_0 = 0$, final position $x = L$, sideways acceleration $a_x = 0$, and $v_{0x} = v_m \cos\theta$, we get $L = (v_m \cos\theta)t$. Solve for t to get $t = \frac{L}{(v_m \cos\theta)}$, the time at which the ball is directly over the wall. Substituting this t into Eq. (3) yields

$$\begin{aligned} y &= d + (v_m \sin\theta)t - \frac{1}{2}gt^2 \\ &= d + (v_m \sin\theta)\frac{L}{v_m \cos\theta} - \frac{1}{2}g\left(\frac{L}{v_m \cos\theta}\right)^2 \\ &= d + L\tan\theta - \frac{gL^2}{2v_m^2 \cos^2\theta}, \end{aligned} \tag{3'}$$

the same expression we got for H in part **a)**, except with θ in place of 60°. So, we could have jumped right to Eq. (3′). The only variables in this equation are y and θ. So, we are done with step 2.

Step 3: Differentiate

Take the derivative of y with respect to θ, and set it to 0. If you are short on time, you could just write in words, "set $\frac{dy}{d\theta} = 0$ and solve for θ." That would get you partial credit for part **c)**. Indeed, unless you have time left over at the end of the exam, you should not waste lots of time completing the math in a 5-point problem! But for the record, here is how the calculus works out:

$$\frac{dy}{d\theta} = L\sec^2\theta - \frac{gL^2}{2v_m^2}\left[\left(\frac{-2}{\cos^3\theta}\right)(-\sin\theta)\right] = 0.$$

Remembering that $\sec\theta = \frac{1}{\cos\theta}$, we can multiply the equation through by $\cos^2\theta$, and divide through by L, to get $1 - \frac{gL}{v_m^2}\tan\theta = 0$. Simplify to get $\tan\theta = \frac{v_m^2}{gL}$, and hence $\theta = \tan^{-1}\left(\frac{v_m^2}{gL}\right)$. You can confirm that this angle is slightly bigger than the firing angle that would make the ball reach its peak directly above the wall.

ANSWER 3

a) Usually in multi-block problems, we are given all the masses and asked for the acceleration (or some quantity related to acceleration). Here, we are given the acceleration and asked for a mass. We want to find m_3 such that the blocks do not move. In other words, we want the "shared" acceleration of the blocks to be $a = 0$. Intuitively, this means that the tendency of block 3 to fall down must balance the tendency of block 1 to slide down its ramp. Using this insight, you could immediately jump to the equation $m_1 g\sin\theta = m_3 g$. But if you did not see this shortcut, let me solve the long way by implementing the usual multi-block strategy.

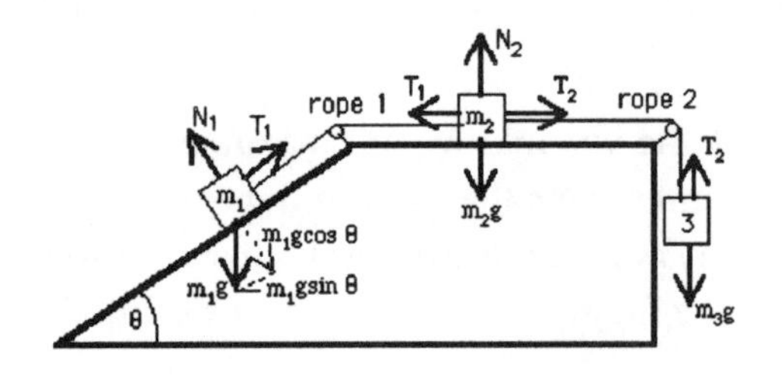

Step 1: Free-body diagram for each block separately.
Step 2: Newton's 2nd law, along direction of motion, for each separate mass

Let me pick down-the-ramp as positive for block 1. This choice commits me to calling leftward positive for block 2, and upward positive for block 3, because all three blocks move together (when they move at all). If you chose up-the-ramp as positive for block 1, that is fine, provided you were consistent. Looking only at the direction of motion, and letting x denote the along-the-ramp direction for block 1, we get

$$\text{Block 1:} \quad \sum F_x = m_1 g\sin\theta - T_1 = m_1 a = 0 \tag{1}$$

$$\text{Block 2:} \quad \sum F_x = T_1 - T_2 = m_2 a = 0 \tag{2}$$

$$\text{Block 3:} \quad \sum F_x = T_2 - m_3 g = m_3 a = 0. \tag{3}$$

Step 3: Solve for the quantity of interest

Usually, we would use these three equations to solve for a, the acceleration. But here, as noted above, we are given $a = 0$, and we are solving for m_3. We can do so with the above three equations, because they contain only three unknowns (m_3, T_1, and T_2).

To complete the algebra efficiently, solve Eq. (1) for T_1 to get $T_1 = m_1 g \sin\theta$. Plug that into Eq. (2) to get $T_2 = m_1 g \sin\theta$. Substitute that into Eq. (3) to get $m_1 g \sin\theta = m_3 g$, the "shortcut" equation mentioned above. Cancel the g's and you get the answer,

$$\begin{aligned} m_3 &= m_1 \sin\theta. \\ &= (2.0 \text{ kg}) \sin 30° \\ &= 1.0 \text{ kg}. \end{aligned}$$

Notice that m_2 does not figure in the answer. Intuitively, this is because the downward pull on block 3 must balance the down-the-ramp pull on mass 1. If this balance is achieved, then block 2 feels no net force, no matter what mass it has.

b) Here again, we can use the standard multi-block strategy. And this time, we are solving for the acceleration; once we know a, we can put it into a kinematic formula to calculate how far block 2 slides in a given time.

Step 1: Free-body diagrams for each block separately

When rope 2 breaks, the tension force T_2 disappears. Block 3 no longer matters.

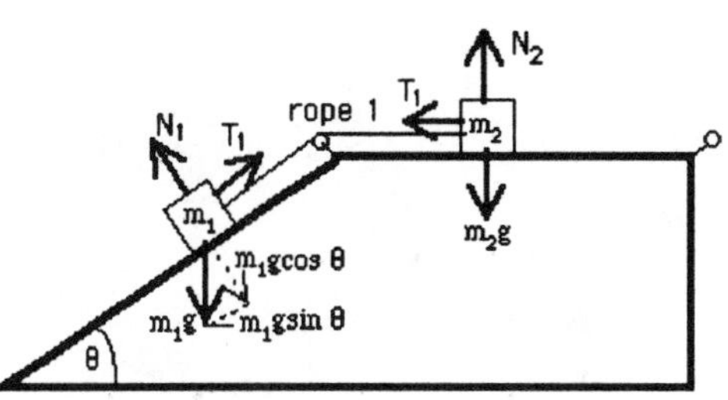

Step 2: Newton's 2nd law, along direction of motion, for each block.

A common mistake is to set $T_1 = m_1 g \sin\theta$. If that equation held, then the net force on block 1 would be 0, in which case it would not accelerate down the ramp. Since block 1 does accelerate, gravity must "overcome" the tension force holding block 1 back: $m_1 g \sin\theta > T_1$.

There is no friction. So, by the usual reasoning,

$$\text{Block 1:} \quad \sum F_x = m_1 g \sin\theta - T_1 = m_1 a \tag{1*}$$

$$\text{Block 2:} \quad \sum F_x = T_1 = m_2 a \tag{2*}$$

Step 3: Solve for the acceleration

We have two equations in two unknowns (T_1 and a). To complete the algebra quickly, add those equations to get $m_1 g \sin\theta = (m_1 + m_2)a$. Isolate a to get

$$\begin{aligned} a &= \frac{m_1}{m_1 + m_2} g \sin\theta \\ &= \frac{2.0 \text{ kg}}{2.0 \text{ kg} + 1.0 \text{ kg}} (9.8 \text{ m/s}^2)(\sin 30°) \\ &= 3.27 \text{ m/s}^2 . \end{aligned}$$

Step 4: Kinematics or Newton's 2nd law to find the quantity of interest

Now that we know the acceleration of the blocks, we can find the distance $\Delta x = x - x_0$ traveled by block 2 in time $t = 1.0$ s. The constant-acceleration kinematic formula relating position to time is

$$\begin{aligned} x &= x_0 + v_{0x}t + \frac{1}{2}a_x t^2 \\ &= 0 + 0 + \frac{1}{2}\left(\frac{m_1}{m_1 + m_2} g \sin\theta\right) t^2 \\ &= \frac{1}{2}(3.27 \text{ m/s}^2)(1.0 \text{ s})^2 \\ &= 1.6 \text{ m}. \end{aligned}$$

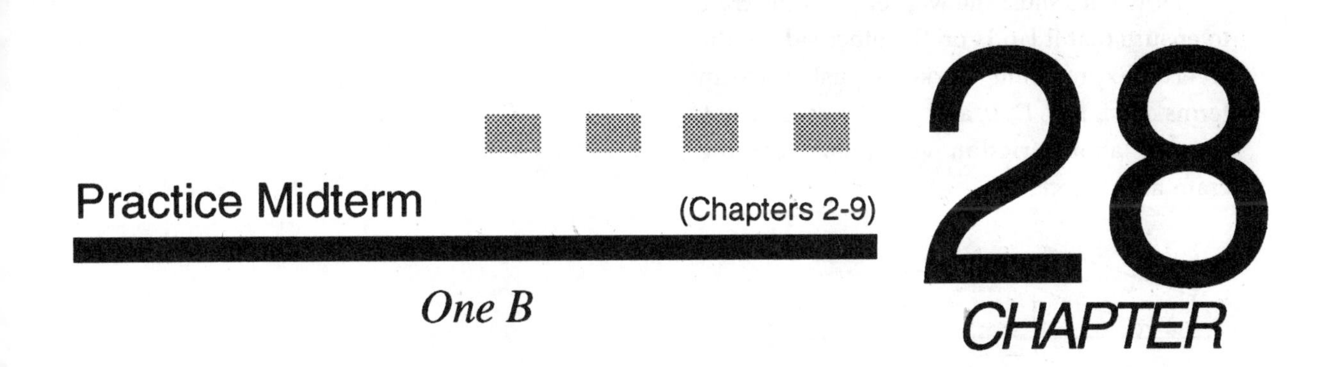

Practice Midterm (Chapters 2-9)

One B

◆ *Three questions, worth 100 points total. You have 80 minutes.*

1. *(30 points)* Mary, anxious to sample the casinos of Reno, drives along a straight California highway toward the Nevada border. Mary is exceeding the speed limit. Indeed, her speed is constant at $v_m = 30$ m/s, about 72 mph.

Hidden behind a billboard a distance $D = 500$ meters from the border is a police car waiting for speeders. At the moment Mary passes the police car, the police car starts chasing Mary. The police car's acceleration is constant. Even though she is being chased, Mary continues driving at constant speed.

The police car started from rest.

a) *(5 points)* What is the smallest acceleration the police car must have to catch Mary before she crosses the border? Express your answer in terms of v_M and D before plugging in the numbers.

b) *(15 points)* Suppose the turbo-boosted police car's acceleration is $a_p = 4.0$ m/s^2. How fast will the police car be moving at the moment it catches up with Mary? Express your answer in terms v_M, D, and a_p before plugging in numbers.

c) *(10 points)* Another car on the road, driven by Jacques, travels at non-constant velocity. Indeed, Jacques' velocity as a function of time looks like this.

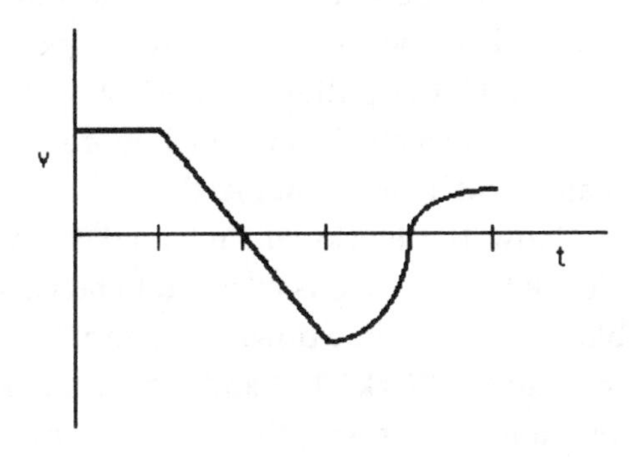

Make a rough sketch of Jacques' position vs. time, and a rough sketch of his acceleration vs. time. You may assumehis initial position is 0.

2. *(35 points)* A rescue worker must get a heavy crate (mass M) of supplies across a river. She plans to do so by pushingthe crate off a ramp hard enough so that it flies across the river and lands on the other side. (The supplies are not fragile.) The ramp has length L and angle θ, and the river has width D.

The crate starts from rest from the bottom of the ramp. The rescue worker applies a constant force until the crate leaves the ramp. The direction of her force is parallel to the surface of the ramp. Because the crate has rollers on the bottom, its coefficient of friction μ is small enough that the worker can accelerate the crate up the ramp.

How hard must the worker push the crate to ensure that it lands on the other side of the river? Express the worker's push force in terms of M, L, θ, D, μ, and g. (If you have not learned about friction yet, then treat the crate as frictionless.)

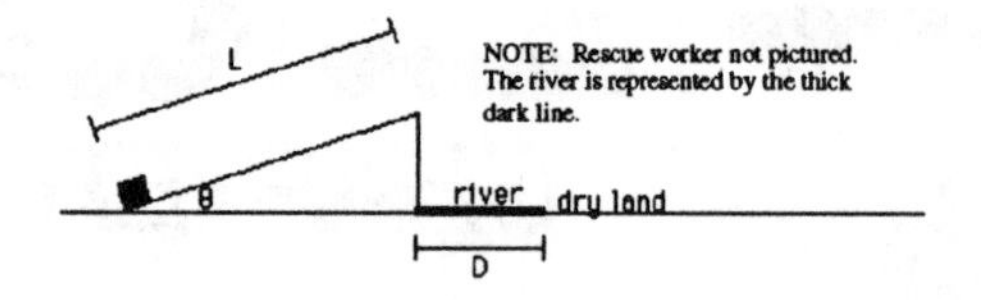

3. *(35 points)* Three blocks are initially at rest in the configuration shown. The pulley is attached to the ramp. The rope goes from block 1 to the pulley to block 2. Block 3, despite its small size, is much more massive than blocks 1 and 2 put together.

Now the blocks are released from rest. Notice that no rope is attached to block 3. As blocks 2 and 3 slide down the ramp, they stay in contact. Blocks 1, 2 and 3 have mass m_1, m_2, and m_3, respectively. The ramp is frictionless, and the rope and pulley are massless.

Express your answers in terms of m_1, m_2, m_3, g, and θ.

a) *(15 points)* What is the tension in the rope?

b) *(20 points)* What is the "contact force" exerted by block 3 on block 2?

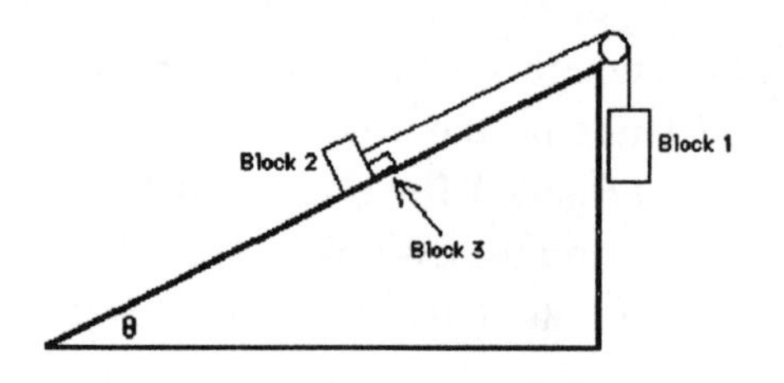

ANSWER 1

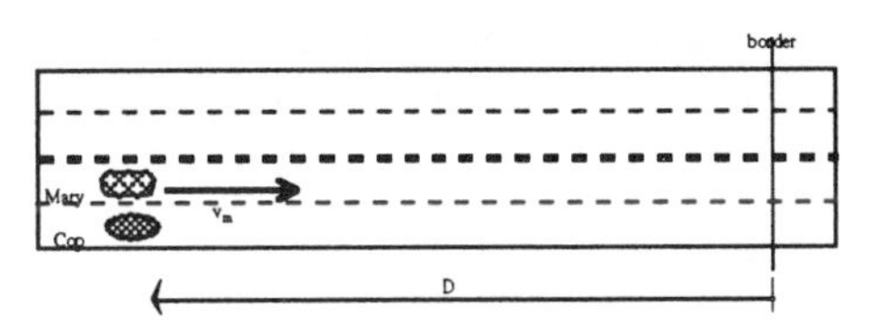

Start with a diagram of what is happening. This "snapshot" is when Mary passes the (stationary) police car, 500 meters from the border.

a) If the cop accelerates as slowly as she can get away with, then she catches Mary *just* as Mary reaches the border. In other words, the police car catches Mary just as Mary reaches $x = D = 500$ m. Therefore, the police car travels $D = 500$ meters in the same time Mary travels $D = 500$ meters.

This suggests a two-part strategy:

(1) Figure out how much time Mary takes to reach the border (i.e., how much time she takes to travel $D = 500$ meters). Call that time t.

(2) Calculate what acceleration the police car must have to move $D = 500$ meters in time t.

I will now implement this strategy.

Step 1: Mary's time to border.

Since we know Mary's initial velocity and her acceleration ($a = 0$, because her velocity is constant), and we want to find how much time she takes to travel distance D, use

$$\text{(MARY)} \qquad x = x_0 + v_0 t + \frac{1}{2}at^2$$
$$D = 0 + v_M t + 0.$$

Solve for t to get $t = \frac{D}{v_M} = \frac{500 \text{ m}}{30 \text{ m/s}} = 16.67$ s. Notice that, since Mary's acceleration is $a = 0$, her equation of motion reduces to the constant-velocity formula $x = v_M t$.

Step 2: Police car's acceleration

Since we now know that the police car must travel distance D in time $t = \frac{D}{v_M} = 16.67$ s, and since we know its initial velocity is $v_0 = 0$, we can again use the kinematic formula

$$\text{(POLICE)} \qquad x = x_0 + v_0 t + \frac{1}{2}at^2$$
$$D = 0 + 0 + \frac{1}{2}a\left(\frac{D}{v_M}\right)^2.$$

Solve for a to get $a = \frac{2v_M^2}{D} = \frac{2(30 \text{ m/s})^2}{500 \text{ m}} = 3.6 \text{ m/s}^2$. That is the smallest acceleration the cop must have to catch Mary before she escapes over the border. Fortunately for Mary, many cars cannot achieve this acceleration.

b) This is a standard "meeting" problem. When two objects meet (or crash or whatever), they have the same position. Therefore, we can invoke the following strategy:

(1) Using kinematics or whatever techniques are available, write x_1 and x_2 (i.e., the positions of object 1 and object 2) as functions of time.

(2) The meeting occurs when $x_1 = x_2$. So, set $x_1(t) = x_2(t)$, and solve for t, the time at which the meeting happens.

(3) Use that t to find whatever you need, in this case v_p, the police car's velocity when they meet.

Let us do it.

Step 1: Write positions as functions of time

For both Mary and the police car, we can use $x = x_0 + v_0 t + \frac{1}{2}at^2$:

$$\text{(MARY)} \qquad x_M = 0 + v_M t + 0$$

$$\text{(POLICE CAR)} \qquad x_p = 0 + 0 + \frac{1}{2}a_p t^2.$$

Step 2: Set Mary's position equal to the police car's position

$$x_M = x_p$$

$$v_M t = \frac{1}{2}a_p t^2.$$

Solve for t to get $t = \frac{2v_M}{a_p} = \frac{2(30\,\text{m/s})}{4\,\text{m/s}^2} = 15\text{ s}$, less than the t found in part **a)**. Therefore, the police car catches Mary before she reaches the border.

Step 3: Use kinematics to solve for the quantity of interest

We want to know the police car's velocity v_p at time $t = \frac{2v_M}{a_p} = 15\text{ s}$. So, we can use

$$\begin{aligned} v_p &= v_0 + a_p t \\ &= 0 + a_p \frac{2v_M}{a_p} \\ &= 2v_M \\ &= 60\text{ m/s}. \end{aligned}$$

The police car, when it catches up with Mary, is moving at double her speed.

c) Let us first sketch position vs. time. During the first time interval, the velocity is positive and constant. So, the car is moving forward at a steady pace. This corresponds to a steadily increasing position, i.e., a straight upward-sloped line on the x vs. t graph.

During the second time interval, the velocity is positive but decreasing. This does *not* mean Jacques goes backwards. It means he continues to move forward, but at a slower and slower rate. So, the x vs. t graph continues increasing, but less and less "steeply"; it levels off. Right when Jacques hits $v = 0$, the slope of the x vs. t graph hits 0.

During the third time interval, Jacques' velocity is negative; he is driving backwards. So, x decreases with time. Furthermore, since his negative velocity is getting bigger and bigger, Jacques position decreases at a faster and faster rate. Mathematically, this means the slope of the x vs. t graph gets more and more negative (steeper and steeper).

During the fourth time interval, Jacques' velocity is still negative. So, Jacques is still going backwards. But his backwards velocity is decreasing. In other words, the rate at which he is going backwards gets smaller and smaller. So, on the x vs. t graph, x continues getting smaller, but at a lesser and lesser rate (slope). The slope hits 0 when $v = 0$.

During the fifth time interval, the velocity is positive. So, the x vs. t graph goes "up" again. Furthermore, the velocity is increasing. So, x increases at a higher and higher rate; the slope of x vs. t increases.

Putting all this together, I get

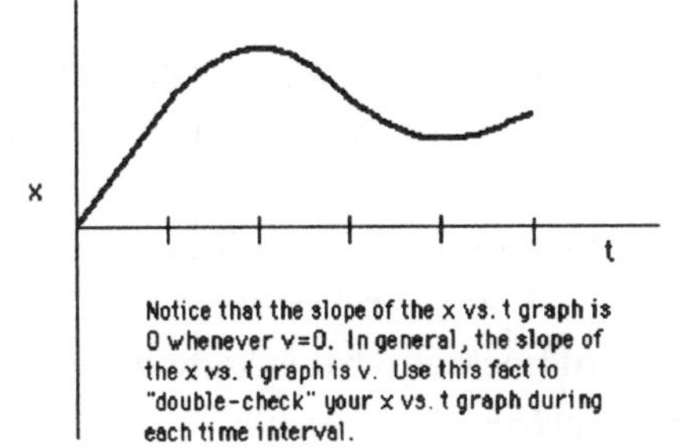

Notice that the slope of the x vs. t graph is 0 whenever v=0. In general, the slope of the x vs. t graph is v. Use this fact to "double-check" your x vs. t graph during each time interval.

Now let us graph acceleration vs. time. The acceleration is the rate at which the velocity changes, i.e., the slope of the v vs. t graph.

During the first time interval, the velocity is constant, and hence $a = 0$.

During the second time interval, the velocity is steadily decreasing. This corresponds to a negative acceleration, i.e., Jacques is moving forward but slowing down. Since the slope of the v vs. t graph is constant, this negative acceleration is constant.

During the third time interval, the velocity continues to get lower, in this case more and more negative (instead of less and less positive). The rate of change of velocity is still constant and negative–indeed, the same slope corresponding to the second time interval.

During the fourth time interval, the slope of the v vs. t graph is positive, and hence, a is positive. Physically, the velocity is "increasing," i.e., getting less and less negative. Furthermore, the rate at which the velocity changes (i.e., the slope of v vs. t) increases. So, in that region, the acceleration is positive and increasing.

By contrast, during the fifth time interval, the velocity continues to increase, but at a slower and slower rate. So, the acceleration continues to be positive, but gets smaller and smaller.

Putting all this together, I get

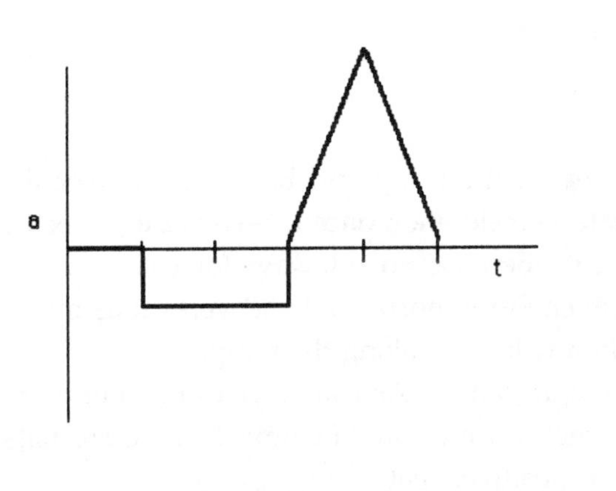

ANSWER 2

Because this difficult multistep problem is arguably the hardest one, you should have considered saving it for last.

Before diving into details, formulate an overall strategy, breaking the problem into manageable "subproblems."

The worker makes the crate speed up as it ascends the ramp. When the crate leaves the ramp, it becomes a projectile, launched at angle θ. Intuitively, whether the crate clears the river depends on its "initial" speed v_0, the speed with which it leaves the ramp. Crucially, the "initial" speed of the projectile is the crate's "final" speed at the top of the ramp.

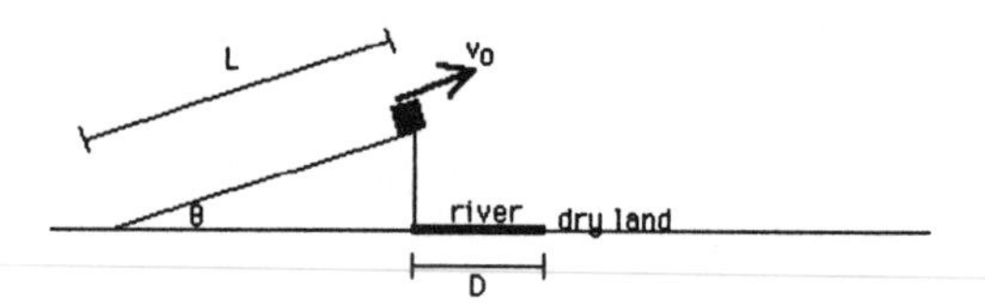

This suggests a way to break up the overall problem:

Subproblem 1: Using projectile reasoning, figure out what "initial" speed v_0 the crate must have (when it leaves the ramp) to clear the river.

Subproblem 2: Using Newton's 2nd law and kinematics, figure out what force the worker must apply to the crate so that it acquires the necessary speed v_0 by the time it reaches the top of the ramp.

Many students reverse the order of these two subproblems. That is o.k., provided you keep careful track of what is known and what you are solving for. I will do subproblem 1 first.

Subproblem 1: Projectile

We are solving for the initial speed v_0 that the "projectile" must possess in order to travel a horizontal distance D by the time it lands, i.e., by the time it falls through a vertical distance $h = L\sin\theta$. On the picture, the curvy line represents the path of the projectile.

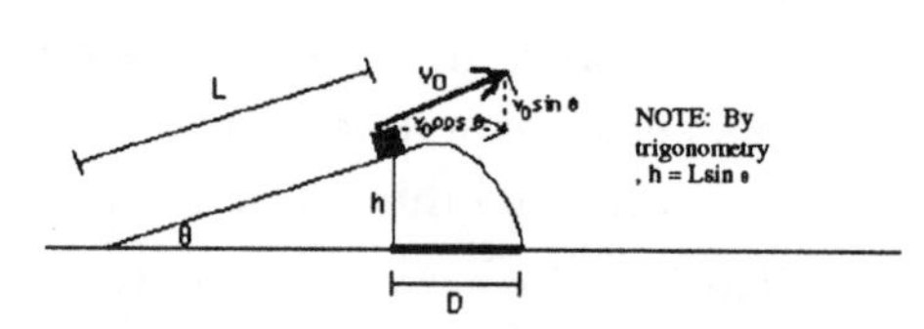

So, a reasonable place to start is

$$\begin{aligned} x &= x_0 + v_{0x}t + \frac{1}{2}a_x t^2 \\ D &= 0 + (v_0\cos\theta)t + 0. \end{aligned} \tag{1}$$

In Eq. (1), a_x is the *horizontal* acceleration *after* the crate leaves the ramp and becomes a projectile, i.e., *after* the worker no longer pushes on the crate. The crate's acceleration once it becomes a projectile *differs* from its acceleration while the worker was pushing it. Because no sideways forces act on the crate once it is in mid-air, $a_x = 0$. In this subproblem, I am choosing horizontal and vertical as my x- and y-direction. (By contrast, in subproblem 2, the x-direction will point along the ramp.)

In Eq. (1), we know the launch angle θ, but not the flight time t. As usual, we can gather more information by considering the vertical aspect of the projectile's motion. In time t, the crate falls through vertical distance $h = L\sin\theta$. So, choosing upward as positive, I get

$$\begin{aligned} y &= y_0 + v_{0y}t + \frac{1}{2}a_y t^2 \\ 0 &= L\sin\theta + (v_0\sin\theta)t - \frac{1}{2}gt^2. \end{aligned} \tag{2}$$

At this point, we have two equations in two unknowns, v_0 and t. So, it is just a matter of algebra to solve for v_0. Remember, we need v_0 so that, in the next subproblem, we can figure out how hard the worker must push in order to make the crate attain that speed. *If you do not have time to complete the algebra, just say in words that you can do so with Eqs. (1) and (2).*

Algebra starts here. From Eq. (1), $t = \frac{D}{v_0 \cos\theta}$. Put that into Eq. (2) to get

$$0 = L\sin\theta + (v_0 \sin\theta)\frac{D}{v_0\cos\theta} - \frac{1}{2}g\left(\frac{D}{v_0\cos\theta}\right)^2.$$
$$= L\sin\theta + D\tan\theta - \frac{1}{2}g\left(\frac{D}{v_0\cos\theta}\right)^2.$$

Multiply through by v_0^2 and isolate v_0 to get

$$v_0 = \sqrt{\frac{gD^2}{2\cos^2\theta(L\sin\theta + D\tan\theta)}}.$$

End of algebra.

Now we can proceed to subproblem 2.

Subproblem 2: Forces on the block while it is on the ramp

We will now find the force with which the rescue worker must push the block to ensure that it leaves the ramp with speed v_0 as given above. So, in this subproblem, v_0 is the crate's *final* velocity as it leaves the ramp.

How can we relate forces to that velocity? Well, invoking the usual force-problem strategy, we can relate the worker's push to the block's acceleration. Then, using kinematics, we can relate the acceleration to the "final" velocity v_0.

In this subproblem, "tilt" the coordinate axes so that the x-direction lies along the ramp. As always, start with a force diagram.

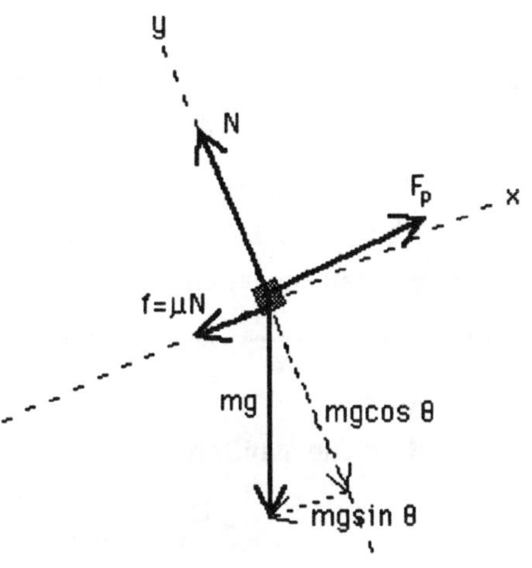

I have let F_p denote the worker's force.

Since the block moves along the ramp, it has no motion in the y-direction: $a_y = 0$. Therefore, the y-forces cancel:

$$\begin{aligned}\sum F_y &= Ma_y \\ &= N - Mg\cos\theta \\ &= 0,\end{aligned}$$

and hence $N = mg\cos\theta$. Now that we know the normal force, we can immediately calculate the frictional force and deal with the x-directed forces. I am choosing up-the-ramp as positive:

$$\begin{aligned}\sum F_x &= Ma_x \\ F_p - Mg\sin\theta - \mu N &= Ma_x \qquad (3) \\ F_p - Mg\sin\theta - \mu Mg\cos\theta &= Ma_x.\end{aligned}$$

We are solving for F_p. But this force equation contains a second unknown, the acceleration. From subproblem 1, however, we know the acceleration must be exactly big enough to make the crate speed up from initial speed 0 to "final" speed v_0 over a distance $x = L$. To calculate this acceleration, use a constant-acceleration kinematic formula:

$$\begin{aligned} v^2_{\text{top of ramp}} &= v^2_{\text{bottom of ramp}} + 2a_x \Delta x \\ v_0^2 &= 0 + 2a_x L. \end{aligned} \tag{4}$$

Remember, "v_0" from subproblem 1 denotes the crate's speed at the *top* of the ramp.

Solve Eq. (4) for a_x to get $a_x = \frac{v_0^2}{2L}$. Substitute that into force Eq. (3) to get

$$\begin{aligned} F_p - Mg\sin\theta - \mu Mg\cos\theta &= M\frac{v_0^2}{2L} \\ &= M\frac{gD^2}{4L\cos^2\theta(L\sin\theta + D\tan\theta)}, \end{aligned}$$

where in the last step I used the expression for v_0 we derived in subproblem 1. Now we just have to isolate F_p to get

$$F_p = M\frac{gD^2}{4L\cos^2\theta(L\sin\theta + D\tan\theta)} + Mg\sin\theta + \mu Mg\cos\theta.$$

That is the answer!

You probably did not have time to complete the algebra. But you can get plenty of partial credit by obtaining equations (1) through (4) and describing in words how to solve for F_p.

ANSWER 3

Use the standard multi-block strategy:

(1) Draw a free-body diagram *for each mass individually*, including only the forces acting *directly* on that mass. For instance, the rope does *not* act directly on block 3. Block 3 is pushed by block 2.

(2) Write Newton's 2nd law along the direction of motion for each mass individually. Since the blocks are attached, they all move together. Therefore, they share the same acceleration.

(3) Use the equations generated in step 2 to solve for the "shared" acceleration.

(4) Substitute that acceleration into kinematic equations, or into a Newtons' 2nd law equation from step 2, to figure out what you need to know.

I will now implement the strategy.

Step 1: Force diagrams

As always, when the pulley is massless, the tension "on both ends" of a rope is the same. I will call it T. Also, block 3 exerts a direct "contact" force on block 2. Call that force F_{23}. By Newton's third law, block 2 must exert an equal force on block 3, in the opposite direction. So, as block 3 pushes block 2 down the ramp, block 2 "pushes back" on block 3, resisting its motion.

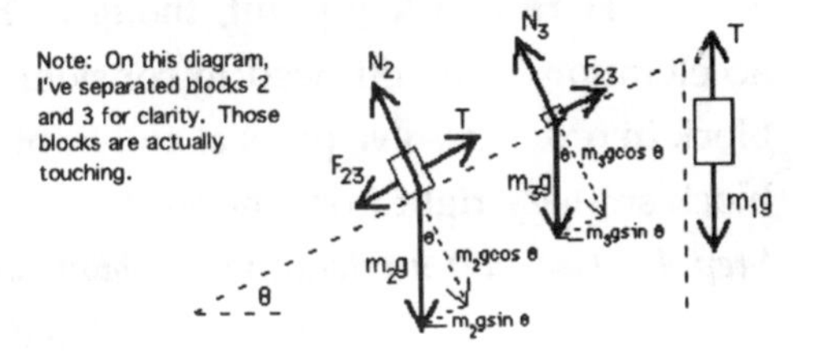

In this problem, everything is frictionless.

Step 2: Newton's 2nd law along direction of motion for each block

First, I must choose which direction counts as positive for each block. I will pick the "down-the-ramp" direction as positive for blocks 2 and 3. I am therefore committed to saying "up" is positive for block 1. Why? Because when blocks 2 and 3 slide down the ramp, block 1 is dragged upward; and my choice of positives must be consistent with the fact that all three blocks move together.

As always, start by writing Newton's 2nd law along the direction of motion. Sometimes, you will also need to consider the perpendicular direction. But here, the normal forces are not important, because there is no friction. Since the blocks move together, they share the same "a."

$$\text{Block 1:} \quad \sum \mathrm{F} = T - m_1 g = m_1 a \tag{1}$$

$$\text{Block 2:} \quad \sum \mathrm{F} = F_{23} - T + m_2 g \sin\theta = m_2 a \tag{2}$$

$$\text{Block 3:} \quad \sum \mathrm{F} = m_3 g \sin\theta - F_{23} = m_3 a. \tag{3}$$

Step 3: Solve for the acceleration

We now have three equations in three unknowns (a, T, and F_{23}). Therefore, it is just a matter of algebra to solve for T and F_{23}. The easiest way to complete the algebra is first to solve for the acceleration a, by adding the three equations. On the left-hand sides, the T's and F_{23}'s cancel, leaving us with

$$-m_1 g + m_2 g \sin\theta + m_3 g \sin\theta = (m_1 + m_2 + m_3)a\ , \tag{*}$$

and hence $a = \frac{-m_1 g + m_2 g \sin\theta + m_3 g \sin\theta}{m_1 + m_2 + m_3}$.

We also could have obtained Eq. (*) by considering the three-block system as a whole. In that case, you use $\sum F_{\text{external}} = Ma$, where $M = m_1 + m_2 + m_3$ is the mass of the three-block system, and "external forces" are those exerted on the blocks by "outside" forces such as gravity, friction, etc. By contrast, "internal" forces are those exerted by the blocks on each other, either directly or via a rope. Here, T and F_{23} are internal forces. The only relevant external forces are the components of gravity along the direction of motion. That is why the left-hand side of Eq. (*) contains only gravity terms.

Here is a key point, though: Even if you considered the system as a whole to obtain the acceleration, you still need to consider the separate Newton's 2nd law equation for each individual block in order to solve parts **a)** and **b)** of this problem. For this reason, you might as well use the multi-block strategy right from the start.

Step 4: Use that acceleration to obtain what you need.

Now we can address parts **a)** and **b)**.

a) Solve Eq. (1) for T, and then substitute in the acceleration we just found:

$$\begin{aligned} T &= m_1(g+a) \\ &= m_1\left(g + \frac{-m_1 g + m_2 g\sin\theta + m_3 g\sin\theta}{m_1+m_2+m_3}\right). \end{aligned}$$

b) To obtain F_{23}, use Eq. (2) or (3) above. I will solve Eq. (3) for F_{23}, and then substitute in our above answer for a, to get

$$\begin{aligned} F_{23} &= m_3(g\sin\theta + a) \\ &= m_3\left(g\sin\theta + \frac{-m_1 g + m_2 g\sin\theta + m_3 g\sin\theta}{m_1+m_2+m_3}\right). \end{aligned}$$

The direction of the contact force exerted by block 3 on block 2 is down the ramp.

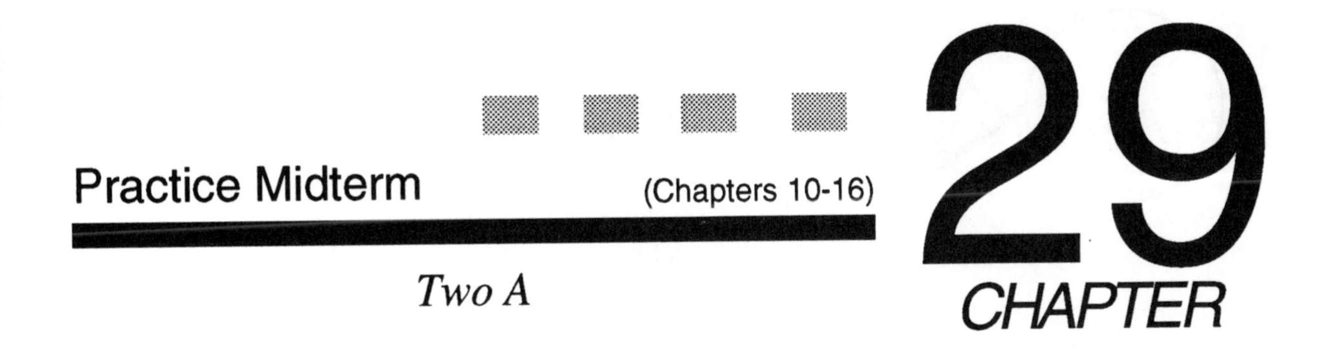

Practice Midterm (Chapters 10-16)

CHAPTER 29

Two A

- *Reminder: Review Problem Set Number 2, in Chapter 24, consists entirely of practice midterm questions on this material. Do them.*
- *Three problems, worth 100 points total. You have 70 minutes.*

1. *(35 points)* A clay ball, of mass m_1, is attached to the end of a massless string of length L. The other end of the string is attached to the ceiling, making a pendulum. The pendulum is released from rest from the essentially horizontal configuration shown below. Meanwhile, a clay block of mass m_2 slides along the frictionless floor. The ball and block collide right at the moment the pendulum reaches the bottom of its swing. The pendulum and block then stick together.

a) *(10 points)* Immediately before the ball and block collide, what was the tension in the string? (The string was essentially vertical at that point.) Answer in terms of the quantities listed above, and g.

b) *(25 points)* To what height off the floor does the ball/block system swing after the collision? Hint: During the collision, no appreciable external force acts on the ball/block system.

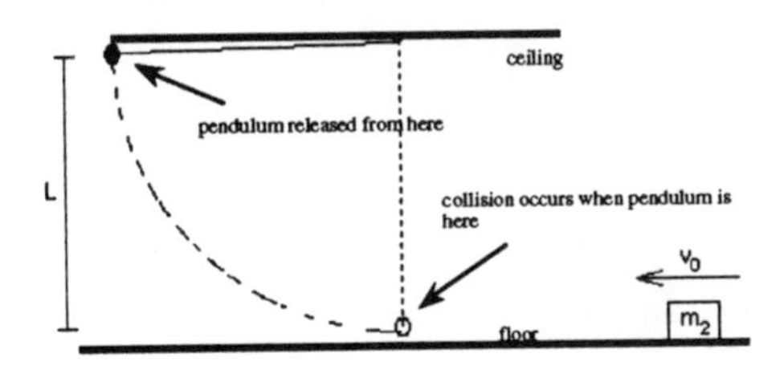

2. *(30 points)* A hollow spherical ball (mass M, radius R) starts from rest at the left end of the track and rolls without slipping until it rolls off the right end. As a result, the ball shoots straight up into the air. The left end of the track is a height H above the right end. How high above the right end of the track does the ball reach (at its peak)?

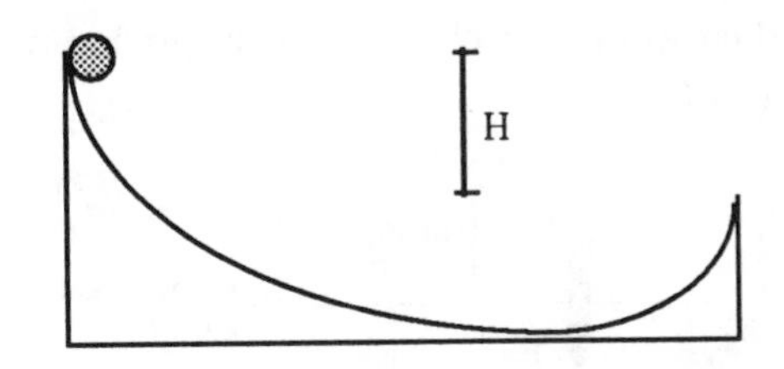

3. *(35 points)* A wheel of radius R has rotational inertia I around its axle, which passes through the center of the wheel. Initially, the wheel is not rotating. At time $t = 0$, a stream of water is shot against the rim of the wheel, at the angle ϕ shown. This stream of water exerts a force F on the rim of the wheel. The direction of this force is the same as the direction of the stream of water. The stream is turned off after a short time t_1.

Express your answers in terms of R, I, ϕ, F, and t_1.

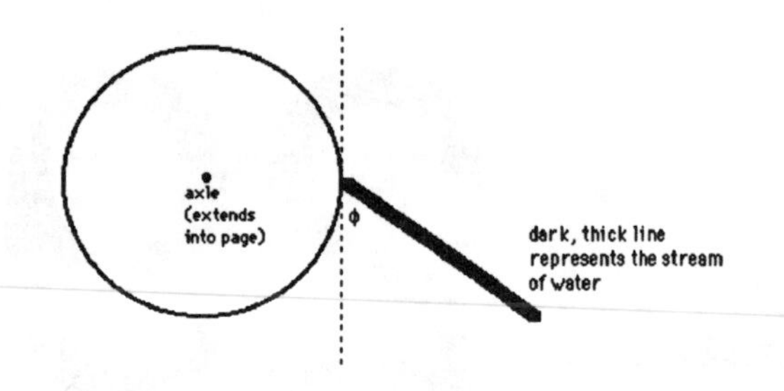

a) *(10 points)* At time t_1, what is the angular velocity of the wheel? Give a direction as well as a magnitude.

b) *(10 points)* How many revolutions did the wheel complete between $t = 0$ and t_1?

c) *(15 points)* At time t_1, Monica decides she wants to slow the wheel down to *one-third* of its current angular velocity. She plans to accomplish this by attaching a long, thin, uniform rod to the same axle on which the wheel is spinning. The end of the rod, not the center of the rod, will be attached to the axle. Monica has many rods to choose from. They all have length d, but they all have different masses. What must be the mass of the rod Monica chooses? Answer in terms of d and I.

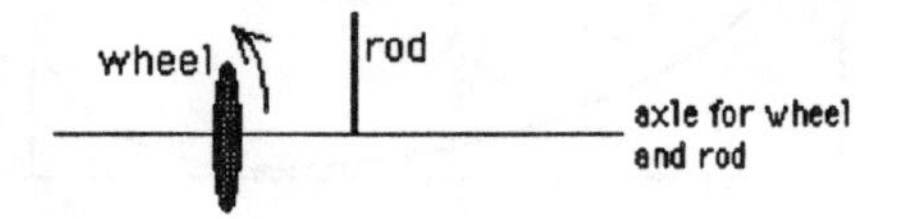

ANSWER 1

a) Since you are trying to find a force, it makes sense to start with a force diagram. We are interested in the tension when the pendulum reaches the bottom of its swing.

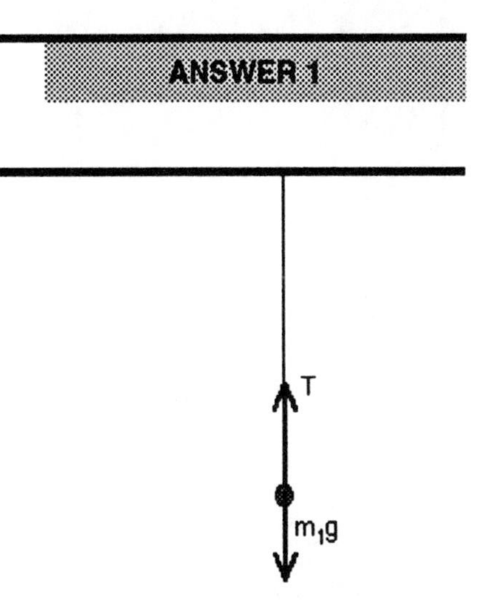

A common mistake is to set the net force equal to 0, in which case $T = m_1 g$. But the pendulum is not motionless at that point; it is still swinging in a circle. An object moving in a circle undergoes radial acceleration $a_{\text{radial}} = \frac{v^2}{r}$. The center of the circle is the top of the string (where it is attached to the ceiling). Therefore, when the pendulum reaches the bottom of its swing, the radial direction is upward. So, by Newton's 2nd law, if we let v_1 denote the ball's speed at the bottom of its swing,

$$\sum F_{\text{radial}} = ma_{\text{radial}}$$
$$T - m_1 g = m_1 \frac{v_1^2}{L}, \quad (1)$$

since the radius of the circle traced out by the ball is L, the string's length.

We cannot yet solve for T, because Eq. (1) contains a second unknown, namely v_1.

To find v_1, use conservation of energy. The "initial" configuration is when the pendulum is released from rest from height L off the ground. "Stage 1" is immediately before the collision, when the pendulum is near floor level, moving at speed v_1. For diagrams, see part **b)**. Energy conservation tells us that

$$E_0 = E_1$$
$$K_0 + U_0 = K_1 + U_1 \quad (2)$$
$$0 + m_1 g L = \frac{1}{2} m_1 v_1^2 + 0.$$

Isolate v_1 to get $v_1 = \sqrt{2gL}$. Now solve Eq. (1) for T, and substitute in this expression for v_1, to get

$$T = m_1 \frac{v_1^2}{L} + m_1 g$$
$$= m_1 \frac{2gL}{L} + m_1 g$$
$$= 2m_1 g + m_1 g$$
$$= 3m_1 g.$$

Notice that $2m_1 g$ is "devoted" to keeping the ball moving in a circle, while the other $m_1 g$ is "devoted" to counterbalancing gravity.

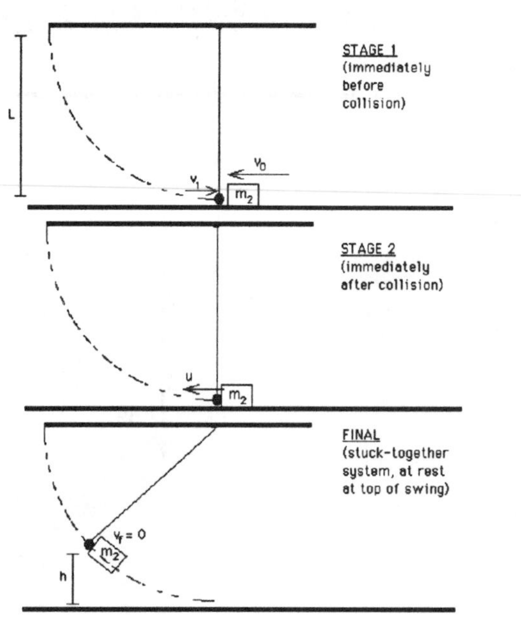

b) This is a typical multi-step problem in which you must use momentum and/or energy conservation to work from one stage to the next. As always, start with pictures. The "INITIAL" situation is drawn in the problem.

Here is the strategy. In part **a)**, using conservation of energy, we already found v_1, the pre-collision speed of the ball. With collision reasoning (conservation of momentum, and if necessary, conservation of energy), we can calculate the post-collision speed of the stuck-together system. I am calling that speed "u." Then, we can again use conservation of energy to find how high the stuck-together system swings. We do not know ahead of time whether it swings left or right. I have drawn left.

Stage 1 → stage 2: Collision reasoning

Since the collision is inelastic, heat gets generated, and therefore, some kinetic energy gets "lost." Because we are not told how much kinetic energy gets lost, conservation of energy cannot help us find u, the post-collision velocity. To find u, start with *linear* momentum conservation.

By the way, you could use conservation of angular momentum instead. I will show you how below. But since the pendulum is not a rigid pivoted object–the string is "loose"–we can use linear momentum. What you *cannot* do is combine linear and angular momenta into the same equation, because those quantities have different units. And if the string were a solid rod, you would *have* to use angular momentum instead of linear momentum conservation, due to the external force exerted by the pivot on the rod during the collision.

I will call leftward positive. Since the pre-collision and post-collision velocities all lie along a straight line, we need not worry about vector components. Setting the momentum at stage 1 equal to the momentum at stage 2, I get

$$p_1 = p_2$$
$$m_2 v_0 - m_1 v_1 = (m_1 + m_2)u.$$

Solve for u, and substitute in our result from part **a)**, $v_1 = \sqrt{2gL}$, to get

$$u = \frac{m_2 v_0 - m_1\sqrt{2gL}}{m_1 + m_2}.$$

Before proceeding to the next step, let me re-derive this expression for u using angular momentum conservation. I will let small "l" denote angular momentum, so as not to confuse it with the string length L. The "origin" is the pivot point where the string touches the ceiling. Immediately before the collision, the block is a point mass moving perpendicular to the relevant "**r**" vector (i.e., the vector from the pivot point to the block). Here, $r = L$. So, using the "special" formula for the angular momentum of a point mass derived in Question 16-5, $l = mv_\perp r$, I get $l_{1,\text{ block}} = m_2 v_0 L$. By similar reasoning, $l_{1,\text{ ball}} = -m_1 v_1 L$. I have implicitly chosen clockwise as positive. That is why the ball's initial angular momentum is negative.

After the collision, the ball and block "share" the same velocity u. We can treat the "blob" as a point mass of mass $m_1 + m_2$. So, again using the point mass formula for angular momentum, we get $l_2 = (m_1 + m_2)uL$.

Putting all this together, we see that

$$l_1 = l_2$$
$$m_2 v_0 L - m_1 v_1 L = (m_1 + m_2)uL.$$

Cancel the L's, and this reduces to the *same equation obtained above using linear momentum conservation*. But that would not be true if the loose string were replaced by a solid rod. In that case, angular and linear momentum conservation would "disagree," because linear momentum is not conserved. You would have to use angular momentum to get the right answer.

Stage 2 → "Final": Energy conservation

As the system swings up, gravity slows it down. More precisely, gravity supplies an external force that acts also as an external torque. For this reason, neither linear nor angular momentum is conserved. We must use energy.

Immediately after the collision, the system has kinetic but no potential energy. At the "peak" of its swing, however, the system is motionless ($v_f = 0$) but carries potential energy, because it is a height h off the ground. We are solving for h.

Expressing the previous paragraph in terms of equations gives me

$$E_2 = E_f$$
$$K_2 + U_2 = K_f + U_f$$
$$\frac{1}{2}(m_1 + m_2)u^2 + 0 = 0 + (m_1 + m_2)gh.$$

Solve for h, and substitute in the above expression for u, to get

$$h = \frac{u^2}{2g}$$
$$= \frac{\left(\frac{m_2 v_0 - m_1\sqrt{2gL}}{m_1 + m_2}\right)^2}{2g}.$$

You should *not* spend time simplifying this expression.

To write K_2, I used the formula for translational kinetic energy, $K_{\text{trans}} = \frac{1}{2}mv^2$. But because the system is a point mass, I equally well could have used the formula for rotational kinetic energy, $K_{\text{rot}} = \frac{1}{2}I\omega^2$. As shown in Chapter 16, you can always write the kinetic energy of a point mass as either translational *or* rotational, but not both. This is because, for a point mass, K_{rot} and K_{trans} are two different ways of saying the same thing:

$$K_{\text{rot}} = \frac{1}{2}I\omega^2$$
$$= \frac{1}{2}(mr^2)\left(\frac{v}{r}\right)^2$$
$$= \frac{1}{2}mv^2 = K_{\text{trans}},$$

where in the second step I used $I_{\text{point mass}} = mr^2$ and also $v = r\omega$.

This is true *only* for point masses. An "extended" (non-point-like) object can have translational or rotational kinetic energy, *or both*. For instance, if an extended object is "pivoted down," i.e., constrained to rotate about some fixed pivot point, then the kinetic energy is purely rotational. But when an extended object is free to move as well as to rotate, it can have both translational *and* rotational kinetic energy. We saw this for balls rolling down hills.

ANSWER 2

Let h denote the ball's peak height. A common error is to conclude that $h = H$, the initial height. Before diving into math, let us see intuitively why that answer is wrong.

If the ball slid like an ice cube instead of rolling, it *would* shoot upward to its initial height. Here is why: Its initial potential energy converts into kinetic energy at the bottom of the track. When the sliding ball shoots up, that kinetic energy converts back into potential energy. At its peak, *all* the kinetic energy has converted back into potential energy. So, the ball's final potential energy equals its initial potential energy, and hence $h = H$. For the sliding ball, this conclusion holds because *all* of the ball's kinetic energy converts into potential energy at the peak.

When the ball rolls, things are different. At the moment the ball leaves the track, it is both moving *and* spinning. When it shoots into the air, it is rotating with some angular speed ω. As the ball rises, it *keeps* spinning. To see this, throw a ball with spin. The ball's angular speed *stays constant* as it flies through the air. (That is because no torque acts on the ball in mid-air. Gravity acts on the center of the ball, and therefore exerts no torque.) Here is the point: When the ball reaches its peak, *it is still spinning*. So, it still has kinetic energy. Therefore, not all of the ball's kinetic energy has converted back into potential energy. As a result, the ball ends up with less potential energy–and hence, less height–than if it had not been spinning.

Let me re-express this conclusion with a formula. Even at its peak, where $K_{\text{trans}} = 0$, the ball still has rotational kinetic energy. Let U_0 denote the ball's initial potential energy at the top of the track, and let subscript "f" refer to the peak. By energy conservation, $U_0 = U_f + K_{f\,\text{rot}}$. From this equation, we see that the final potential energy is less than the initial potential energy. So, while flying upward, the ball never reaches its initial height.

The most common error leading to the mistaken conclusion that $h = H$ is to reason that, since the ball rolls without slipping, its linear and angular speed are related by

$$v = R\omega \quad \text{ROLLING WITHOUT SLIPPING CONDITION}.$$

At its peak, the ball's speed is $v = 0$. Therefore, according to the rolling-without-slipping condition, its angular speed is $\omega = 0$ at the peak.

Here is why the above reasoning fails: *Once in mid-air, the ball no longer rolls without slipping*. Its angular speed becomes "decoupled" from its linear speed. So, the rolling without slipping condition no longer applies. It applies only when the ball is on the track. By contrast, the ball at its peak has nonzero ω, even though $v = 0$.

This qualitative discussion suggests a problem-solving strategy. As noted above, after the ball leaves the ramp, it keeps all of its angular speed. So, we can find the angular speed at the "launch" point, and then put it into the above energy conservation equation. Alternatively, we could find the linear speed at the take off point, and then treat the ball as a projectile. (The fact that it is spinning in no way affects its projectile motion, as you can confirm by throwing a ball with and without spin, provided you do not throw a "curve ball." A curve ball curves due to the ball's interaction with the surrounding air.)

Whichever strategy we choose, we need to find the ball's "take off" speed or angular speed.

Subproblem 1: Find ω and/or v at the "take off" point

Since the ball rolls without slipping along the track, it has both translational and rotational kinetic energy. Furthermore, its linear and angular speed are related by the above rolling without slipping condition. Let us measure our heights from the "take off" point on the right end of the track. So, the ball's initial height is H and its "take off" height is 0.

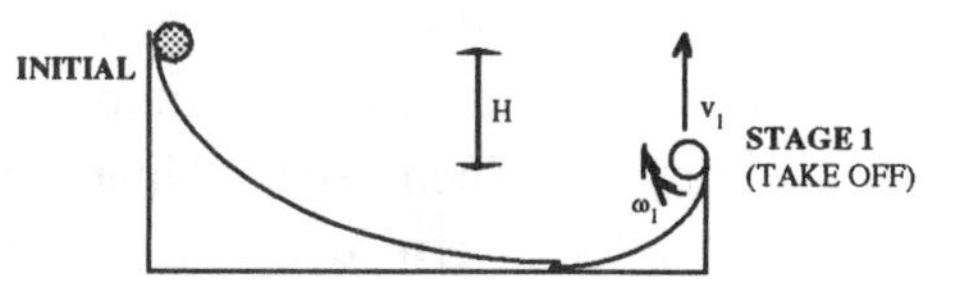

Substituting all this into energy conservation yields

$$\begin{aligned} E_0 &= E_1 \\ U_0 + K_0 &= U_1 + K_1 \\ MgH + 0 &= 0 + \frac{1}{2}Mv_1^2 + \frac{1}{2}I\omega_1^2 \\ &= \frac{1}{2}M(R\omega_1)^2 + \frac{1}{2}\left(\frac{2}{3}MR^2\right)\omega_1^2, \end{aligned}$$

where in the last step I used the rolling without slipping condition ($v = R\omega$) and the rotational inertia of a spherical shell, $I = \frac{2}{3}MR^2$. Solve for ω_1 to get

$$\omega_1 = \sqrt{\frac{6gH}{5R^2}}, \text{ and hence } v_1 = R\omega_1 = \sqrt{\frac{6gH}{5}}.$$

Subproblem 2: Solve for h using conservation of energy or projectile reasoning

Projectile reasoning: The ball is "thrown" upward with initial velocity $v_{\text{initial}} = v_1$. Its final speed at the peak is $v_f = 0$, and we are solving for the vertical distance $\Delta y = h$. So, we can use the constant-acceleration kinematic formula $v_f^2 = v_{\text{initial}}^2 + 2a\Delta y$. Solve for Δy to get

$$\Delta y = h = \frac{v_f^2 - v_{\text{initial}}^2}{2a} = \frac{0 - v_1^2}{2(-g)} = \frac{\frac{6gH}{5}}{2g} = \frac{3}{5}H.$$

Let me re-solve this subproblem using

Energy conservation: The initial energy equals the energy at stage 1, which equals the final energy (at the peak). So, we can set $E_0 = E_f$, or $E_1 = E_f$. I will choose the first option. Remember, at its peak, although the ball has speed $v_f = 0$, it is still spinning with angular velocity $\omega_f = \omega_1$.

$$\begin{aligned} E_0 &= E_f \\ U_0 + K_0 &= U_f + K_f \\ MgH + 0 &= Mgh + \frac{1}{2}I\omega_1^2 \\ &= Mgh + \frac{1}{2}\left(\frac{2}{3}MR^2\right)\frac{6gH}{5R^2}, \end{aligned}$$

where in the last step I invoked our above expression for ω_1 from subproblem 1. Notice that all the M's, g's, and R's cancel, leaving us with $H = h + \frac{2}{5}H$. So, $h = \frac{3}{5}H$, in agreement with the answer found by projectile reasoning.

ANSWER 3

a) In a linear analog to this problem, we would be given the force with which (say) a desk gets pushed across the room for a known time, and we would be asked for the final velocity. To solve that linear problem, we would use a standard strategy: First, set $\sum F = ma$ to find the desk's acceleration. Then use the kinematic formula $v = v_0 + at$.

Here, we can use the rotationally analogous strategy. First, invoke the rotational form of Newton's 2nd law, $\sum \tau = I\alpha$, to solve for the angular acceleration α. Then substitute that angular acceleration into the rotational analog of $v = v_0 + at$, namely $\omega = \omega_0 + \alpha t$.

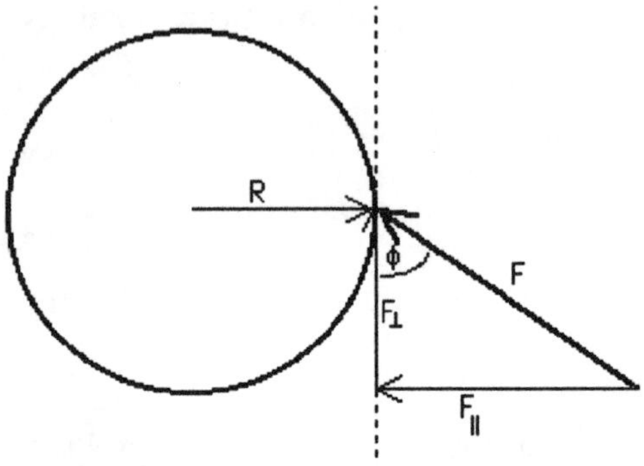

So first, we need to find the torque τ exerted by the water stream on the wheel. We must break up that force into its components parallel and perpendicular to **R**, the radius vector from the center to the contact point.

So, the torque is $\tau = F_{\perp}r = (F\cos\phi)R$. Notice that we got $\cos\phi$ instead of $\sin\phi$. That is because ϕ is not the angle between **F** and **R**. (The angle between **F** and **R**, when they are placed tail to tail, is $90° + \phi$.)

Since this is the only torque acting on the wheel, Newton's 2nd law gives

$$\sum \tau = I\alpha$$
$$(F\cos\phi)R = I\alpha,$$

and hence $\alpha = \frac{FR\cos\phi}{I}$.

Now that we have the angular acceleration, we can use rotational kinematics. Let ω_1 denote the angular speed at time t_1:

$$\omega_1 = \omega_0 + \alpha t_1$$
$$= 0 + \frac{FR\cos\phi}{I}t_1.$$

What is the direction of this angular velocity? From the diagram, the water makes the wheel spin counterclockwise. Twist the fingers on your right hand so that your fingers point counterclockwise. Your thumb sticks up toward you. So, the direction of the angular velocity is up out of the page, perpendicular to the plane of the wheel.

b) Since we already obtained the angular acceleration in part **a)**, we can jump straight to kinematics. Use the rotational analog of $x = x_0 + v_0 t + \frac{1}{2}at^2$, namely

$$\theta = \theta_0 + \omega_0 t + \frac{1}{2}\alpha t^2$$
$$= 0 + 0 + \frac{1}{2}\left(\frac{FR\cos\phi}{I}\right)t_1^2.$$

By dimensional analysis, you can confirm that this answer is dimensionless, i.e., in radians. Since one revolution corresponds to 2π radians,

$$\#\text{ of revolutions} = \left\{\frac{1}{2}\left(\frac{FR\cos\phi}{I}\right)t_1^2\right\} \times \frac{1\text{ rev}}{2\pi\text{ rad}} = \frac{1}{4\pi}\left(\frac{FR\cos\phi}{I}\right)t_1^2\text{ revs.}$$

c) The key physical insight you must have is this: When the rod gets attached to the axle, no external torque acts on the system. The only torques experienced by the rod, axle, and wheel are the torques they exert on each other. So, the angular momentum of the overall system is conserved. Intuitively, we expect that attaching the rod slows down the system. Let subscript "1" refer to time t_1, and subscript "f" refer to the system immediately after the rod is attached.

When the rod and wheel "share" the same axle, they automatically share the same angular speed, because they have no choice but to rotate together. We want this final angular speed to be one third of the angular speed at time t_1: $\omega_f = \frac{\omega_1}{3}$. Given all this, let us see where angular momentum conservation takes us. Recall that the given "I" is the wheel's rotational inertia.

$$L_1 = L_f$$
$$I_{\text{wheel}}\omega_1 = (I_{\text{wheel}} + I_{\text{rod}})\omega_f$$
$$I\omega_1 = (I + I_{\text{rod}})\frac{1}{3}\omega_1 \qquad \left[\text{since } \omega_f = \frac{1}{3}\omega_1\right]$$
$$I\omega_1 = \left(I + \frac{1}{3}md^2\right)\frac{1}{3}\omega_1 \qquad \left[\text{since } I_{\text{rod}} = \frac{1}{3}ml^2\right].$$

I used the rotational inertia of a rod pivoted at its end, not pivoted at its center.

We are done! We can immediately solve for m, the rod's mass, in terms of d and the wheel's rotational inertia, I. Notice that the ω_1's cancel. Multiply through by 9, divide through by d^2, and then isolate m to get $m = \frac{6I}{d^2}$.

Practice Midterm (Chapters 10-16)

30 CHAPTER

Two B

- *4 questions, worth 100 points total. You have 75 minutes.*

1. *(20 points)* A block of mass m is pushed up a ramp by Anne. Anne's push is directed at an angle ϕ above the surface of the ramp, as drawn below. The ramp angle is θ, and the coefficient of kinetic friction between the block and ramp is μ. The force exerted by Anne is F_A. Points A and B are a distance L apart.

   Express your answers in terms of the quantities listed above, and g.

   a) *(10 points)* How much work does Anne do pushing the block from point A to point B?

   b) *(10 points)* How much *total* work is done on the block as it is pushed from point A to point B?

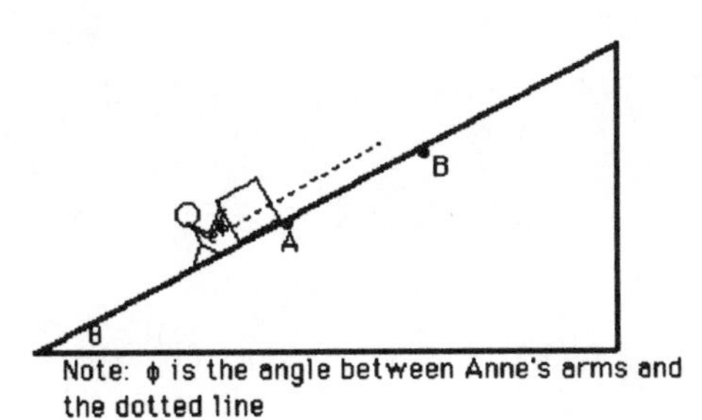

Note: ϕ is the angle between Anne's arms and the dotted line

2. *(25 points)* Consider the oddly shaped ramp (mass m_2) drawn below. Attached to the right end of the ramp is a massless, horizontal spring of spring constant k, at its equilibrium length. The spring is a distance h below the top of the ramp. Taped to the top of the ramp is a small block of mass m_1. Initially, this whole system slides leftward across the *frictionless* floor with speed v_0.

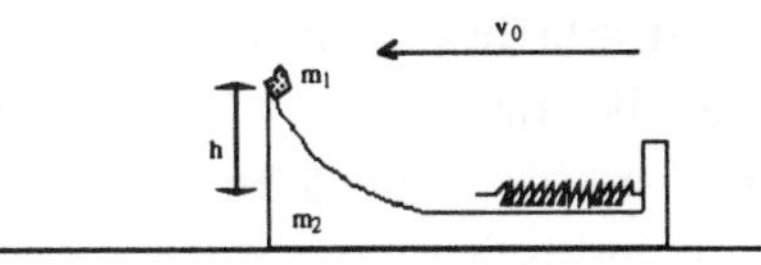

At time $t = 0$, the tape breaks. So, the block slides down the ramp and compresses the spring.

   a) *(10 points)* What is the block's speed with respect to the floor immediately before the block reaches the spring? Answer in terms of m_1, m_2, h, v_0, and g. You will get 9 points out of 10 for setting up the relevant equation or equations, even if you do not complete the algebra.

   b) *(15 points)* Let s denote the maximum distance by which the spring gets compressed from its equilibrium length. Solve for s in terms of m_1, m_2, h, v_0, and g.

3. *(30 points)* A "rotating condiment tray" (mass M, radius R) is a thin solid disk mounted on frictionless bearings so that it can rotate frictionlessly around its center. The disk is parallel to the ground.

   The tray is rotating clockwise such that a point on the edge of the disk has speed v_0. Suddenly, a spider of mass m drops straight down and lands on the tray a distance $\frac{R}{3}$ from the center.

   Express all answers in terms of M, R, and v_0.

a) *(10 points)* Immediately after the spider lands on the tray, what is the rotational speed of the tray?

b) *(10 points)* Now suppose the spider starts "running" in a clockwise circle of radius $\frac{R}{3}$. What must be the spider's (linear) speed so that the tray is no longer rotating (with respect to the floor)?

c) *(10 points)* Now the spider again stops moving with respect to the tray. So, the spider is sitting a distance $\frac{R}{3}$ from the center, and the tray has the same angular speed it had in part **a)**. The Spider covers itself with silk and attaches itself to the tray. Because of the silk, air resistance exerts a drag force on the spider. The drag force is constant, with magnitude F_D.

How much time will the tray take to stop rotating? Answer in terms of F_D and the quantities listed above.

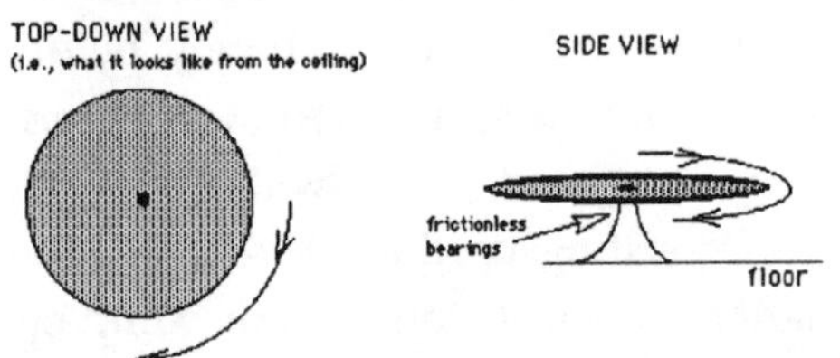

4. *(25 points)* A solid sphere (mass M, radius R, moment of inertia $\frac{2}{5}MR^2$) starts at rest on the top of a semicircular track, as drawn below. The semicircular track has radius s. The sphere rolls without slipping along the track, until it falls off the track at point B. Point A is a vertical distance H above the ground, but you are not given the height of point B.

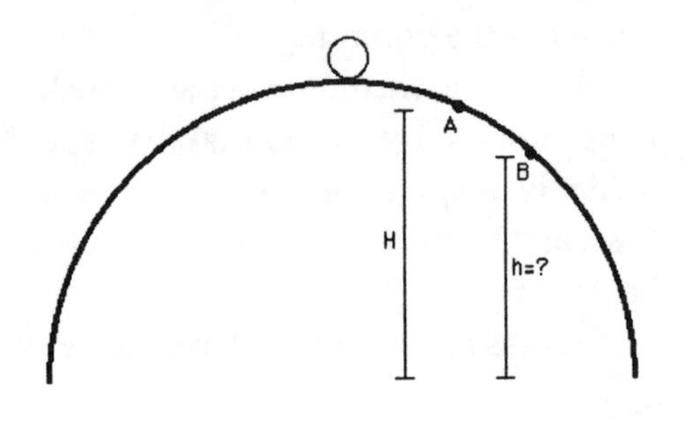

Express youranswers in terms ofM, R, H, and g. Since $s >> R$, you may neglect the fact that the ball's center of mass is slightly above the semicircular track.

a) *(10 points)* What is the angular speed ω of the sphere when it reaches point A?

b) *(15 points)* Set up, but do not solve, equations that would enable you to solve for the height h at which the sphere falls off the track, in terms of M, R, H, s, and g.

ANSWER 1

Let us start with a free-body diagram.

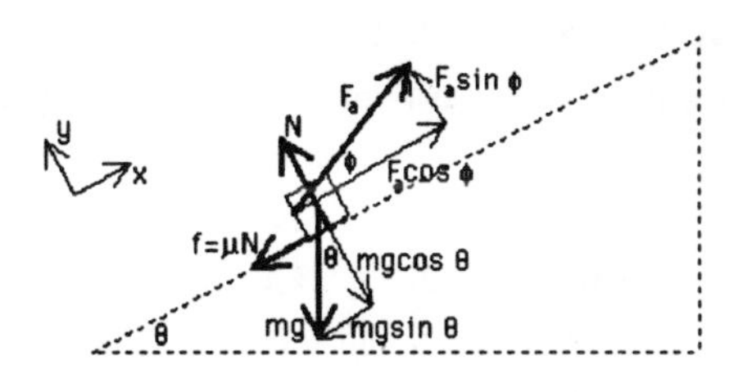

a) The work done *by a single force* is given by the dot product of that force with the displacement $\mathbf{s}$. So, the work done by Anne is $W_A = \mathbf{F}_A \cdot \mathbf{s} = F_{A\parallel}L$, where $F_{A\parallel}$ is the component of Anne's force along the direction of motion, i.e., parallel to the ramp. From the diagram, $F_{A\parallel} = F_A \cos\phi$. So $W_A = F_{A\parallel}L = (F_A \cos\phi)L$. The most common error is to confuse the work done by *Anne* with the net work done on the block by all the forces.

b) To find the total (net) work done on the block, we must either sum up the work done by each force individually, or else find the work performed by the net force. These methods give the same answer, because they are really two different ways of saying the same thing. Whichever technique you choose, you must start with a force diagram, breaking the forces into their components parallel and perpendicular to the direction of motion, as drawn above.

Only force components along the direction of motion perform work. Intuitively, this is because only a force along the direction of motion, either forward or backward, can "help" or "resist" the motion. So,

$$W_{\text{total}} = \mathbf{F}_{\text{net}} \cdot \mathbf{s} = F_{x\,\text{net}}L. \tag{1}$$

From the free-body diagram, the net force in the x-direction is

$$F_{x\,\text{net}} = \sum F_x = F_A \cos\phi - mg\sin\theta - \mu N, \tag{2}$$

The only unknown in Eq. (2) is N, the normal force. We can find N by considering the y-directed forces. Because the block neither jumps off nor burrows into the ramp, its y-ward acceleration is $a_y = 0$. So

$$\sum F_y = ma_y$$
$$N + F_A \sin\phi - mg\cos\theta = 0.$$

Solving for N yields $N = mg\cos\theta - F_A \sin\phi$. Notice that, since Anne "helps" the normal force to resist gravity's y-ward pull, the normal force does not need to push as hard as it usually does. That is why N is less than its "usual" value of $mg\cos\theta$.

In any case, we can substitute this expression for N into Eq. (2) to get the net x-directed force, and then substitute that force into Eq. (1) to find the total work done on the block:

$$\begin{aligned} W_{\text{total}} &= F_{x\,\text{net}}L \\ &= [F_A \cos\phi - mg\sin\theta - \mu N]L \\ &= [F_A \cos\phi - mg\sin\theta - \mu(mg\cos\theta - F_A \sin\phi)]L. \end{aligned}$$

ANSWER 2

To solve this, you must use both energy *and* momentum conservation. As the block slides down and compresses the spring, energy is conserved, and no heat gets generated. You might think that momentum is not conserved, because the external force of gravity pulls the block. But in parts **a)** and **b)**, we are interested only in the horizontal component of the momentum. No *horizontal* external force acts on the system. The only horizontal forces experienced by the block, ramp, and spring are the forces they exert on each other. So, horizontal momentum is conserved throughout the whole process.

a) As the block slides down the ramp, the ramp does *not* stay at speed v_0. The block exerts a normal force on the ramp that tends to speed up the ramp. So, we cannot just "switch reference frames," find the block's speed with respect to the ramp to be $v_{\text{rel}} = \sqrt{2gh}$, and then say the block's speed with respect to the ground is $\sqrt{2gh} - v_0$. This answer is wrong because the ramp speeds up from initial speed v_0, *and* because the block's speed with respect to the ramp (immediately before hitting the spring) is not $\sqrt{2gh}$. That relative speed was calculated assuming that the block "carries" all the kinetic energy of the system. But it does not, not even according to an observer moving leftward at v_0. As just noted, even that observer "sees" the ramp speeding up, and therefore gaining kinetic energy. We have to use energy conservation more carefully.

To do so, start with pictures. Stage 1 is right before the block reaches the spring. Let v_1 and u_1 denote the velocity of the block and ramp at this point. I am guessing that v_1 is rightward. If I am wrong, I will get a negative answer for v_1. As discussed above, we expect that u_1 is bigger than v_0, the initial speed of the ramp and block. All these velocities are *with respect to the ground.*

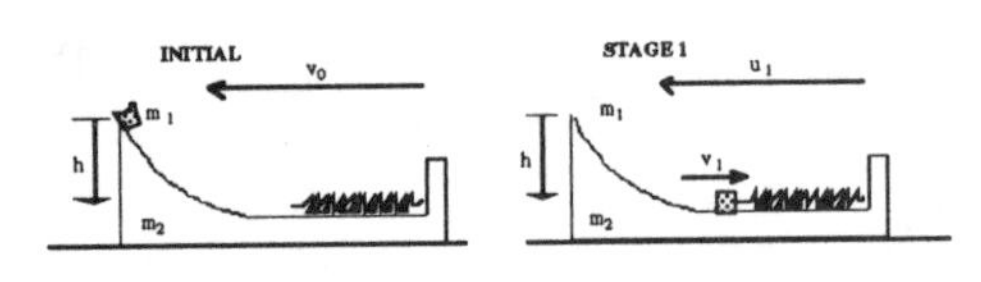

Neither the ramp's gravitational potential energy nor the spring's elastic potential energy changes between the initial configuration and stage 1. So, we only have to worry about the block's potential energy. Keep in mind that initially, the block and ramp move together at speed v_0, since the ramp carries along the block.

$$\begin{aligned} E_0 &= E_1 \\ U_0 + K_0 &= U_1 + K_1 \\ m_1 gh + \frac{1}{2}m_1 v_0^2 + \frac{1}{2}m_2 v_0^2 &= 0 + \frac{1}{2}m_1 v_1^2 + \frac{1}{2}m_2 u_1^2. \end{aligned} \tag{1}$$

We cannot yet solve Eq. (1) for v_1, because it contains a second unknown, the ramp's speed u_1. We need another equation. To get it, use momentum conservation. I will pick leftward as positive.

$$\begin{aligned} p_0 &= p_1 \\ m_1 v_0 + m_2 v_0 &= -m_1 v_1 + m_2 u_1. \end{aligned} \tag{2}$$

Now that we have two equations in the two unknowns (v_1 and u_1), it is just a matter of algebra to solve for v_1. Unfortunately, the algebra yields a nasty quadratic equation. *It is not worth the time you would spend solving that equation to earn 1 extra point.* So, you should skip the algebra and move on to part **b)**. But if your instructor weights algebra more heavily, then you should do it.

b) We can recycle the strategy of part **a**). A possible error is to set the block's final speed (when the spring reaches maximum compression) equal to 0. It is true that when the spring reaches maximum compression, the block momentarily comes to rest *with respect to the ramp*. But the ramp itself is still sliding leftward with some final speed v_f. So, the block gets carried along at that same speed.

Why do the block and ramp "share" the same speed when the spring reaches maximum compression? Well, before the spring reaches that point, the block is sliding rightward across the ramp. And after the spring starts "de-compressing," the block slides leftward across the ramp. Maximum compression is the "turning point," the moment when the block "changes its direction" with respect to the ramp. So, right at that moment, the block is motionless with respect to the ramp; the block is "carried along" by the ramp. Hence, the ramp and block share the same speed v_f at that moment.

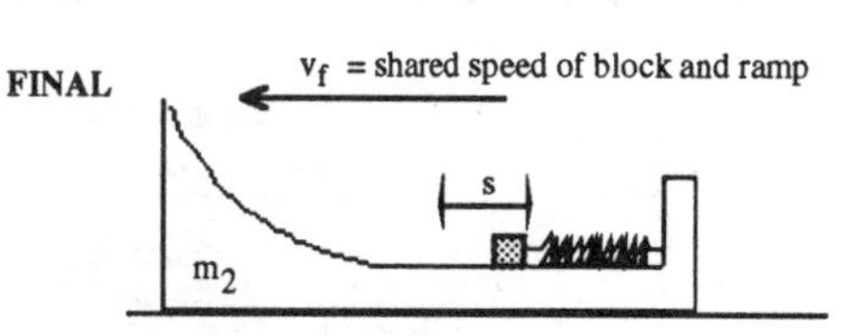

Since energy is conserved for the entire process (with negligible heat generated), $E_0 = E_1 = E_f$. So, we can set the final energy equal to either the initial energy or the stage 1 energy. I will use the initial energy. Since the spring gets compressed a distance s from equilibrium, its "final" potential energy is $U_f = \frac{1}{2}ks^2$.

$$\begin{aligned} E_0 &= E_f \\ U_0 + K_0 &= U_f + K_f \\ m_1gh + \frac{1}{2}m_1v_0^2 + \frac{1}{2}m_2v_0^2 &= \frac{1}{2}ks^2 + \frac{1}{2}m_1v_f^2 + \frac{1}{2}m_2v_f^2. \end{aligned} \tag{3}$$

We cannot yet solve for s, because the equation contains a second unknown, v_f. To find v_f, conserve momentum. Again, since momentum is conserved for the entire process, we can set $p_f = p_0$, or $p_f = p_1$. I will choose the first option:

$$\begin{aligned} p_0 &= p_1 \\ m_1v_0 + m_2v_0 &= m_1v_f + m_2v_f. \end{aligned} \tag{4}$$

From Eq. (4), $v_f = v_0$! Intuitively, as the block slides down the ramp, the ramp speeds up. But as the block compresses the spring, the spring pushes rightward on the ramp, slowing it down. At maximum compression, the ramp has slowed down to its initial speed.

In any case, when you set $v_f = v_0$ in Eq. (3), all the kinetic energy terms cancel, leaving us with $m_1gh = \frac{1}{2}ks^2$. Solve for s to get $s = \frac{2m_1gh}{k}$. This is the same answer we would have obtained if the ramp had been nailed down. But you could not just jump to that conclusion, because it is not always true. Here is why:

Suppose the floor is nearly but not completely frictionless. Then it will not be the case that $m_1gh = \frac{1}{2}ks^2$, because some of the initial potential energy gets converted into heat energy due to frictional rubbing. My point is that the spring's compression *does* in general depend on whether the ramp is allowed to slide. Only in this *particular* problem do we happen to get the same answer as if the ramp were nailed down.

Read the following paragraph only if you tried to switch reference frames and set $\frac{1}{2}m_1v_{\text{rel}}^2 = \frac{1}{2}ks^2$, where v_{rel} is the block's speed with respect to the ramp, in stage 1. This equation is wrong for a subtle reason. If the ramp picked out an inertial frame of reference, then $\frac{1}{2}m_1v_{\text{rel}}^2$

would indeed equal $\frac{1}{2}ks^2$. But during the spring's compression, the ramp accelerates, because the compressed spring pushes on it. Since the ramp accelerates, the frame of reference defined by the ramp before the compression *differs* from the reference frame defined by the ramp after the compression. Therefore, the right-hand and left-hand sides of $\frac{1}{2}m_1 v_{rel}^2 = \frac{1}{2}ks^2$ are written with respect to *different* frames of reference, which is not physically valid.

ANSWER 3

a) You should recognize this to be a "rotational collision" problem. If we consider the tray and the spider to be our "system," then angular momentum is conserved, because the only torques experienced by the tray and by the spider are the torques they exert on each other. During and after the collision, no "outside" torque acts on the system, until part c).

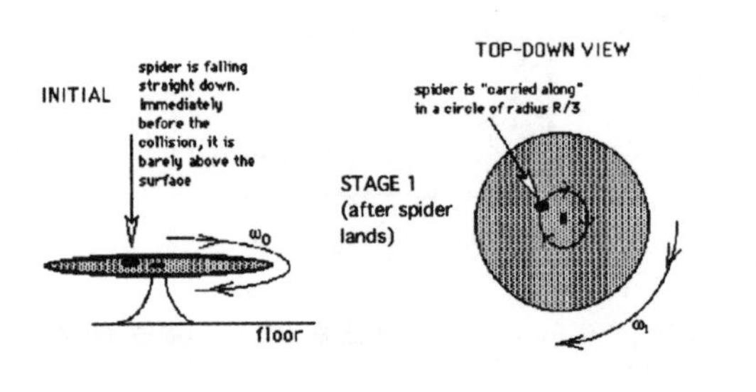

As always in conservation problems, picture the relevant stages.

After it lands, the spider gets "carried along" at the tray's new angular speed, ω_1. We are solving for ω_1. Initially, while falling, the spider has no angular momentum, even though it is moving. That is because it travels parallel to the axis of rotation, an imaginary vertical line through the center of the tray. Intuitively, the falling spider does not rotate around that axis; no component of its motion is tangential to an imaginary circle drawn around the axis of rotation. So, the spider's initial angular momentum is 0.

(Remember, because linear and angular momentum have different units, you cannot combine them together into a generalized "momentum conservation" equation.)

The rotational inertia of a solid disk (i.e., a thin solid cylinder) is $I_{disk} = \frac{1}{2}MR^2$. Since the spider is a point mass, its rotational inertia is $I_{spider} = mr^2$. After the spider lands, its distance from the pivot point is $r = \frac{R}{3}$. Putting all this together, we get

$$\begin{aligned}
L_0 &= L_1 \\
I_{disk}\omega_0 &= [I_{disk} + I_{spider}]\omega_1 \\
\frac{1}{2}MR^2\omega_0 &= \left[\frac{1}{2}MR^2 + m\left(\frac{R}{3}\right)^2\right]\omega_1 \\
\frac{1}{2}MR^2\frac{v_0}{R} &= \left[\frac{1}{2}MR^2 + m\left(\frac{R}{3}\right)^2\right]\omega_1.
\end{aligned} \tag{1}$$

In the last step, I took advantage of the fact that the tangential velocity of a point on a rotating object is $v = r\omega$, where r is the point's distance from the pivot. The problem tells us that initially, a point on the edge of the disk ($r = R$) moves at speed v_0. Therefore, $v_0 = R\omega_0$.

(Notice that we could have used the "special" formula for a point mass' angular momentum, $L_{point\ mass} = mv_{\perp}r$. After the spider lands, $r = \frac{R}{3}$ and $v_{\perp} = \frac{\omega_1 R}{3}$. So, according to this formula, the spider's stage 1 angular momentum is $L_{spider} = m(\omega_1 \frac{R}{3})\frac{R}{3}$, in agreement with the above expression obtained using $L_{spider} = I_{spider}\omega_1$.)

In any case, we can immediately solve Eq. (1) to get

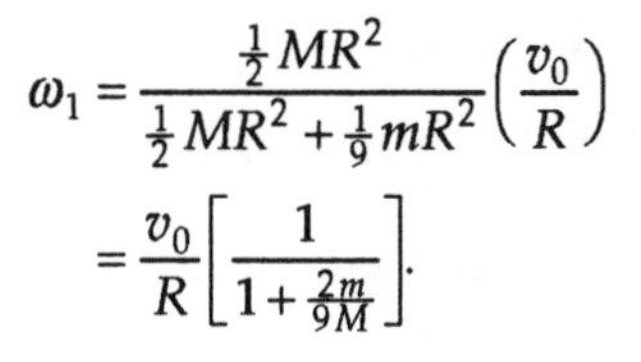

$$\omega_1 = \frac{\frac{1}{2}MR^2}{\frac{1}{2}MR^2 + \frac{1}{9}mR^2}\left(\frac{v_0}{R}\right) = \frac{v_0}{R}\left[\frac{1}{1+\frac{2m}{9M}}\right].$$

b) Angular momentum conservation again applies, because the only torques experienced by the spider and tray are the torques they exert on each other. Intuitively, when the spider starts to run clockwise, its little legs push "backwards" against the tray, exerting a backward (counterclockwise) torque that stops the tray. (Once the spider reaches its final constant speed, it no longer exerts a torque.) Here is the situation when the spider has achieved speed v_2, the velocity needed to stop the disk.

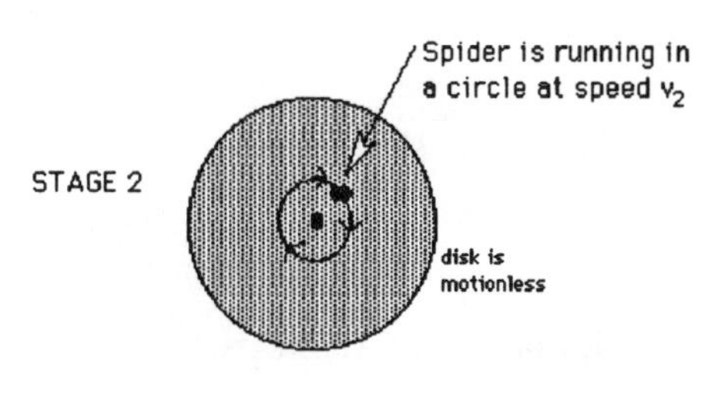

I will use the special point mass angular momentum formula: $L_{2\text{ spider}} = mv_{2\perp}r = \frac{mv_2R}{3}$. The disk is motionless at this stage: $L_{2\text{ disk}} = 0$.

Because angular momentum is conserved throughout the process (until the spider spins silk), we can set the stage 2 angular momentum equal to either the stage 1 or the initial angular momentum: $L_0 = L_1 = L_2$. I will use the initial angular momentum:

$$L_0 = L_2$$
$$I_{\text{disk}}\omega_0 = L_{2\text{ spider}} + L_{2\text{ disk}}$$
$$\frac{1}{2}MR^2\frac{v_0}{R} = mv_2\frac{R}{3} + 0.$$

Remember, v_0 is the initial speed of a point on the edge of the tray. Solve for v_2 to get $v_2 = \frac{3M}{2m}v_0$.

c) Angular momentum is no longer conserved, because wind resistance (the drag force) generates an external torque on the system. This torque slows down the system until it stops.

If this were a linear problem, a known force would decelerate a car, and we would need to find how much time the car takes to slow down from its initial speed to $v_f = 0$. To solve that standard force problem, we would first set $\sum F = ma$ to calculate the acceleration. Then we would use the kinematic formula $v_f = v_i + at$ to calculate the "stopping time." So here, we can use the rotational analog of this strategy. First, set $\sum \tau = I\alpha$ to find the angular acceleration α. Then use the rotational analog of $v_f = v_i + at$, namely $\omega_f = \omega_i + \alpha t$, to find the stopping time τ.

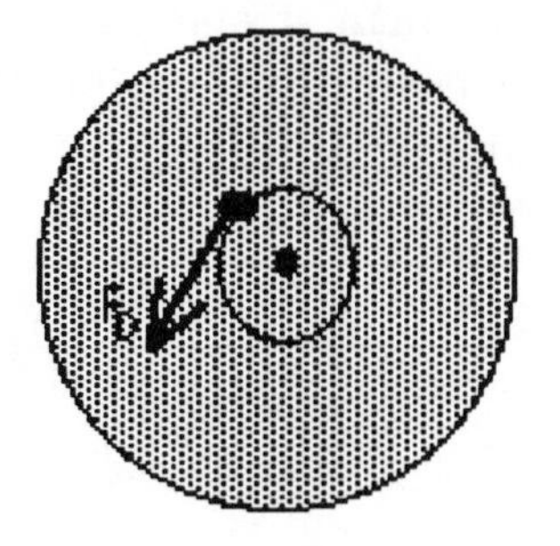

As always, diagram the forces.

The only relevant force on the system is F_D, the drag force. A drag force always acts "backwards" compared to the system's direction of motion. Notice that $\mathbf{F}_D$ acts entirely perpendicular to the radius vector connecting the pivot point to the "contact point" where the force acts. I will pick the direction of rotation (clockwise) as positive. So, from $\tau = F_\perp r$ and Newton's 2nd law, we get

$$\sum \tau = I\alpha$$
$$-F_D r = \left[I_{\text{disk}} + I_{\text{spider}}\right]\alpha$$
$$-F_D r = \left[\frac{1}{2}MR^2 + m\left(\frac{R}{3}\right)^2\right]\alpha.$$

We must use the rotational inertia of the whole system, spider included, because the whole system decelerates due to the drag force. Divide through by R^2 and isolate α to get

$$\alpha = -\frac{F_D}{3R[\frac{1}{2}M + \frac{1}{9}m]}.$$

The minus sign indicates that the system *decelerates*.

Now that we know α, we can calculate the time it takes the system to stop spinning. The final angular speed is $\omega_f = 0$. The "initial" angular speed, in this part of the problem, is the "final" angular speed from part **a)**, i..e, the system's angular speed after the spider lands. In part **a)**, we found that angular speed to be $\omega_1 = \frac{v_0}{R}\left[\frac{1}{1+\frac{2m}{9M}}\right]$. So, we can start with $\omega_f = \omega_1 + \alpha t$, and solve for t to get

$$t = \frac{\omega_f - \omega_1}{\alpha} = \frac{0 - \frac{v_0}{R}\left[\frac{1}{1+\frac{2m}{9M}}\right]}{-\frac{F_D}{3R[\frac{1}{2}M+\frac{1}{9}m]}} = \ldots\text{algebra}\ldots = \frac{3v_0 M}{2F_D}.$$

ANSWER 4

a) Since the slope of the track keeps changing, the torque on the rolling sphere is not constant. Therefore, the angular acceleration is not constant. We cannot use our constant-acceleration kinematic formulas. Instead, we must use conservation of energy.

As always, diagram the relevant stages. An "initial" picture is provided in the problem. At point A, the sphere is moving *and* rotating.

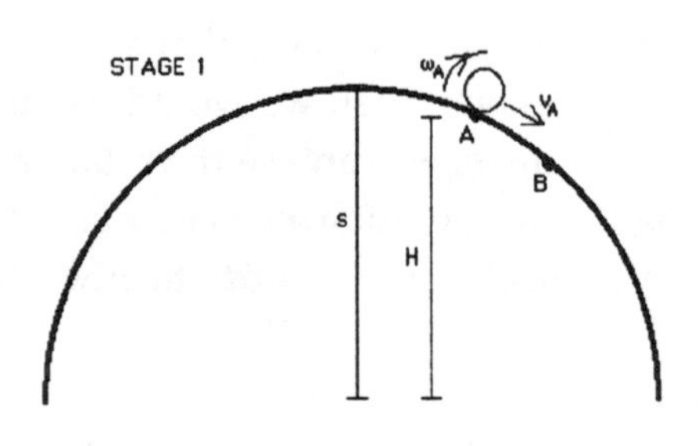

The sphere starts off a distance s above the ground, since s is the radius of the semicircular track. So, if we measure our heights from ground level, the initial potential energy is $U_0 = Mgs$. At point A, the sphere is a distance H above the ground. So, it still has some potential energy, though less than it started with: $U_A = MgH$. This lost potential energy converts into the translational *and* rotational kinetic energy of the sphere.

$$
\begin{aligned}
E_0 &= E_A \\
U_0 + K_0 &= U_A + K_A \\
Mgs + 0 &= MgH + \frac{1}{2}Mv_A^2 + \frac{1}{2}I\omega_A^2 \\
&= MgH + \frac{1}{2}Mv_A^2 + \frac{1}{2}\left(\frac{2}{5}MR^2\right)\omega_A^2 \\
&= MgH + \frac{1}{2}M(R\omega_A)^2 + \frac{1}{2}\left(\frac{2}{5}MR^2\right)\omega_A^2,
\end{aligned} \tag{1}
$$

where in the last step I used

$v_{\text{cm}} = R\omega$ ROLLING WITHPUT SLIPPING CONDITION.

Here, v_{cm} denotes the center-of-mass velocity, i.e., the velocity of the object as a whole. That is the "v" appearing in $K_{\text{trans}} = \frac{1}{2}mv^2$ and all other *linear* (translational) motion formulas.

We can immediately solve Eq. (1) for ω_A to get $\omega_A = \sqrt{\frac{10g(s-H)}{7R^2}}$.

b) To solve this, we need a physical insight about what is happening when the sphere barely loses contact with the track. Right before reaching point B, the sphere is still "skimming" along the circular track, barely touching the track at all. And at point B, the sphere loses contact with the track entirely. Therefore, the normal force exerted by the track on the sphere at point B is $N = 0$. (In general, when one object "loses contact" with another, $N \to 0$ at the loss-of-contact point.) So, gravity is the only force left acting on the sphere at point B. In the following free-body diagram, I break up gravity into its "radial" and "tangential" components.

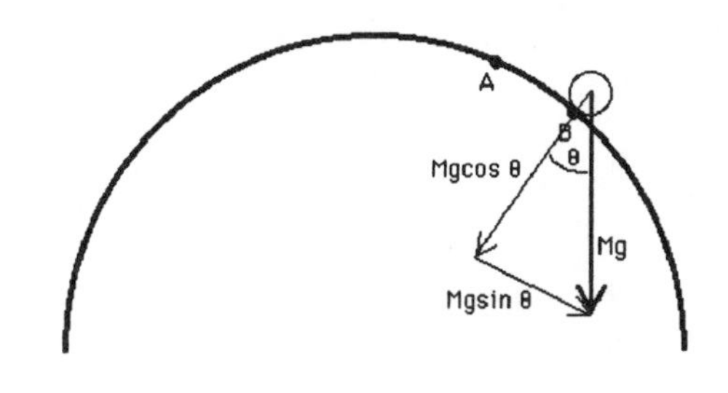

Right at point B, the sphere is still moving along the circular track, i.e., it is still traveling in a circle of radius s. Therefore, it must experience a radial acceleration $a_{\text{radial}} = \frac{v_B^2}{r} = \frac{v_B^2}{s}$, where v_B denotes the ball's velocity at point B. So, by Newton's 2nd law, using the above force diagram, we get

$$
\begin{aligned}
\sum F_{\text{radial}} &= m\frac{v_B^2}{r} \\
Mg\cos\theta &= M\frac{v_B^2}{s}.
\end{aligned} \tag{2}
$$

This equation seems pointless right now, because it does not contain the variable we are seeking, h. By examining the force diagram, however, you can see that h relates to the angle θ. Let me make this more explicit by drawing another diagram. The angle in this new diagram is the *same* θ appearing in the force diagram, because in both cases the hypotenuse points radially towards the center.

From this diagram, we see that

$$h = s\cos\theta. \tag{3}$$

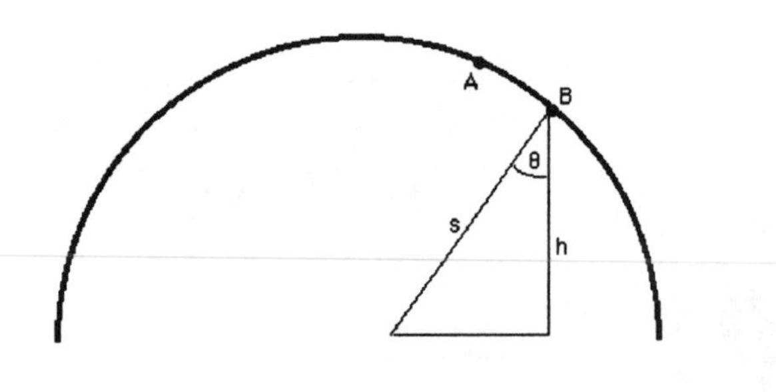

Can we solve for h yet? Not quite. Sure, we know how the θ in Eq. (2) relates to the height we are trying to find. But Eq. (2) contains another unknown, namely v_B. Fortunately, we can obtain an expression for v_B the same way we found ω_A in part **a)** above: energy conservation. Relating the initial energy to the energy at point B, and remembering the rolling without slipping condition ($v = R\omega$), I get

$$\begin{aligned} E_0 &= E_B \\ Mgs &= Mgh + \frac{1}{2}Mv_B^2 + \frac{1}{2}I\omega_B^2 \\ &= Mgh + \frac{1}{2}Mv_B^2 + \frac{1}{2}\left(\frac{2}{5}MR^2\right)\left(\frac{v_B}{R}\right)^2. \end{aligned} \tag{4}$$

Equations (2) through (4) are three equations in three unknowns (h, v_B, and θ). So, after some intense algebra, you can solve for h. You were not supposed to complete the algebra.

Chapter 31

Practice Final Exam (Chapters 1-22)

Number One

- *Be sure to do Chapter 25, which contains nothing but practice final exam questions.*
- *This practice final exam contains 6 questions, each worth the same number of points. You have 3 hours.*

1. Two blocks, of mass m and $3m$, are slung over a pulley by a massless rope, as shown below. The masses are initially at the same height. The pulley is a uniform solid disk of mass $12m$ and radius R. The total length of the rope, including the part that is slung over the pulley, is L. The masses are released at time $t = 0$.

a) How much time does the pulley take to complete one revolution?

b) At what time will the smaller mass crash into the pulley? (You may treat the mass as being much smaller than the pulley.)

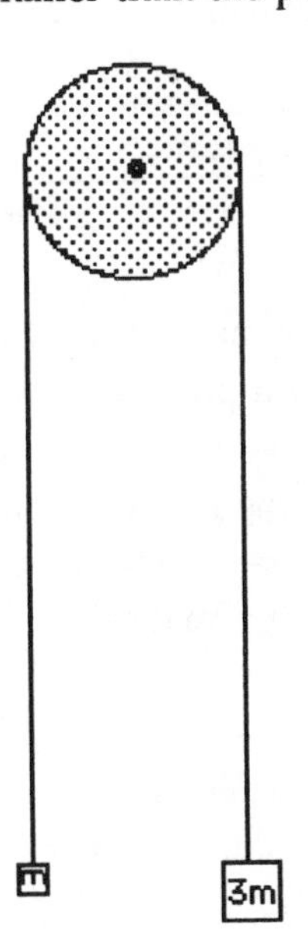

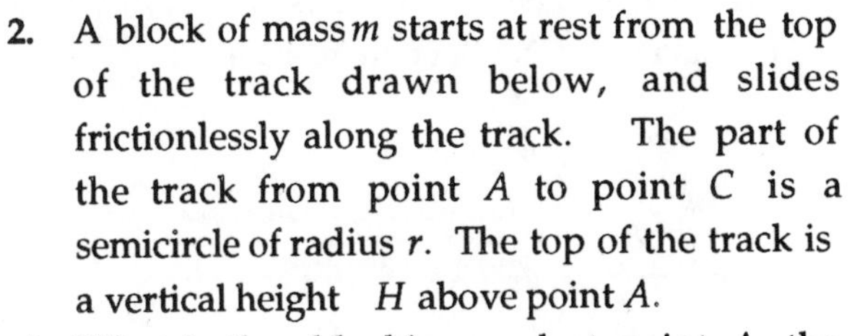

2. A block of mass m starts at rest from the top of the track drawn below, and slides frictionlessly along the track. The part of the track from point A to point C is a semicircle of radius r. The top of the track is a vertical height H above point A.

a) What is the block's speed at point A, the bottom of the track? Express your answer in terms of the quantities listed above.

b) What is the normal force exerted by the track on the block when the block reaches point B? Point B is where the track is vertical.

c) What is the minimum speed v_{min} that the block must have at point A to ensure that it does not fall off the track, but instead stays in contact with the track all the way to point C?

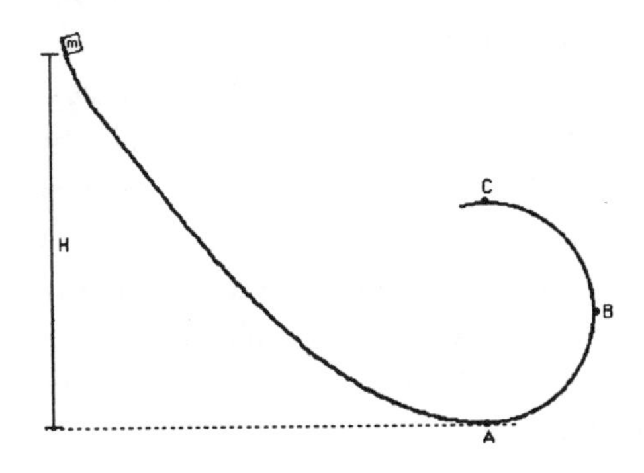

3. In outer space, two clay meteors collide and stick together. The collision occurs a distance D from the center of the Earth. After the collision, the stuck-together meteors end up zooming towards the Earth.

   Before colliding, both meteors had mass m. Observers on Earth notice that, immediately before the collision, meteor 1 had velocity v_1, and meteor 2 had velocity v_2, where v_1 and v_2 are at right angles. (See the diagram below.)

a) What fraction of the meteors' kinetic energy was lost during the collision?

b) How fast are the stuck-together meteors traveling immediately before crashing into the Earth? The Earth's mass is M_E, and its radius is R_E. Neglect air resistance.

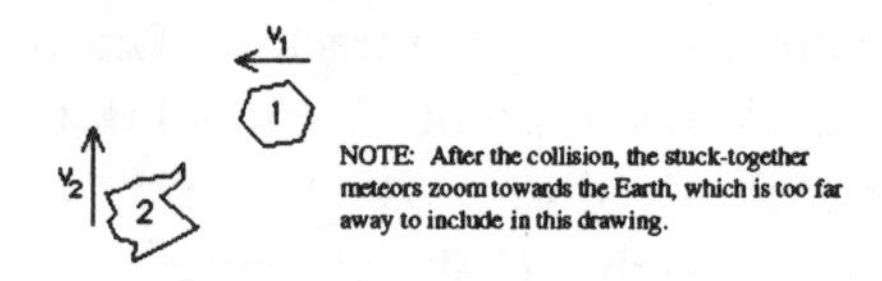

4. The system drawn below is embedded in a truck. Both springs have equilibrium length L and spring constant k. Therefore, when the mass is at $x = 0$, neither spring is stretched or compressed. The block has mass M, and the floor is nearly frictionless.

a) Suppose the truck accelerates rightward with acceleration a_t. The mass will eventually settle at a position behind $x = 0$. How far behind $x = 0$ will the mass settle? Call your answer D, and solve for D in terms of M, k, L, and a_t.

b) Suppose the truck's acceleration began at time $t = 0$. At time $t = t_1$, the truck abruptly stops. What is the amplitude of the resulting oscillations of the block? Neglect friction, and express your answer in terms of D and the constants listed in the problem.

c) What is the period of the resulting oscillations?

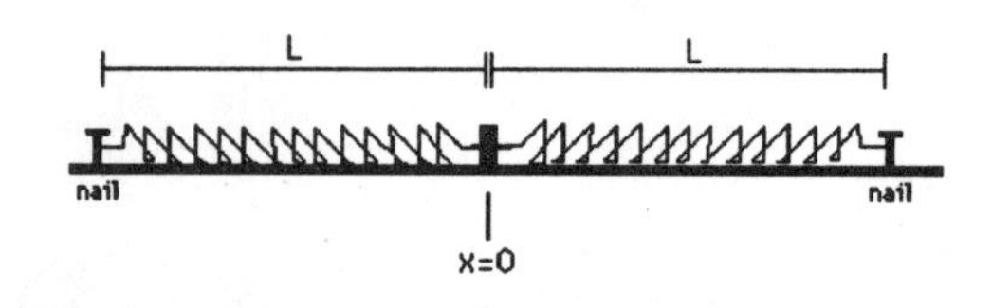

5. Water is pumped into the left end of the pipe drawn below, at pressure $5p_0$, where p_0 is atmospheric pressure. (Strictly speaking, $5p_0$ is the pressure at point K.) The density of water is ρ. The water eventually flows out the right end of the pipe into the air.

   Both sections of pipe are cylindrical, which means the cross-sectional areas are circular. The radius of the thick segment of pipe is r, and the radius of the thin segment is $\frac{r}{2}$. Both segments have length L. (The "kink" between the two segments is very short. Neglect it.)

a) Consider a single droplet of water that travels through the pipes in a straight line. The droplet enters the left end of the pipe at time $t = 0$. At what time does the water droplet leave the right end of the pipe? Answer in terms of the quantities listed in the problem, and g if needed.

b) The mouth (right end) of the pipe is a height h above ground level. What horizontal distance does that droplet travel in the air before landing on the ground?

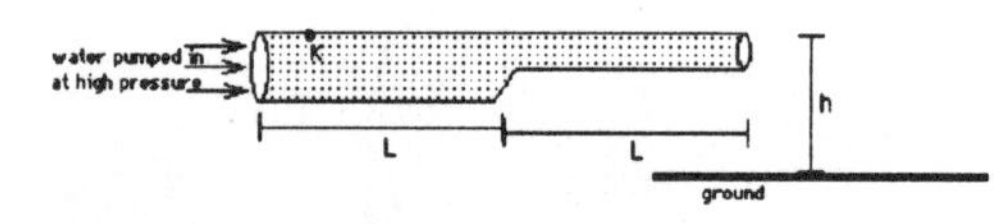

6. A thin rod of mass M and length L is attached to a wall by a pin joint, as drawn below. A rope of mass $m_r << M$ and length $L\sqrt{3}$ is fastened to the rod at its endpoint, as shown. The rod makes a 60° angle with the wall, while the rope makes a 30° angle with the wall.

   Throughout this problem, you may assume that when the rope is plucked or otherwise disturbed, the rod does not "jiggle" significantly.

a) Someone plucks the rope so as to produce the lowest resonant frequency, i.e., the lowest-frequency standing wave. What is that frequency?

b) While the rope is oscillating due to the pluck in part **a)**, which point on the rope undergoes the *largest* oscillations? Explain your answer.

c) Let A denote the amplitude of this standing wave, by which I mean the maximum displacement from equilibrium of the point on the rope that undergoes the largest oscillations. It turns out that the displacement y of this point as a function of time is given by $y = A\sin(2\pi ft)$, where f is the frequency.

During its oscillations, what is the maximum velocity of this point? Express your answer in terms of A and g, along with the masses and lengths given in the problem.

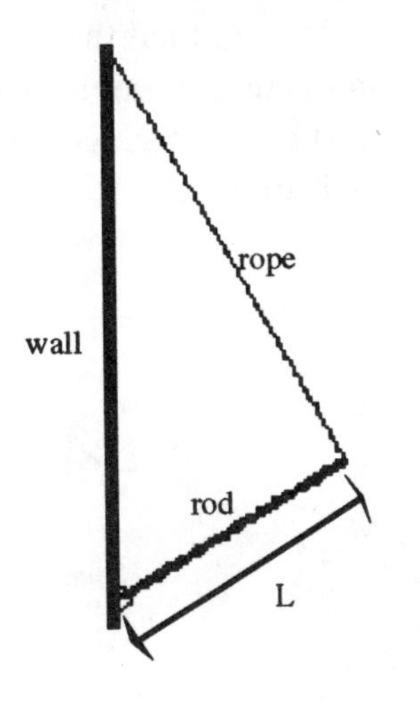

ANSWER 1

a) Since there is no frictional rubbing or other loss of mechanical energy, you can use either the standard multi-block strategy, or energy conservation. I will show both ways.

Method 1: Multi-block strategy

(1) Make free-body diagrams for each individual object,

(2) Write Newton's 2nd law (along the direction of motion) for each object separately, using $\sum\tau = I\alpha$ instead of $\sum F = ma$ for rotating objects.

(3) Algebraically solve for the linear or angular acceleration; and then

(4) Use that acceleration in a kinematic equation to figure out something about the motion. Here, we will use α to calculate the time it takes the pulley to rotate through angle $\theta = 2\pi$ radians.

Now I will implement the strategy.

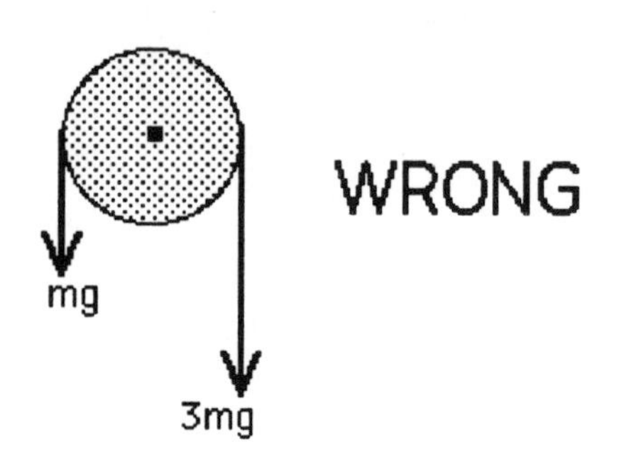

Step 1: Force diagrams for each separate object

The most common mistake is to draw the following force diagram for the pulley.

This is wrong, because the masses do not pull directly on the pulley. Instead, the masses help to create tensions in the rope, and these tensions pull on the pulley. If the tension in the left segment of rope were mg, then the net force on the little mass would be 0, and hence, it would not move. The tension in the left rope must be greater than mg, so that the net force on the little block is upward.

Because the pulley is not massless, the tensions in the two segments of rope are *different*. So, here are the accurate diagrams.

Step 2: Newton's 2nd law for each object

Notice that I have made a consistent choice of positive directions in my diagrams. Because the pulley rotates, we must use $\sum\tau = I\alpha$. Since the blocks move in tandem, they share the same acceleration a.

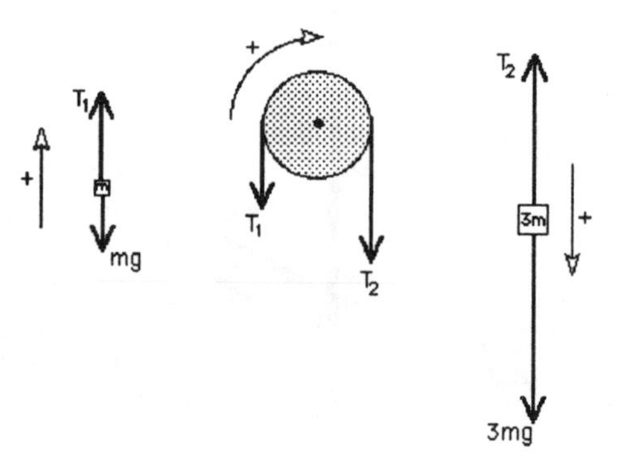

$$\text{SMALL MASS: } \sum F = T_1 - mg = ma \tag{1}$$

$$\text{BIG MASS: } \sum F = 3mg - T_2 = 3ma \tag{2}$$

$$\text{PULLEY: } \sum\tau = T_2R - T_1R = I\alpha = \frac{1}{2}(12m)R^2\alpha. \tag{3}$$

I have plugged in the book value for a solid disk's rotational inertia, remembering that the disk's mass is $M = 12m$.

We want to solve this set of equations for the angular acceleration α, so that we can use rotational kinematics to complete the problem. Buts Eqs. (1) through (3) contain four unknowns (α, a, T_1, T_2). Fortunately, the blocks' linear acceleration relates in a simple way to the pulley's angular acceleration. I explain this in detail in Chapter 15, Question 1. Here is the capsule summary: Since the blocks and rope move together, and since the rope and the rim of the pulley move together, it follows that $a_{\text{blocks}} = a_{\text{tang, rim of pulley}}$. (The subscript "tang" stands for "tangential.") But the tangential acceleration of a point moving in a circle of radius R is $a_{\text{tang}} = \alpha R$. So, since "$a$" denotes the blocks' acceleration, we have

$$a = \alpha R. \tag{4}$$

With these four equations, we now have sufficient information to solve for α.

Step 3: Algebra to solve for the linear or angular acceleration.

We want α. To get it quickly, substitute αR for a in Eqs. (1) and (2). Then divide Eq. (3) through by R. Finally, add these modified equations. (1) through (3). When you do so, the T_1 and T_2 terms cancel, leaving us with $-mg + 3mg = 10mR\alpha$, and hence $\alpha = \frac{g}{5R}$.

Step 4: Kinematics

Now that we know the angular acceleration a, we can solve for the time needed to complete one revolution by using the rotational analog of $x = x_0 + v_0 t + \frac{1}{2}at^2$, namely $\theta = \theta_0 + \omega_0 t + \frac{1}{2}\alpha t^2$. We can set $\theta_0 = 0$, in which case the final angle (1 revolution later) is $\theta = 2\pi$ radians. Since the system starts from rest, the initial angular speed is $\omega_0 = 0$. So, solving for t gives

$$t = \sqrt{\frac{2\theta}{\alpha}} = \sqrt{\frac{2(2\pi)}{\frac{g}{5R}}} = \sqrt{\frac{20\pi R}{g}}.$$

I will now re-derive this answer using a different method.

Method 2: Energy conservation.

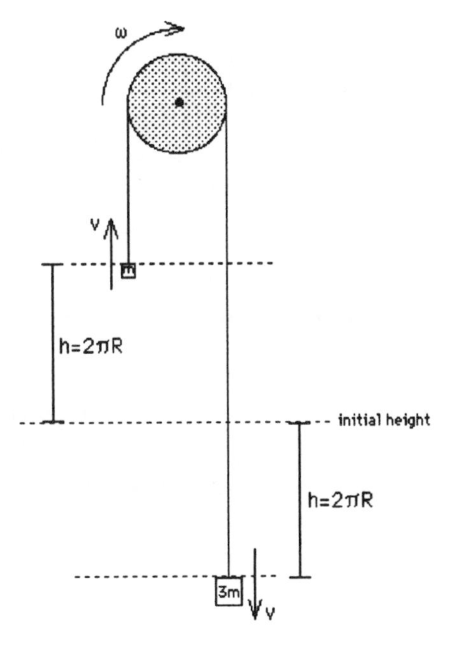

When the pulley completes one revolution, its rim moves a distance $2\pi R$. Therefore, the rope is dragged along by that same distance. Since the blocks are attached to the rope, they also move that same distance. So, during one revolution of the pulley, the small block rises by a distance $h = 2\pi R$, and the big block falls by that same h.

The "final" velocities of the two blocks are the same–call it v–because the blocks move in sync. Using energy conservation, we can solve for v or ω. Then we will use kinematics to solve for t.

By the way, v and ω are related in the same way that a and a are related. Since the blocks, rope, and rim of the pulley all move together, $v_{\text{blocks}} = v_{\text{tang, rim of pulley}}$; and since the tangential speed of a point moving in a circle is $v_{\text{tang}} = \omega R$, we have $v = \omega R$.

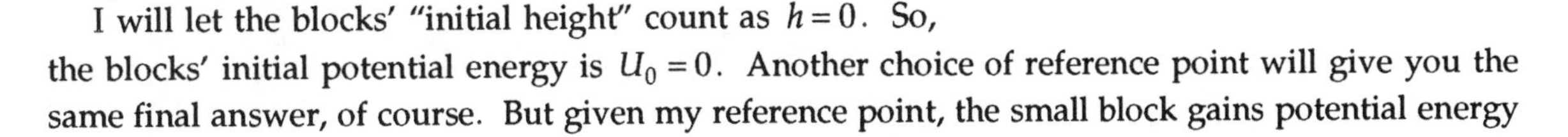

I will let the blocks' "initial height" count as $h = 0$. So, the blocks' initial potential energy is $U_0 = 0$. Another choice of reference point will give you the same final answer, of course. But given my reference point, the small block gains potential energy *mgh*, and the big block *loses* $(3m)gh$. Therefore, since I am setting the big block's initial potential energy to 0, I must make its final potential energy negative, to reflect the fact that it loses potential energy.

The pulley's kinetic energy is rotational. According to conservation of energy,

$$K_0 + U_0 = K_f + U_f$$

$$0+0=\left[\frac{1}{2}mv^2+\frac{1}{2}(3m)v^2+\frac{1}{2}I\omega^2\right]+[mgh+3mg(-h)] \quad (5)$$

$$=\left[\frac{1}{2}m(\omega R)^2+\frac{1}{2}(3m)(\omega R)^2+\frac{1}{2}\left(\frac{1}{2}(12m)R^2\right)\omega^2\right]+[-2mgh],$$

where in the last step I set $v=\omega R$, and substituted in the pulley's rotational inertia. We can immediately solve for ω, the pulley's "final" angular speed after 1 revolution. I will not complete the algebra. But given ω, you could then use $\omega^2=\omega_0^2+2\alpha\Delta\theta$ to solve for the angular acceleration, since we know $\omega_0=0$ and $\Delta\theta=2\pi$. Finally, once we have α, we can find the time it takes the pulley to complete 1 revolution, using kinematics as in step 4 above.

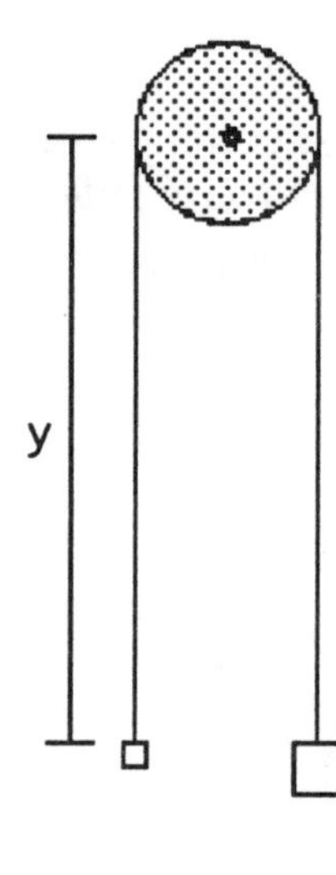

b) To solve this, you must figure out y, the distance traveled by the small block before crashing.

Since the part of the rope wrapped around the top half of the pulley traces out half a circumference, it has length $\frac{1}{2}(2\pi R)=\pi R$. But the whole rope, consisting of two vertical segments of length y and a semicircular segment of length πR, has length L: $2y+\pi R=L$. So, $y=\frac{L-\pi R}{2}$.

Now that you know y, you can combine your results of part **a)** with kinematic reasoning to figure out the corresponding t. For instance, in part a) method 1, we found that the pulley's angular acceleration is $\alpha=\frac{g}{5R}$. Therefore, the blocks' linear acceleration is $a=\alpha R=\frac{g}{5}$. Given this information, you can immediately use $y=y_0+v_0t+\frac{1}{2}at^2$, with $y_0=0$, $v_0=0$, and $y=\frac{(L-\pi R)}{2}$. Solving for t gives

$$t=\sqrt{\frac{2y}{a}}=\sqrt{\frac{5(L-\pi R)}{g}}.$$

ANSWER 2

a) Because the track's slope keeps changing, so does the net force on the block. Therefore, its acceleration is not constant. You cannot use your old constant-acceleration kinematic formulas. Instead, we must use energy conservation. Since it is easy to visualize, I will not bother drawing pictures. There is no heat due to frictional rubbing.

$$E_0 = E_A$$

$$K_0 + U_0 = K_A + U_A$$

$$0 + mgH = \frac{1}{2}mv_A^2 + 0.$$

Solve for v_A to get $v_A=\sqrt{2gH}$.

b) Since we are calculating a force, I will begin as always with a force diagram. We are interested in the forces on the block at point B.

Remember, a normal force always points perpendicular to the surface exerting the force. Since the track at point B is vertical, the normal force is horizontal.

How can we solve for N? You must take advantage of the fact that the block is moving in a *circle* of radius r. Therefore, it undergoes a radial acceleration (towards the center of the circle) of magnitude $a_{\text{radial}} = \frac{v_B^2}{r}$, where v_B denotes the velocity at point B. From the force diagram, the only radial force is N; gravity points tangential to the track. So, Newton's 2nd law tells us that

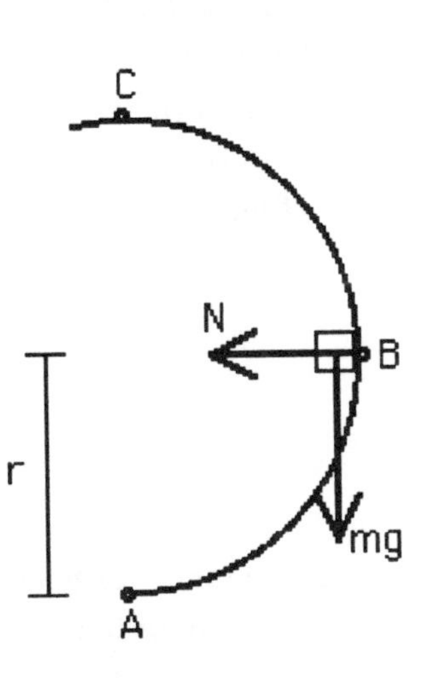

$$\sum F_{\text{radial}} = ma_{\text{radial}}$$
$$N = m\frac{v_B^2}{r}. \tag{1}$$

To finish solving for N, we just need to know the block's velocity at point B. Duplicating the reasoning of part **a)**, we can use energy conservation. Indeed, since $E_0 = E_A = E_B$, we can either set E_B equal to E_A, or we can set E_B equal to E_0. I will choose the second option. At point B, the block is a height r off the ground. So,

$$E_0 = E_B$$
$$K_0 + U_0 = K_B + U_B$$
$$0 + mgH = \frac{1}{2}mv_B^2 + mgr.$$

Isolate v_B to get $v_B = \sqrt{2g(H-r)}$. Put that v_B into Eq. (1) to get

$$N = m\frac{v_B^2}{r} = \frac{2mg(H-r)}{r}.$$

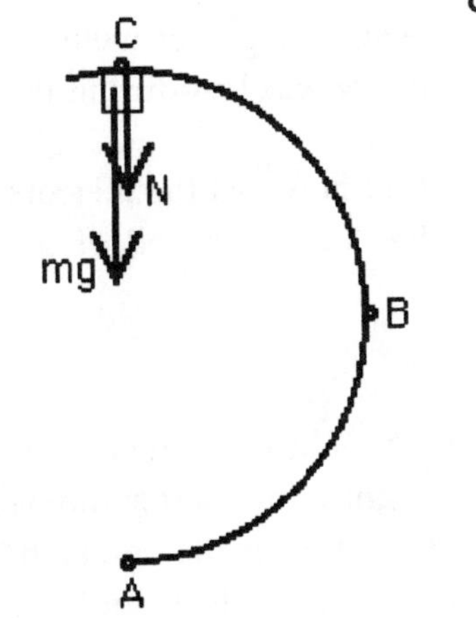

c) If it has enough velocity, the block easily reaches the top of the track. But if it is moving too slowly at point A, it falls off the track somewhere between points B and C–assuming it even reaches point B. We are finding $v_{\min}$, the *smallest* speed it can have at point A such that the block nonetheless reaches point C.

You might assume that we should set $v_C = 0$, in order to encode the assumption that the block *barely* reaches point C. But if the block were moving that slowly at the top of the semicircle, it would have fallen off the track before reaching C. Let me postpone for two paragraphs an intuitive explanation of why this is so. For now, I will ask you to take it on faith that the block must still be moving at point C. So, the question becomes: How *fast* is the block moving at point C, if it is barely staying in contact with the track?

If the block *barely* stays in contact with the track at point C, then the track is barely touching the block, and vice versa. In other words, the normal force is $N \approx 0$ at point C.

Nonetheless, at point C, the block is still moving in a circle of radius r. So, its radial acceleration is $a_{\text{radial}} = \frac{v_C^2}{r}$. To use this acceleration in Newton's 2nd law, we need to know the forces acting on the block at point C.

As just noted, however, the normal force is $N \approx 0$. So, the only radial force (i.e., the only force pointing towards the center of the circle) is mg:

$$\sum F_{\text{radial}} = ma_{\text{radial}}$$

$$mg = m\frac{v_C^2}{r}.$$

Solve for v_C to get $v_C = \sqrt{gr}$.

Given all this, I can explain why the block must still be moving at point C. If it were moving too slowly, or not moving at all, then $\frac{mv^2}{r}$ would be very small. In fact, the radial force (due to gravity) would be bigger than $\frac{mv^2}{r}$. As a result, the block would no longer keep moving in a circle, but would "spiral in." Only when the net radial force *equals* $\frac{mv^2}{r}$ does the object continue moving in a perfect circle.

Now that we know how fast the block must be moving at point C, we can use energy conservation to calculate the speed it must have possessed at point A, to ensure that it reaches point C. At point C, the block is a height $h = 2r$ off the ground.

$$E_A = E_C$$

$$K_A + U_A = K_C + U_C$$

$$\frac{1}{2}mv_{\text{min}}^2 + 0 = \frac{1}{2}mv_C^2 + mg2r$$

$$= \frac{1}{2}m(gr) + mg2r \qquad \left[\text{since } v_C = \sqrt{gr}, \text{ from above}\right].$$

Cancel the m's, and solve for v_{min} to get $v_{\text{min}} = \sqrt{5gr}$.

ANSWER 3

You must break up this problem into two parts:

Subproblem 1: Momentum conservation to find the post-collision speed of the stuck-together meteors. Call this speed v_3. Once you know v_3, you can calculate how much kinetic energy was lost during the collision.

Subproblem 2: Gravity considerations, specifically conservation of energy, to find how fast the meteors are moving right before impact with the Earth. In this subproblem, v_3 is the "initial" speed of the two-meteor blob when it is a distance D from the center of the Earth.

Let us go for it.

Subproblem 1: Conservation of momentum

You could not use energy conservation here, because an unknown amount of heat gets generated during the inelastic collision. We must use momentum. At first glance, you might think momentum is not conserved, because an external force, Earth's gravity, acts on the system during the collision. Strictly speaking, this is true. But since the collision takes so little time, Earth's gravity does not have time to speed up the meteors very much during the collision. Put another way, during the few hundredths of a second over which the collision takes place, the "internal" forces exerted by the meteors on each other are much bigger than the external gravitational force. Therefore, to excellent approximation, momentumis conserved during the collision. For this reason, we can use momentum conservation to address *any* "quick" (impulsive) collision.

Momentum conservation would be easy to apply, except that the meteors' momenta do not all lie along a straight line. Therefore, *we must worry about vector components* . We cannot just say that

$$m_1 v_1 + m_2 v_2 = (m_1 + m_2) v_3 \qquad \textbf{WRONG,}$$

because that equation fails to distinguish x-directed motion from y-directed motion. (Similarly, you cannot say that someone who walks 3 miles north and 4 miles west ends up 7 miles from her starting point. Vectors do not add like numbers.)

Instead, we must *separately* conserve the x- and y-component of momentum. By doing so, we can separately calculate the x- and y-component of v_3, the post-collision velocity. Then, we can "Pythagoras together" those components to find the total post-collision speed.

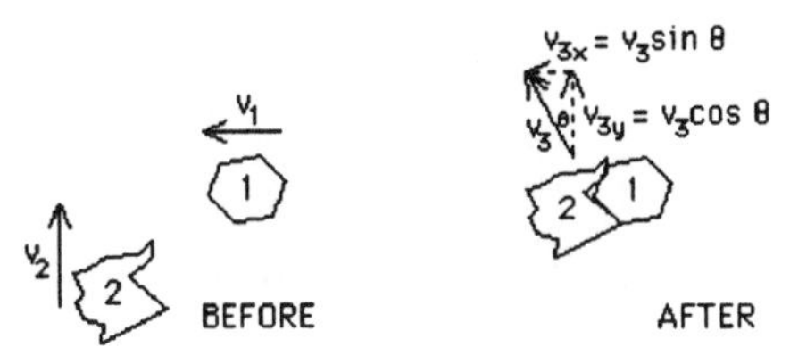

After the meteors stick together, they have total mass $2m$.

Let us start with x-momentum. I will pick leftward as positive. Initially, only meteor 1 has x-momentum.

$$p_{x\text{ before}} = p_{x\text{ immediately after collision}}$$
$$mv_1 = (2m)v_{3x}$$

Solve for v_{3x} to get $v_{3x} = \frac{v_1}{2}$.

By similar reasoning, we can find the y-component of the post-collision velocity:

$$p_{y\text{ before}} = p_{y\text{ immediately after collision}}$$
$$mv_2 = (2m)v_{3x}.$$

Solve for v_{3y} to get $v_{3y} = \frac{v_2}{2}$.

Now that we know the x- and y-components of $\mathbf{v}_3$, we can find the total speed v_3:

$$v_3 = \sqrt{v_{3x}^2 + v_{3y}2} = \sqrt{\left(\frac{1}{2}v_1\right)^2 + \left(\frac{1}{2}v_2\right)^2} = \frac{1}{2}\sqrt{v_1^2 + v_2^2}.$$

Do not draw any general conclusions from the simple form of this answer. Usually in two-dimensional momentum problems, things gets messier.

a) Now that we know the post-collision speed, we can calculate what fraction of kinetic energy was lost during the collision. The fraction of kinetic energy lost is $\frac{K_{\text{lost}}}{K_0} = \frac{K_0 - K_f}{K_0}$. Why? Well, suppose the system starts with $K_0 = 100$ joules of kinetic energy, but ends up with only $K_f = 60$ joules. Then the system lost $K_{\text{lost}} = K_0 - K_f = 40$ joules. So, the fraction of kinetic energy lost was $\frac{K_{\text{lost}}}{K_0} = \frac{K_0 - K_f}{K_0}$, which is 40%.

In this case, the initial kinetic energy is $K_0 = \frac{1}{2}mv_1^2 + \frac{1}{2}mv_2^2$, while the post-collision kinetic energy is

$$
\begin{aligned}
K_f &= \frac{1}{2}(2m)v_3^2 \\
&= \frac{1}{2}(2m)\left(\frac{1}{2}\sqrt{v_1^2 + v_2^2}\right)^2 \\
&= \frac{1}{2}(2m)\left(\frac{1}{4}\right)\left(v_1^2 + v_2^2\right) \\
&= \frac{1}{4}mv_1^2 + \frac{1}{4}mv_2^2.
\end{aligned}
$$

So, the fraction of kinetic energy lost is

$$
\begin{aligned}
\text{fraction of } K \text{ lost} &= \frac{K_{\text{lost}}}{K_0} = \frac{K_0 - K_f}{K_0} \\
&= \frac{\left[\frac{1}{2}mv_1^2 + \frac{1}{2}mv_2^2\right] - \left[\frac{1}{4}mv_1^2 + \frac{1}{4}mv_2^2\right]}{\frac{1}{2}mv_1^2 + \frac{1}{2}mv_2^2} \\
&= \frac{1}{2}.
\end{aligned}
$$

During the collision, exactly 50% of the initial kinetic energy gets converted into heat and other forms of "non-mechanical" energy.

b) To solve part **b)**, let us proceed to

Subproblem 2: Gravity/energy to find speed at impact

In this subproblem, the relevant "initial" situation is immediately after the collision, when the two-meteor blob is a distance D from the center of the Earth, moving at speed v_3. As the blob "falls" toward Earth, it speeds up. In this subproblem, we are trying to find v_f, its speed right before impact with the Earth.

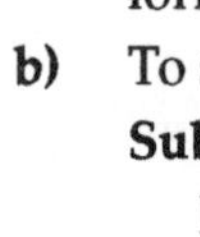

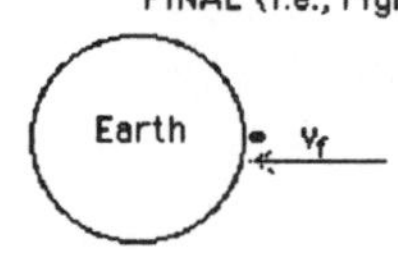

The meteor's "final" distance from the center of the Earth is $r = R_E$. Using the general expression for gravitational potential energy, $U = -\frac{GMm}{r}$, where r is measured to the centers of the relevant objects, and remembering that the two-meteor blob has mass $2m$, I get

$$
\begin{aligned}
E_0 &= E_f \\
\frac{1}{2}(2m)v_3^2 - \frac{GM_E(2m)}{D} &= \frac{1}{2}(2m)v_f^2 - \frac{GM_E(2m)}{R_E}.
\end{aligned}
$$

The $2m$'s cancel. Solve for v_f to get

$$
\begin{aligned}
v_f &= \sqrt{v_3^2 + \frac{2GM_E}{R_E} - \frac{2GM_E}{D}} \\
&= \sqrt{\frac{1}{2}\left(v_1^2 + v_2^2\right)^2 + \frac{2GM_E}{R_E} - \frac{2GM_E}{D}},
\end{aligned}
$$

where in the last step I invoked our result from subproblem 1 above, $v_3 = \frac{1}{2}\sqrt{v_1^2 + v_2^2}$.

ANSWER 4

a) In disguised form, this is a standard force problem. When the block "settles," it is motionless with respect to the truck. Therefore, the block's total acceleration (i.e., its acceleration *with respect to the ground*) is a_t, because the block gets "carried along" by the truck. It follows from Newton's 2nd law that the net force on the block must be $F_{net} = ma_t$. The problem then reduces to figuring out where the block must be so that the springs push and pull it with net force ma_t. Remember, " ma_t " is not a "truck force" acting on the block. The truck does not exert a direct forward force on the block. Rather, the block's acceleration is the *result* of the forces supplied by the springs.

Before diving into details, let us picture what happens when the truck starts accelerating. The block gets "thrown backward." But no force actually *pushes* the block backward. Rather, the block "tries" to sit still while the truck accelerates forward. So, the block moves backwards with respect to the truck. Eventually, it settles into the position drawn below, an unknown distance D behind $x = 0$. As just noted, at this point, the block is not moving with respect to the truck, i.e., it is no longer sliding along the floor.

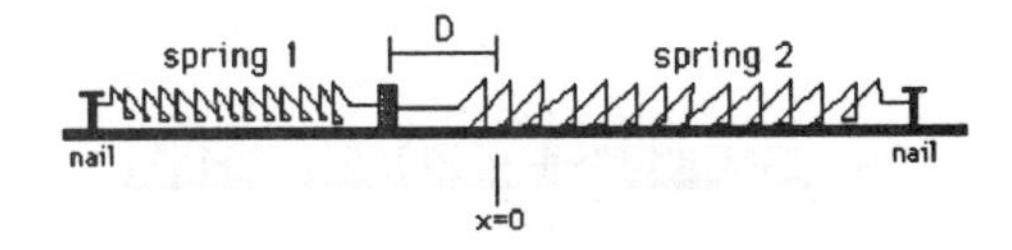

To solve for D, draw a force diagram and use Newton's 2nd law. Since spring 2 is stretched a distance $x = D$, it pulls the block rightward with force $F = kx = kD$. Since spring 1 is compressed by distance $x = D$, it *pushes* the block rightward, with force $F = kD$. Notice that, instead of trying to sort out the minus signs, I used physical reasoning to figure out the directions of these forces.

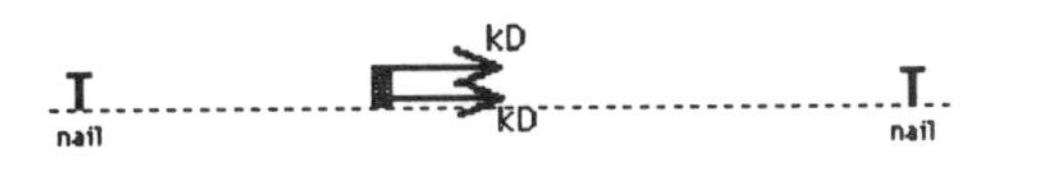

Because all the forces and accelerations in this problem point along the same straight line, we need not worry about vector components. As mentioned above, since the block is motionless with respect to the truck, but the truck has acceleration a_t, the block's acceleration is a_t. So, Newton's 2nd law *applied to the block, not the truck,* demands that

$$\sum F = ma$$
$$kD + kD = Ma_t.$$

Solve for D to get $D = \frac{Ma_t}{2k}$. Notice that the equilibrium length of the springs is irrelevant.

Intuitively speaking, the springs provide the forward force needed to make the block "keep up" with the accelerating truck. When you ride a car, this forward force is provided by the back of your seat.

b) At first glance, you might think the block's amplitude of oscillation is D. To see why this wrong, you must picture what is happening physically. Let v_1 denote the truck's speed right before it stops. Since the block "keeps up" with the truck, v_1 is also the block's speed immediately before the truck stops. By Newton's 1st law, an object in motion tends to stay in motion. Therefore, when the truck stops, the block *keeps moving forward* with initial speed v_1. (Similarly, during a car crash, the passengers keep moving forward towards the windshield, unless they are wearing seat belts.) When the truck suddenly stops, it essentially "throws" the block forward at speed v_1. Of course, the block does not keep moving at that speed forever, because the springs exert forces on it.

Here is the point: This problem is the same as if, on a motionless floor, we would pulled the block back a distance D, and then given it initial forward speed v_1. We are looking for A, the resulting amplitude of oscillation. This A will be bigger than D, because the block was "thrown."

First, I will figure out v_1. Remember, v_1 is the truck's *final* speed right before it stops. Since the truck's initial speed was $v_0 = 0$, its final speed is

$$\begin{aligned} v_1 &= v_0 + at \\ &= a_t t_1. \end{aligned}$$

To find A, you could use oscillation kinematic reasoning. But that turns out to be complicated, due to phase factors. It is easier to conserve energy. As always, make "initial" and "final" diagrams. By "initial," I mean right when the truck stopped, i.e., right when the block was thrown forward at speed v_1. As my "final" configuration, I will consider the block when it has moved forward to its maximum displacement, $x = A$. Remember, $x_{max} = A$ is the "turning point" at which the block momentarily comes to rest before starting to move back towards equilibrium. So, the block's velocity at that point is $v_f = 0$.

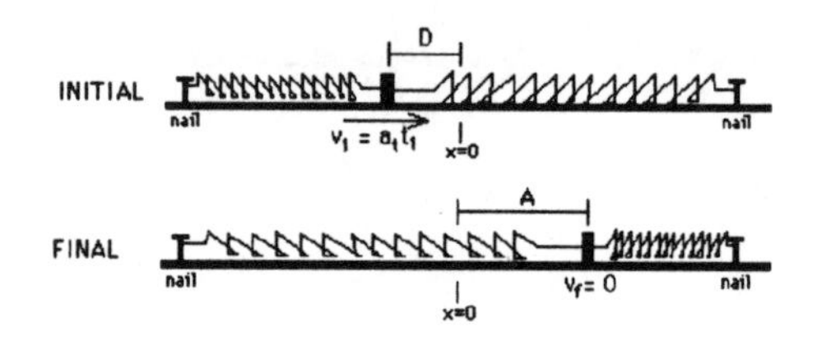

Since no frictional heat gets generated, conservation of energy tells us that

$$\begin{aligned} E_0 &= E_f \\ K_0 + U_{0,\text{ spring }1} + U_{0,\text{ spring }2} &= K_f + U_{f,\text{ spring }1} + U_{f,\text{ spring }2} \\ \frac{1}{2}Mv_1^2 + \frac{1}{2}kD^2 + \frac{1}{2}kD^2 &= 0 + \frac{1}{2}kA^2 + \frac{1}{2}kA^2. \end{aligned}$$

Solve for A, and set $v_1 = a_t t_1$ from above, to get

$$A = \sqrt{\frac{\frac{1}{2}Ma_t^2 t_1^2 + kD^2}{k}}.$$

c) Here is a fundamental property of all simple harmonic motion: *The period of oscillation in no way depends on the amplitude or the initial velocity. It depends only on the forces that act on the system when it is displaced from equilibrium.* To demonstrate this principle, use your shoe as a pendulum. You will see that small oscillations take the same time as large oscillations, provided the "large" oscillations are not so big that the small angle approximation ($\sin\theta \approx \theta$) breaks down. So, even if you could not solve part **b)**, you can address this problem.

To find the period, we must find the angular frequency ω, and then use $T = \frac{2\pi}{\omega}$. To calculate the angular frequency ω, you must use $\omega = \sqrt{\frac{\text{stuff}}{M}}$. Remember, the restoring force on a harmonically oscillating object is $F = -(\text{stuff})x$. For a review of all this material, see Answer 20-2, in which I lay out the following strategy for finding "stuff":

<u>Stuff-finding strategy:</u>

(1) Draw a force diagram of the system when it is displaced an arbitrary distance x from equilibrium. (Do not draw the system at equilibrium, because then the net force on it will be 0.)

(2) Write down the net force on the system along the direction of motion.

(3) By writing that net force in the form $F_{net} = -(\text{stuff})x$, figure out stuff.

I will now implement this strategy. Remember, the point of all this is to find "stuff" so that we can substitute it into $\omega = \sqrt{\frac{\text{stuff}}{M}}$ to obtain the angular frequency. (And once we know ω, we can easily calculate the period T.)

Stuff-finding step 1: Force diagram of system displaced from equilibrium.

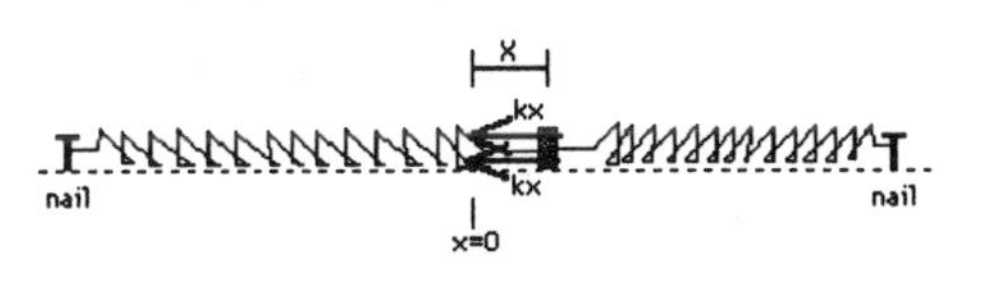

Notice that both springs exert forces in the same direction, because the compressed spring *pushes* while the stretched spring *pulls*. I omit the normal and gravitational forces, because they cancel out, and besides, we care only about the forces along the direction of motion.

Stuff-finding step 2: Write net force along direction of motion. $F_{net} = -2kx$.

Stuff-finding step 3: Express F_{net} *in the form* $F_{net} = -(\text{stuff})x$, *and read off stuff.*

With no further effort, we see that $\text{stuff} = 2k$.

End of stuff-finding strategy.

So, $\omega = \sqrt{\frac{\text{stuff}}{M}} = \sqrt{\frac{2k}{M}}$, and hence $T = \frac{2\pi}{\omega} = 2\pi\sqrt{\frac{M}{2k}}$. Here, we perhaps could have "guessed" that $\omega = \sqrt{\frac{2k}{M}}$. But I wanted to review the stuff-finding strategy, because you will need it to address more complex examples of simple harmonic motion.

ANSWER 5

a) To solve this, we need to know the speed of water in both segments of pipe. Call those speeds v_0 (in the thick section of pipe) and v_1 (in the thin section). Once we find those velocities, we should be able to figure out the "travel time" using kinematic reasoning.

Subproblem 1: Find the water speed in both segments of pipe.

As often happens in fluid flow problems, we must invoke both Bernoulli's law and flux conservation to find what we are looking for. When using those equations, we must pick two points to compare. Well, we know the pressure at point K. The other "chosen" point should live inside the thin segment of pipe. I will pick the mouth of the pipe (point D), because we know the pressure there.

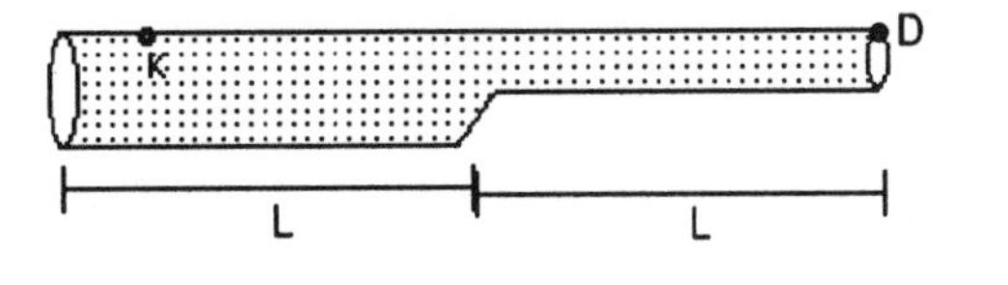

At point D, the water pressure must be "equalized" with the surrounding air pressure: $p_D = p_0$. Also, points K and D are both a height h off the ground. So, from Bernoulli's law,

$$5p_0 + \frac{1}{2}\rho v_0^2 + \rho gh = p_0 + \frac{1}{2}\rho v_1^2 + \rho gh. \tag{1}$$

The ρgh terms cancel. Nonetheless, Eq. (1) contains two unknowns, v_0 and v_1. To get a second equation, consider flux conservation, according to which the volume per second of water entering the pipe equals the volume per second of water leaving the pipe:

$$v_0 A_{\text{thick section of pipe}} = v_1 A_{\text{thin section of pipe}}$$

$$v_0 \pi r^2 = v_1 \pi \left(\frac{r}{2}\right)^2, \tag{2}$$

where I have used the area of a circle. Now we have two equations in two unknowns (v_0 and v_1). So, only algebra remains.

Algebra starts here. From Eq. (2), cancelling out the π's and r^2's, we get $v_1 = 4v_0$. Water travels four times faster in the thinner segment of pipe. Substitute that expression for v_1 into Eq. (1), and cancel the ρgh terms, to get $5p_0 + \frac{1}{2}\rho v_0^2 = p_0 + \frac{1}{2}\rho(4v_0)^2$. Isolate v_0 to get $v_0 = \sqrt{\frac{8p_0}{15\rho}}$, and hence $v_1 = 4v_0 = 4\sqrt{\frac{8p_0}{15\rho}}$.

End of algebra.

Subproblem 2: Kinematics to get the travel time t

Now that we know the velocities in both segments of pipe, we can figure out the droplet's total "travel time" by separately solving for the travel time in the thick segment, call it t_0, and the travel time in the thin segment, call it t_1. The total time is then $t = t_0 + t_1$.

A common mistake is to say the average velocity is $\bar{v} = \frac{v_0+v_1}{2}$, and then use $\bar{v} = \frac{\Delta x}{\Delta t} = \frac{2L}{\Delta t}$. But the average velocity is not $\frac{(v_0+v_1)}{2}$, because the droplet does not spend equal *time* moving at both velocities. Since it traverses the thick segment more slowly, it spends more time in that segment. So, the average velocity is "weighted" more towards v_0 than towards v_1. For a detailed discussion of this, see Chapter 2, Question 2-4(b).

Since the water in the thick segment flows at constant velocity, we can use the constant-velocity formula $v = \frac{\Delta x}{\Delta t}$. The droplet travels at speed v_0 over a distance $\Delta x = L$ in time t_0. So, $v_0 = \frac{L}{t_0}$, and hence $t_0 = \frac{L}{v_0}$. By equivalent reasoning, the droplet's travel time in the thin segment of pipe (also of length L) is $t_1 = \frac{L}{v_1}$. Putting all this together, and substituting in the values obtained above for v_0 and v_1, gives

$$
\begin{aligned}
t &= t_0 + t_1 \\
&= \frac{L}{v_0} + \frac{L}{v_1} \\
&= L\left[\frac{1}{v_0} + \frac{1}{v_1}\right] \\
&= L\left[\sqrt{\frac{15\rho}{8p_0}} + \frac{1}{4}\sqrt{\frac{15\rho}{8p_0}}\right] \\
&= L\left[\frac{5}{4}\sqrt{\frac{15\rho}{8p_0}}\right].
\end{aligned}
$$

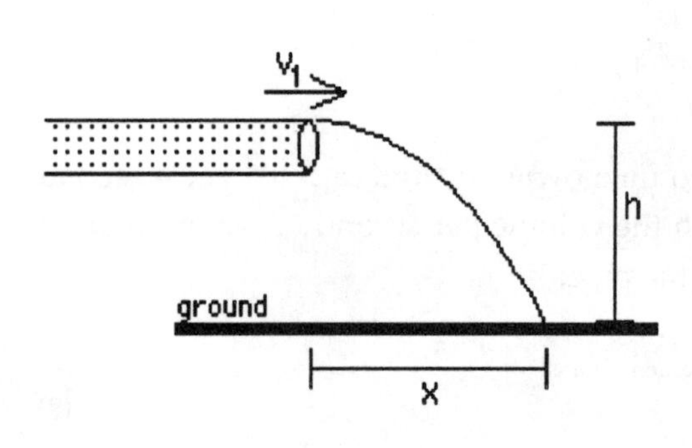

b) Once the droplet leaves the mouth of the pipe, it becomes a projectile.

As the diagram shows, the droplet's "initial" velocity is v_1 in the horizontal direction; and its initial height off the ground is $y_0 = h$.

We are solving for x, the horizontal distance covered. Let t denote the amount of time the droplet spends in the air. *This t differs from the t you obtained in part **a**), which was the amount of time the droplet spends in the pipe.* Let us start with the horizontal displacement equation:

$$x = x_0 + v_{0x}t + \frac{1}{2}a_x t^2 \\ = 0 + v_1 t + 0. \tag{3}$$

Remember, since the droplet feels no *horizontal* forces while in mid-air, its horizontal acceleration is $a_x = 0$. (This is true in general of projectiles.)

To solve Eq. (3) for x, we need to know the flight time t. As usual in projectile problems, you can extract more information by considering the vertical (as well as the horizontal) aspect of the motion. Let me pick upward as positive. The droplet starts at $y_0 = h$ and ends at $y = 0$. It has no initial *vertical* velocity, because it gets shot out of the pipe sideways. So, the vertical version of Eq. (3) gives us

$$y = y_0 + v_{0y}t + \frac{1}{2}a_y t^2 \\ 0 = h + 0 - \frac{1}{2}gt^2. \tag{4}$$

We now have enough information to find x. First, solve Eq. (4) for t to get $t = \sqrt{\frac{2h}{g}}$, and substitute that time into Eq. (3) to get

$$x = v_1 t \\ = v_1\left(\sqrt{\frac{2h}{g}}\right) \\ = 4\sqrt{\frac{8p_0}{15\rho}}\left(\sqrt{\frac{2h}{g}}\right),$$

where in the last step I substituted in our result for v_1 from part **a)**, subproblem 1.

You could not use conservation of energy to solve this part. Energy conservation would have enabled you to find the final *velocity*, but not the horizontal distance covered.

ANSWER 6

a) The hardest part of this problem is getting started. We are looking for the frequency of the lowest-frequency (i.e., "most lackadaisical") standing wave, the 1-humped mode. But with standing waves, the "easy" quantity to calculate is wavelength, not frequency.

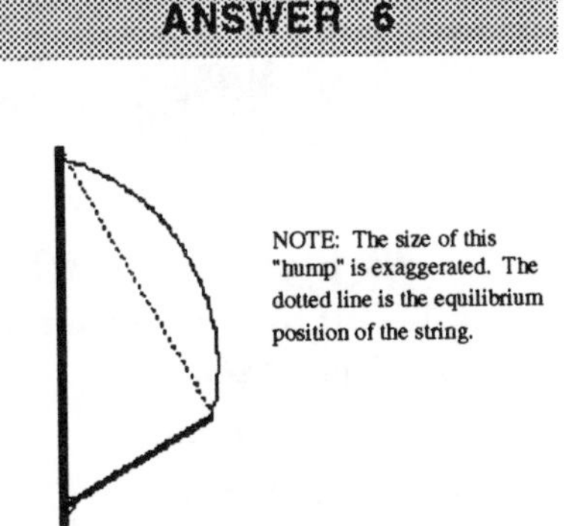

From this picture, we see that half a wavelength "fits" on the string, and hence, the full wavelength is *twice* the rope's length: $\lambda = 2(\text{rope length}) = 2(L\sqrt{3})$. You can reach this same conclusion using $\lambda_n = \frac{2(\text{string length})}{n}$, where n is the number of humps. Here, $n = 1$.

So, we know the wavelength and want the frequency. The connection is given by the general wave "kinematic" relation

$$v = \lambda f. \tag{1}$$

If we can somehow figure out the velocity of waves on this rope, then we can put it into Eq. (1) and solve for f. So, the problem reduces to finding v.

Well, a fundamental fact about waves is that their speed v depends *only* on properties of the medium, specifically, the "tautness" and "heaviness." For a one-dimensional medium such as a rope, this gets expressed as

$$v = \sqrt{\frac{\text{Tension}}{\mu}}, \tag{2}$$

where μ is the linear mass density, i.e., the mass per length. Since we know the mass and length of the rope, we can easily calculate μ. So, if we can find the tension, then we can substitute it into Eq. (2) to get v, which we can then put into Eq. (1) to obtain f.

The problem now reduces to finding the tension in the rope. But that is a standard statics question! Before actually finding the tension, let me summarize how that statics subproblem fits into the overall problem. We are going to

(1) Use statics reasoning to find the tension in the rope,

(2) Use that tension to calculate the speed of waves on the rope; then

(3) Use that speed to figure out the frequency of the standing wave.

Let us do it:

Subproblem: Statics reasoning to find tension in rope

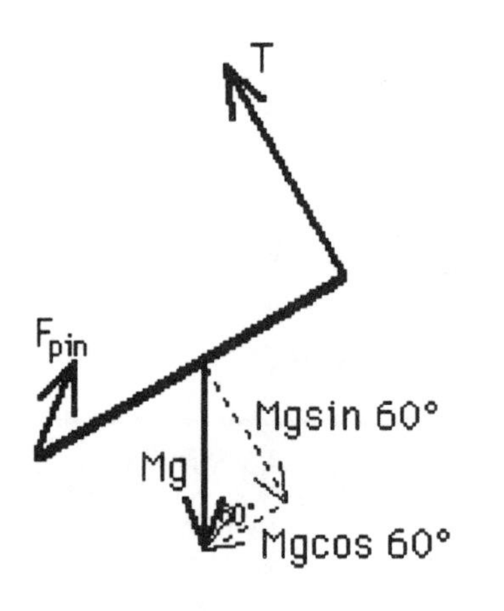

Statics step 1: Force diagram

Since the sum of a triangle's angles is 180°, the angle between the rope and the rod must be 90°. So, here is a force diagram of the rod. Let T denote the tension in the rope. Remember, gravity always acts on the center of mass, which here is a distance $\frac{L}{2}$ from the pin joint.

I am guessing about the direction of the pin joint's force. Fortunately, as we will soon see, F_{pin} plays no role here.

Statics step 2: Choose reference pivot point.

I will choose the pin joint, i.e., the bottom left end of the rod. Given this choice, the pin joint exerts no torque; since F_{pin} acts right on the pivot point, the "r" in $\tau = F_{\perp} r$ is 0.

Statics step 3: Torque and/or force equations: $\sum \tau = 0$, $\sum F_x = 0$, $\sum F_y = 0$.

As usual, I will start with torques. The component of gravity perpendicular to the relevant radius vector (from the bottom to the middle of the rod) is $Mg \sin 60°$. So, $\sum \tau = TL - (Mg \sin 60°)\frac{L}{2} = 0$. We can immediately solve for T, without invoking any force-balancing equations. (Sometimes you get lucky.) Cancel the L's and isolate T to get $T = \frac{1}{2} Mg \sin 60°$.

End of statics subproblem.

Now that we know the tension in the rope, we can calculate the wave velocity v. The rope's mass per length is $\mu = \frac{m_r}{L\sqrt{3}}$. So, the wave velocity is

$$v = \sqrt{\frac{\text{Tension}}{\mu}} = \sqrt{\frac{\frac{1}{2} Mg \sin 60°}{\frac{m_r}{L\sqrt{3}}}} = \sqrt{\frac{L\sqrt{3} Mg \sin 60°}{2m_r}}.$$

And now that we know the wave speed v, we can use $v = \lambda f$ to obtain the frequency. Remember from above that the wavelength of the $n = 1$ standing wave is twice the rope's length: $\lambda = 2(L\sqrt{3})$. So, solving $v = \lambda f$ for f yields

$$f = \frac{v}{\lambda} = \frac{\sqrt{\frac{L\sqrt{3}Mg\sin 60^\circ}{2m_r}}}{2L\sqrt{3}},$$

which simplifies down to $f = \frac{1}{4}\sqrt{\frac{Mg}{Lm_r}}$.

b) As the above "hump" sketch shows, the lowest resonant frequency corresponds to the longest possible standing wave, which is one big hump. That hump is highest in the middle of the rope. So, the midpoint of the rope undergoes the largest oscillations.

Notice that you could solve this part solely by drawing pictures; no math was needed.

c) You might think that the answer is the v we found in part **a)**. But the wave velocity v is the speed with which crests and troughs propagate along the rope. An individual point on the rope is not "carried down the rope" by the wave–even for traveling waves. Instead, for "transverse" waves, a point on the rope just oscillates back and forth perpendicular to the rope. You can see this by making standing waves on a stretched slinky. An individual "coil" of the slinky oscillates back and forth perpendicular to the direction in which the slinky is stretched.

Let me call the direction along which the rope is stretched the x-direction. Then, individual points on the rope oscillate in the y-direction.

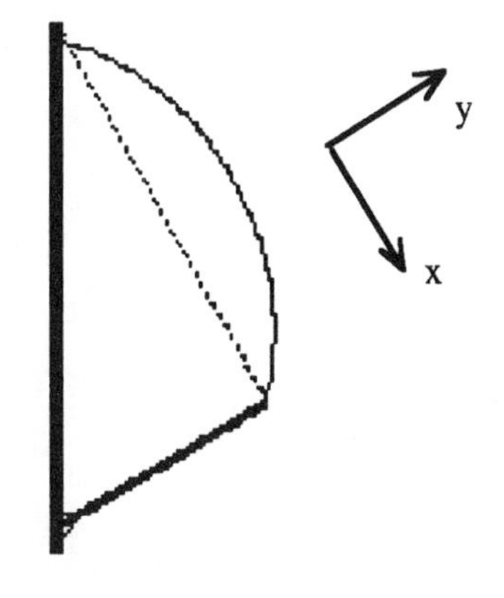

Since a point on the rope moves solely in the y-direction, the speed of a point on the rope is the rate at which its y-position changes: $v_y = \frac{dy}{dt} = \frac{d[A\sin(2\pi ft)]}{dt} = 2\pi fA\cos(2\pi ft)$. Since cos(...) oscillates between +1 and −1, v_y reaches its maximum value when $\cos(2\pi ft) = 1$. So, set $\cos(2\pi ft) = 1$ to get $v_{\text{max}} = 2\pi fA = 2\pi A\left(\frac{1}{4}\sqrt{\frac{Mg}{Lm_r}}\right)$, where I have invoked my part **a)** answer for f.

You can confirm that the midpoint of the rope reaches v_{max} right at the moment the rope passes through its equilibrium position, i.e., right when the "hump" momentarily disappears as the standing wave transitions from being an "upward-pointing hump" to being a "downward-pointing hump." You have seen this before. A simple harmonic oscillator, such as a mass on a string, reaches maximum speed as it passes through equilibrium. This is not coincidence. A point on a standing wave *is* a simple harmonic oscillator. Its position as a function of time, after all, is given by $y = A\sin(2\pi ft) = A\sin(\omega t)$, where the angular frequency is $\omega = 2\pi f$.

Practice Final Exam (Chapters 1-22)

Number Two

32 CHAPTER

◆ *6 questions, each worth the same number of points. You have 3 hours.*

1. Consider the manually operated water cannon drawn below. The thin section of the cannon has cross-sectional area A, while the thick section has cross-sectional area $10A$.

   The water cannon operates as follows: The whole cannon is filled with water (density ρ_w). Then, the piston plugging the thick section of the cannon is pushed rightward *with constant force and constant speed*, forcing water through the cannon. As a result, water spurts out the right end of the cannon into the air.

   Kim stands near the top of a cliff. In the valley below, a small bush has just caught fire. Kim aligns her water cannon horizontally, and prepares to shoot water at the bush. The bush is a distance D to the right of the mouth of the water cannon, and a distance s below the mouth of the water cannon.

   In this problem, you may take air pressure to be given, p_{air}. Your answers may also contain A, ρ_w, D, s, and g.

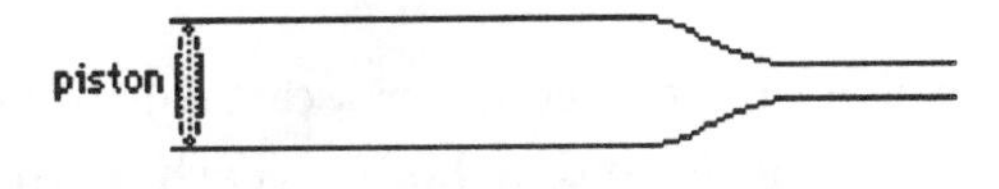

a) With what speed must water leave the water cannon in order to hit the bush? Neglect air resistance.

b) With what force must Kim push on the piston in order for water to hit the bush? If you did not solve part **a)**, let "v" denote the answer to part **a)**, and solve in terms of v and the given constants.

c) With what power is she pushing?

2. Pretend the solar system consists only of the Sun and the Earth, as drawn below. Neglect the moon and the planets. Some data: Earth's mass $= M_E$, Sun's mass $= M_S$, Earth's radius $= R_E$, Sun's radius $= R_S$, Sun-to-Earth distance (from center of Sun to center of Earth) $= R_{SE}$.

a) Assuming the Earth's orbit is approximately circular, how much time does the Earth take to complete one orbit around the Sun? Express your answer in terms of the constants listed above, and any universal physical constants you need.

b) In the picture below, the Earth orbits the Sun counterclockwise. A cannon ball is fired straight up from point A (i.e., along the Earth's direction of motion around the Sun). The ball's initial speed relative to the Earth is v_0, which is so large that the ball eventually gets *very* far from the solar system, an effectively infinite distance away. When the ball is very far away, what is its speed v *with respect to the Sun?*

c) Let us fancifully pretend that the cannon ball is so massive that the act of firing it causes the Earth to slow down to 80% of its current speed. Draw a rough sketch of Earth's new orbit around the Sun. Then, set up (but do not solve) the equation or equations that would enable you to find r_{min}, the closest distance the Earth gets to the Sun in that new orbit. In this part of the problem, neglect the cannon ball.

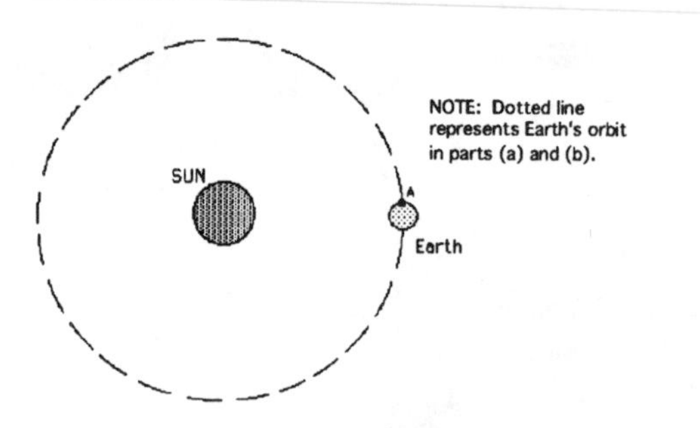

3. In this problem, consider a cart consisting of four wheels and a body. Each wheel is approximately a uniform solid disk of mass $m_w = 5.0$ kg and radius $R = 0.10$ meter. The body, which is essentially a flat metal sheet, has mass $m_b = 3.0$ kg.

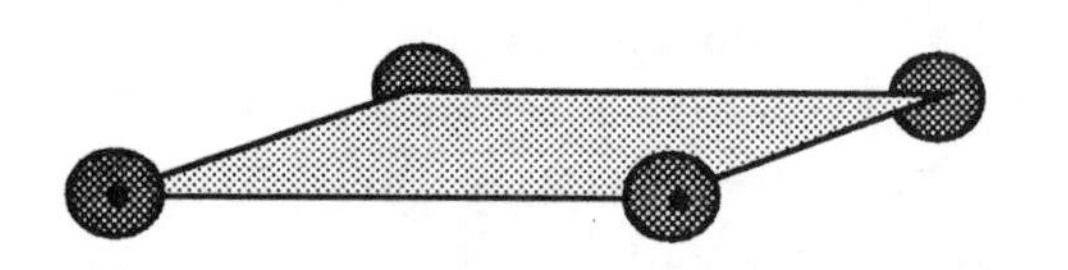

Unfortunately, this cart has a design flaw. The wheels are attached insecurely to the body. In fact, when a wheel spins at a rate of 30 revolutions/secondor more, it falls off the body.

This cart is released from rest from the top of a huge ramp that makes a 30° angle with the horizontal. The cart rolls down the ramp (i.e., the wheels roll without slipping).

a) How far along the ramp will the cart roll before falling apart?

b) After the wheels fall off, it is observed that the body of the cart slides along the ramp for a distance $d = 60$ meters before coming to rest. During the body's slide, how much work did gravity do on the cart?

c) What is the coefficient of kinetic friction between the body and the ramp?

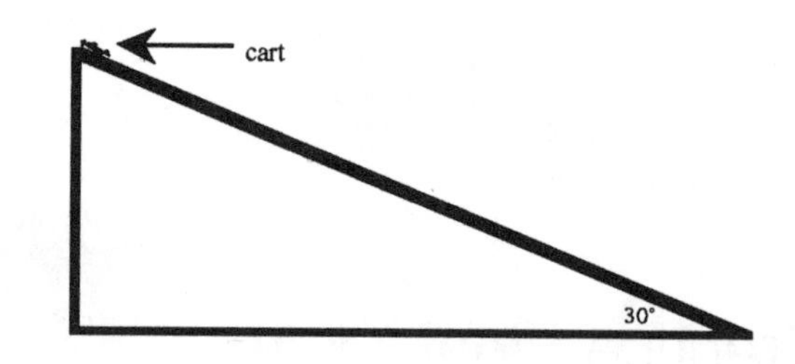

4. A block of mass m, sitting on a *frictionless* ramp of ramp angle θ, is attached to a rope, the other end of which is wrapped around a pulley of mass M_p and radius R, as drawn below. The pulley is approximately a uniform solid disk pivoted at its center. The ramp is nailed to the floor. The block is released from rest, a distance L from the bottom of the ramp.

After the block slides smoothly off the ramp (without "bouncing" or losing any speed), it slides frictionlessly along the floor until it encounters a horizontal spring of spring constant k. The other end of the spring is attached to a wall. The spring is initially at its equilibrium length.

a) How far will the block compress the spring from equilibrium?

Hint: When the block slides off the bottom of the ramp, the rope goes slack, and therefore no longer affects the block's motion.

b) Let $t = 0$ denote the time at which the block first touches the spring. At what time will the spring be fully compressed?

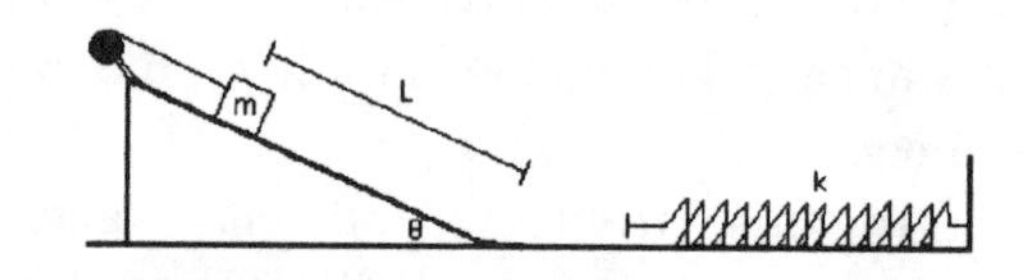

5. Consider a long thin rod with a small solid sphere attached to its end. The rod has mass m and length L. The sphere has mass M. The sphere is *very small* compared to the rod.

Suppose this system is stored in the corner of a room, as drawn below. The rod makesa 60° angle with the floor. The essentially massless rope is stretched horizontally and is attached to a point two thirds of the way up the rod.

a) What is the tension in the rope?

b) The rope breaks. What is the rod's angular speed immediately before it crashes into the floor?

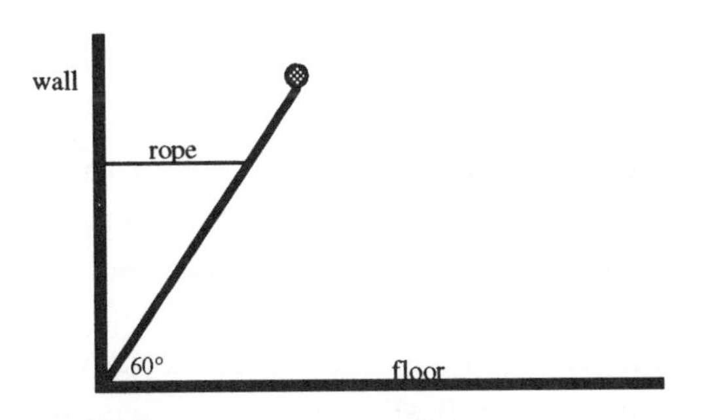

6. Consider a computer hard disk spinning inside a disk drive. The disk is uniformly solid and has mass M and radius R.

The disk is old. As a result, the disk has developed four little (nearly massless) protrusions sticking out from the edge of the disk, as shown below. The protrusions are equally spaced around the disk. Each time one of these protrusions hits point A on the casing, a click is produced. When the disk spins fast, you can hear the fast-paced clicking as a high-pitched whine.

Suppose the disk is spinning frictionlessly at angular speed ω_0. (You may assume that the disk does not slow down appreciably as a result of the protrusions' hitting the casing.) Suddenly, a piece of dirt, which has been building up on the ceiling of the casing, drops straight down onto the computer disk, landing and sticking a distance $\frac{R}{2}$ from the center of the disk. The dirt has mass m. In this problem, you may take the speed of sound as given, v_s.

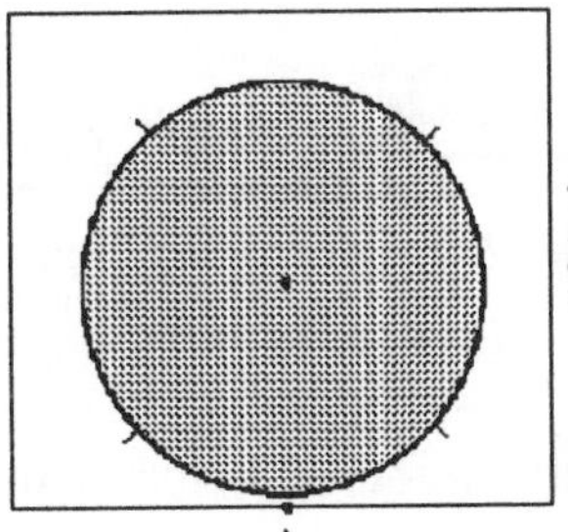

Top-down view of disk inside its rectangular casing (box). Notice the four protrusions.

a) Before the dirt falls onto the disk, what is the wavelength of sound produced by the clicking? Answer in terms of the constants given above.

b) After the dirt sticks to the disk, what is the wavelength of sound produced by the clicking? Would a person hear the pitch get higher or lower after the dirt lands?

ANSWER 1

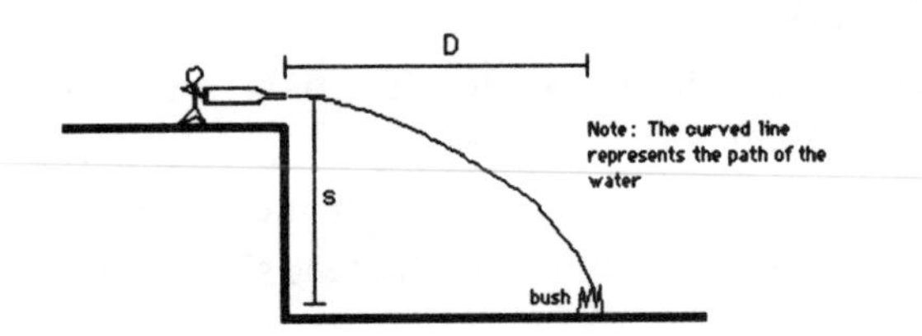

a) If you make a good sketch, this turns into a standard projectile problem. The bush sits a distance D to the right of, and a distance s below, the mouth of the water cannon.

In this problem, we are solving for v_0, the speed with which water must leave the cannon to travel horizontal distance D in the same time it falls through vertical distance s. Notice that v_0 is entirely horizontal, because the cannon is held horizontally. So, $v_{0x} = v_0$, while $v_{0y} = 0$.

As usual in projectile motion problems, we must consider both the horizontal and the vertical aspect of the motion, in order to gather enough information. Specifically, in this case, we can use vertical information to find how long the water stays in the air; and then we can substitute this "flight time" into a horizontal equation relating the water's initial speed to its horizontal distance traveled.

Picking upward as positive, we see that the water starts at height $y_0 = s$, and ends at $y = 0$. Its initial *vertical* speed, as noted above, is $v_{0y} = 0$. And its vertical acceleration is due to gravity, $a_y = -g$. So, from the vertical displacement formula, we get

$$y = y_0 + v_{0y}t + \frac{1}{2}a_y t^2.$$

$$0 = s + 0 - \frac{1}{2}gt^2.$$

Solve for the "flight time" t to get $t = \sqrt{\frac{2s}{g}}$. Now we can put that time into the horizontal displacement equation. Remember, a projectile's *horizontal* acceleration is $a_x = 0$, because only *vertical* forces act on it (i.e., gravity).

$$x = x_0 + v_{0x}t + \frac{1}{2}a_x t^2.$$

$$D = 0 + v_0 t + 0$$

$$= v_0\sqrt{\frac{2s}{g}},$$

and hence $v_0 = D\sqrt{\frac{g}{2s}}$.

b) We need to calculate what rightward force F_p Kim must exert on the piston to make water shoot out the mouth of the cannon at speed $v_0 = D\sqrt{\frac{g}{2s}}$.

As always when dealing with forces, start with a force diagram. This diagram shows the forces acting *on the piston.*

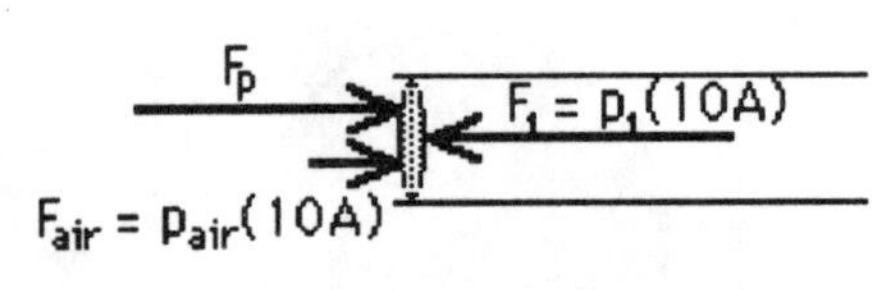

Here, p_1 denotes the fluid pressure at the piston, and F_1 is the corresponding force. I have used the definition of pressure, $p = \frac{\text{force}}{\text{area}}$, and hence $F = pA$. It is easy to forget that air pressure "helps" Kim push the piston.

How can we solve for F_p? Usually, Newton's 2nd law helps. Since the piston moves at constant velocity, neither speeding up nor slowing down, its acceleration is 0. So,

$$\sum F_x = ma$$
$$F_p + p_{\text{air}}(10A) - p_1(10A) = 0,$$

and hence

$$F_p = (p_1 - p_{\text{air}})10A. \tag{1}$$

To solve Eq. (1) for F_p, we must find p_1. So, this reduces to a fluid flow problem.

As usual, I will start by labeling some points. We are trying to find p_1. From part **a)**, we know that water exits the pipe at speed $v_2 = "v_0" = D\sqrt{\frac{g}{2s}}$. We also know the fluid pressure at point 2. Since water "touches" air at that point, the water pressure must be "equalized" to the surrounding air pressure: $p_2 = p_{\text{air}}$. Given this information, we can invoke our usual fluid flow "tools," namely the flux conservation (continuity) equation and Bernoulli's law. See Chapter 22.

Flux conservation: $(10A)v_1 = Av_2,$

Bernoulli: $p_1 + \frac{1}{2}\rho_w v_1^2 + \rho_w g s = p_{\text{air}} + \frac{1}{2}\rho_w v_2^2 + \rho_w g s.$

(In Bernoulli's law, you could have set $h = 0$ instead of $h = s$. It does not matter; those terms cancel.) The only unknowns are p_1 and v_1. So, we have enough information to solve for p_1.

Algebra starts here. Solve the flux conservation equation for v_1, and substitute in our part **a)** answer, to get $v_1 = \frac{1}{10}v_2 = \frac{1}{10}D\sqrt{\frac{g}{2s}}$. The water in the thick segment of pipe flows ten times slower. Put that expression into Bernoulli's law, cancel the $\rho_w g s$ terms, and solve for p_1 to get

$$\begin{aligned}
p_1 &= p_{\text{air}} + \frac{1}{2}\rho_w v_2^2 - \frac{1}{2}\rho_w v_1^2 \\
&= p_{\text{air}} + \frac{1}{2}\rho_w (v_2^2 - v_1^2) \\
&= p_{\text{air}} + \frac{1}{2}\rho_w \left(\left[D\sqrt{\frac{g}{2s}} \right]^2 - \left[\frac{1}{10} v_1^2 D\sqrt{\frac{g}{2s}} \right]^2 \right) \\
&= p_{\text{air}} + \frac{1}{2}\rho_w \left(1 - \frac{1}{100}\right)\frac{D^2 g}{2s} \\
&= p_{\text{air}} + \frac{99\rho_w D^2 g}{400s}.
\end{aligned}$$

End of algebra.

Now that we know the water pressure at the piston, substitute it into Eq. (1) above to get

$$F_p = (p_1 - p_{\text{air}})10A$$
$$= \left(p_{\text{air}} + \frac{99\rho_w D^2 g}{400s} - p_{\text{air}}\right)10A.$$
$$= \frac{99\rho_w D^2 g}{400s}10A.$$

c) We are seeking the power expended *by Kim*. Power is work (or energy) per time. Here, we mean the work done *by Kim*, which is $W_{\text{by Kim}} = \mathbf{F}_p \cdot \Delta\mathbf{x}$. Fortunately, Kim's push force $\mathbf{F}_p$ is entirely along the direction of motion, and hence the dot product reduces to a "regular" product: $W_{\text{by Kim}} = F_p \Delta x$. So, $P = \frac{W_{\text{by Kim}}}{\Delta t} = \frac{F_p \Delta x}{\Delta t} = F_p v$, where in the last step I used $v = \frac{\Delta x}{\Delta t}$. Here, v is the piston's speed.

Do we know v? Well, water flows through the cannon *because* the piston pushes it through. So, the speed of water in the thick part of pipe, v_1, equals the piston's speed. For instance, if the piston is pushed at 2 centimeters per second, then the water in front of the piston also flows at 2 centimeters per second.

Above, in the midst of the algebra, we found that $v_1 = \frac{1}{10}D\sqrt{\frac{g}{2s}}$. So, Kim's power is

$$P = F_p v_1$$
$$= \left[\frac{99\rho_w D^2 g}{400s}10A\right]\left[\frac{1}{10}D\sqrt{\frac{g}{2s}}\right].$$

ANSWER 2

a) We are looking for the period T of the Earth's orbit. In other words, we are trying to express "1 year" in terms of the given constants. First, by using "centripetal force" reasoning, we can calculate the Earth's speed around the Sun. Then we can figure out how the Earth's speed relates to its period.

Since the Earth traces a circle around the Sun of radius R_{SE}, it must experience a radial acceleration (towards the Sun) $a_{\text{radial}} = \frac{v^2}{R_{SE}}$. The only force acting on the Earth, namely the Sun's gravitational pull, is entirely radial. So, by Newton's 2nd law,

$$\sum F_{\text{radial}} = M_E a_{\text{radial}}$$
$$\frac{GM_S M_E}{R_{SE}^2} = M_E \frac{v^2}{R_{SE}}, \quad (1)$$

which we can immediately solve for v to get $v = \sqrt{\frac{GM_S}{R_{SE}}}$.

But how does the Earth's speed relate to its period T? Intuitively, the faster the Earth whizzes around the Sun, the shorter its period, i.e., the less time it takes to complete a full revolution. Let is formalize this intuition. In time T, the Earth completes one "circumference" around the Sun, a distance $2\pi R_{SE}$. Since the Earth's speed is constant, we have

$$v = \frac{\Delta x}{\Delta t} = \frac{2\pi R_{SE}}{T}. \quad (2)$$

So, we can immediately solve for T, and then substitute in our expression for v found above.

$$T = \frac{2\pi R_{SE}}{v} = \frac{2\pi R_{SE}}{\sqrt{\frac{GM_S}{R_{SE}}}} = \sqrt{\frac{4\pi^2}{GM_S}} R_{SE}^{\frac{3}{2}}.$$

Notice that, by algebraically rearranging this equation, you get $\frac{R_{SE}^3}{T^2} = \frac{GM_S}{4\pi^2}$, which is Kepler's 3rd law: The ratio of r^3 to T^2 for any planet is the same, because that ratio equals a constant that depends only on the Sun's mass, not on the planet's mass. That constant is $\frac{GM_S}{4\pi^2}$. Eqs (1) and (2) are how you derive Kepler's 3rd law.

b) As the cannon ball gets farther from the Earth and Sun, the gravitational forces on it decrease. Therefore, its acceleration is not constant. You cannot use your old kinematic formulas. As always when forces get too messy, use conservation, in this case, energy conservation. Start with diagrams of the relevant stages of the process.

Do not set the final velocity equal to 0, because the ball's initial speed v_0 may exceed the escape velocity. In that case, when the ball gets very far away, it is still moving. We are solving for how *fast* it is still moving.

Since the ball escapes from both the Earth's and the Sun's gravitational pull, we must take into account the ball's potential energy due to the Earth and also its potential energy due to the Sun. Fortunately, since energies have no direction (i.e., they are "scalars," not vectors), we can just add the potential energies.

A big stumbling block in this problem is relative velocities. We want the ball's final speed with respect to the Sun. So, both sides of our energy conservation equation had better be written with respect to the Sun. But the given initial speed, v_0, is relative to the Earth. How can we "switch reference frames" to re-express that initial velocity relative to the Sun?

To explore this question, keep in mind that the Earth circles the Sun at 70000 mph. So, even before getting fired, the cannon ball is already moving at 70000 mph relative to the Sun. Let is say the ball gets fired along the Earth's direction of motion at 1000 mph. Then its speed relative to the Sun is even bigger than 70000 mph, namely 71000 mph. It is like throwing a ball forward at 1 mph from a car cruising at 70 mph. The ball's speed relative to the street is 71 mph.

From these intuitive examples, we see that

$$v_{\text{ball relative to Sun}} = v_{0\ \text{ball relative to Earth}} + v_{\text{Earth relative to Sun}}$$
$$= v_0 + \sqrt{\frac{GM_S}{R_{SE}}},$$

where I have invoked our part **a)** expression for the Earth's speed around the Sun, found from Eq. (1) above. In summary, the ball's initial speed relative to the Sun is $v_0 + \sqrt{\frac{GM_S}{R_{SE}}}$, not v_0.

Gravitational potential energy is $U_{\text{grav}} = -\frac{GMm}{r}$, where r is measured from the *centers* of spherical objects. So initially, the cannon ball is a distance $r = R_E$ from the center of the Earth, and $r = R_{SE}$ from the center of the Sun. (The Sun's radius plays no role here.) Given all this, we can invoke energy conservation:

$$E_0 = E_f$$

$$K_0 + U_{0\text{ due to Earth}} + U_{0\text{ due to Sun}} = K_f + U_{f\text{ due to Earth}} + U_{f\text{ due to Sun}}$$

$$\frac{1}{2}m\left(v_0 + \sqrt{\frac{GM_S}{R_{SE}}}\right)^2 + \left(-\frac{GM_E m}{R_E}\right) + \left(-\frac{GM_S m}{R_{SE}}\right) = \frac{1}{2}mv^2 + \left(-\frac{GM_E m}{\infty}\right) + \left(-\frac{GM_S m}{\infty}\right)$$

$$= \frac{1}{2}mv^2 + 0 + 0.$$

The m's cancel, because the cannon ball's mass does not matter. Solve for the final speed v to get

$$v = \sqrt{\left(v_0 + \sqrt{\frac{GM_S}{R_{SE}}}\right)^2 - \frac{2GM_E}{R_{SE}} - \frac{2GM_S}{R_{SE}}}.$$

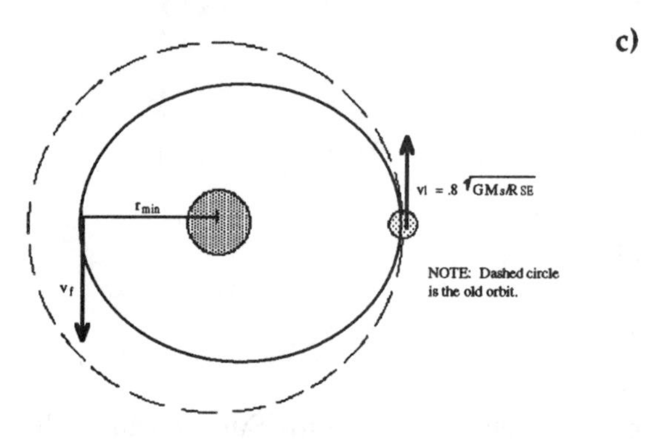

c) All closed orbits caused by gravitational attraction are either circular or elliptical. The Earth gets "knocked out" of its circular orbit, into an elliptical one. The orientation of the ellipse is counterintuitive; see Question 19-4. But it makes sense that the new orbit is "inside" the old orbit, because the Earth slowed down. Since the new orbit is closed, the Earth eventually "returns" to its starting point.

We are solving for r_{min}. Immediately after the Earth gets slowed down, it is still a distance R_{SE} from the Sun, as the diagram shows. At that point, its speed is 80% of its old circular-orbit speed from part **a)**: $v_1 = .8\sqrt{\frac{GM_S}{R_{SE}}}$. Its "final" speed at r_{min} differs from v_1, because the Earth speeds up as it "falls" closer to the Sun. So, the Earth goes fastest at r_{min}, and slowest at its farthest point from the Sun.

We can start with energy conservation. We are talking about the Earth's energy, gravitationally interacting with the Sun (*not* the cannon ball).

$$E_1 = E_f$$

$$K_1 + U_1 = K_f + U_f$$

$$\frac{1}{2}mv_1^2 + \left(-\frac{GM_S M_E}{R_{SE}}\right) = \frac{1}{2}mv_f^2 + \left(-\frac{GM_S M_E}{r_{min}}\right).$$

Though we know v_1, we cannot yet solve for r_{min}, because this equation contains a second unknown, namely v_f. As just discussed, v_f is bigger than v_1. But how much bigger? To obtain more information, you must notice that the only force acting on the Earth as it orbits the Sun is the Sun's gravitational pull. This force points radially toward the Sun; no component of the force is perpendicular to the "radius vector," an imaginary line connecting the Sun to the Earth. Therefore, the force exerts no torque. Since no external torque acts on the Earth as it orbits the Sun, angular momentum is conserved.

Because the Earth is tiny compared to the distance between the Earth and Sun, we can treat Earth as a point mass. Recall from Question 16-5 that the angular momentum of a point mass is $L_{\text{point mass}} = mv_\perp r$, where $v_\perp$ is the component of v perpendicular to the radius vector. When it is closest to and farthest from the Sun, the Earth moves entirely perpendicular to the radius vector; see the above diagram. So, we need not break the velocities into components. Angular momentum conservation gives

$$L_1 = L_f$$
$$M_E v_1 R_{SE} = M_E v_f r_{\min}.$$

Now we have two equations in the two unknowns, v_f and $r_{\min}$. So, you can solve for $r_{\min}$. You were not required to complete the algebra.

ANSWER 3

a) We are told that the wheels fall off when they reach angular speed 30 rev/s. Before continuing, we had better convert to radians per second, or else some of our rotation formulas will not work: $\omega_f = (30\ \frac{\text{rev}}{\text{s}})(\frac{2\pi\,\text{rad}}{\text{rev}}) = 60\pi$ rad/s $= 188.5$ rad/s. We are asked how much distance the cart has covered when its wheels reach angular speed $\omega_f = 188.5$ rad/s.

Intuitively, as the cart rolls farther and farther down the hill, its wheels spin faster and faster. To find the exact relation between how fast the wheels spin and how far the cart has rolled, we can use either forces and torques (and kinematics), or energy conservation. As usual, energy conservation gets us there faster.

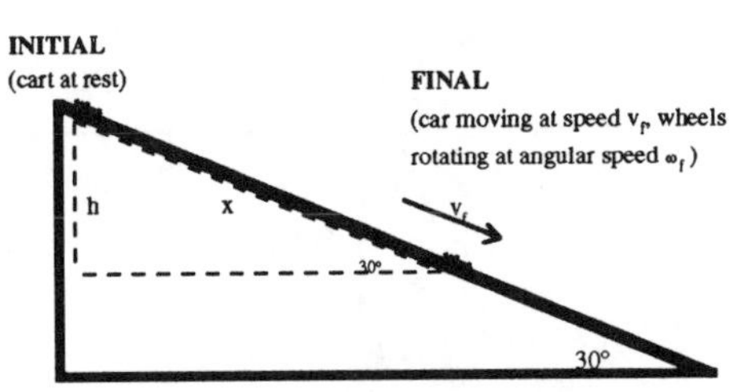

Start by picturing the relevant stages.

We are solving for x, the distance along the ramp that the cart rolls before falling apart. Notice from the diagram that the cart falls through height $h = x\sin 30°$.

I will choose the cart's "final" position as my zero point for measuring heights. So, the final potential energy is $U_f = 0$, while the initial potential energy is $U_0 = (m_b + 4m_w)gh$. (The total mass consists of the body plus four wheels.)

Initially, the cart has no kinetic energy. But at its final position, things get complicated. The body and each wheel has *translational* kinetic energy, because each of those parts moves at (unknown) velocity v_f. In addition, the wheels–but *not* the body–have *rotational* kinetic energy. Specifically, each wheel has $K_{\text{rot}} = \frac{1}{2} I_w \omega_f^2$. The total rotational kinetic energy is four times this value, since the cart has four wheels. Putting all this together, and remembering that a solid disk has rotational inertia $I_w = \frac{1}{2} m_w R^2$, we get

$$E_0 = E_f$$
$$K_0 + U_0 = K_{f\,\text{trans}} + U_f$$
$$0 + (m_b + 4m_w)gh = \frac{1}{2}(m_b + 4m_w)v_f^2 + 4\left[\frac{1}{2} I_w \omega_f^2\right] + 0 \qquad (1)$$
$$(m_b + 4m_w)g(x\sin 30°) = \frac{1}{2}(m_b + 4m_w)v_f^2 + 4\left[\frac{1}{2}\left(\frac{1}{2} m_w R^2\right)\omega_f^2\right],$$

where in the last step I substituted in $h = x\sin 30°$ (from the above diagram). Can we solve for x? We know that the wheels fly off when $\omega_f = 188.5$ rad/s. But Eq. (1) still contains another unknown besides x, namely v_f. Fortunately, since the wheels roll without slipping, their linear and angular speeds relate in a simple way:

$$v = R\omega, \quad \textbf{ROLLING WITHOUT SLIPPING CONDITION}$$

and hence $v_f = R\omega_f$. Substitute this expression for v_f into Eq. (1) to get

$$(m_b + 4m_w)g(x\sin 30°) = \frac{1}{2}(m_b + 4m_w)(R\omega_f)^2 + 4\left[\frac{1}{2}\left(\frac{1}{2}m_w R^2\right)\omega_f^2\right].$$

Now x is the only unknown. Solve for it to get

$$\begin{aligned} x &= \frac{\frac{1}{2}m_b R^2\omega_f^2 + \frac{1}{2}4m_w R^2\omega_f^2 + 4(\frac{1}{2})(\frac{1}{2})m_w R^2\omega_f^2}{(m_b + 4m_w)g\sin 30°} \\ &= \frac{[\frac{1}{2}m_b + 3m_w]R^2\omega_f^2}{(m_b + 4m_w)g\sin 30°} \\ &= \frac{[\frac{1}{2}(3\text{ kg}) + 3(5\text{ kg})](0.1\text{ m})^2(188.5\text{ s}^{-1})^2}{[(3\text{ kg}) + 4(5\text{ kg})](9.8\text{ m/s}^2)\sin 30°} \\ &= 52\text{ m}. \end{aligned}$$

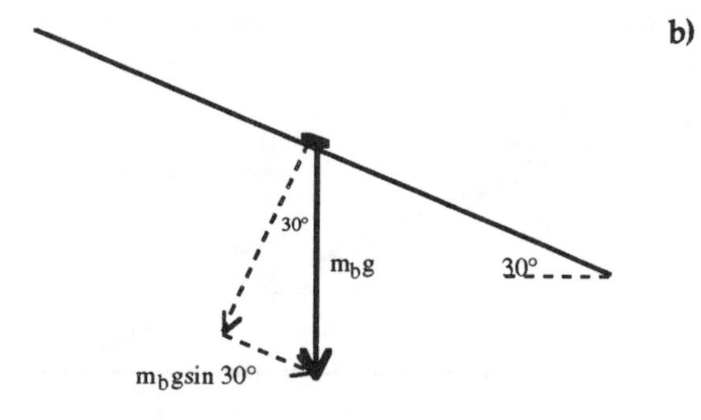

b) This problem tests whether you know the difference between the work done by a single force, and the net work. You are asked only for the work done by gravity: $W_{\text{by grav}} = \mathbf{F}_{\text{grav}} \cdot \Delta\mathbf{x} = F_{\text{grav}\,\parallel}\Delta x$, where $F_{\text{grav}\,\parallel}$ is the component of gravity parallel to the direction of motion, i.e., along the ramp. To find this component of gravity, break up the relevant vector.

The component of gravity along the direction of motion (i.e., parallel to the ramp) is $m_b g\sin 30°$. Use m_b, not $m_b + 4m_w$, because the problem asks for the work done by gravity on the *body*, not on the wheels, as it slides a distance $\Delta x = d = 60$ meters. So,

$$\begin{aligned} W_{\text{by grav}} &= F_{\text{grav}\,\parallel}\Delta x \\ &= (m_b g\sin 30°)d \\ &= (3\text{ kg})(9.8\text{ m/s}^2)(\sin 30°)(60\text{ m}) \\ &= 880\text{ J}. \end{aligned}$$

To solve this problem, you did not need to consider any forces other than gravity. If you did so, you may have been finding the net work.

c) Here, you *do* need a full force diagram. Before drawing it, however, let is formulate an overall strategy. We know that the body slides a distance d. Its "initial" speed is how fast it was going when the wheels fell off, i.e., its final speed from part **a)**. From this kinematic information, we can figure out the body's deceleration during its slide. From that deceleration and a force diagram, we can calculate the frictional force, and hence, the coefficient of friction.

(You can also figure out the net force on the body using the work-energy energy plus the definition of work. But once you find that net force, you will still need a detailed force diagram to relate it to friction.)

Let us start with the kinematics. From part **a)**, remembering the rolling without slipping condition, the cart's speed when the wheels fall off is $v_f = R\omega_f$, where ω_f is the "fall-apart" angular speed of the wheels, $\omega_f = 30$ rev/s $= 188.5$ rad/s. So,

$$"v_0" = R\omega_f = (.10 \text{ m})(188.5 \text{ s}^{-1}) = 18.8 \text{ m/s}.$$

I am calling this velocity v_0, because the cart's speed when the wheels fall off is the initial speed with which the body slides down the ramp. At this point, we know the body's initial speed, and also the distance it slides. Therefore, to find the acceleration, we can use $v^2 = v_0^2 + 2a\Delta x$. The body reaches final speed $v = 0$ after sliding through distance $\Delta x = d$. So, solving for acceleration gives

$$\begin{aligned} a &= \frac{v^2 - v_0^2}{2d} \\ &= \frac{0 - (18.8 \text{ m/s})^2}{2(60 \text{ m})} \\ &= -2.96 \text{ m/s}^2. \end{aligned}$$

The minus sign indicates that the body slows down instead of speeding up.

Now that we know the body's acceleration along the ramp, a_x, let is draw the force diagram.

Since the body undergoes no y-motion, $a_y = 0$:

$$\begin{aligned} \sum F_y &= m_b a_y \\ N - m_b g \cos 30° &= 0, \end{aligned}$$

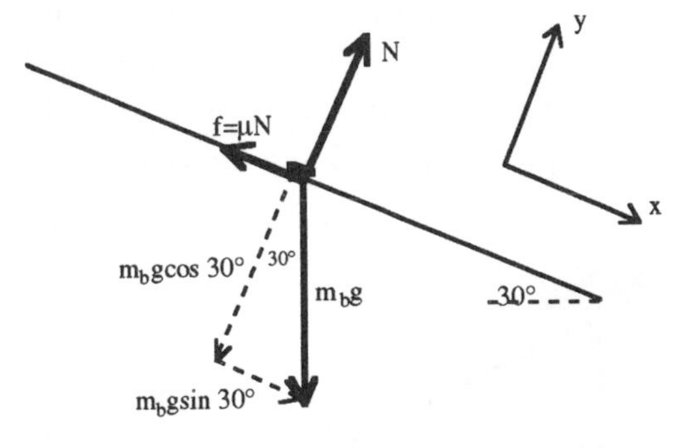

and hence $N = m_b \cos 30°$. Substitute that normal force into the x-force equation, picking down-the-ramp as positive, to get

$$\begin{aligned} \sum F_x &= m_b a_x \\ m_b g \sin 30° - \mu N &= m_b a_x \\ m_b g \sin 30° - \mu(m_b g \cos 30°) &= m_b a_x. \end{aligned}$$

Solve for μ, and substitute in the value for a_x obtained above. Remember, a_x is negative.

$$\begin{aligned} \mu &= \frac{g \sin 30° - a_x}{g \cos 30°} \\ &= \frac{9.8 \text{ m/s}^2 \sin 30° - (2.96 \text{ m/s}^2)}{9.8 \text{ m/s}^2 \cos 30°} \\ &= 0.93, \end{aligned}$$

which is very high. (Remember, μ ranges from $0 =$ very slippery to $1 =$ very rough.) The body "scrapes" along the ground.

ANSWER 4

Let us think things through physically, to formulate a strategy. The block accelerates down the ramp. After it leaves the ramp, the block slides along the floor at constant speed, because the rope no longer pulls on it, as stated in the part **a)** hint. Therefore, whatever speed the block has at the bottom of the ramp is the speed with which it crashes into the spring. In other words, the block's "ramp exit" velocity is also its "initial" velocity in a spring problem. This suggests a way to break the problem into two manageable subproblems:

Subproblem 1 : A standard "multi-block" force problem to find the block's speed at the bottom of the ramp. Call that speed v_1.

Subproblem 2: Use v_1 as the "initial" velocity in a spring-compression problem. In part **a)**, we can find the maximum compression using energy conservation. In part **b)**, we must use oscillation kinematics to find the relevant time. (More on this later.)

Now I will do both subproblems.

Subproblem 1: Multi-block

I will not review the whole multi-block strategy in detail. If you have trouble with this part, work through Question 15-1. Since everything is frictionless, and therefore no heat gets generated, you could use either the multi-block force-and-torque method, *or* energy conservation. Although energy conservation turns out to be quicker, I will review the multi-block method.

Multi-block step 1: Draw separate force diagrams for each relevant object

On the pulley, I will leave out the downward force of gravity and also the force supplied by the "support bar" holding the pulley in place. All the forces on the pulley must cancel, because the pulley does not move. But the pulley *spins*, due to the torque provided by tension. (Gravity and the support-bar force exert no torque, because they act on the center.)

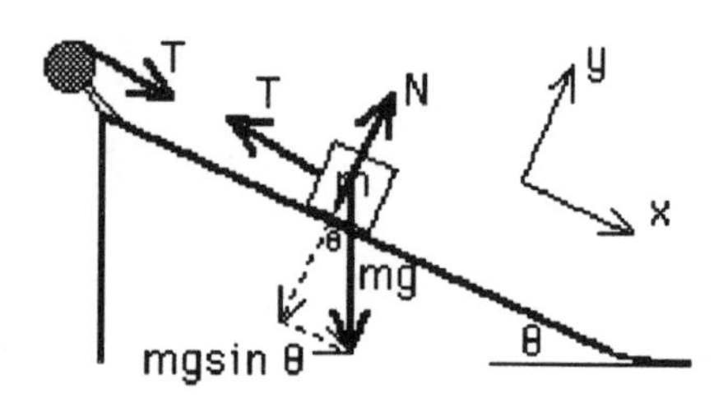

Multi-block step 2: Newton's 2nd law for each separate object

We must use the rotational form of Newton's 2nd law, $\sum \tau = I\alpha$, on the pulley. I will choose down-the-ramp as positive for the block. This commits me to calling clockwise positive for the pulley:

BLOCK: $\sum F_x = ma$

$$mg\sin\theta - T = ma \tag{1}$$

PULLEY: $\sum \tau = I\alpha$

$$TR = \frac{1}{2}M_p R^2 \alpha. \tag{2}$$

I have used the book value for a solid disk's (cylinder's) rotational inertia.

We are trying to solve for the block's acceleration a, so that we can use it to calculate the block's velocity at the bottom of the ramp. Unfortunately, these two equations contains three unknowns: a, T, and α. But as we have seen before, since the block, the rope, and the rim of the pulley move together, it follows that $a_{\text{block}} = a_{\text{tang, rim of pulley}}$. (The subscript "tang" stands for "tangential.") And the tangential acceleration of a point moving in a circle of radius R is $a_{\text{tang}} = \alpha R$. So,

$$a = \alpha R. \tag{3}$$

Multi-block step 3: Solve for the acceleration or force you want

We have three equations in three unknowns (a, T, and α). To quickly solve for acceleration, notice from Eq. (3) that $\alpha = \frac{a}{R}$. Put this into Eq. (2), and then divide Eq. (2) through by R. Now add Eq. (1) to the modified Eq. (2). The $+T$ in Eq. (1) cancels the $-T$ in Eq. (2), leaving us with $mg\sin\theta = (m + \frac{1}{2}M_p)a$, and hence

$$a = \frac{mg\sin\theta}{m + \frac{1}{2}M_p}.$$

Step 4: Kinematics

Now that we know the block's acceleration along the ramp, we can use an old constant-acceleration kinematic equation to find how fast it is traveling after sliding a distance L along the ramp. Call the "final" speed at the bottom of the ramp v_1. Then,

$$\begin{aligned} v_1^2 &= v_0^2 + 2a\Delta x \\ &= 0 + 2\left(\frac{mg\sin\theta}{m + \frac{1}{2}M_p}\right)L, \end{aligned}$$

and hence

$$v_1 = \sqrt{2\frac{mg\sin\theta}{m + \frac{1}{2}M_p}L}.$$

As mentioned above, you could have obtained this result using energy conservation, provided you were careful to include the pulley's rotational kinetic energy.

Now that we know v_1, this problem reduces to the following: A block of mass m slides with speed v_1 toward a spring.

STAGE 1 (block only, since rope and pulley are no longer relevant)

FINAL

Subproblem 2: Spring reasoning

a) We are looking for the amplitude A, i.e., the maximum compression reached by the spring. As usual, energy conservation provides the easiest way to find A.

At maximum compression ($x = A$), the block momentarily comes to rest: $v_f = 0$. A spring's potential energy is $U = \frac{1}{2}kx^2$, where x is the distance by which it has been stretched or compressed from equilibrium. So, energy conservation tells us that

$$\begin{aligned} E_{1\,(\text{block only})} &= E_2 \\ K_1 + U_1 &= K_f + U_f \\ \frac{1}{2}mv_1^2 + 0 &= 0 + \frac{1}{2}kA^2. \end{aligned}$$

Solve for A to get

$$A = \sqrt{\frac{mv_1^2}{k}} = \sqrt{\frac{m}{k}}v_1 = \sqrt{\frac{m}{k}}\sqrt{2\frac{mg\sin\theta}{m+\frac{1}{2}M_p}L},$$

where I have used our expression for v_1 from subproblem 1 above.

b) I see two related ways to answer this. The easiest is to realize that compressing the spring from equilibrium ($x = 0$) to maximum displacement ($x = +A$) corresponds to exactly one quarter of a full oscillation. (A full oscillation would go $x = 0 \to x = +A \to x = 0 \to x = -A \to x = 0$.) So, we could figure out the period T using $T = \frac{2\pi}{\omega}$, and then divide by 4.

Alternatively, we could use oscillation kinematics, starting with $x = A\sin(\omega t)$. (If you use cosine instead of sine, you will need to introduce a 90° phase constant. If we use sine, the phase constant goes to zero. You can confirm this by noticing that, at $t = 0$, $x = A\sin(\omega 0) = 0$.) We are looking for the time at which $x = A$. So, we could set $x = A\sin(\omega t) = A$, and solve for t.

Whichever of these techniques you use, you must first find ω, the angular frequency of the mass-and-spring system. In general, you would have to derive "stuff" and set $\omega = \sqrt{\frac{\text{stuff}}{m}}$. See Question 16-2 for an example. But here, we are dealing with a single spring, which exerts a force $F = -kx$. So, for a spring, stuff $= k$. Therefore, $\omega = \sqrt{\frac{k}{m}}$. That is what we need to know.

To complete the problem, use either of the above-mentioned techniques. I will use the first:

$$\begin{aligned} t &= \frac{1}{4}T \\ &= \frac{1}{4}\left(\frac{2\pi}{\omega}\right) \\ &= \frac{1}{4}\left(\frac{2\pi}{\sqrt{\frac{k}{m}}}\right) \\ &= \frac{\pi}{2}\sqrt{\frac{m}{k}}. \end{aligned}$$

Interestingly, this time does not depend on how fast the block was moving before it reached the spring. If v_1 was large, the spring compresses through a large distance, but it covers that distance relatively quickly. If v_1 was small, the spring compresses through a small distance, but it covers that distance slowly. In either case, the spring takes the same time to compress. Indeed, in general, a harmonic oscillator's period of oscillation does not depend on the amplitude.

ANSWER 5

a) Since this is a standard statics problem, I will use the usual strategy.

Step 1: Force diagram

I plan to choose the bottom end of the rod as my pivot point. So, I will break up most of the forces into their components parallel and perpendicular to the rod, because all the relevant "r" vectors will lie along the rod.

N_1 is the normal force exerted by the floor on the rod, and N_2 is the normal force exerted by the wall on the rod. It turns out that neither of those forces plays a role here. Gravity acts on the center of mass, halfway up the rod. Notice that gravity also acts on the sphere. (Alternatively, you could draw a *single* gravity arrow emanating from the overall center of mass of the rod-and-sphere system. Since I am too lazy to figure out where that center of mass is located, I drew separate gravity arrows for the rod and the sphere.)

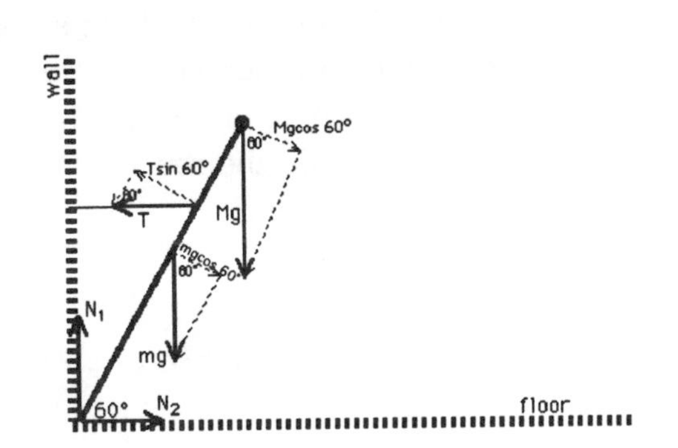

Step 2: Choice of reference pivot point

As mentioned above, I am choosing the bottom of the rod as my reference pivot point. That is why I did not break up $\mathbf{N}_1$ and $\mathbf{N}_2$; those forces act directly on the pivot point, and therefore exert no torques.

Step 3: Torque and force balancing equations

As usual, I will start with torques. For gravity acting on the rod, the **r** vector runs along the rod and has length $r = \frac{L}{2}$. The component of mg perpendicular to this **r** vector is $mg\cos 60°$. So, the torque due to gravity on the rod is $\tau_{\text{grav, rod}} = F_{\perp}r = (mg\cos 60°)\frac{L}{2}$. By similar reasoning, the torque due to gravity on the small sphere is $\tau_{\text{grav, sphere}} = Mg\cos 60°\, L$. Both of these gravitational torques are clockwise. By giving those torques plus signs, I have implicitly picked clockwise as positive.

The component of tension perpendicular to the rod is $T_{\perp} = T\sin 60°$. Since the rope's "contact point" is two thirds of the way up the rod, $r = \frac{2}{3}L$. Crucially, the torque due to tension is counterclockwise. So, $\tau_{\text{tension}} = -T\sin 60°\frac{2}{3}L$.

Since the system does not rotate, the net torque must be zero:

$$\sum \tau = mg\cos 60°\frac{L}{2} + Mg\cos 60°\, L - T\sin 60°\frac{2}{3}L = 0.$$

Often, we would need to write force-balancing equations to acquire more information. But here, we can solve the torque-balancing equation for T, because the equation contains no other unknowns. Doing so yields $T = \frac{3}{2}(\frac{1}{2}m + M)g\cot 60°$.

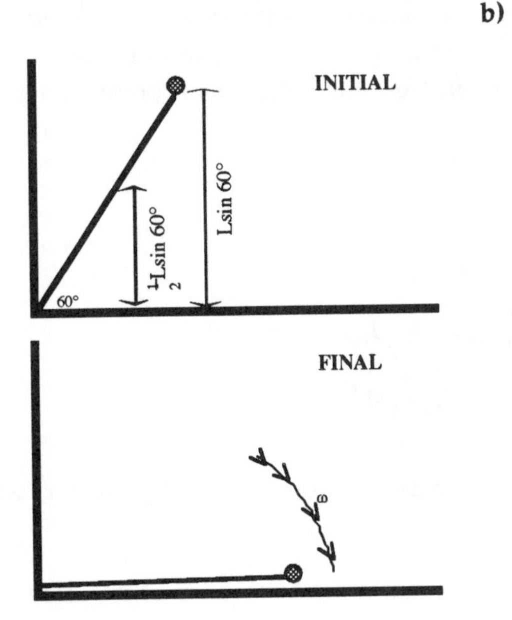

b) Your first inclination might be to use torques and angular acceleration, along with kinematics. Unfortunately, this technique does not work, because the angular acceleration is not constant. Here is why: As the rod falls, the angle it makes with the vertical keeps changing. Since gravity always points vertically, the angle between the rod and the gravitational force vector keeps changing. Therefore, the torques due to gravity keep changing, because those torques depend on angle. (See the force diagram above.) Because the torques are not constant, the angular acceleration is not constant. You cannot use constant-acceleration kinematic formulas.

As usual, when the forces or torques are not constant, use conservation. Gravity exerts an external torque on the system, and hence angular momentum is not conserved. But energy *is* conserved. Just before the system crashes into the floor, the sphere and rod are moving together with the same angular speed, ω. We are solving for ω.

Let is measure our heights from floor level. Then the rod and sphere both end up with essentially no potential energy. As the diagram indicates, the sphere starts out at height $h = L\sin 60°$, while the rod's center of mass starts off at *half* that height. Remember, when calculating potential energies, you must measure your heights to the center of mass. So, the initial potential energy is

$$U_0 = U_{0\text{ sphere}} + U_{0\text{ rod}} = Mg(L\sin 60°) + mg\left(\frac{1}{2}L\sin 60°\right).$$

By the way, you could also calculate U_0 by finding the center-of-mass height of the whole sphere-and-rod system, h_{cm}. Then, $U_0 = (M+m)gh_{cm}$. This technique gives the *same* expression for U_0 as you get using $U_0 = U_{0\text{ sphere}} + U_{0\text{ rod}}$. Personally, I find it easier to calculate potential energies by treating the system as a bunch of "pieces"–here, a rod and sphere–and adding together the separate potential energy terms for each piece. But some people prefer working with the overall center of mass. Take your pick.

Let is talk about the final kinetic energy. Since the rod is not a point mass, and since it rotates around a fixed pivot point, its kinetic energy is purely rotational. By contrast, the sphere is small enough to be considered a point mass. Therefore, we can treat its kinetic energy as either translational or rotational, but not both. (See Question 16-5 for a review of these considerations.) I will express the point mass' kinetic energy rotationally, partly because the sphere and rod "share" the same angular speed ω.

A common mistake in this problem is to use $I = \frac{2}{5}MR^2$ for the sphere. But as your textbook shows, that is the rotational inertia of a sphere *spinning about its center*. So, $\frac{2}{5}MR^2$ would be the right "I" to use if the sphere were spinning in place. But here, the sphere as a whole is rotating around an outside pivot point, the bottom of the rod. You could calculate its new "I" using the parallel axis theorem. But as you can confirm, if the sphere's radius is much much less than the rod's length, then the rotational inertia you get from the parallel axis theorem is extremely close to the rotational inertia of a point mass, $I_{\text{point}} = Mr^2$, where r is the distance from the pivot to the point mass. Here, $r = L$, the rod's length. So, you can treat the sphere as a point mass, with rotational inertia $I_{\text{point mass}} = ML^2$.

Putting all this together, energy conservation gives us

$$\begin{aligned} E_0 &= E_f \\ K_0 + U_{0\text{ sphere}} + U_{0\text{ rod}} &= K_f + U_f \\ 0 + Mg(L\sin 60°) + mg\left(\frac{1}{2}L\sin 60°\right) &= \frac{1}{2}I_{\text{rod}}\omega^2 + \frac{1}{2}I_{\text{point mass}}\omega^2 + 0. \\ &= \frac{1}{2}\left(\frac{1}{3}mL^2 + ML^2\right)\omega^2. \end{aligned}$$

Notice that you must add the rotational inertias, *not* the masses. The rod gets a "$\frac{1}{3}$" factor, while the point mass does not.

We can immediately solve for ω to get

$$\omega = \sqrt{\frac{2[mg\frac{L}{2}\sin 60° + MgL\sin 60°]}{\frac{1}{3}mL^2 + ML^2}} = \sqrt{\frac{g\sin 60°(m+2M)}{L(\frac{1}{3}m+M)}}.$$

ANSWER 6

a) We are looking for λ, the wavelength of sound produced by the clicking, given v_s, the speed of sound. So, it makes sense to try

$$v_s = \lambda f_{\text{sound}} = \frac{\lambda}{T_{\text{sound}}} \quad (1)$$

where f_{sound} and T_{sound} are the frequency and period of the sound waves. (I am using subscript "sound" to distinguish the frequency of sound waves from the rotational frequency of the disk.) Remember, frequency is cycles per second, while period is the time needed to complete one cycle. That is why $f = \frac{1}{T}$. If we can somehow find f_{sound} or T_{sound}, we can easily solve Eq. (1) for λ.

To find f_{sound}, you cannot just plug in $\omega_0 = 2\pi f = \frac{2\pi}{T}$. Here is why: In this problem, ω_0 is a *rotational* (angular) speed, while f_{sound} is a *wave* frequency. The formula $\omega = \frac{2\pi}{T} = 2\pi f$ is valid only if both sides of the equation refer to rotational motion, or if both sides refer to wave motion, or if both sides refer to simple harmonic motion. Otherwise, you are equating apples to oranges.

Nonetheless, the disk's angular speed *should* relate in some simple way to f_{sound}. Intuitively, the faster the disk spins, the more frequently sound waves get produced. Let is see if we can formalize this intuition into an equation.

Each click corresponds to a sound wave getting generated. Every time the disk spins through 90° (i.e., a quarter of a full rotation), it clicks. Since we know ω_0, we can calculate the time between clicks. But the time between clicks is the time between sound waves, i.e., the *period* of the sound waves. In symbols, $t_{\text{between clicks}} = T_{\text{sound}}$. Given ω_0, I see two closely related methods for calculating this time:

Method 1: Going from ω_0 to T_{sound} using the definition of angular speed

By definition, angular speed is the change in angle per time. Since the angular speed is constant, we can write $\omega = \frac{\Delta\theta}{\Delta t}$. We want to know the time between clicks, which is the time needed for the disk to cover $90° = \frac{\pi}{2}$ radians. Since the time between clicks is $\Delta t = T_{\text{sound}}$, we have

$$\omega_0 = \frac{\Delta\theta}{\Delta t} = \frac{\frac{\pi}{2}}{T_{\text{sound}}},$$

and hence $T_{\text{sound}} = \frac{\pi}{2\omega_0}$.

Method 2: Going from ω_0 to T_{sound} using the rotational period

As you can derive using $\omega = \frac{\Delta\theta}{\Delta t}$, angular speed is related to the rotational period T_{rot} by $T_{\text{rot}} = \frac{2\pi}{\omega}$. Since a click gets produced every fourth of a rotation, the period with which sound waves are produced is

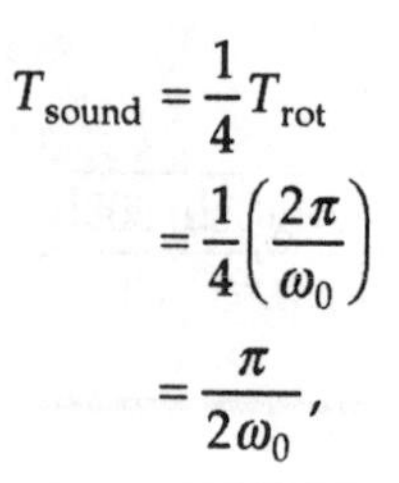

$$T_{\text{sound}} = \frac{1}{4}T_{\text{rot}} = \frac{1}{4}\left(\frac{2\pi}{\omega_0}\right) = \frac{\pi}{2\omega_0},$$

the same answer obtained by method 1.

In any case, now that we know the period of the clicking, we can solve Eq. (1), $v_s = \frac{\lambda}{T_{\text{sound}}}$, for wavelength: $\lambda = v_s T_{\text{sound}} = v_s \frac{\pi}{2\omega_0}$.

b) In part **a)**, we just found that the wavelength of sound waves produced by the clicking is related to the disk's angular speed by

$$\lambda = \frac{v_s \pi}{2\omega}. \tag{2}$$

So, once we find the disk's new angular speed (after the dirt lands), we can easily put that ω into Eq. (2) to obtain the new wavelength.

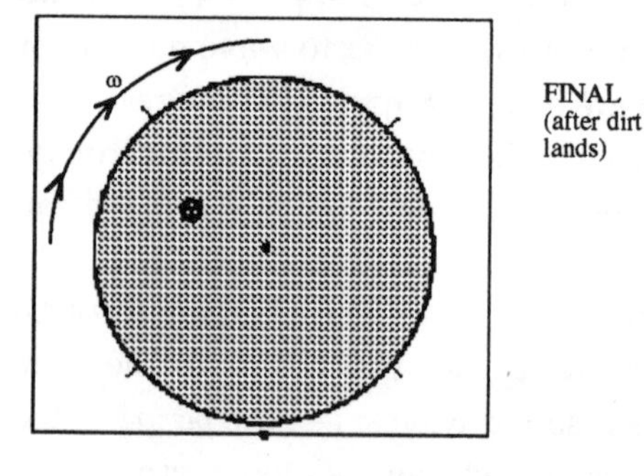

To find the new angular speed, use conservation of angular momentum. Intuitively, when the dirt lands, the disk slows down, resulting in a lower frequency of sound waves; **you would hear the pitch get lower.**

After the dirt lands, it gets "carried along" by the disk at the new angular speed, ω.

Initially, the dirt has no angular momentum in the relevant "direction," because it falls straight down onto the disk. (Remember, the angular momentum of a point mass is proportional to $v_\perp$, the component of the velocity perpendicular to **r** and *along the plane of rotation*. Since the plane of rotation here is horizontal, a vertically-moving object has no angular momentum in the relevant direction). So,

$$L_0 = L_f$$
$$I_{\text{disk}}\omega_0 = (I_{\text{disk}} + I_{\text{dirt}})\omega. \tag{3}$$

The rotational inertia of a solid disk is $I_{\text{disk}} = \frac{1}{2}MR^2$. The rotational inertia of a point mass (e.g., a dirt speck) is $I_{\text{point mass}} = mr^2$. Since the dirt sits a distance $r = \frac{R}{2}$ from the pivot point, $I_{\text{dirt}} = m(\frac{R}{2})^2$. Substitute these rotational inertias into Eq. (3), and solve for ω, to get

$$\omega = \omega_0 \frac{I_{\text{disk}}}{I_{\text{disk}} + I_{\text{dirt}}} = \omega_0 \frac{\frac{1}{2}MR^2}{\frac{1}{2}MR^2 + m(\frac{R}{2})^2} = \omega_0 \frac{M}{M + \frac{1}{2}m}.$$

As expected, the angular speed decreases. Substitute this new angular speed into Eq. (2) to get the new wavelength of sound:

$$\lambda = \frac{v_s \pi}{2\omega} = \frac{v_s \pi}{2\omega_0}\left(\frac{M + \frac{1}{2}m}{M}\right).$$